Greener Synthesis of Organic Compounds, Drugs and Natural Products

Greener Synthesis of Organic Compounds, Drugs and Natural Products

Edited by
Ahindra Nag

CRC Press
Taylor & Francis Group
Boca Raton London New York

CRC Press is an imprint of the
Taylor & Francis Group, an **informa** business

First edition published 2022
by CRC Press
6000 Broken Sound Parkway NW, Suite 300, Boca Raton, FL 33487-2742

and by CRC Press
2 Park Square, Milton Park, Abingdon, Oxon, OX14 4RN

Library of Congress Cataloging-in-Publication Data
Names: Nag, Ahindra, editor. Title: Greener synthesis of organic compounds / edited by Ahindra Nag. Description: First edition. | Boca Raton : CRC Press, 2022. | Includes bibliographical references and index. Identifiers: LCCN 2021045720 | ISBN 9780367544034 (hardback) | ISBN 9780367544089 (paperback) | ISBN 9781003089162 (ebook) Subjects: LCSH: Organic compounds--Synthesis. | Green chemistry. Classification: LCC QD262 .G666 2022 | DDC 547/.2--dc23/eng/20220104 LC record available at https://lccn.loc.gov/2021045720

ISBN: 978-0-367-54403-4 (hbk)
ISBN: 978-0-367-54408-9 (pbk)
ISBN: 978-1-003-08916-2 (ebk)

DOI: 10.1201/9781003089162

Typeset in Times
by SPi Technologies India Pvt Ltd (Straive)

Contents

Preface vii
Editor ix
Contributors xi

1. **Green Chemistry and Green Catalysts** 1
   Ahindra Nag and Himadri Sekhar Maity

2. **New Greener Developments in Direct Amidation of Carboxylic Acids** 23
   Andrea Ojeda-Porras and Diego Gamba-Sánchez

3. **Greener Methods for Halogenation of Aromatic Compounds** 41
   Paola Acosta-Guzmán and Diego Gamba-Sánchez

4. **Microwave as a Greener Alternative in the Synthesis of Organic Compounds** 57
   Paola Acosta-Guzmán

5. **Photochemical Reactions as a Useful and Easy to Implement and Scale Up, New Method for the Synthesis of Chemicals** 75
   Angelo Albini

6. **Biocatalysis in Green Biosolvents** 89
   Margherita Miele, Laura Ielo, Vittorio Pace, and Andrés R. Alcántara

7. **Palladium-Catalyzed Suzuki–Miyaura Cross-Coupling in Continuous Flows** 119
   Remi Nguyen, Virinder S. Parmar, and Christophe Len

8. **Synthesis of Bioactive Heterocyclic Compounds** 137
   Athar Ata and Samina Naz

9. **The Use of Small Particle Catalysts in Pursuit of Green and Sustainable Chemistry** 151
   John A. Glaser

10. **Greener Organic Transformations by Plant-Derived Water Extract Ashes** 177
    Bipasa Halder and Ahindra Nag

11. **Application of Starch in the Synthesis of *N*-substituted Pyrroles by a Simple and Green Route** 191
    K. Arabpourian and Farahnaz K. Behbahani

12. **Greener Synthesis of Potential Drugs** 195
    Renata Studzińska, Renata Kołodziejska, and Daria Kupczyk

13. **Selected Green Efforts to Utilization of Carbohydrates** 229
    Michela I. Simone

14. **Greener Synthesis of Natural Products** 241
    Renata Kołodziejska, Renata Studzińska, Hanna Pawluk, and Alina Woźniak

15. **The Prelude of Green Syntheses of Drugs and Natural Products** ... 289
Leonardo Xochicale-Santana, C. C. Vidyasagar, Blanca M. Muñoz-Flores, and Víctor M. Jiménez Pérez

16. **Biosynthesis of Natural Products** ... 305
Athar Ata, Samina Naz, and Kenneth Friesen

Problems and Answers ... **319**

Index ... **347**

Preface

The 21st century is recognized as the new era of green chemistry where more and more emphasis has been placed on protecting the earth against human devastation and it provides us a proactive track for the sustainable progress of future science and technology. In 1992, after assessing the impact of manufacturing processes of active pharmaceutical ingredients (APIs) on the degree of environmental pollution, it was found that the chemical and pharmaceutical industry generates very high Environmental Factor (E factor) (kg waste/kg product) of higher than 100, indicating a huge amount of residues. Anastas and Warner first developed the 12 principles of "Green Chemistry" in order to promote green and sustainable chemical manufacturing processes. After that, green chemistry focuses on the invention, design and application of environmentally friendly chemical products and processes to reduce or eliminate negative impacts on human health and the environment by avoiding the use and generation of hazardous substances. The investigations in sustainable technology such as photochemistry, microwave technology and nanotechnology, microbe biotransformation have been processed to achieve the ultimate goal of waste-free and energy-efficient syntheses in both industry and the academic world.

In the new millennium, the diversity-oriented synthesis (DOS) for the library of "privileged medicinal structure or scaffolds" of heterocyclic architectures via one-pot multi-component reactions (MCRs) using a natural catalyst in the presence of water acting as a greener reaction medium has received significant importance which is discussed in the book from the standpoint of green chemistry. A greener, efficient and economic approach for the synthesis of several types of symmetric trisubstituted methane derivatives (TRSMs) such as bis(indolyl) methane, bis-coumarins, bis-pyrones, bis(pyrazolyl) methanes, bis-lawsones and bis-dmedones via either electrophilic substitution or cascade Knoevenagel–Michael-type reaction of one molecule of aldehyde with two molecules heterocycles or 1,3 cyclic diketone systems has been discussed by using "water extract of tamarind seed ash (WETSA)" as a renewable and eco-friendly catalyst.

There are specialized books available in green chemistry but the subject material in most is presented in a diffused or highly specialized form such as non-detailed discussions of the theoretical and experimental processes, and students and researchers have to go through or search for various textbooks, journals and pharmacopeias. The major objectives of writing this book are to cater to the present needs of greener synthesis for students and researchers in a lucid, condensed and cohesive form.

I express my indebtedness and gratitude to all the professors who have contributed chapters to the book. I also acknowledge my indebtedness to my wife Jayita Datta and my sons Aritra and Anindya for their encouragement and sustained cooperation. I am also thankful to my research scholars Dr. Himadri Sekhar Maity and Dr. Bipasa Halder for their assistance in completing the book. The cooperation of publishers Mrs. Renu Upadhya and Mrs. Jytosna Jangra, an Taylor and Francis groups, UK, in bringing out the book is very much appreciated.

It is our hope that this book will focus on new directions of greener synthesis and also fulfill the demand of all those who study the subjects either in academic courses (graduate and post-graduate) or through research applications. I gladly invite constructive suggestions from professors, students, scientists and researchers for further their need and improvement in the text.

Ahindra Nag

Editor

Professor **Ahindra Nag**, B.Tech, M.Sc., Ph.D., MICHE, FIC, DEM, has teaching and research experiences of 34 years in the Chemistry Department, Indian Institute of Technology, Kharagpur, India. He has published 100 research papers, 12 textbooks and 3 patents. He has guided 17 research scholars and was invited as a Visiting Professor to Rome (Italy), Academia Sinica (Taiwan) and Tennessee (USA) institutes. He is an editorial member of different international journals and is also the secretary of the International Green Chemistry Society.

Contributors

Paola Acosta-Guzmán
Laboratory of Organic Synthesis, Bio and Organocatalysis
Chemistry Department
Universidad de los Andes
Bogotá, Colombia

Angelo Albini
Department of Chemistry
University of Pavia
Pavia, Italy

Andrés R. Alcántara
Department of Chemistry in Pharmaceutical Sciences
Complutense University
Madrid, Spain

K. Arabpourian
Department of Chemistry, Karaj Branch
Islamic Azad University
Karaj, Iran

Athar Ata
Department of Chemistry
Richardson College for the Environmental and Science Complex
The University of Winnipeg
Winnipeg, Manitoba, Canada

Farahnaz K. Behbahani
Department of Chemistry, Karaj Branch
Islamic Azad University
Karaj, Iran

Kenneth Friesen
Department of Chemistry
Richardson College for the Environmental and Science Complex
The University of Winnipeg
Winnipeg, Canada

Diego Gamba-Sánchez
Laboratory of Organic Synthesis, Bio and Organocatalysis
Chemistry Department
Universidad de los Andes
Bogotá, Colombia

John A. Glaser
United States Environmental Protection Agency, Office of Research & Development
Center for Environmental Solutions and Emergency Response
Cincinnati, Ohio

Bipasa Halder
Chemistry Department
Indian Institute of Technology Kharagpur
Kharagpur, India

Laura Ielo
Department of Chemistry
University of Torino
Torino, Italy

Víctor M. Jiménez Pérez
Facultad de CienciasQuímicas
Universidad Autónoma de Nuevo León
Av. Universidad S/N, Ciudad Universitaria, San Nicolás de Los Garza Nuevo León
Nuevo León, México

Renata Kołodziejska
Nicolaus Copernicus University in Toruń
Collegium Medicum in Bydgoszcz
and
Department of Medical Biology and Biochemistry
Bydgoszcz, Poland

Daria Kupczyk
Nicolaus Copernicus University in Toruń
Collegium Medicum in Bydgoszcz
Department of Medical Biology and Biochemistry
Bydgoszcz, Poland

Christophe Len
Chimie ParisTech
PSL Research University
CNRS, Institute of Chemistry for Life and Health Sciences
Paris, France

Himadri Sekhar Maity
Chemistry Department
Indian Institute of Technology
Khargpur, India

Margherita Miele
Department of Pharmaceutical Chemistry
University of Vienna
Vienna, Austria

Blanca M. Muñoz-Flores
Facultad de CienciasQuímicas
Universidad Autónoma de Nuevo León
Av. Universidad S/N, Ciudad Universitaria, San Nicolás de Los Garza Nuevo León
Nuevo León, México

Ahindra Nag
Chemistry Department
Indian Institute of Technology
Kharagpur, India

Samina Naz
Department of Chemistry
The University of Winnipeg
Winnipeg, Manitoba, Canada

Remi Nguyen
Chimie ParisTech, PSL Research University
CNRS, Institute of Chemistry for Life and Health Sciences
Paris, France

Andrea Ojeda-Porras
WESTCHEM, School of Chemistry
University of Glasgow
and
Loudon Laboratory
University Avenue
Glasgow, United Kingdom

Vittorio Pace
Department of Chemistry
University of Torino
Torino, Italy

Virinder S. Parmar
Chimie ParisTech, PSL Research University
CNRS, Institute of Chemistry for Life and Health Sciences
Paris, France

Hanna Pawluk
Department of Medical Biology and Biochemistry
Nicolaus Copernicus University in Toruń
Collegium Medicum in Bydgoszcz
Bydgoszcz, Poland

Michela I. Simone
Discipline of Chemistry
University of Newcastle, University Drive
Callaghan, Australia

Renata Studzińska
Faculty of Pharmacy, Department of Organic Chemistry
Nicolaus Copernicus University in Toruń
Collegium Medicum in Bydgoszcz
Bydgoszcz, Poland

C. C. Vidyasagar
School of Basic Sciences and Research in Chemistry
Rani Channamma University
Godihal, India

Alina Woźniak
Department of Medical Biology and Biochemistry
Nicolaus Copernicus University in Toruń
Collegium Medicum in Bydgoszcz
Bydgoszcz, Poland

Leonardo Xochicale-Santana
Facultad de CienciasQuímicas
Universidad Autónoma de Nuevo León
Av. Universidad S/N, Ciudad Universitaria, San Nicolás de Los Garza Nuevo León
Nuevo León, México

1

Green Chemistry and Green Catalysts

Ahindra Nag and Himadri Sekhar Maity

CONTENTS

1.1 Green Chemistry 1
1.1.1 Water as a Greener Solvent 2
1.1.2 Photochemistry 3
1.1.3 Microwave-assisted Synthesis 3
1.1.4 Tandem Reaction 3
1.1.5 Click Reactions 3
1.1.6 Multicomponent Reactions 4
1.1.7 Flow Chemistry Reactions 4
1.1.8 Versatile, Small and Biologically Active Molecules with Diverse Functionality 6
1.1.9 Phenolic Compounds 6
1.1.10 Heterocyclic Compounds 7
1.2 Green Catalysts 7
1.2.1 Lipase, Esterase and Yeast as Biocatalysts 8
1.2.1.1 Lipase 8
1.2.1.2 Esterase 12
1.2.1.3 Yeast 12
1.2.2 Plant as Biocatalyst 14
1.2.3 Waste Feedstock as Green Catalyst 15
1.2.3.1 Biomass Waste as Catalyst (Homogenous or Heterogeneous) 15
1.2.4 Heterogeneous Catalysts from Waste Materials 16
1.2.5 Green Nanoparticles as Heterogeneous Catalyst 17
1.2.6 Ecocatalyst as Heterogeneous Catalyst from Plant Parts 18
1.2.7 Carbon Nanoparticles as Heterogeneous Catalysts 18
References 18

1.1 Green Chemistry

The 21st century is recognized as the new era of green chemistry which overlaps with all sub-disciplines of chemistry. It is a swiftly developing field that provides us a proactive track for the sustainable progress of future science and technology.[1] Green chemistry is a practical as well as philosophical concept which significantly focuses on the invention, design and application of environmentally friendly chemical products and processes to reduce or to eliminate negative impacts on human health and the environment by avoiding the use and generation of hazardous substances.[2] Its goal is to improve the quality of life and the competitiveness of industry, by developing safer and more eco-friendly chemistry. Paul T. Anastas for the first time in 1991 coined the term green chemistry.[3] In 1998, Paul T. Anastas and John C. Warner[4] provided a set of principles known as the "Twelve Principles of Green Chemistry" to guide chemists in achieving this goal via the practice of green chemistry. The 12 principles address a range of ways to reduce the environmental and health impacts of chemical production and also guide research priorities for the development of green chemistry technologies. For this, Paul T. Anastas is regarded as the father of green chemistry. In 1995, Paul T. Anastas also helped to persuade US President Bill Clinton to launch the Presidential Green Chemistry Challenge, which encourages still by offering president's environmental youth award (PEYA) of five citations each year to best youth scientists of companies as well as academics who have done an outstanding job of implementing the principles.[3] From the perspective of green chemistry as a central issue in both academic and industrial research in the 21st century and considering the increase of environmental pollution and its intensive impact on living systems, the world is revolving around the sustainable development of environmentally benign, clean and economically feasible organic syntheses using green reagents, eco-friendly catalysts, benign reaction mediums and greener reaction conditions (*e.g.* microwave heating, ultrasound irradiation, infrared radiation, flow chemistry, electrolysis, grinding method and twin screw extrusion) to meet the fundamental scientific challenges of shielding the environment. Environmental factor (E-factor) is directly related to greener synthesis. E-factor is

DOI: 10.1201/9781003089162-1

the actual amount of all waste materials [Kgs (raw materials) – Kgs (product)] formed in the process including solvent losses and waste from energy production.[5]

$$\text{E} - \text{factor} = [\text{Kgs (raw materials)} - \text{Kgs (product)}] / [\text{Kgs (product)}]$$

Higher E-factor means more waste and negative environmental impact. For example, the first laboratory synthesis of its anti-impotence drug sildenafil citrate (Viagra) by drugmaker Pfizer had an E-factor of 105. Then, Pfizer's researchers cut Viagra's E-factor to 8 by eliminating hazardous chlorinated solvents, hydrogen peroxide and oxalyl chloride.[3] After that success, Peter Dunn, the leader of the Viagra synthesis team, became the head of the more systematic green-chemistry drive started by Pfizer in 2001. Pfizer reduced the E-factor of the anticonvulsant pregabalin (Lyrica) from 86 to 9 and modified similar improvements for the antidepressant sertraline and the non-steroidal anti-inflammatory drug celecoxib. "These three products alone have eliminated more than half a million metric tons of chemical waste", commented Dunn.[3]

1.1.1 Water as a Greener Solvent

From the perspective of green chemistry, the use of environmentally friendly reaction mediums such as water, ionic liquids, polyethylene glycol (PEG), supercritical fluids (especially supercritical carbon dioxide (scCO_2)), organic carbonate solvents, perfluorinated solvents and *glycerol* instead of hazardous organic solvents is one of the most fundamental contents of green chemistry.[6] Nature's own reaction medium, i.e. water plays an essential role in life processes as well as organic syntheses. From the standpoint of green chemistry, water as a reaction solvent has gained significant attention for many organic transformations because water is considered as non-toxic, abundantly available, cheap, safe for handing, non-flammable and environmentally benign compared to other organic solvents.[7] In addition, water not only increases the rate and yield of reactions but also enhances unique enantioselectivity in a chiral synthesis which is not observed for reactions in organic solvents.[8] Water has emerged as a greener solvent by the significant accelerating effect on versatile organic transformations owing to its high polarity, a network of hydrogen bonds, hydrophobic interaction, trans-phase interaction, high surface tension and high specific heat capacity.[6,9] Furthermore, water-mediated reactions offer the key advantage of insolubility of the final products, which facilitates their isolation by a simple filtration method. For instance, "in/on water reactions at the surface/interface" and "phase-transfer" techniques are the major platforms used in advanced synthetic chemistry for easy isolation of the products and catalyst from the aqueous reaction medium.[10]

The first example of organic synthesis of indigo was described by Baeyer and Drewsen in 1882 (Figure 1.1).[11] In the synthesis, a suspension of *o*-nitrobenzaldehyde in aqueous acetone was treated with a solution of sodium hydroxide. There was immediate formation of the characteristic blue color of indigo and the product subsequently precipitated.

Breslow and co-workers reported[12] in 1980 that an acceleration of the Diels–Alder reaction under "in water" condition was achieved at very high dilution to dissolve the reactants and also observed that the cycloaddition of cyclopentadiene (0.4 mM) and butanone (25.5 mM) was 740 times faster in water than in isooctane (Figure 1.2). The increased selectivity could be obtained with water (endo/exo = 21.4) compared to the same reaction in cyclopentadiene (endo/exo = 3.85) and the same results were obtained in protic solvent (ethanol and methanol) and hydrocarbons.

Sharpless and co-workers used[13] "on water" condition under which substantial rate enhancement was noticed when the organic reactants were insoluble in the aqueous phase. The "on water" protocol provided not only a better reaction in terms of yield and rate but also offered a green separation which was the cycloaddition reaction of quadricyclane and dimethyl azo-dicarboxylate (Figure 1.3).

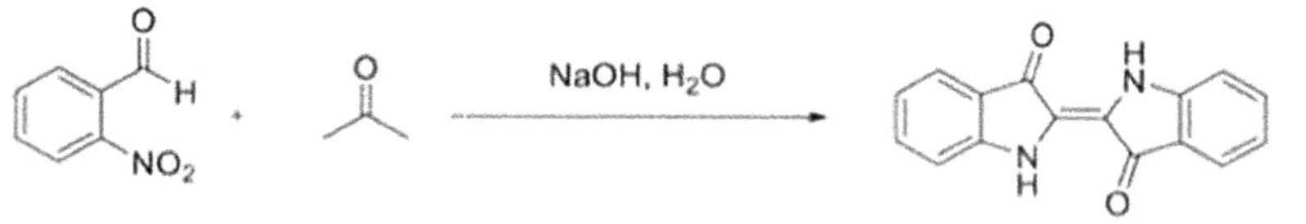

FIGURE 1.1 Synthesis of indigo in aqueous medium.

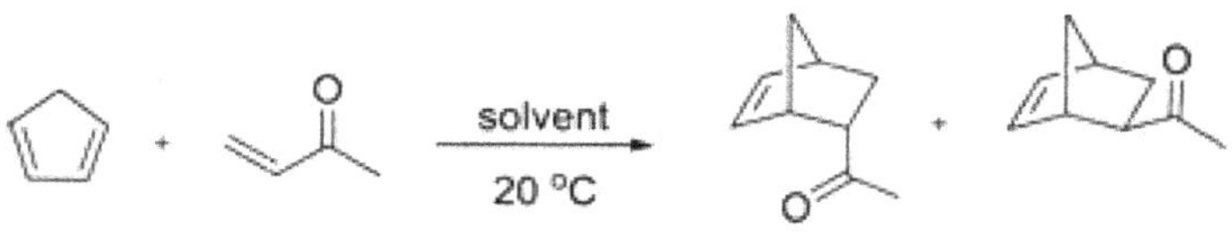

FIGURE 1.2 Diels–Alder reaction in aqueous medium.

FIGURE 1.3 Cycloaddition reaction of quadricyclane and dimethyl azo-dicarboxylate.

Water is considered as an eco-friendly solvent for the development of versatile C–C, C–N, C–S, C–O and C–P bond-forming reactions such as Diels–Alder reaction, aldol condensation, Knoevenagel condensation, Claisen rearrangement, Michael addition, Mannich reaction, coupling reactions (Suzuki–Miyaura, Heck, Stille, Sonogashira, Kumada and Negishi), click reactions, allylation reactions, benzoin condensation, multi-component reactions, epoxidation reaction, radical addition reaction, Grignard-type additions and many others.[14,15] These reactions provide a significant tool for the construction of many important bioactive compounds. Among the various water-mediated reactions, multi-component reactions, click reactions and tandem reactions via one-pot strategy have received much attention for the synthesis of the structurally diverse, small and privileged scaffold of natural products as well as drug or drug-like molecules discussed below.

1.1.2 Photochemistry

Photochemistry, an important branch of chemistry, has been used recently to perform a chemical reaction caused by absorption of ultraviolet[16] (wavelength from 100 to 400 nm). There are two basic laws to complete photochemical reactions.[16] The first law is the Grotthuss–Draper law which states that light must be absorbed by a compound in order for a photochemical reaction to take place. The second law is the Stark–Einstein law which states that for each photon of light absorbed by a chemical system, only one molecule is activated for subsequent reaction. The efficiency of the photochemical process is given by its quantum yield (Φ), which is defined as the number of moles of a stated reactant disappearing or the number of moles of a stated product produced. The common example is the formation of thymine dimer, a ring product formed by the reaction of two thymine molecules. Thymine dimer is an important class of products formed in DNA upon UV irradiation as an example of cycloaddition of unsaturated molecules[17] (Figure 1.4).

1.1.3 Microwave-assisted Synthesis

This technique is green as well as clean chemistry reaction where microwave irradiation is used as a source of heat for chemical reaction.[18] It has the advantages of enhanced reaction rates, higher yields, greater selectivity and economics for the synthesis of organic and inorganic molecules, peptide synthesis, polymer synthesis, nanotechnology and also in drug discovery. This process produces high yields and lower quantities of side-products. Purification of products is easier and, in some cases, selectivity is modified.[19] Indeed, new reactions and conditions that cannot be achieved by conventional heating can be performed using microwaves. It is generally solvent-free reaction as the absence of organic solvents in reactions leads to a clean, efficient and economical technology and safety is increased significantly.

1.1.4 Tandem Reaction

In recent years, one-pot reactions as an attractive synthetic concept are well-known for improving overall process efficiency and reducing production wastes.[20] A tandem reaction, also known as a domino reaction or cascade reaction, is a chemical process which combines multi-step reactions in a single reaction vessel and acts as an efficient alternative as it simplifies the synthesis route, avoids the isolation of intermediates, lowers the operation cost, minimizes the waste output as well as energy consumption and reduces the use of solvents.[21] The one-pot synthesis of benzylidenemalononitriles through tandem deacetalization–Knoevenagel condensation reaction or oxidation–Knoevenagel condensation approach is an example of tandem reactions.[20,22]

1.1.5 Click Reactions

Click chemistry encompasses a group of powerful linking reactions that produce diverse molecular entities in medicinal chemistry, biotechnology, materials science and polymer science through one-pot strategy from readily available potential building blocks under simple reaction conditions.[23] The term "click chemistry" was introduced by Sharpless.[24] The catchy term "click" indicates reactions that are modular in approach, occur irreversibly resulting in high yield of a single product, are efficient, selective and versatile in nature, can be performed in benign and easily removable solvent (like water) without requiring chromatographic purification, and proceed with high reaction specificity (in some cases, with both regio-specificity and stereo-specificity).[24–26] Huisgen's 1,3-dipolar azide-alkyne cycloaddition (AAC) is considered to be an important click reaction for the formation of two regio-isomeric five-membered nitrogen-containing heterocycles, i.e. the 1,4- and 1,5-isomers of substituted 1,2,3-triazoles under thermal conditions.[27] Among all the click reactions, Cu(I)-catalyzed azide-alkyne 1,3-dipolar cycloaddition (CuAAC) for the synthesis of 1,4-disubstituted-1,2,3-triazole via exclusively as well as in a regioselective manner is the jewel in the crown.[28] American chemist K.B. Sharpless has referred to this cycloaddition as "the cream of the crop" of click chemistry and "the premier example of a click reaction".[23,28] Analogous ruthenium-catalyzed

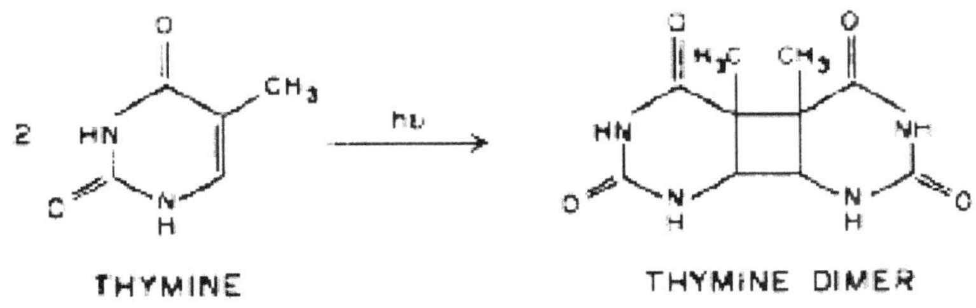

FIGURE 1.4 Formation of thymine dimer.

azide-alkyne cycloaddition (RuAAC) reaction reported by the Sharpless group in 2005 is another example of click reaction for the selective formation of 1,5-disubstituted-1,2,3-triazole in the presence of ruthenium, instead of copper catalyst.[29]

1.1.6 Multicomponent Reactions

With the advent of sustainable and green chemistry practices, implementation of several transformations in a single manipulation through multicomponent reaction (MCR) strategy has emerged as powerful and ideal bond-forming tools in organic, combinatorial and medicinal chemistry for the fabrication of biologically important scaffolds.[30] MCRs are those reactions in which three or more reactants react simultaneously to provide the product in a single step under appropriate reaction conditions. MCRs offer a handful of advantages such as operational simplicity, structural complexity and diversity, atomic and structural economy, high selectivity, step economy, high convergence, high bond-forming index, one-pot reaction, high yield, shorter reaction time, avoidance of time consuming protection and deprotection processes, low costs, minimization of waste, labor, energy and reduction of expensive purification technique making the process green and more eco-friendly.[31] At present, there are many well-known MCRs such as Strecker reaction, Hantzsch reaction, Biginelli reaction, Asinger reaction, Mannich reaction, Passerini reaction, Ugi reaction, Kabachnik–Fields reaction, Prins reaction and Gewald reaction.[32] These reactions are valuable assets for the construction of the novel, complex and structurally diverse molecular entities with attractive drug or drug-like features.[29] The process of the MCR has been discussed in Chapter 10.

1.1.7 Flow Chemistry Reactions

Flow chemistry reactions are continuous flow reactions where reactant components are pumped in a tube or pipe at a controlled temperature to complete the reactions.[33] The process is robust, control and stability inherent in steady state operation of continuous process. Pharmaceutical industries generally rely on manufacturing of pharmaceutical ingredients in multipurpose batch or semi-batch reactors, but in the present system that interest among researchers is arising toward continuous flow manufacturing of organic molecules, including highly functionalized and chiral compounds. Automated flow-based techniques enable optimization and determination of the kinetics of chemical mechanisms at the milligram scale. The advantages of the reactions are as follows:

(a) Cleaner products with excellent selectivity. Greener operation with reduced solvent consumption. (b) Safer reactions as the less probability hazardous intermediates (30a) Faster reactions as the reactions are proceeded with controlled temperature and flow reactors are easily pressurized ultimately it enhances heat and mass transfer. That means this reaction allows temperatures 100°C–150°C above their normal boiling point, therefore creating reaction rates that are 10000s of times faster. This process is called superheating. The high surface area to volume ratio is 1000 times greater than those of a bath reactor, which enables almost immediate heating or cooling and ultimately therefore temperature control. (c) Ability to operate cryogenic processes at higher temperatures to enable gas–liquid reactions. (d) To carry out hazardous chemistry (*e.g.* hazardous reaction). (e) To stimuli multistep reaction sequences in an undisrupted and automated fashion which requires no intermediate handling. Thus, this typical advantageous technology can be combined with microwave irradiation, supported reagents or catalysts, photochemistry, inductive heating, electrochemistry, new solvent systems, 3D printing or microreactor technology. (f) Integrated synthesis analysis[33] either by FTIR or LCMS after dilution of the sample and injection on it. (g) Easier and well-defined scale up routes from laboratory to pilot production. Hence the process is gaining interest in industries not only for chemical transformations and separations, but also crystallizations, drying and formulation are all integrated into one single, fully automated continuous process.

In flow chemistry reaction, the reactor unit is typically either a tube or microstructure device (micro reactor) which is generally made of copper, stainless steel, tantalum, zirconium or per fluorinated polymer. Commercial units in the laboratory scale integrate all the compact units that require the users to put the reactant components in the feed tank with optimized reaction temperature and check the quality of the product at regular intervals. There are four types of reactions occurring in the reactors[34] (Figure 1.5)

(a) In the first case, no catalyst is added. (b) In the second case, the reaction mixture is mixed with support reagent and no catalyst is used (30a). (c) In the third type of reaction, homogenous catalyst is mixed with the reactant and the product is formed with catalyst. (d) In the fourth type of reaction, support catalyst is previously incorporated into the reactor and the mixture of reactants is passed through the reactor to get only product. The diagrammatic view of the flow chemistry equipment is shown in Figure 1.6.

Kirschning and co-workers reported[35] newly developed multistep continuous flow synthesis process of olanzapine using inductive heating (IH) technology which reduces reaction times and increases process efficiency based on the induction

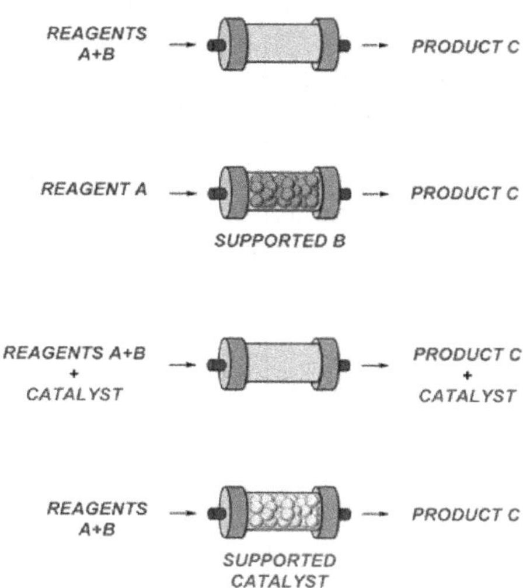

FIGURE 1.5 Reactors applied in flow chemistry.

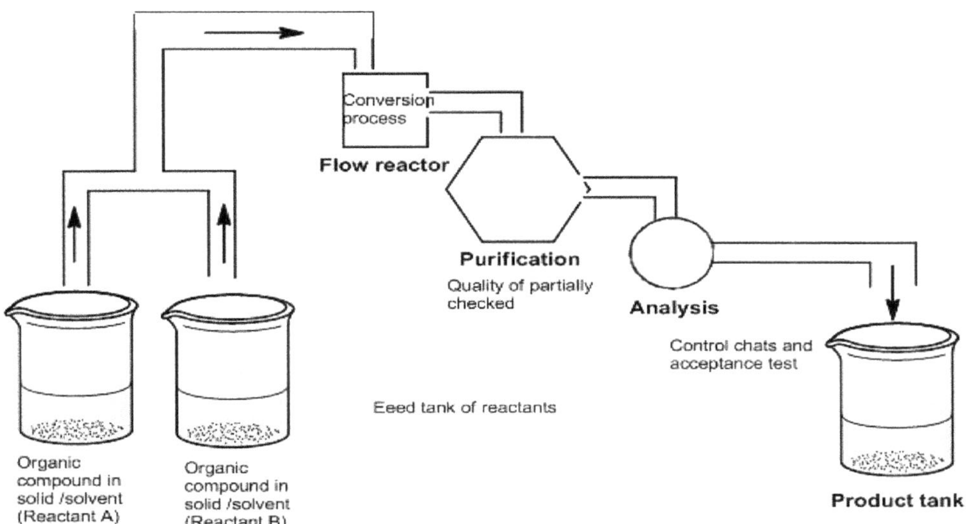

FIGURE 1.6 Diagrammatic view of flow chemistry equipment.

of an electromagnetic field (at medium or high frequency depending on nanoparticle sizes). The synthetic step consisted of coupling aryl iodide and aminothiazole using Pd_2dba_3 as catalyst and xantphos as ligand and ethyl acetate as solvent. After quenching with distilled water and upon in-line extraction in a glass column, the crude mixture was passed through a silica cartridge in order to remove Pd catalyst. Nitroaromatic compound was then subjected to reduction with Et_3SiH into a fixed bed reactor containing Pd/C at 40°C to get aniline. The product from the reactor was mixed with HCl (0.6 M methanol solution) and heated at 140°C to get the product with an overall yield of 88% (Figure 1.7).

Jamison and co-workers reported[36] synthesis of ibuprofen where continuous flow Friedel–Crafts acylation of isobutylbenzene with propionyl chloride was performed with aluminum chloride as a Lewis acid using a 250 μL PFA reactor coil heated at 87°C with a residence time of 60 seconds. The outlet was treated with aqueous HCl, and the crude mixture was extracted using an inline membrane separator operating at 200 psi. Aryl ketone was formed with 95% yield and then mixed with a DMF solution of trimethyl orthoformate (TMOF) and subjected to 1,2-aryl oxidative migration using ICl as promoter. Oxidation to ester took place in a 900 μL PFA reactor coil at 90°C with a residence time of 60 seconds. The final

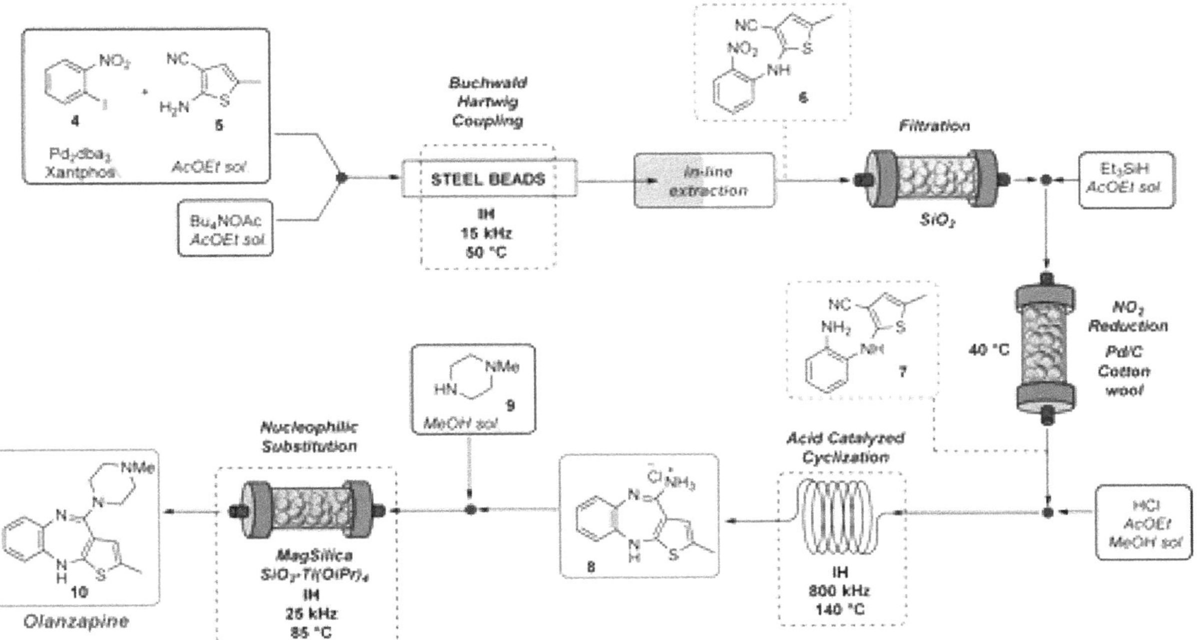

FIGURE 1.7 Continuous flow synthesis of olanzapine. (Flow Chemistry: Recent Developments in the Synthesis of Pharmaceutical Products; R. Porta et al. *Org. Process Res. Dev*. 20, 1, 2–25, 2016. (Permission from ACS https://pubs.acs.org/doi/abs/10.1021/acs.oprd.5b00325?src=recsys))

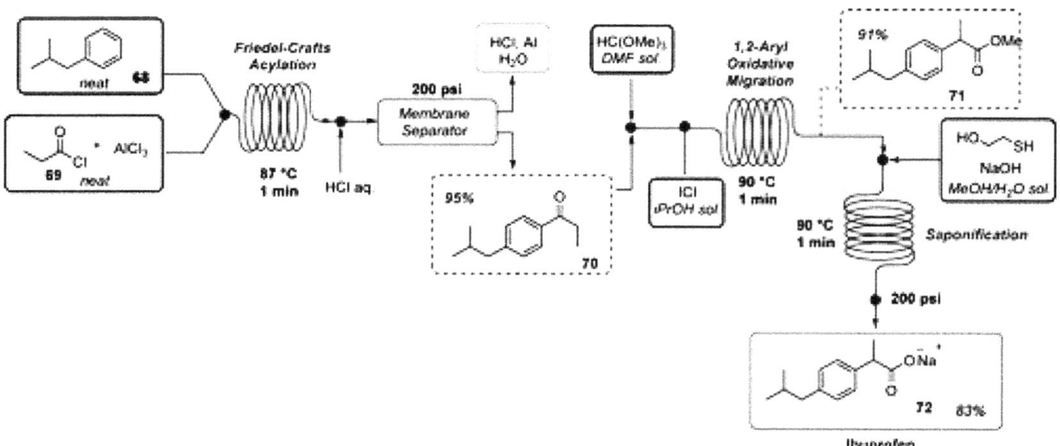

FIGURE 1.8 Continuous flow synthesis of ibuprofen. (Flow Chemistry: Recent Developments in the Synthesis of Pharmaceutical Products; R. Porta et al. *Org. Process Res. Dev*. 20, 1, 2–25, 2016 (Permission from ACS https://pubs.acs.org/doi/abs/10.1021/acs.oprd.5b00325?src=recsys))

synthetic step involved a simultaneous quench of excess ICl and saponification of ester. Crude mixture exiting from oxidation reactor was combined with a water/methanol mixture containing sodium hydroxide and 2-mercaptoethanol.Then the mixture was heated at 90°C for 60 seconds in a 3.9 mL PFA reactor coil. After quenching with aqueous hydrochloric acid and extraction with hexanes, pure compound ibuprofen is obtained with 83% yield (Figure 1.8).

This process principle and procedure has also been discussed in Chapter 7.

1.1.8 Versatile, Small and Biologically Active Molecules with Diverse Functionality

Nowadays, the prominent theme of diversity-oriented synthesis (DOS) of structurally versatile molecular libraries in water medium, aims to generate small molecules with diverse functionally which possess skeletons found in natural products, drug-like molecules and materials.[37]

1.1.9 Phenolic Compounds

Phenolic compounds are a large and diverse group of molecules produced by plants as secondary metabolites.[38] The term "phenolic" or "polyphenol" can be defined chemically as a substance which possesses at least one aryl ring containing a minimum of one hydroxyl group. Phenolic compounds may be classified into simple phenols (*e.g.* phenol and cresol), phenolic acids (*e.g.* gallic, protocatechuic, vanillic and syringic acids), aldehyde forms of phenolic acids (*e.g.* vanillin and syringaldehyde), hydroxycinnamic acids (*e.g.* ferulic acid, caffeic acid and *p*-coumaric acid), hydroxyl cinnamyl alcohols (*e.g.* coniferyl, sinapyl, syringyl and *p*-coumaryl alcohols), phenylacetic acids, acetophenones, hydroxystyrenes, stilbenes, flavonoids, tannins and lignins.[39] Some representative examples of phenolic compounds are summarized in Figure 1.9. Phenolic compounds exhibit a broad range of biological activities, such as antioxidant, antimicrobial, anti-inflammatory, antiallergenic, anti-artherogenic, anti-thrombotic, anti-cancer and so on.[38,39]

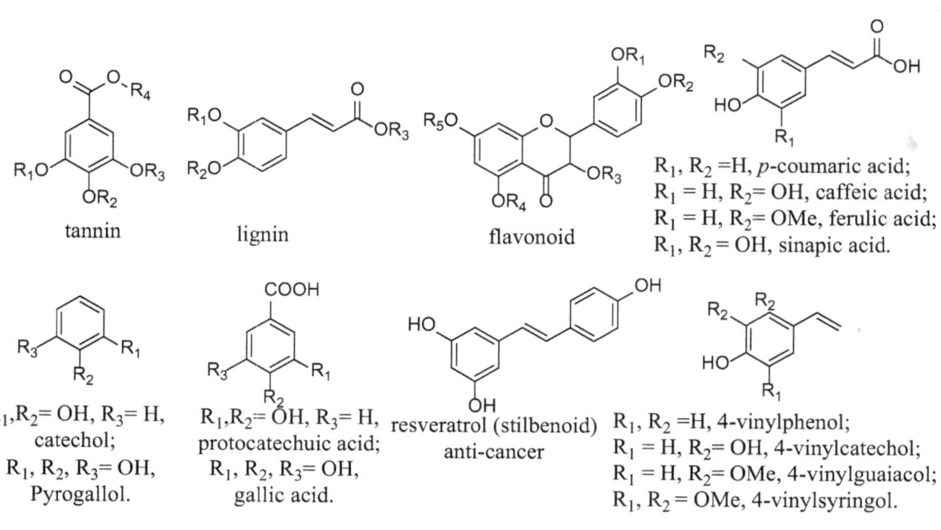

tannin

lignin

flavonoid

R_1, R_2 =H, *p*-coumaric acid;
R_1 = H, R_2= OH, caffeic acid;
R_1 = H, R_2= OMe, ferulic acid;
R_1, R_2 = OH, sinapic acid.

R_1,R_2= OH, R_3= H, catechol;
R_1, R_2, R_3= OH, Pyrogallol.

R_1,R_2= OH, R_3= H, protocatechuic acid;
R_1, R_2, R_3= OH, gallic acid.

resveratrol (stilbenoid) anti-cancer

R_1, R_2 =H, 4-vinylphenol;
R_1 = H, R_2= OH, 4-vinylcatechol;
R_1 = H, R_2= OMe, 4-vinylguaiacol;
R_1, R_2 = OMe, 4-vinylsyringol.

FIGURE 1.9 Some representative examples of phenolic compounds.

1.1.10 Heterocyclic Compounds

In the past decades, most of the scientists have been interested in the synthesis of heterocyclic unit-containing molecules due to their important medicinal and pharmaceutical applications and their use as precursors for the library of natural products and drug or drug-like molecules.[40] In organic chemistry, heterocycle units are present in nature more than 70% as promising bioactive and drug molecules which are essential for mankind by boosting the quality of life.[41] Among the various heterocycles, much attention is paid to nitrogen and oxygen-containing heterocycles due to their biological significances.[42] Heterocyclic compounds can be usefully classified based on their electronic structure. A variety of five- and six-membered heterocyclic compounds such as pyrazole, imidazole, isoxazole, oxazole, pyridazine, phthalazine (benzopyridazine), pyrimidine, pyrazine, uracil, triazole, tetrazole, pyran, benzopyran, triazine and phthalide are present largely in nature owing to the stability of five- and six-member ring systems.[43,44] Among the six-membered heterocycles, pyran or oxine with molecular formula C_5H_6O is a significant class of oxygen-containing heterocyclic compounds. Pyran is non-aromatic due to the presence of five carbon atoms and one oxygen atom along with two double bonds. 2*H*-pyrans and 4*H*-pyrans are the two isomers of pyrans which vary by the position of the double bonds. 2*H*-pyran is named like that due to the presence of the saturated carbon at position 2 whereas, in 4*H*-pyran, the saturated carbon is present at position 4. The classification of pyran-based heterocyclic molecules is determined by the presence of either 2*H*- or 4*H*-pyran scaffold (Figure 1.10).

Pyran derivatives represent the key moiety of various naturally occurring and synthetic products such as benzopyrans, coumarins, flavonoids, xanthones, chromone, sugars and naphthoquinones displaying diverse medicinal and biological activities.[41,44]

Benzopyrans, commonly known as chromenes, are the polycyclic oxygenated and elite class of heterocyclic compounds resulting from the fusion of a benzene and pyran moiety, and are extensively present in edible fruits and vegetables as natural alkaloids, flavonoids, tocopherols and anthocyanins.[45] The chromene skeletons are mainly observed as either 2*H*-chromene or 4*H*-chromene. 2*H*-chromenes and 4*H*-chromenes are well-known key pharmacophores for many naturally occurring as well as synthetic drugs or drug-like molecules displayed with a wide spectrum of potent biological and medicinal activities including antioxidant, antimicrobial, antibacterial, antifungal, antiviral, anticoagulant, anti-vascular, anti-tumor, anti-proliferative, anti-inflammatory, anti-cancer, anti-rheumatic, antidepressant, anti-tuberculosis, anti-hepatitic, anti-parasitic, antibiotic, anti-infertility and many others.[45] The glimpse of structures of some biologically active drugs or drug-like 2*H*- or 4*H*-chromene derivatives are summarized in Figure 1.11.

In many cases, chromene frameworks act as drugs because of its key feature of lipophilic nature which easily helps to cross the cell membrane. Over the last few years, many pre-clinical and clinical trials are done on the compounds with 2*H*-chromene and 4*H*-chromene moiety.[42] According to the SAR studies, the substitution with specific groups on the chromene nucleus enhances the ability of molecules in preventing the disorders. This process principal and procedure has been discussed in detail in Chapter 8.

1.2 Green Catalysts

Among the "Twelve Principles of Green Chemistry", safer syntheses, use of renewable feedstock as a catalyst, utilizing safer solvents (use of green solvent or solvents elimination), and design for energy efficiency can be regarded as the most important ones for the synthetic chemists.[46] These must be designed aiming not only their immediate applications but also incorporating their footprints for future generations. In this regard, the investigations of sustainable, efficient and economically feasible alternative catalyst systems for chemical transformations using natural, renewable, biodegradable and bio-based feedstock as ecocatalyst are extensively acknowledged owing to their easy access, ease of biodegradability, cheap price, low toxicity and non-flammable properties.[47]

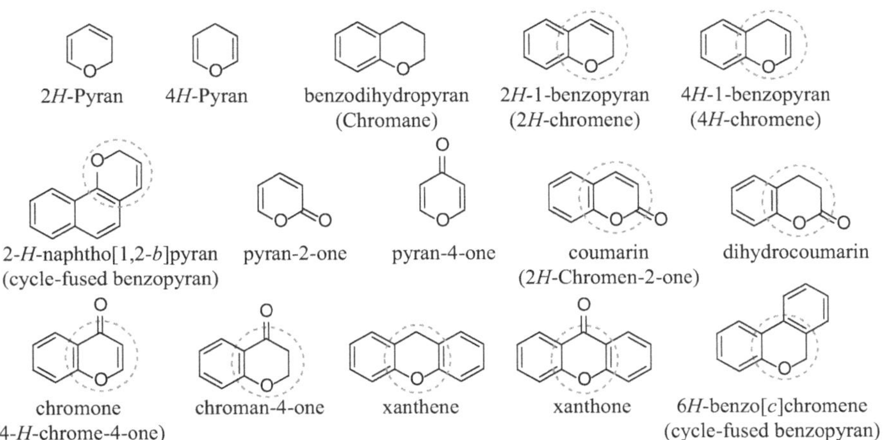

FIGURE 1.10 Heterocycles with the core structure of pyran and benzopyran unit.

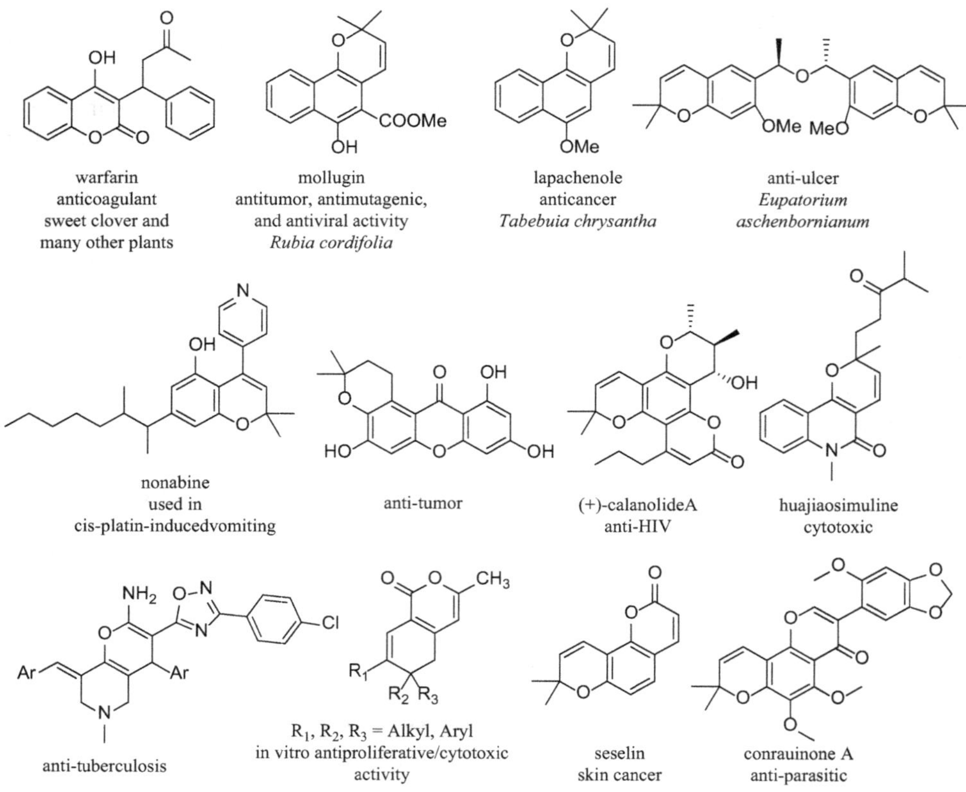

FIGURE 1.11 Representative examples of some natural and synthetic potent drugs or drug-like 2*H*- or 4*H*-chromene derivatives.

1.2.1 Lipase, Esterase and Yeast as Biocatalysts

Biocatalysts involve the use of catalysts to catalyze and speed up chemical reactions in biocatalytic processes. The system is considered as a biotransformation process using catalysts such as lipase and esterase enzymes, yeasts, natural catalysts, etc. It can catalyze novel small molecule transformations that may be difficult or impossible using classical synthetic organic chemistry. Utilizing natural or modified enzymes to perform organic synthesis is termed as chemoenzymatic synthesis.

1.2.1.1 Lipase

Lipases are a class of enzymes which catalyze[48] the hydrolysis of long-chain triglycerides and biotransformations are now accepted as a common methodology for the preparation of organic compounds, natural products and chiral pharmaceuticals. It will react in mild reaction conditions and with a high degree of selectivity. The uses of lipases are as follows:

a. Lipases or triacylglycerol acyl hydrolases are the most frequently used biocatalysts in organic synthesis. Lipases are able to discriminate between enantiotopic groups and between the enantiomers of a racemate. Use of such biocatalysts helps to perform enantioselective hydrolytic reactions and the formation of ester bond. They efficiently catalyze various other reactions such as amidation, aminolysis, thiotransesterification and oximolysis.
b. Lipases can be used in asymmetric synthesis like the kinetic resolution of racemic alcohols, acids, esters or amines and in the desymmetrization of prochiral compounds.
c. Lipases find use in the stereoselective biotransformations to carry out the kinetic resolution (KR) of mixtures and enzymatic desymmetrization (EED) of prochiral compounds.
d. Lipases in organic solvent can be used for the enantioselective preparation of intermediates for the synthesis of chiral drugs (single isomer) by aminolysis, transesterification or enzymatic hydrolysis.

Lipases are distributed among higher animals, microorganisms and plants where they fulfill the key role in the turnover of lipids. Of the plentiful enzymes used, over a half is from fungi and yeast and over a one-third is from bacteria with the remainder divided between animal (8%) and plant (4%) sources.[49] Although pancreatic lipases have been traditionally used for various purposes, it is now well established that microbial lipases are preferred for commercial applications due to their multifold properties, easy extraction procedures and unlimited supply. Microbes are preferred to plants and animals as sources of enzymes because of the following reasons:

a. They are generally cheaper to produce.
b. The raw materials used for enzyme production are easily available and its quality and composition can be easily identified and controlled.

c. Microbial lipases are produced by methods which can be scaled up easily to meet the current market demand.
d. The level of contaminants is much less in microbial enzymes compared to that of plant and animal tissues.

Among the various lipase-producing microorganisms (bacteria, fungi, yeasts and actinomyces) *Candida*, *Pseudomonas*, *Rhizomucor* and *Rhizopus* species stand out nowadays as sources of most commercially available enzyme preparations.[50]

Lipases belong to the class of serine hydrolases which is composed of Ser-Asp-His and in some lipases Glu is present instead of Asp. Another pentapeptide consensus sequence found around the active site of serine is Gly-X_1-Ser-X_2-Gly (X_1 = histidine, X_2 = glutamic or aspartic acid).[51] In most lipases, a part of the enzyme molecule covers the active site with a short α-helix called "lid" or "flap". The side of the α-helical lid facing the catalytic site, as well as the protein chains surrounding the catalytic site, is composed mainly of hydrophobic side chains.[51] The phenomenon of interfacial activation is often associated with the reorientation of this α-helical lid structure in the vicinity of the active site making the active site accessible which otherwise remains buried inside.[51]

The mechanism for hydrolysis or esterification is same for both lipases and it comprises of four steps:[52]

Step 1: Substrate is bound to the active serine resulting in an intermediate which is stabilized by the histidine and aspartic acid residues of the protein molecule.

Step 2: Release of alcohol and formation of an acyl enzyme complex.

Step 3: Attack of nucleophile (water in hydrolysis, alcohol or ester in esterification and transesterification).

Step 4: Yield of product and free enzyme after resolution.

Most lipases act at a specific position on the glycerol backbone of a lipid substrate and converts triglyceride substrates to monoglycerides and free fatty acids as shown below in Figure 1.12.

Nag et al.[53] synthesized biosurfactants used as food emulsifiers in food industries by treatment of different types of carbohydrates with fatty acids in the presence of *Candida rugosa* lipase at normal room temperature (Figure 1.13).

Nag et al.[54] reported resolution of DL-menthol by immobilized lipases and the resolution depended on different factors such as polarity and nonpolarity of solvents, different temperatures and specific activity of different lipases.

Nag et al.[55] produced different terpene esters by reaction of terpenes with different fatty acids in the presence of *Candia anatarctica* lipase. The reaction conversion of the reaction was monitored by proton NMR (Figure 1.14).

Aliquot: 100 µL each time/analyzed by ^{1}H NMR Spectroscopy (400 MHz $CDCl_3$)

Quantification of ester content in the reaction mixture % Ester conversion $= [E/(A + E)]/x\,100$

E = Integration value of methylene protons of ester
A = Integration value of methylene protons of alcohol

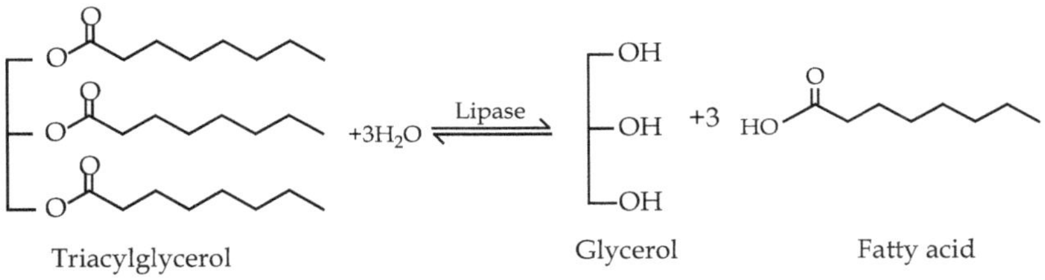

FIGURE 1.12 The catalytic action of lipase.

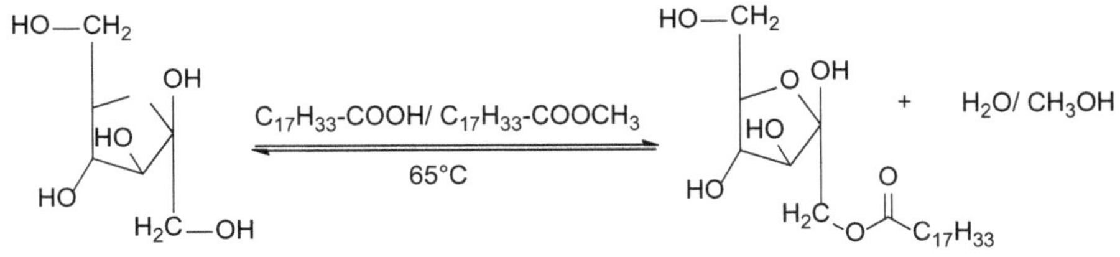

FIGURE 1.13 Lipase-catalyzed acylation to produce biosurfactants.

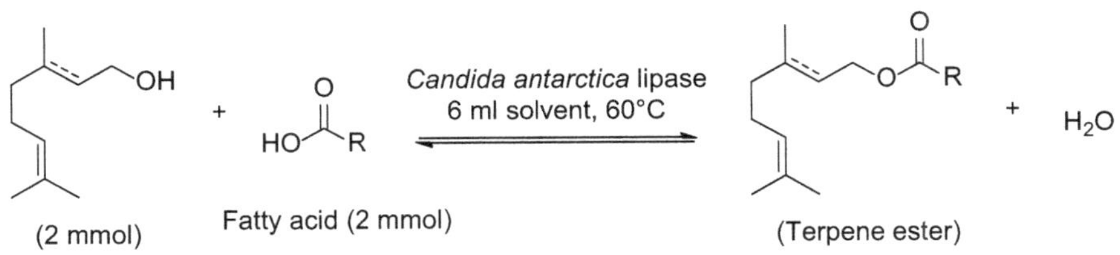

FIGURE 1.14 Lipase-catalyzed conversion of terpene esters.

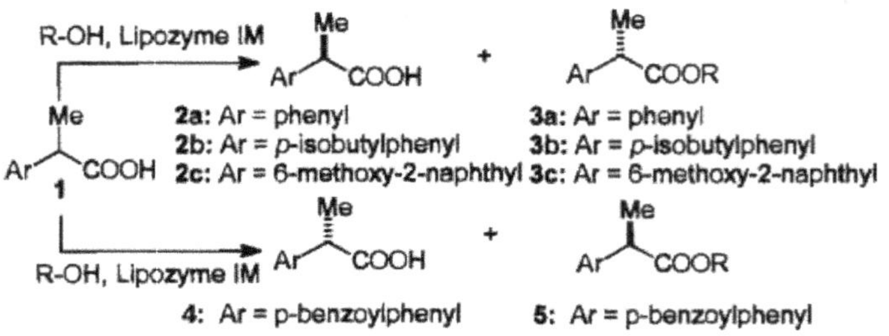

R-OH, Lipozyme IM
Me
Ar COOH
+
Me
Ar COOR
Me
Ar COOH
1
2a: Ar = phenyl
2b: Ar = p-isobutylphenyl
2c: Ar = 6-methoxy-2-naphthyl
3a: Ar = phenyl
3b: Ar = p-isobutylphenyl
3c: Ar = 6-methoxy-2-naphthyl
R-OH, Lipozyme IM
Me
Ar COOH
+
Me
Ar COOR
4: Ar = p-benzoylphenyl
5: Ar = p-benzoylphenyl

FIGURE 1.15 Enantioselective esterification of 2-arylpropionic acids catalyzed by immobilized *Rhizomucor miehei* lipase.

Recently there has been tremendous demand for chiral drug substances to focus on single stereoisomer, instead of racemic mixtures. For this reason, kinetic resolution processes or enzymatic desymmetrization of prochiral compounds have special relevance in pharmaceutical industries.

The 2-arylpropionic acids, bearing a single stereogenic tertiary center, constitute an important class of non-steroidal anti-inflammatory agents that relieve inflammation by inhibiting cyclooxygenase, thereby regulating the arachidonic acid cascade. Since the pharmacological activity of the (S)-isomer of 2-arylpropionic acid is reported[56] to be stronger than that of the (R)-isomer, development of an efficient, enantioselective synthetic route to the (S)-isomer has received considerable attention (Figure 1.15). The lipase selectivity, however, was not sufficient for preparative scale resolution, and the water content strongly affected the reaction rate. Trinidad López-Belmonte et al.[56] reported the *Rhizopus miehei* lipase-catalyzed resolution of the racemic mixture but selectivity was extremely low.

Park et al.[57] suggested that enantioselectivity in lipase-catalyzed resolution of racemic ketoprofen was mainly dependent on the sources of lipase, alcohol moiety, organic solvent and water content. Ethanol was used as the alkyl donor and the optimum water content required for highly efficient enzymatic resolution was determined to be 0.1–0.15% (v/v), which was maintained using salt hydrates such as $Na_2SO_4 \cdot 10H_2O$. (*S*)-Ketoprofen could be obtained with high enantioselectivity (E = 15) in *n*-hexane supplemented with ethylene dichloride (20% (v/v)) using commercially available *Candida antarctica* lipase.

Basak and Nag et al.[58] anticipated the following strategies to synthesize anti-inflammatory pure chiral drugs. The methyl ester of ibuprofen, naproxen or flurbiprofen was reduced with sodium borohydride in methanol. The mixture was allowed to stir for 4 hr after which the reaction volume was reduced to 5 mL. Water was added and the mixture was extracted with ethyl acetate. The combined organic layers were dried over anhydrous sodium sulphate and then evaporated. The product was purified by silica gel chromatography using hexane: ethyl acetate = 3:1.The alcohol was then acetylated with acetic anhydride, triethylamine and a catalytic amount of DMAP using dichloromethane as solvent. The mixture was allowed to stir for 1 hr at room temperature and then washed with water and extracted with dichloromethane. The organic layer was dried over Na_2SO_4 and then evaporated to leave a white solid from which the product was isolated by silica gal chromatography using hexane: ethyl acetate = 7:1. The acetyl derivative was hydrolyzed with the help of the enzyme (PPL) in phosphate buffer (pH 8.0) and acetone. After 50% conversion (estimated by TLC), the mixture was extracted with ethyl acetate, dried over Na_2SO_4 and solvent was evaporated. The residue, upon chromatography on silica gel, furnished the product which was isolated from hexane: ethyl acetate (1:1). The absolute configuration was determined by comparison with the literature values for S-isomer. The oxidation of the (+) alcohols were carried out with PDC using DMF as solvent. After 72 hr, the mixture was partitioned between ethyl acetate and aqueous $NaHCO_3$. The aqueous layer was adjusted to pH 2 and then re-extracted with ethyl acetate. The organic layer was dried with Na_2SO_4, filtered and evaporated to leave the acid as a white solid (yield ~50%) having identical ^1H NMR with the authentic racemic sample (Figure 1.16).

Kato et al.[59] reported an efficient synthetic procedure for obtaining optically active ketoprofen using the *Mucor javanicus* lipase, one of nine commercially available hydrolytic enzymes, showed good enantioselectivity (E = 50) for racemic ketoprofen trifluoroethyl ester in phosphate buffer (pH 7.0) containing 30% acetone. Lipase immobilized on Toyonite 200-A showed the best selectivity (E = 55) and reactivity. Moreover, the lipase could be recycled at least five times (Figure 1.17).

(-) Paroxetine hydrochloride is a selective serotonin (5-HT) reuptake inhibitor used as an antidepressant. (-) Paroxetine alcohol intermediate can also be prepared by the enzymatic hydrolysis of the corresponding ester derivative or the enzymatic acylation of the primary alcohol in the reaction by *Candida antarctica* lipases CAL-A and CAL-B by Fernandez et al.[48] Both enzymes gave very good results in terms of yield and enatioselectivities. In the reaction process, the two *Candida antarctica* lipases showed opposite stereochemical preference. CAL-A catalyzes the acylation of the (3S, 4R) alcohol, whereas CAL-B prefers the (3R, 4S) enantiomer. In the last case, the remaining alcohol posses the correct absolute correct for the synthesis of (-) Paroxetine (Figure 1.18).

Citalopram is a very selective inhibitor of serotonin and an efficient human antidepressant. It can be obtained by the resolution of the corresponding cyanodiol with a primary and a tertiary hydroxyl group. Again, to get optically pure citalopram, on testing with several lipases CAL-B, vinyl acetate and acetonitrile gave the best results (E = 70) by Fern´andez-Solares et al.[60] *Candida antarctica* lipase B (CAL-B) catalyzes the enzymatic acetylation of the primary benzylic alcohol with high enantioselectivity at the quaternary stereogenic center. The enzymatic enantioselective hydrolysis of the 3-acetyloxymethyl derivative catalyzed by CAL-B is also possible (Figure 1.19).

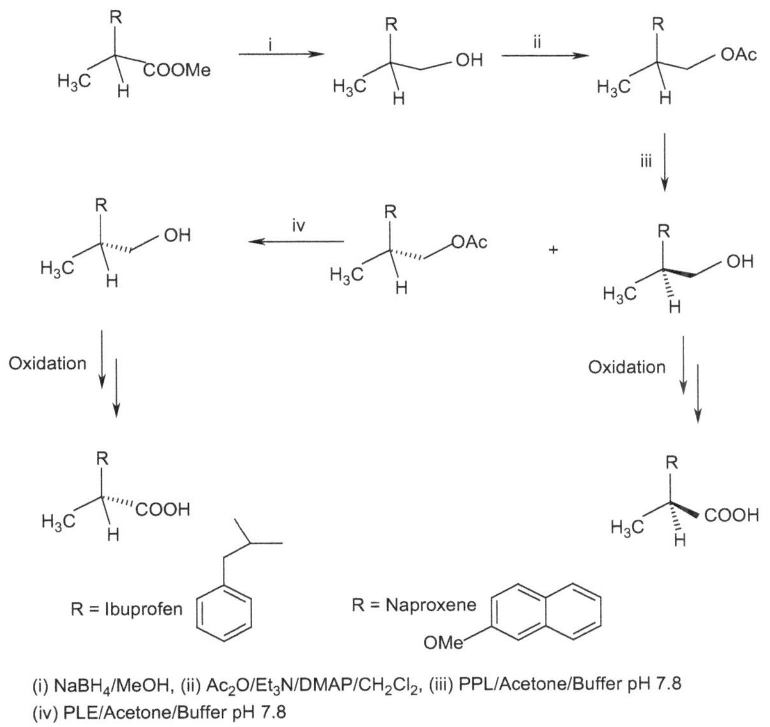

(i) $NaBH_4$/MeOH, (ii) Ac_2O/Et_3N/DMAP/CH_2Cl_2, (iii) PPL/Acetone/Buffer pH 7.8
(iv) PLE/Acetone/Buffer pH 7.8

FIGURE 1.16 Lipase-catalyzed synthesis of anti-inflammatory drugs.

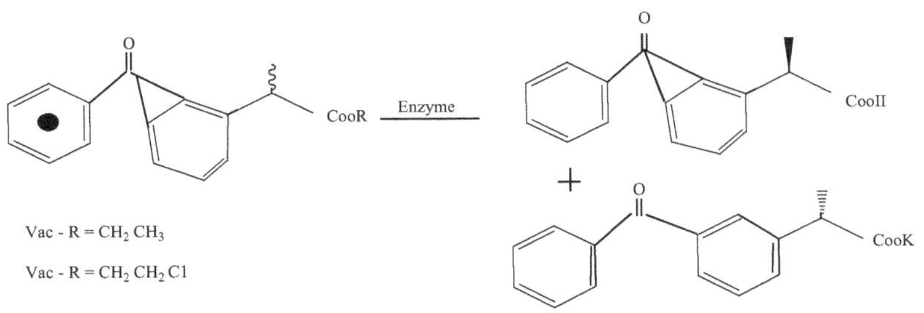

FIGURE 1.17 Enzymatic-catalyzed enantioselective hydrolysis of ketoprofen monochloroethyl ester and trifluroethyl ester.

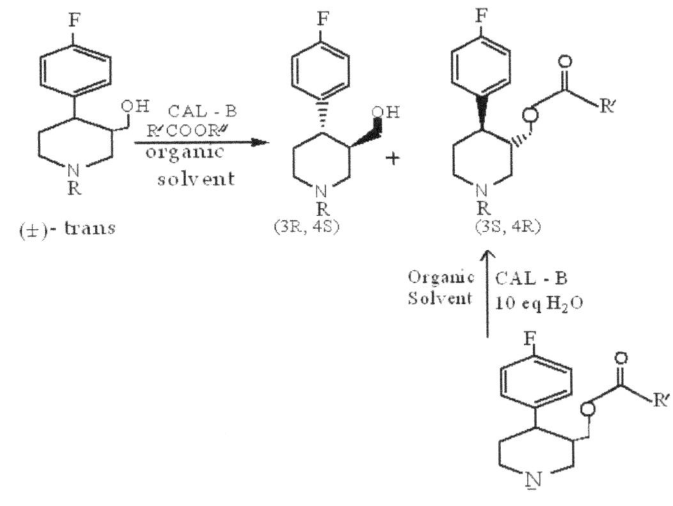

FIGURE 1.18 Lipase-catalyzed preparation of (-) Paroxetine precursor.

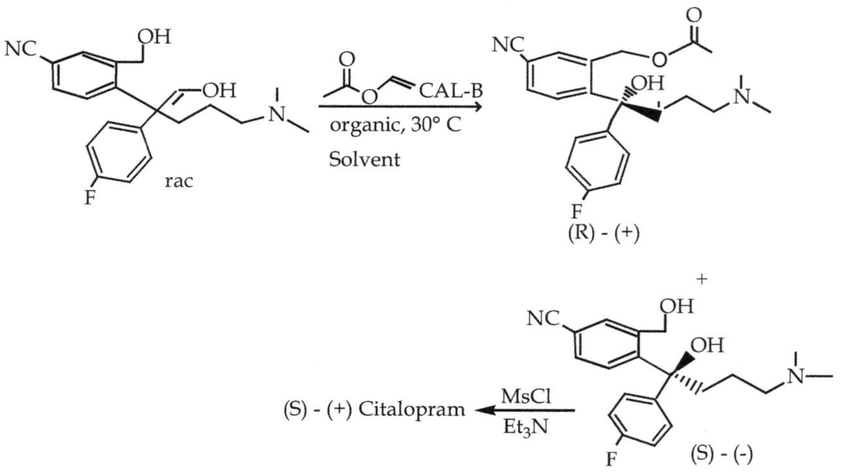

FIGURE 1.19 Chemoenzymatic synthesis of (s) – (+) citalopram.

1.2.1.2 Esterase

An esterase is a hydrolase enzyme that splits esters into an acid and an alcohol in a chemical reaction with water called hydrolysis. It differs from lipase that esterases[61] preferentially break ester bonds of shorter-chain fatty acids while lipases can break long-chain fatty acids which are typically insoluble or at least poorly soluble in water.

On the basis of sources generally varied nomenclature are reported in the literature, (*e.g.* carboxylesterase, cholinesterase, acetyl xylan esterase, aryl esterase, phosphotriesterase, phenolic esterase, pig liver esterase (PLE), acetylcholine esterase, cholesterol esterase, ferulic acid esterase, tannin esterase, etc.). Animal esterase, PLE is generally isolated as acetone powder from fresh pig liver by Nag et al.[61] The specific activity of esterase is found out by spectrophotometrically using p-nitrophenyl acetate. Microbial esterases can be used in the form of cells from both bacteria and fungi to catalyze the reactions. But the problem is to produce the microbial cells in a suitable medium. Sometimes controlling the reaction becomes more complex; however, the selectivity is more than the isolated enzyme.

Thromboxane A_2 is an exceptionally potent proaggregatory and vasoconstrictor substance produced by the metabolism of arachidonic acid in blood platelets and other tissues.[61] Together with the potent antiaggregatory and vasodilator, it is thought to play a role in the maintenance of vascular homeostasis and to contribute to the pathogenesis of a variety of vascular disorders. Compound 1 is an intermediate for thromboxane synthetase inhibitor (compound 2).

An attempt to synthesize 1 as given in (Figure 1.20) is difficult as the conversion of 3 to 4 needs a selective transformation of saturated ester over α, ß unsaturated or aromatic esters.

Ester groups attached with α-unsaturated group or aromatic ring are much less prone to be hydrolyzed by normal hydrolytic conditions like NaOH/CH $_3$OH or LiOH/THF due to the formation of electronically unstable tetrahedral transition state. The selective transformation of saturated ester was not useful under these conditions due to the formation of multiple products. When compounds having both saturated and unsaturated aromatic ester functionalities (acetone/buffer pH 7.8; PLE), Nag et al.[61] found that the saturated ester functionality to be smoothly hydrolyzed leaving behind unsaturated one, the reaction time ranging from 4 hr to 20 hr with the yield of saturated acid 95% or higher (Figure 1.21). The degree of hydrolysis was measured by ^{1}H NMR. The presence of unsaturated acid was never observed in NMR. This was a new finding and the selectivity achieved by PLE.

Paracetamol is one of the safest analgesic drugs in the world. But the following difficulties are countered during preparation of paracetamol from p-aminophenol which is an unstable compound and undergoes oxidation to p-quinone and related compounds, thus the isolation in the pure form may be tricky. To overcome the above difficulties, Nag et al.[62] have selected p-nitrophenol rather p-aminophenol to prepare N,O-diacetyl-p-aminophenol which was resistant to oxidation. Selectively, hydrolysis (Figure 1.22) of O-acetyl in preference over N-acetyl group was achieved successfully using PLE by Nag et al.[62]

Asakawa et al.[63] described the preparation of a new chiral building block containing a benzylic quaternary stereogenic center via the highly enantioselective PLE-mediated hydrolysis of dimethyl 2-(2-chloro-5-methoxyphenyl)-2-methylmalonate, as well as the absolute configuration of the new chiral building block, which has been elucidated through the formal total synthesis of (_)-physostigmine (Figure 1.23).

1.2.1.3 Yeast

Yeasts have been used as biocatalysts, mainly the dehydrogenases. Particularly, dehydrogenases are classified as oxidoreductases and have been widely used for the reduction of carbonyl groups of aldehyde or ketones and carbon–carbon double bonds. These enzymes catalyze the reactions involving direct transfer of hydride anion either to oxidized coenzyme or from reduced coenzyme to a substrate.[64] The coenzymes in the oxidized forms are nicotinamide adenine dinucleotide (NAD^+) and nicotinamide adenine dinucleotide phosphate ($NADP^+$). The reduced forms are represented as NADH and NADPH (Figure 1.24). Both nicotinamides are present in the living

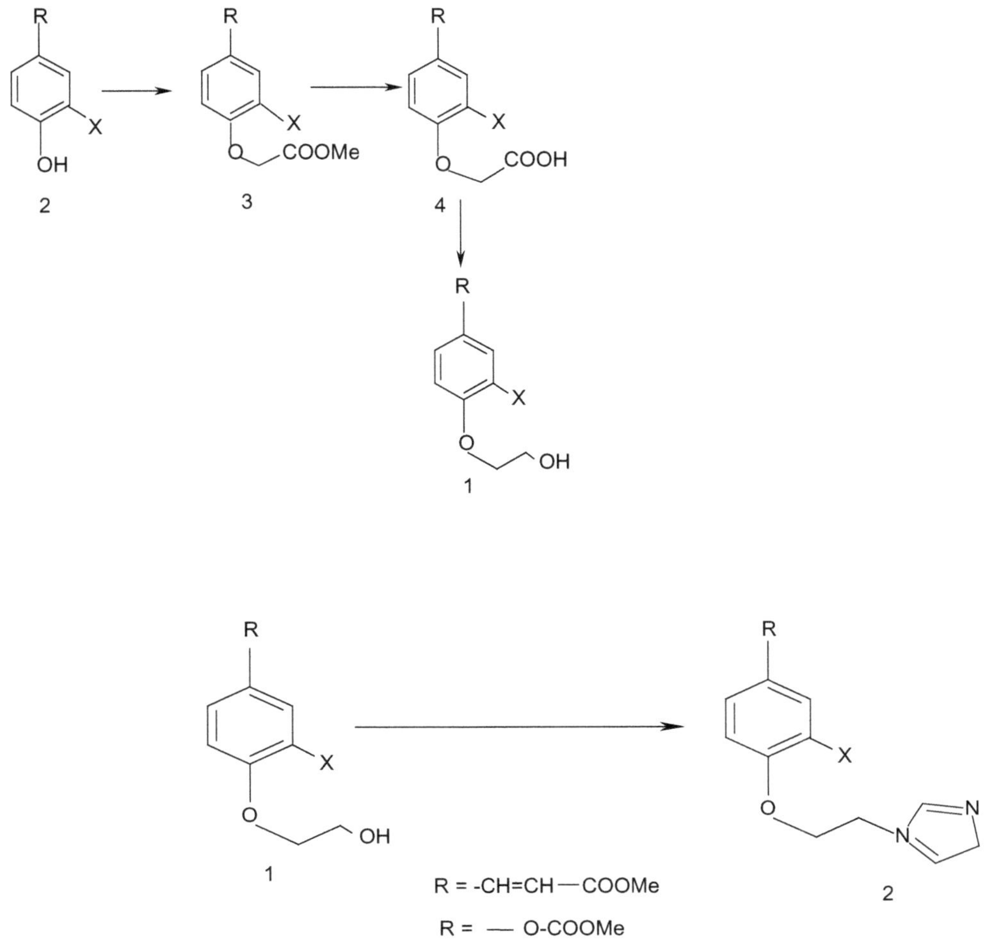

R = -CH=CH—COOMe

R = — O-COOMe

FIGURE 1.20 Esterase synthesis of thromboxane synthetase inhibitor.

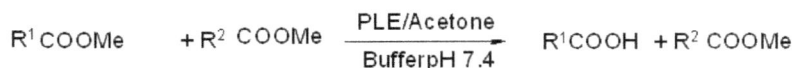

$R^1COOMe + R^2COOMe \xrightarrow[\text{Buffer pH 7.4}]{\text{PLE/Acetone}} R^1COOH + R^2COOMe$

FIGURE 1.21 Selective transformation of saturated esters by esterase.

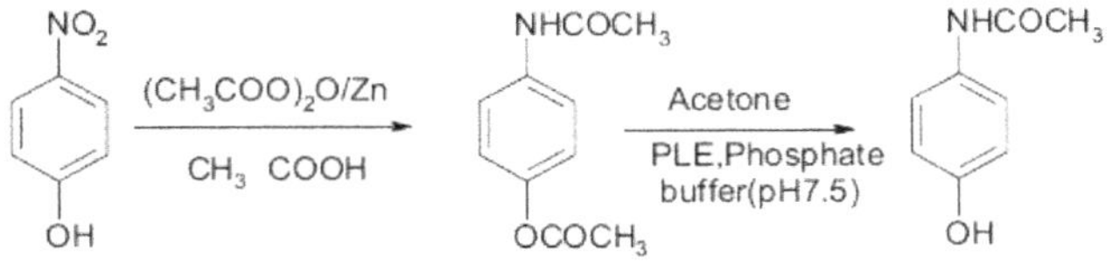

FIGURE 1.22 Esterase synthesis of paracetamol.

cells and representations of their structures[64] are provided in Figure 1.25.

Generally, the dehydrogenases may stereoselectively catalyze a reduction of a non-natural pro-chiral ketone, due to their conformational flexibility, producing alcohol with prevalence of one enantiomer, when both enantiomers are formed. For example, Figure 1.26 shows a reduction of ketone **3** mediated by an alcohol dehydrogenase isolated from *Thermoanaerobium brockii* giving a mixture of alcohols in 95% of (*R*)-**4** and 5% of (*S*)-4.[65]

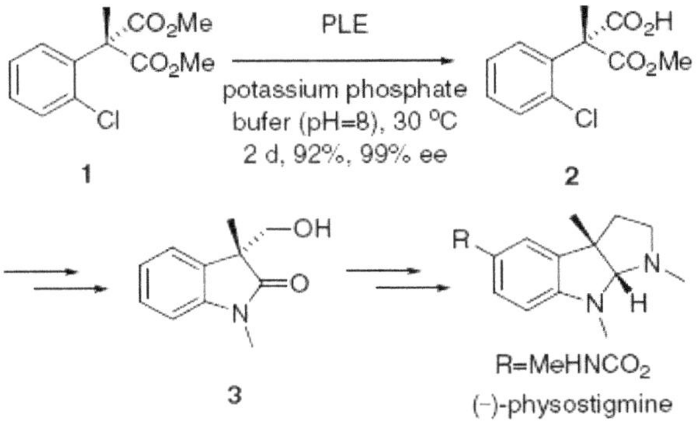

FIGURE 1.23 Enantioselective PLE-mediated hydrolysis of dimethyl 2-(2-chlorophenyl)-2-methylmalonate 1 and a formal total synthesis of (_)-physostigmine.

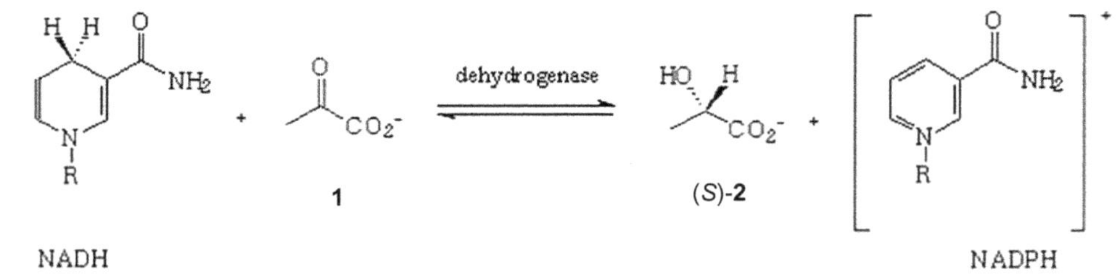

FIGURE 1.24 Conversion of NADH and NADPH.

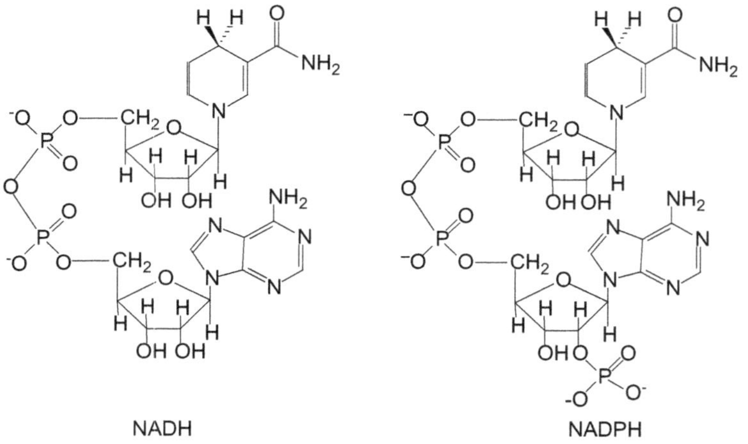

FIGURE 1.25 NADH and NADPH structures.

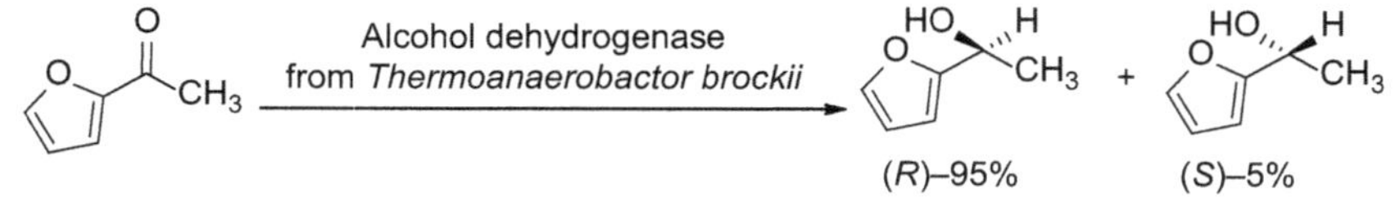

FIGURE 1.26 Reduction of ketone compound by *Thermoanaerobium brockii*.

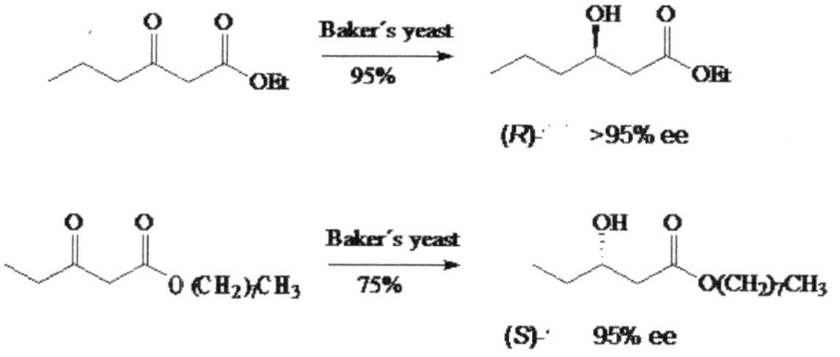

FIGURE 1.27 Reduction of β-ketoesters by baker's yeast.

Reduction of β-ketoesters mediated by baker's yeast shows ethyl 3-oxo-hexanoate reduction where the bulk of acidic moiety was enlarged in it would give with *R* instead of *S* configuration (Figure 1.27). On the other hand, if the bulk of alcohol moiety of the alkyl β-ketoester is enlarged as in n-octyl 2-oxo-pentanoate, the baker's yeast reduction product will have *S* configuration.[66] Therefore, the desired configuration of the product is obtained by switching the relative size of substituent. This divergent behavior is not due to an alternative fit of the substrates in a single enzyme, but rather due to the presence of different dehydrogenases, possessing opposite stereochemical preferences, which compete for the substrate.[67]

1.2.2 Plant as Biocatalyst

Nowadays, organic syntheses using plants as biocatalyst are the challenging goals of green catalytic chemistry. There are many reports in literature for a variety of bio-organic transformations such as reduction of carbonyl compounds, oxidation of alcohols, Knoevenagel condensation, Friedel–Crafts reaction, MCRs, coupling reactions and many more by using edible plants, plant tubers (*e.g.* wild carrot, beet and radish), extract of plant leaves (*e.g.* banana and maize), seeds (*e.g.* soaked *Phaseolus aureus L.*), natural oil (corn oil), fruit juices (*e.g.* lemon, apple, pineapple, tamarind, coconut water and star fruit) and aqueous extract of plant parts (*e.g. Acacia concinna* pods, *Sapindustrifolistus L.* and banana peels) which play roles as natural green catalysts.[68]

Nag et al.[69] have shown that cucumber juice (*Cucumis sativus L.*) acted as a green and highly efficient alternative for versatile organic synthesis (Figure 1.28).

Nag et al. reported[70] that reduction and decarboxylation of aromatic aldehydes and aromatic acids, which are single step reactions, both have been performed by the juice of coconut (*Cocos nucifera L.*) abbreviated as ACC and cucumber juice abbreviated as CSJ (*Cucumus sativus L.*). From this idea, they decided to apply[71] methodology in MCRs for the synthesis of substituted diols such as biscoumarins (**3a-j**), bispyrones (**3k-o**) and bis(pyrazolyl)methanes (**3p-t**) using one molecule of aromatic aldehyde and two molecules of 4-hydroxycoumarin (**2a**), 4-hydroxy-6-methyl-2-oxo-pyran (**2b**) and 3-methyl-1-phenyl-5-pyrazolone (**2c**) (Figure 1.30) and substituted 4*H*-pyrans like 2-amino-5-oxo-4-aryl-4*H*, 5*H*-pyrano[3,2-*c*] chromene-3-carbonitrile (**5a-l**), 2-amino-7-methyl-5-oxo-4-aryl-4*H*, 5*H*-pyrano[4,3-*b*]pyran-3-carbonitrile (**5m-q**), 6-amino-3-methyl-1-phenyl-4-aryl-1, 4-dihydropyrano[2,3-*c*] pyrazole-5-carbonitrile (**5r-v**) using one molecule of aromatic aldehyde, one molecule of active methylene nitrile and one molecule of 4-hydroxycoumarin (**2a**), 4-hydroxy-6-methyl-2-oxo-pyran (**2b**) and 3-methyl-1-phenyl-5-pyrazolone (**2c**) respectively (Figure 1.29) in the presence of ACC.

In order to extend the scope of this present protocol, Nag et al.[71] introduced active methylene nitriles to form different 4*H*-pyrans (Table 1.1) *via* the one-pot reaction among active methylene nitriles, different C–H enolizable carbonyl compounds and various aromatic aldehydes using ACC juice (Figure 1.30).

As a synthetic procedure, active methylene nitrile (**4**, 1 mmol) was added to the mixture of C–H enolizable carbonyl compounds (**2a/2b/2c**, 1 mmol) and aromatic aldehyde (**2**, 1 mmol) in 5 mL ACC. The reaction mixture was stirred at 40°C maximum for 24 hr under an inert atmosphere. All the reactions proceeded smoothly and gave higher yields in case of malononitrile (**4a**) compared to ethyl-2-cyanoacetate (**4b**) (Table 1.1). All the compounds were purified by only a simple filtration method and characterized by ^{1}H, ^{13}C NMR and mass spectroscopy analysis.

In conclusion, a simple, eco-friendly and novel procedure was demonstrated for the synthesis of bisenols and 2-amino-4*H*-pyrans using natural feedstock coconut juice. The major importance of using ACC juice is higher yields, no work-up and no column chromatography.

1.2.3 Waste Feedstock as Green Catalyst

1.2.3.1 Biomass Waste as Catalyst (Homogenous or Heterogeneous)

Recently, on complying with the greener perspectives, waste feedstocks as natural catalysts have also been used in either homogenous or heterogeneous phase for versatile organic syntheses. Several organic transformations such as coupling reactions (Suzuki–Miyaura cross-coupling, Sonogashira cross-coupling), Dakin reaction, MCRs and oxidative dimerization–cyclization reaction have been carried out using natural waste feedstocks, for example, water extract of ash of plant parts (water extract of rice straw ash, i.e. WERSA, water extract of banana peel ash, i.e. WEB), biomass-derived compounds (lignosulfonic acid as an organic waste generated from

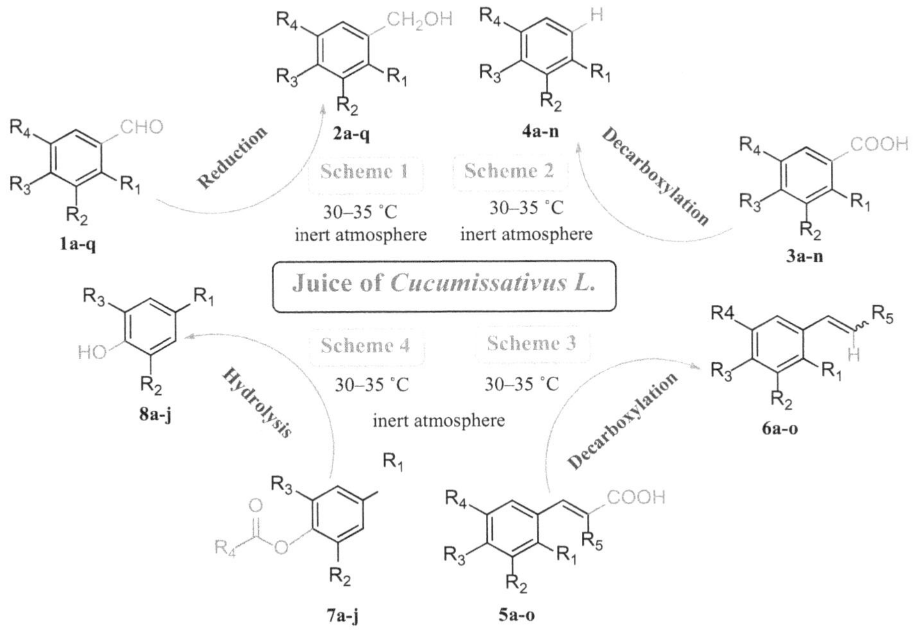

FIGURE 1.28 Reduction, decarboxylation and hydrolysis in the presence of cucumber juice.

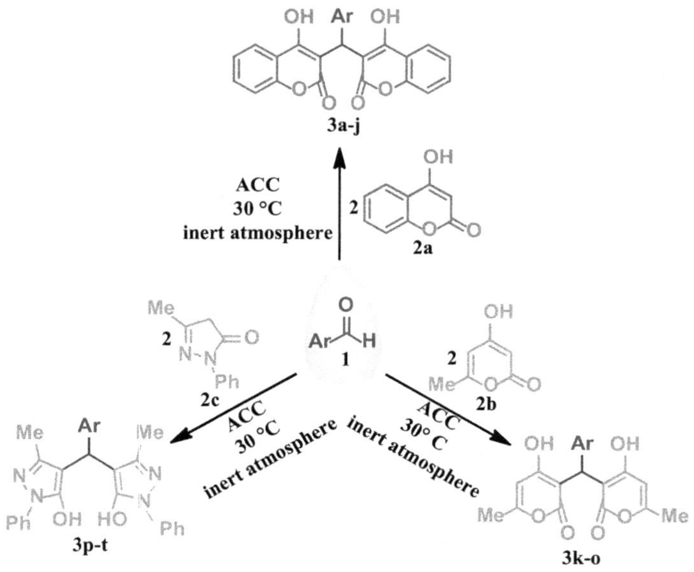

FIGURE 1.29 Synthesis of substituted diols derivatives in the presence of ACC.

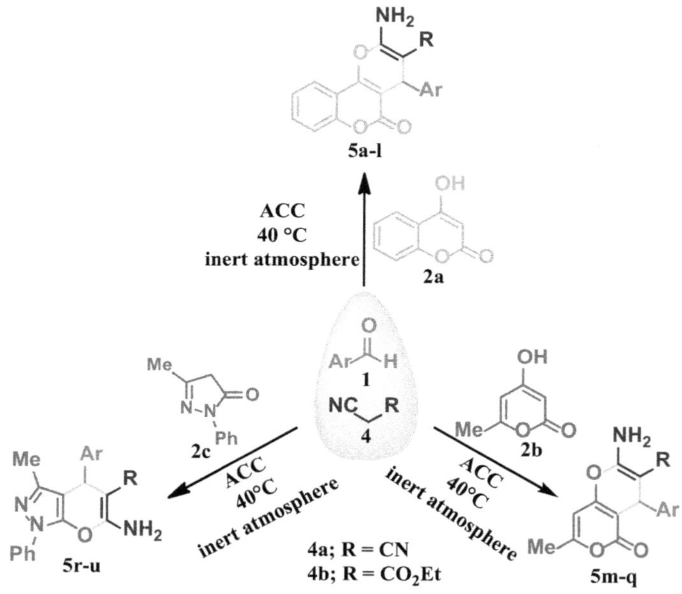

FIGURE 1.30 Synthesis of substituted 2-amino-4H-pyran derivatives in the presence of ACC.

the pulp and papermaking industry) and crude enzyme from waste materials (crude peroxidase from onion solid waste).[72]

1.2.4 Heterogeneous Catalysts from Waste Materials

In the current industrial scenario, heterogeneous catalysis plays a significant role in versatile organic syntheses over homogeneous catalysts conforming to the goals of green and sustainable chemistry. The application of heterogeneous catalysts in liquid phase provides several advantages over homogeneous ones, including operational simplicity, enhanced stability, non-volatility, efficiency, non-corrosiveness, economical, environmentally friendliness, easy recyclability and recoverability.[73] Many organic reactions such as multicomponent reactions, coupling reactions, aldol condensation reaction and click reactions are performed by using natural heterogeneous catalysts such as clay, zeolite, dolomite, apatite, kaolin, natural phosphates, white marble, silica from agricultural wastes and animal waste parts (bone, bovine tendons, bovine serum albumin, hog pancreas, chitosan and calcined eggshell power).[74] This process has been discussed in detail in Chapter 10.

In addition, the heterogenization of homogeneous catalysts has gained much attention in recent years. The fabrication of natural as well as waste feedstock (biomass) to solid heterogeneous acid catalysts by functionalization of solid surface (*e.g.* carbon, silicates, alumina, cellulose and polymer) with strong liquid acids (*e.g.* chlorosulfonic acid, sulfuric acid and phosphoric acid) shows a good response in future for organic

TABLE 1.1

Substrate Scope for the Synthesis of Diverse 4*H*-pyran Derivatives[a], [b]

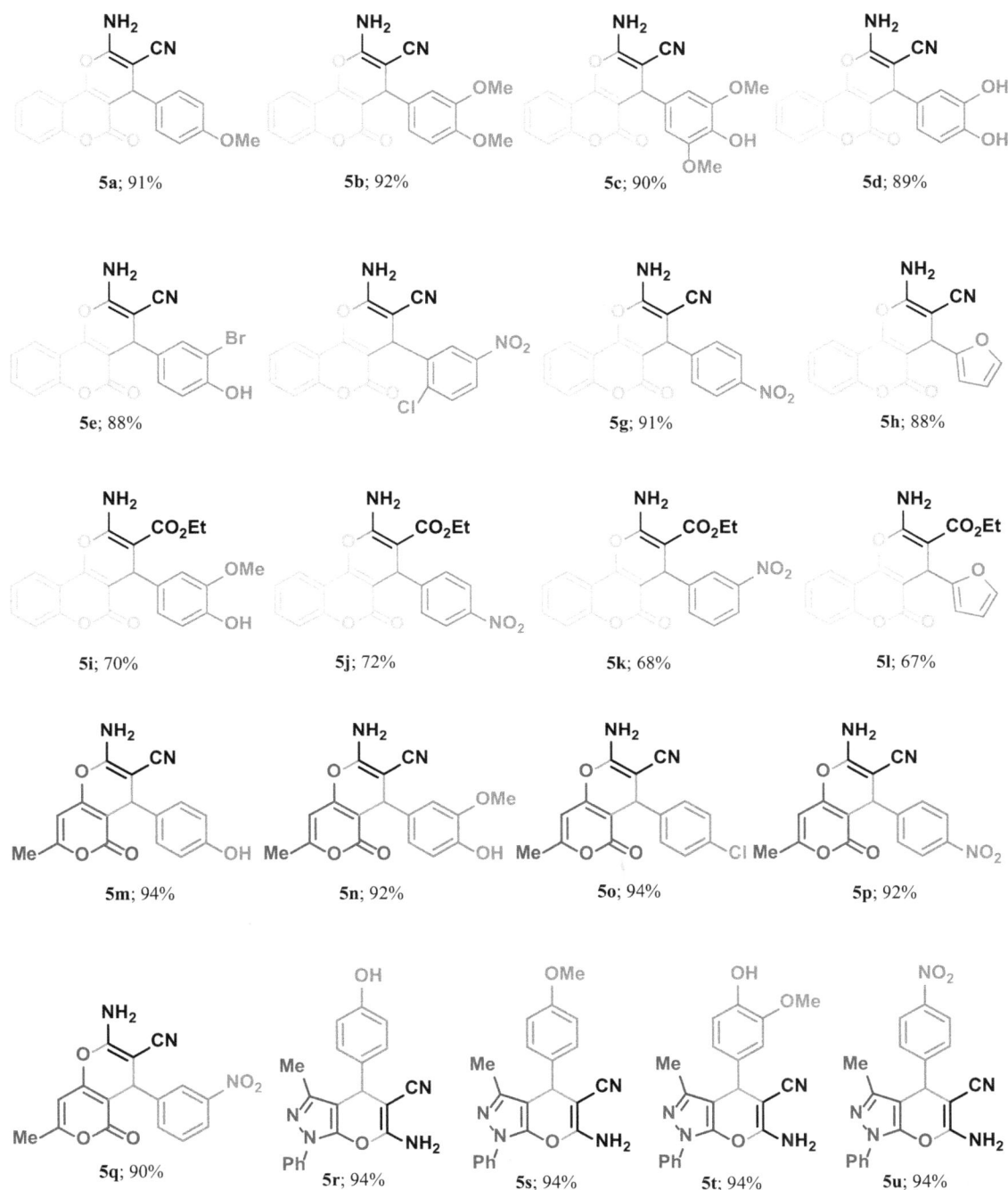

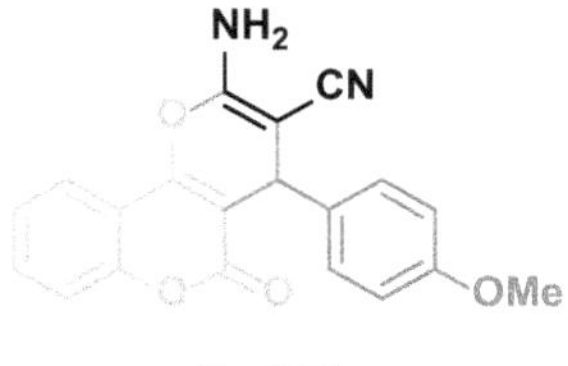

5a; 91% 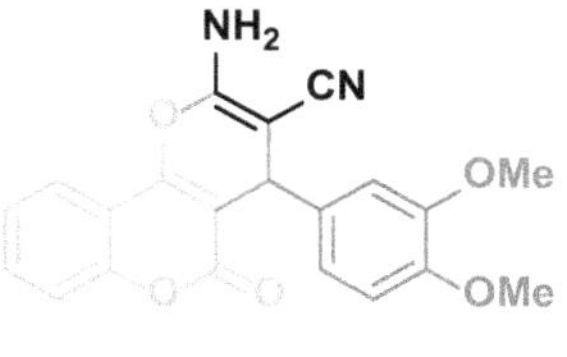 **5b**; 92% 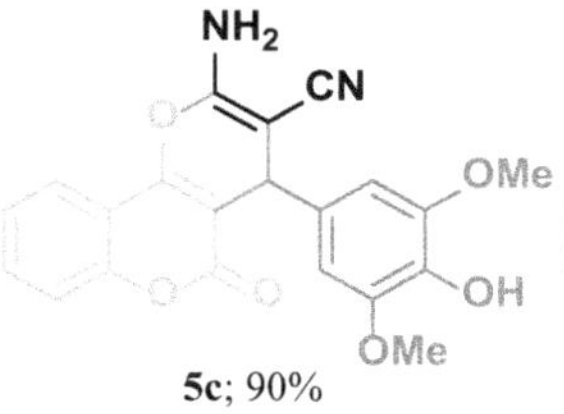 **5c**; 90% 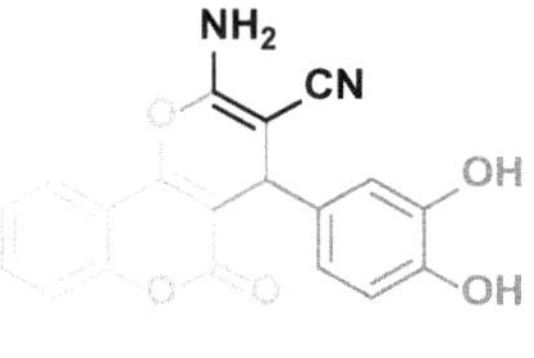 **5d**; 89%

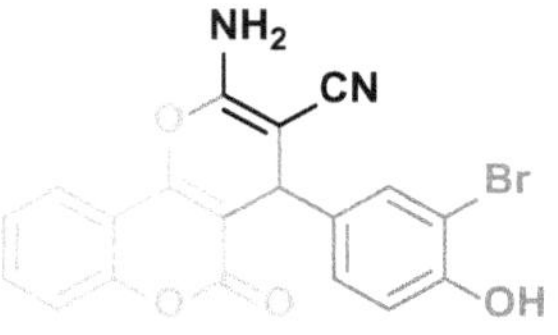

5e; 88% **5f** **5g**; 91% 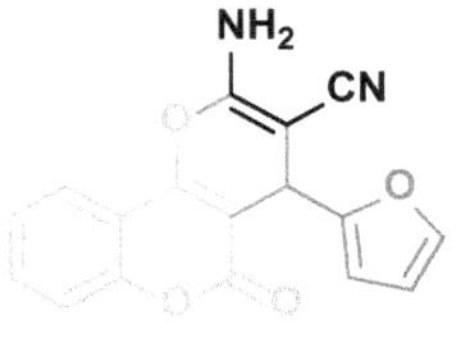 **5h**; 88%

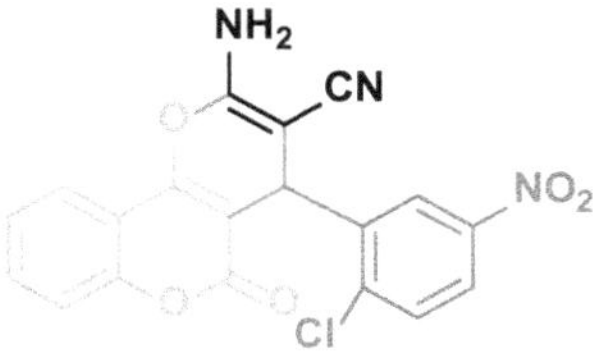

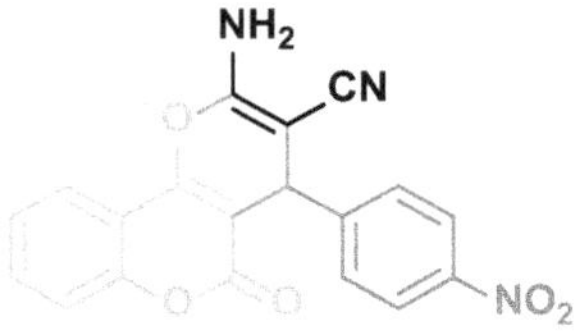

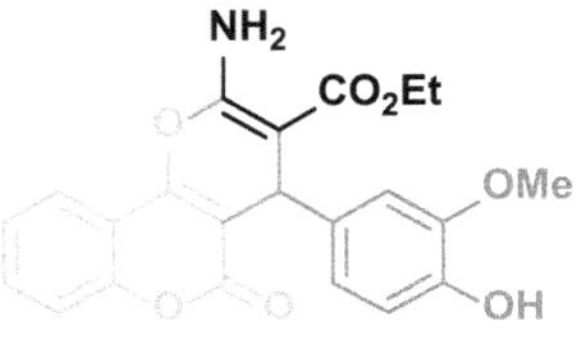

5i; 70% 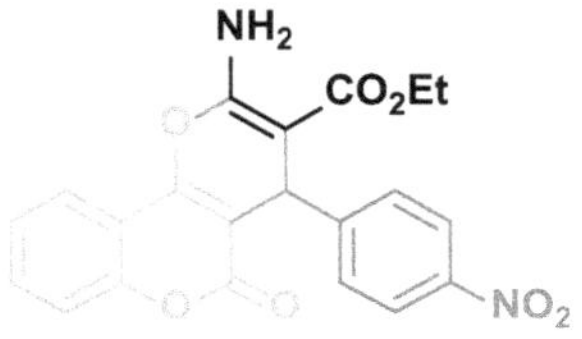 **5j**; 72% 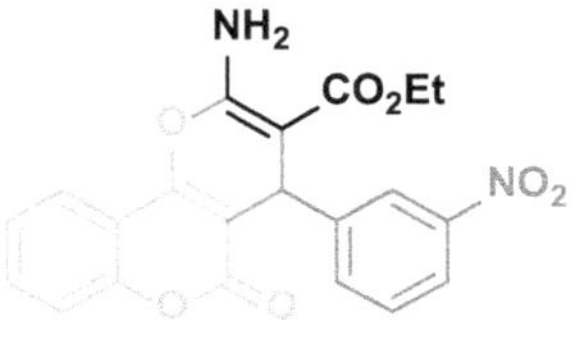 **5k**; 68% 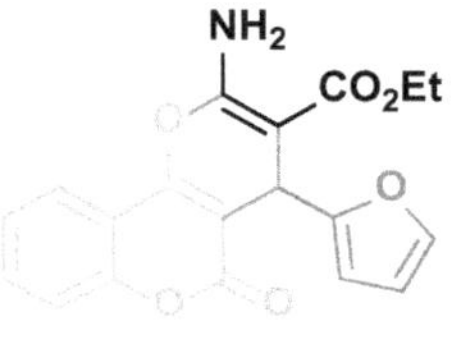 **5l**; 67%

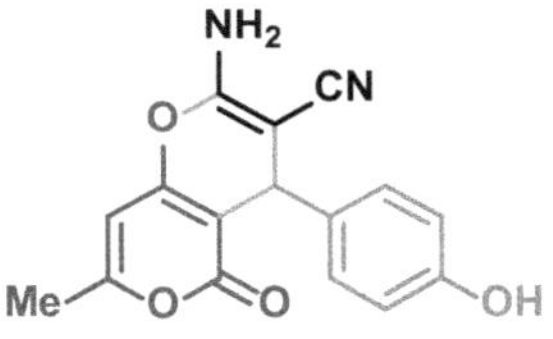

5m; 94% 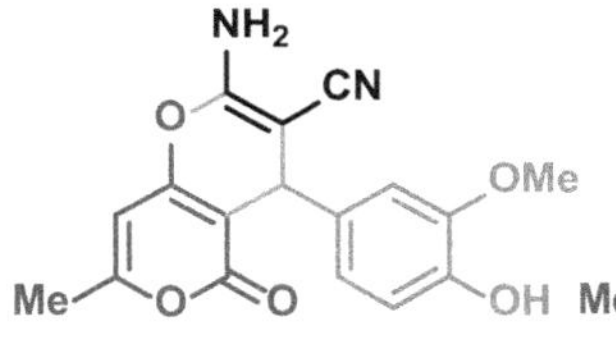 **5n**; 92% 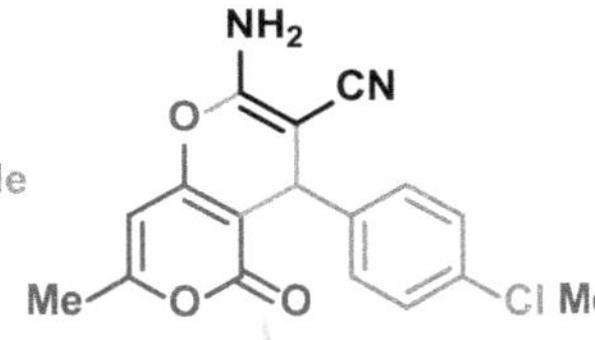 **5o**; 94% 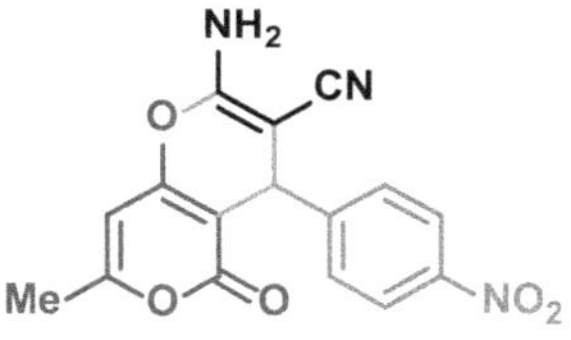 **5p**; 92%

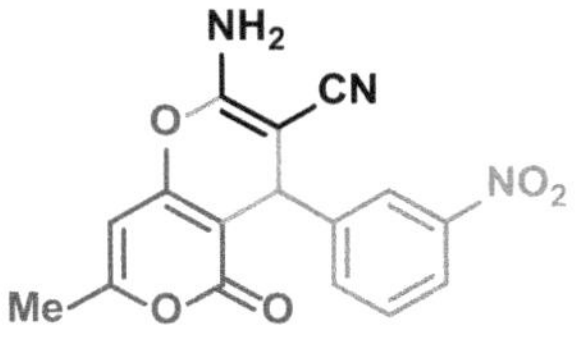

5q; 90% 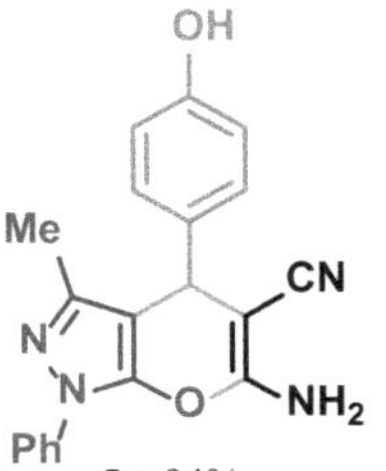 **5r**; 94% 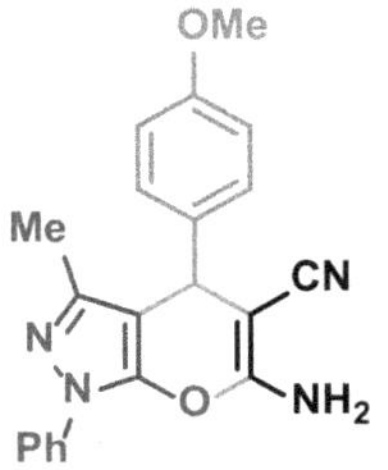 **5s**; 94% 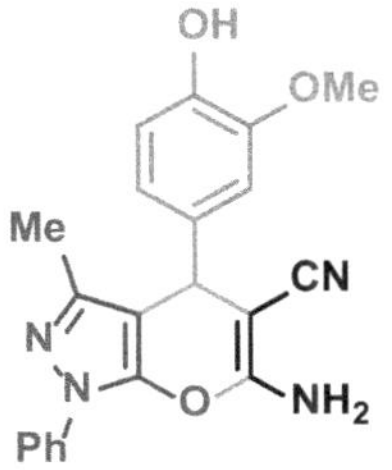 **5t**; 94% 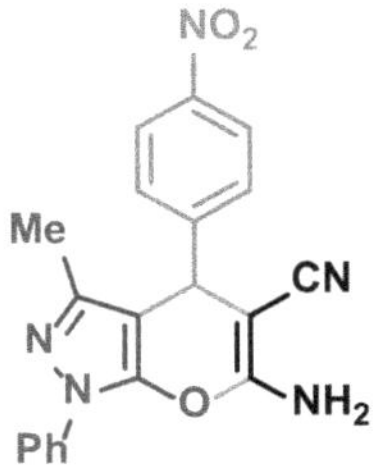 **5u**; 94%

[a] Reaction conditions: aromatic aldehyde (**1**; 1 mmol), different C–H enolizable carbonyl compound (2a/2b/2c; 1 mmol), active methylene nitrile (**4**, 1 mmol) in 5 mL ACC stirred at 40°C maximum for 24 hr under inert atmosphere.

[b] Isolated yields.

transformations, in particular from the green synthetic chemistry point of view.[75] There are several reported examples of solid heterogeneous acid catalysts such as sulfonated rice husk ash (RHA-SO_3H), sulfonated wool (wool-SO_3H), sawdust sulphonic acid (SD-OSO_3H), cellulose sulfuric acid, xanthan sulfuric acid, starch sulfuric acid and many others for synthesis of different types of organic compounds.[75,76]

1.2.5 Green Nanoparticles as Heterogeneous Catalyst

A "strong marriage" between nanotechnology and green chemistry has emerged as an essential avenue because it signifies an innovative way to build an environment sustainable society. Nanoparticles as heterogeneous catalysts have been extensively used in organic synthesis owing to their fundamental size- and shape-dependent properties.[77] This process principal has been discussed in detail in Chapter 9.

Sustainable synthesis and surface fabrication of nanomaterials using green chemistry have been attracted mounting interest in recent years due to their outstanding catalytic activities, easy handling, low cost, reusability and biocompatibility.[78] Recently, the trends of green synthesis of metal (Pd, Cu, Ni, Ag, Fe and Zn) nanoparticles (NPs) using plant extracts of animal wastes (agro-industrial waste) such as papaya peels, banana pulp and peel, beetroot, oyster shell powders, plant shoots of *Anisopappuschinensis*, root of *Salvadorapersica L.*, leaf of *soybean* (*Glycine max*), rice husks, miswak, broth of *Cinnamomum camphora* leaf, bark of *Cinnamom zeylanicum*, sun-dried tulsi leaves, *Pulicaria glutinosa and many more as* bioreductants have garnered significant popularity for versatile organic transformations like oxidation of alcohols, reduction of nitro groups, coupling reaction (Suzuki cross-coupling, Stille coupling and Sonogashira coupling) and click reactions.[79] This process has been discussed in detail in Chapter 9.

1.2.6 Ecocatalyst as Heterogeneous Catalyst from Plant Parts

In addition, several natural waste feedstocks (biomass) are the natural reservoir of transition metals which are very precious in organic transformation.[80] In other words, wastes have become useful and innovative chemical tools. In many plants, the metallophytes are able to hyperaccumulate several transition metals under the form of metallic cations such as Pd^{2+}, Cu^{2+}, Ni^{2+}, Zn^{2+}, Mn^{2+}, Ag^{+} and Co^{2+}.[81] After utilization of these metals via either phytoextraction or rhizofiltration, these transition metals are known as ecocatalyst named Eco-M (where M is the metal predominantly accumulated by the plant), such as Eco-Pd®, Eco-Cu®, Eco-Ni®, Eco-Zn® and Eco-Mn® which are directly used in many organic transformations such as reduction of nitro groups, Knoevenagel condensation reaction, bromination reaction, reductive amination of ketones, Ullmann reaction, coupling Reaction (Suzuki cross-coupling, Stille coupling, Heck–Mizoroki coupling), hydrogenation, Friedel–Crafts alkylation and acylation.[80–82] This process is discussed in detail in Chapter 10.

1.2.7 Carbon Nanoparticles as Heterogeneous Catalysts

Nano carbon materials have emerged as promising alternatives to conventional metal-based catalysts in recent years.[83] The use of heterogeneous carbon materials for the transformation or synthesis of organic or inorganic substrates is termed as carbocatalysis.[84] Nano carbon catalysts including carbon nanotubes (CNTs), onion-like carbon (OLC), nano diamonds, carbon quantum dots (CQDs), graphite/graphene oxide and mesoporous carbon have attracted great attention as potential green catalysts for various organic transformations such as acid/base reactions (hydrolysis of cellulose and Knoevenagel condensation reaction), oxidation reactions (oxidation of alkane, alcohol, thiol, amine and sulphide), reduction reactions (selective hydrogenation and reduction of nitrophenol), polymerization of various olefin monomers, nucleophilic addition of alcohol to an epoxide, esterification reaction, acetalization of aldehyde, Michael addition reaction, MCRs, condensation reactions (*e.g.* Claisen–Schmidt reaction), Friedel–Crafts reaction and many others.[83–85] In addition, there are a few reports of fluorescent CQDs as an effective green and recyclable carbocatalyst.[86] The use of carbon catalysts is attractive in this regard because of their low cost, non-toxic and natural abundance.

Using the literature survey and investigation of a variety of organic transformations such as reduction, decarboxylation, deacetylation, debenzoylation, electrophilic substitution, tandem reaction, Knoevenagel condensation, Michael addition reaction and click reaction for the synthesis of a glimpse of structurally diverse compounds such as phenolic compounds, aryl cyano derivatives, symmetric trisubstituted methane derivatives (TRSMs), 2-amino-4*H*-pyrans, 2-amino-4*H*-chromenes, 2-amino-spiro-oxindole-chromenes, 1*H*-pyrazolo[1,2-*b*] phthalazine-5,10-dione derivatives, 3-hydroxy-2-oxyindoles and 3-substituted phthalides using eco-friendly green catalysts, biocatalysts and natural feedstocks will direct newer synthesis of organic compounds, drugs and natural products.

REFERENCES

1. Clark, J. H. Chapter 1: Green and Sustainable Chemistry: An Introduction, in *Green and Sustainable Medicinal Chemistry: Methods, Tools and Strategies for the 21st Century Pharmaceutical Industry*, pp. 1–11, 2016.
2. Lu, C. W.; Wang, F.; Li, J. J.; Yu, S. J.; Au, Y.; *Res. Chem. Intermed.*, 2018, 44, 1035–1044.
3. Anastas, P. T.; Warner, J. C.; *Green Chemistry: Theory and Practice*, Oxford University Press, New York, pp. 30, 2018.
4. Anastas, P. T.; Zimmerman, J. B.; *Environ. Sci. Technol.*, 2003, 37, 94A.
5. Sheldon, R. A.; *Green Chem.*, 2017, 19, 18–43.
6. (a) Safaei, H. R.; Shekouhy, M.; Rahmanpur, S.; Shirinfeshan, A.; *Green Chem.*, 2012, 14, 1696–1704.
7. (a) Maleki, B.; Nasiri, N.; Tayebee, R.; Khojastehnezhad, A.; Akhlaghi, H. A. *RSC Adv.*, 2016, 6, 79128–79134.
8. Hasaninejad, A.; Shekouhy, M.; Zare, A.; Ghattali, S. M. S. H.; Golzar, N. J.; *Iran. Chem. Soc.*, 2011, 8, 411–42.
9. Maleki, B. *Org. Prep. Proced. Int.*, 2016, 48, 303–318.
10. Chen, Y.; Zhang, W.-Q.; Yu, B.-X.; Zhao, Y.-M.; Gao, Z.-W.; Jiana, Y.-J.; Xu, L.-W.; *Green Chem.*, 2016, 18, 6357–6366.
11. Baeyer, A.; Drewsen, V.; *Ber.*; 1882, 15, 2856.
12. Rideout, D. C.; Breslow, R.; *J. Am. Chem. Soc.*, 1980, 102, 7816–7817.
13. Narayan, S.; Muldoon, J.; Finn, M. G.; Fokin, V. V.; Kolb, H. C.; Sharpless, K. B.; *Angew. Chem. Int. Ed.*, 205, 44, 3275–3279.
14. (a) Safaei, H. R.; Shekouhy, M.; Rahmanpur, S.; Shirinfeshan, A.; *Green Chem.*, 2012, 14, 1696–1704 (b) Lindström, U. M.; *Chem. Rev.*, 2002, 102, 2751–2772.
15. (a) Hooshmand, S. E.; Heidari, B.; Sedghi, R.; Varma, R. S.; *Green Chem.*, 2019, 21, 381–405. (b) Baruah, D.; Konwar, D.; *Catal. Commun.*, 2015, 69, 68–71 (c) Khan, M.; Kuniyil, M.; Shaik, M. R.; Khan, M.; Adil, S. F.; Al-Warthan, A.; Alkhathlan, H. Z.; Tremel, W.; Tahir, M. N.; Siddiqui, M. R. H.; *Catalysts*, 2017, 7, 20 (d) Li, C.; *Chem. Rev.*, 1993, 93, 2023–2035 (e) Zhang, H.; Han, M.; Chen, T.; Xu, L.; Yu, L.; *RSC Adv.*, 2017, 7, 48214–48221 (f) Beare, K. D.;

McErlean, C. S. P.; *Tetrahedron Lett.*, 54, 1056–1058, 2013. (g) Banitaba, S. H.; Safari, J.; Khalili, S. D.; *Ultrason. Sonochem.*, 2013, 20, 401–407.
16. Coxon, J. M.; Halton, B.; *Organic Photochemistry*, Cambridge University Press, 1987.
17. Gilbert, A.; Baggott, J.; *Essentials of Molecular Photochemistry*, Blackwell Science Ltd, 1990.
18. Horikoshi, S.; Serpone, N.; *General Introduction to Microwave Chemistry*. Wiley-VCH Verlag GmbH & Co. KGaA, USA, 1–28, 2015.
19. Polshettiwar, V.; Varma, R. S.; *Chem. Soc. Rev.*, 2008, 37, 1546–1557.
20. Toyao, T.; Saito, M.; Horiuchi, Y.; Matsuoka, M.; *Catal. Sci. Technol.*, 2014, 4, 625–628.
21. (a) Wang, H.; Wang, C.; Yang, Y.; Zhao, M.; Wang, Y.; *Catal. Sci. Technol.*, 2017, 7, 405–417 (b) Li, P.; Yu, Y.; Liu, H.; Cao, C. Y.; Song, W. G.; *Nanoscale.*, 2014, 6, 442–448 (c) Merino, E.; Verde-Sesto, E.; Maya, E. M.; Iglesias, M.; Sánchez, F.; Corma, A.; *Chem. Mater.*, 2013, 25, 981–988.
22. (a) Shiju, N. R.; Alberts, A. H.; Khalid, S.; Brown, D. R.; Rothenberg, G.; *Angew. Chem. Int. Ed.*, 2011, 50, 9615–9619 (b) Xu, L.; Li, C. G.; Zhang, K.; Wu, P. *ACS Catal.* 2014, 4, 2959–2968.
23. (a) Thirumurugan, P.; Matosiuk, D.; Jozwiak, K.; *Chem. Rev.*, 2013, 113, 4905–4979 (b) Chandrasekaran, S. *Click Reactions in Organic Synthesis*, 1st Edition. Wiley-VCH Verlag GmbH & Co. KGaA, 2016 (c) Zhao, Z.; Yao, Z.; Xu, X.; *Curr. Org. Chem.* 2017, 21, 2240–2248.
24. Kolb, H. C.; Finn, M. G.; Sharpless, K.
25. (a) Reddy, K. R.; Maheswari, C. U.; Rajgopal, K.; Kantam, M. L.; *Synth. Commun.*, 2008, 38, 2158–2167 (b) Lal, S.; Alez, S. D. *Org. Chem.* 2011, 76, 2367–2373.
26. Huisgen, R. In *1,3-Dipolar Cycloaddition Chemistry*, Ed. A. Padwa, John Wiley & Sons, New York, 1984, 1, 11–76.
27. Rostovtsev, V. V.; Green, L. G.; Fokin, V. V.; Sharpless, K. B.; 2002, *Angew. Chem. Int. Ed.*, 41, 2596–2599.
28. Kolb, H. C.; Sharpless, B. K.; *Drug Discov. Today* 2003, 8, 1128–1137.
29. Zhang, L.; Xue, P.; Sun, H. H. Y.; Williams, I. D.; Sharpless, K. B.; Fokin, V. V.; Jia, G.; *J. Am. Chem. Soc.* 2005, 127, 15998–15999.
30. (a) Rotstein, B. H.; Zaretsky, S.; Rai, V.; Yudin, A. K.; *Chem. Rev.*, 2014, 114, 8323–8359 (b) Dömling, A.; Wang, W.; Wang, K. *Chem. Rev.*, 2012, 112, 3083–3135. (c) Cioc, R. C.; Ruijter, E.; Orru, R. V. A. *Green Chem.* 2014, 16, 2958–2975.
31. (a) Haji, M.; *Beilstein J. Org. Chem.*, 12, 1269–1301, 2016 (b) Ganem, B.; *Acc. Chem. Res.*, 2009, 42, 463–472. (c) Dekamin, M. G.; Eslami, M.; Maleki, A.; *Tetrahedron.*, 2013, 69, 1074–1085.
32. (a) Domling, A.; Ugi, I. *Angew. Chem. Int. Ed.*, 2000, 39, 3169–3210. (b) Pham, K.; Zhang, Z.; Shen, S.; Ma, L.; Hu, L.; *Tetrahedron*, 2013, 69, 10933–10939 (c) Asinger, F.; Thiel, M.; *Angew. Chem.*, 1958, 70, 667–683.
33. Sato, T.; Uozumi, Y.; Yoichi, M. A. Y.; *ACS Omega*, 2020, 5, 4126938–26945.
34. Porta, R.; Benaglia, M.; Puglisi, A.; *Org. Process Res. Dev.*, 216, 20, 1, 2–25.
35. Tsubogo, T.; Oyamada, H.; Kobayashi, S.; *Nature*, 2015, 520, 329.
36. Snead, D. R.; Jamison, T. F. *Angew. Chem. Int. Ed.* 2015, 54, 983.
37. (a) Kumar, A.; Maurya, R. A.; Saxena, D.; *Mol. Divers.*, 2010, 14, 331–341 (b) Burke, M. D.; Schreiber, S. L.; *Angew. Chem. Int. Ed.*, 2004, 43, 46–58 (c) Spring, D. R.; *Org. Biomol. Chem.* 2003, 1, 3867–3870.
38. Kumar, H.; Choudhary, N.; Varsha; Kumar, N.; Seth, S.; Seth, R.; *J. Food Sci. Technol.*, 2014, 46–59.
39. (a) Kadiri, O.; *Int. J. Food Prop.*, 2017, 20, S798–S809 (b) Alu'datt, M. H.; Rababah, T.; Alhamad, M. N.; Al-Mahasneh, M. A.; Almajwal, A.; Gammoh, S.; Ereifej, K.; Johargy, A. *Int. Food Chem.*, 2017, 218, 99–106.
40. (a) Hinman, J.; Hoeksema, H.; Caron, E. L.; Jackson, W. G.; *J. Am. Chem. Soc.*, 1956, 78, 1072–1074 (b) Poupaert, J.; Carato, P.; Colacino, E.; *Curr. Med. Chem.*, 2005, 12, 877–885.
41. (a) Laskar, S.; Brahmachari, G.; *Org. Biomol. Chem.*, 2014, 2, 01–05 (b) Brahmachari, G.; Banerjee, B.; *ACS Sustain. Chem. Eng.*, 2014, 2, 411–422.
42. (a) Martins, M. A. P.; Frizzo, C. P.; Moreira, D. N.; Buriol, L.; Machado, P.; *Chem. Rev.*, 2009, 109, 4140–4182 (b) Das, V. K.; Borah, M.; Thakur, A. J. *J. Org. Chem.*, 2013, 78, 3361–3366.
43. Kalaria, P. N.; Karad, S. C.; Raval, D. K.; *Eur. J. Med. Chem.* 2018, 158, 917–936.
44. (a) Guo, R.-Y.; An, Z.-M.; Mo, L.-P.; Wang, R.-Z.; Liu, H.-X.; Wang, S.-X.; Zhang, Z.-H.; *ACS Comb. Sci.*, 2013, 15, 557–563 (b) Albadi, J.; Mansournezhad, A.; *Res. Chem. Intermed.* 2016, 42, 5739–575.
45. Fan, X.; Feng, D.; Qua, Y.; Zhang, X.; Wang, J.; Loiseau, P. M.; Andrei, G.; Snoeck, R.; Clercq, E. D.; *Bioorg. Med. Chem. Lett.*, 2010, 20, 809–813 (b) Niknam, K.; Khataminejad, M.; Zeyaei, F.; *Tetrahedron Lett.*, 2016, 57, 361–365. (c) Rueping, M.; Sugiono, E.; Merino, E.; *Chem. Eur. J.*, 2008, 14, 6329–6332 (d) Kumar, A.; Maurya, R. A.; Sharma, S. A.; Ahmad, P.; Singh, A. B.; Bhatia, G.; Srivastava, A. K.; *Bioorg. Med. Chem. Lett.*, 2009, 19, 6447–6451 (e) Alvey, L.; Prado, S.; Saint-Joanis, B.; Michel, S.; Koch, M.; Cole, S. T.; Koch, M.; Cole, S. T.; Tillequin, F.; Janin, Y.L.; *Eur. J. Med. Chem.*, 2009, 44, 2497–2505.
46. Anastas, P. T.; Heine, L. G.; Williamson, T. C.; *Green Chemical Syntheses and Processes: Introduction*, American Chemical Society: Washington, DC, Chapter 1, pp. 1–6, 2000.
47. Saikia, B.; Borah, P.; Barua, N. C.; *Green Chem.* 2015, 17, 4533–4536.
48. Fernandez, G.-V., Brieva R.; Gotor V.; *J. Mol. Enzymatic Catalysts* 2006, 40, 111–120.
49. Das, M.; PhD thesis, Enzymatic polishing of germinated brown rice, 2006, Indian Institute of Technology, Kharagpur, West Bengal, India.
50. Costas, M.; Deive, F.J.; Longo, M.A.; *Process Biochem.*, 2004, 39, 2109–2114.
51. Saxena, B.K.; Davidson, W.K.; Sheoran, A.; *Process Biochem.*, 2003, 39, 239–247.
52. Zaidi, A., Gainer, J.L., Carta, G., Mrani, A., Kadiri, T., Belarbi, Y., Mir, A.; *J. Biotechnol.*, 2002, 93, 209–216.
53. Nag, A.; Patil, D.; Leonardis, A.D.; *J. Food Biochem.*, 2010, 32, 45–72.
54. Nag, A.; Lee, G.; Shaw, J.; Shaw, F.; *J. Am. Agri. Food Chem.*, 2002, 50, 2, 262–266.
55. Nag, A.; Das, D.; Patil, D.; *Chemical Papers* 2011, 65, 1, 9–15.

56. López-Belmonte, M. T.; Alcántara, A. R.; Sinisterra, J. V.; *J. Org. Chem.*, 1997, 62, 6, 1831–1840.
57. Park, H.J.; Choi, W.J.; Huh, E.C.; Lee, C.Y.; *J. Biosci. Bioeng.*, 1997, 87, 1831–1840.
58. Basak, A.; Nag, A.; Bhattacharya, G.; Mandal, S.; Nag, S.; *Tetrahedron Asymmetry* 2000, 11, 2403–2407.
59. Kato, K.; Tanaka, S.; Gong, Y.; *J. Biosci. Bioeng.*, 1999, 87, 76–81.
60. Fernadez-Solares, R.; Brieva, M.; Quiros, I.; *Tetrahedron Asymmetry*, 2004, 2, 341–345.
61. (a) Fozan, P.; Jonson, P.H.; Peteson, M.T.; Peterson, S.B.; *Biochimie*, 2000, 82, 1033–1041 (b) Basak, A.; Bhattacharjee, G.; Nag, A.; *Biotech Letter*, 1993, 15, 1, 22–23.
62. Nag, A., Nag, S.; Patil, D.; Basak, A.; *J. Pharmaceutical Chem.*, 2008, 42, 8, 281–284.
63. Asakawa, K.; Noguchi, N., Takashima, S., Nakada, M.; *Tetrahedron Asymmetry*, 2008, 19, 2304–2309.
64. Nelson, D. L.; Cox, M. M.; *Lehninger Principles of Biochemistry*, 4[a] ed., W.H. Freeman and Company, New York, 2005.
65. Drueckhammer, D. G.; Barbas, C. F.; Nozaki, K.; Wong, C. H.; Wood, C. Y.; Ciufolini, M. A.; *J. Org. Chem.*, 1988, 53, 1607–16.
66. Dillon, M. P.; Hayes, M. A.; Simpson, T. J.; Sweeney, J. B.; Fermenting baker's yeast reduction of ethyl 2-oxohexanoate. In *Biocatalyse for Fine Chemicals Synthesis*, John Willey & Sons, Chichester, 2, 1999.
67. Faber, K.; *Biotransformations in Organic Chemistry*, 5[a] ed., Springer-Verlag, Berlin, 2004.
68. (a) Saha, A.; Jana, A.; Choudhury, L. H.; *New J. Chem.*, 2018, 42, 17909–17922 (b) Patil, M. A.; Ubale, P. A.; Karhale, S. S.; Helavi, V. B.; *Der. Chemica. Sinica.*, 2017, 8, 1, 198–205 (c) Vekariya, R. H.; Patel, K. D.; Patel, H. D.; *Res. Chem. Intermed.*, 42, 7559–7579, 2016 (d) Luna, H.; Vázquez, L. H.; Reyo, A.; Arias, L.; Manjarrez, N.; Ocana, A. N.; *Ind. Crops Prod.*, 2014, 59, 105–108 (e) Chavan, H. V.; Bandgar, B. P.; *ACS Sustainable Chem. Eng.*, 2013, 1, 8, 929–936 (f) Menges, N.; Şahin, E.; *ACS Sustainable Chem. Eng.*, 2, 226–230, 2014 (g) Kumaraswamy, G.; Ramesh, S.; *Green Chem.*, 2003, 5, 306–308 (h) Pal, R.; *Int. J. Org. Chem.*, 201, 3, 136–142 (i) Fonseca, A. M.; Monte, F. J. Q.; Oliveira, M. D. C. F. D.; Mattos, M. C. D.; Cordell, G. A.; Braz-Filho, R.; Lemos, T. L. G.; *J. Mol. Catal. B Enzym.*, 2009, 57, 78–82 (j) Pavokovic, D.; Buda, R.; Andrašec, F.; Roje, M.; Bubalo, M. C.; Redovnikovic, I. R. *Tetrahedron Asymmetry*, 2017, 28, 730–733 (k) Bagull, S. D.; Rajput, J. D.; Bendre, R. S.; *Environ. Chem. Lett.*, 2017, 15, 725–731 (l) Pal, R.; *Int. J. Adv. Chem.*, 2014, 2, 27–33.
69. Misra, K.; Maity, H.S.; Nag, A.; *Mol. Catal. B. Enzym.*, 2012, 82, 92–95.
70. Maity, H.S.; Mishra, K.; Mahata, T.; *Rsc Adv.*, 2016, 6, 24446–24450.
71. Halder, B.; Maity, H. S.; Nag, A.; *Curr. Organocatal.*, 2019, 6, 20–27.
72. (a) Chen, W.; Peng, X. W.; Zhong, L. X.; Li, Y.; Sun, R. C.; *ACS Sustainable Chem. Eng.*, 2015, 3, 1366–1373 (b) Boruah, P. R.; Ali, A. A.; Saikia, B.; Sarma, D.; *Green Chem.*, 2015, 17, 1442–1445 (c) Moussouni, S.; Saru, M. L.; Ioannou, E.; Mansour, M.; Detsi, A.; Roussis, V.; Kefalas, P.; *Tetrahedron Lett.*, 2011, 52, 1165–1168 (d) Dewan, A.; Sarmah, M.; Bora, U.; Thakur, A. J.; *Tetrahedron Lett.*, 2016, 57, 3760–3763 (e) Boruah, P. R.; Ali, A. A.; Chetia, M.; Saikia, B.; Sarma, D.; *Chem. Commun.*, 51, 11489–11492, 2015 (f) Kotzebue, L. R. V.; Oliveira, J. R. D.; Silva, J. B. D.; Mazzetto, S. E.; Ishida, H.; Lomonaco, D.; *ACS Sustainable Chem. Eng.*, 2018, 6, 5485–5494.
73. (a) Khaksar, S.; Tajbakhsh, M.; Gholami, M. C. R.; *Chimie*, 2014, 17, 30–34 (b) Dusi, M.; Mallat, T.; Baiker, A.; *Catal. Rev.*, 2000, 42, 213–278 (c) Bordoloi, A.; Sahoo, S.; Lefebvre, F.; Halligudi, S. B.; *J. Catal.*, 2008, 259, 232–239.
74. (a) Mejdoubi, K. E.; Sallek, B.; Cherkaoui, H.; Chaair, H.; Oudadesse, H.; *Kinet. Catal.*, 2018, 59, 290–295 (b) Fallah, A.; Tajbakhsh, M.; Vahedi, H.; Bekhradni, A.; *Res. Chem. Intermed.* 2017, 43, 29–43 (c) Patel, K. G.; Misra, N. M.; Vekariya, R. H.; Shettigar, R. R.; *Res. Chem. Intermed.*, 44, 289–304, 2018 (d) Borhade, A. V.; Uphade, B. K.; Gadhave, A. G.; *Res. Chem. Intermed.* 2016, 42, 6301–6311 (e) Riadi, Y.; Mamouni, R.; Azzalou, R.; Boulahjar, R.; Abrouki, Y.; El Haddad, M.; Routier, S.; Guillaumet, G.; Lazar, S.; *Tetrahedron Lett.*, 2010, 51, 6715–6717 (f) Sangsuwan, R.; Sangher, S.; Aree, T.; Mahidol, C.; Ruchirawat, S.; Kittakoop, P.; *RSC Adv.*, 2014, 4, 13708–13718 (g) He, T.; Zeng, Q. Q.; Yang, D. C.; He, Y. H.; Guan, Z. *RSC Adv.*, 2015, 5, 37843–37852.
75. (a) Seddighi, M.; Shirini, F.; Mamaghani, M.; *RSC Adv.*, 2013, 3, 24046–24053 (b) Pathak, G.; Das, D.; Rajkumari, K.; Rokhum, L.; *Green Chem.*, 2018, 20, 2365–2373.
76. (a) Karhale, S.; Patil, M.; Rashinkar, G.; Helavi, V. *Res. Chem. Intermed.*, 2017, 43, 7073–7086 (b) Shirini, F.; Seddighi, M.; Mamaghani, M.; *Res. Chem. Intermed.* 2015, 41, 8673–8680.
77. (a) Saha, M.; Das, B.; Pal, A. K. C. R.; *Chimie*, 2013, 16, 1079–1085 (b) Tabrizian, E.; Amoozadeh, A.; *Catal. Sci. Technol.*, 2016, 6, 6267–6276.
78. Khan, M.; Al-Marri, A. H.; Khan, M.; Mohri, N.; Adil, S. F.; Al-Warthan, A.; Siddiqui, M. R. H.; Alkhathlan, H. Z., Berger, R., Tremel, W., Tahir, M. N.; *RSC Adv.*, 2014, 4, 24119–24125.
79. (a) Dewan, A.; Sarmah, M.; Thakur, A. J.; Bharali, P.; Bora, U.; *ACS Omega* 2018, 3, 5327–5335 (b) Veerakumar, P.; Muthuselvam, I. P.; Hung, C.T.; Lin, K. C.; Chou, F. C.; Liu, S. B. *ACS Sustainable Chem. Eng.*, 2016, 4, 6772–6782 (c) Singh, J.; Mehta, A.; Rawat, M.; Basu, S.; *J. Environ. Chem. Eng.*, 2018, 6, 1468–1474.
80. G. Clave, C. Garel, C. Poullain, B. L. Renard, T. K. Olszewski, B. Lange, M. Shutcha, M. P. Faucon, C. Grison; *RSC Adv.*, 2016, 6, 59550–59564.
81. Garel, C.; Renard, B. L.; Escande, V.; Galtayries, A.; Hesemann, P.; Grison, C. *Appl. Catal. A Gen.*, 2015, 504, 272–286.
82. (a) Escande, V.; Velati, A.; Garel, C.; Renard, B. L.; Petit, E., Grison, C.; *Green Chem.*, 2015, 17, 2188–2199 (b) Gopiraman, M.; Wei, K.; Zhang, K. Q.; Chung, I. M.; Kim, I. S.; *RSC Adv.*, 2018, 8, 4531–4547 (c) Orha, L.; Tukacs, J. M.; Gyarmati, B.; Szilagyi A.; Kollar, L.; Mika, L. T. *ACS Sustainable Chem. Eng.*, 2018, 6, 5097–5104.
83. (a) Su, D. S.; Perathone, S.; Centi, G.; *Chem. Rev.*, 113, 5782–5816, 2013 (b) Navalon, S.; Dhakshinamoorthy, A.; Alvaro, M.; Garcia, H.; *Chem. Rev.*, 2014, 114, 6179–6212 (c) Duan, X.; Ao, Z.; Li, D.; Sun, H.; Zhou, L.; Suvorova, A.; Saunders, M.; Wang, G.; Wang, S. *Carbon*, 2016, 103, 404–411 (d) Duan, X.; Sun, H.; Wang, S.; *Acc. Chem. Res.*, 2018, 51, 678–687.

84. (a) Gupta, N.; Khavryuchenko, O.; Wen, G.; Wu, K. H.; Su, D.; *Carbon*, 2018, 130, 714–723 (b) Tang, P.; Hu, G.; Li, M.; Ma, D.; *ACS Catal.*, 2016, 6, 6948–6958.
85. (a) Espinosa, J. C.; Navalon, S.; Alvaro, M.; Dhakshinamoorthy, A.; Garcia, H.; *ACS Sustainable Chem. Eng.*, 2018, 6, 5607–5614 (b) Dreyer, D. R.; Jia, H. P.; Bielawski, C. W. *Angew. Chem. Int. Ed.*, 2010, 49, 6813–6816 (c) Kausar, N.; Mukherjee, P.; Das, A. R. *RSC Adv.*, 2016, 6, 88904–88910.
86. (a) Caputo, J. A.; Frenette, L. C.; Zhao, N.; Sowers, K. L.; Krauss, T. D.; Weix, D. J.; *J. Am. Chem. Soc.*, 2017, 139, 4250–4253 (b) Majumdar, B.; Mandani, S.; Bhattacharya, T.; Sarma, D.; Sarma, T. K.; *J. Org. Chem.*, 2017, 82, 2097–2106 (c) Wang, H.; Zhuang, J.; Velado, D.; Wei, Z.; Matsui, H.; S. Zhou; *ACS Appl. Mater. Interfaces* 2015, 7, 27703–27712.

2

New Greener Developments in Direct Amidation of Carboxylic Acids

Andrea Ojeda-Porras and Diego Gamba-Sánchez

CONTENTS

2.1 Introduction ... 23
2.2 Mechanistic Considerations ... 24
2.3 Catalyst-free Reactions ... 25
2.4 Organic Additives ... 28
2.5 Boron ... 31
2.6 Mesoporous Solids as Heterogeneous Catalyst ... 32
2.7 Other Catalytic Systems ... 36
2.8 Conclusion ... 39
References ... 39

2.1 Introduction

Amides, also denoted as carboxamides, are organic compounds characterized by having a carbonyl group directly bonded to a nitrogen atom. Depending on the substituents on the nitrogen, amides can be classified as primary, secondary or tertiary. When the nitrogen is only bonded to the carbon of the carbonyl group, it is a primary amide. Addition of one or two additional carbon substituents led to secondary and tertiary amides, respectively. A lactam is a cyclic amide (Figure 2.1).

The chemical properties of amides are governed by the delocalization of the lone pair of the nitrogen atom on to the carbonyl group, forming a "partial" C–N double bond (Scheme 2.1). As a result, the nitrogen of an amide is less basic than the nitrogen of an amine. Additionally, the N–H bond(s) in amides are more acidic compared to amines since resonance of the lone pair of the conjugated base is possible. This delocalization is also responsible for the limited rotation around the C–N bond and its structural rigidity.

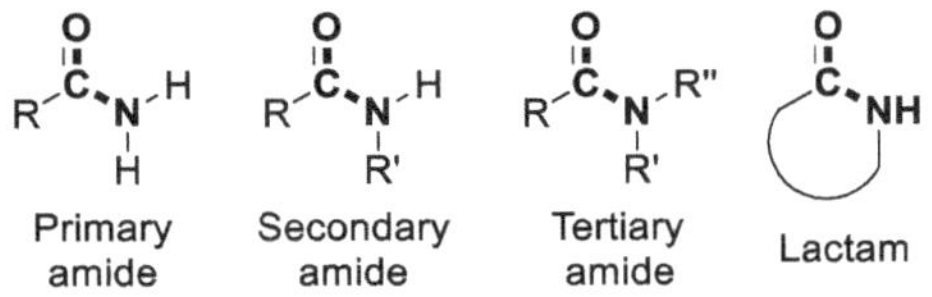

Primary amide Secondary amide Tertiary amide Lactam

FIGURE 2.1 Amide functional group.

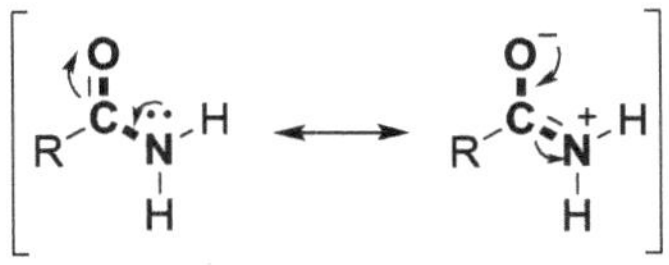

SCHEME 2.1 Delocalization of nitrogen lone pair.

Due to its polarity, high stability and conformational diversity, the amide bond is one of the most important functional groups in organic chemistry. For instance, the amide linkage constitutes the structural base of peptides and proteins by binding together different amino acids.[1] Additionally, the amide functional group is present in several bioactive and pharmaceutical compounds; for example, penicillin, ponatinib, diltiazem and atorvastatin (one of the most sold drugs worldwide) contain an amide moiety in their structure (Figure 2.2).[2] It is estimated that *N*-acylation reactions of amines, in order to form the amide bond, constitute 16% of all reactions carried out in the pharmaceutical industry.[3] Furthermore, approximately 25% of commercially available pharmaceuticals[4] and one-third of the new drug candidates contain an amide bond.[5] Moreover, there are also numerous applications of amides in plastics, pesticides, perfumes, lubricants, polymers, agrochemicals and detergents.[6] Furthermore, amides have been used as precursors in the synthesis of a great variety of high value functional groups such as nitriles and carbonyl compounds and have been successfully employed in cross-coupling reactions.[7] Aromatic amides are also frequently used in the synthesis of heterocyclic compounds.

Ideally, the amide bond can be formed by reacting a carboxylic acid with either an amine or ammonia releasing water as the only by-product. However, this reaction has been traditionally considered of little synthetic value due to the demanding conditions required. The use of activated carboxylic acid derivatives, such as acid chlorides or anhydrides, as well as the use of coupling reagents has emerged as an alternative to drive the reaction (Scheme 2.2).[2] Even though these methods can be applied to a wide variety of substrates affording the desired amides in excellent yields, under mild conditions, the high amount of waste, toxicity of the coupling reagents and

DOI: 10.1201/9781003089162-2

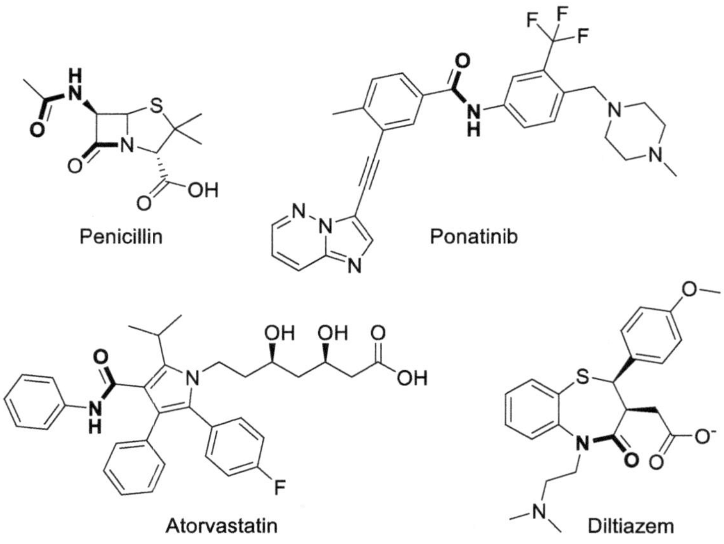

FIGURE 2.2 Penicillin and artovastatin.

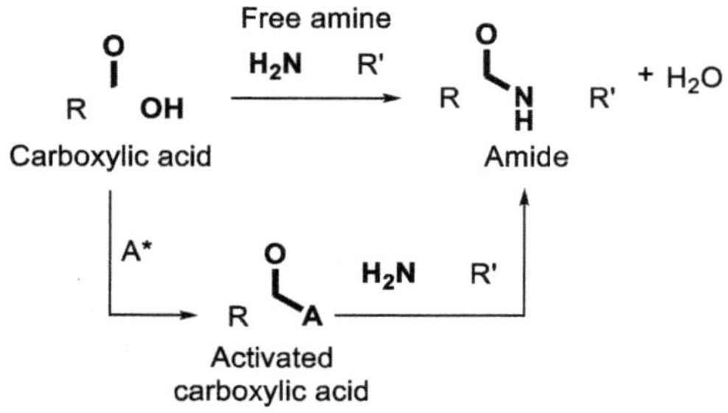

SCHEME 2.2 General amide synthesis.

low atom economy are the major drawbacks to an efficient and eco-friendly way of amide synthesis.

Due to the importance of the amide functional group and in the context of researching the development of new sustainable synthetic methods, the ACS Green Chemistry Institute and the global pharmaceutical corporations identified "amide formation avoiding poor atom economy reagents" as one of the top challenges in organic chemistry in 2007[8] and in 2018.[9] This chapter presents a general overview of environmentally friendly methods toward the synthesis of amides by direct amidation of carboxylic acids, emphasizing the concept development. Starting with a brief description of the reaction mechanism, several alternatives using organic additives, boron-based compounds and mesoporous solids as catalysts are then described. Applications in the synthesis of natural products and bioactive compounds are presented at the end of the chapter.

2.2 Mechanistic Considerations

Even though amide synthesis by direct amidation of carboxylic acids has been reported since 1958,[10] it was not until 2010 that a detailed mechanistic study of the reaction was performed.[11] Traditionally, the use of elevated temperatures[12] for this transformation was assumed due to the formation of the ammonium carboxylate when mixing a carboxylic acid with a free amine. It was also assumed that this unreactive carboxylate was immediately formed due to the acid-base reaction and could not lead to the desired amide at room temperature unless a thermal amidation over 100°C was performed (Scheme 2.3). Traditionally, the scope of the reaction tended to be limited to the substrate's thermal stability given the high temperatures required.

In 2011, Whiting and coworkers[11a] performed a series of experiments in order to understand the mechanism of thermal amidation between carboxylic acids and amides in non-polar aprotic solvents. This study revealed that the formation of the carboxylate depends on the chemical properties of both

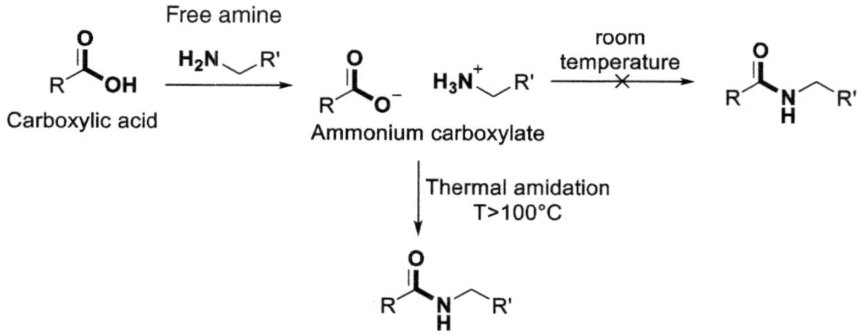

SCHEME 2.3 Formation of ammonium carboxylate in amide synthesis.

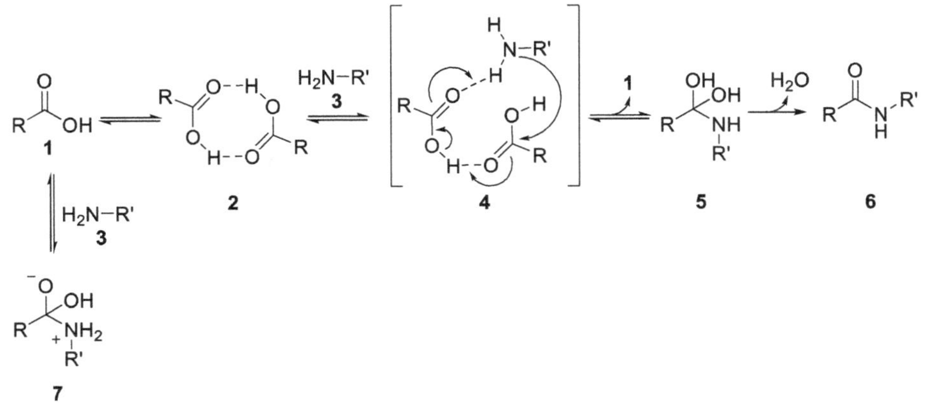

SCHEME 2.4 Proposed mechanism for the direct amidation of carboxylic acids.

reagents such as pKa, basicity, steric and electronic effects, and does not prevent condensation from happening. As shown is Scheme 2.4, further computational studies proved dimerization of the carboxylic acid through hydrogen bonding. Nucleophilic attack from amine **3** to dimer **2** leads to the formation of neutral tetrahedral intermediate **5** and releases one molecule of acid *via* transition state **4**. Formation of zwitterionic species **7** is not relevant since it rapidly collapses to the starting materials and has been proven to be highly unstable. Active removal of water plays a key role in the reaction rate, for example, addition of molecular sieves or the use of a Dean–Stark trap accelerates the amidation. Quantum chemical simulations have proven that the presence of water directly affects the concentration of active dimer **2** by hydration of the hydrogen bond.[13] An excess of water produces separation between carboxylic acid molecules to such an extent dimer **2** is not formed and they are no longer activated. Consequently, nucleophilic attack by the amine is no longer possible.

Even though the mechanism showed in Scheme 2.4 is widely accepted, the presence of additives and catalysts as well as the use of non-conventional heating and energy sources might drive the reaction through a different pathway. Although the following sections are focused in different approaches toward the amide synthesis in a greener way, a brief discussion of the mechanism reaction will be presented when relevant.

2.3 Catalyst-free Reactions

N-formylation of amines is one of the most used transformation in organic chemistry as formamides are employed as catalysts in several reactions[14] and are intermediates in the synthesis of biologically active compounds such as oxazolidinones,[15] aryl imidazoles,[16] 1,2 dihydroquinolines[17]and fungicides.[18] Nowadays, the use of toxic formylating agents and/or expensive catalysts are considered the main drawbacks of this transformation. In 2011, Bhanage and coworkers reported a catalyst-free *N*-formylation of amines using formic acid **8** or ethyl formate **9**.[19] Beyond being a solvent-free method, this work evidences the effect of the formylating agent in the yield and time of the reaction. As shown in Scheme 2.5a, when using ethyl formate instead of formic acid with aniline **10a**, *o*-toluidine **10b** and 2-fluoroaniline **10c**, a slight decrease of the yield was observed with a considerable increase in the reaction time. Moreover, while formylation of **10d** and **10e** with formic acid yielded desired product in good yields; poor yields were achieved when attempting this transformation using **9**, independent of the reaction time. Interestingly, this methodology was also applied to aliphatic and aromatic secondary amines (such as **12** and **14**). Overall, formic acid proved to be a better formylating agent than ethyl formate, releasing water as by-product. However, the reaction media is considerably more acidic.

Even though the chemical properties of the starting materials play a key role in the outcome of any reaction, the effect of the solvent cannot be underestimated. The type of solvent employed determines the solubility of the starting materials, products and reaction rate among others. For example, the amidation of oleic acid **16** was possible due to the use of ethanol as the reaction solvent.[20] In this catalyst-free methodology, the reaction was carried out in a sealed autoclave at 60°C using 1 equivalent of acid and 1.2 equivalents of amine. Interestingly, even though a wide variety of amides were isolated in moderate to excellent yields, NMR analysis of the crude revealed traces of ethyl oleate **16a**. Moreover, when **16** was subjected to the same solvothermal conditions in the absence of amine, **16a** was isolated as the major compound. Additionally, the presence of different esters was observed in the ^{1}H NMR crude of the reaction when methanol and 2-propanol were used as solvents instead. Moreover, no reaction was observed under solvent-free conditions. This suggests ethanol reacts with **16** generating ester **16a**, which after nucleophilic attack of the amine affords the amide **17** as the only product (Scheme 2.6a). Oleic acid afforded the desired amides **17a** and **17b** in good yields when reacting with primary amines and the enantiopurity of (*R*)-1-phenylethylamine was preserved in amide **17c**. Reaction with ammonia afforded amide **17d** in low yield (35%) but with high conversion (76%), thus suggesting purification difficulties. Secondary **17e** and aromatic amines **17f** were also successfully employed, as well as different primary aliphatic and aromatic acids (products **18** and **19**). Strangely, isomerization of the *trans* to the *cis* double bond was only observed when pyrrolidine was used **17e**. The selectivity of this reaction is also worth mentioning. Isolating **20** in 78% yield proves that aliphatic primary amines are more reactive than aromatic amines and even though the ethanolic ester has been proven to

a) Formylation of aromatic amines

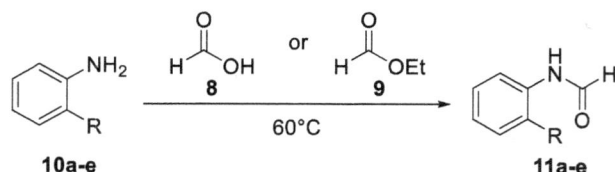

10a-e → (8 or 9, 60°C) → 11a-e

Amine	10a	10a	10b	10b	10c	10c	10d	10d	10e	10e
R	H	H	CH_3	CH_3	F	F	Cl	Cl	$COCH_3$	$COCH_3$
Formylating agent	8	9	8	9	8	9	8	9	8	9
Time	5 min	6h	15 min	18h	40 min	6h	15 min	24h	2.5h	48h
Yield	93	89	97	89	92	84	92	42	78	40

b) Formylation of secondary amines

12 → (Conditions) → 13

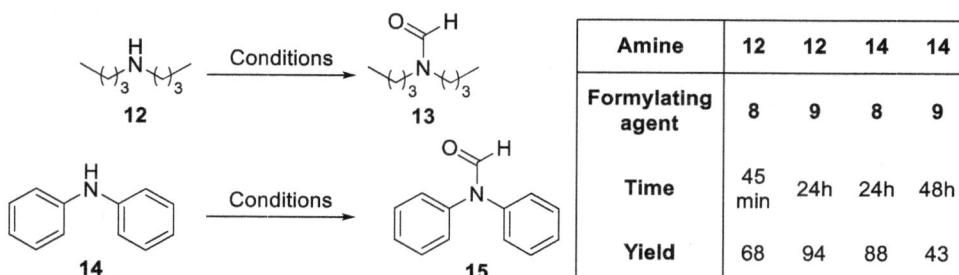

14 → (Conditions) → 15

Amine	12	12	14	14
Formylating agent	8	9	8	9
Time	45 min	24h	24h	48h
Yield	68	94	88	43

SCHEME 2.5 Formylating agent's effect in yield and reaction time.

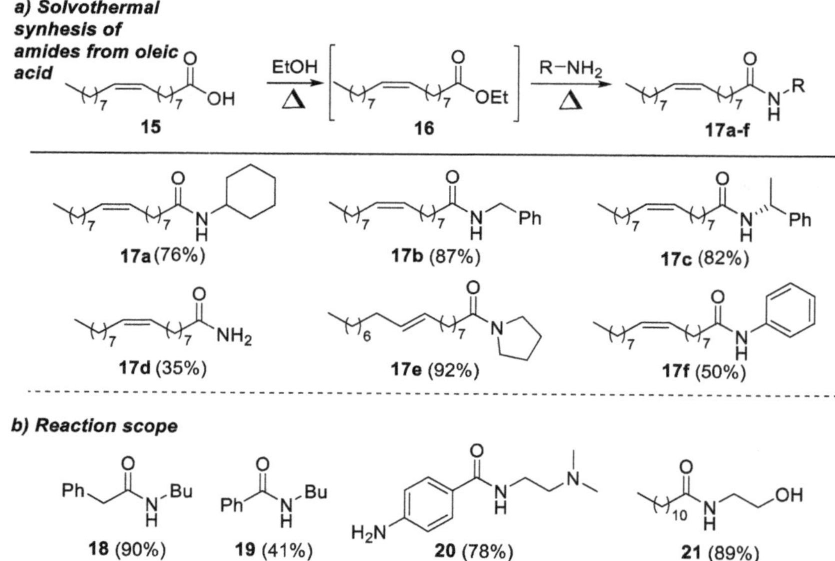

SCHEME 2.6 Solvothermal synthesis of amides.

be an intermediary, in the presence of an amine, the amide is the isolated product (no esterification was observed with ethanolamine but **21** was obtained in 89% yield). Even though this work is not a direct amidation of a carboxylic acid, the formation of the ester and its amidation is a one-pot reaction and does not require any catalyst but the use of an alcohol as solvent, the activation energy is reached under solvothermal conditions (Scheme 2.6).

Non-conventional heating methods have also been employed in the synthesis of amides. In 2008, You and coworkers

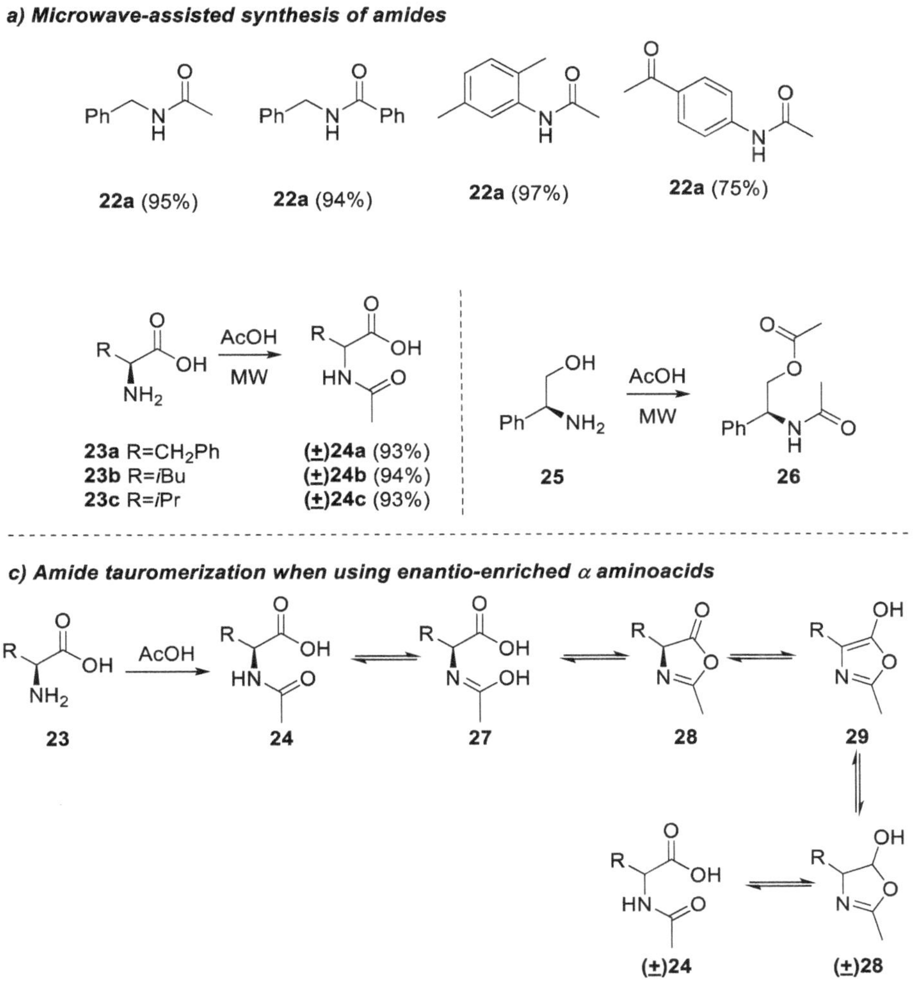

a) Microwave-assisted synthesis of amides

22a (95%) 22a (94%) 22a (97%) 22a (75%)

23a R=CH$_2$Ph (±)24a (93%)
23b R=iBu (±)24b (94%)
23c R=iPr (±)24c (93%)

AcOH, MW

25 26

c) Amide tauromerization when using enantio-enriched α aminoacids

23 24 27 28 29 (±)24 (±)28

SCHEME 2.7 Microwave-assisted synthesis of amides.

described a catalyst-free synthesis of primary amides under microwave irradiation.[21] As shown in Scheme 2.7a, this methodology was successfully applied to aliphatic primary and aromatic amines with acetic and benzoic acid, yielding the desired products **22a-d** in almost quantitative yield. Furthermore, additional chemoselective experiments were completed (Scheme 2.7b). When performing the reaction with enantiopure α-amino acids **23**, it proved to be selective toward amine acetylation. Even though this could be used as an amine's selective protection method, racemization of the products **24a-c** is a big drawback and must not be dismissed. Isolation of enantiopure ester **26** reveals no selectivity between the hydroxyl and the amino group but suggests tautomerization after amide formation when enantioenriched amino acids are employed. As shown in Scheme 2.7c, after amide formation, dehydration of **27** affords **28** and thus equilibrium between **28** and **29** is presumed to be responsible for the racemization, and this important feature is commonly observed in amino acid chemistry. Remarkably, all reactions took place in less than 15 minutes, being much faster than traditional synthetic methods.

As mentioned before, amide synthesis by condensation of carboxylic acids and amines is the ideal transformation in terms of atom economy. Nevertheless, as was discussed in previous examples, it is also a highly problematic reaction; it usually needs for special conditions or as it will be discussed later the use of additives or catalysts is imperative. Nevertheless, additional green alternatives to the direct condensation must not be dismissed. A catalyst-free decarboxylative amidation of α-ketoacids **30** promoted by light (23 W household bulbs) has been recently described.[22] Control and trapping experiments allowed the authors to confirm the generation of singlet oxygen by irradiation (Scheme 2.8a). Mechanistic studies suggest the formation of α-imino acid **32** as a key intermediary. Singlet oxygen abstraction of one electron from **32** generates **33**. Decarboxylation of **33** followed by reaction with water affords enol **35** *via* radical **34**. Amide is finally obtained after tautomerization of **35**. In order for the reaction to proceed, visible light, water and oxygen must be present in the media. This methodology was successfully applied to aromatic and aliphatic α-keto acids with primary and secondary amines. As shown in Scheme 2.8b, the reaction is selective toward α-keto acids in the presence of carboxylic acids and toward amines in the presence of alcohols (compounds **37** and **38** were isolated in 60% and 51% yield, respectively).

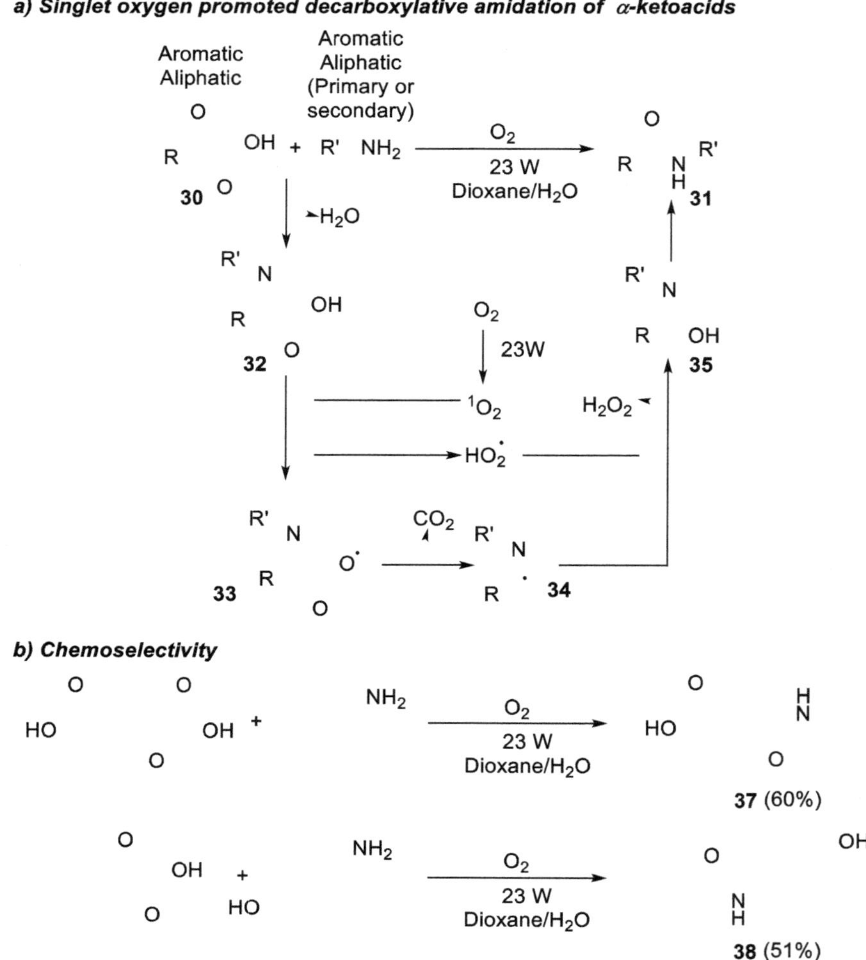

SCHEME 2.8 Singlet oxygen promoted decarboxylative amidation of α-ketoacids

2.4 Organic Additives

Organocatalysis has played a major role in organic chemistry, especially in the asymmetric synthesis of complex natural products. The use of environmentally friendly metal-free catalysts is highly encouraged in green chemistry due to its availability, low toxicity and low cost. As mentioned before, the direct amidation of carboxylic acids can be achieved by using light; this process can be accelerated with organic compounds (such as dyes) catalysts in several photo redox reactions, increasing yield and scope while diminishing reaction time. In 2018, Singh and coworkers reported a photoredox amidation reaction of carboxylic acids and amines in the presence of eosin Y **39** as catalyst.[23] As shown in Scheme 2.9a, the desired amides were isolated in good to excellent yields in less than three hours in toluene at room temperature. Interestingly, this sustainable approach was applied to both aromatic and aliphatic carboxylic acids with aliphatic primary and secondary amines, with chemoselectivity toward the aliphatic primary amine observed in the presence of aromatic **40c** or heteroaromatic amines **40b**. Highly hindered amines **40a** as well as further functionalized carboxylic acids **40d** were also successfully synthesized. A plausible mechanism is illustrated in Scheme 2.9b. The catalytic cycle starts with the photoexcitation of eosin Y to eosin Y*. This reactive species is reduced by the carboxylate anion generating radical **41** and radical anion eosin Y.—. Eosin Y is regenerated by oxygen while intermediate **42** is obtained from **41** by a diradical coupling reaction. Nucleophilic attack on the carbonyl group at **42** by the amine yields one equivalent of the desired amide and one equivalent of peracid salt **43** which is also transformed in the amide. Due to the regeneration of Eosin Y, the single electron transfer that enables the reaction can be done with 2 mol% catalyst.

Among the most desired characteristics of a green process, are the use of water and alternative sources of energy; the coupling of organo- and electrocatalysis proved successful in amide synthesis. Recently, an electrochemical *N*-acylation in the presence of tetrabutylammonium bromide (TBAB) under aqueous conditions was explored.[24] In this case, an equimolar solution of carboxylic acid, amine, base (Cs_2CO_3) and TBAB in water were subjected to a current of 40 mA over 30 minutes (Scheme 2.10). Diverse aromatic acids with electron-withdrawing **44a** and electron-donating groups **44b** were effectively coupled to primary **44c** and secondary aliphatic amines **44d** as well as heteroaromatic amines **44e** and hydrazine hydrate **44f**.

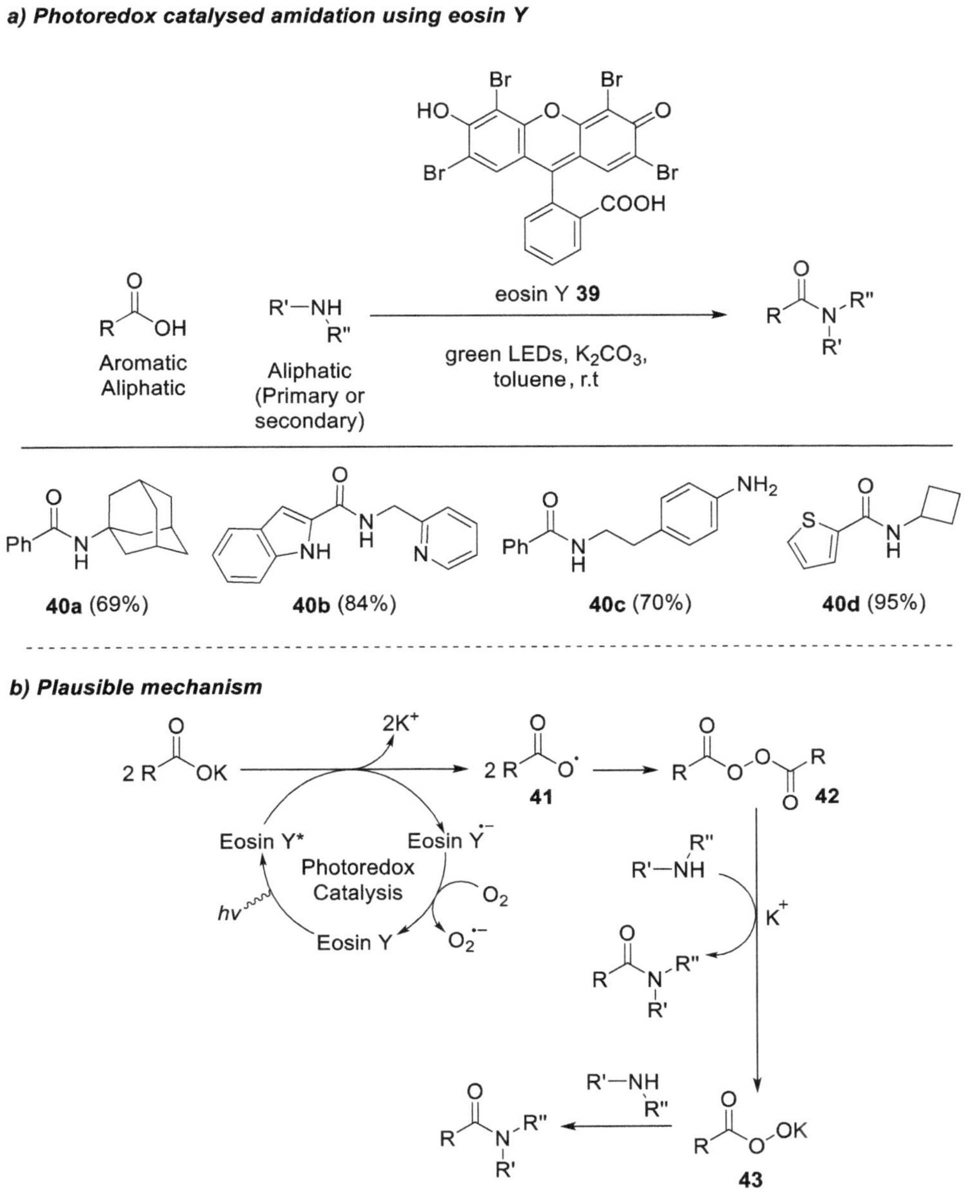

SCHEME 2.9 Photoredox-catalyzed amidation using eosin Y.

Furthermore, melatonin **44g**, the hormone that regulates the sleep–wake cycle, was successfully synthesized in 85% yield. The presence of TBAB plays a key role in the reaction as the source of bromide ions. As presented in Scheme 2.10c, intermediate **45c** is obtained from carboxylic salt **45a** via **45b**. After a homolytic cleavage of **45c**, radical **45d** is formed. Bromide ions are regenerated *in situ* at the cathode from the reduction of bromine radicals. Once radical **45e** is formed, a final oxidation to the desired amide takes place at the anode. This green methodology avoids the use of toxic reagents while showing a broad substrate scope using water as solvent.

As shown in Scheme 2.10c, activation of the carboxylic acid by *in situ* generation of acyl bromide **45c** plays a key role in the reaction. Even though the use of activating agents and coupling reagents is highly discouraged, several efforts have been proposed to make this approach greener. Esters are one of the most employed activated derivatives of carboxylic acids and their formation *in situ* has been widely explored despite its low atom economy. From this point of view, 2,4,6-trichloro-1,3,5-triazine (TCT) **46** offers an interesting alternative by using a 1:3 ratio to the carboxylic acid in addition to being commercially available, stable, non-volatile and non-toxic. Nowadays, direct amidation of carboxylic acids via triazine ester is considered an environmentally friendly alternative. In 2018, a mechanochemical synthesis of primary amides using TCT and NH_4SCN as an ammonia source was reported.[25] Aromatic **47a-e** and aliphatic **47f-g** amides were successfully synthesized in excellent yield just after 5 minutes of grinding the reaction mixture. Interestingly, no Michael addition product was observed when synthesizing **47h**. The presence of Fmoc as a protecting group was also tolerated **47i**. Additionally, TCT has also been used in the synthesis of secondary and tertiary

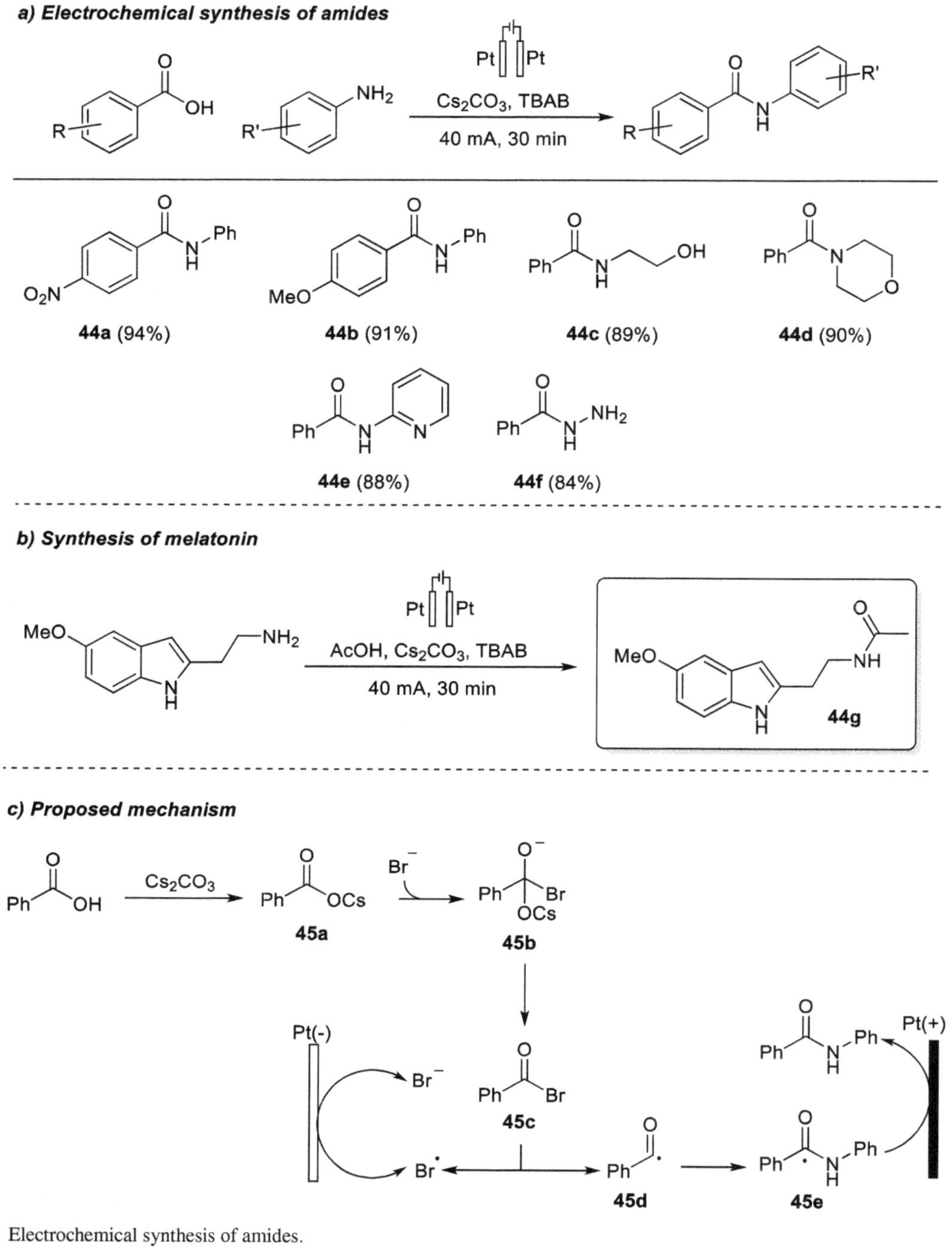

a) Electrochemical synthesis of amides

Pt | Pt

Cs_2CO_3, TBAB

40 mA, 30 min

44a (94%) 44b (91%) 44c (89%) 44d (90%)

44e (88%) 44f (84%)

b) Synthesis of melatonin

Pt | Pt

AcOH, Cs_2CO_3, TBAB

40 mA, 30 min

44g

c) Proposed mechanism

Cs_2CO_3

45a 45b 45c 45d 45e

Pt(-) Pt(+)

SCHEME 2.10 Electrochemical synthesis of amides.

amides. In 2019 choline chloride/urea was employed as a biodegradable and eco-friendly solvent for amidation reaction at room temperature.[26] In this case, primary **48a-c** and secondary **48d-e** amines were successfully coupled to benzoic acid. The use of aliphatic and additional aromatic carboxylic acids with electron-withdrawing and electron-donating groups was also explored (Scheme 2.11).

Besides, activation of carboxylic acids as esters can be performed under acidic conditions. In 2012, Mandal and coworkers reported the mild activation of *N*-protected amino acids **49** as benzyl esters **50** under microwave irradiation using benzyl alcohol as solvent.[27] After treatment with an amine, the desired amides **51** were isolated in moderate to good yields (Scheme 2.12). The E-factor of the reaction (ratio of the mass of waste generated per mass of product obtained) can be significantly reduced through the addition of ethanol to precipitate the product, negating the need for chromatography. As such, the remaining reaction mixture can be recycled. Even though the formation of the benzyl ester decreases the atom economy of this transformation, activation by microwave irradiation and significant reduction of the waste generated make this procedure a green eco-friendly alternative in large-scale synthesis of amides. Additionally, organic additives containing silicon have been recently employed as coupling reagents[28] and reducing agents in the amide synthesis.[29]

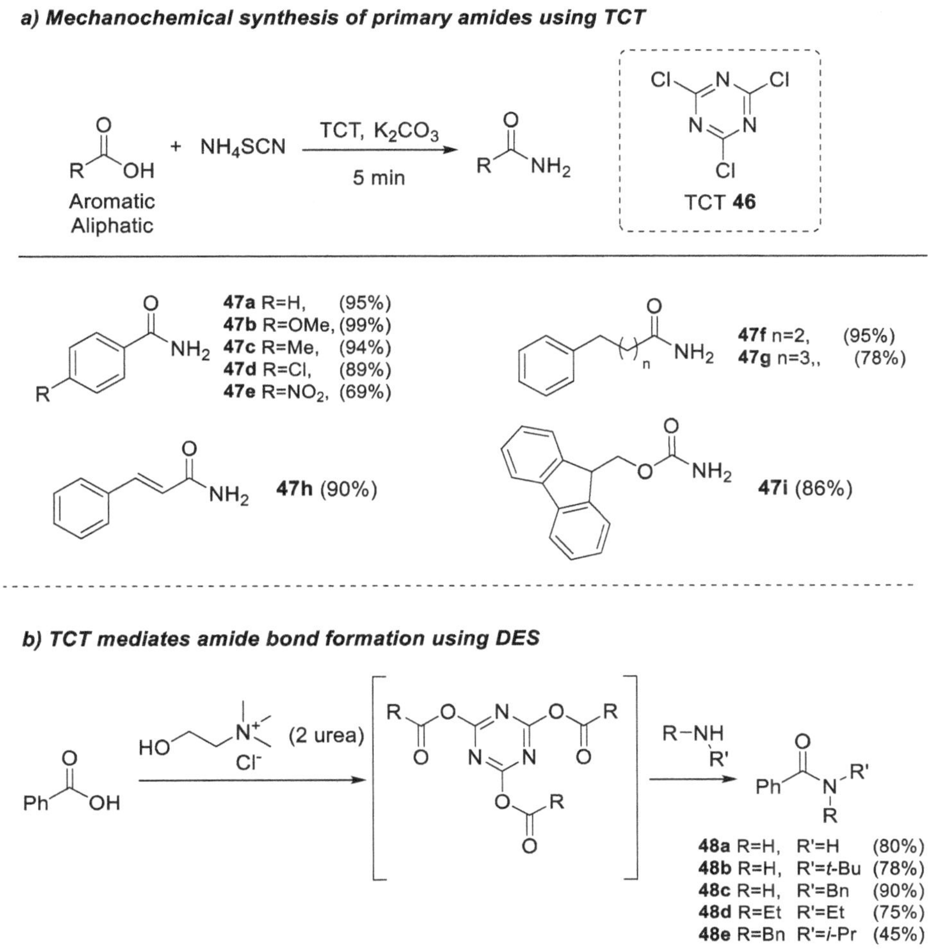

a) Mechanochemical synthesis of primary amides using TCT
+ NH4SCN
TCT, K2CO3
5 min
Aromatic
Aliphatic
TCT 46
47a R=H, (95%)
47b R=OMe, (99%)
47c R=Me, (94%)
47d R=Cl, (89%)
47e R=NO2, (69%)
47f n=2, (95%)
47g n=3,, (78%)
47h (90%)
47i (86%)
b) TCT mediates amide bond formation using DES
(2 urea)
48a R=H, R'=H (80%)
48b R=H, R'=t-Bu (78%)
48c R=H, R'=Bn (90%)
48d R=Et R'=Et (75%)
48e R=Bn R'=i-Pr (45%)

SCHEME 2.11 Mechanochemical synthesis of amides.

p-TsOH (0.3 equiv)
BnOH (2 equiv)
MW, 100°C, 50W, 5min
49
50
(2 equiv)
40°C
Crude Mixture
EtOH
Recycling of the
reaction mixture
51a 68%
51a 68%
51a 68%
51

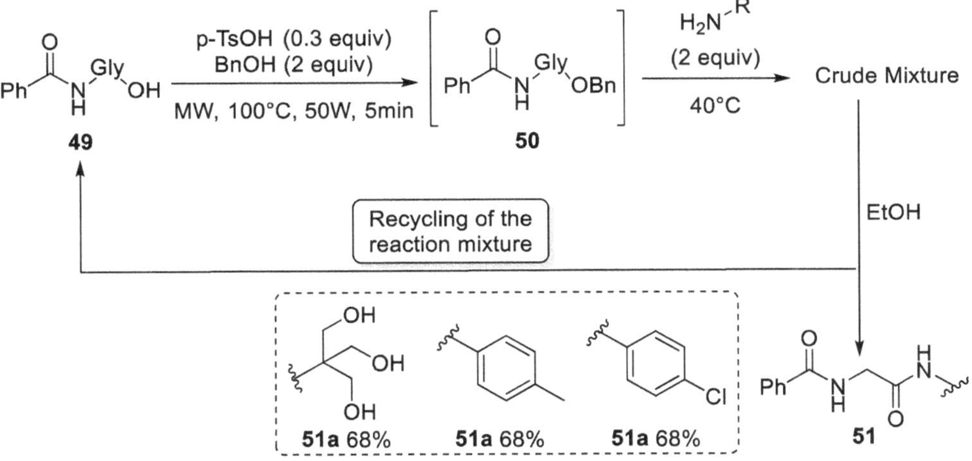

SCHEME 2.12 E-factor reduction in amide synthesis from *N*-protected amino acids.

Even though organic additives represent a remarkable improvement toward the development of green methodologies in amide and peptide synthesis; the waste generated, long reaction times and high temperatures are drawbacks that need further optimization.

2.5 Boron

Boron-containing compounds have been widely applied in rearrangements, Suzuki couplings, allylboration reactions and hydroboration-oxidation processes.[30] The first amidation

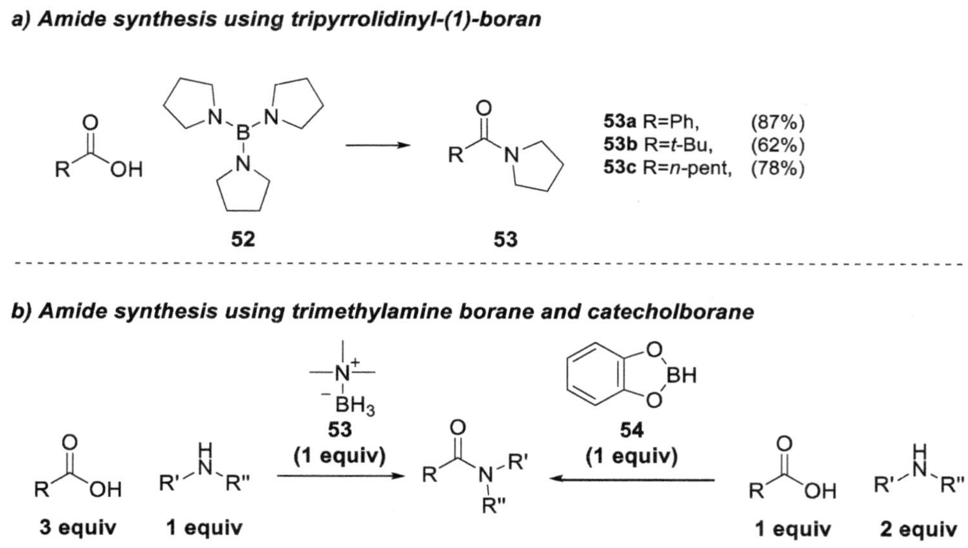

a) Amide synthesis using tripyrrolidinyl-(1)-boran

53a R=Ph, (87%)
53b R=*t*-Bu, (62%)
53c R=*n*-pent, (78%)

b) Amide synthesis using trimethylamine borane and catecholborane

SCHEME 2.13 Amide synthesis using stoichiometric amounts of boron reagents.

reaction of carboxylic acids in the presence of aminoborane compounds was reported in 1965 (Scheme 2.13).[31] In this solvent-free approach, the desired amides were obtained in good yields when mixing a carboxylic acid with one equivalent of tripyrrolidinyl-(1)-boran **52**. Since only one of the amino groups of the borane compound is present in the final product, the poor atom economy of this transformation remains a major disadvantage. Additional methodologies that required the use of stoichiometric amounts of borane trimethylamine complex **54a**[32] or catecholborane **54b**[33] and an excess of either the carboxylic acid or the amine were explored in the following years (Scheme 2.13b).

In 1996, Yamamoto reported the use of electron-poor arylboronic acids **55** and **56** as catalysts in the direct amidation of carboxylic acids.[34] These Lewis acids exhibited an increased reactivity while tolerating acidic, basic and aqueous conditions. Even though this was the first approach toward a greener methodology of amide synthesis using boron-based compounds; the higher temperatures, low yields and long reaction times were not ideal. It was not until 2008 when a waste-free amidation procedure at room temperature was described by Hall and coworkers.[35] A comparison between **55** and **56** with **57** showed a significant increase of the yield when catalytic amounts of **57** were used in the direct amidation of phenylacetic acid with benzylamine. Attractively, the use of equimolar amounts of amines and carboxylic acids facilitated the product purification and allowed the catalyst to be easily recovered and reused. Furthermore, the mild reaction conditions allowed the synthesis of biologically interesting highly functionalized compounds. Protected serotonin **58a**, indomethacin **58b** and **58c** and ibuprofen **58d** and **58e** amide derivates were successfully synthesized in good to excellent yields when applying this methodology (Scheme 2.14).

Encouraged by the low racemization observed in ibuprofen derivatives **58d** and **58e**, peptide synthesis catalyzed by boronic acids emerged as the next target. Further studies conducted by Hall proved second-generation catalyst **59** to be a better catalyst compared to **57**.[36] Higher yields in shorter reaction times were obtained by direct amidation of aliphatic and heteroaromatic carboxylic acids with aliphatic amines when molecular sieves were added to the reaction mixture. As shown in Scheme 2.15, double *N*-protected α-phthaloyl amino acids **60** were successfully coupled to C-protected α-amino acids **61** using 25 mol% of **59** at 40°C. Direct amidation of α-azido acids **63** and C-protected α-amino esters **64** were also achieved. Although no epimerization was observed and the desired compounds were easily isolated, the low reactant concentration and the number of molecular sieves required to drive the reaction forward are considered the main drawbacks of this methodology.

The mechanism of the reaction in the presence of boronic acids differs from those discussed above. In this case, computational studies suggest the activation of the carboxylic acid as acyloxyboronic species **66**. Even though the reaction mechanism is still under consideration, two different pathways are currently accepted (Scheme 2.16). In earlier studies, attack of the amine to the tetravalent boron intermediate **66** followed by dehydration affords tetrahedral intermediate **67**.[37] Finally, intermediate **67** can rearrange, regenerating the catalyst and forming the amide product. In the second and more recent approach, acyl boronate **68** is formed after dehydration of intermediate **66**. Nucleophilic attack of the amine affords the amide *via* zwitterion intermediate **69**.[38] The successful use of boronic esters in amide synthesis and especially in peptide coupling is a result of the mild conditions employed. Nowadays, the use of boronic acids as catalysts is one of the most attractive green alternatives in the synthesis of amides to the point statistical analysis of experiments that are being conducted in order to give a better understanding of the reaction and its further application in industry.[39]

2.6 Mesoporous Solids as Heterogeneous Catalyst

In green chemistry, the use of heterogeneous catalysts is highly encouraged due to its ease of recovery and separation of the reaction mixture. Nowadays, silica stands as a groundbreaking

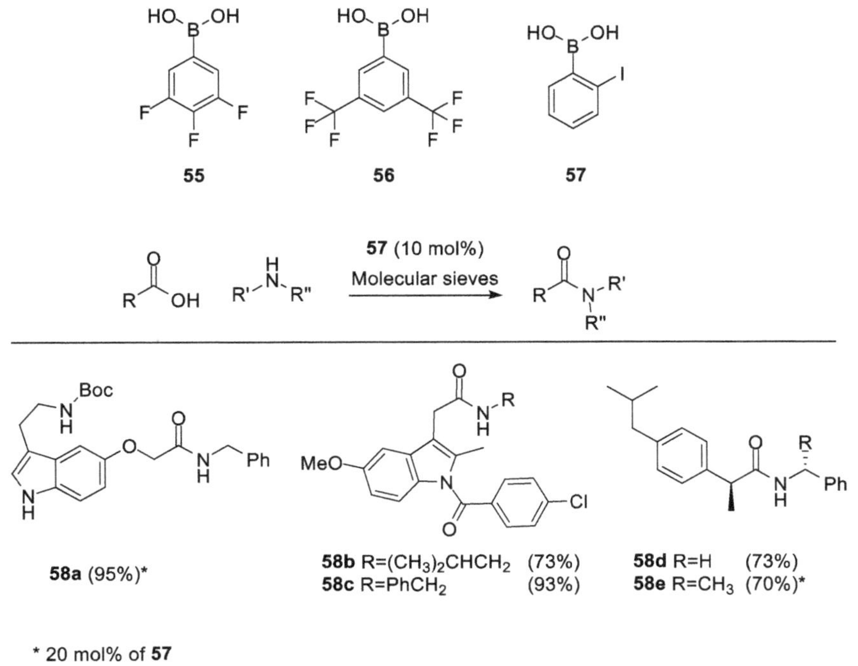

55 **56** **57**

57 (10 mol%)
Molecular sieves

58a (95%)*

58b R=$(CH_3)_2CHCH_2$ (73%)
58c R=$PhCH_2$ (93%)

58d R=H (73%)
58e R=CH_3 (70%)*

* 20 mol% of **57**

SCHEME 2.14 Boronic acids as catalysts in amide formation.

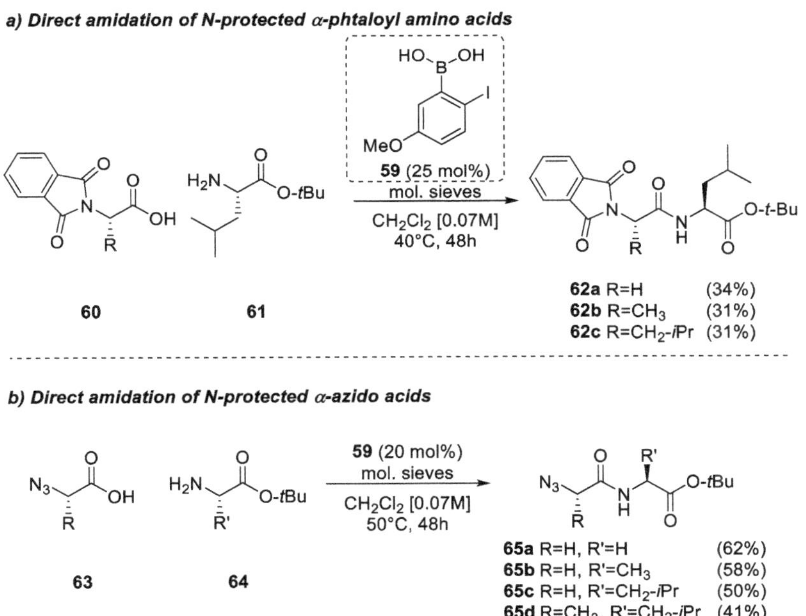

SCHEME 2.15 Amidation of amino and azido acids catalyzed by boronic acid **59**.

solid support thanks to its availability, low cost, abundance, weak acidity and hydrophobicity. One of the first examples in the use of silica as a green heterogeneous catalyst in the direct amidation of carboxylic acids was described in 2009 by Clark and coworkers (Scheme 2.17).[40] In this approach, several aliphatic carboxylic acids were successfully amidated in good to excellent yields when pre-activated silica was employed. Unfortunately, low yields were observed when aromatic acids were used instead. Even though an equimolar mixture of acid and amine can be employed, experiments conducted proved that silica must be pre-activated at 700°C and water needs to be removed by a Dean–Stark tramp or by the addition of molecular sieves and thus increasing the amount of waste generated. Still, it was proved that silica could be recovered and a preliminary assay was achieved to demonstrate the feasibility of a flow process. A more recent study allowed the formylation of primary, secondary and heteroaromatic amines at room temperature under solvent-free conditions (Scheme 2.17b).[41] Interestingly, the catalyst could be reused at least four times without deactivation, and excellent chemoselectivity

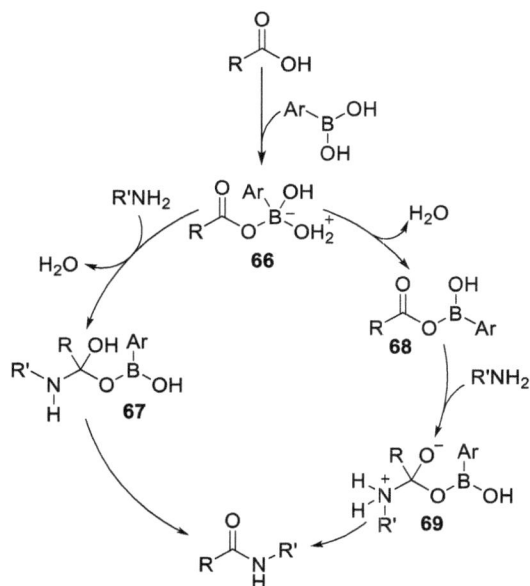

SCHEME 2.16 Mechanism of amide coupling using boronic acids as catalyst.

for substrates with hydroxy functionalities was demonstrated. Although the E-factor was reduced, the use of an excess of formic acid as well as the two-step synthesis of the functionalized mesoporous silica (Des/SBA-15) employed as a catalyst might be considered a drawback in the path toward an environmentally friendly synthesis of amides.

In 2015, we published a novel green methodology for the synthesis of amides under microwave irradiation using silica gel as a solid support.[42] Even though a microwave-assisted synthesis of amide had been described by Loupy[43] restricted to aliphatic acids with primary aliphatic amines we successfully demonstrated (Scheme 2.18) the combination of MW heating and the use of silica gel as solid support to be optimal for the amidation of aliphatic and aromatic carboxylic acids with primary and secondary amines in addition to aromatic amines. Several experiments to prove the selectivity of the reaction were performed. Firstly, reaction optimization was conducted using cinnamic acid as starting material. Isolation of **71** in 94% yield showed no traces of the Michael addition to the double bond. Secondly, the selectivity between different types of amines was demonstrated by two additional experiments showing primary aliphatic amines to be more reactive than secondary and aromatic amines (compound **72** was isolated as the only product in excellent yields when benzoic acid was reacted with butylamine in the presence of either pyrrolidine or aniline). Lastly, selective *N*-acylation in the presence of O and S was also showed. Despite the numerous advantages of this methodology, such as short reaction time, easy work up and purification of the desired product, this methodology cannot be applied in peptide synthesis.

Parallel to our work, an additional methodology using mesoporous silica SBA-15 was developed[44] (Scheme 2.19). Although the reaction is performed under reflux and is strongly dependent on the temperature, it was successfully applied to *N*-Boc protected alanine **75**, affording benzylated compound **76** in 87% yield. In addition, the recovery and reutilization of the catalyst was evaluated in the synthesis of procainamide **80**, a widely used pharmaceutical as an arrhythmic agent. Direct amidation of *p*-nitrobenzoic acid **77** with *N,N*-diethyl ethylenediamine afforded intermediate **79** in 95% isolated yield. The catalyst was recovered by filtration and activated by calcination

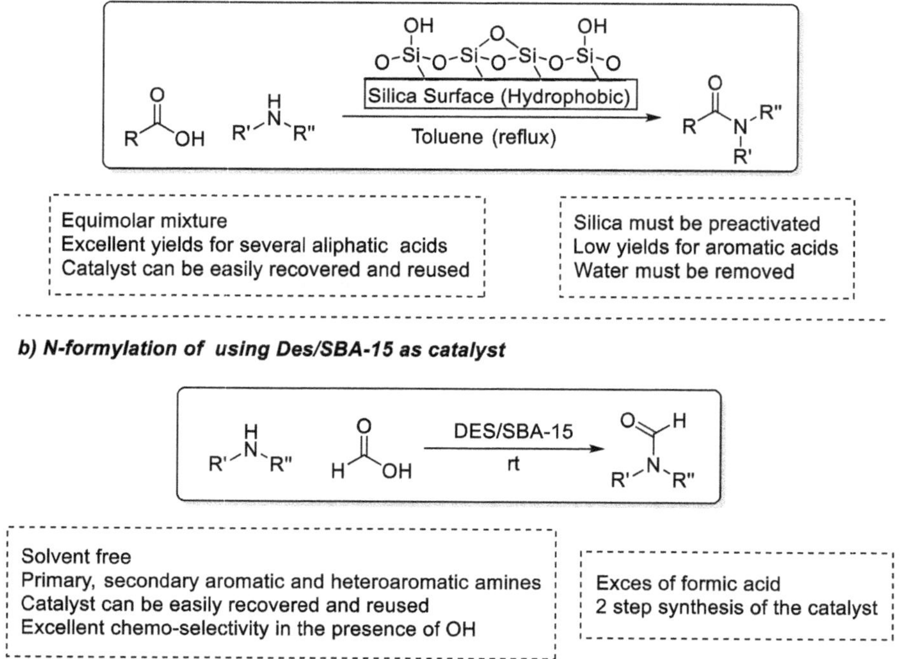

SCHEME 2.17 Amidation of carboxylic acids using silica as a heterogeneous catalyst.

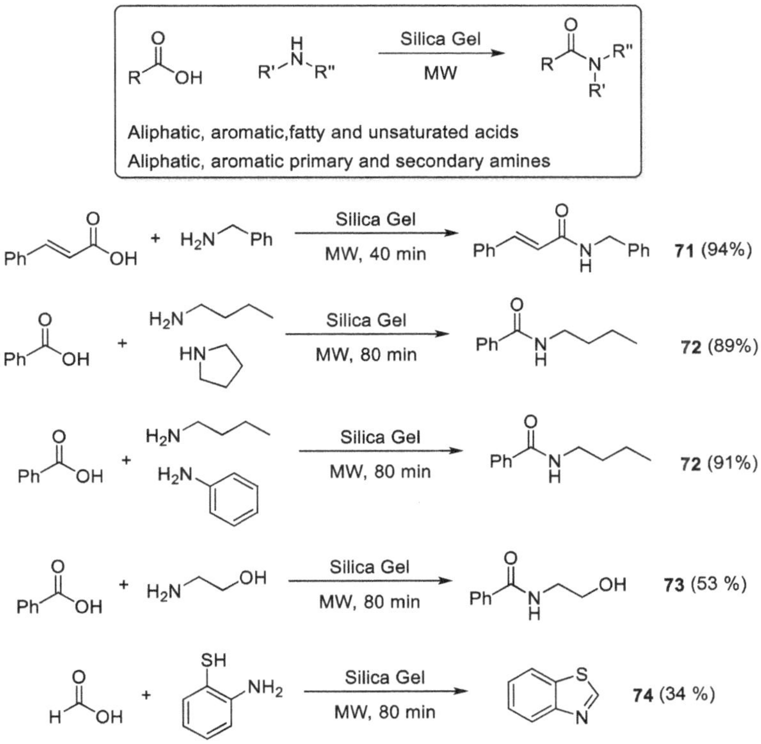
Silica Gel
MW
Aliphatic, aromatic,fatty and unsaturated acids
Aliphatic, aromatic primary and secondary amines
Silica Gel
MW, 40 min
71 (94%)
Silica Gel
MW, 80 min
72 (89%)
Silica Gel
MW, 80 min
72 (91%)
Silica Gel
MW, 80 min
73 (53 %)
Silica Gel
MW, 80 min
74 (34 %)

SCHEME 2.18 Direct amidation under MW irradiation using silica gel as solid support.

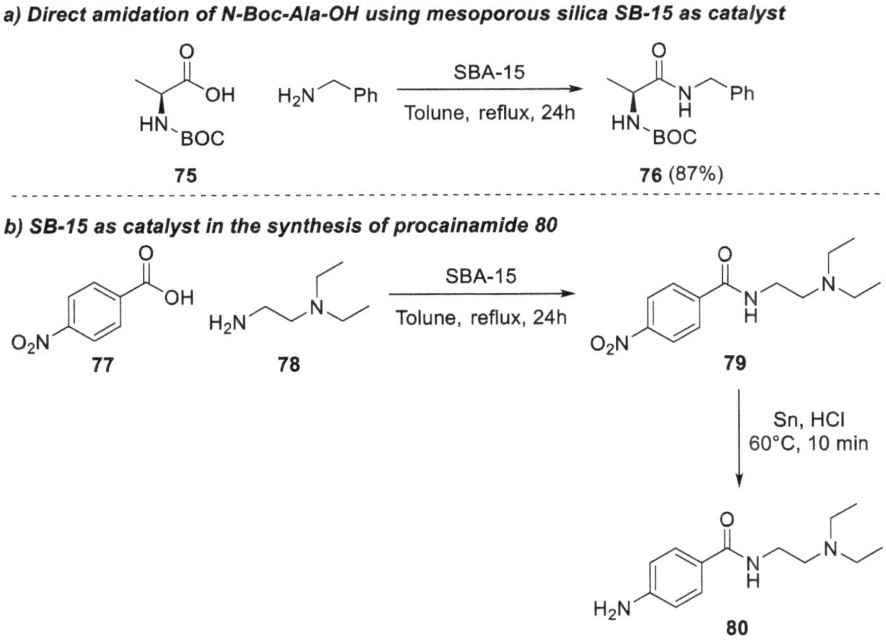
a) Direct amidation of N-Boc-Ala-OH using mesoporous silica SB-15 as catalyst
SBA-15
Tolune, reflux, 24h
75
76 (87%)
b) SB-15 as catalyst in the synthesis of procainamide 80
SBA-15
Tolune, reflux, 24h
77
78
79
Sn, HCl
60°C, 10 min
80

SCHEME 2.19 Direct amidation using mesoporous silica SB-15 as catalyst.

at 550°C, over five hours, then it was successfully employed in four additional cycles. Gas chromatography demonstrated the activity of the catalyst did not decline since the yields were over 99% in all cases. A heterogeneous reduction in the presence of Sn afforded **80** in 90% yields over two steps.

In addition to silica, other mesoporous materials have been employed as catalyst. A solvent-free formylation method using a NaY functionalized zeolite under mild conditions afforded excellent yields in short reaction times (Scheme 2.20).[45] Even though the catalyst can be easily recovered and reused, an excess amount of formic acid must be employed and thus a larger amount of waste is generated. Additionally, the use of activated alumina balls[46] and a mesoporous carbonaceous solid acid (Starbon® acid)[47] have also been reported. In both

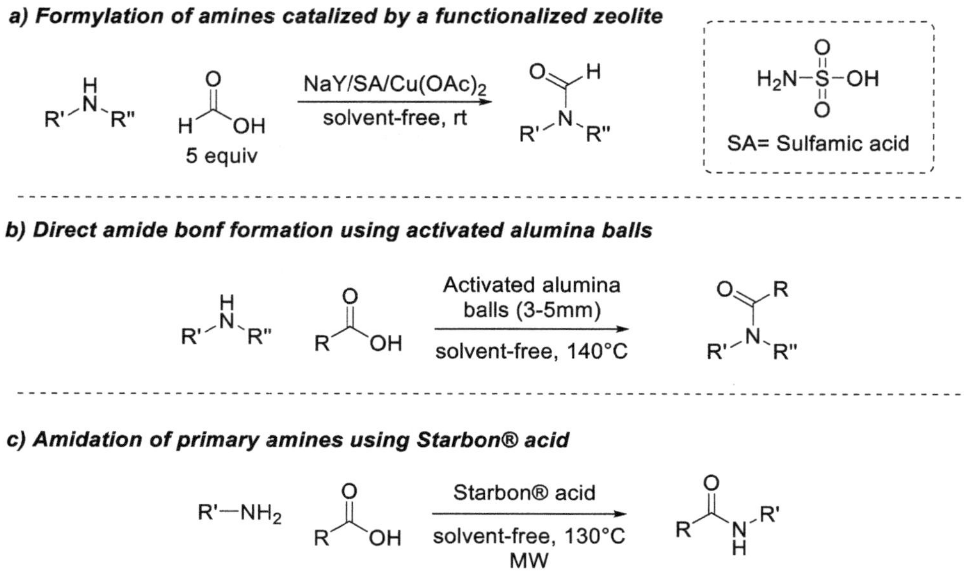

a) Formylation of amines catalized by a functionalized zeolite

NaY/SA/Cu(OAc)2, solvent-free, rt; 5 equiv; SA= Sulfamic acid

b) Direct amide bonf formation using activated alumina balls

Activated alumina balls (3-5mm), solvent-free, 140°C

c) Amidation of primary amines using Starbon® acid

Starbon® acid, solvent-free, 130°C, MW

SCHEME 2.20 Direct amidation catalyzed by (a) zeolite, (b) activated alumina balls, (c) Starbon®acid.

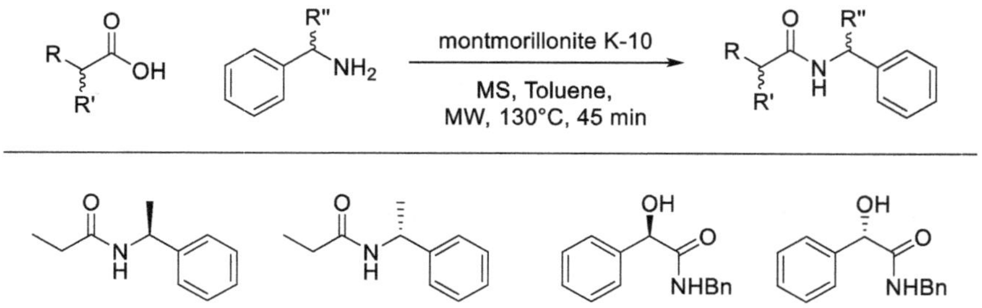

montmorillonite K-10, MS, Toluene, MW, 130°C, 45 min

81a (88%) **81b** (90%) **81c** (76%) **81d** (79%)

SCHEME 2.21 Montmorillonite K-10 as catalyst in the synthesis of α-branched amides.

cases, an equimolar mixture of acids and amines can be used as the starting material and the catalyst can be reused without decreasing its activity (Schemes 2.20b, c). The mild reaction conditions, ease work-up and the displacement of the reaction toward the products by the absorption of water by the catalyst makes these methodologies extremely useful.

Lastly, we would like to mention the use of commercially available montmorillonite K-10 in the amidation of sterically hindered α-branched carboxylic acids and amines and its application in the synthesis of sterically hindered enantio-pure compounds.[48] In this case, the integrity of pre-existing chiral centers was conserved after performing the reaction in toluene and under MW irradiation. Compounds **81a-d** were successfully isolated in good yields after 45 minutes when using molecular sieves to remove the produced water (Scheme 2.21). Isolation of **81c** and **81d** proved the chemoselectivity of the reaction since no traces of the ester were observed.

2.7 Other Catalytic Systems

Several transition metals have been employed as homogeneous and heterogeneous catalysts in the direct amidation of carboxylic acids due to its stability, conductivity and great variety in the oxidation state. The synthesis of formamides catalyzed by $ZnCl_2$[49] or ZnO[50] and the use of a sulfated tungstate[51] or $FeCl_3$[52] have been reported in the past few years. In 2018, an air stable bis(pentamethylcyclopentadienyl) zirconium perfluorooctanesulfonate **82** was successfully employed as a recyclable catalyst in the synthesis of *N*-substituted amides.[53] As shown in Scheme 2.22, paracetamol **83** was successfully isolated in 88% yield after 5 hours. Interestingly, moclobemide **84**, a commercially available medicament used in the treatment of depression, was effectively synthesized in a 10 mmol scale.

As described above, transition metals are excellent catalysts in the synthesis of amides. Even though they are usually employed as salts, oxides and coordination complexes, its presence in nanoparticles has been recently encouraged. For more than 20 years, nanoscience and nanotechnology have experienced an increased interest, particularly in the synthesis of novel and greener nano catalysts.[54] Due to their high surface area and unique physical, chemical, electronic, magnetic and thermic properties, transition metal nanoparticles have been widely employed as recyclable catalysts in the synthesis of formamides under solvent-free conditions.[55] By being in the interphase of homogeneous and heterogeneous catalysts, their removal from the reaction mixture is easily achieved by filtration, centrifugation and even by application of an external

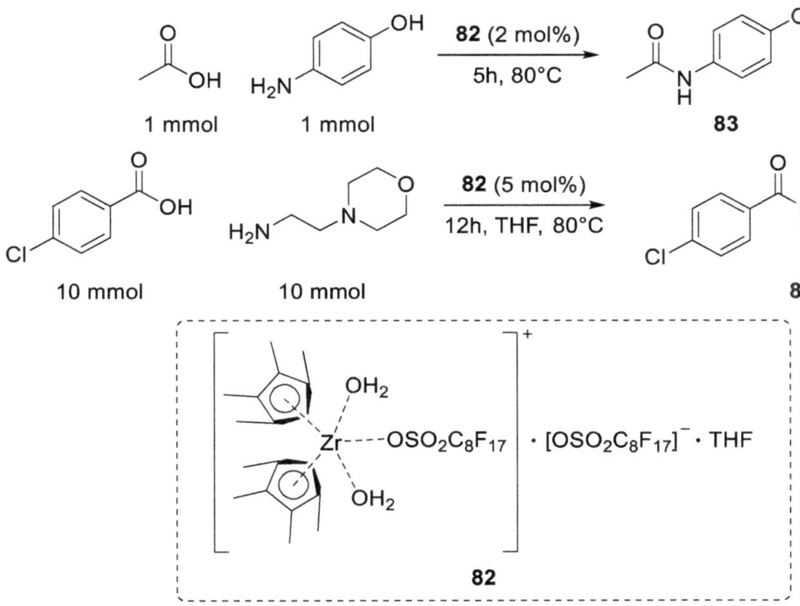

SCHEME 2.22 Synthesis of biologically active compounds using **82** as catalyst.

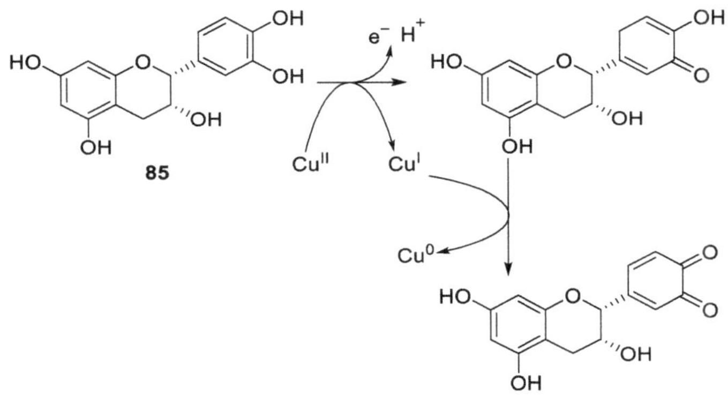

SCHEME 2.23 In situ green synthesis of Cu nanoparticles.

magnetic field, facilitating the reaction work-up and product purification. The synthesis and characterization of these above-mentioned nanoparticles have been widely reported and standardized; however, the use of several surfactants and tetraethyl orthosilicate still represents a drawback in terms of atom economy. In 2015, Nasrollahzadeh reported the synthesis of copper nanoparticles using *Anthemis xylopoda* flowers and their applications for *N*-formylation of amines at room temperature.[56] As shown in Scheme 2.23, the presence of several antioxidant flavonoid glycosides (such as luteolin glucoside **58**) in the aqueous extract of the flowers is responsible for the reduction of the Cu(II) species and the in situ formation of the nanoparticles. However, despite the short reaction times, high yields and reusability of the nanoparticles observed, to our knowledge, there is only one report of amide synthesis using diverse acids instead of amine formylation.[57]

Additionally, ionic liquids have been employed in the synthesis of amides, showing different advantages such as low toxicity, thermal stability and low vapor pressure. In 2018, the synthesis of several benzamides in the presence of diatomite earth@IL/$ZrCl_4$ **86** under ultrasonic irradiation was described (Scheme 2.24).[58] Interestingly, the reactions were completed in less than an hour and the catalyst was reused at least five times without observing a significant decrease in the yield. Remarkably, Yuan and coworkers reported the use of a heteropolyanion-based ionic liquid **87** under solvent-free conditions in the synthesis of *N*-formyl α-amino acid derivatives.[59] As shown in Scheme 2.24b, racemization was not observed by HPLC when compounds **88a-b** were isolated in excellent yields.

Finally, as illustrated previously, amides are present in the structural core of several natural occurring molecules and are frequently used as stating materials in the synthesis of biological active compounds. In this section, we present a few selected examples involving the condensation of carboxylic acids with amines in the green synthesis of diverse moieties of current interest. For example, oxindoles **90** are aromatic heterocyclic organic compounds that have recently being studied due to its pharmaceutical properties.[60] In 2005, several *N*-substituted oxindoles were achieved through a MW-assisted amidation of 2-halo-arylacetc acids **89** followed by an intramolecular amidation catalyzed by palladium.[61] When alkylamines are employed

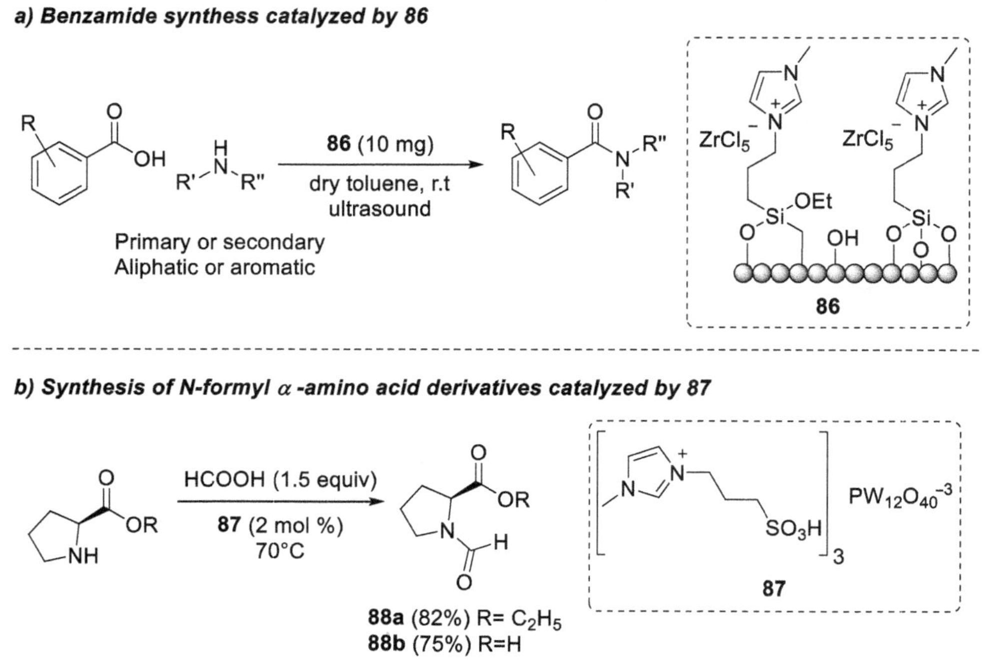

b) Synthesis of N-formyl α-amino acid derivatives catalyzed by 87

SCHEME 2.24 Ionic liquids as catalyst in the synthesis of amides.

as a starting material, isolation of the amide intermediate is not necessary (Scheme 2.25).

Pyrimidine-based heterocyclic compounds have been widely studied due to their antiviral, antibiotic, and anti-inflammatory properties.[62] As shown in Scheme 2.26, a green multi-component approach involving an *N*-formylation of **91** followed by a Knoevenagel condensation allows the synthesis of several barbituric and thiobarbituric acid derivatives **92** in a catalyst free-procedure.[63] Once the *N*-formylation of the amine has taken place, a nucleophilic attack of **93** to the carbonyl of the amide followed by the release of water results in the desired compound.

Finally, we would like to mention that even though the use of coupling agents is highly discouraged due to its toxicity, poor

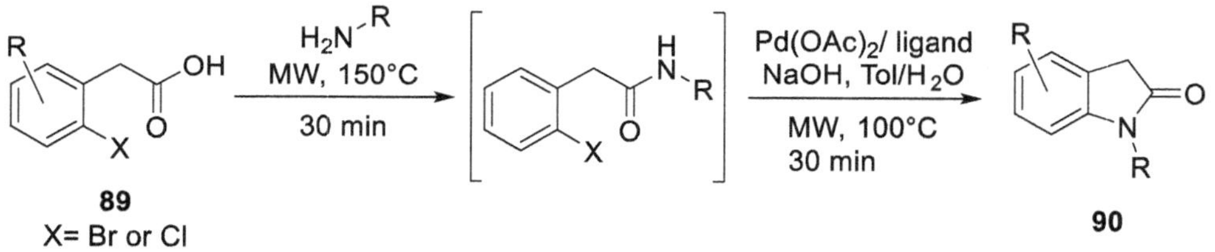

SCHEME 2.25 Microwave-assisted synthesis of oxindoles **90**.

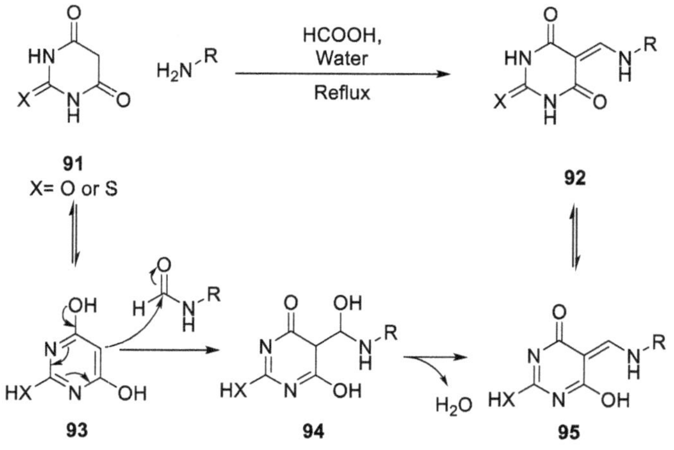

SCHEME 2.26 Green synthesis of barbituric and thiobarbituric acid derivatives.

atom economy and use of non-environmentally friendly solvents such as DMF and DCM, several efforts have been made to make this approach greener.[64]

2.8 Conclusion

As the reader should notice at this stage, the direct amidation of carboxylic acids with amines is one of the most powerful and promising strategies for the synthesis of bioactive compounds. The results and current methods are still far away for massive industrial applications in compounds like peptides; however, several teams worldwide are pursuing hard research trying to achieve this goal.

The activation of the carboxylic acid is still a significant problem, and even if its dual reactivity (nucleophile-electrophile) has been used, it seems like stronger activation is required. Further details about the reaction mechanisms, principally with solid supports and catalysts, will be highly appreciated in order to better understand the reactions pathways and to develop new catalytic systems or processes.

In conclusion, even if conceptually simple, direct amidation of carboxylic acids is still a big challenge for the scientific community and making it greener is one of the most important future aims.

REFERENCES

1. Humphrey, J. M.; Chamberlin, A. R. *Chemical Reviews* 1997, 97, 2243–2266.
2. Valeur, E.; Bradley, M. *Chemical Society Reviews* 2009, 38, 606–631.
3. Narendar Reddy, T.; Beatriz, A.; Jayathirtha Rao, V.; de Lima, D. P. *Chemistry – An Asian Journal* 2019, 14, 344–388.
4. Ghose, A. K.; Viswanadhan, V. N.; Wendoloski, J. J. *Journal of Combinatorial Chemistry* 1999, 1, 55–68.
5. Carey, J. S.; Laffan, D.; Thomson, C.; Williams, M. T. *Organic & Biomolecular Chemistry* 2006,4, 2337–2347.
6. Reddy, T. N.; de Lima, D. P. *Asian Journal of Organic Chemistry* 2019, 8, 1227–1262.
7. Bourne-Branchu, Y.; Gosmini, C.; Danoun, G. *Chemistry – A European Journal* 2019, 25, 2663–2674.
8. Constable, D. J. C.; Dunn, P. J.; Hayler, J. D.; Humphrey, G. R.; Leazer, J. J. L.; Linderman, R. J.; Lorenz, K.; Manley, J.; Pearlman, B. A.; Wells, A.; Zaks, A.; Zhang, T. Y. *Green Chemistry* 2007, 9, 411–420.
9. Bryan, M. C.; Dunn, P. J.; Entwistle, D.; Gallou, F.; Koenig, S. G.; Hayler, J. D.; Hickey, M. R.; Hughes, S.; Kopach, M. E.; Moine, G.; Richardson, P.; Roschangar, F.; Steven, A.; Weiberth, F. J. *Green Chemistry* 2018, 20, 5082–5103.
10. (a) Dunlap, F. L. *Journal of the American Chemical Society* 1902, 24, 758–763; (b) Mitchell, J. A.; Reid, E. E. *Journal of the American Chemical Society* 1931, 53, 1879–1883.
11. (a) Charville, H.; Jackson, D. A.; Hodges, G.; Whiting, A.; Wilson, M. R. *European Journal of Organic Chemistry* 2011, 2011, 5981–5990; (b) Charville, H.; Jackson, D.; Hodges, G.; Whiting, A. *Chemical Communications* 2010, 46, 1813–1823.
12. (a) Cossy, J.; Pale-Grosdemange, C. *Tetrahedron Letters* 1989, 30, 2771–2774; (b) Jursic, B. S.; Zdravkovski, Z. *Synthetic Communications* 1993, 23, 2761–2770.
13. Chocholoušová, J.; Vacek, J.; Hobza, P. *The Journal of Physical Chemistry A* 2003, 107, 3086–3092.
14. (a) Kobayashi, S.; Nishio, K. *The Journal of Organic Chemistry* 1994, 59, 6620–6628; (b) Kobayashi, S.; Yasuda, M.; Hachiya, I. *Chemistry Letters* 1996, 25, 407–408.
15. Lohray, B. B.; Baskaran, S.; Srinivasa Rao, B.; Yadi Reddy, B.; Nageswara Rao, I. *Tetrahedron Letters* 1999, 40, 4855–4856.
16. Chen, B.-C.; Bednarz, M. S.; Zhao, R.; Sundeen, J. E.; Chen, P.; Shen, Z.; Skoumbourdis, A. P.; Barrish, J. C. *Tetrahedron Letters* 2000, 41, 5453–5456.
17. Kobayashi, K.; Nagato, S.; Kawakita, M.; Morikawa, O.; Konishi, H. *Chemistry Letters* 1995, 24, 575–576.
18. (a) Grant, H. G.; Summers, L. A. *Australian Journal of Chemistry* 1980, 33, 613–617; (b) Schwan, M.; Bobylev, M. *The FASEB Journal* 2007, 21, A998–A998.
19. Dhake, K. P.; Tambade, P. J.; Singhal, R. S.; Bhanage, B. M. *Green Chemistry Letters and Reviews* 2011, 4, 151–157.
20. Dalu, F.; Scorciapino, M. A.; Cara, C.; Luridiana, A.; Musinu, A.; Casu, M.; Secci, F.; Cannas, C. *Green Chemistry* 2018, 20, 375–381.
21. Wang, X. J.; Yang, Q.; Liu, F.; You, Q. D. *Synthetic Communications* 2008, 38, 1028–1035.
22. Xu, W.-T.; Huang, B.; Dai, J.-J.; Xu, J.; Xu, H.-J. *Organic Letters* 2016, 18, 3114–3117.
23. Srivastava, V.; Singh, P. K.; Singh, P. P. *Tetrahedron Letters* 2019, 60, 40–43.
24. Ke, F.; Xu, Y.; Zhu, S.; Lin, X.; Lin, C.; Zhou, S.; Su, H. *Green Chemistry* 2019, 21, 4329–4333.
25. Jaita, S.; Phakhodee, W.; Chairungsi, N.; Pattarawarapan, M. *Tetrahedron Letters* 2018, 59, 3571–3573.
26. Salimiyan, K.; Saberi, D. *Chemistry Select* 2019, 4, 3985–3989.
27. Thalluri, K.; Nadimpally, K. C.; Paul, A.; Mandal, B. *RSC Advances* 2012, 2, 6838–6845.
28. Sayes, M.; Charette, A. B. *Green Chemistry* 2017, 19, 5060–5064.
29. Hamstra, D. F. J.; Lenstra, D. C.; Koenders, T. J.; Rutjes, F. P. J. T.; Mecinović, J. *Organic & Biomolecular Chemistry* 2017, 15, 6426–6432.
30. Dimitrijević, E.; Taylor, M. S. *ACS Catalysis* 2013, 3, 945–962.
31. Nelson, P.; Pelter, A. *Journal of the Chemical Society* (Resumed) 1965, 5142–5144.
32. Trapani, G.; Reho, A.; Latrofa, A. *Synthesis* 1983, 1983, 1013–1014.
33. Collum, D. B.; Chen, S.-C.; Ganem, B. *The Journal of Organic Chemistry* 1978, 43, 4393–4394.
34. Ishihara, K.; Ohara, S.; Yamamoto, H. *The Journal of Organic Chemistry* 1996, 61, 4196–4197.
35. Al-Zoubi, R. M.; Marion, O.; Hall, D. G. *Angewandte Chemie International Edition* 2008, 47, 2876–2879.
36. Fatemi, S.; Gernigon, N.; Hall, D. G. *Green Chemistry* 2015, 17, 4016–4028.
37. Marcelli, T. *Angewandte Chemie International Edition* 2010, 49, 6840–6843.
38. Wang, C.; Yu, H.-Z.; Fu, Y.; Guo, Q.-X. *Organic & Biomolecular Chemistry* 2013, 11, 2140–2146.
39. Arnold, K.; Batsanov, A. S.; Davies, B.; Whiting, A. *Green Chemistry* 2008, 10, 124–134.
40. Comerford, J. W.; Clark, J. H.; Macquarrie, D. J.; Breeden, S. W. *Chemical Communications* 2009, 2562–2564.

41. Azizi, N.; Edrisi, M.; Abbasi, F. *Applied Organometallic Chemistry* 2018, 32, e3901.
42. Ojeda-Porras, A.; Hernández-Santana, A.; Gamba-Sánchez, D. *Green Chemistry* 2015, 17, 3157–3163.
43. Perreux, L.; Loupy, A.; Volatron, F. *Tetrahedron* 2002, 58, 2155–2162.
44. Tamura, M.; Murase, D.; Komura, K. *Synthesis* 2015, 47, 769–776.
45. Kazemi, S.; Mobinikhaledi, A.; Zendehdel, M. *Chinese Chemical Letters* 2017, 28, 1767–1772.
46. Ghosh, S.; Bhaumik, A.; Mondal, J.; Mallik, A.; Sengupta, S.; Mukhopadhyay, C. *Green Chemistry* 2012, 14, 3220–3229.
47. (a) Luque, R.; Budarin, V.; Clark, J. H.; Macquarrie, D. J. *Green Chemistry* 2009, 11, 459–461. (b) Jursic, B. S.; Zdravkovski, Z. *Synthetic Communications* 1993, 23, 2761–2770.
48. Kumar, M.; Sharma, S.; Thakur, K.; Nayal, O. S.; Bhatt, V.; Thakur, M. S.; Kumar, N.; Singh, B.; Sharma, U. *Asian Journal of Organic Chemistry* 2017, 6, 342–346.
49. Chandra Shekhar, A.; Ravi Kumar, A.; Sathaiah, G.; Luke Paul, V.; Sridhar, M.; Shanthan Rao, P. *Tetrahedron Letters* 2009, 50, 7099–7101.
50. Hosseini-Sarvari, M.; Sharghi, H. *The Journal of Organic Chemistry* 2006, 71, 6652–6654.
51. Chaudhari, P. S.; Salim, S. D.; Sawant, R. V.; Akamanchi, K. G. *Green Chemistry* 2010, 12, 1707–1710.
52. Basavaprabhu; Muniyappa, K.; Panguluri, N. R.; Veladi, P.; Sureshbabu, V. V. *New Journal of Chemistry* 2015, 39, 7746–7749.
53. Li, N.; Wang, L.; Zhang, L.; Zhao, W.; Qiao, J.; Xu, X.; Liang, Z. *ChemCatChem* 2018, 10, 3532–3538.
54. Grunes, J.; Zhu, J.; Somorjai, G. A. *Chemical Communications* 2003, 2257–2260.
55. (a) Reddy, M. B. M.; Ashoka, S.; Chandrappa, G. T.; Pasha, M. A. *Catalysis Letters* 2010, 138, 82–87; (b) Das, V. K.; Devi, R. R.; Raul, P. K.; Thakur, A. J. *Green Chemistry* 2012, 14, 847–854; (c) Kooti, M.; Nasiri, E. *Journal of Molecular Catalysis A: Chemical* 2015, 406, 168–177; (d) Ahmadi Meleh Amiri, K.; Saadati, Z.; Vafayi Bagheri, Z. *Silicon* 2019, 11, 2117–2125.
56. Nasrollahzadeh, M.; Sajadi, S. M.; Hatamifard, A. *Journal of Colloid and Interface Science* 2015, 460, 146–153.
57. Das, V. K.; Devi, R. R.; Thakur, A. J. *Applied Catalysis A: General* 2013, 456, 118–125.
58. Ahmadi, M.; Moradi, L.; Sadeghzadeh, M. *Research on Chemical Intermediates* 2018, 44, 7873–7889.
59. Chen, Z.; Fu, R.; Chai, W.; Zheng, H.; Sun, L.; Lu, Q.; Yuan, R. *Tetrahedron* 2014, 70, 2237–2245.
60. Kaur, M., Chapter 6 - Oxindole: A Nucleus Enriched With Multitargeting Potential Against Complex Disorders. In *Key Heterocycle Cores for Designing Multitargeting Molecules*, Silakari, O., Ed. Elsevier: 2018; pp. 211–246.
61. Poondra, R. R.; Turner, N. J. *Organic Letters* 2005, 7, 863–866.
62. Dhorajiya, B. D.; Bhakhar, B. S.; Dholakiya, B. Z. *Medicinal Chemistry Research* 2013, 22, 4075–4086.
63. (a) Dhorajiya, B. D.; Dholakiya, B. Z. *Research on Chemical Intermediates* 2015, 41, 277–289 (b) Dhorajiya, B. D.; Dholakiya, B. Z. *Green Chemistry Letters and Reviews* 2014, 7, 1–10.
64. (a) Wilson, K. L.; Murray, J.; Jamieson, C.; Watson, A. J. B. *Organic & Biomolecular Chemistry* 2018, 16, 2851–2854 (b) MacMillan, D. S.; Murray, J.; Sneddon, H. F.; Jamieson, C.; Watson, A. J. B. *Green Chemistry* 2013, 15, 596–600.

3

Greener Methods for Halogenation of Aromatic Compounds

Paola Acosta-Guzmán and Diego Gamba-Sánchez

CONTENTS

3.1 Introduction 41
3.2 Brominations 42
3.3 Chlorinations 44
3.4 Iodinations 50
3.5 Fluorinations 51
3.6 Conclusions 53
References 55

3.1 Introduction

The halogenation of aromatic compounds is one of the most antique reactions in organic chemistry; even before the enouncement of the theory of aromaticity and the famous solution to the problem of benzene's structure, proposed by Friedrich August Kekulé in 1865,[1] chemists already knew how to substitute one or more hydrogens of an aromatic compound by a chlorine, bromine or iodine. Kekulé recognized that using the right reagents chemists could produce mono or polyhalogenations of aromatic compounds like benzene; even more interesting, at that time the proportion of isomers obtained in a bis-halogenation were easily quantified. Some years later (in 1887), Henry E. Armstrong announced what can be seen as the first approach to describe a reaction mechanism, even if the concept was developed some years later.[2]

These historic data show the significance and tradition of the halogenation of aromatic compounds by electrophilic aromatic substitution (EAS); a very important reaction whose mechanism (for halogenations) was studied in detail during the decades of 1950s[3] and 1960s,[4] establishing the well-known and accepted mechanism that all we know now (Scheme 3.1).

It is noteworthy that the EAS is by far the most used reaction to introduce halogens in aromatic rings; however, it is not the only way. Particularly, the iodination and fluorination reactions have always been more demanding than the brominations or chlorinations; a traditional iodination must be accomplished using an oxidant to prevent the equilibrium from displacing the reagents, and the lack of efficient electrophilic fluoride sources was a strong limitation for many years. Recently, many methods to introduce halogens to an aromatic ring (directly or by steps) have been developed, mainly motivated by the need of efficient synthetic methods to produce halogenated aromatics, since they are present in an important number or drugs, drug-like molecules and because they are also among the most significant synthetic intermediates (Figure 3.1) for examples of important molecules containing halogens).

As mentioned before, fluorine has attracted the attention of researchers in medicinal chemistry, consequently, fluorinated drugs are among the most sold all around the world, some examples are fluconazole,[5] used for the treatment of several fungal infections, and olaparib[6] which is used principally in women to treat ovarian, fallopian tube and breast cancer. However, there are also examples of drugs containing the other halogens on aromatic rings: aripiprazole[7] (antipsychotic) and chloroquine[8] (antimalarial) for chlorine; amiodarone[9] (an antiarrhythmic agent) and triiodothyronine[10] for iodine and bromo-dragonfly[11] and brimonidine[12] (for the treatment of ocular hypertension)

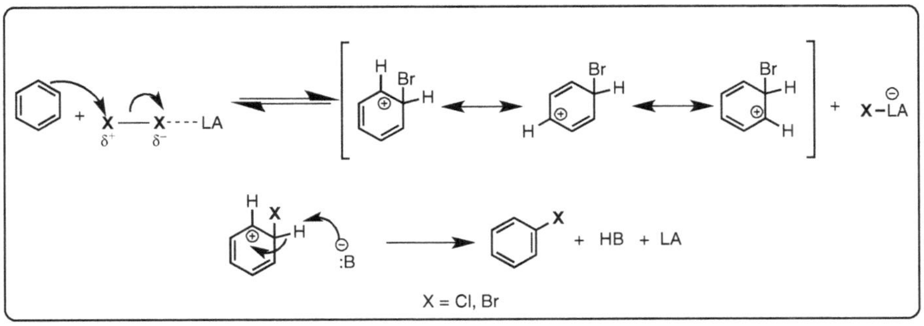

X = Cl, Br

SCHEME 3.1 General mechanism for EAS (halogenations).

DOI: 10.1201/9781003089162-3

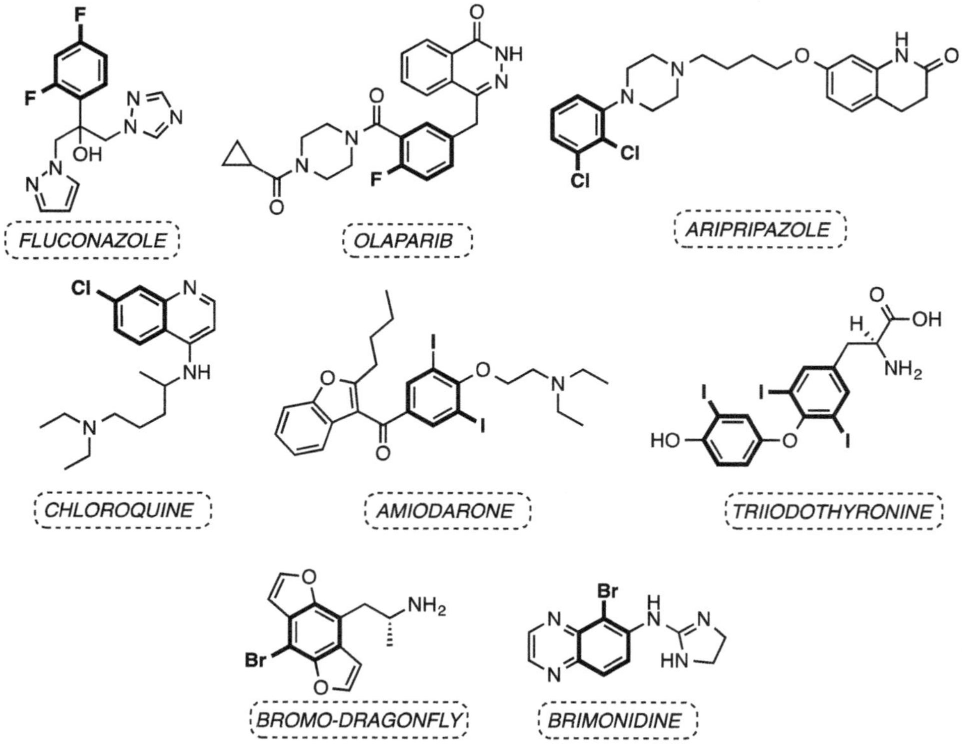

FIGURE 3.1 Selected examples of important molecules containing halogens.

for bromine are examples of drugs and drug-like molecules containing halogens.

This chapter is focused on some selected recent examples of eco-friendly halogenations of aromatics, we will emphasize the concept development and the proposed solutions for environmental problems, such as: the reduction of waste, the use of alternative methods of heating, the use of cheaper and more active catalysts, and when appropriate, we will also highlight the disadvantages and problems of the cited methods.

We will start with the more commonly used halogenations (brominations and chlorinations), then we will turn our attention to the more demanding iodinations, finishing with fluorinations. The order of the discussion was selected based on the importance of halogenated aromatics, in other words, we decided to treat the fluorinations at the end of the chapter since we think the fluorinated compounds are by far the most important (between the four types treated here) in medicinal chemistry, mainly because of the fluorine atom properties which have attracted the attention of many researchers all around the world.

It is noteworthy to mention that we will mix the synthetic methods and not all of them are EASs, when appropriate, we will introduce other mechanisms in order to make the reading experience comfortable.

3.2 Brominations

Aryl bromides are fundamental building blocks in organic synthesis, principally in cross-coupling reactions, consequently they are used in the synthesis of more complex organic molecules; additionally, a significant number of aryl bromide compounds have been isolated from marine organisms, and some of them have shown interesting biological properties. The synthetic methods to access aryl bromides have been recently reviewed in literature, with the focus on the development of general synthetic methods, but not on their environmental implications.[13]

As we mentioned before, the classic bromination of aromatics is an old known reaction, it is typically accomplished using bromine and a Lewis acid, and the major problems are the same for all EASs, leading to problems of regioselectivity of secondary brominations or brominations of substituted aromatics. In addition, when the substituent is an electron-withdrawing group (EWG), the reaction time is increased substantially or requires very harsh conditions, making it incompatible with highly functionalized molecules; a good example is the method described by Saiganesh and co-workers in 2007.[14] They used a mixture of *N*-bromosuccinimide (NBS) and concentrated H_2SO_4 and were able to accomplish the bromination of deactivated aromatics at a relatively low temperature (60°C); unfortunately, this method is clearly useful for simple aromatics (Scheme 3.2), but unsuitable for complex polyfunctionalized molecules since the use of concentrated sulfuric acid will cause undesired side reactions. However, it has some green chemistry advantages since it is used as a Brønsted acid, and consequently, the residues are water soluble salts, and the succinimide can be isolated or separated from the final product.

Since the former method was applied with EWGs, or deactivating groups, bromination is made exclusively at the meta position; however, sometimes we need to prepare the ortho-brominated substrate, where it is (to the best of our

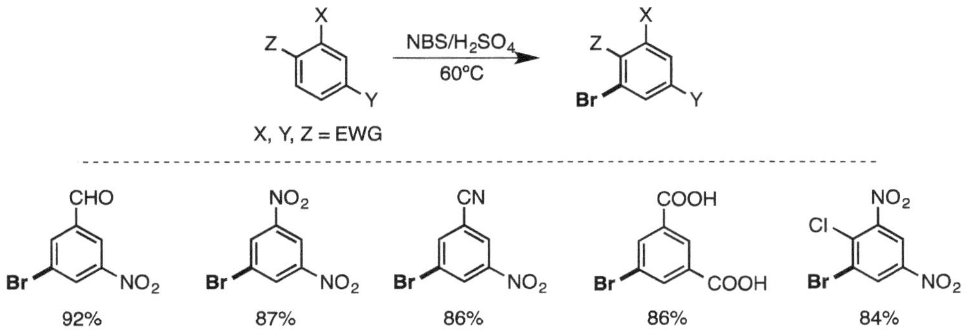

SCHEME 3.2 Bromination of deactivated rings.

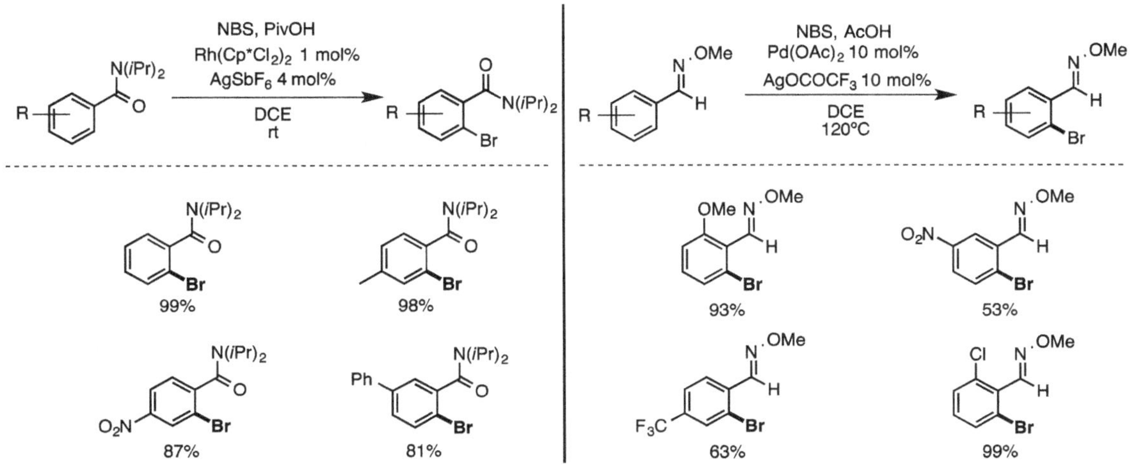

SCHEME 3.3 Ortho-selective bromination using directing groups and metal catalysis.

knowledge) impossible to make this substitution using classic EASs. Transition metal catalysis provides a useful alternative route to accessing the ortho-substituted aromatics, examples of that are the use of Rh or Pd catalysts and ortho-directing groups published by Glorius[15] and Fabis,[16] respectively. It must be highlighted that the use of expensive catalysts is an important drawback of these methodologies; however, the activity of the catalysts permits the use of low charges and both reactions have a great substrate scope (Scheme 3.3).

It should be clear by now that the mechanisms of these methods are not EASs, both are based on C-H activations promoted by the transition metal, and the steps of the mechanism proceed through that of a usual cross-coupling reaction, that means oxidative addition followed by reductive elimination (Scheme 3.4), both methods use directing groups (carbonyls, amides or oximes to mask aldehydes), so the ortho-regioselectivity is excellent.

Some other authors have used sterically hindered bromide sources as an alternative to accomplish brominations at the para position,[17] unfortunately, the use of toxic metals like tin makes this method incompatible with the green chemistry principles (Scheme 3.5).

Classic brominations have also been improved in order to make them well-suited to green chemistry. The most common approach is the use of a bromide source complemented with an oxidant; the main advantage is that many oxidants and bromide sources (salts) are water soluble, thus making those methods feasible in water or alcoholic mixtures which are less polluting than classic organic solvents. Besides, the production of waste is also reduced. The sole problem is that the conditions are so mild that they are only applicable to highly activated rings; however, the methods are very easy to use even in laboratories

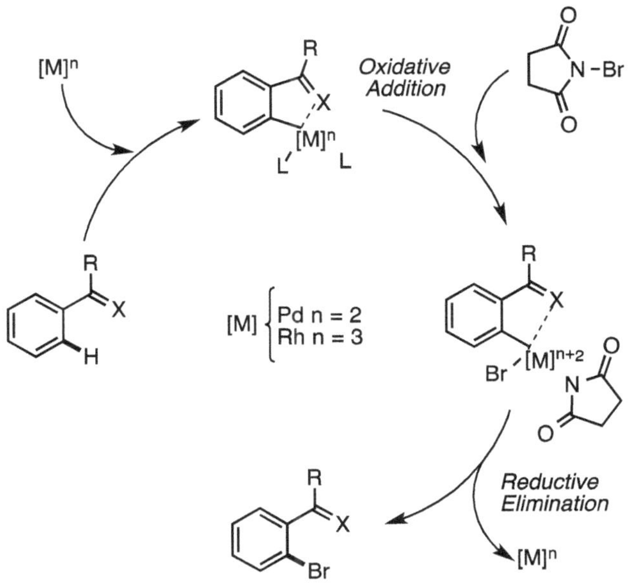

SCHEME 3.4 Mechanism for ortho-selective brominations.

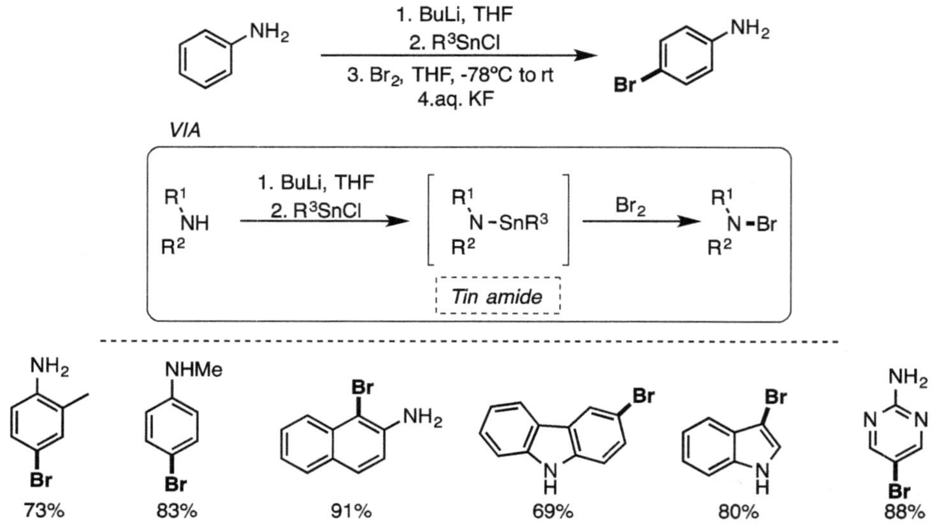

SCHEME 3.5 Bromination with N-Br hindered reagents.

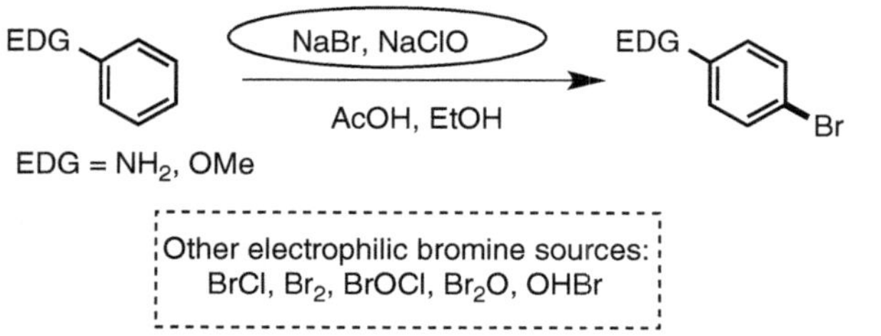

SCHEME 3.6 Bromination with NaBr/NaClO and alternative electrophilic bromide sources.

for undergraduate students as demonstrated by Luong in 2012[18] using NaBr as the bromide source and bleach (NaOCl) as the oxidant. The reaction is performed in water and ethanol, producing very good results (Scheme 3.6). It is noteworthy to mention that some authors declare that the oxidation of Br^- produces Br_2 exclusively, however, it has been demonstrated that other Br-OR species can be formed (especially in chlorinated water), that can act as sources of electrophilic bromine for the bromination reaction.[19]

Depending on the oxidant, reaction conditions and substrate, the reaction may take a different pathway generating radicals, thus making it unsuitable with substrates with alkyl chains; however, some authors have taken advantage of the radical generation, developing a method of tandem cyclization-bromination in water and with eco-friendly oxidants. In this case the radicals react with the aromatic ring because it is activated *in situ*, otherwise the reaction will be done on the lateral alkyl chain (depending on the substrate) as is clearly shown in the mechanism of Scheme 3.7.[20]

Heterogenous catalysis is one of the most useful strategies in green chemistry, and brominations have not been overlooked, as demonstrated by a recent work using graphene oxide (GO) as heterogeneous catalyst and Br_2. The GO acts as the oxidant, in fact, one epoxide present in the GO structure reacts with Br_2 producing the brominating species GO-Br, in other words, a new source of electrophilic bromine.[21] The bromide residue reacts, producing Br_2 and Br-OH, two additional sources of electrophilic bromine; thus the global reaction is very clean since no wastes are generated, and the GO may be reused even with NBS as the bromine source (Scheme 3.8).

The production of highly functionalized aromatic bromides is still a challenge, we may cite one example of an alternative method that may be used to produce functionalized bromides, particularly brominated boronic acids – molecules that are important not only because of their potential use in cross-coupling reactions but also by their own biological properties. The use of silver salts proved convenient to accomplish brominations of boronic acids, selectively in the ortho position (Scheme 3.9).[22]

The reaction starts with the formation of the ate complex by the reaction of the boronic acid with EtOH, then the activation of Br_2 by $AgSO_4$ leads to the addition of an electrophilic bromine to the ate complex and the concomitant formation of AgBr. The intermediate species has a more electrophilic bromide, which can react easily at the ortho-position yielding the final product.

3.3 Chlorinations

Despite the huge occurrence of chlorinated aromatics and heteroaromatics, the most common reagents for chlorination – Cl_2, *t*-BuOCl and SO_2Cl_2 – are still regularly used due to the high reactivity; however, their low selectivity and significant volumes of waste produced make them less than ideal from a green chemistry standpoint. Alternative reagents like NCS do give some good results but its lower reactivity can have an impact on the overall yield, and in some cases, no reaction at all.

Having this in mind, some teams have developed alternative reagents that perform electrophilic chlorinations with well-defined advantages, such as better reactivity, better selectivity, recyclability of the reagent, milder reaction conditions, among others. We will cite two important examples, selected due to

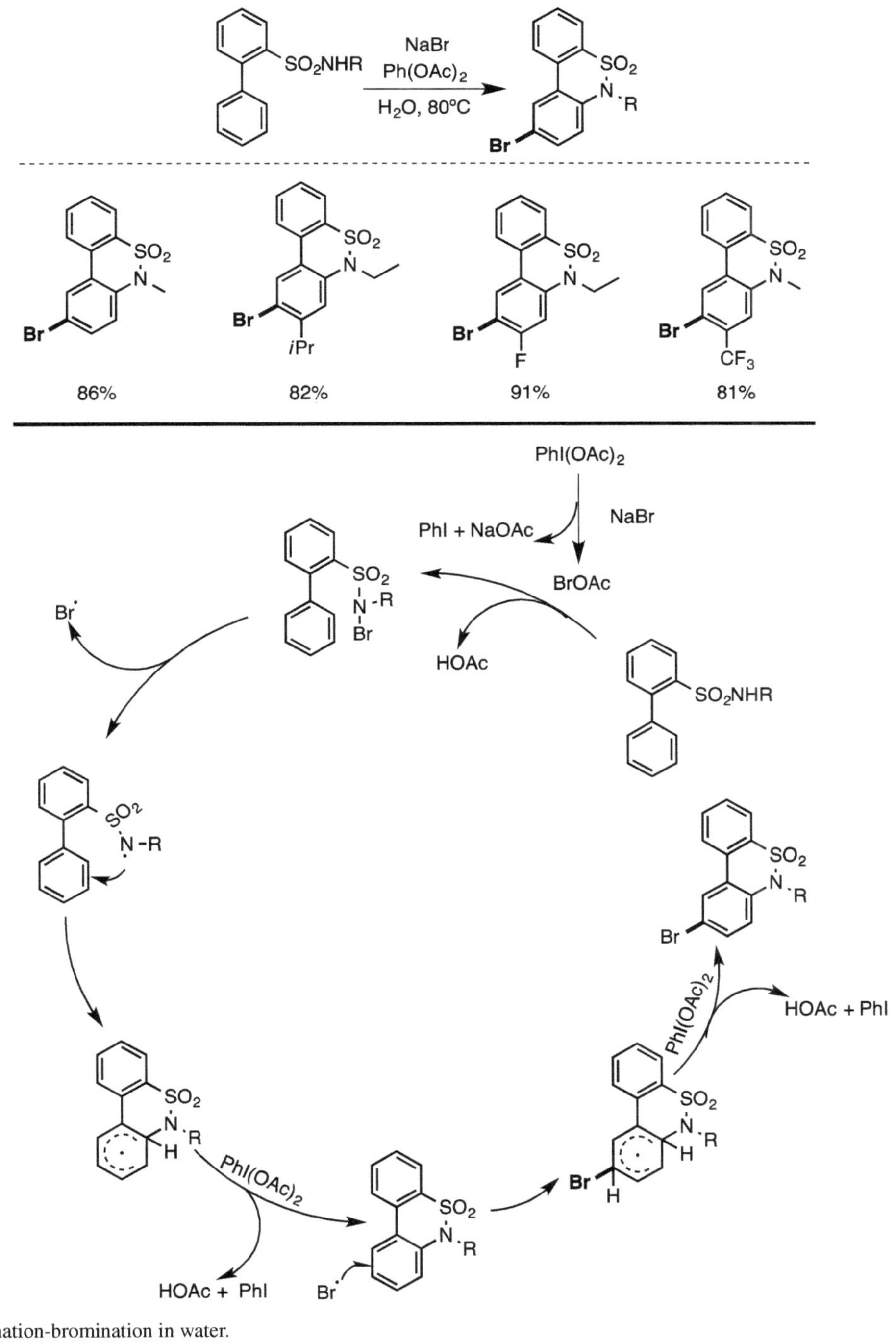

SCHEME 3.7 Amination-bromination in water.

their scope and applicability. The first one was reported by Baran and co-workers in 2014,[23] the Palau`chlor or CBMG is a guanidine-based chlorinating reagent which exhibits very good chlorination activity toward heteroaromatics, it is inexpensive, can be prepared in deca-gram scale and is useful under soft reaction conditions. In most cases it produces better results than other chorine sources, and most importantly, it can be used with other aromatics even in late-stage transformations. Scheme 3.10 shows a comparison between CBMG and other chlorine sources for the chlorination of the same molecule, and some examples of other successful chlorinations. It must be highlighted that CBMG reacts by a classic EAS, thus the more activated (nucleophilic ring) will be selectively chlorinated under the presence of other aromatics.

Other researchers have also been interested in developing new reagents for chlorination or in the application of reagents known for other transformations in electrophilic chlorination of aromatics. Hypervalent iodine regents have shown a good scope as electrophile sources, and hypervalent chlorine is not an exception. Xue and co-workers[24] described a good method of chlorination using 1-chloro-1,2-benziodoxol-3-one (CBDO), a known reagent that showed excellent reactivity

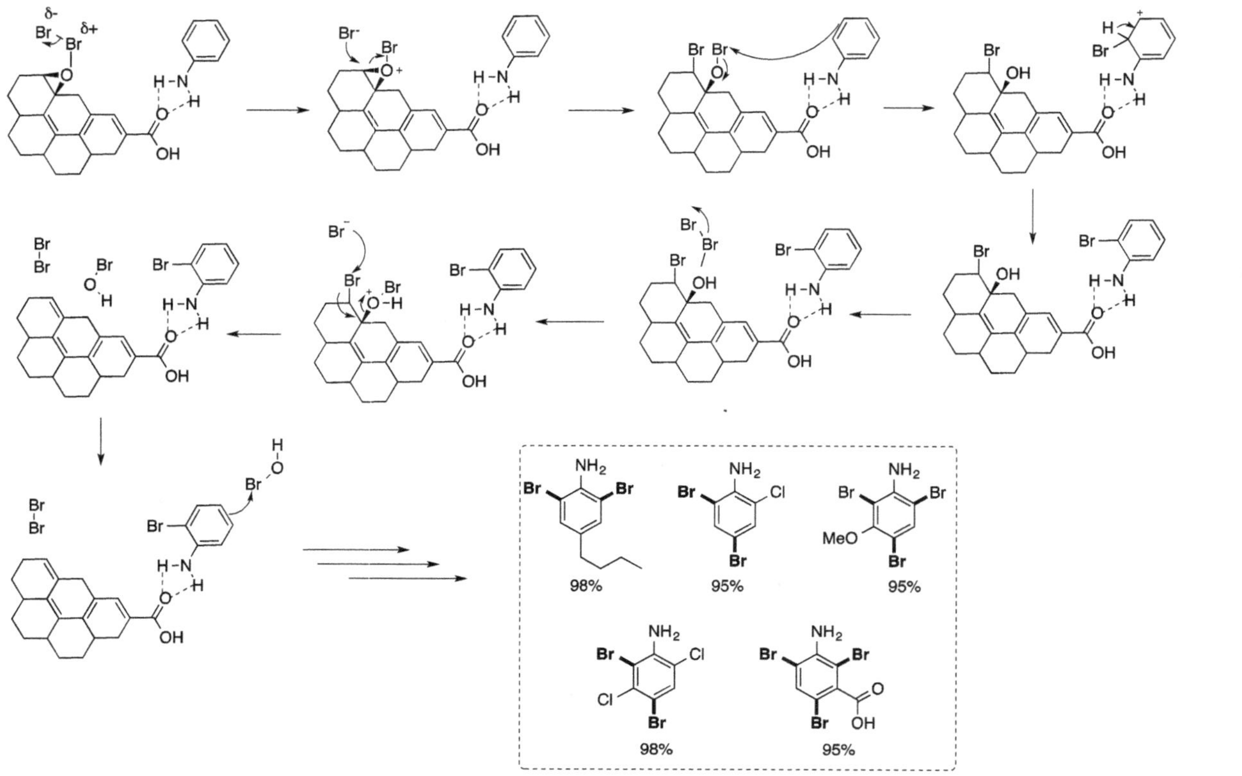

SCHEME 3.8 NBS as the bromine source.

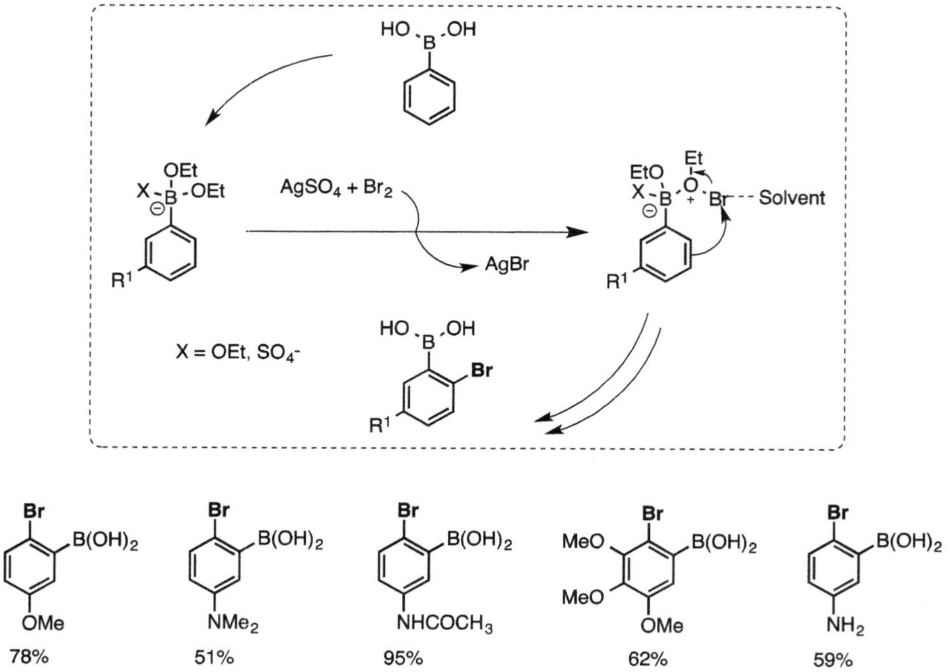

SCHEME 3.9 Bromination of boronic acids catalyzed by Ag.

with heteroaromatics and activated carbocycles. The reaction is performed at room temperature and the reagents can be easily regenerated as shown in the Scheme 3.11. The sole problem of this method is the use of toxic DMF as the solvent, and alternative polar solvents remain to be tested.

Besides the development of new reagents, the introduction of a metal catalysis is still an alternative to increase the reactivity of all unreactive reagents. As we mentioned before, NCS is unsuitable for demanding chlorinations; however, its activity can be improved by the introduction of a Fe catalysis, notably

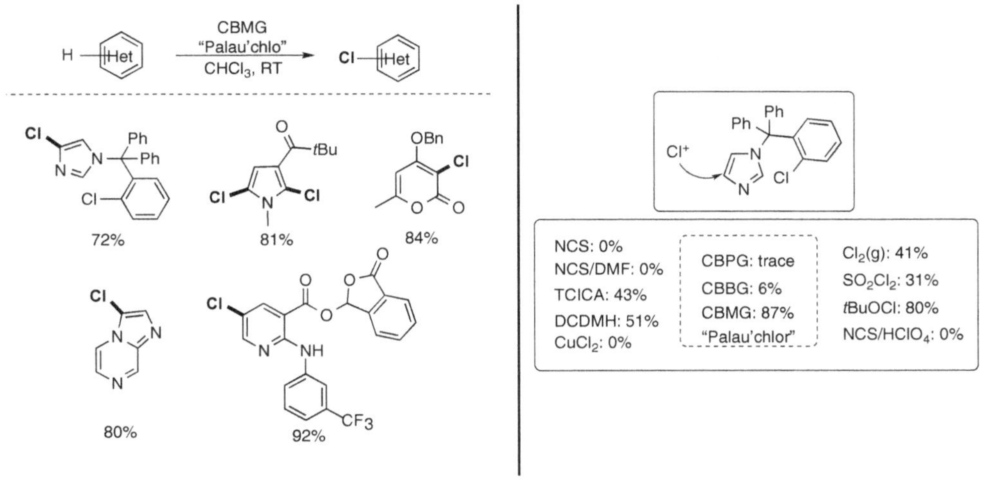

SCHEME 3.10 Chlorination of aromatic with Palau'chlor.

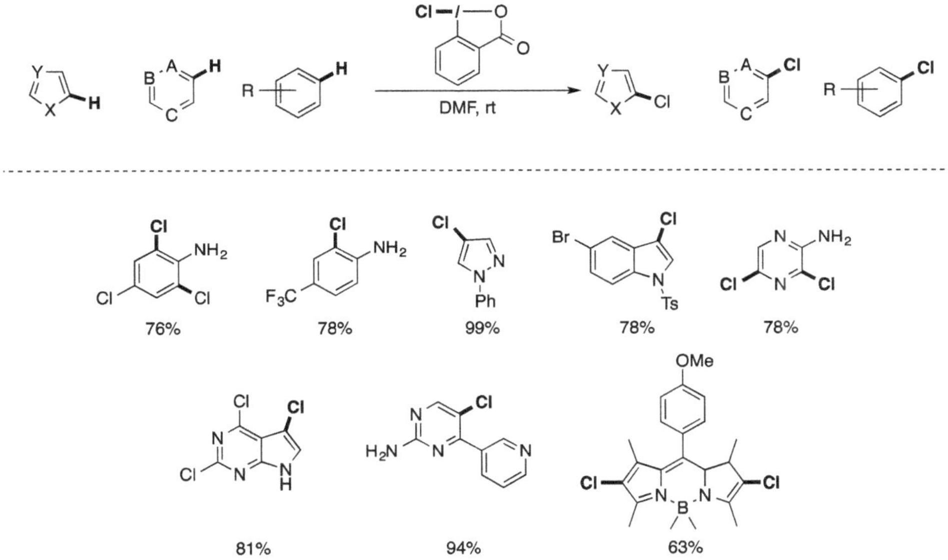

SCHEME 3.11 Chlorination with hypervalent iodine reagent.

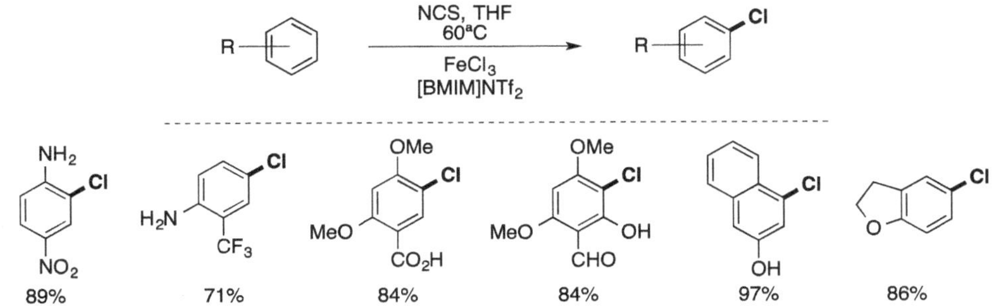

SCHEME 3.12 Chlorination of aromatics with $Fe(NTf_2)_3$ as catalyst.

the use of $Fe(NTf_2)_3$, proved adequate for chlorinations using NCS at low temperatures, unfortunately, the reaction is limited to the use of activated aromatics, but para-selectivity is very high[25] (Scheme 3.12).

As usual in electrophilic substitutions, the regioselectivity is an important issue and the chlorination of activated aromatics yields mainly para- and ortho-chlorinated products. Yu and coworkers[26] developed an amazing alternative to obtain the meta-chlorinated product. The method is based on the use of a Pd(II) catalyst and directing groups. As shown in Scheme 3.13 the directing group reacts with the catalyst generating an ortho-Pd derivative, which constitutes at the same time a "protection" of the ortho position and an activation of the meta-position. The insertion of the ligand in position 2 and the Pd(II)

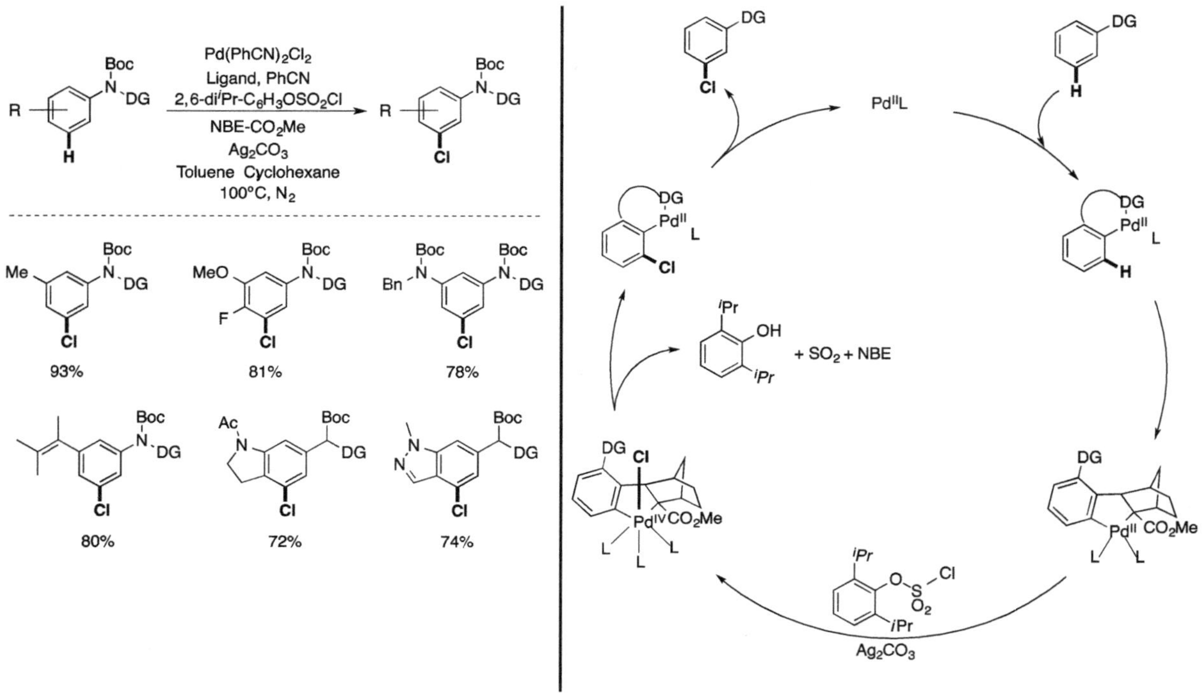

SCHEME 3.13 Meta-chlorination of aromatic amines.

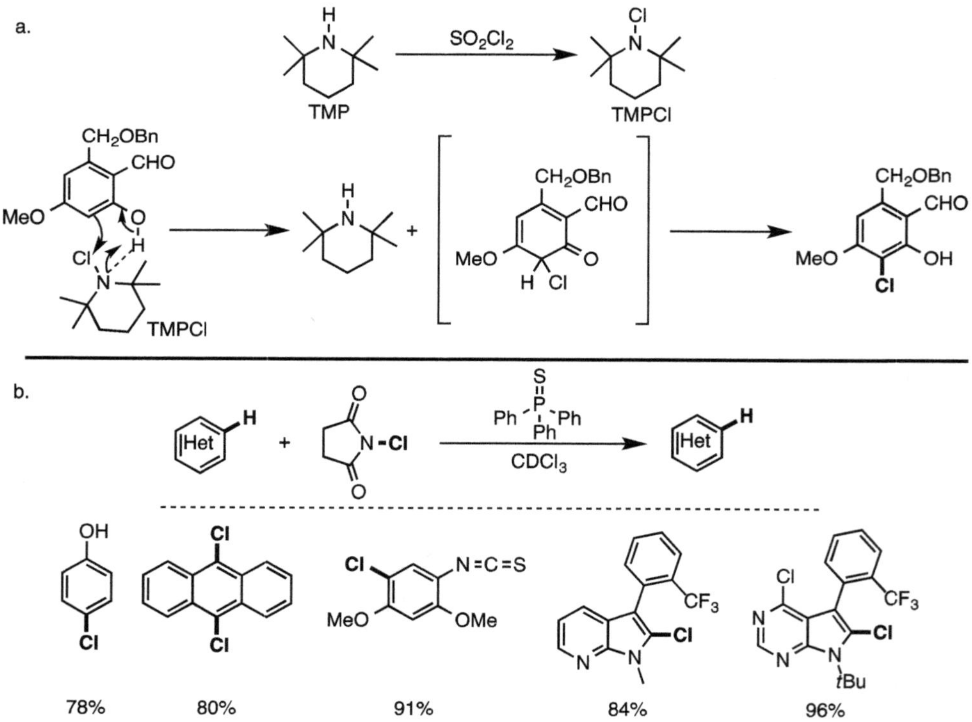

SCHEME 3.14 Alternative chlorinations with classic reagents.

in position 3 affords an intermediate which after oxidation will produce the species with a chlorine ready to be inserted between the carbon and Pd, thus generating the chlorinated product and regenerating the catalyst.

Another interesting alternative in green chemistry is the use of organocatalysis, since typically the residue causes less pollution compared to metals, besides the use of green solvents is usually accompanied by organic catalysts. However, this is not always the case and even if the organocatalysis is a good alternative, in some cases, it has several problems. Examples of these issues are the works exemplified in Schemes 3.14a and 3.14b. Both authors used traditional sources of chlorine;

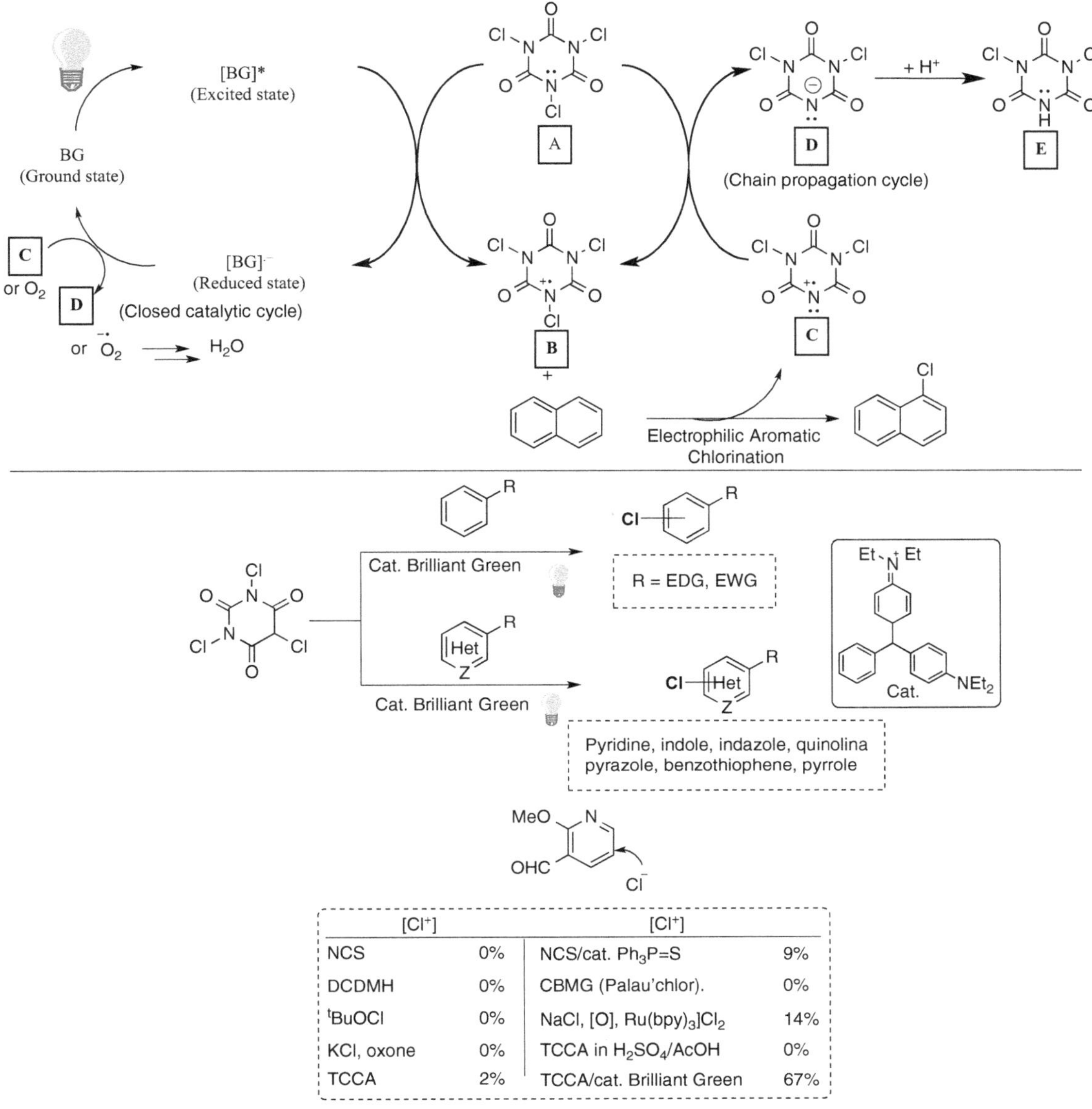

[Cl⁺]		[Cl⁺]	
NCS	0%	NCS/cat. $Ph_3P{=}S$	9%
DCDMH	0%	CBMG (Palau'chlor).	0%
^tBuOCl	0%	NaCl, [O], $Ru(bpy)_3Cl_2$	14%
KCl, oxone	0%	TCCA in H_2SO_4/AcOH	0%
TCCA	2%	TCCA/cat. Brilliant Green	67%

SCHEME 3.15 Photocatalytic chlorination.

in the first case SO_2Cl_2[27] and in the second case NCS[28], besides the residue in both cases maybe causes more pollution than transition metals; nevertheless, these two reactions have some advantages. In Scheme 3.14a, the method is ortho-selective and can be carried out under mild conditions; in Scheme 3.14b, the method is mild and may be applied in functionalized substrates.

Fortunately, organocatalysis provides a huge spectrum of possibilities and nowadays, combined with some other areas, like photochemistry or photocatalysis is a very important source of effective methods for halogenations, particularly for chlorinations. An exciting example is the combination of visible light with an organic catalyst to transfer energy and achieve chlorinations that are otherwise unfeasible with classic or even more modern methods. The reaction is based (as usual with photocatalysis) on the use of a molecule which can absorb visible light and this molecule is excited to a higher level of energy, providing a more active chlorine source, thus making the reaction applicable to less reactive substrates (Scheme 3.15).[29] This method can be applied to substituted aromatics and heteroaromatics. Besides, Scheme 3.15 also shows a comparison of this method with other known chlorination methods, proving its superiority.

Finally, organocatalysis has also provided a way for regiodivergent differentiation. Using two different catalysts, Gustafson was able to selectively produce ortho- or para-chlorinated phenols.[30] Selectivity is achieved presumably by the difference in transition state energies and provides a method with a magnificent potential, since future developments will be certainly focused on the reactions of less activated rings. Scheme 3.16 shows the general reaction and a couple of examples of the regioselective chlorination.

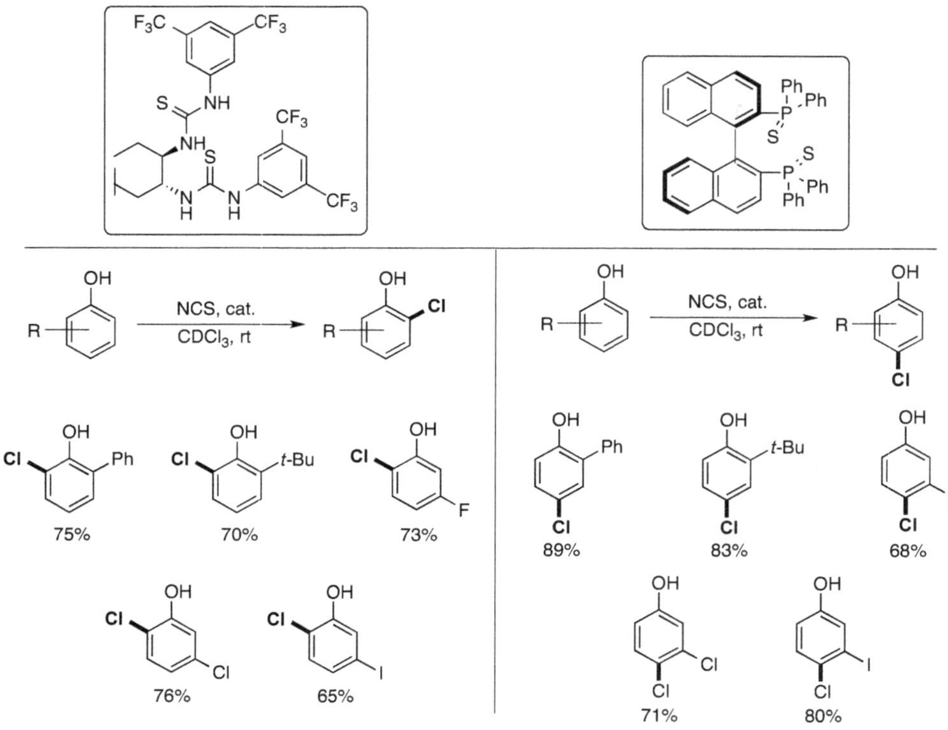

SCHEME 3.16 Organocatalytic chlorination.

3.4 Iodinations

Iodinations by EAS typically uses an oxidant to transform the HI into I_2 shifting the equilibrium to the products, meaning special conditions are often used. However, some of the methods described for chlorination or bromination reactions are also useful for iodinations. Examples are the methods described by Sutherland and co-workers described in Scheme 3.12, virtually the same method can be applied for iodinations using Fe[31] or Ag[32] catalysts.

A slight modification of the method described in Scheme 3.2 (use of NBS/H_2SO_4) is the use of a mixture of NIS and trifluoroacetic acid for iodination of deactivated aromatics.[33] Finally, the use of hypervalent iodine was described for chlorinations in Scheme 3.11; a different system (IBX/I_2) was used successfully in the iodination of activated aromatics.[34]

In addition, some methods have been developed using transition metal catalysts, organocatalysts and hypervalent iodine among others, in the same line that those discussed in previous sections. We can mention the use of Pd(II) salts, since it was demonstrated that the catalyst can be recovered and reused at least five times without loss of activity. Both methods described by the team of Jin-Quan Yu are based on the use of a combination of $Pd(AcO)_2$ and I_2, as mentioned before the major advantage is the recyclability of the catalyst. Besides, the reactions were performed on phenylacetic acids[35] or phenylacetic amides,[36] both are very important structures in the chemistry of bioactive compounds, and the iodination is selectively accomplished in the ortho position. Interestingly, iodination of simple activated rings like anisole did not work, suggesting the reaction does not proceed via an EAS even if authors suggest the presence of electrophilic iodine sources (Scheme 3.17).

Other metals have been used and Co/Ag is a good example of synergic contribution of two metals to produce a more reactive catalytic system.[37] It was applied to the ortho-iodination of amides and the reaction is based on the mechanism described in Scheme 3.18, basically the process starts with a redox reaction between Co(II) and Ag(I) generating a complex of Co(III) with the substrate as a ligand and Ag(0) as residue. The reaction follows by the activation of the ortho position by the insertion of Co, followed by metal-halogen exchange, forming the C-I bond, then the product is liberated by oxidation of Ag(0) and reduction of Co(III) regenerating the catalytic system. This reaction produces very low amounts of waste fitting with the green chemistry concept.

Turning our attention to organocatalytic iodinations, we found an interesting method using 1,3-diiodo-5,5-dimethylhydantoin (DIH) and disulfides as organocatalysts.[38] The advantages are the use of low amounts of catalysts and an easy to handle and less toxic source of iodine. Unfortunately, the reaction only works with activated aromatics and π-excessive heterocycles. The proposed mechanism is described in Scheme 3.19 and is based on the interaction between the hydantoin and the disulfide to increase the electrophilic character of the iodine promoting a normal EAS.

A very interesting method, recently described, is the one-pot oxidation-iodination of Fisher–Borsche products using polyethylene glycol 400 (PEG 400).[39] The method is performed in water and uses H_5IO_6 as the oxidant, generating *in situ* the electrophilic source of iodine, taking advantage of the PEG structure; in fact, this method used PEG as an alternative to the more polluting and extremely toxic crown ethers, as can be observed in the reaction mechanism on Scheme 3.20.

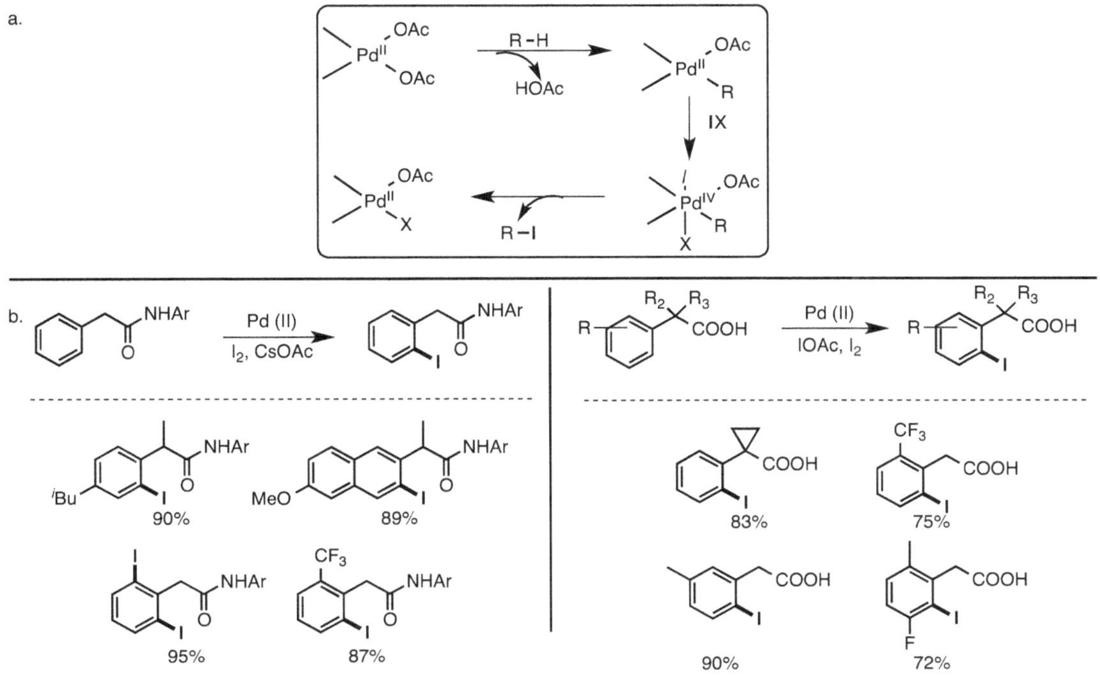

SCHEME 3.17 Iodination of phenylacetic acids and amides with Pd(II) catalysts.

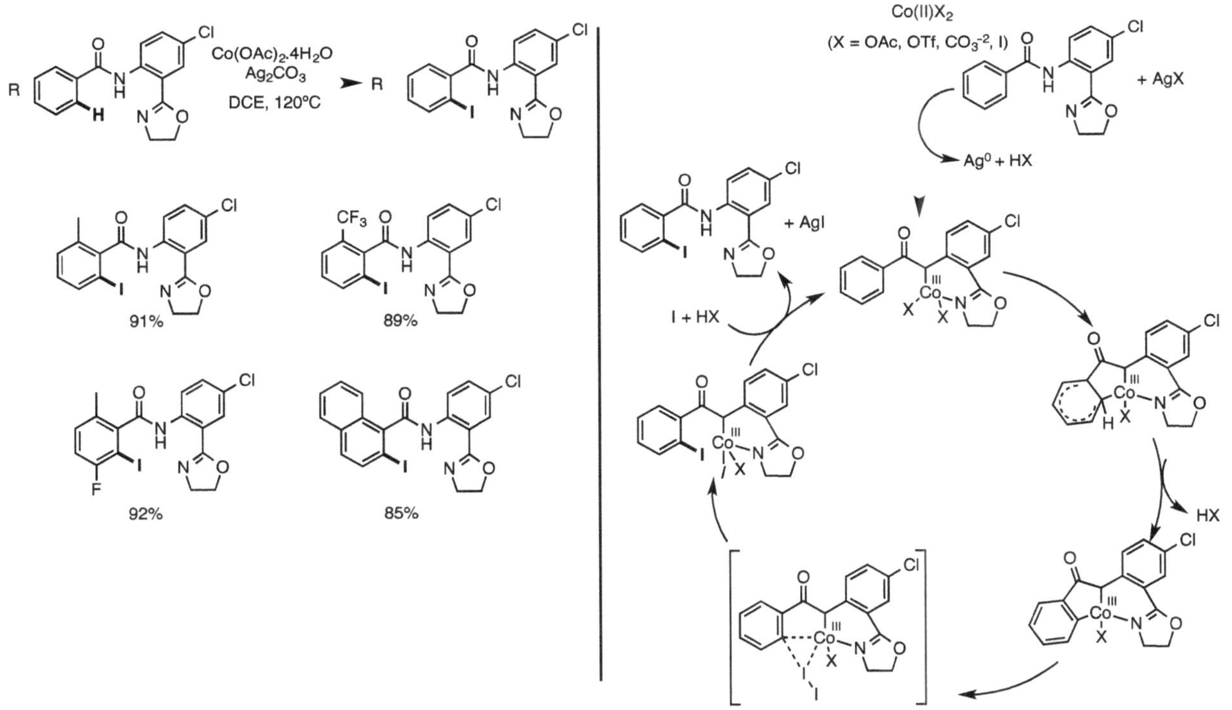

SCHEME 3.18 Ortho-iodination of amides with Co catalyst.

3.5 Fluorinations

Fluorine has attracted the attention of many organic chemists, principally because of its special properties; for instance, introducing a fluorine atom in the correct position of a bioactive molecule will improve the metabolic resistance, and it may change the lypo/hydrosolubility improving the bioavailability and thus increasing the biological power of the active molecule. These are a few examples of advantages or improvements achieved by insertion of a fluorine into a bioactive molecule. Unfortunately, the classic EASs do not work with fluorine, principally because of the reactivity of F_2; for many years, the goal was to find an appropriate electrophilic fluorine source to properly achieve EASs, thus the most common way to introduce an electrophilic fluorine into aromatics is the use of

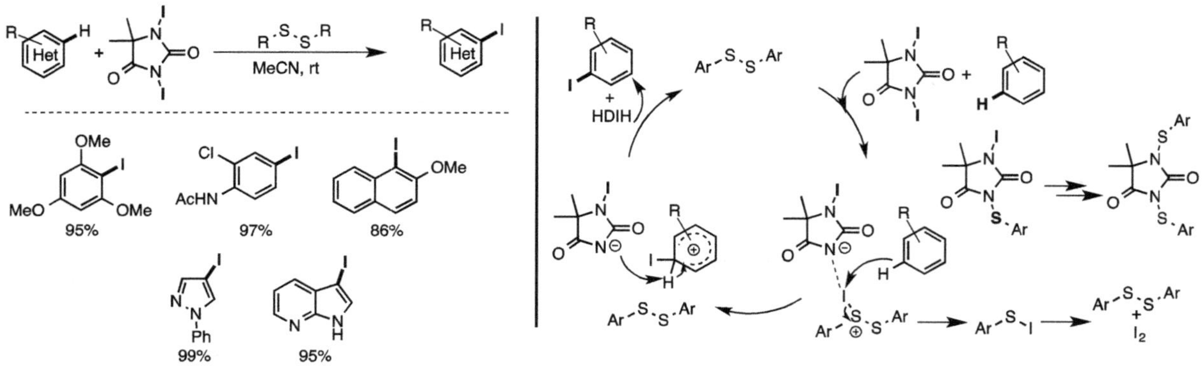

SCHEME 3.19 Organocatalytic iodination using disulfides.

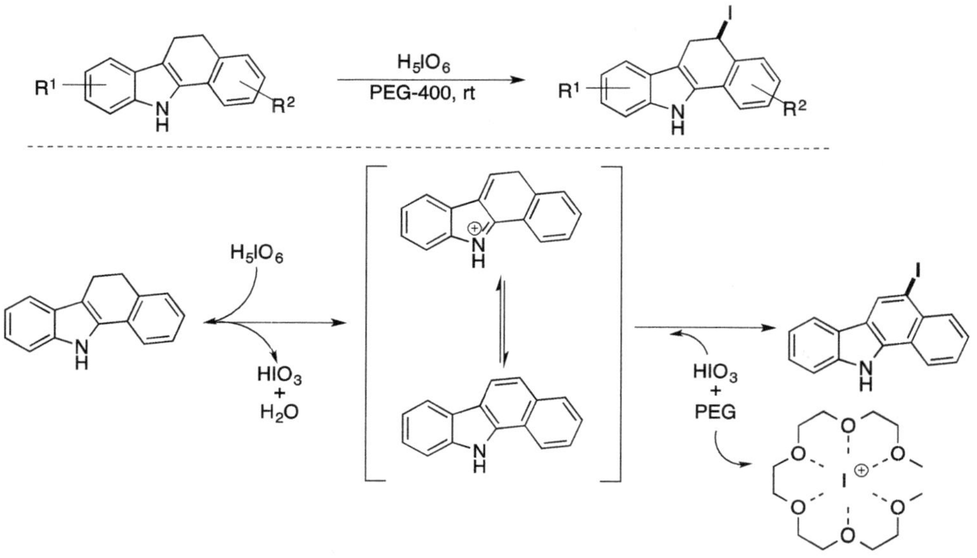

SCHEME 3.20 PEG-400 promoting iodination of aromatics.

N-fluorobenzenesulfonimide (NFSI), but many other reagents are now available. Direct fluorination has suffered from other problems (besides the electrophilic reagent), such as high temperatures, toxicity of the used fluorinating agent and in some cases the need of special equipment compatible with HF or F^-, among others. Consequently, researchers have developed an impressive number of protocols for fluorinations of aromatics, however, most of those methods are indirect, meaning that the aromatic compound should be previously functionalized and a substitution by a fluorine is then accomplished. The most famous process of this kind is the so-called Balz–Shiemann reaction, which is based on the generation of a diazonium salt with a BF_4^- as the counterion (it is the most common, but examples with other counterions can be found in literature), followed by heating or irradiation produces the corresponding fluorinated product through substitution of the N_2 with fluorine. The concept is simple, molecular nitrogen is maybe the best leaving group, so the C-N bond can be cleaved producing a cationic intermediate, which is trapped by a nucleophilic fluoride. Therefore, the reaction of aromatics bearing good leaving groups may be the essence for other fluorine substitutions; other researchers in modern chemistry have used this concept, using iodonium salts,[40] triflates,[41] iodine,[42] among others[43] with the help of transition metals like Cu, Pd, Ni, Ag, etc. On the other hand, sequences of stannilation-fluorination or borination-fluorination are among the most used methods to introduce fluorine in aromatics, also using transition metals as catalysts, we can mention the work of Sanford,[44] Ritter[45] and Hartwig[46] as the most representative researchers in this field; however, as those are indirect methods we are not going to present them here in detail, even if amazing achievements have been described, besides, the use of tin or boron are not within the principles of green chemistry, and because the atom economy for most of them is far away from the goal in green chemistry.

In consequence, we are going to describe some direct methods for the fluorination of aromatics, and readers will notice that work from the researchers mentioned above also fits with this description.

Fluorinations by EASs suffer from the same problems that we have mentioned in previous sections, principally regioselectivity; however, an interesting approach for para-fluorination of

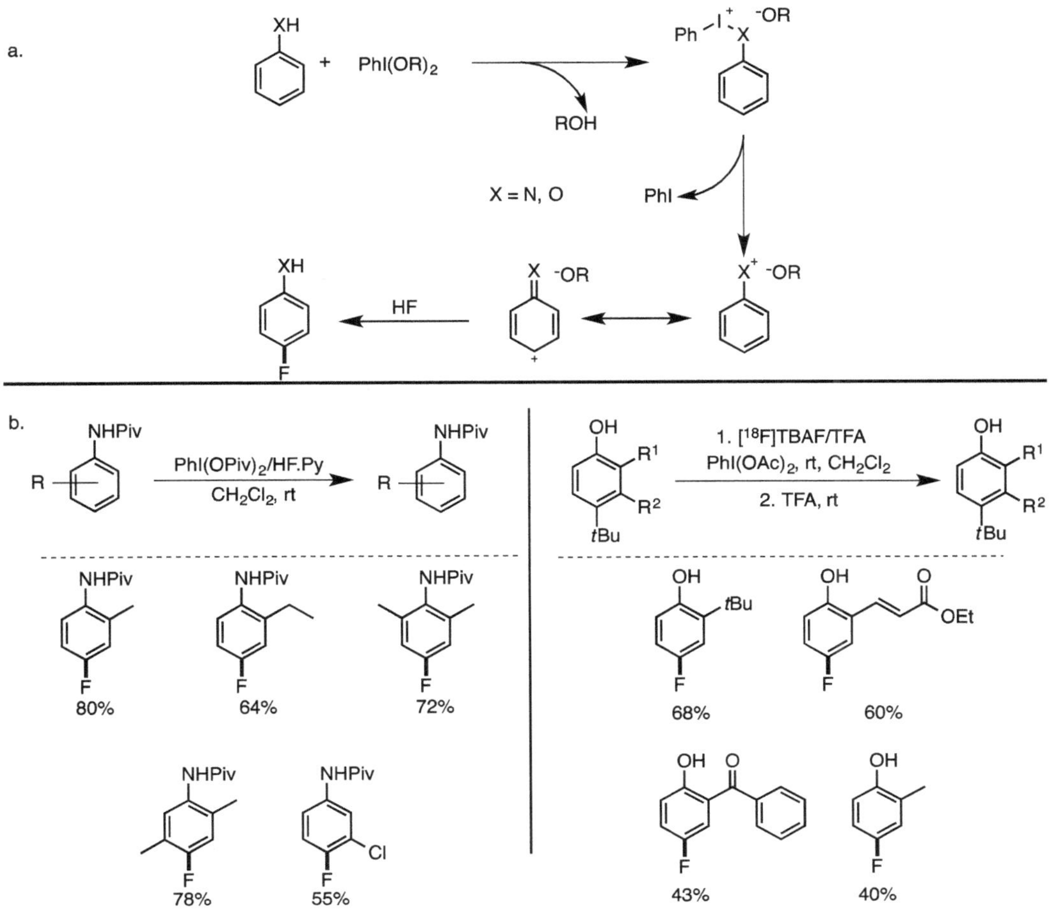

SCHEME 3.21 Direct fluorinations with nucleophilic fluoride.

electron-rich aromatics (phenols and anilines) is the so-called aryl umpolung. Typically, an aromatic ring with electron-donating groups will act as a nucleophile, but its reaction with hypervalent iodine reagents will produce a cationic intermediate (electrophile) which will react with a source of nucleophilic fluoride (Scheme 3.21). This method was applied successfully to several activated aromatics.[47]

The fluorination of deactivated aromatics is certainly more difficult and examples with both electrophilic and nucleophilic fluoride sources can be found. Most of these methods use directing groups and transition metal catalysts, and ortho-directed substitution and para-directed are described in the literature.

We will first mention a method that uses quinoline as the directing group and auxiliary to accomplish ortho-fluorinations (mono and difluorination) of deactivated aromatics. The most important feature of this method is the use of nucleophilic fluoride from AgF, even if there is a clear proposal about the reaction mechanism, the reaction scope, the reagents and the possibility for further functionalization make this method suitable from a green chemistry perspective.[48] (Scheme 3.22).

Other methods are based on the use of ortho-directing groups and sources of electrophilic fluorine, typically accompanied with Pd catalysts, and each method has advantages and weaknesses, so we make a list of ortho-directing groups and fluoride sources in Figure 3.2.[49]

Even if all those methods use electrophilic fluorine, none of them proceed with EAS, instead the directing group serves to activate a position and to induce the insertion of the metallic catalysts, increasing the reactivity of one single position, consequently the regioselectivity of those methods is excellent, as well as the functional groups' tolerance.

The main problem is the atom economy; for application in the synthesis of complex active molecules the directing group has to be cleaved and typically discarded; therefore, Sorensen and Chen[50] developed a method that uses the ortho-directing group as catalysts, unfortunately the reaction is limited to the use of aromatic aldehydes, but is still a significant improvement and certainly an inspiration for other researchers. See Scheme 3.23.

3.6 Conclusions

We have described herein a select group of examples for halogenation of aromatics. Clearly the time for simple electrophilic aromatic substitutions seems to have passed, and the time for

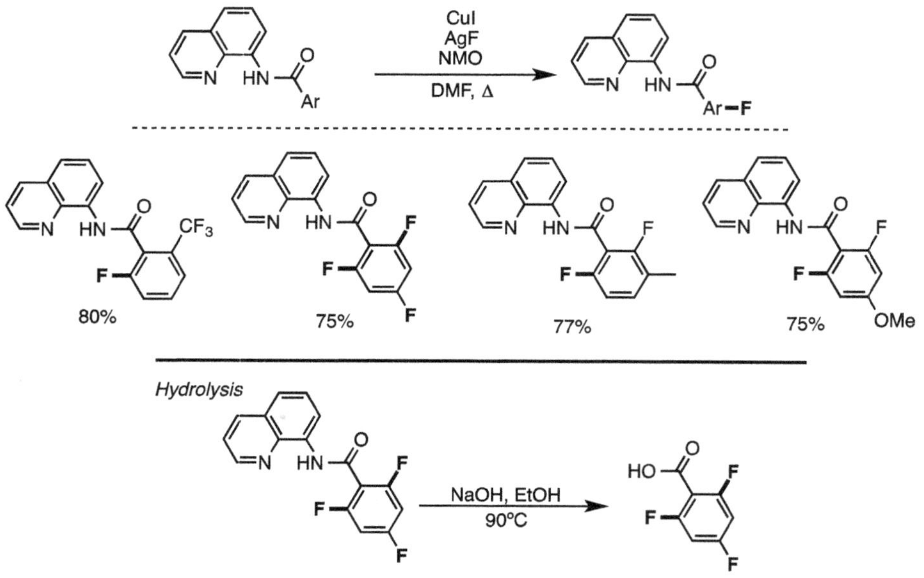

SCHEME 3.22 Fluorination with quinoline as the directing group.

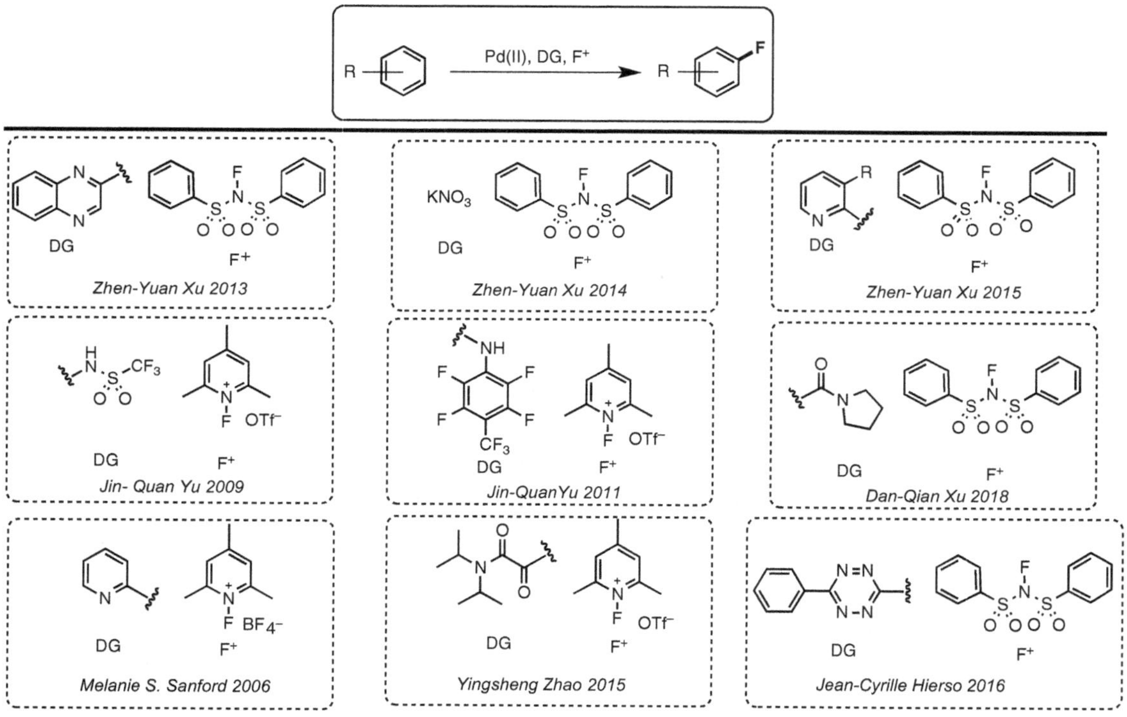

FIGURE 3.2 Sources of electrophilic fluorine.

more selective and mild halogenations has arrived. The aromatic (aryl) umpolung opens the possibility to use sources of nucleophilic halogens without previous functionalization and several approaches have been recently described.

The use of electrophilic sources of halogen is still an alternative and the recent methods are mainly based on the use of metal catalysts and directing groups; however, organocatalytic approaches are continuously appearing, reducing the waste residue and using fewer polluting reagents.

Despite the great achievements and continuous growth of this field, halogenation of aromatics still needs to fit better with the green chemistry principles. The use of alternative heating sources combined with organocatalysis or heterogeneous catalysis needs more effort from the scientific community.

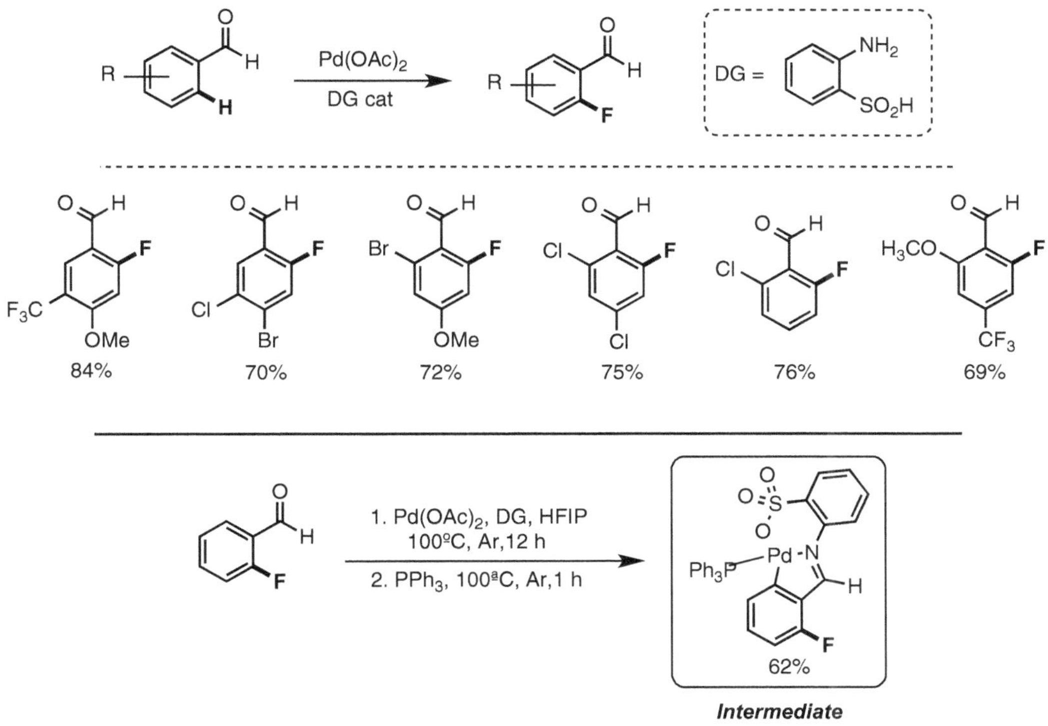

SCHEME 3.23 Fluorination using catalytic directing auxiliary.

REFERENCES

1. Kekulé, A., *Bulletin de la Societé Chimique de France* 1865, 3, 98–110.
2. Armstrong, H. E., *Journal of the Chemical Society* 1887, 51, 258–268.
3. (a) Andrews, L. J.; Keefer, R. M., *Journal of the American Chemical Society* 1959, 81, 1063–1067; (b) Mason, S. F., *Journal of the Chemical Society* (Resumed) 1959, 1233–1239; (c) Robertson, P. W., *Journal of the Chemical Society* (Resumed) 1954, 1267–1270; (d) Stock, L. M.; Brown, H. C., *Journal of the American Chemical Society* 1959, 81, 5615–5620; (e) Keefer, R. M.; Ottenberg, A.; Andrews, L. J., *Journal of the American Chemical Society* 1956, 78, 255–259; (f) Robertson, P. W.; de la Mare, P. B. D.; Swedlund, B. E., *Journal of the Chemical Society* (Resumed) 1953, 782–788.
4. (a) Keefer, R. M.; Andrews, L. J., *Journal of the American Chemical Society* 1960, 82, 4547–4553; (b) Olah, G. A.; Kuhn, S. J.; Hardie, B. A., *Journal of the American Chemical Society* 1964, 86, 1055–1060.
5. Emami, S.; Ghobadi, E.; Saednia, S.; Hashemi, S. M., *European Journal of Medicinal Chemistry* 2019, 170, 173–194.
6. Griguolo, G.; Dieci, M. V.; Miglietta, F.; Guarneri, V.; Conte, P., *Future Oncology* 2020, 16, 717–732.
7. Frampton, J. E., *Drugs* 2017, 77, 2049–2056.
8. Mushtaque, M. S., *European Journal of Medicinal Chemistry* 2015, 90, 280–295.
9. Mujović, N.; Dobrev, D.; Marinković, M.; Russo, V.; Potpara, T. S., *Pharmacological Research* 2020, 151, 104521.
10. Parmentier, T.; Sienaert, P., *Journal of Affective Disorders* 2018, 229, 410–414.
11. O'Connor, R. E.; Keating, J. J., *Drug Testing and Analysis* 2014, 6, 658–667.
12. Sharma, S.; Trikha, S.; Perera, S.; Aung, T., *Clinical Ophthalmology* 2015, 9, 2201–2207.
13. Voskressensky, L. G.; Golantsov, N. E.; Maharramov, A. M., *Synthesis* 2016, 48, 615–643.
14. Rajesh, K.; Somasundaram, M.; Saiganesh, R.; Balasubramanian, K. K., *The Journal of Organic Chemistry* 2007, 72, 5867–5869.
15. Schröder, N.; Wencel-Delord, J.; Glorius, F., *Journal of the American Chemical Society* 2012, 134, 8298–8301.
16. Dubost, E.; Fossey, C.; Cailly, T.; Rault, S.; Fabis, F., *The Journal of Organic Chemistry* 2011, 76, 6414–6420.
17. Smith, M. B.; Guo, L.; Okeyo, S.; Stenzel, J.; Yanella, J.; LaChapelle, E.; *Organic Letters* 2002, 42321–2323.
18. Cardinal, P.; Greer, B.; Luong, H.; Tyagunova, Y., *Journal of Chemical Education* 2012, 89, 1061–1063.
19. Sivey, J. D.; Bickley, M. A.; Victor, D. A., *Environmental Science & Technology* 2015, 49, 4937–4945.
20. Li, L.; Li, Y.; Zhao, Z.; Luo, H.; Ma, Y.-N., *Organic Letters* 2019, 21, 5995–5999.
21. Ghorpade, P. V.; Pethsangave, D. A.; Some, S.; Shankarling, G. S., *The Journal of Organic Chemistry* 2018, 83, 7388–7397.

22. Al-Zoubi, R. M.; Hall, D. G., *Organic Letters* 2010, 12, 2480–2483.
23. Rodriguez, R. A.; Pan, C.-M.; Yabe, Y.; Kawamata, Y.; Eastgate, M. D.; Baran, P. S., *Journal of the American Chemical Society* 2014, 136, 6908–6911.
24. Wang, M.; Zhang, Y.; Wang, T.; Wang, C.; Xue, D.; Xiao, J., *Organic Letters* 2016, 18, 1976–1979.
25. Mostafa, M. A. B.; Bowley, R. M.; Racys, D. T.; Henry, M. C.; Sutherland, A., *The Journal of Organic Chemistry* 2017, 82, 7529–7537.
26. Shi, H.; Wang, P.; Suzuki, S.; Farmer, M. E.; Yu, J.-Q., *Journal of the American Chemical Society* 2016, 138, 14876–14879.
27. Saper, N. I.; Snider, B. B., *The Journal of Organic Chemistry* 2014, 79, 809–813.
28. Maddox, S. M.; Nalbandian, C. J.; Smith, D. E.; Gustafson, J. L., *Organic Letters* 2015, 17, 1042–1045.
29. Rogers, D. A.; Bensalah, A. T.; Espinosa, A. T.; Hoerr, J. L.; Refai, F. H.; Pitzel, A. K.; Alvarado, J. J.; Lamar, A. A., *Organic Letters* 2019, 21, 4229–4233.
30. Maddox, S. M.; Dinh, A. N.; Armenta, F.; Um, J.; Gustafson, J. L., *Organic Letters* 2016, 18, 5476–5479.
31. Racys, D. T.; Warrilow, C. E.; Pimlott, S. L.; Sutherland, A., *Organic Letters* 2015, 17, 4782–4785.
32. Racys, D. T.; Sharif, S. A. I.; Pimlott, S. L.; Sutherland, A., *The Journal of Organic Chemistry* 2016, 81, 772–780.
33. Bergström, M.; Suresh, G.; Naidu, V. R.; Unelius, C. R., *European Journal of Organic Chemistry* 2017, 2017, 3234–3239.
34. Moorthy, J. N.; Senapati, K.; Kumar, S., *The Journal of Organic Chemistry* 2009, 74, 6287–6290.
35. Mei, T.-S.; Wang, D.-H.; Yu, J.-Q., *Organic Letters* 2010, 12, 3140–3143.
36. Wang, X.-C.; Hu, Y.; Bonacorsi, S.; Hong, Y.; Burrell, R.; Yu, J.-Q., *Journal of the American Chemical Society* 2013, 135, 10326–10329.
37. Kommagalla, Y.; Chatani, N., *Organic Letters* 2019, 21, 5971–5976.
38. Iida, K.; Ishida, S.; Watanabe, T.; Arai, T., *The Journal of Organic Chemistry* 2019, 84, 7411–7417.
39. Ghom, M. H.; Naykode, M. S.; Humne, V. T.; Lokhande, P. D., *Tetrahedron Letters* 2019, 60, 1029–1031.
40. Ichiishi, N.; Canty, A. J.; Yates, B. F.; Sanford, M. S., *Organic Letters* 2013, 15, 5134–5137.
41. Watson, D. A.; Su, M.; Teverovskiy, G.; Zhang, Y.; García-Fortanet, J.; Kinzel, T.; Buchwald, S. L., *Science* 2009, 325, 1661–1664.
42. Fier, P. S.; Hartwig, J. F., *Journal of the American Chemical Society* 2012, 134, 10795–10798.
43. Lee, E.; Hooker, J. M.; Ritter, T., *Journal of the American Chemical Society* 2012, 134, 17456–17458.
44. (a) Ye, Y.; Schimler, S. D.; Hanley, P. S.; Sanford, M. S., *Journal of the American Chemical Society* 2013, 135, 16292–16295; (b) Ye, Y.; Sanford, M. S., *Journal of the American Chemical Society* 2013, 135, 4648–4651.
45. (a) Tang, P.; Furuya, T.; Ritter, T., *Journal of the American Chemical Society* 2010, 132, 12150–12154; (b) Furuya, T.; Strom, A. E.; Ritter, T., *Journal of the American Chemical Society* 2009, 131, 1662–1663; (c) Mazzotti, A. R.; Campbell, M. G.; Tang, P.; Murphy, J. M.; Ritter, T., *Journal of the American Chemical Society* 2013, 135, 14012–14015; (d) Furuya, T.; Ritter, T., *Organic Letters* 2009, 11, 2860–2863.
46. Fier, P. S.; Luo, J.; Hartwig, J. F., *Journal of the American Chemical Society* 2013, 135, 2552–2559.
47. (a) Gao, Z.; Lim, Y. H.; Tredwell, M.; Li, L.; Verhoog, S.; Hopkinson, M.; Kaluza, W.; Collier, T. L.; Passchier, J.; Huiban, M.; Gouverneur, V., *Angewandte Chemie International Edition* 2012, 51, 6733–6737; (b) Tian, T.; Zhong, W.-H.; Meng, S.; Meng, X.-B.; Li, Z.-J., *The Journal of Organic Chemistry* 2013, 78, 728–732.
48. Truong, T.; Klimovica, K.; Daugulis, O., *Journal of the American Chemical Society* 2013, 135, 9342–9345.
49. (a) Lou, S.-J.; Chen, Q.; Wang, Y.-F.; Xu, D.-Q.; Du, X.-H.; He, J.-Q.; Mao, Y.-J.; Xu, Z.-Y., *ACS Catalysis* 2015, 5, 2846–2849; (b) Wang, X.; Mei, T.-S.; Yu, J.-Q., *Journal of the American Chemical Society* 2009, 131, 7520–7521; (c) Ning, X.-Q.; Lou, S.-J.; Mao, Y.-J.; Xu, Z.-Y.; Xu, D.-Q., *Organic Letters* 2018, 20, 2445–2448; (d) Hull, K. L.; Anani, W. Q.; Sanford, M. S., *Journal of the American Chemical Society* 2006, 128, 7134–7135; (e) Chen, C.; Wang, C.; Zhang, J.; Zhao, Y., *The Journal of Organic Chemistry* 2015, 80, 942–949; (f) Lou, S.-J.; Xu, D.-Q.; Xia, A.-B.; Wang, Y.-F.; Liu, Y.-K.; Du, X.-H.; Xu, Z.-Y., *Chemical Communications* 2013, 49, 6218–6220; (g) Testa, C.; Gigot, É.; Genc, S.; Decréau, R.; Roger, J.; Hierso, J.-C., *Angewandte Chemie International Edition* 2016, 55, 5555–5559; (h) Lou, S.-J.; Xu, D.-Q.; Xu, Z.-Y., *Angewandte Chemie International Edition* 2014, 53, 10330–10335; (i) Chan, K. S. L.; Wasa, M.; Wang, X.; Yu, J.-Q., *Angewandte Chemie International Edition* 2011, 50, 9081–9084.
50. Chen, X.-Y.; Sorensen, E. J., *Journal of the American Chemical Society* 2018, 140, 2789–2792.

4

Microwave as a Greener Alternative in the Synthesis of Organic Compounds

Paola Acosta-Guzmán

CONTENTS

4.1 Introduction ... 57
4.2 Microwave Effects ... 58
4.3 Applications in the Synthesis of Active Compounds ... 59
4.3.1 Drugs Synthesis ... 59
4.3.2 Chromanes ... 63
4.3.3 Indoles ... 65
4.3.4 Quinolines ... 66
4.3.5 Cumarines ... 67
4.3.6 Pyrimidines and Derivatives ... 69
4.3.7 Benzotriazoles ... 70
4.3.8 Triazinas ... 70
4.4 Conclusions ... 71
References ... 71

4.1 Introduction

Nowadays, the main goal of modern organic chemistry, in academia and industry, is the creation of molecular diversity and complexity combining economic and environmental features to reach the criteria of sustainable chemistry. Therefore, it is important to have synthetic processes based on transformations with multiple bond formations; it is also important to use alternative energy sources in order to simplify the whole process of discovery of innovative organic compounds.

In this context, microwave irradiation has been used since the beginning of the 20th century as an alternative source of energy inorganic reactions. Originally, microwaves were applied to heat foods; however, Percy Spencer used them during the 1940s with the purpose of accelerating a chemical reaction. The first report of the use of microwaves as an alternative source of energy in chemical reactions appeared in 1986[1] that the first report of the heating using microwaves in a synthetic process.

The first tests were carried out in domestic microwave ovens using glass or Teflon containers, but without any control of the pressure or temperature supplied to the system, which on several occasions led to explosions due to the rapid heating of the organic solvents. However, during the first 15 years, domestic microwaves were still used until the appearance of specially designed reactors that allowed the use of this type of unconventional heating in synthetic methodologies in a safer and more controlled manner.

Microwaves have shown many advantages when they are applied in organic syntheses; some of them are: (i) better yields and higher purity of the reaction products because of a minor quantity of byproducts obtained and consequently, the purification step is faster and easier; (ii) shorter reaction times because microwaves interact directly with the molecules increasing the rate of heating, which is supported by experimental data showing that microwaves may enhance the rates of chemical reactions 1000 times compared to those under conventional heating; (iii) the reactions can be performed at higher temperatures; this is because the solvent can be overheated under microwave irradiation; thus, the reactions are completed in a few minutes instead of hours. An example is the synthesis of fluorescein which usually takes about 10 hours for conventional heating; however, it can be achieved in only 35 minutes using microwave irradiation.[2]

As a result of the advantages of the use of microwave irradiation in synthetic procedures, this technology has been included within the precepts of green chemistry which are defined as a set of principles aimed at reducing or eliminating the use or generation of dangerous chemical substances that may intervene in chemical processes (Figure 4.1). Since this technique represents a clean methodology, reproducible, that allows reactions to be carried out under solvent-free conditions, it implies lower energy costs as the reaction times are shorter and allows selectivity in the products obtained.[3]

Thereby, the applications of microwave irradiation on a laboratory scale are very well documented; unfortunately, their industrial applications have several drawbacks due to the big volume of reagents. In fact, the microwave effect is limited by the "penetration factor"; when large volumes are irradiated, microwaves lose their energy as they have to pass through the

DOI: 10.1201/9781003089162-4

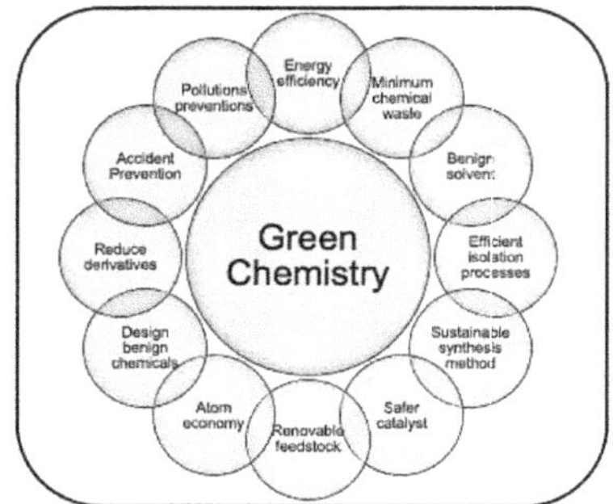

FIGURE 4.1 Principles of green and sustainable organic chemistry.

surface to the interior. A plausible solution is flow heating[4] wherein the system is heated while the reagents circulate in a closed reactor.

This chapter discusses some selected recent examples of microwave-assisted reactions applied in the synthesis of drug and drug-related molecules, some drug derivatives and heterocyclic compounds with promising biological applications. The emphasis is on the importance of the molecule and how microwaves help in the synthesis of different compounds and also on the advantages shown by microwaves over conventional heating, in terms of reaction times, yield and reduction of the number of steps.

The chapter starts with some description of the theoretical aspects in order to understand how microwaves act and which is the microwave effect that promote the organic reactions and allows shorter reaction times, also how using microwaves the desired product is obtained in better conditions. Then, the chapter turns to the discussion of some of the applications of microwaves in the synthesis of medicines and heterocyclic compounds. The aim of this chapter is to demonstrate the significance of microwaves in organic synthesis and in procedures that improve the preparation of existing drugs as well as the discovery of promising compounds for the treatment of different diseases.

4.2 Microwave Effects

Chemical reactions often require thermal activation. Traditionally, heating is done using oil baths or water baths and an appropriate solvent al reflux. Since the appearance of microwave reactors dedicated to organic synthesis in 2000, the situation has gradually changed and Microwave Radiation-Assisted Organic Synthesis (SOARM) has firmly established as a powerful technology for the thermal activation of chemical reactions, showing enormous advantages such as shorter reaction times – from hours to minutes or even seconds – and reduction of collateral reactions involving the formation of by-products and therefore increase of reaction yields. In some cases, the diastereoselectivity of target molecules is increased, even though diastereoselectivity is not a concept that is favored by heating.

Microwave irradiation is electromagnetic irradiation in the frequency range of 0.3–300 GHz; therefore, it lies between

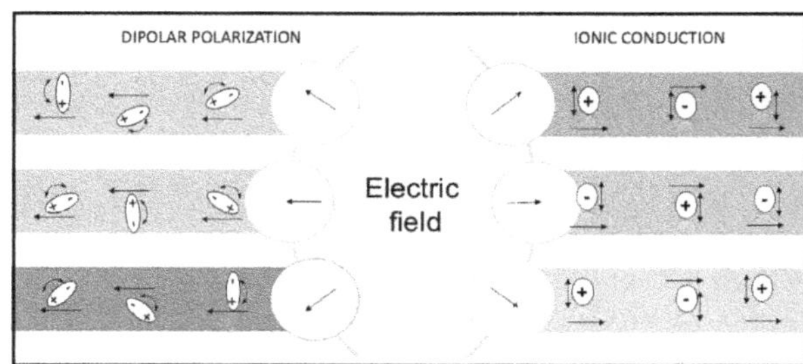

FIGURE 4.2 Different mechanisms of heating by an electric field.

infrared and radio frequencies. These electromagnetic waves are formed by two components, an electric field and a magnetic field, with the electric field being the most important for wave–material interaction, while in some cases, such as metals, the magnetic field interaction may be the most relevant. The electric component is responsible of heating by two mechanisms: dipolar polarization and ionic conduction (Figure 4.2).

Dipolar polarization is an interaction between the electric field and the polar molecules. The molecules try to align themselves while the electric field of the microwave radiation is oscillating. As the consequence, this rotational movement produces an energy transfer by molecular friction. On the other hand, ionic conduction takes place when free ionic species are in the solution, which under the influence of the electric field of the radiation tries to orient themselves analogously to the dipole rotation. The result is instantaneous localized superheating.[5]

Microwave-induced chemistry is based on the efficient heating of materials; however, the heating characteristics of a particular material under microwave irradiation depend on the dielectric properties of the material. For this reason, the materials can be classified into three categories based on their interactions with microwaves. The first one corresponds to high dielectric materials which lead to strong absorption of microwaves and consequently to rapid heating of the medium. They are the most important class of materials for microwave-induced synthesis, for example, polar solvents or aqueous solutions. The second class is composed of materials that are transparent to microwaves, exhibiting only small interactions with the penetrating microwaves; examples are borosilicate glass, ceramic, Teflon, fused quartz, among others. The last category is composed of materials that reflect microwaves, there is no, or only small, coupling of energy into the system. In this case, the temperature increases in the material only marginally, for example, metals.

The ability of some materials to convert electromagnetic energy into heat energy at a given frequency and temperature is calculated using the equation ($\varepsilon''/\varepsilon = \tan \delta$), where δ is the dissipation factor (often called the loss tangent), ε'' is the dielectric loss, which measures the efficiency with which heat is generated from the electromagnetic irradiation and ε is the dielectric constant of the material. The ability to absorb microwaves of a given material is directly proportional to its dissipation factor. Likewise, it is proposed that the interfacial polarization may also contribute to the heating effect when the conducting particles are in contact with a non-conducing medium such as a heterogeneous reaction.

Hence, the larger the dielectric constant, the greater the interaction with the solvent. So, solvents such as water, DMF, methanol, acetone, ethyl acetate, acetic acid, chloroform and dichloromethane are heated when they are irradiated with microwaves, while solvents like toluene, hexane, CCl_4 and diethyl ether do not interact; in other words, they are more transparent to microwaves.

In addition, solvents can be classified as high ($\tan \delta > 0.5$), medium ($\tan \delta = 0.1–0.5$) and low microwave-absorbing solvents ($\tan \delta < 0.1$). Other common solvents without a permanent dipolar moment, such as carbon tetrachloride or benzene ordioxane, are more or less microwave-transparent.[6]

Despite having the theoretical arguments mentioned above that allow knowing the basic understanding of high-frequency electromagnetic irradiation and microwave–matter interactions, the exact reasons why and how microwaves improve the conditions for a chemical reaction to take place is still a matter of debate in the scientific community manage to improve the chemical reactions; however, a series of proposals on the existence of the so-called specific microwave effects, or also known as non-thermal, which cannot be achieved by conventional heating have been mentioned. These microwave effects can be regarded as thermal and non-thermal. The first ones result from microwave heating, which may result in a different temperature regime, whereas non-thermal effects are specific effects resulting from non-thermal interactions between the substrate and the microwaves. Some examples of this microwave effect are: (i) the superheating effect of solvents at atmospheric pressure, (ii) the selective heating, for example, strongly microwave-absorbing reagents in a less polar reaction medium and (iii) the elimination of wall effects caused by inverted temperature gradients.

The conventional heating is comparatively slow and energy-inefficient because the energy is transferred from the walls to the interior of the reaction mixture. Consequently, a large portion of the energy is lost to the environment through the conduction of materials and convection currents. In this case, the heating effect is heterogeneous and is dependent on some properties of the materials such as conductivity, specific heat and density, which can lead to higher surface temperatures and therefore non-uniform sample temperatures as well as higher thermal gradients.

In contrast, microwave irradiation results in efficient internal heating by the direct transfer of microwave energy to dipoles or ions that are present in the reaction mixture. Also, the temperature profile shows that microwave heating allows a rapid increase of solvent temperature and quick cooling as well. As a consequence, this direct "in-core" heating, results in inverted temperature gradients compared to a conventionally heated system.[7]

Finally, the literature mentions that the main difference between reaction rates and selectivity under microwave and conventional heating has been explained by the thermal effects mentioned before, summarized as rapid heating, superheating, hot spots, selective heating and simultaneous cooling.

4.3 Applications in the Synthesis of Active Compounds

4.3.1 Drugs Synthesis

Aspirin is a non-steroidal, anti-inflammatory drug (NSAID). It was the first drug of this class to be discovered. It has many applications, including relieving pain and swelling, managing various conditions and reducing the risk of cardiovascular events in people with high risk. People also use it as an anti-inflammatory or a blood thinner. Recently, microwave-induced heating has been employed in the synthesis of some analgesic drugs demonstrating its advantages in terms of purity, yield and reaction time. A good example is a method described by Montes and coworkers[8]; they proposed the synthesis of aspirin from salicylic acid and acetic anhydride under microwave irradiation, thus making an evaluation of different catalysts; the evaluation included some acids (H_2SO_4, H_3PO_4, $MgBr_2.OEt_2$, $AlCl_3$) and bases ($CaCO_3$, NaOAc, Et_3N, DMAP) (Scheme 4.1a). The study demonstrated that the best catalysts, in terms of reaction time, were the Brønsted acids. Additionally, under basic conditions, aspirin was obtained with better yield, high purity and without polymer formation. This was because the calcium carbonate ($CaCO_3$) worked very well under microwaves and it did not work under conventional heating.

On the other hand, the comparison of the results when the reaction is performed without any catalyst seems very interesting. Under traditional heating, no reaction was observed; however, under microwave irradiation, the reaction was completed in 10–13 minutes without polymer formation producing the highest yield of >90%. This decrease in reaction time using microwave irradiation is attributed to an increase in the polarity of the medium or a charge separation in the reaction transition state (Scheme 4.1b); as the transition state is stabilized, the activation energy decreases.[9]

Commonly prescribed worldwide as a highly effective analgesic and anti-inflammatory agents, the NSAIDs including ibuprofen, naproxen and indomethacin, among others, have demonstrated a mechanism of action involving inhibitions of prostaglandin synthesis, particularly in swollen tissue by inhibiting the enzyme cyclooxygenase (COX),[10] but the big problem is that of two distinct cyclooxygenase isoforms known as COX-2 and COX-1. The inhibition of the enzyme COX-2 leads to anti-inflammatory action, while the inhibition of COX-1 results in side effects such as affection of the renal function, gastric mucosa problems, vascular hemostasis and the autocrine response to circulating hormones.[11] In this context, the development of a reliable and selective anti-inflammatory agent with low or null side effects is a long-standing medicinal chemistry problem with significant social implications. Some efforts to improve the gastrointestinal tolerance of NSAIDs by masking the free carboxylic group temporarily have included the preparation and evaluation of numerous ester and amide prodrugs.[12] With the intention of making an interesting contribution to this problem, Hall and coworkers[13] reported the synthesis of novel NSAID bis-conjugates with acetaminophen and

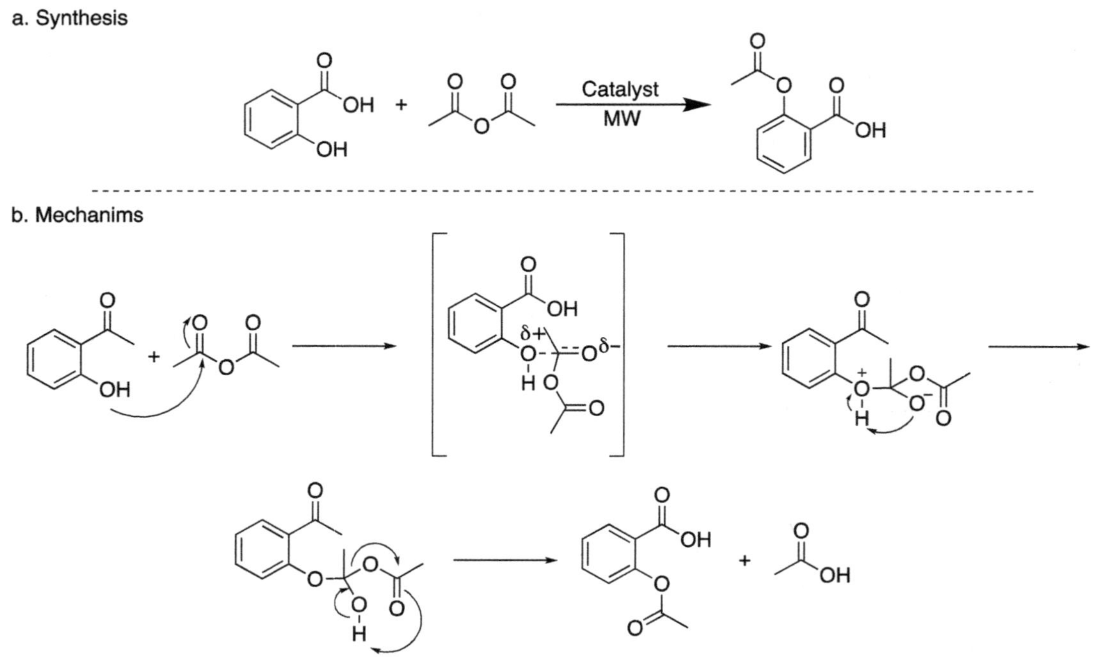

a. Synthesis

Catalyst
MW

b. Mechanims

SCHEME 4.1 (a,b) Synthesis of aspirin promoted by microwaves.

amino acid linkers using benzotriazole chemistry, the synthesis started with the coupling of the Boc-aminoacylbenzotriazoles with acetaminophen using DMAP in THF under microwave irradiation to obtain the Boc-protected amino acid acetaminophen conjugates. Then, Boc-protected amino acid–acetaminophen conjugates were deprotected with dioxane–HCl solution; the crude product was used in the next step without any purification. Finally, the desired compounds were obtained by coupling the unprotected amino acid–acetaminophen conjugates with NSAID-benzotriazolides in the presence of K_2CO_3 in DMF under microwave irradiation at 50 W, 70°C (Scheme 4.2).

All the compounds obtained using this methodology were evaluated to determine their anti-inflammatory activity. The results showed that all the synthesized compounds exhibited more potent anti-inflammatory activity than their parent drugs (ibuprofen, naproxen, indomethacin); based on the observed results, the authors established structure–activity relationships (SARs) and proposed: the alkyl/arylalkyl function of the amino acid residue controls the anti-inflammatory potency of the compounds generally being the compound with benzyl group in the amino acid function the most potent agent between the alkyl functions. Additionally, the compounds were evaluated to determine the ulcer index values and the study showed that none of the anti-inflammatory analogs caused ulcers, lesions or erosions in rat's gastric mucosa.

The neurocognitive disorder can occur when some virus infiltrates the central nervous system and causes a neuroinflammatory cascade; scientists have estimated that 15–50% of individuals with HIV have HIV-associated neurocognitive disorder (HAND). Even when the disease is well-managed, the latent viral reservoir can cause persistent inflammation. Additionally, some alterations in brain structure and function accompany HAND, with diffuse degradation in gray and white matter volume, as well as changes in brain function.[14] As a contribution to this health problem, the investigations have aimed at the synthesis of new compounds that increase liposolubility, since there exists a relationship between lipophilicity, membrane permeability and CNS penetration. These studies involved modifications of the nucleoside base with lipophilic functional groups[15] or the phosphate groups of nucleotides as well as the modification of the 5'-hydroxyl group of the parent anti-HIV nucleosides, such as zidovudine, stavudine, 3'-azido-2',3'-dideoxyuridine[16] and abacavir.[17] In this sense, Sriram contributed to this investigation,[18] with the synthesis of Schiff bases of abacavir modifying the 2-amino group. He used microwave irradiation to achieve the condensation between the drugs and some aldehydes and ketones under the presence of acetic acid. The products were obtained in short times (1–3 min) compared to the traditional heating (3 hours); also the products were obtained with high purity after recrystallization from EtOH/$CHCl_3$ (Scheme 4.3).

All the synthesized compounds were evaluated to establish the inhibitory effect on the replication of HIV-1 in CEM cell lines. The values of EC_{50} (effective concentration of compound (lM) achieving 50% protection in MT-4 cell lines against the cytopathic effect of HIV-1) and CC_{50} (cytotoxic concentration of compound (lM) required to reduce the viability of mock infected CEM cells by 50%) demonstrated excellent anti-HIV activity. The compound with a methyl group in the aromatic ring was the most potent (EC_{50} of 0.05 μM) in comparison to the parent drug (EC_{50} of 1.6 μM). Additionally, the big relevance of the present study is that the Schiff bases synthesized have higher lipophilicity (log P) than the corresponding drugs; consequently, their permeation properties through viral cell membranes were improved.

Continuing with the studies on the synthesis of drugs to treat CNS disease, the other critical problem of commercial drugs is

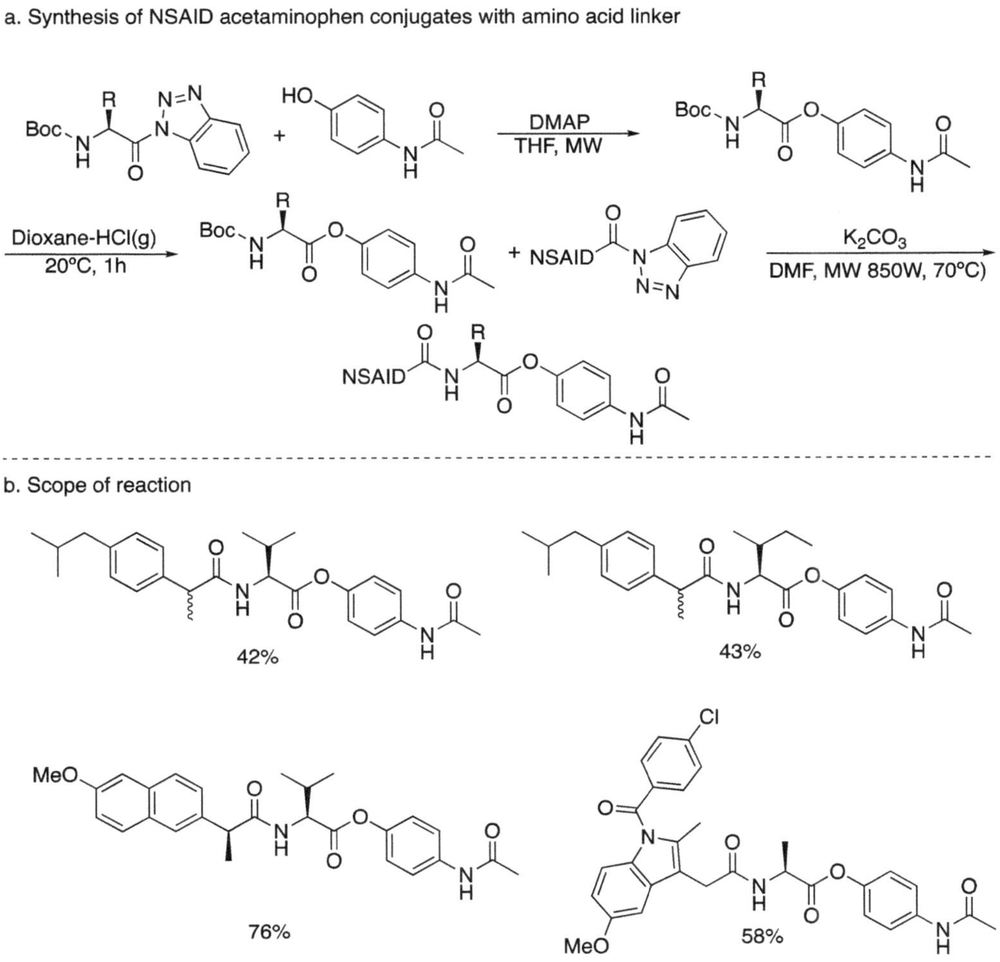

a. Synthesis of NSAID acetaminophen conjugates with amino acid linker

DMAP, THF, MW

Dioxane-HCl(g), 20°C, 1h

K_2CO_3, DMF, MW 850W, 70°C)

b. Scope of reaction

42% 43% 76% 58%

SCHEME 4.2 Synthesis of novel NSAID bis-conjugates with acetaminophen.

the low brain bioavailability; thus, new strategies like blood–brain barrier (BBB) shuttles[19] and molecular Trojan horses[20] have also been explored for brain delivery *via* passive diffusion or active transport. A desirable BBB shuttle should have physicochemical properties consistent with CNS-active drugs, including a limited number of hydrogen-bond donors (HBDs), low molecular weight and sufficient lipophilicity. Since the shuttle will ultimately be linked to the parent drug, the shuttle itself should be smaller and have fewer HBDs than a typical CNS drug. Based on those criteria, *N*-benzylamide derivatives are ideal shuttle candidates since they are hydrophobic, small, aromatic and readily synthesized.

Marder[21] predicted the reason why the NSAIDs have exhibited little to no clinical therapeutic efficacy against AD; proposing that this is due to extensive plasma-protein binding. To give a solution to this problem, the researchers proposed blocking the carboxylic acid moiety introducing the function ester[22]; this strategy should reduce the number of HBDs; as a consequence, it increases the lipophilicity and prevents ionization of the carboxyl group; additionally, this modification increases the percentage of unbound drugs in circulation and potentially reduces gastric side effects common in chronic NSAID use.

To contribute to improving the brain bio-availabilities of known NSAIDs, Young and coworkers designed a *N*-benzylamine shuttle system. The compounds were synthesized using a three-step strategy. First, the formation of acid chloride using classic conditions ($(COCl)_2$ and DMF), after a nucleophilic acyl substitution reaction was performed between the acid chloride and benzylamine. Finally, the authors included microwave irradiation to promote the nucleophilic substitution reaction between the NSAID carboxylate (Scheme 4.4). Unfortunately, the synthetic procedure has an issue of chemoselectivity. In fact, the substitution and elimination products are in competing. Authors tried different conditions (solvents and temperatures); nevertheless, they only obtained the desired products in yields ranging between 17–25% for the final step. All the compounds synthesized using this methodology exhibited physicochemical properties such as molecular weights, calculated log of the octanol–water partition coefficient, topological polar surface areas (TPSA) and number of HBDs and hydrogen-bond acceptors (HBAs) in the range of other CNS drugs. In addition, exploring physicochemical features relevant to CNS delivery, the stabilities of these NSAID *N*-benzylamide conjugates were tested. First, the stabilities of some of the conjugates at 37°C in phosphate-buffered saline (PBS; pH 7.4) were assessed, they exhibit long half-lives in this medium (>48 h) and short-to-moderate half-lives in the human plasma.

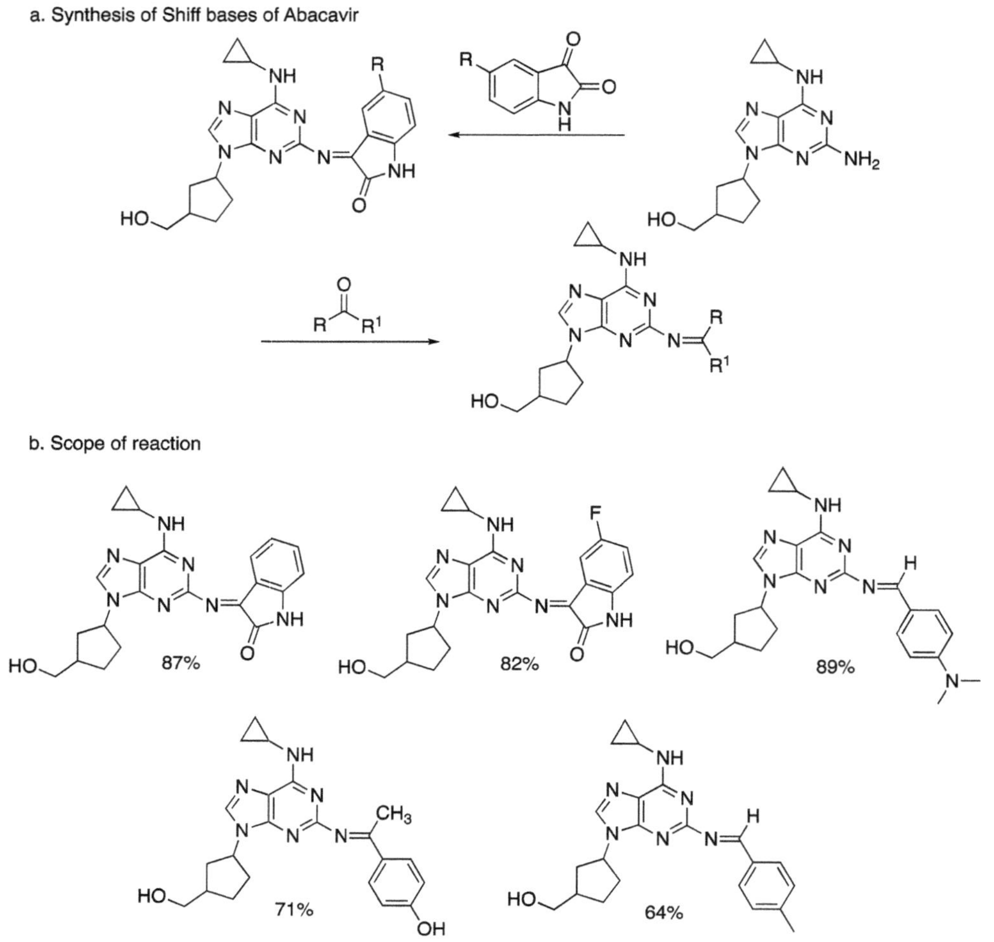

a. Synthesis of Shiff bases of Abacavir

b. Scope of reaction

87% 82% 89% 71% 64%

SCHEME 4.3 Bases of Shiff derivatives of abacavir.

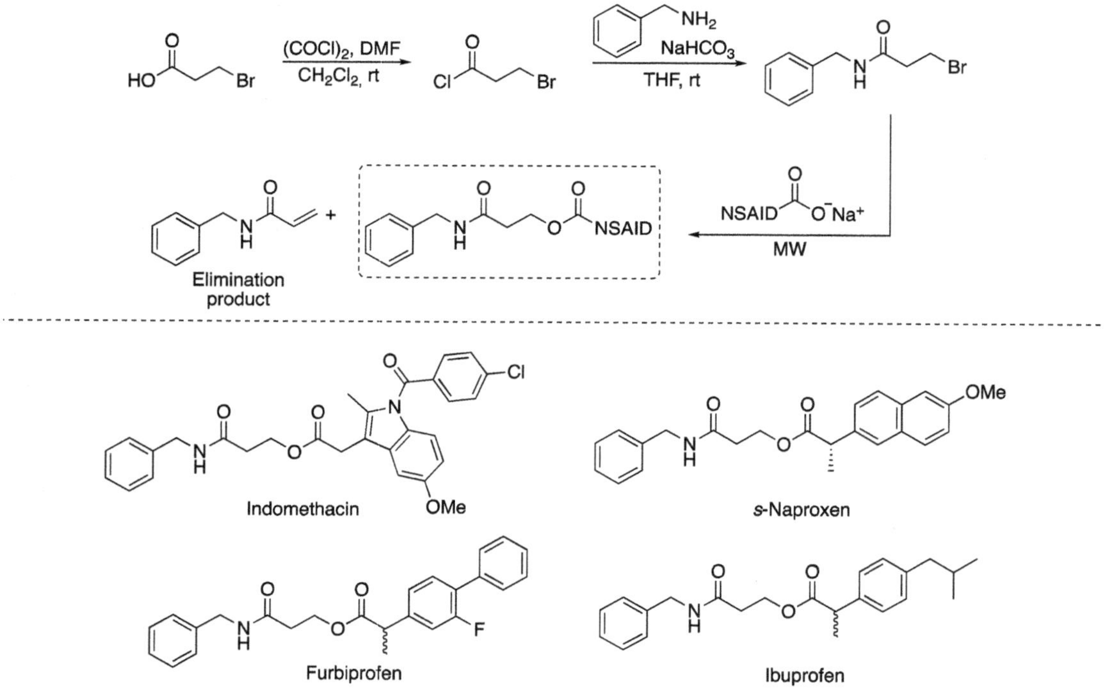

(COCl)$_2$, DMF, CH$_2$Cl$_2$, rt; NaHCO$_3$, THF, rt; NSAID, MW

Elimination product

Indomethacin; s-Naproxen; Furbiprofen; Ibuprofen

SCHEME 4.4 Synthesis of NSAID *N*-benzylamide conjugates.

With these results, the authors suggested that *N*-benzylamide NSAID conjugates are promising CNS.[23]

Diabetes is one of the World's most concerning health problems and millions of patients are using antidiabetic drugs (ADDs) in order to control blood glucose. Normally, metformin is used for the treatment of this disease. This medication is used to decrease hepatic (liver) glucose production, decrease GI glucose absorption and increase the target cell insulin sensitivity.[24] Nevertheless, metformin has slow and incomplete absorption from the upper intestine after oral administration because it is a strong base (like all bisguanidines) and highly water-soluble. Additionally, in minimal doses, the metformin has gastrointestinal adverse effects like diarrhea, vomit, abdominal pain and nausea, among others.[25] To provide a solution to this problem, lipophilic prodrugs were studied with the aim of increasing permeability and passive diffusion across the cell membranes. In this way, sulfenamides have proven to be promising candidates as more lipophilic prodrugs than metformin, since the powerful cyclization behavior can be avoided by this prodrug approach.[26] Generally, sulfenamide prodrugs are rapidly bio-converted to their parent drugs by endogenous free thiols, like glutathione and cysteine, and free thiol-containing proteins.[27] In this context, Huttenen and coworkers proposed a fast and convenient microwave-assisted one-pot synthesis to prepare more lipophilic sulfenamide prodrugs of metformin. First, the procedure implied the liberation of the basic metformin by stirring metformin hydrochloride with sodium hydroxide 1 M for 30 min. After that, basic metformin, silver nitrate and disulfide were dissolved in anhydrous MeOH in a sealed pressure-resistant glass tube and irradiated with microwaves until reaching a temperature of 80°C for 30 min. Finally, the reaction mixture was filtered and the filtrate was treated with acetic acid or hydrochloric acid 1M; this last step was necessary because sulfenamides seemed to be somewhat unstable in their basic form. The desired compounds were obtained with moderate yield; however, this microwave-assisted synthesis is faster than other methods (Scheme 4.5). The prepared prodrugs had significantly increased lipophilicity; the shortest alkyl group had only a modest improvement on lipophilicity while the longer alkyl groups seemed to increase the amount of the prodrugs in the octanol phase, especially at pH values between 6.5 and 7.4. Therefore, these prodrugs have a great potential to improve passive intestinal absorption of metformin, since the pH of the intestine varies typically between 4.5 and 8.0. Furthermore, the rates of bioconversion of the sulfenamide prodrugs were determined in 1 mM reduced glutathione and cysteine solutions at 37°C (pH 7.4). The compound with butylthio moiety was the only sulfenamide prodrug that degraded and released metformin quantitatively. Finally, according to the preliminary *in vivo* studies, the octylthio prodrug was also absorbed mostly intact after oral administration in rats.[28]

Diseases that affect bone health such as osteoporosis, Paget's disease and hypercalcemia are a significant public health problem that contributes substantially to morbidity in aged world population. However, beginning in the 1960s, it was discovered that bisphosphonates have a high affinity for bones and are effective for the treatment of this diseases associated with bone mineralization. For this reason, several members of this group of agents are either in clinical use or well advanced in clinical trials.

The bisphosphonates contain a phosphate-carbon-phosphate (P-C-P) core structure that targets them to bone and renders them resistant to enzymatic degradation. The possible inhibition of bone resorption *in vitro* by bisphosphonate was demonstrated by initial observations that the bisphosphonate structure had a high affinity for bone and inhibited the degradation of hydroxyapatite crystals. Due to the importance of this type of compound, different efforts are being made to improve their obtention.[29] In this sense, McKenna presents a new strategy to synthesize bisphosphonate drugs (risedronate, zoledronate, pamidronate, alendronate and eridronate) by a one-pot synthesis employing microwave irradiation. The reaction was carried out by the addition of the respective carboxylic acid to phosphoric acid, then, oxalyl chloride was added and the mixture was irradiated for 3–7 min. Finally, the reaction was quenched with water and the hydrolysis led to bisphosphonic acid (Scheme 4.6). Thus, this protocol has effectively reduced the reaction time from at last 7.5 h to less than 20 min, while maintaining a good yield of product in comparison to the traditional method for the preparation of these compounds.[30]

4.3.2 Chromanes

The chromene system is an important structural class of oxygen heterocycles fused to aromatic rings; this class of heterocyclic compounds having diverse biological and chemical

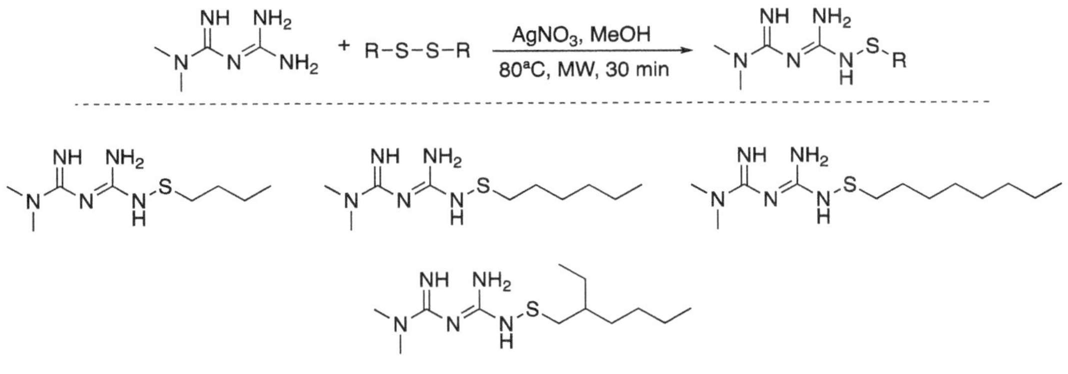

NH NH$_2$ + R–S–S–R $\xrightarrow[80°C,\ MW,\ 30\ min]{AgNO_3,\ MeOH}$ NH NH$_2$ S R

SCHEME 4.5 Sulfenamide prodrugs of metformin obtained under microwave.

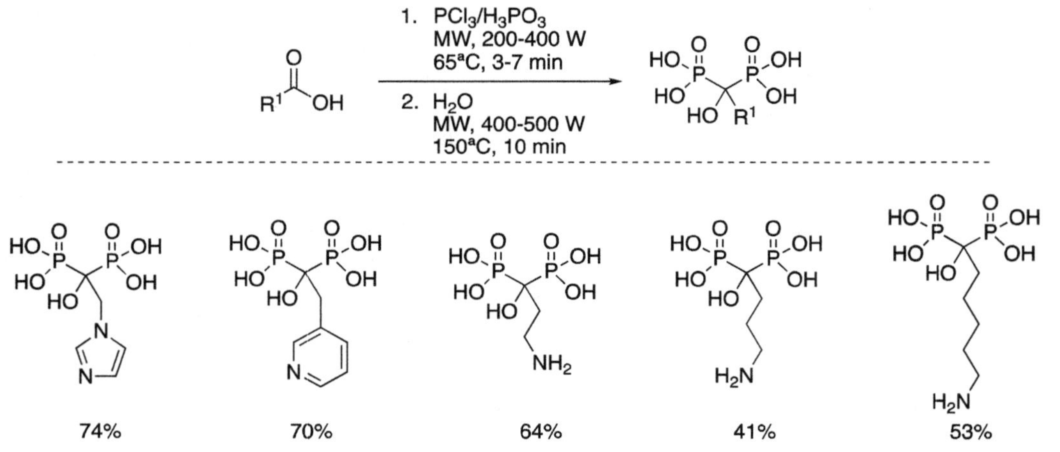

SCHEME 4.6 Synthesis of bisphosphonates.

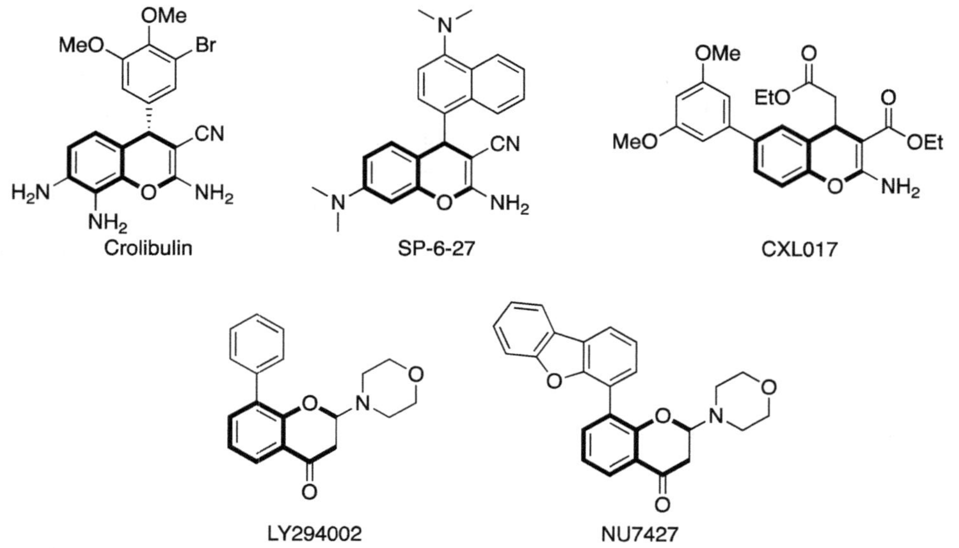

FIGURE 4.3 Biologically active chromenes.

importance, shows diverse biological activities such as antioxidant, antibacterial, antifungal, anti-inflammatory and antitumor effects. Also, the chromene skeleton is found in a myriad of biologically and chemically important natural products; an example of an unnatural analog is the Cromakalim that is a potassium channel opener.

In 2010, some synthetic chromene analogs emerged as potent anticancer agents. Crolibulin A chromene analog, is in Phase II of clinical screening for aplastic thyroid cancer with the National Cancer Institute (Figure 4.3). Additionally, some bicyclic 4*H*-chromene analogs have shown promising anticancer activity for various cancer cell lines. These new chromenes have shown strong cytotoxicity against human cancer cell lines through various pathways, including microtubule depolymerization.[31] Because of the predominance of the chromene scaffold, the scope of chromene research is multifarious, extending from rather simple to highly complex molecules; for this reason, considerable efforts in developing new and general methods to prepare them have been made for many years. In this sense, multicomponent reactions and microwave (high-speed) synthesis are considered as one of the most important tools to produce biologically and chemically important chromenes in a time-sensitive manner. However, environmentally friendly, efficient and economical methods for the synthesis of biologically important chromenes remain a significant challenge in synthetic chemistry.[32]

Due to the importance of the chromene system and in the context of researching the development of new sustainable synthetic methods. The group of Patil reported the microwave-assisted synthesis of a potent anticancer agent (*in vitro*) by a tricomponent reaction among phenol, aldehyde and malononitrile (Scheme 4.7). All the chromenes synthesized under these conditions showed activity in the nanomolar range (IC_{50}: 7.4–640 nM) in two melanoma, three prostate and four glioma cancer cell lines. Early studies on the mechanism of action suggested that these novel chromenes interact with the colchicine-binding site in tubulin.[33]

Shah and coworkers[34] synthetized novel pyrazole-substituted 4*H*-chromenes with good antibacterial activity against

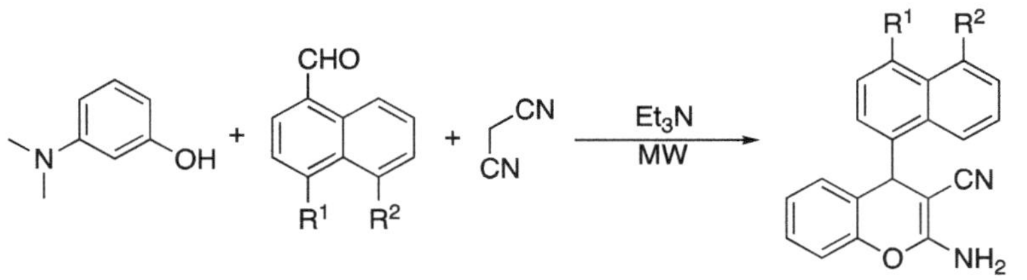

SCHEME 4.7 Synthesis of 4*H*-chromene analogs.

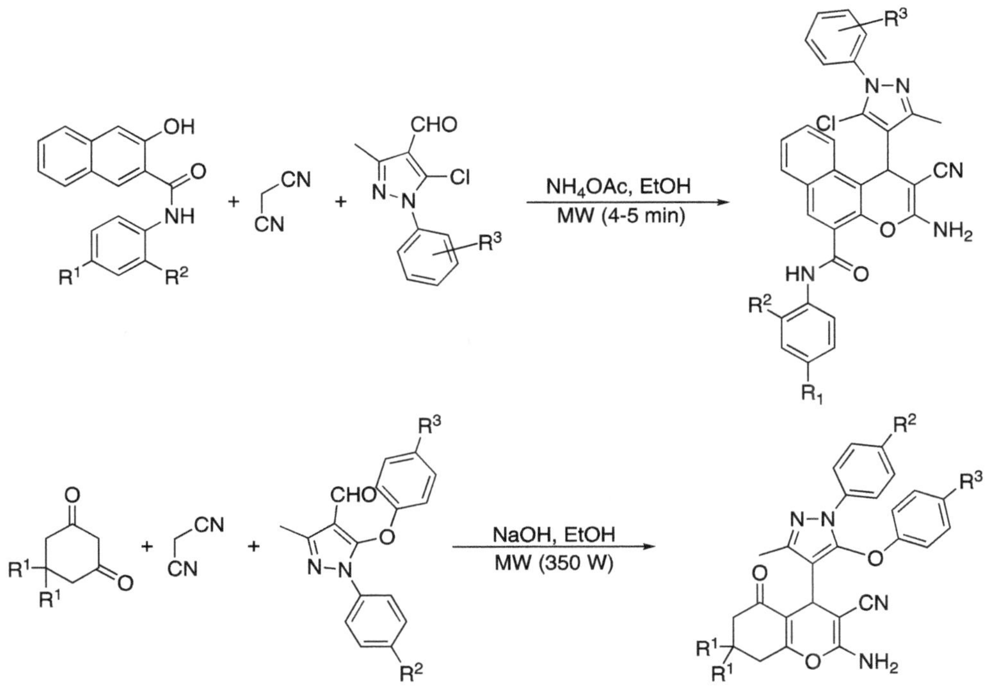

SCHEME 4.8 (a,b) Synthesis of 4-heteroarly-chromene derivatives.

Escherichia coli compared to standard drug ampicillin. They obtained the compounds after a one-pot reaction between various substituted 5-chloro-3-methyl-1-aryl-4,5-dihydro-1*H*-pyrazole-4-carbaldehydes, 2-naphthols and malononitrile using microwaves under the presence of catalytic amounts of ammonium acetate (Scheme 4.8a); the products were obtained with excellent yields (>85%). Then, Sangani and his group,[35] continued to work on the development of new pyrazole-substituted 4*H*-chromenes to find possible SARs. In this case, substituted 5-phenoxypyrazole-4-carbaldehydes reacted with dimedone and malononitrile under the presence of NaOH and microwave irradiation (Scheme 4.8b). The new compounds were screened against three Gram-positive bacteria (*Streptococcus pneumoniae, Clostridium ani* and *Bacillus subtilis*) and three Gram-negative bacteria (*Salmonella typhi, Vibrio cholerae* and *E. coli*). After the synthesis, they established SARs among the 4*H*-chromenes. The SAR revealed that a gem dimethyl group on the benzopyrane ring with either a chlorine or methyl substituent on the *O*-phenyl ring of the pyrazole moiety, methyl group on the *N*-phenyl ring of the pyrazole moiety as well as a methyl group on the *N*-phenyl ring of the pyrazole moiety are essential to obtain good antibacterial activities.

4.3.3 Indoles

The indole structure represents an essential structural moiety in drug discovery. In recent years, many synthetic pathways leading to indole derivatives have been developed. On the other hand, 3-substituted indoles are common components of drugs and are generally found to be of pharmaceutical interest in a variety of therapeutic areas.[36] This basic skeleton is present in the essential amino acid tryptophan and the key neurotransmitter serotonin. Additionally, this heterocycle is present in several commercial drugs of the triptan family that were developed for the treatment of migraine headaches, for example, frovatriptan, zolmitriptan and sumatriptan (Figure 4.4). Other well-known drugs such as the antibiotic indolmycin and the NSAID Indomethacin also have an indole moiety.[37]

Since 1866, when Bayer made the first synthesis of indole, a large number of synthetic routes have been published. Nonetheless, microwave synthesis has emerged as a valuable tool in the search of novel medicinally relevant indoles. This technology offers the opportunity of synthesizing pharmacologically important indoles in a time-sensitive manner. In the literature exist several classic name reactions to build this important system such as Ficher, Reissert, Gassmann, Batcho–Leimgruber, Bischler-Möhlau, Hegedus, Nenitzescu

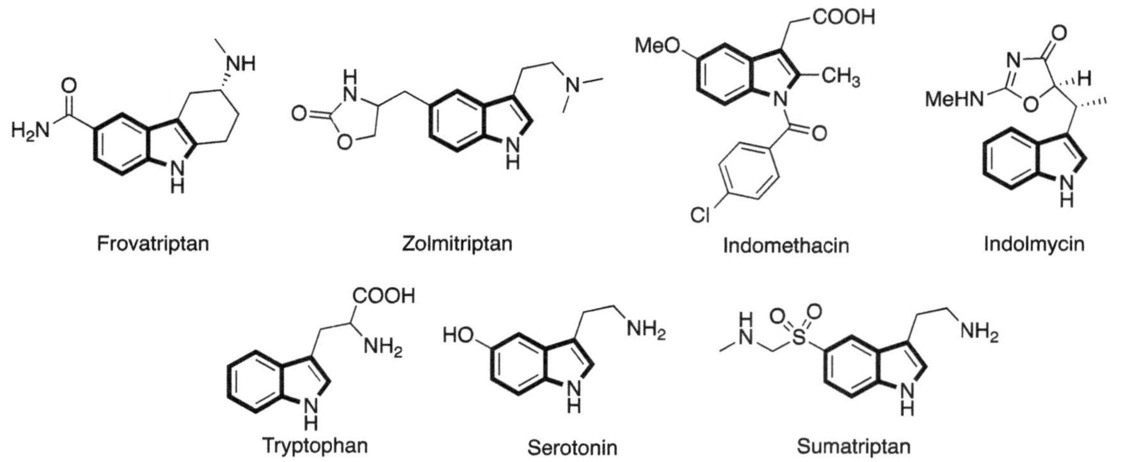

Frovatriptan Zolmitriptan Indomethacin Indolmycin

Tryptophan Serotonin Sumatriptan

FIGURE 4.4 Commercial drugs with indole system in their structure.

and Madelung reaction.[38] However, the Ficher syntheses have maintained their versatility until today to produce a large number of indole analogs compared to any other established route. This method allows formation of indoles from the reaction of arylhydrazines and carbonyl compounds (ketones and aldehydes) under acid conditions *via* the cyclization of hydrazine intermediate.

In this context, Barbieri and coworkers[39] adopted the Ficher indole synthesis to obtain carbazoles; this nitrogen-containing tricyclic scaffold presents a wide variety of biological activities including anti-Alzheimer, anti-Parkinson, antimicrobial and anticancer activities.[40] They reported the use of microwave irradiation in the one-pot synthesis of tetrahydro carbazole and tetrahydro-7*H*-pirido[*a*]carbazoles (Scheme 4.9); they optimized different variables such as temperature, power and time to obtain good-to-excellent yields in the final product. The main advantages of the microwave methodology were shortening of the synthetic pathway from a two-step convectional method to a one-pot synthesis, cleaner reaction products, higher yields and shorter reaction times.

Other researchers have also been interested in developing new indole systems through the application of microwave irradiation. Lipinska[41] described a good method to prepare 2-heteroaryl-5-methoxyindole as the key intermediate in the preparation of indolo[2,3-*a*]quinolizine alkaloids. This approach allows to easily replace the traditional harsh conditions (180–200°C heating) with microwave-induced solid-supported (MK10/$ZnCl_2$) Fischer indolization (Scheme 4.10).

4.3.4 Quinolines

Quinolines are benzo-fused pyridine heterocycles and this motif is one of the most prevalent heterocyclic scaffolds since it is present in many natural products. Furthermore, quinoline moiety is an essential pharmacophore and a crucial functionality because of its wide variety of pharmacological activities which include anticancer,[42] anti-HIV,[43] antibacterial,[44] antioxidant,[45] anti-inflammatory,[46] antimalarial,[47] antifungal,[48] antiproliferative,[49] antihypertensive,[50] antitubercular properties,[51] among others.

Having this in mind, Jassem and coworkers[52] decided to study the quinoline system as a potential antibreast cancer drug because this type of cancer is a malignant and deadly disease around the world since the current treatments are limited by

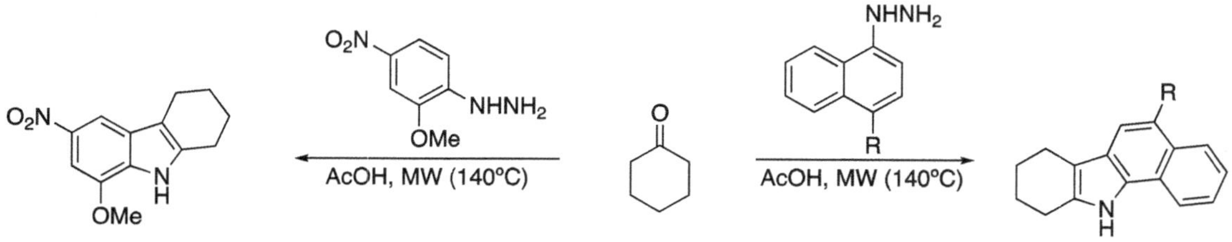

AcOH, MW (140°C) AcOH, MW (140°C)

SCHEME 4.9 One-pot synthesis of tetrahydro-7*H*-pyrido[*a*]carbazoles.

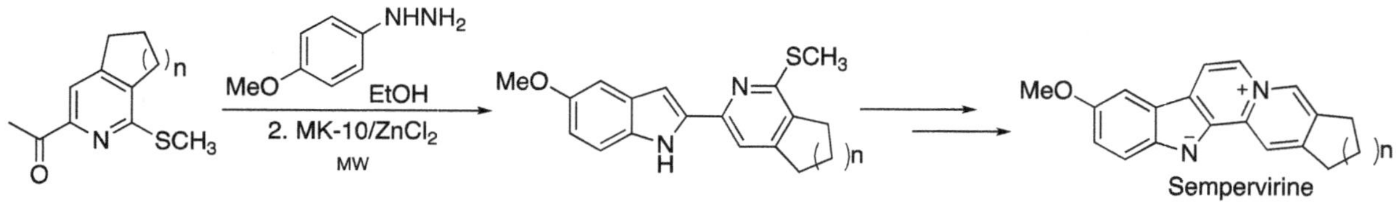

EtOH 2. MK-10/$ZnCl_2$ MW

Sempervirine

SCHEME 4.10 Synthesis of 2-heteroaryl-5-methoxyindole intermediate in the pathway synthesis of sempervirine.

the emergence of cure-resistant cancer cells. Nowadays, in the targeted therapy of breast cancer, human epidermal growth factor receptor 2 (HER2) is being considered as a promising route to design novel drugs because it plays an important role in the regulation of cyclic cell signaling pathways involving cell proliferation and cellular replication.[53] In their work, the authors reported two novel *N*-substituted quinolone derivatives; the synthesis involved the obtention of a key intermediate from the reaction between 2,5-difluoro-4-(pyrrolidin-1-yl) benzoyl chloride and ethyl 3-(diethylamino) acrylate; however, under conventional heating, the product could not be obtained, no matter the solvent or the base used. Therefore, on switching over to microwave irradiation, the coupling reaction was feasible; the next step was a transaminolysis with cyclopropylamine using DBU as an efficient catalyst to yield quinolone derivatives (Scheme 4.11). This method is an innovative tool for the improvement of some cyclization reactions without any difficulties to achieve a nucleophilic ring closure. Additionally, docking studies have been used to determine any interaction between the new compounds and the residues where the activity lies in the active site cavity of HER2. This study exhibited hydrogen bonds, polar and Van der Waals interactions. Also, the binding energy and hydrogen bonding interactions showed that the synthesized compounds are considered to have potential activity against breast cancer. Besides, the new quinoline derivatives were evaluated against HIV-1 and HIV-2 founding potent activity against HIV-1 replication.

Another example of how microwaves can be used in the synthesis of quinolines was reported by Olayenca and coworkers.[54] They utilized a microwave-assisted approach as a green methodology to access a series of 2-propylquinoline-4-carbohydrazides hydrazones derivatives. It involved a four-step synthesis where the last step was promoted by microwaves. Interestingly, the targeted compounds were obtained with good-to-excellent yields within short reaction times (1–3 min) in an eco-friendly manner (Scheme 4.12). The *in vitro* screening of the synthesized compounds and gentamicin as a reference compound was carried out on *Pseudomonas aeruginosa*, *Staphylococcus aureus*, *E. coli*, *Proteus vulgaris*, *Bacillus lichen* form is and Micrococcus variant, using the agar diffusion method. Compounds with the pyridine ring have emerged as the most potent antimicrobial hydrazides and hydrazones with the lowest MIC value of $0.39 \pm 0.02 - 1.56 \pm 0.02$ mg/mL. These significant antibacterial activities may be explained based on the site of action of hydrazones and hydrazides, where they interact with bases of the DNA of the organisms, and thus, inserts between the stacked bases of helix.[55]

4.3.5 Cumarines

Coumarin and its derivatives are some of the most abundant class of compounds found widely in plants such as *Euphorbiaceae, Rutaceae, Orchidaceae and Asteraceae* and they are formed in the metabolic pathway of the shikimic acid.[56] Also, coumarins are products from the secondary metabolism of some microorganisms, fungi and animals. These compounds have shown exceptional biological activity, usually associated with low toxicity. Some of them are reported as promising for use in anti-inflammatories, antioxidants, anticancer, antivirals, or antimutagenic, among others.[57] Recent studies revealed that several biopharmacological activities of this nucleus are manifested in interactions with receptors such as xanthine oxidase, monoamine oxidase, cholinesterase and aromatase. Furthermore, lactone ring (present in the structure of coumarin) can make strong polar contacts allowing to acylate target proteins, which is very important for the covalent mechanism of inhibition of some receptors.

Hosamani and coworkers[58] decided to make a contribution, developing an efficient and effective new chemotherapy agent by incorporating a hybrid multifunctional molecule which may help in lowering the toxicity, as well as better selectivity toward the cancerous cells based on coumarin and maltol.

The hybrid was prepared by the initial obtention of 4-bromomethylcoumarins using the Pechman cyclization of phenol with 4-bromoethylacetophenone, after the target products were obtained by the reaction between maltol and anhydrous K_2CO_3 with 4-bromomethylcoumarins obtained in the former step with DMF as solvent under both conventional and microwave irradiation (Scheme 4.13). According to those results, it was clear that microwaves proved to be extremely fast providing good-to-excellent yields (84–94%) compared to the conventional method (57–72%). The synthesized compounds were evaluated for their *in-vitro* anticancer activity against two human cancer cell lines viz., A-549 (human lung carcinoma) and HeLa (human cervical cancer). Among the tested compounds, those with methyl groups as substituents and methoxy

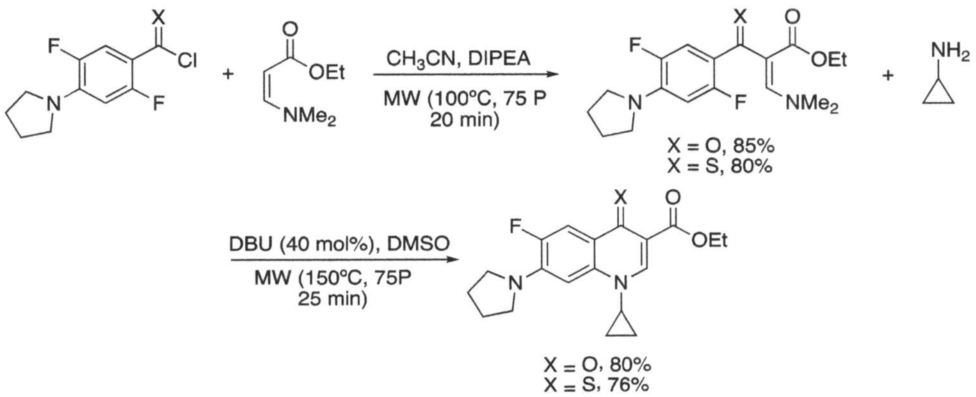

CH_3CN, DIPEA, MW (100°C, 75 P 20 min)

X = O, 85%
X = S, 80%

DBU (40 mol%), DMSO, MW (150°C, 75P 25 min)

X = O, 80%
X = S, 76%

SCHEME 4.11 Synthesis of some drug-like quinolone derivatives.

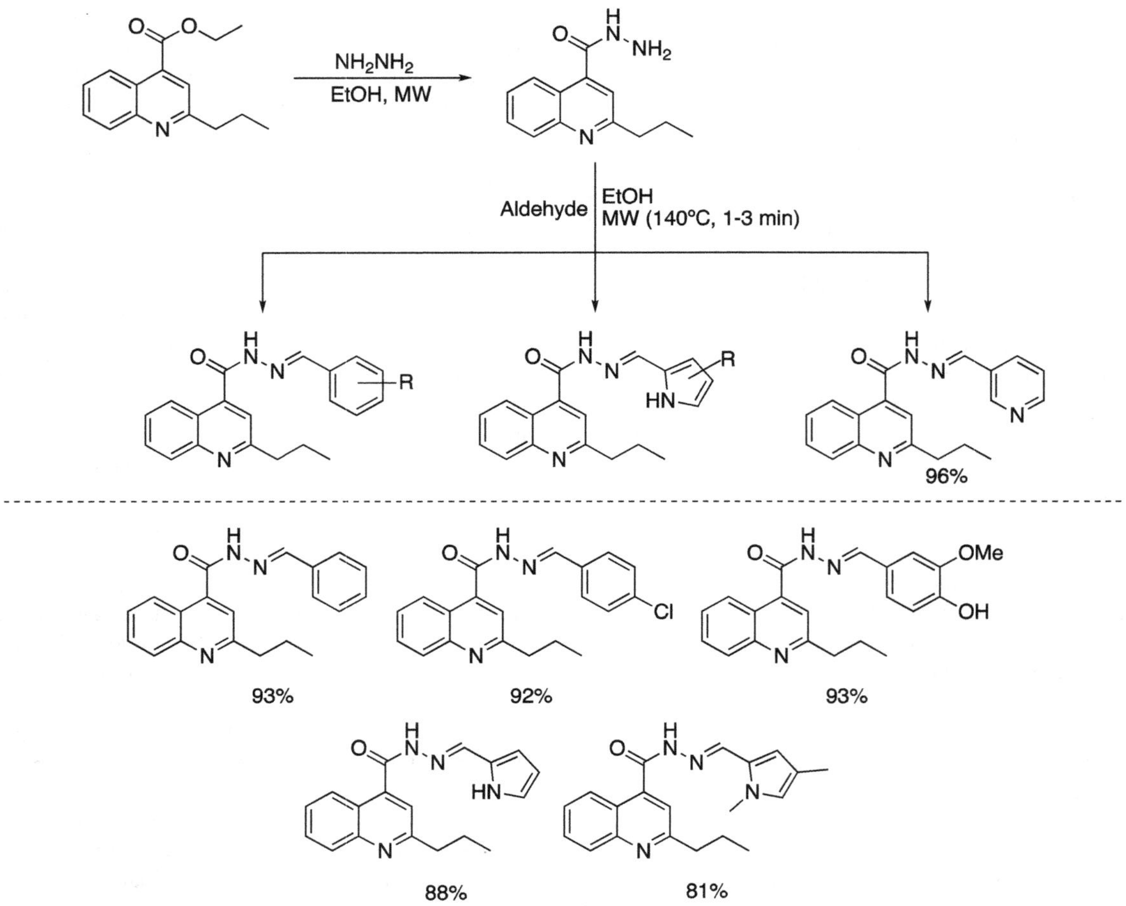

NH2NH2
EtOH, MW
Aldehyde
EtOH
MW (140°C, 1-3 min)
96%
93%
92%
93%
88%
81%

SCHEME 4.12 Microwave-assisted synthesis of 2-propylquinoline-4-carbohydrazide hydrazone derivatives.

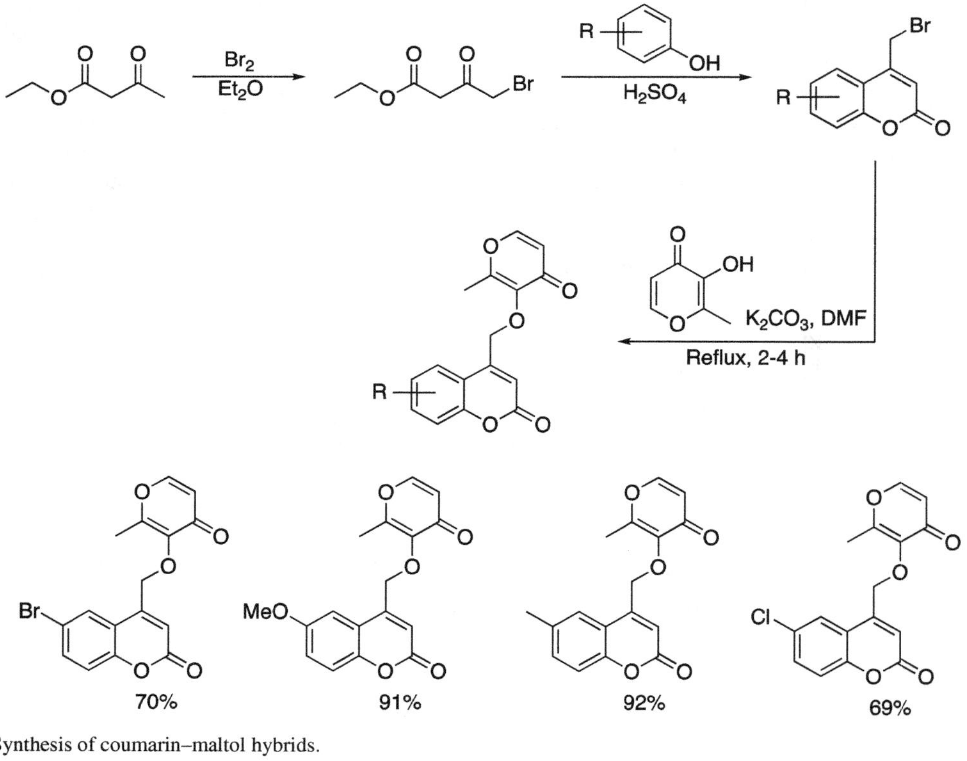

Br_2
Et_2O
H_2SO_4
K_2CO_3, DMF
Reflux, 2-4 h
70%
91%
92%
69%

SCHEME 4.13 Synthesis of coumarin–maltol hybrids.

groups in position 6were found to have potent cytotoxicity with IC_{50} values in the range of 2.47–4.26 μM on A-549 and HeLa cancer cells.

4.3.6 Pyrimidines and Derivatives

Six-member rings containing at least one nitrogen are very important in heterocyclic chemistry; for this reason, chemists have been interested in finding different synthetic strategies to optimize this kind of nucleus. Normally the literature reported the use of conventional heating in the procedures but sometimes the results are somehow poor. In this context, microwaves have been proposed as useful promoters of different organic reactions. In the group of Mohan, they decided to employ this unconventional heating source in the design and synthesis of a series of 1,2,3,4-tetrahydropyrimidene-5-carbonitrile derivatives based on bioisosteric similarities with isoniazid.

This isoniazid (INH) belongs to the group of the first line of antitubercular drugs. INH is believed to kill mycobacteria by inhibiting the biosynthesis of mycolic acids, a critical component of the cell wall.[59] In agreement with the above, they thought to replace the pyridine ring of INH with its pyrimidine bioisostere. For this aim, they propose a multicomponent reaction which involves a one-pot procedure involving two important organic reactions (Knoevenagel condensation and Michael addition) using an arylaldehyde, cyanoacetate and urea/thiourea with potassium carbonate and finally ethanol as the solvent (Scheme 4.14). The synthesis was also carried out under conventional heating, but the results were less efficient. Additionally, a docking study demonstrated that most of the target compounds occupied energetically more favorable positions in the active site cavity than the isoniazid. However, none of the compounds was found to be better in their *in vitro* antimycobacterial activity compared to isoniazid.[60]

Jain and coworkers reported a simple protocol for the efficient preparation of aryl- and heteroaryl-substituted dihydropyrimidinones under the same three-component protocol but changing urea/thiourea for guanidine nitrate. Their synthesis was also promoted by microwave irradiation and was carried out under solvent-free conditions which also provided improved selectivity, enhanced reaction rates, cleaner products and manipulative simplicity compared to other one-pot procedures previously reported that required strong protic or Lewis acids, prolonged reaction times and high temperatures (Scheme 4.15).[61]

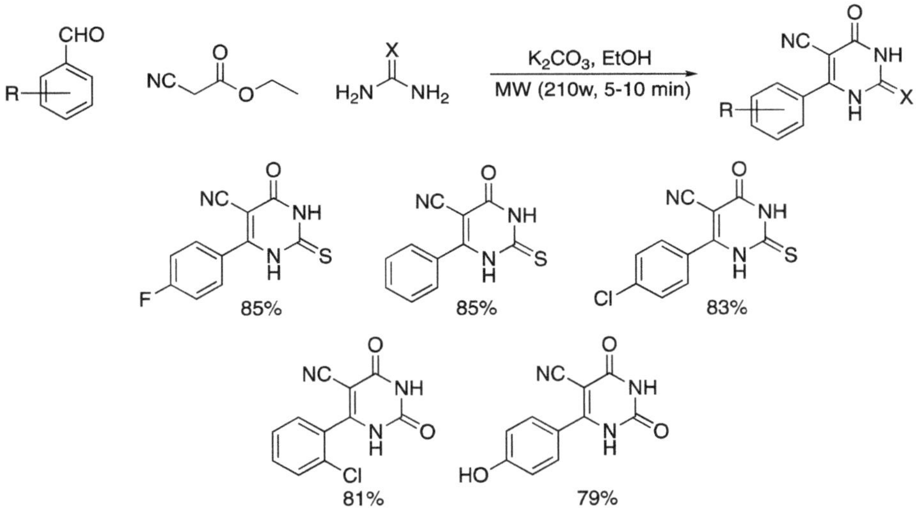

SCHEME 4.14 Synthesis of series of 1,2,3,4-tetrahydropyrimidine-5-carbonitril derivatives similar to isoniazid.

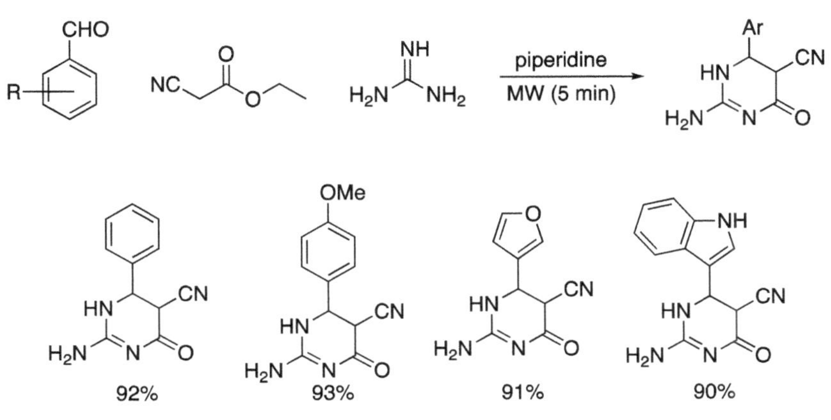

SCHEME 4.15 Microwave-assisted synthesis of dihydropyrimidinone derivatives.

4.3.7 Benzotriazoles

Azoles are one of the crucial structural nitrogen-containing heterocycles. In particular, the five-membered ring triazoles including 1,2,3-triazole, 1,2,4-triazole and benzofused triazoles, as well as their derivatives usually have remarkable pharmacological properties such as anticancer, antifungal, antibacterial, antiviral and antitubercular activities. A good example for the last one is the study about new 2-oxo-4-substituted aryl-azetidine derivatives of benzotriazole reported by Dubey and coworkers. Their synthetic strategy was aimed at synthesizing highly biologically active heterocycles containing benzotriazole and azetidinone moieties. They obtained a mixture of the 1- and 2-substituted isomers of benzotriazole derivatives in all steps (Scheme 4.16). However, the 1-substituted compound was obtained in higher yields. All the reactions were carried out under both conventional and microwave heating. Comparing the results some reactions required 5.5 to 8 h with conventional heating and were completed within 2–5 min under microwave irradiation, yields were improved from 60–72% to 84–92%.

All the synthesized compounds were screened against *Mycobacterium tuberculosis* and some microorganisms for their antimicrobial activities. The results showed that generally, compounds possessing electron-withdrawing groups have good antibacterial and antitubercular activity. They also conducted madecytotoxicity analysis and found that none of the compounds was toxic.[62]

4.3.8 Triazinas

Among the heterocyclic compounds that have been significantly benefited by the use of microwave irradiation, we found the triazine system. They have been promising scaffolds which are mainly used as novel anticonvulsant drugs such as Lamotrigine.[63] The above impelled Irannejad and coworkers to propose the synthesis of new *S*-substituted 1,2,4-triazine-3-thols by microwave-assisted green synthesis.

The synthetic pathway implied the obtention of 5,6-bis-aryl-1,2,4-triazine-3-thiols by the condensation of 1,2-diarylketones with thiosemicarbazide and a catalytic amount of hydrochloric acid under microwave irradiation at 120°C for 10 min using different ratios of water and ethanol; however, the crucial step for the reaction optimization was choosing the right amount of water and ethanol and found (2:1, 1:1 and 1:2) as the best mixtures, as the yield depends on the lipophilicity (LogP) or solubility of the starting material and final compound in the selected solvent (Scheme 4.17a).

Furthermore, based on previous studies, they proposed a mechanism to explain the obtention of the desired compounds (Scheme 4.17b). The first step is the feasible and easy formation of thiosemicarbazone intermediate in the presence of acid. The next step demands much more energy provided by the high energy of microwave heating; this step implied that the weak nucleophilic nitrogen of thioamide attacks the protonated carbonyl group to form the triazine system. Moreover, compared to the conventional synthetic methods, this approach has

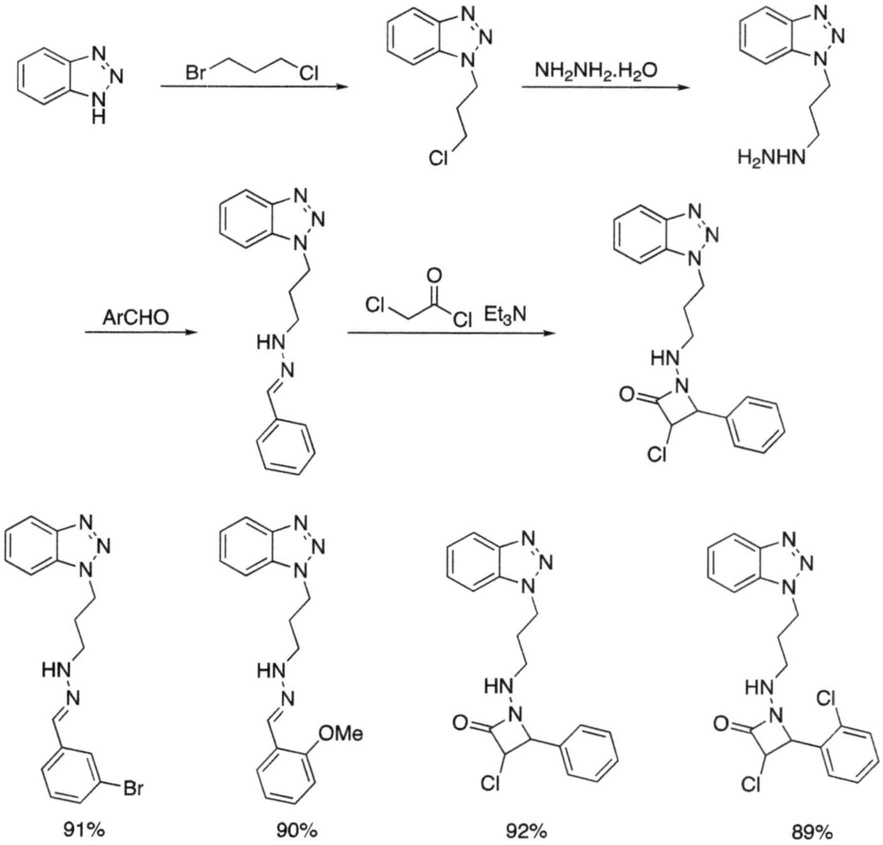

Br Cl

$NH_2NH_2.H_2O$

ArCHO

Cl Cl Et_3N

91% 90% 92% 89%

SCHEME 4.16 Synthesis of 2-oxo-4-substituted aryl-azetidine derivatives of benzotriazole.

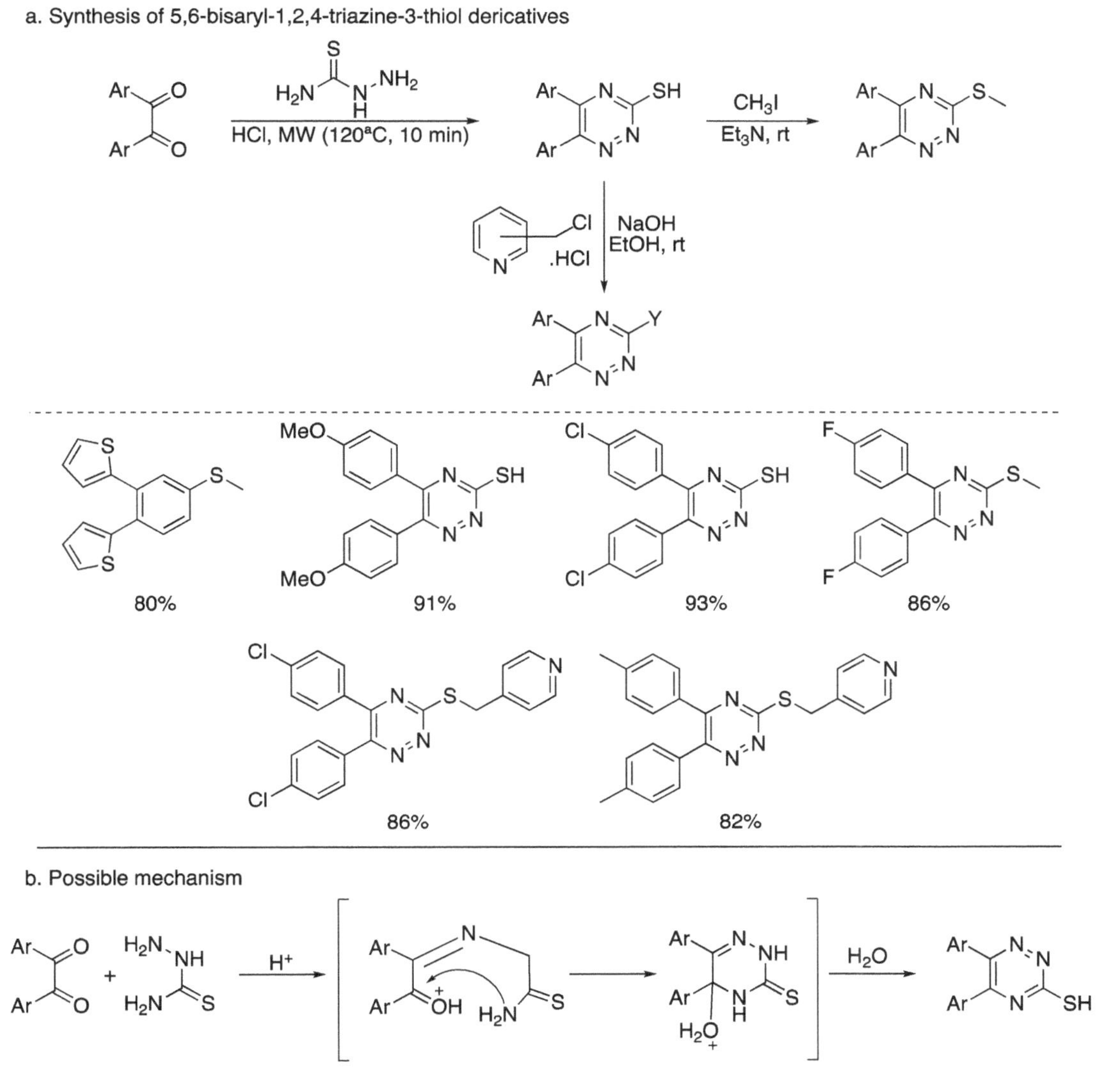

a. Synthesis of 5,6-bisaryl-1,2,4-triazine-3-thiol dericatives

HCl, MW (120°C, 10 min)

CH_3I Et_3N, rt

NaOH EtOH, rt

80% 91% 93% 86%

86% 82%

b. Possible mechanism

SCHEME 4.17 Green procedure for the synthesis of 5,6-bisaryl-1,2,4-triazine-3-thiol derivatives.

some advantages: it is faster as well as nontoxic and they used environmentally benign solvents. All the compounds were evaluated in order to determine the anticonvulsant activity. The study demonstrated that compound with 4-pyridylmethylthio moiety on the triazine presents the highest protection in both electroshock- and PTZ-induced seizures tests.[64]

4.4 Conclusions

I have described herein the theoretical aspects that allow understanding how microwaves act, and showed a select group of examples for drug synthesis, drug derivatives and biologically active compounds promoted by microwave irradiation. At this point, the readers of this chapter should recognize microwave irradiation as a powerful alternative to conventional heating. Microwaves are also one of the most powerful and promising techniques for the synthesis of bioactive compounds and they allow to even improve the currently known synthesis for some medicaments. All of these are due to microwave irradiation not only allowing to reduce reaction times but also because the desired product obtained is cleaner and with better selectivity, which leads to a more sustainable synthesis and is therefore included within the principles of green chemistry. Nevertheless, the results and current methods are still far away from massive industrial applications because microwave irradiations have some limitations with big volumes; however, scientists are working on some strategies and are designing different reactors to allow using this technique for industrial applications.

REFERENCES

1. Gedye, R.; Smith, F.; Westaway, K.; Ali, H.; Baldisera, L.; Laberge, L.; Rousell, J., *Tetrahedron Lett.* 1986, 27, 279–282.
2. (a) Grewal, A.; Kumar, K.; Redhu, S.; Bhardwaj, S., *Int. Res. J. Pharm. Appl. Sci.* 2013, 3, 278–285; (b) Nain, S.; Singh, R.; Ravichandran, S., *Adv. J. Chem. A* 2019, 2, 94–104.

3. Polshettiwar, V.; Varma, R. S., *Chem. Soc. Rev.* 2008, 37, 1546–1557.
4. Priecel, P.; Lopez-Sanchez, J. A., *ACS Sustain. Chem. Eng.* 2019, 7, 3–21.
5. (a) General Introduction to Microwave Chemistry. In *Microwaves in Catalysis*, 2015; pp. 1–28; (b) Horikoshi, S.; Serpone, N., *General Introduction to Microwave Chemistry*. 2015; pp. 1–28.
6. (a) Gude, V. G.; Patil, P.; Martinez-Guerra, E.; Deng, S.; Nirmalakhandan, N., *Sustain. Chem. Process* 2013, 1, 1–31; (b) Microwave Theory. In *Microwaves in Organic and Medicinal Chemistry*, 2012; pp. 9–39.
7. (a) Kappe, C. O., 2004, 43, 6250–6284; (b) Das, R.; Mehta, D.; Bhardawaj, H., *Int. J. Res. Dev. Pharm. Life Sci.* 2012, 1, 32–39.
8. Montes, I.; Sanabria, D.; García, M.; Castro, J.; Fajardo, J., *J. Chem. Educ.* 2006, 83, 628–631.
9. Lidström, P.; Tierney, J.; Wathey, B.; Westman, J., *Tetrahedron* 2001, 57, 9225–9283.
10. Jain, H. K.; Mourya, V. K.; Agrawal, R. K., *Bioorg. Med. Chem. Lett.* 2006, 16, 5280–5284.
11. Moreau, A.; Praveen Rao, P. N.; Knaus, E. E., *Bioorg. Med. Chem.* 2006, 14, 5340–5350.
12. Huang, Z.; Velázquez, C. A.; Abdellatif, K. R. A.; Chowdhury, M. A.; Reisz, J. A.; DuMond, J. F.; King, S. B.; Knaus, E. E., *J. Med. Chem.* 2011, 54, 1356–1364.
13. Tiwari, A. D.; Panda, S. S.; Girgis, A. S.; Sahu, S.; George, R. F.; Srour, A. M.; Starza, B. L.; Asiri, A. M.; Hall, C. D.; Katritzky, A. R., *Org. Biomol. Chem.* 2014, 12, 7238–7249.
14. Hall, S. A.; Lalee, Z.; Bell, R. P.; Towe, S. L.; Meade, C. S., *Prog. Neuropsychopharmacol. Biol. Psychiatry*. 2021, 104, 110040–110051.
15. Sriram, D.; Yogeeswari, P.; Gopal, G., *Eur. J. Med. Chem.* 2005, 40, 1373–1376.
16. Chu, C. K.; Bhadti, V. S.; Doshi, K. J.; Etse, J. T.; Gallo, J. M.; Boudinot, F. D.; Schinazi, R. F., *J. Med. Chem.* 1990, 33, 2188–2192.
17. McGuigan, C.; Harris, S. A.; Daluge, S. M.; Gudmundsson, K. S.; McLean, E. W.; Burnette, T. C.; Marr, H.; Hazen, R.; Condreay, L. D.; Johnson, L.; De Clercq, E.; Balzarini, J., *J. Med. Chem.* 2005, 48, 3504–3515.
18. Sriram, D.; Yogeeswari, P.; Myneedu, N. S.; Saraswat, V., *Bioorg. Med. Chem. Lett.* 2006, 16, 2127–2129.
19. Oller-Salvia, B.; Sánchez-Navarro, M.; Giralt, E.; Teixidó, M., *Chem. Soc. Rev.* 2016, 45, 4690–4707.
20. Pardridge, W. M., *Nat. Rev. Drug* 2002, 1, 131–139.
21. Marder, K., *Curr. Neurol. Neurosci.* 2010, 10, 336–337.
22. Halen, P. K.; Murumkar, P. R.; Giridhar, R.; Yadav, M. R., *Mini. Rev. Med. Chem.* 2009, 9, 124–139.
23. Eden, B. D.; Rice, A. J.; Lovett, T. D.; Toner, O. M.; Geissler, E. P.; Bowman, W. E.; Young, S. C., *Bioorg. Med. Chem. Lett.* 2019, 29, 1487–1491.
24. Huttunen, K. M.; Mannila, A.; Laine, K.; Kemppainen, E.; Leppänen, J.; Vepsäläinen, J.; Järvinen, T.; Rautio, J., *J. Med. Chem.* 2009, 52, 4142–4148.
25. Wathey, B.; Tierney, J.; Lidström, P.; Westman, J., *Drug Discov. Today* 2002, 7, 373–380.
26. Huttunen, K. M.; Leppänen, J.; Vepsäläinen, J.; Sirviö, J.; Laine, K.; Rautio, J., *J. Pharm. Sci.* 2012, 101, 2854–2860.
27. (a) Nti-Addae, K. W.; Laurence, J. S.; Skinner, A. L.; Stella, V. J., *J. Pharm. Sci.* 2011, 100, 3023–3027; (b) Nti-Addae, K. W.; Stella, V. J., *J. Pharm. Sci.* 2011, 100, 1001–1008.
28. Huttunen, K. M.; Leppänen, J.; Laine, K.; Vepsäläinen, J.; Rautio, J., *Eur. J. Pharm. Sci.* 2013, 49, 624–628.
29. Grey, A.; Reid, I. R., *Ther. Clin. Risk Manag.* 2006, 2, 77–86.
30. Mustafa, D. A.; Kashemirov, B. A.; McKenna, C. E., *Tetrahedron Lett.* 2011, 52, 2285–2287.
31. (a) Kemnitzer, W.; Kasibhatla, S.; Jiang, S.; Zhang, H.; Zhao, J.; Jia, S.; Xu, L.; Crogan-Grundy, C.; Denis, R.; Barriault, N.; Vaillancourt, L.; Charron, S.; Dodd, J.; Attardo, G.; Labrecque, D.; Lamothe, S.; Gourdeau, H.; Tseng, B.; Drewe, J.; Cai, S. X., *Bioorg. Med. Chem. Lett.* 2005, 15, 4745–4751; (b) Patil, S. A.; Wang, J.; Li, X. S.; Chen, J.; Jones, T. S.; Hosni-Ahmed, A.; Patil, R.; Seibel, W. L.; Li, W.; Miller, D. D., *Bioorg. Med. Chem. Lett.* 2012, 22, 4458–4461; (c) Das, S. G.; Srinivasan, B.; Hermanson, D. L.; Bleeker, N. P.; Doshi, J. M.; Tang, R.; Beck, W. T.; Xing, C., *J. Med. Chem.* 2011, 54, 5937–5948.
32. Patil, S. A.; Patil, S. A.; Patil, R., *Future Med. Chem.* 2015, 7, 893–909.
33. Pfeffer, S. R.; Seibel, W. L.; Pfeffer, L. M.; Miller, D. D.; 2014, 11, 400–406.
34. Shah, N. K.; Shah, N. M.; Patel, M. P.; Patel, R. G., *J. Chem. Sci.* 2013, 125, 525–530.
35. Sangani, B. C.; Shah, M. N.; Patel, P.; Patel, G. R., 2012, 77, 1165–1174.
36. El Sayed, M. T.; Ahmed, K. M.; Mahmoud, K.; Hilgeroth, A., *Eur. J. Med. Chem.* 2015, 90, 845–859.
37. Patil, S. A.; Patil, R.; Miller, D. D., *Curr. Med. Chem.* 2011, 18, 615–637.
38. Gribble, G. W., *Contemp. Org. Synth.* 1994, 1, 145–172.
39. Barbieri, V.; Ferlin, M. G., *Tetrahedron Lett.* 2006, 47, 8289–8292.
40. Głuszyńska, A., *Eur. J. Med. Chem.* 2015, 94, 405–426.
41. Lipińska, T., *Tetrahedron Lett.* 2004, 45, 8831–8834.
42. Zablotskaya, A.; Segal, I.; Geronikaki, A.; Shestakova, I.; Nikolajeva, V.; Makarenkova, G., *Pharmacol. Rep.* 2017, 69, 575–581.
43. Zhong, F.; Geng, G.; Chen, B.; Pan, T.; Li, Q.; Zhang, H.; Bai, C., *Org. Biomol. Chem.* 2015, 13, 1792–1799.
44. Sun, N.; Du, R.-L.; Zheng, Y.-Y.; Huang, B.-H.; Guo, Q.; Zhang, R.-F.; Wong, K.-Y.; Lu, Y.-J., *Eur. J. Med. Chem.* 2017, 135, 1–11.
45. Murugavel, S.; Jacob Prasanna Stephen, C. S.; Subashini, R.; Anantha Krishnan, D., *J. Photochem. Photobiol. B* 2017, 173, 216–230.
46. Pinz, M. P.; Reis, A. S.; de Oliveira, R. L.; Voss, G. T.; Vogt, A. G.; Sacramento, M. D.; Roehrs, J. A.; Alves, D.; Luchese, C.; Wilhelm, E. A., *Regul. Toxicol. Pharmacol.* 2017, 90, 72–77.
47. Vijayaraghavan, S.; Mahajan, S., *Bioorg. Med. Chem. Lett.* 2017, 27, 1693–1697.
48. Ben Yaakov, D.; Shadkchan, Y.; Albert, N.; Kontoyiannis, D. P.; Osherov, N., *J. Antimicrob. Chemother.* 2017, 72, 2263–2272.
49. Nathubhai, A.; Haikarainen, T.; Koivunen, J.; Murthy, S.; Koumanov, F.; Lloyd, M. D.; Holman, G. D.; Pihlajaniemi, T.; Tosh, D.; Lehtiö, L.; Threadgill, M. D., *J. Med. Chem.* 2017, 60, 814–820.

50. Kumar, H.; Devaraji, V.; Joshi, R.; Jadhao, M.; Ahirkar, P.; Prasath, R.; Bhavana, P.; Ghosh, S. K., *RSC Adv.* 2015, 5, 65496–65513.
51. Bodke, Y. D.; Shankerrao, S.; Kenchappa, R.; Telkar, S., *Russ. J. Gen. Chem.* 2017, 87, 1843–1849.
52. Jassem, A. M.; Dhumad, A. M.; Almashal, F. A.; Alshawi, J. M., *Med. Chem. Res.* 2020, 29, 1067–1076.
53. Dent, S.; Oyan, B.; Honig, A.; Mano, M.; Howell, S., *Cancer Treat. Rev.* 2013, 39, 622–631.
54. Ajani, O. O.; Iyaye, K. T.; Aderohunmu, D. V.; Olanrewaju, I. O.; Germann, M. W.; Olorunshola, S. J.; Bello, B. L., *Arab. J. Cem.* 2020, 13, 1809–1820.
55. Ajani, O. O.; Obafemi, C. A.; Nwinyi, O. C.; Akinpelu, D. A., *Bioorg. Med. Chem.* 2010, 18, 214–221.
56. Nazari, Z. E.; Iranshahi, M., 2011, 25, 315–323.
57. (a) Mamidala, S.; Peddi, S. R.; Aravilli, R. K.; Jilloju, P. C.; Manga, V.; Vedula, R. R., *J. Mol. Struct.* 2021, 1225, 129114–129128; (b) Milanović, Ž. B.; Dimić, D. S.; Avdović, E. H.; Milenković, D. A.; Marković, J. D.; Klisurić, O. R.; Trifunović, S. R.; Marković, Z. S., *J. Mol. Struct.* 2021, 1225, 129256–129268.
58. Koparde, S.; Hosamani, K. M.; Barretto, D. A.; Joshi, S. D., *Chem. Data Collect.* 2018, 15–16, 41–53.
59. Ahmad, S.; Mokaddas, E., *Respir. Med. CME* 2010, 3, 51–61.
60. Mohan, S. B.; Ravi Kumar, B. V. V.; Dinda, S. C.; Naik, D.; Prabu Seenivasan, S.; Kumar, V.; Rana, D. N.; Brahmkshatriya, P. S., *Bioorg. Med. Chem. Lett.* 2012, 22, 7539–7542.
61. Bhatewara, A.; Jetti, S. R.; Kadre, T.; Paliwal, P.; Jain, S., *Int. J. Med. Chem.* 2013, 2013, 197612–197618.
62. Dubey, A.; Srivastava, S. K.; Srivastava, S. D., *Bioorg. Med. Chem. Lett.* 2011, 21, 569–573.
63. Kaushik, D.; Khan, S. A.; Chawla, G., *Eur. J. Med. Chem.* 2010, 45, 3960–3969.
64. Irannejad, H.; Naderi, N.; Emami, S.; Ghadikolaei, R. Q.; Foroumadi, A.; Zafari, T.; Mazar-Atabaki, A.; Dadashpour, S., *Med. Chem. Res.* 2014, 23, 2503–2514.

5

Photochemical Reactions as a Useful and Easy to Implement and Scale-Up, New Method for the Synthesis of Chemicals

Angelo Albini

CONTENTS

5.1 Introduction 75
5.2 Excited States Chemistry 75
5.3 How to Carry Out Satisfactory Photochemical Reactions 78
5.4 Photochemical Reactions with Oxygen 80
5.5 Photoinitiated Processes 80
5.6 Energy Transfersensitization 80
5.7 Photoinitiated Processes: Generation of Radicals 81
5.8 Choice of Apparatus 81
5.9 Cost Issue 83
5.10 Conclusion and Outlook 87
Notes 88
References 88

5.1 Introduction

Photochemical reactions offer often a convenient, but not rarely virtually the only alternative for many important synthetic transformations; yet, the insertion of photochemical steps in synthetic procedures remain limited (about 15% of reported procedures, and only a tiny fraction of the patented ones), and of the further applications of photochemistry, artificial photocatalysis as an alternative way for obtaining energy from natural sources as well as for recovering polluted waters likewise remain minoritarian (<10%). Certainly, such percentages demonstrate a significant improvement with respect to the minimal contribution given in the 19th century, but remain far from achieving the potential of such reactions, in particular, because of the unusual intrinsic 'green' character of such procedures. The purpose of the ensuing presentation is to attempt to clarify the cause of such a situation and possibly some convenient way to obtain synthetically useful reactions. When the first light-induced reactions were reported in the literature (end of the 19th century and beginning of the 20th century), scientists were puzzled particularly because these results seemed to flatly refute the basic tenets of statistic chemical kinetics, a triumph of the chemistry of that time. According to that theory, the fraction of the molecules that reached the energy of the 'activated complex would react. Enhancing the temperature would lead to a flattening of the Gaussian curve of the energy distribution and to a shift toward the right-hand side (see Figure 5.1). The number of such complexes increased and so did the rates of the reactions. Photochemical reaction, however, had virtually negligible temperature coefficient and the molecules generated in this way had large energy, not comparable with that absorbed with a photon (Schemes 5.1 and 5.2). All of such prejudices were lifted when it was understood the reactions reported involved a step through the electronically excited states when considering the entire transformation; the path was diabatic, it occurred in part in the excited, in part in the ground state, while in the ground state any path was adiabatic (see Figures 5.2 and 5.3).[1]

5.2 Excited States Chemistry

The examination of the UV–visible spectrum and experience suggest which may be the structure, and thence the reactivity of S_1 and T_1, as well as the effect of solvent or any further parameter. The way the chemistry is rationalized is the same as that of the ground state, with often participation of radical species, as one may expect due to the fact that energetic states are expected since large energies easily lead to the cleavage of bonds.

Schemes 5.3 and 5.4 illustrate some typical photochemical reactions from a variety of 'chromophores', that is a moiety to which the most important orbitals are located, and named by indicating first the starting, then the arriving MOs, e.g., p-p* double bonds, n-p* for non-bonding MOs. The photochemistry observed is rationalized in the same way as it is usually done with ground states (see Scheme 5.1), for the monomolecular geometrical isomerization of X = Y double bonds, rearrangement of benzene derivatives, electrocyclic processes, 2+2 intramolecular photocycloadditions, cleavage of s bonds, intramolecular reactions of aromatic nitroaromatic, and, respectively, the bimolecular photoreactions (H abstraction by carbonyls and nitroaromatics, carbonyl-alkene and

DOI: 10.1201/9781003089162-5

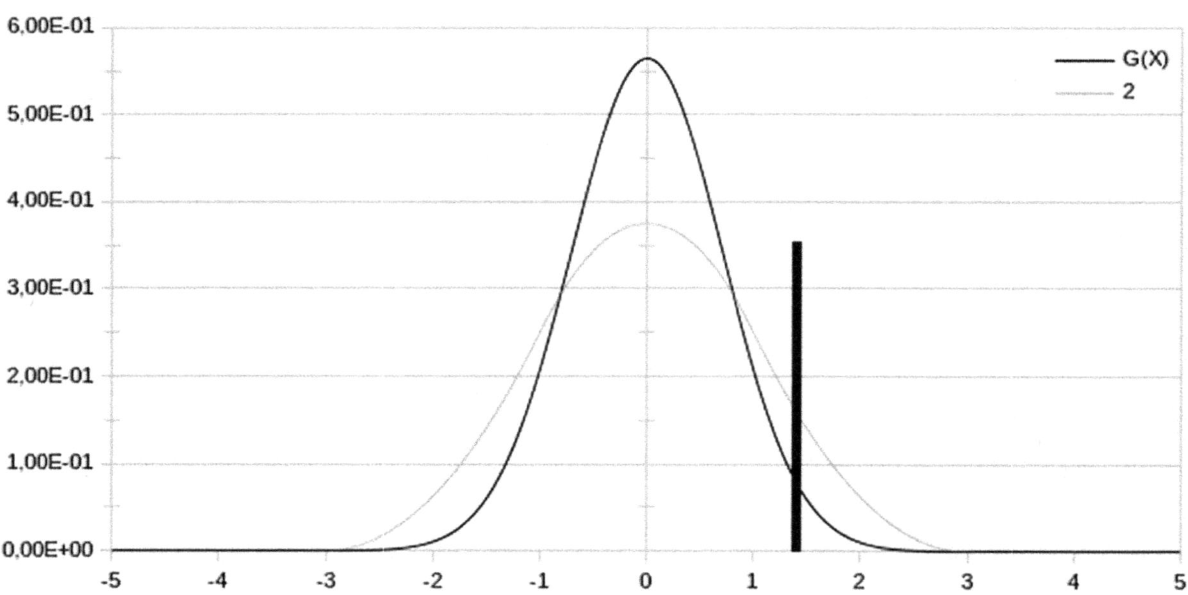

FIGURE 5.1 Energy is distributed among particles as indicated by the Gaussian curves. The fraction laying above the energy of the activated complex (see the vertical line in the figure) react. Upon enhancing the temperature, the curves flatten somewhat and the fractions of particles reacting increase.

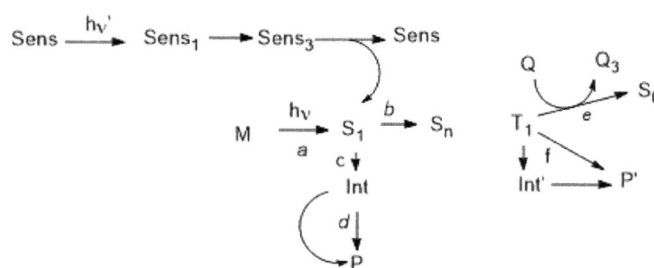

SCHEME 5.1 Reactions via the singlet and the triplet states (compare with Sec. 7).

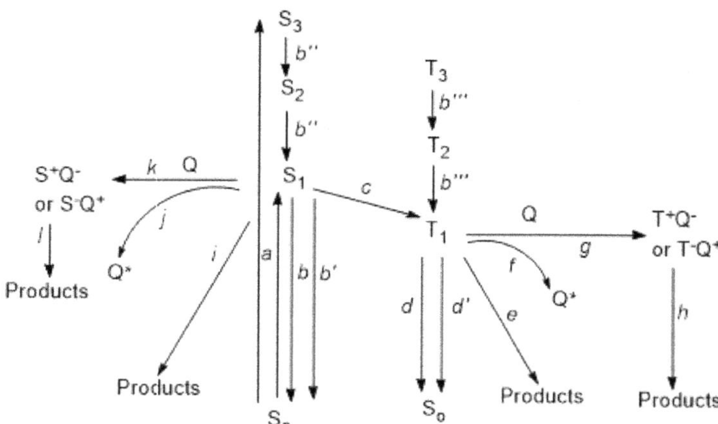

SCHEME 5.2 The vocabulary of photochemistry: absorption of a photon (a), and internal conversion from higher-lying states to S_1 (b) is followed the by-products formation (i) in competition with bimolecular quenching (k, l), or by accessing the triplet state via intersystem crossing (c) to a triplet and again IC from T_n to T_1 (b''')and products formation (e) or quenching (g, h); usually, bimolecular reactions were more significant in this case due to the longer lifetime of this state. Emission of light from either S_1 (fluorescence, path b) or the triplet state (phosphorescence, path d) further competes, as may do an internal conversion from S_1 (b') or intersystem crossing from T_1(d') to S_0.

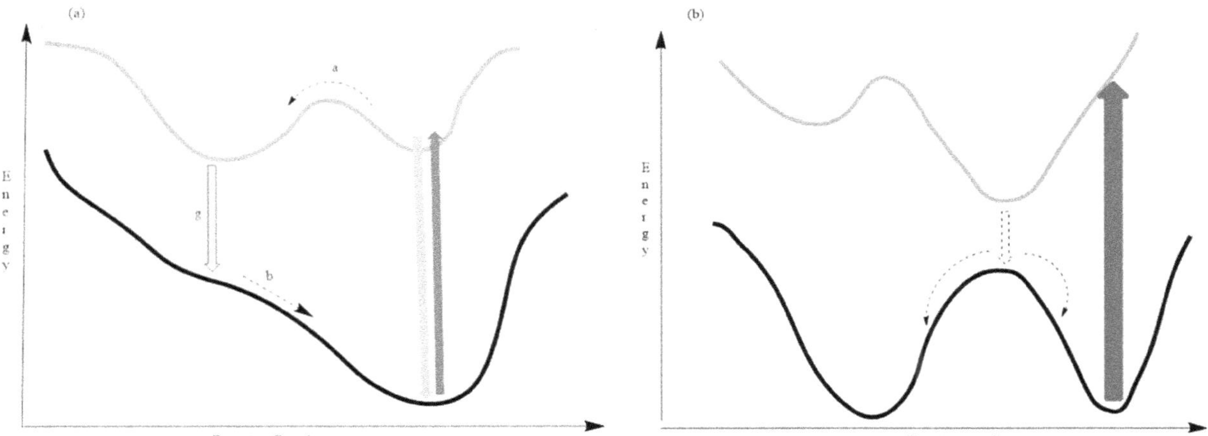

FIGURE 5.2 Schematic representation of the processes occurring on an excited state surface (PES) by absorption of a quantum of light, represented as a red vertical arrow, since the nuclei have no time to change their configuration during the fast (ca. 10^{-14} s^{-1}) absorption process. The configuration reached may be close to a minimum in the excited state surface, and has the time to reach the equilibrium with the environment (case a) or be close to a steep drop so that a pre-dissociation situation is reached (case b, any quantum above a limit causes the cleavage of the molecule). Figure 5.2b also introduces the case of 'no reasonably closely lying' maximum in the ground state. The molecule reaches a local minimum from which it funnels in approximately 1 to 1 yield toward both sides (avoided crossing). Reproduced from *Yuri V. Il'ichev, YV, Johnson & Johnson Photochemistry. Theoretical Concepts and Reaction Mechanisms*, visited on March, 13, 2021.

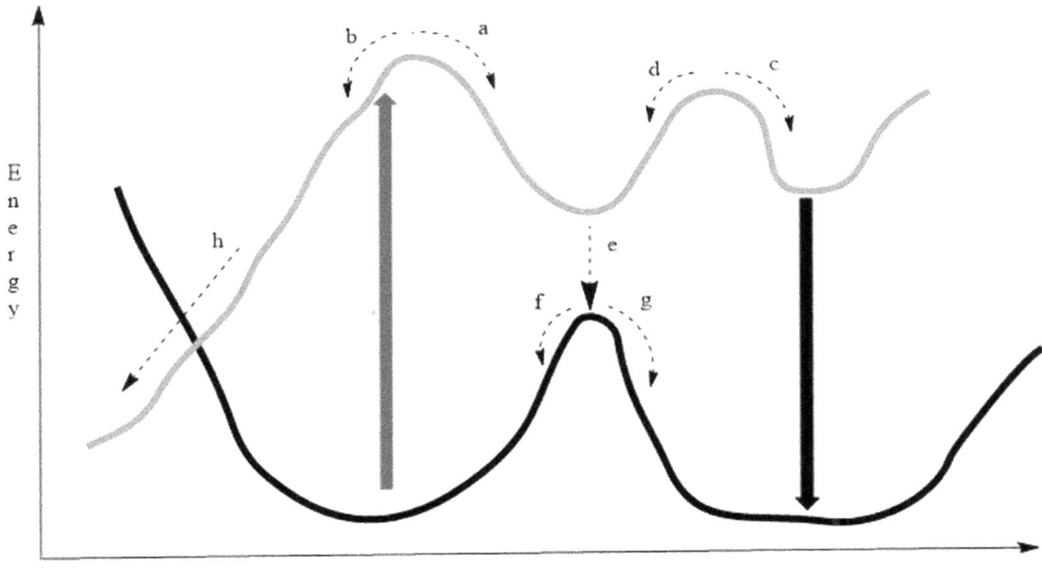

FIGURE 5.3 A photochemical reaction occurs in part on the excited-state potential energy surface PES before returning to the ground state PES. Different cases are presented, after the absorption of a quantum of light: (a) a local minimum is reached, but then two paths are likely, (b), leading to a ground state maximum and from them to two products, paths (c) and (d), or to avoided crossing, path f and then directly to a product. Reproduced from *Yuri V. Il'ichev, YV, Johnson & Johnson, Photochemistry. Theoretical Concepts and Reaction Mechanisms*, visited on March, 13, 2021.

enone-alkene 2+2 cycloaddition, 2+2, 3+2 and 4+2 cycloaddition of benzene derivatives and larger (hetero)aromatics, the addition of alkenes, alkynes, (hetero)aromatics to halogenated benzene derivatives via an intermediate, formally via a triplet phenyl cation).

As for the classing of photoreactions, it has been hinted above that 'chemistry of electronical excited states' would be a better name for the reactions we are discussing presently, and the fact that the difference between the energy of ground versus excited states and UV/visible radiation is in a sense, a cindicence. The safest choice is registering the UV/visible spectrum. Side reactions are no frequent occurrence because the short lifetime of electronically excited states makes it quite difficult to overcome even a small activation barrier; thus, in

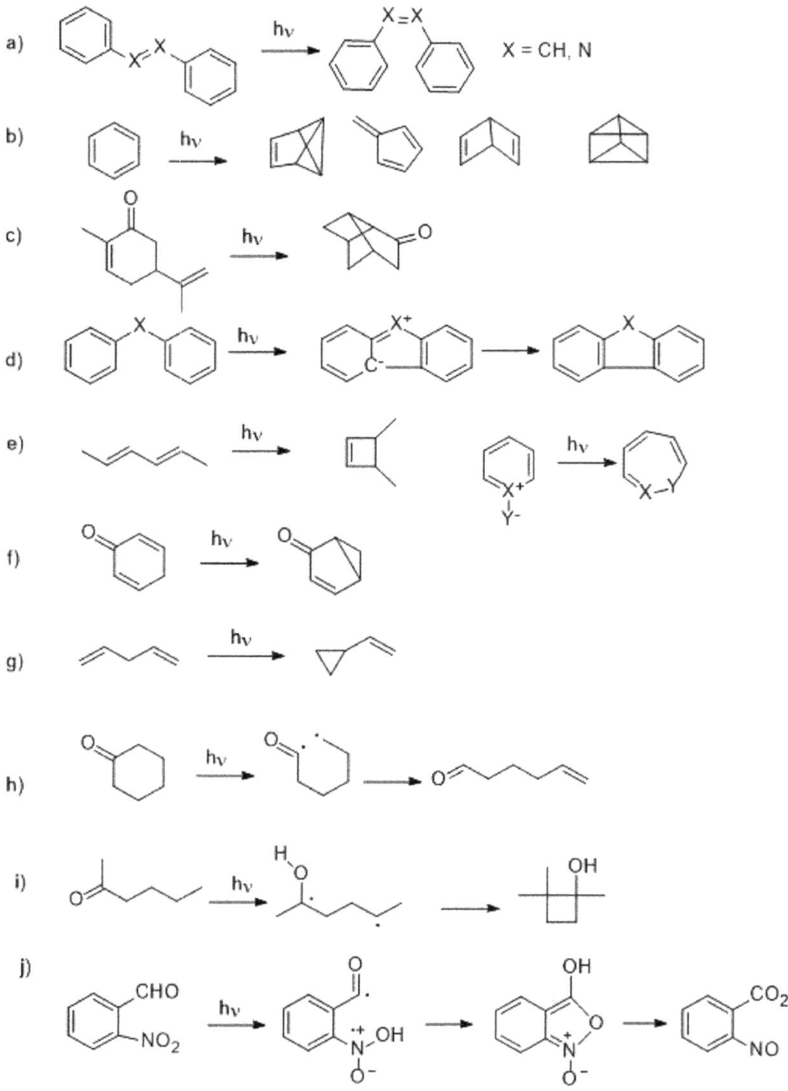

SCHEME 5.3 Examples of unimolecular photoreactions.

general, there is only a single accessible hill. Rather, selectivity limits may result from over-irradiation, when the primary products do absorb significantly in the same region as the starting material, but this is a general issue in chemical synthesis.

5.3 How to Carry Out Satisfactory Photochemical Reactions

There is something that is called 'dark' photochemistry (with no photons); this is no contradiction in terms, in the sense that aromatic compounds smoothly add oxygen likewise cleaves back and liberate singlet oxygen. Several molecules based on the generation and cleavage of cyclic peroxide esters have been accurately described and are used also for reactions used in forensic applications, as is the case for luminol, see Scheme 5.5.

These cleave under appropriate conditions by retro 2+2photocycloaddition to yield a pair of carbonyl functions that in these cases correspond to highly fluorescent (hetero) aromatic molecules that are formed in the excited state and emit. In other cases, a chemical reaction occurs.

By definition, light has to be adsorbed to cause a chemical effect (and this is alluded to, in what is called the first law of photochemistry). Although the interest by scholars for the light/matter interconversion never slackened, with a consistent interest for themes such as the bleaching of dyed fabric molecules when exposed to light. The first law of photochemistry, attributed to Grotthus and Draper,[1] has indeed formulated much earlier, in the frame of the continue interest by scientists for phenomena such as the bleaching of dyed fabric, of flowers or biological samples). Therefore, in order to carry out a satisfactory photochemical reaction, one should first check that light is adsorbed and further that it is the molecule chosen for the study, rather than by the solvent or the material by which is made the container. In general, there is no competing photoreaction since it is not frequent that there are several paths accessible during the short lifetime of the excited states. However, it would be reasonable to carry out the irradiation both in air

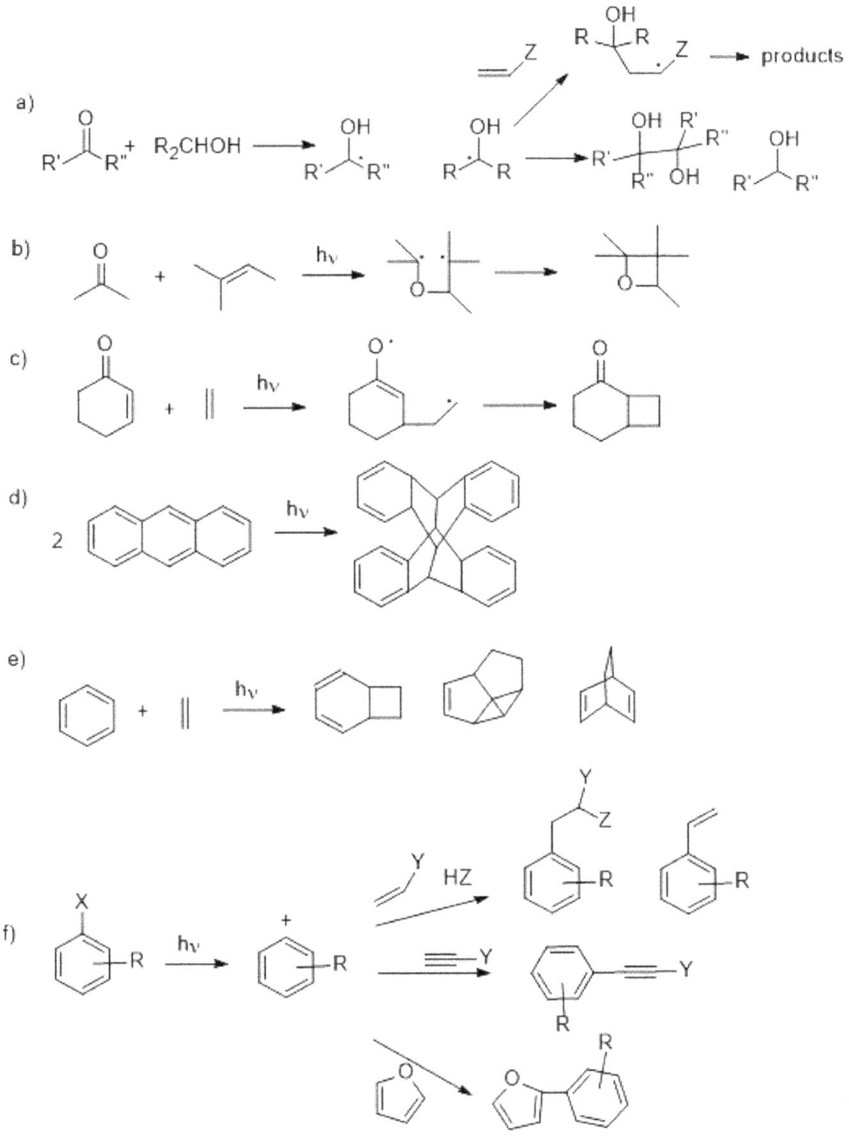

SCHEME 5.4 Examples of bimolecular photoreactions.

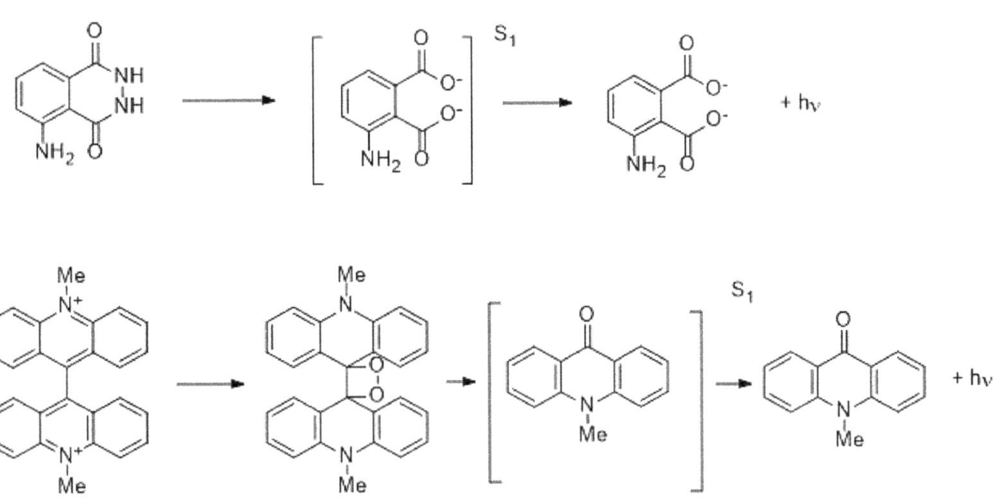

SCHEME 5.5 Generation of chemiluminescent peroxides.

equilibrated and in nitrogen, or better argon-equilibrated (= simply flush for 5 min the solution), for comparison and guidance during a future experience.

5.4 Photochemical Reactions with Oxygen

Oxygen has a leading role in photochemistry. The dioxygen molecule O_2 has, contrary to organic molecules, a triplet ground state, while the singlet states are of only slightly higher energy since they have the same electronic occupancy (S_1 vs T_0 energy gap is only 22.4 kcal mol^{-1}, see Scheme 5.6).

In-depth studies have demonstrated that ground-state O_2 has a weakly diradical character and thus, while generally unreactive, fully traps radicals. On the other hand, the low energetic gap between singlet and triplet O_2 makes accessible an energy transfer from any excited molecule, and thus produces a singlet molecule in any case; singlet O_2 is a strong electrophile and easily reacts with a large number of molecules (Scheme 5.7). When the distinction of the two processes is difficult or not necessary, the oxygen reactions may be simply indicated as an activated oxygen, reacting oxygen species (ROS).

5.5 Photoinitiated Processes

As indicated in Section 5.2, a photochemical reaction involves electronically excited states. However, this is not always the case, and the much larger concept of photoinitiated processes is required for a full presentation. First of all, not always the molecule absorbing light and that reacting are identical. As an example, in Section 5.2, it is mentioned that triplet states cannot be reached by the direct excitation of many molecules. However, under suitable conditions, the choice of another molecule (Sens) and if this has an efficient ISC to ^{3}Sens and higher energy than ^{3}M; then it may transfer its energy to M and generate ^{3}M indirectly (this again is a step of various efficiency, but nowadays can be credibly predicted).

5.6 Energy Transfersensitization

Not necessarily the molecule that absorbs the quantum of light reacts. In fact, it has been stated in Section 5.2 that ISC may be slow or even negligible: however, when a further molecule is present that absorbs a photon and then undergoes an effective ISC to its own triplet (^{3}Sens) and this happens to be higher in energy than the triplet considered ^{3}M), then triplet processes become apparent (see Scheme 5.8, energy transfer sensitization).

Another highly relevant process that is photoinitiated but does not proceed via the excited states, is electron transfer photosensitization (Scheme 5.9). We are not accustomed to mentioning electron transfer as a likely path with organic molecules unless strongly donating or accepting inorganic reagents are used, say sodium metal in the Birch naphthalene reduction, or when electrochemical activation is applied. Quite different is the reaction of excited states. Now, one has to transfer an electron, not to the LUMO of the acceptor, but to the hole created by excitation from the HOMO, or, in the other words, it is not the abstraction of an electron from the HOMO, but transfer from the single electron promoted into the LUMO by excitation (see Scheme 5.10).

An electron is transferred to the oxidizing cation M^{n+}, or the same step takes place in the cathodic compartment by electrolysis, or to an excited acceptor.

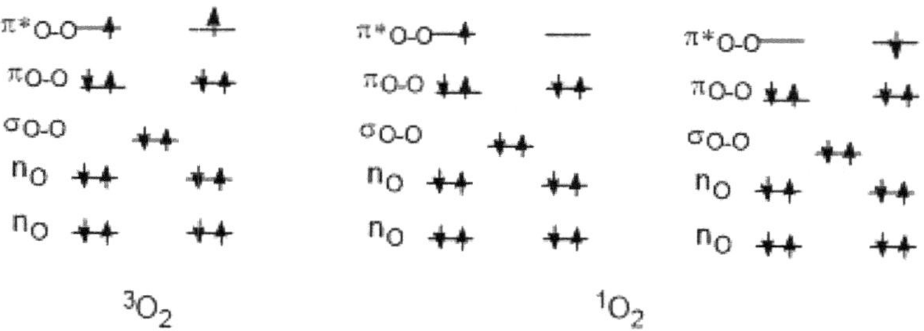

SCHEME 5.6 Electronic structure of the dioxygen, ground state triplet and excited singlet state (a doubly degenerate state).

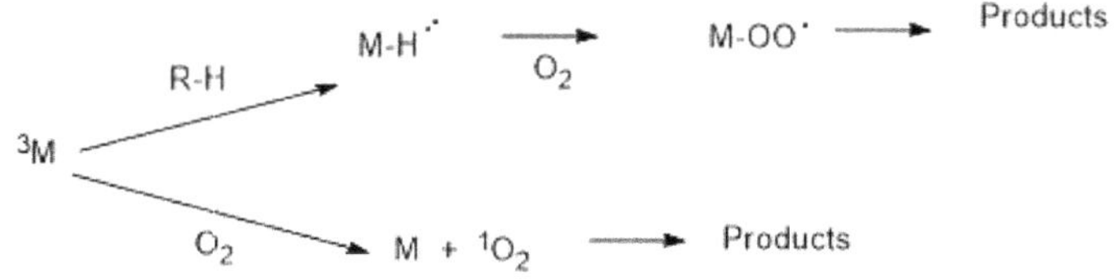

SCHEME 5.7 Paths leading to oxygen addition.

A-B ⟶ A· —RH⟶ AH + R·

SCHEME 5.8 A photoinitiated reaction. Photoinduced cleavage of A-B produces radicals A, than may enter in a chain reaction.

Sens → ^{1}Sens → ^{3}Sens →(M) Sens + ^{3}M → Products

SCHEME 5.9 A sensitizer undergoes an efficient intersystem crossing to its own triplet (^{3}Sens) and this happens to be higher in energy than ^{3}M, energy transfer to M with simultaneous spin flipping to give ^{3}M occurs and a new chemistry may take place in such a state.

D M* A

D+· M* A-·

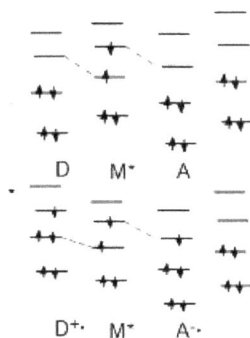

SCHEME 5.10 Electron transfer reactions involving excited states are rare in organic chemistry, due to the fact that molecules are rarely strong enough acceptors or donors. However, electronic excitation makes electron transfer between organic molecules a much faster process.

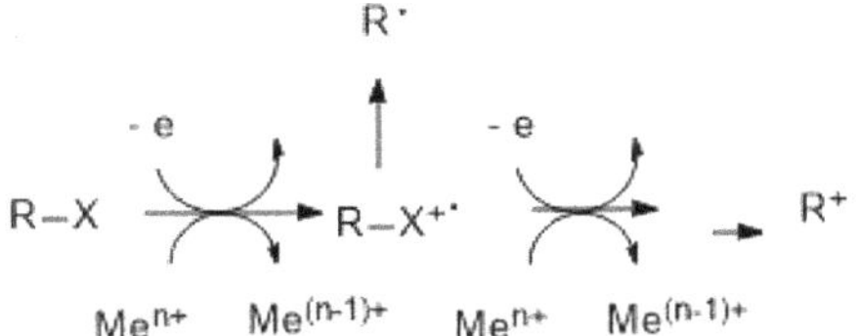

SCHEME 5.11 Sensitized electron transfer activation.

$$A + D \rightleftarrows A^{\delta-}\text{--}D^{\delta+}$$

SCHEME 5.12 A photoreaction may not also involve full electron transfer, but a charge transfer complexation via a complex existing only in the excited state (in short, 'exciplexes').

Scheme 5.11 shows the course of single-electron transfer reactions in the presence of a suitable oxidizer. When this contains a good electrofugal group, X+, the radical cation cleaves and yields a radical (R+.→leads to radical R). This is the actual chemical intermediate justifying the observed chemistry and of course in the reaction occurring in a specular way if the first-formed radical anion contains a suitable nucleofugal group that is expelled to give radical R (RX+.→ + R.+X+, just as R-X-.→ R . X-).

Such reactions via radical ions are well differentiated from reactions via electron donor–acceptor complexes (or exciplexes), which have a great potential in the photochemistry and photophysics of excited states, particularly among aromatic derivatives (see Scheme 5.12).

5.7 Photoinitiated Processes: Generation of Radicals

The oldest type of photochemical reaction, and still the most useful one used for the generation of commodities, is the photogeneration of highly reactive intermediates, in particular,

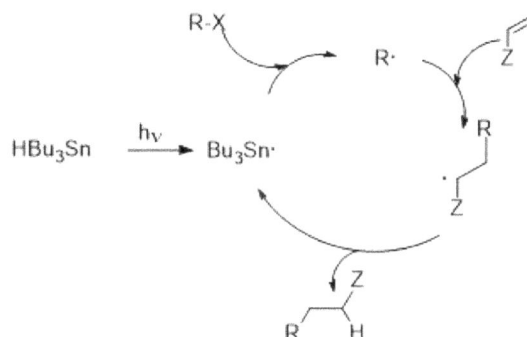

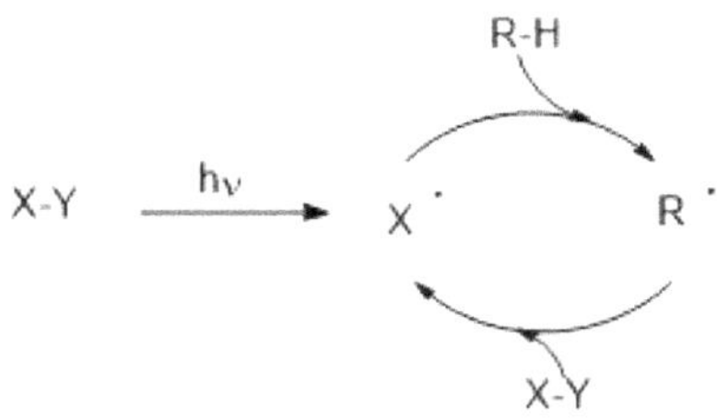

R-X
R·
Z
HBu$_3$Sn
hν
Bu$_3$Sn·
R
Z
Z
R H

SCHEME 5.13 Radicals are formed in a chain reaction, via two- or three-link chains (and thus, two or three different radicals, respectively, are present at the same time and react selectively. A photon is required only in the initiation step, and thus quantum yields may skyscraper to infinite, or at least reach very high values (10^3 or more).

radicals. In a purist theoretical approach, the 'true' photochemical reaction ends there, since the ground state PES has been reached, but from the synthetic point of view, the nice part is yet to come. The difference between the two concepts may be recognized, and with radicals, the intervening of chain in processes is expected, so that quantum yields of 10^3 and above are quite common. Classic cases are the halogenation of alkanes and the alkylation of a,b-unsaturated ketones, esters and related derivatives (Giese method) (Scheme 5.13).

5.8 Choice of Apparatus

We refer here to the liquid phase, although both solid-state and gas-phase photochemistry have been making fantastic advancements in recent times. Chemical synthesis in the lab requires very little sophistication. Let us take a moment for understanding where rays go in the apparatus we are planning. If using freely available sun irradiation, take into account that this is a rather weak source, particularly in the UV. The key objection would always be that because of the continuously changing weather conditions, one cannot have reliable irradiation times. In more sunny countries, it has been demonstrated *ad abundantiam* that any photochemical reaction is revealed by passing the solution into a tube placed in the focus of a concentrating mirror (see Figure 5.4).

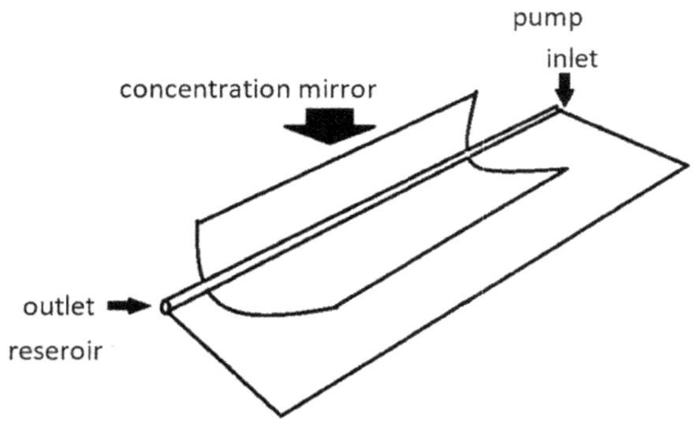

FIGURE 5.4 Photochemical reaction carried out by passing the solution in a tube placed on the focus of a concentrating mirror, applied to the use under solar irradiation conditions.

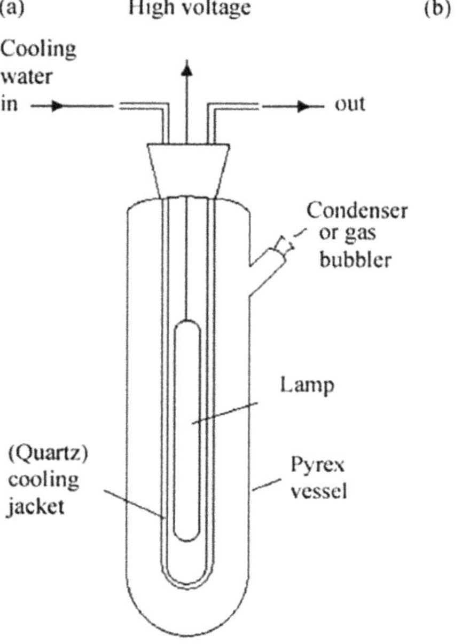

FIGURE 5.5 Immersion well apparatus for medium-pressure lamps.

The most widely used lamps are low-pressure (10^{-3} to 10^{-5} atm under operating conditions) mercury arcs. They consume a low electrical power of 6–16W and are often identified as germicidal lamps (disinfecting the environment by killing bacteria is, in fact, their main application) or mercury resonance lamps. These are supplied as quartz (or rather fused silica, a synthetic amorphous SiO_2) tubes of various lengths, typically 20–60 cm (although lamps longer than 1 m are available), and with 1.0–2.4 cm diameter (see Figure 5.6). In these lamps, >80% of the emission occurs at 254 nm (and a fraction at 185 nm, a wavelength at which the common quartz is not transparent and thus is available only if a high-purity, UV-grade quartz is used). Under these conditions, the excitation of most classes of organic compounds (including many solvents) is ensured. It must be taken into account that given the large size of the lamp, the number of photons emitted per surface unity is low. Therefore, these lamps are most useful for external irradiation by using (quartz) tubes for the irradiated solutions. On the other hand, heating under operating conditions is modest with these lamps; thus there is no heat to be adsorbed.

A key issue with these lamps is the possibility to add a phosphor on the outer surface. These (inorganic) materials absorb the irradiation and emit a chemiluminescent radiation at a longer wavelength. The most used of such phosphors emit at ca. 310 nm, in the UV-B, at ca. 350 nm (UV-A, Wood light or black light) and at ca. 405 nm, UV and violet light. The emitted light has about the same intensity as the light absorbed.

Passing to longer wavelengths has, first of all, the advantage that Pyrex glassware can be used since this material does not absorb over 310 nm, although some loss of transparency develops with the aging of the lamps. In medium-pressure (also called high-pressure) mercury arcs (under operating conditions, close to 1 atm), it can be further noticed that an unbreakable material, such as poly(methylacrylate) often used in biological laboratories are used (obviously only for aqueous solutions). In principle, this glassware should absorb in the same range as Pyrex but are a poor choice for (prolonged) use with photobiological experiments. These lamps are small ampules, of ca. 1–5 cm length; under these conditions, the resonance line at 254 nm is fully reabsorbed and the emission is a continuum with stronger emissions close to the emission

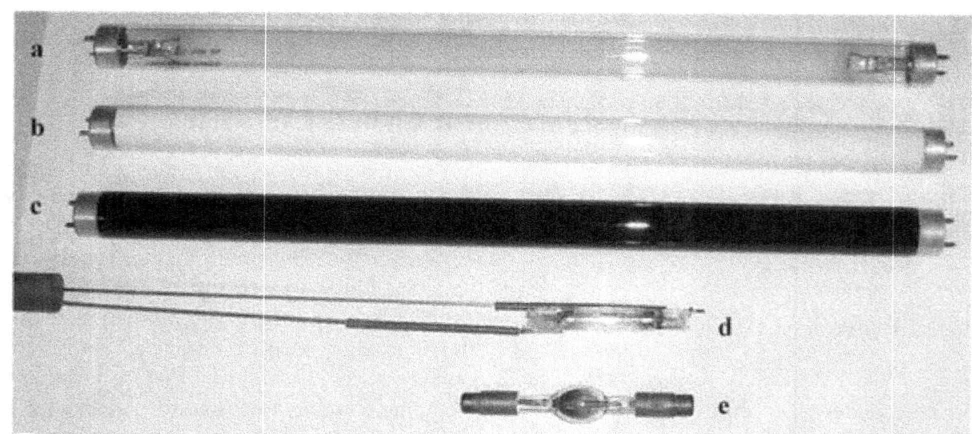

FIGURE 5.6 A picture of different lamps used for photochemical synthesis. (Reproduced by permission from ref 2 see text).

of Hg exciplexes; considerable heat is evolved, but a gentle stream of water is sufficient to cool the apparatus. With the small dimension of these lamps, the optimal configuration is the immersion well apparatus, with a stream of cooling water circulating (see Figure 5.5), better not to leave the experiment unattended, however. Should it happen and the water ceases to circulate, usually no major fire results, since the contacts of the lamp melt and the ampule smoothly lands on the well' tip).

A convenient alternative is the rapidly developing LEDs. These are semiconductors that emit incoherent electroluminescence over a short wavelength range. These small devices are fitted with optics that shape the emission (parallel or with an angle), and are available in a large variety of monochromatic emitting types (actually over a narrow range, typically 20 nm) for almost any wavelength. The conversion of absorbed power (which is in the hundred mW range) into light depends on the wavelength, and LEDs that are reasonably effective light sources are available over the whole of the visible spectrum. LEDs emitting in the same range as Pyrex, UVA, down to ca. 320 or 310 nm are also available, although their emission is much less intense (one-tenth or less compared to those emitting in the visible), and they are somewhat more expensive (a couple of bucks, at any rate). The advantages are that such sources are cheap and long-lived (>10^4 h), and although a single LED is too weak (but more powerful sources are expected to become available in a few years), a set of perhaps 20 LEDs would be adequate. Mounting these light sources on a cylindrical surface, at the center of which test tubes containing the solution can be accommodated, might represent an efficient approach to carrying out photochemical reactions with small volumes (2–30 mL) and will most likely become increasingly common.

Finally, the high (also called very high) mercury lamps are small quartz ampules, where the lines from mercury atoms are reabsorbed and an essential continuous emission over the UV and visible spectrum results. These arcs operate at a pressure of ca. 10^2 atm and develop much heat. Their lifetime is limited and the risk of explosion is high. Thus, these are mounted only in a steel container with cooling fins and a mirror at the back and a lens in the front, so that an almost parallel ray impinges on a cuvette containing the solution results. This is a convenient mounting for the precise measurement of the quantum yield at a precise wavelength, by using appropriate filters, but certainly too expensive for a synthetic explorative study. Another high-pressure lamp that is conveniently used with similar results is the argon arc, likewise emitting over the entire UV and visible range (similar to the black-body emission) (Figure 5.7).

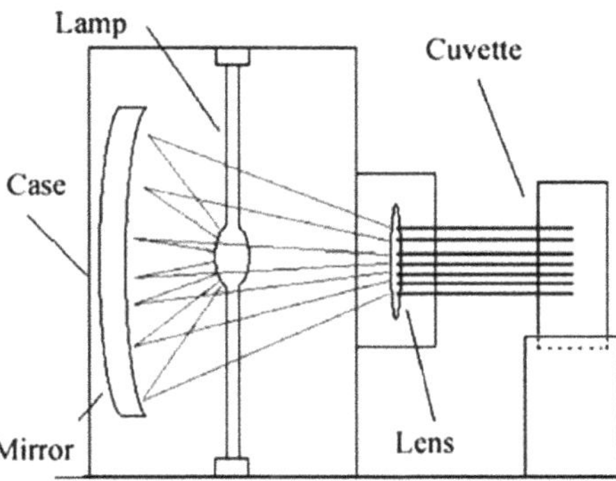

FIGURE 5.7 Mounting of very-high-pressure arcs on an optical bench, an excellent solution for measuring the quantum yield. The dependence of the product's yield on temperature (if any) is conveniently determined by measuring the temperature of the cooling solution circulating in the cuvette, or, if one prefers to work on a larger scale, by measuring the temperature of the cooling medium in the immersion well reactor (see Figure 5.5).

5.9 Cost Issue

An important barrier to the introduction of a photochemical step is of psychological origin since many synthetically interested chemists appear to think that photochemical reactions are never clean, and always a mess, and, at any rate, require complex and delicate instrumentation. I can but repeat unambiguously that on the contrary, photochemical reactions are very clean and any new reaction of synthetic value is always welcome, particularly by physical chemist friends. It is certainly possible that a (serendipitous) discovery, will again contribute to widening the borders of (industrial) research on photochemistry (Figure 5.8). As a matter of fact, photochemical reactions are scarcely considered in the industry. There has been a time where a process appeared in some way promising, while not always a satisfactory solution was arrived at [3], or, in some cases, it seemed too expensive complex to devote a reactor to a photochemical step, while the likelihood that introducing such a step would lead to a delay was obviously disturbing. Indeed, this is what could be demanded from the industry, although this was contrary to the letter, and more so to the spirit of green chemistry (check that the appropriate conditions are met in Scheme 5.14).

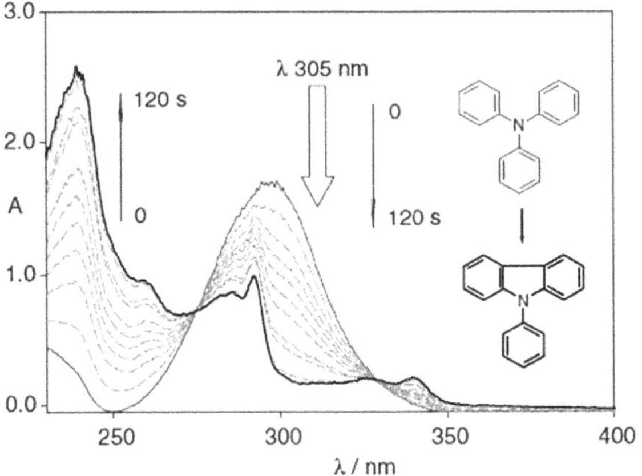

FIGURE 5.8 Photochemical conversion of triphenylamine into *N*-phenylcarbazole. At 305 nm, the photoproduct has only a modest absorption and irradiation at that wavelength leads to the regular conversion that can be continued up to completion. This would not be the case if the irradiation were carried out at a shorter wavelength (because the inner filter effect by the photoproduct would progressively slow down the reaction; at any rate, conserving the isosbestic is no information about the fact the *per se* no indication that a single product is formed, but only that the same products mixture is formed consistently). Irradiations carried out by using a solvent that absorbs competitively, such as acetone, likewise take place in the regular way.

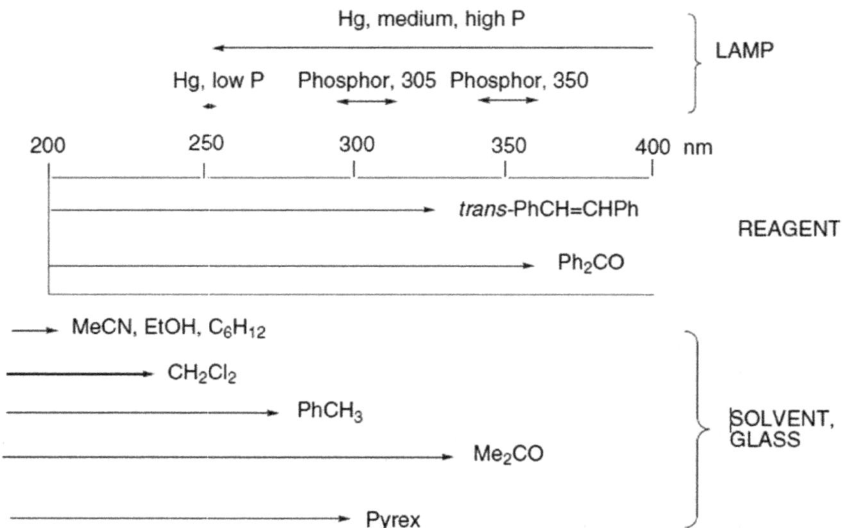

SCHEME 5.14 Checking that the conditions for a successful photochemical reaction are met. First of all, light must be absorbed by the molecule chosen. (Reprinted by permission from ref 2).

The economic aspects were not, however, determining to assess whether a photochemical process could substitute a thermal one, may be no simple matter, even when the required data were available. Such evaluations have been carried out and it appears that no unsurmountable barrier to extend it to a larger application of photochemistry exists. Particular attention has been given to the choice of lamps. It may well be that some prefer to use high-voltage lamps (high-pressure Hg or Ar lamps), but it is, in general, too expensive to use them, and is only to be done when it is aimed to measure photodecomposition at close-lying wavelengths, and thus having a relatively uniform emission is imperative. Staying on explorative studies, it is much better to make recourse either to the sun, the fraction of the emission relative to UV is high enough to identify relatively low quantum yield reactions (this has been called the 'window chemistry'), or to low-pressure mercury lamps that offer much better light emitted–absorbed power ratio, are long-lived and consume very little. These are easily found as germicidal lamps (254 nm), or, when a phosphor transmits longer wavelengths and do not forget LEDs (see Section 5.7). When lamps and a pair of tubes containing the solution are ready and a light is passed in the solution on the on signal any photoreaction occurring will be observed. At any rate, one should not be discouraged from studying low quantum yield reactions, since a molecule not reacting under these conditions simply waits that another quantum hits the molecule, and any problem means only that what is required for completing the synthesis is irradiating for a longer time, unless both the initial compound and the products formed absorb in the same and are in a photochemical equilibrium (a typical case is that of previtamin D), certainly not to be chosen as a first topic of research.[2–4]

As for the industrial application, a photochemical synthesis with a reaction sequence of a reagent leading to the same product was quite difficult to perform. The step economies, the complexity of the purifications or even the yields of the two pathways were often neglected in the cost analyses, and the mere cost of the photochemical equipment was usually the only parameter considered. Oddly enough, the economic aspects were not determined to assess whether a photochemical process could be developed on an industrial scale. The reason is that photochemical processes were not competitive enough in terms of process design to stand a real chance against the well-understood thermal processes, whether this may be more from tradition, or intellectual laziness is unmaterial, the fact is that for the momenta limited choice of photochemical reactions was examined, practically only those that had no sensible thermal alternative has been used by the industry – those that were not thermally feasible and those that underwent a trivial and straight forward scale-up (Scheme 5.15).

One of the first reactions successfully developed at the industrial scale was the photooximation of saturated hydrocarbons (NOCl → NO + Cl), first serendipitously discovered in 1919 when a solution of nitrosyl chloride in heptane stand in the sunlight and noted an unexpected discoloring of the solution was observed. Shortly afterward, the demand for polyamide fibers skyrocketed. By1975, Toyo Rayon (now Toray Industries Inc.) fruitfully produced 150.000 tons per year of caprolactam. Applying the same reaction, Aquitaine-Organico (now Arkema S.A.) produced 4.000 tons per year of lauryl lactam for Nylon 12-manufacturing.[10] As early as 1987, researchers at the New Hampshire University proposed a conceptual project for the solar powering of the industrial photooximation of cyclohexane.

Photooximation processes reached remarkable success despite their low quantum yield. At the beginning of industrial photochemistry, industries tended to rely more confidently on free-radical reactions. The high quantum yield of this class of transformations (up to 10^3) allowed to cut the costs connected to the low light output of mercury vapor lamps, then limited to a few hundred watts. With a quantum yield of ca. 2500, the most notable process was the photochlorination of benzene to give g-hexachlorocyclohexane (lindane, see structure in Scheme 5.16), an insecticide and rodenticide, the photoinitiated version of which, by the mid-1950s, replaced the older

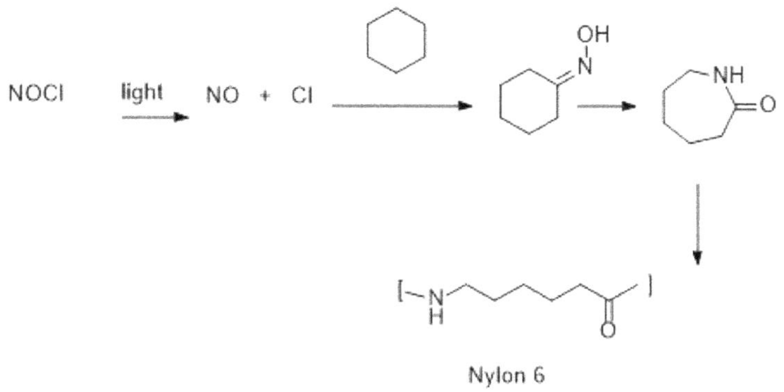

NOCl light NO + Cl → cyclohexane oxime → caprolactam → Nylon 6

SCHEME 5.15 Photochemical synthesis of cyclohexanoxime, the monomer of Nylon 6.[5–9]

TABLE 5.1

Suggestions for Choosing the Bet Irradiation Time for Obtaining the Full Conversion of A Light-Absorbing Chemical into Products[i]

Lamp	Irradiation (nm)	Einstein (min $^{-1}$ cm $^{-2}$)[h]	Volume Irradiation (mL)	Time Taken to Convert a 5 × 10^{-2} M Solution (Φ = 1) (h)
Low-pressure Hg	External, 254[a]	~4 × 10^{-6}	10–100	0.25
Low-pressure Hg arc phosphor–coated	External, 305[b]	~1 × 10^{-6}	10–100	1
Low-pressure Hg arc phosphor–coated	External, 350[c]	~8 × 10^{-7}	10–100	1.2
medium-pressure Hg arc	Immersion 300–400[d]	1 × 10^{-5}	80–1000	0.1
LED, UV	External 310[e]	~2 × 10^{-8}	1–10	40
LED, blue	External 400[f]	~1 × 10^{-6}	1–10	1
Solar light	External 330–400[g]	~1 × 10^{-7}	Any	10

[a] Quartz lamp, 5–25 cm long, check the tube, mounted parallelly is likewise made of quartz.
[b] Quartz lamp, 5–25 cm long, coated with a white phosphor, check the tube, mounted parallelly is likewise made of quartz.
[c] Quartz lamp, 5–25 cm long coated with a black phosphor the tube with the solution may be of Pyrex.
[d] Quartz lamp, 2–10 cm long, the inner compartment has to be in quartz, while the inner immersion well is in Pyrex.
[e] Six LED, 0.3–2 cm^2, around the tube with the solution.
[f] Six LED, 0.3–2 cm^2, around the tube with the solution.
[g] Based on results obtained when using cylindrical Pyrex vessels, 10–100 nm.
[h] The Einstein is an Avogadro number of photons; the use of this measure unity rather than the number of photons corresponds to the use of moles rather than of molecules and has been introduced by Bodenstein as best fitting with the chemist's habits (Bodenstein, M, Wagner, C, Z *Physic Chem*, 1929, 3(B), 456–458).
[i] It is advised to check in every case how much the studied chemical adsorbs light by comparing the solution with the neat solvent.

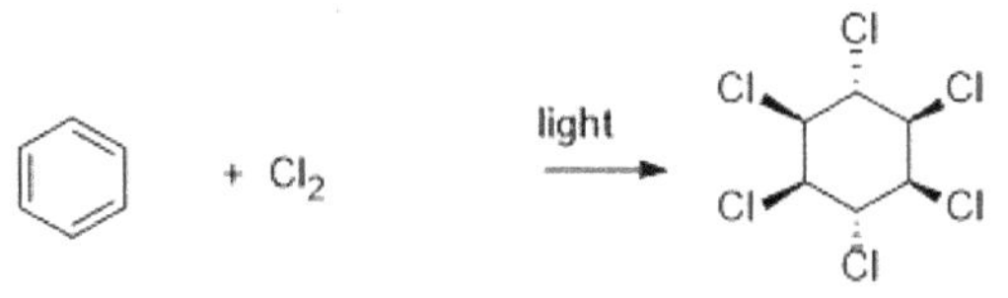

benzene + Cl_2 light → hexachlorocyclohexane

SCHEME 5.16 Structure of the γ-lindane.[10, 11]

thermal chlorination. Despite the unsatisfactory chemical yield (15%), the lower temperatures required in the new process granted a higher yield for the desired g-isomer (one out of the six possible stereoisomers).

In 2009, lindane production has been banned due to environmental concerns; back in 1973, however, its global production was 70 million dollars worth. Furthermore, other photochlorinations found their way to the industry: the production of alkyl chlorides from methane, ethane and toluene was reported and exploited by several companies by the end of the 1970s.[10–12]

Similarly, light-driven sulfochlorinations and sulfoxidations found their space in the industry. These radical chain reactions, with a quantum yield around 2000, were used for the industrial synthesis of surfactants and emulsifiers.[12] The first process to be developed was sulfochlorination; however, unbranched alkanes were necessary to avoid regioselectivity issues and low conversions (30–50%) were required to limit polysulfochlorination, as both processes led to undesirable products. To avert these limitations, by the mid-1960s, sulfochlorination was replaced by sulfoxidation in the detergent industry. Hoechst AG (then Aventis Deutschland, now Sanofi-Aventis) developed an efficient industrial sulfoxidation process, and, in 1976, photochemically produced ca. 50.000 tons per year of alkanesulfonates for preparing detergents.

In the 1960s, Dragoco (now Symrise AG) and Firmenich SA first implemented a photosensitized process toward the synthesis of RoseOxide) used as perfume. Rose Bengal is used as a sensitizer, allowing the photooxidation of citronellol to yield

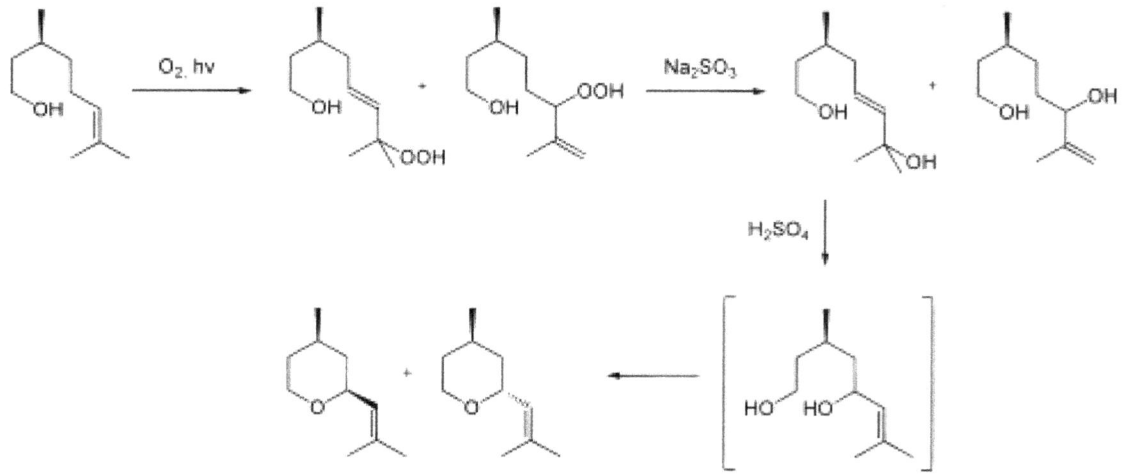

SCHEME 5.17 Oxygen addition to citronellol to yield rose oxide.

two regioisomeric hydroperoxides, and then reduced to alcohols by sulfite (Scheme 5.17).[13]

The major product is then converted to a mixture of rose oxide stereoisomers by sequential allylic rearrangement and cyclization. Another photosensitized process found is in the industrial synthesis of Vitamin A via Wittig olefination by BASF: a stereoisomeric mixture of 11-cis and all-trans vitamin A acetate is irradiated in the presence of tetraphenylporphinatozinc or chlorophyll to achieve the complete isomerization to the all-trans product.[14] A similar approach has been patented by Hoffmann-LaRoche. Most interestingly, at the same time, an elegant biomimetic approach was introduced within the industry for the synthesis of previtamin D_3.[15]

As within skin cells, precursor 7-dehydrocholesterol (achieved in four steps from cholesterol) generates previtamin D_3 by ring-opening of the cyclohexadiene moiety upon absorption of light. On mild heating, this intermediate rearranges to previtamin D_3 via a sigmatropic 1,7-hydrogen shift. A thorough investigation of the irradiation conditions has been required since previtamin D_3 can easily rearrange to undesired tachysterols and lumisterols upon irradiation. Moreover, mercury lamp emission and 7-dehydrocholesterol absorption spectra overlap poorly, leading to great radiant energy waste. Thus, proper light filters and a two-step irradiation sequence are usually adopted,[15] see Scheme 5.18.

Despite the efforts, as of today, the overall yield for the industrial preparation of Vitamin D_3 is still lower than 20%.

As a *continuous* flow device is scale-independent, a single reactor can, in principle, be used to process a few milligrams of substrate up to nearly a kilogram per day (see below), with a range of advantages, such as the following: i) due to much shorter path lengths, high-concentration solutions can be irradiated effectively; ii) mirror wise, however, large volumes of very-low-concentration solutions can be irradiated. This is particularly useful for reactions with competing *inter*molecular side reactions, e.g., dimerization and polymerization, and

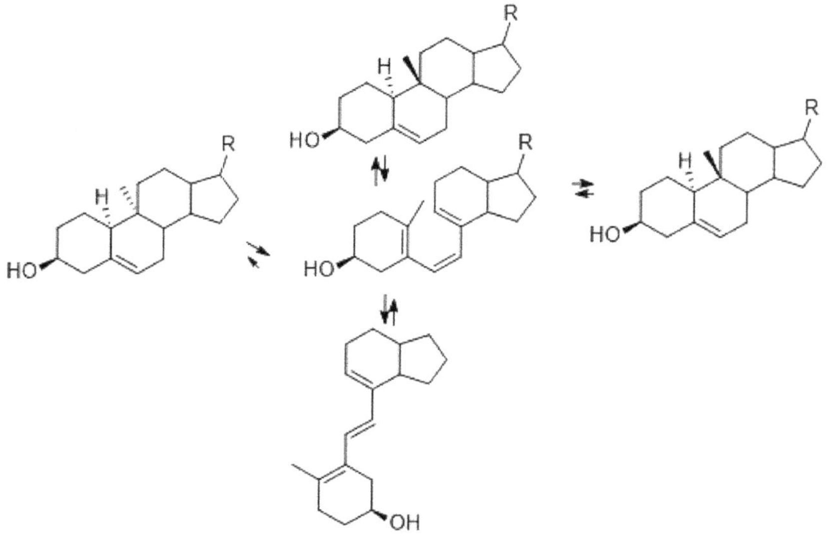

SCHEME 5.18 Irradiation of 7-dehydrocholesterol causes thermal and photochemical equilibria to be established, which makes it impossible to obtain 100% yield of previtamin D_3.

iii) the photolyzed solution can be concentrated by continuous evaporation and the solvent is recycled for new irradiation of the starting material. This can dramatically cut down the solvent footprint, particularly in dilute reactions where large volumes of solvent would be required to process large quantities of the substrate. It is true that by allowing the bulk solution to be kept remote from the lamp, only a minimal amount of flammable solvent comes close to a potential ignition source at any one time, and was a large improvement of safety conditions, but despite these safety advantages, what followed preceded freezing of photochemical research, according to the idea that that topic was a research theme for good times, or at least not to be seriously compared with thermal methods until the price continued to drop. New successes included the re-discovery of the Paterno synthesis of oxetanes, as well as the research of extra-fuels that were obtained by generating small ring compounds. In the 1990s, most chemists believed that preparative photochemistry had no future. A renewed interest in large-scale preparative organic photochemistry sprouted in the last decade of the 20th century.

The renaissance of photochemistry came along with the introduction of high-power LEDs, flow techniques, mini- and microreactors. Adopting LED sources featuring a narrow emission band may be the winning modification when it is difficult to irradiate a single compound and requires no expensive cooling apparatus; thus resulting in a cheaper alternative to the old irradiation methods, as it is the case for the industrial synthesis of vitamin A in the natural all-trans configuration.[16]

The successes of microreactors in thermal processes have been more than duplicated in photochemistry. Mini- and micro(-flow) reactors allow a simpler control of irradiation time and a more efficient penetration of light in the solution. In the case of flow reactors, scalability would be much easier as it would imply the serial coupling of multiple reactors, without changing the optimized conditions for each one of them. In addition to these technical considerations, it should be noted that on the laboratory scale, huge advancements in the field of photochemistry have been made. Novel and once-elusive transformations can now be achieved through photoredox catalysis. The use of the appropriate photocatalysts allows, in principle, to extend photochemical processes to non-absorbing starting materials, broadening the scope of the photochemical synthetic aspects.

This renewed interest in light-driven transformations is well exemplified by two recent reports from the industry. In 2011, researchers at Sanofi-Aventis designed a novel pathway toward the synthesis of the alkaloid artemisinin (see Scheme 5.19), an antimalaria drug formerly extracted from Artemisia annual leaves starting from artemisinin acid that can be biosynthetically produced in yeast; by 2014, Sanofi was able to achieve 60 tons per year of Artemisinin with high purity and yield using just three synthetic steps (see Scheme 5.19).

Sanofi's synthesis of artemisinin still relies on well-known reactor design and old mercury vapor lamps as the illumination source, while the first industrial synthesis of camptothechin derivatives that employed modern microreactors was developed at Hereus and started from artemisinic acid that can be biosynthetically produced in yeast. By 2014, Sanofi was able to achieve 60 tons per year of Artemisinin with high purity and yield using just three synthetic steps (see Scheme 5.17).[17]

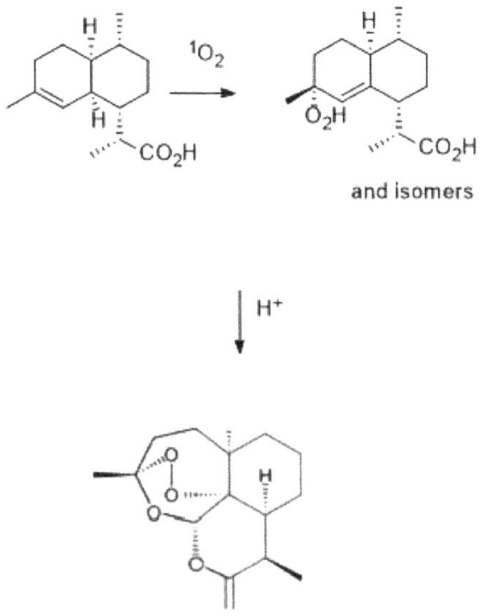

SCHEME 5.19 Oxygen addition applied the synthesis of artemisinin.

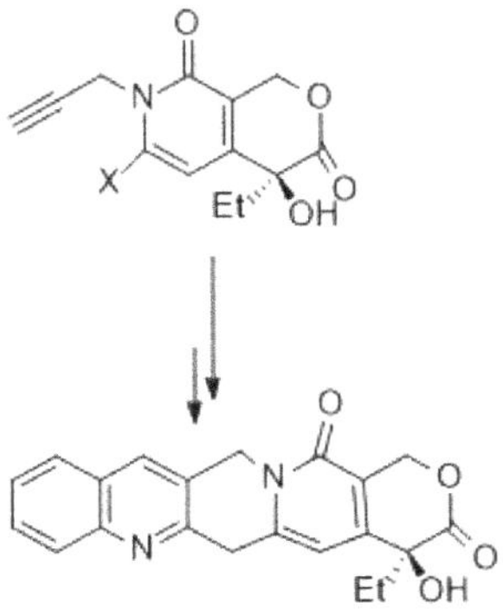

SCHEME 5.20 Photoinitiated radical synthesis of camptotechin.

For this purpose, 12 identical reactors were operated in parallel, illuminated by their own individual light sources (250 W high-pressure Hg lamp, 350–400 nm). 10-Hydroxycamptothecin and 7-ethyl-10-hydroxycamptothecin are obtained with high yields and conversions from a reduction of the respective *N*-oxides (see Scheme 5.20). The microsystem array operated at 0.6 weight % concentration, with respect to the 0.1 weight% concentration of the classical batch system, affording 2 kg per day of 10-hydroxycamptothecin with 90% yield at 95% conversion (with respect to a 50% yield at 85% conversion for the batch system by using a multi-lamps apparatus fitted with six 15W lamps.

5.10 Conclusion and Outlook

The interest of the chemical industry in photochemistry seems to undergo a really variable course with high peaks when the price of oil increases very much. Common citizens have no access to sensible predictions on when the next (final) energy crisis will hit but everybody knows that scientists will have to restart chemistry a step earlier, and leave behind such a convenient, if smelling, starting material as oil that organisms have kindly stored for us a long, long time ago, under the Earth's surface.[18–20] If such a begin from the start has to begin anew will take place, however, it will require that the tenets of green chemistry have to be respected from the beginning. If this is correct, then photochemistry has to be among the most important issues, in view of the fact that there is no need to store aggressive chemicals, or at most, in a limited amount, the minimal heat evolution (attention, however, don't forget to carry out the thermochemical balance of the reaction), the much-limited risk of fugitive reactions, the easy application with 'novel' reactors and the simple scale-up.[2] My wish would be that this time, the beginning is free from skeletons in the cupboards, contrary, for example, to the case of lead tetraethyl that the car-production industry imposed when it was already proven that this excellent anti-shock agent that regularized fuel combustion was toxic. Thus, is likely that such convenient light sources ad Hg arcs will have to be abandoned, but others have been, or are being, or will be developed.

If I am allowed a personal note, do not worry too much about (low) quantum yields or to achieve the dream photoreactor, with perfectly uniform irradiation everywhere, while there is a large number of new photoreactions to discover, first.

NOTES

1 Another fact that generated some misunderstanding between physics and chemists was that in his earlier papers Einstein had not discussed in detail the application of his equivalence law, and only later clarified that he was referring to the generation of electronically *excited states* (quantum yield, $\Phi = 1$), while chemists were accustomed to measure, and could not ignore the fact that the quantum yield of *end products* varied from $<10^{-3}$ to $>10^{3}$.

2 Light-induced reactions have been felt from the beginning as a way to obtain chemicals in a smoother way, closer to what nature had been establishing.[18–20]

REFERENCES

1. Some remarks on the first law of photochemistry, *Photochem. Photobiol. Sci.*, 2016, **15**, 319–324.
2. Albini, A, Germani, L, Photochemical methods, In Albini, A, Fagnoni, M, *Handbook of Synthetic Photochemistry*, Wiley, Weinheim, 2009.
3. Albini, A, Fagnoni, M, Green chemistry and photochemistry were born at the same time, *Green Chem.*, 2004, **6**, 1–6.
4. Gu, GD, Okamura, WH, Synthesis of vitamin D, *Chem. Rev.*, 1995, 95, 1877–1952.
5. Lynn, EV, A new reaction of paraffinic hydrocarbons, *J. Amer. Chem. Soc.*, 1919, 41, 368–370.
6. Talukdar, T, Wong, EHS, Mathur, VK, Caprolactam production by direct solar flux, *Sol. Energy*, 1992, 47, 165–171.
7. Mosher, MW, Bunce, NJ, The free radical oximation of alkanes by nitrosyl chloride, *Can. J. Chem.*, 1971, 49, 28–34.
8. Griffiths, OM, Ruggeri, M, Baxendale, IR, Photochemical flow oximation of alkanes, *Synlett*, 2020, 31, 1908–1912.
9. Talukdar, T, Wong, EHS, Mathur, VK, Caprolactam production by direct solar flux, *Sol. Energy*, 1992, 47, 165–171.
10. Fischer, M, Industrial applications of photochemical syntheses, *Angew. Chem. Int. Ed. Engl.*, 1978, 17, 16–26.
11. Pferner, KH, Photochemistry in industrial synthesis, *J. Photochem. Photobiol. A Chem.*, 1990, 51, 81–86.
12. Braun, AM, Pesch, G, Oliveros, E, Industrial Photochemistry, In Griesbeck, A, Olgemoller, M, Ghetti, F (Eds), *Handbook of Photochemistry and Photobiology*, 3rd edition, 2012, CRC, Orlando, Chapter 1.
13. Alsters, PL, Jary, W, Nardello-Rataj, V, Aubry, JM, "Dark" Singlet Oxygenation of β-Citronellol: A Key Step in the Manufacture of Rose Oxide, *Org. Process Res. Dev.*, 2010, 14, 1, 259–262.
14. Zaghdoudi, K, Ngomo, O, Vanderesse, R, Arnoux, P, Myrzakhmetov, B, Frochot, C et al. Extraction, identification and photo-physical characterization of Persimmon (*Diospyros kaki* L.), *Food*, 2017, 6, 4.
15. Kopecki, D, Leveque, F, Seeberger, PH, A continuous-flow process for the synthesis of Artemisinin, *Chem. A Eur. J.*, 2013, 19, 5450–5456.
16. Basso, A, Capurro, P, Recent applications of photochemistry on large scale synthesis, *Photochemistry*, 2021, 48, 293–324.
17. Turconi, J, Griolet, F, Guevel, R, Oddon, D, Villa, R, Geatti, A et al, Semisynthetic Artemisinin, the Chemical Path to Industrial Production, *Org. Prep. Res. Dev.*, 2014, 18, 417-R.
18. Paternò, E, Nuovi orizzonti della sintesi in fotochimica orgzanica, *Gazz. Chim. Ital.*, 1909, 39, 213–220.
19. Ciamician, G, Actions chimiques de la lumiere, *Bull. Soc. Chim. France*, IV Ser III, i–xxvii.
20. Stobbe, H, Photochemieorganischer Verbindungen, *Zh. Electrochem Angew. Phys. Chem.*, 1908, 33, 73.

6

Biocatalysis in Green Biosolvents

Margherita Miele, Laura Ielo, Vittorio Pace, and Andrés R. Alcántara

CONTENTS

6.1 Introduction ... 89
6.2 Methyltetrahydrofuran (2-MeTHF) ... 91
6.2.1 Lipase-Catalyzed Acylation of Alcohols in 2-MeTHF ... 98
6.2.2 Enzyme-Catalyzed Hydrolysis Using 2-MeTHF as (Co)solvent ... 100
6.2.3 Bioreductions Using 2-MeTHF as (Co)solvent ... 101
6.2.4 Use of Lyases and 2-MeTHF as (Co)solvent ... 101
6.2.5 Biocatalytic Cascades Using 2-MeTHF as (Co)solvent ... 101
6.2.6 Recent Examples of Biocatalysis in 2-MeTHF ... 102
6.3 Glycerol and Glycerol-Derived Solvents ... 107
6.4 Gamma-Valerolactone (GVL) ... 107
6.5 Dihydrolevoglucosenone ... 108
6.6 Conclusion ... 108
References ... 109

6.1 Introduction

A well-known paradigm inside Green Chemistry states that "the best solvent is no solvent".[1,2] Nevertheless, most of the known organic transformations do require the use of solvents, as they are crucial not only for running a reaction, but also for the subsequent work-up (separation and/or purification of the final products). Consequently, it has been assessed that solvents (annual industrial-scale production higher than 20 million metric tons[3]) typically constitute around 80% of all materials required for the effective completion of an archetypal synthetic procedure,[3,4] and this value approaches 90% when the synthesis of active pharmaceutical ingredients (APIs) is intended.[1,5] This is not a very sustainable scenario and turns even worse bearing in mind that the most commonly used organic solvents, usually displaying high toxicities, are generally derived from fossil non-renewable resources; accordingly, this fact causes severe environmental and economic concerns when facing large-scale chemical processes.[6,7]

Therefore, modern green chemistry is focused on the smart replacement of these solvents by more sustainable alternatives.[7] In this sense, a "green solvent" must fulfill some physicochemical characteristics,[2,8–11] such as low vapor pressure, high boiling point and high biodegradability under environmental conditions. Additionally, it must be odorless, easily recyclable and should possess an almost universal capability for dissolving as many chemicals as possible; finally, yet importantly, it must be reasonably priced and should derive from renewable resources.

Inside the denomination of "green solvents", we can include a great variety of chemical structures; obviously, water has been traditionally considered as the greenest possible solvent, attending to its cost, ready availability and its non-toxic, non-polluting and non-flammable behavior.[12] Anyhow, as most organic compounds are not soluble in water, it had not commonly been used for synthetic purposes due to the commonly accepted axiom (*corpora non agunt nisi soluta*) that substances should be dissolved "in water" to react; nevertheless, organic reactions involving water-insoluble compounds can be efficiently conducted "on water", that is, at the water interface.[13] This concept has fostered the use of water for organic reactions. Additionally, other commonly cited green solvents are ionic liquids (ILs),[14–18] supercritical fluids[19,20] (mainly supercritical CO_2[21,22]), deep-eutectic solvents (DES)[23–34] and, last but not least, biomass-derived solvents, usually abbreviated as "biosolvents".[35] These solvents, which share similar properties with those derived from fossil resources, satisfy several of the criteria required for being cataloged as a green solvent, such as accessibility, biodegradability, little toxicity and affordable prices.[36] For this reason, their use is increasingly being considered as a valuable and sustainable alternative compared to petrol-derived counterparts.[11,36,37]

Biosolvents are some of the compounds which could be obtained in a biorefinery, that is, a refinery that converts biomass to bioenergy (biofuels, power and/or heat) and other beneficial bio-based products (food, feed, chemicals, materials and solvents).[38,39] Most of the biorefineries actually working (or planned to start working)[40–42] are based on lignocellulosic material,[43] consisting of high recalcitrance phenolic lignin and polymeric carbohydrates such as cellulose and hemicelluloses. From these ones, different carbohydrate molecules possessing 5 (xylose, arabinose) or 6 carbon atoms (glucose, mannose, galactose and rhamnose) can be obtained, serving as starting

DOI: 10.1201/9781003089162-6

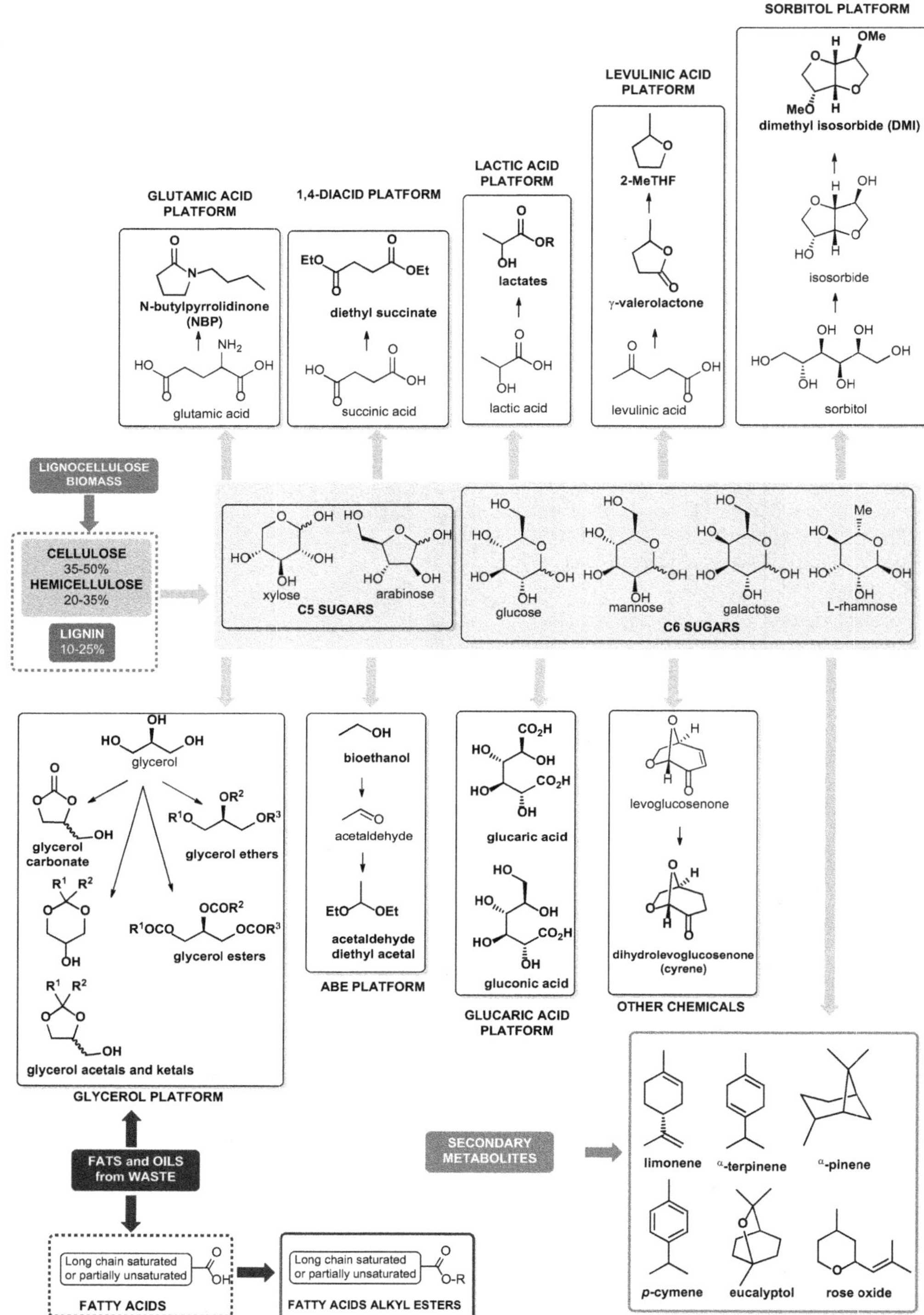

SCHEME 6.1 A resume of the most common biosolvents and their procedence.

materials for the generation (*via* biological or chemical conversions) of different building blocks classified into different platforms,[43] which in turn can subsequently be converted in valuable chemicals, materials and solvents, as presented in Scheme 6.1.

As can be seen, most of the main biosolvents used in organic synthesis today are covered in Scheme 6.1. The most obvious choice is bioethanol,[44–47] which is the starting building block for the production of acetaldehyde diethyl acetal (1,1-diethoxyethane), an attractive biosolvent.[48] Some other biosolvents are glycerol, its ether, acetals, carbonates and esters[49–55]; several low-melting mixtures of carbohydrates[56,57]; esters of lactic acid[58,59] and gluconic acid[60–63]; 2-methyltetrahydrofuran (2-MeTHF),[64–66] dihydrolevoglucosenone (commercial name Cyrene),[67,68] or γ-valerolactone (GVL).[69–76] More recently, some other biomass-derived chemicals have been proposed as green solvents, such as isosorbide dimethyl ether (Me_2Isos or DMI)[77–80] and diethyl succinate (Et_2Suc),[81–84] both derived from cellulose, or *N*-butylpyrrolidinone (NBP), a dipolar aprotic solvent[85–91] which can be obtained from glutamic acid.[92] Finally, we cannot forget mentioning that such a very common solvent, such as ethyl acetate (EtOAc), is readily available from biomass[93,94] and is often disregarded in the context of biosolvents.

Another important family of biosolvents comes from secondary metabolites of plants, accessible through biorefineries derived from lignocellulose waste.[95] In this sense, most commonly used terpenes acting as biosolvents are limonene[96–102] or *p*-cymene[3,103–105]; some new alternatives, such as γ-terpinene,[48,106,107] α-pinene,[48] eucalyptol[48,108–110] or (+)-rose oxide[48] (4-methyl-2-(2-methylprop-1-en-1-yl)tetrahydro-2*H*-pyran) have been recently reported. Additionally, from waste fats and oils, apart from glycerol also available from C5-C6 sugars, it is possible to obtain fatty acids that, upon esterification, would furnish fatty acid alkyl esters. These methyl esters (FAME), known as biodiesel, although mainly used as biofuels, can be also used as biosolvents,[111–114] with methyl soyate being one of the most reported.[115,116]

Thus, to illustrate the use of these green biosolvents, different examples would be presented hereinafter, considering their applicability in classical organic chemistry as well as in biocatalyzed procedures.

6.2 Methyltetrahydrofuran (2-MeTHF)

Although originally considered as a biofuel,[117] 2-MeTHF is a renewable alternative to THF and other organic solvents,[65,66,118] as it can be synthesized from xylose or glucose, through furfural and levulinic acid valorization[119] (see Scheme 6.1). Remarkably, following the furfural route, a reduction of 97% solvent emissions compared to non-renewable THF production has been reported[120]; furthermore, applying acidic pathways starting from levulinic acid (via γ-valerolactone (GVL) intermediate), it is possible to generate a dual high-value stream of 2-MeTHF and GVL.[66] A detailed Life Cycle Assessment (LCA, a tool to assess the environmental impacts and resources used throughout a product's life cycle, that is, from raw material acquisition, via production and use phases, to waste management[121] for the production of 2-MeTHF from three biomass sources (corn stover, sugar cane bagasse and rice straw) has been described,[122] showing that the energy demand and environmental damage caused by crop production far prevailed over that of biomass processing, reporting an energy consumption of around 0.2 MJ/kg.

Regarding its chemical–physical properties, 2-MeTHF shows lower water miscibility, higher stability and lower volatility compared to THF (2-MeTHF has melting and boiling points of −136°C and 80.2°C, respectively, whereas THF values are −108.4°C and 66°C, respectively). Likewise, it has been reported that 2-MeTHF displays low toxicity, possess neither mutagenicity nor genotoxicity characteristics,[123,124] and the human Permitted Daily Exposure (PDE) limit (below which there would be negligible safety concerns for patients exposed to 2-MeTHF) is 6.2 mg/day.[125] Very recently, a No Observed Adverse Effect Level (NOAEL) of 250 mg/kg/day has been reported, therefore supporting the safe use of 2-MeTHF in the pharmaceutical industry.[126] Albeit 2-MeTHF is generally considered to be readily degradable,[66,127] though literature focusing on the degrading pathways is still scarce. Anyhow, regardless of its biogenic origin, the CHEM21 selection guide still ranks 2-MeTHF as problematic due to its high flammability,[128] even considering that its flash point (−11°C) is higher than that from hexane (−30°C).[119]

2-MeTHF had been briefly benchmarked to other solvents in some previous publications[129,130]; nevertheless, the work conducted by Simeó et al.[131] describing the regioselective acylation of several nucleosides catalyzed by lipase B from *Candida antarctica*, to render either 3′or 5′-esters depending on the substrate, can be considered as the real starting point for the modern use of 2-MeTHF as an alternative for biotransformations. Subsequently, many other examples have been reviewed elsewhere by Alcántara and coworkers[64,66]; some of them are shown in Table 6.1.

TABLE 6.1
Some Previously Reported Examples of Biotransformation Conducted in 2-MeTHF

Entry	Reaction Type Lipase-Catalyzed Acylation of Lineal Alcohols	Reference
#01	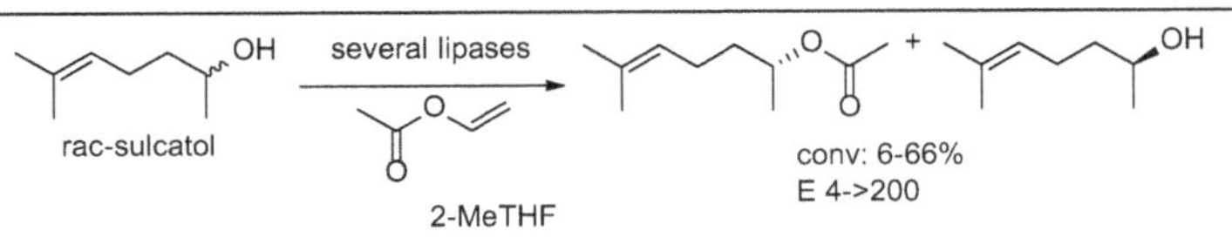 rac-sulcatol; several lipases; 2-MeTHF; conv: 6-66%; E 4->200	Belafriekh et al.[132]
#02	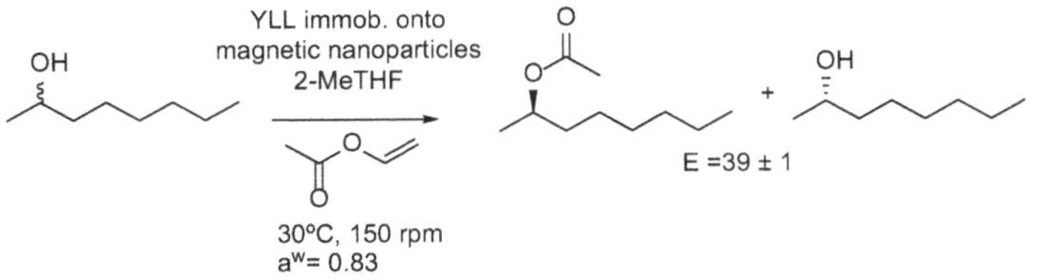 YLL immob. onto magnetic nanoparticles; 2-MeTHF; 30°C, 150 rpm; a^W= 0.83; E =39 ± 1	Liu et al.[133]

(Continued)

TABLE 6.1 (Continued)

Entry	Reaction Type: Lipase-Catalyzed Acylation of Lineal Alcohols	Reference
#03	*Pseudomonas stutzeri* lipase; Shvo´s catalyst; Trifluoroethyl butyrate; 2-MeTHF, 55 °C, Ar atm. — 48 h; 85 % conv.; > 99 % eep	Hoyos et al.[134]
#04	*Pseudomonas stutzeri* lipase immobilized on Accurel MP101; *metal-associated meso-porous silicates*; Vinyl butyrate; 2-MeTHF, 50 °C — 5 h; 96.8 %conv.; 97 % eep	Nieguth et al.[135]
#05	*Pseudomonas stutzeri* lipase: commercial crude enzyme or covalently immobilized on a porous epoxy-type copolymer; Trifluoroethyl butyrate; 2-MeTHF, 55 °C, Ar atm. Ar = Ph, 4-MeO-C_6H_4, 4-iPr-C_6H_4, 4-EtO-C_6H_4, 3,4-diCl-O-C_6H_3, 2-furyl, 3-thienyl — 6-8 h; 48-50 % conv.; > 99 % eep	Aires-Trapote et al.[136]
#06	*Pseudomonas stutzeri* lipase covalently immobilized on a porous epoxy-type copolymer; *Shvo catalyst*; Trifluoroethyl butyrate; 2-MeTHF, 55 °C, Ar atm.; up to 6 recycles — 24 h; 92% conv.; > 99 % eep	Aires-Trapote et al.[136]
#07	lipase; 4h, 90°C; 2-MeTHF/solvent; RCO$_2$H. R = -$(CH_2)_2$-S-S-$(CH_2)_2$-CO_2H; -$(CH_2)_2$-O-$(CH_2)_2$-CO_2H; -$(CH_2)_n$-CO_2H (*n*=2,3,4,5,6,7,8,9,10); -(CH=CH)-CO_2H (Zand E); *p*-C_6H_4-CO_2H	Schuster et al.[137]
#08	Vinyl acetate, CAL-B; 2-MeTHF / CO_2-exp. 2-MeTHF. Conversion 29 %, E > 200; Conversion 40 %, E > 200; Conversion 19 %, E > 200; Conversion 26 %, E > 200	Hoang et al.[138] Hoang et al.[139]

Entry	Reaction Type: Lipase-Catalyzed Acylation of Lineal Alcohols	Reference
	Lipase-catalyzed acylation of cyclic alcohols	
#09	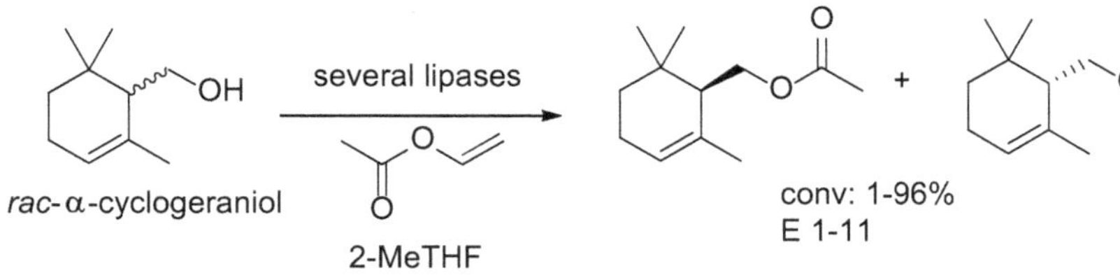 *rac*-α-cyclogeraniol; several lipases; 2-MeTHF; conv: 1-96%; E 1-11	Belafriekh et al.[132]
#10	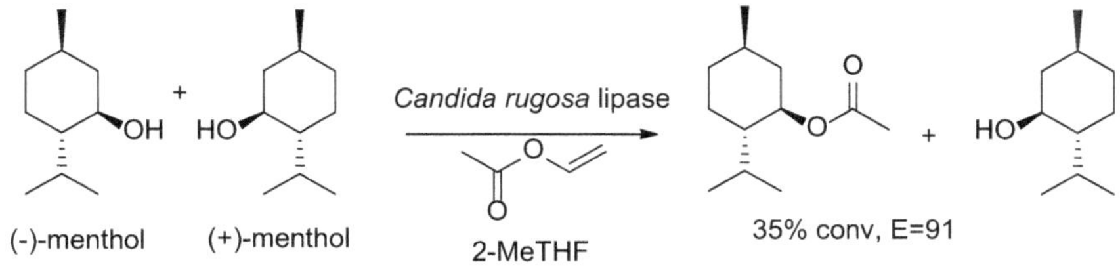 (-)-menthol (+)-menthol; *Candida rugosa* lipase; 2-MeTHF; 35% conv, E=91	Belafriekh et al.[132]
#11	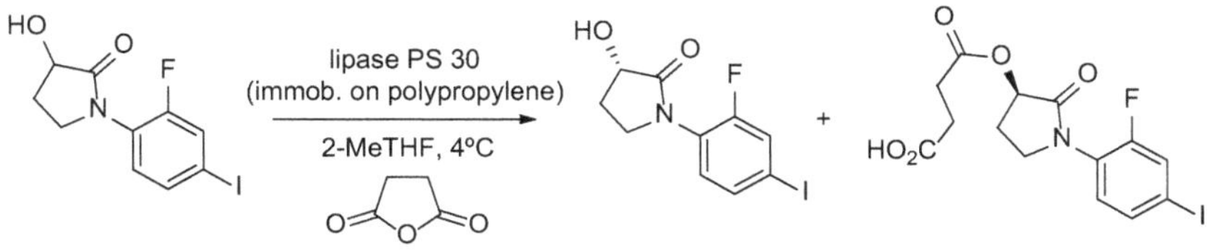 lipase PS 30 (immob. on polypropylene); 2-MeTHF, 4°C	Singh et al.[140]
	Lipase-catalyzed acylation of nucleoside and analogs	
#12	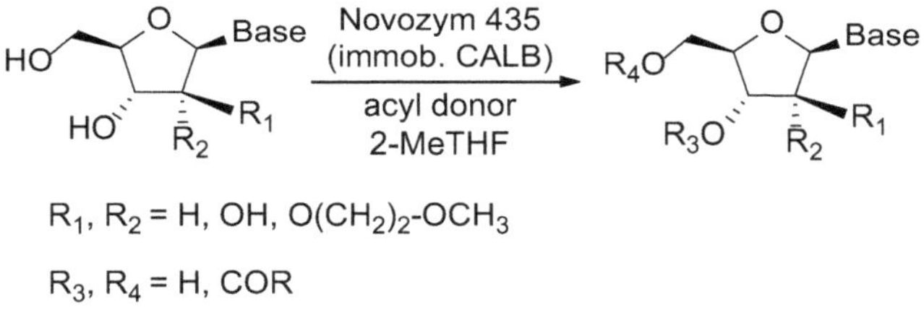 Novozym 435 (immob. CALB); acyl donor; 2-MeTHF; R_1, R_2 = H, OH, $O(CH_2)_2$-OCH_3; R_3, R_4 = H, COR	Simeó et al.[131]
#13	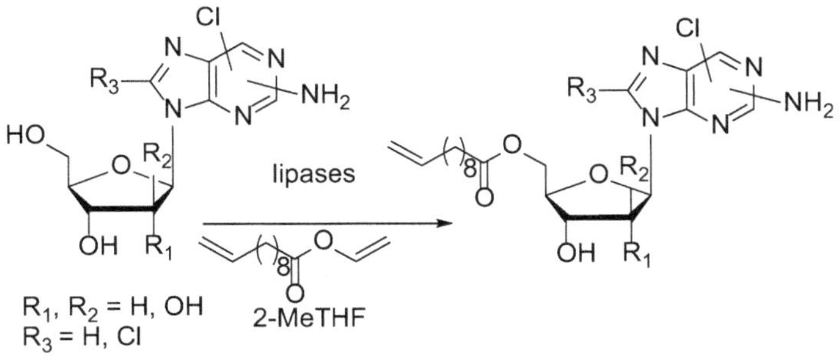 lipases; 2-MeTHF; R_1, R_2 = H, OH; R_3 = H, Cl	Gao et al.[141] Gao et al.[142]
#14	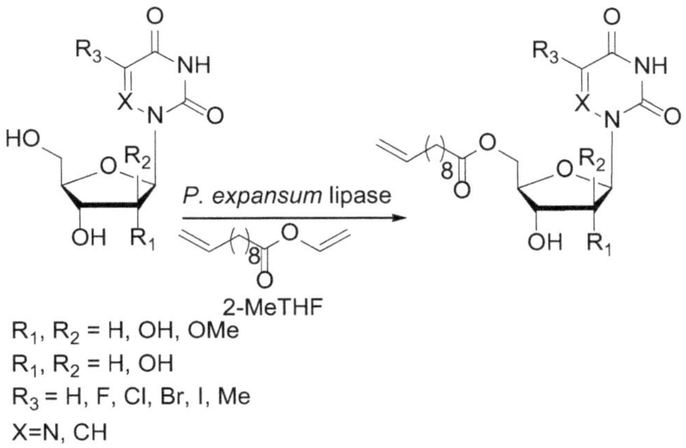 *P. expansum* lipase; 2-MeTHF; R_1, R_2 = H, OH, OMe; R_1, R_2 = H, OH; R_3 = H, F, Cl, Br, I, Me; X=N, CH	Gao et al.[143]

(*Continued*)

TABLE 6.1 (Continued)

Entry	Reaction Type: Lipase-Catalyzed Acylation of Lineal Alcohols	Reference
#15	cordycepin; Novozym 435 (immob. CALB), vinyl acetate, 2-MeTHF	Chen et al.[144] Zhang et al.[145]
	Lipase-catalyzed *O*-acylation of natural products	
#16	polydatin (0.03 mol); lipase TL IM, 2-MeTHF (3 mL), R-CO-O-C(CH$_3$)=CH$_2$ (0.27 mol); R = -(CH$_2$)$_n$-CH$_3$ n=2,4,8,10,12,14; -(CH=CH)$_n$-CH$_3$ n=1,2; -(CH$_2$)$_8$-CH=CH$_2$; → **6-*O*-acyl-polydatin pro-drug**	Wang et al.[146] Wang et al.[147] Wang et al.[148]
#17	gastrodin; lipase, 2-MeTHF/THF mixtures, R-CO-O-C(CH$_3$)=CH$_2$, R = -(CH$_2$)$_8$-CH=CH$_2$; → 7"-ester (MAJOR) + 6'-ester (minor)	Yang et al.[149]
#18	Novozyme 435, *t*-BuOH/2-MeTHF, 1/4 (v/v), 50°C, 200 rpm	Hu et al.[150] Hu et al.[151]
#19	G M G G M G; PSL-C lipase, vinyl acetate, 2-MeTHF/([C$_4$MIm][BF$_4$]) 20/80 v/v 55°C; → G M G G M G	Chen et al.[152]
#20	lipase, 2-MeTHF, 47°C; 53-60% conversion, 98-99% ee	Sundell et al.[153]

Entry	Reaction Type: Lipase-Catalyzed Acylation of Lineal Alcohols	Reference
	Lipase-catalyzed *N*-acylations	
#21	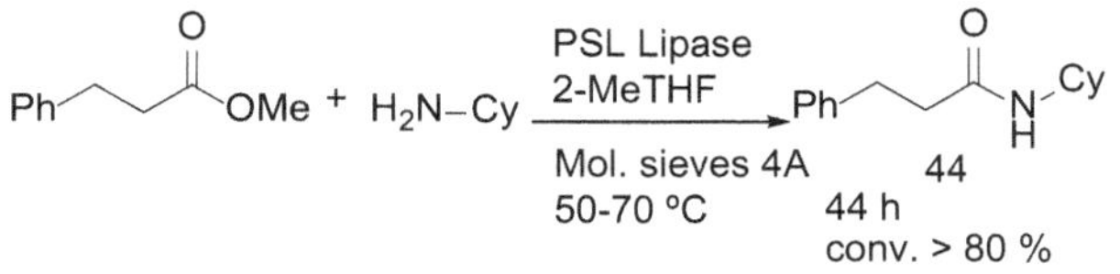 Ph-CH$_2$CH$_2$-CO-OMe + H_2N–Cy → (PSL Lipase, 2-MeTHF, Mol. sieves 4A, 50-70 °C, 44 h) Ph-CH$_2$CH$_2$-CO-NH-Cy 44; conv. > 80 %	van Pelt et al.[154]
#22	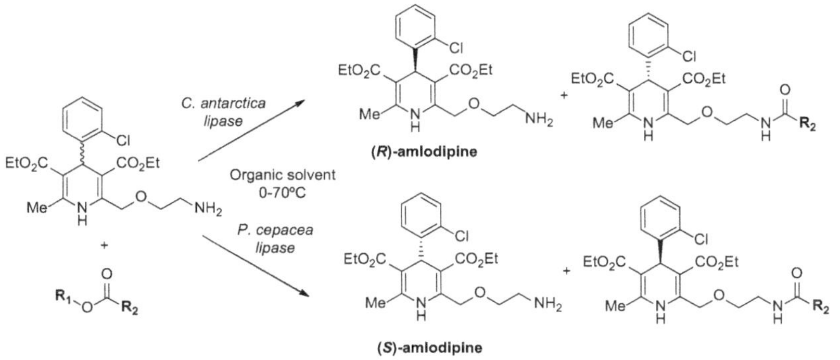 *C. antarctica* lipase / *P. cepacea* lipase; Organic solvent 0-70ºC; (*R*)-amlodipine; (*S*)-amlodipine. R_1 = Me, Et, n-Pr, i-Pr, n-Bu, i-Bu, t-Bu, n-pentyl, n-hexyl, isoamyl, 2-methoxyethyl, Cy, CH_2X, CHX_2or CX_3(wherein X = F, Cl, Br,); R_2 = Me, Et, n-Pr, i-Pr, n-Bu, i-Bu, t-Bu, linear or ramified C5-10 alkyl, Cy, Ph, *o*-, *m*- or *p*-monosubstituted Ph, polysubstituted Ph or heteroaryl	Filipan et al.[155]
#23	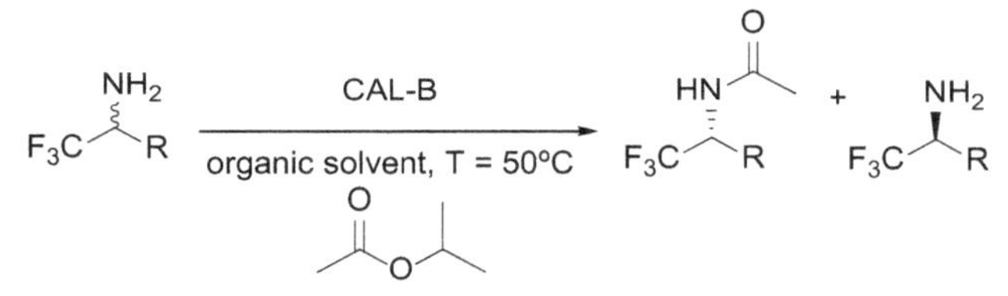 CAL-B, organic solvent, T = 50ºC	Cheng et al.[156]
#24	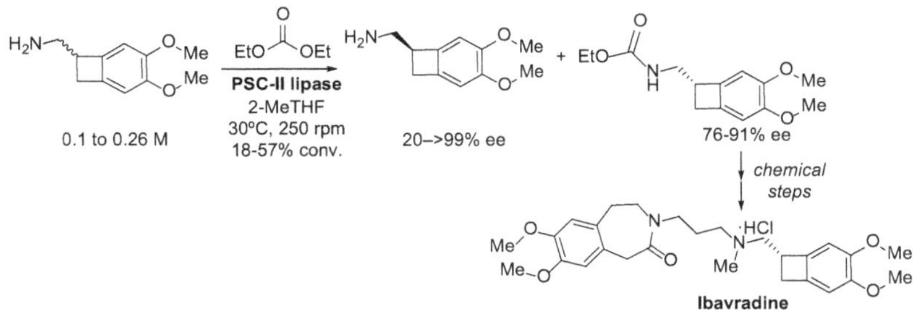 0.1 to 0.26 M; EtO-CO-OEt; **PSC-II lipase**, 2-MeTHF, 30ºC, 250 rpm, 18-57% conv.; 20–>99% ee; 76-91% ee; chemical steps; **Ibavradine**	Pedragosa-Moreau et al.[157]
#25	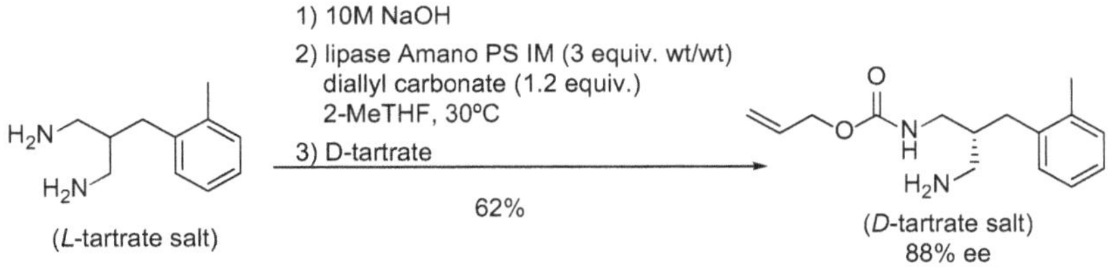 (*L*-tartrate salt); 1) 10M NaOH 2) lipase Amano PS IM (3 equiv. wt/wt) diallyl carbonate (1.2 equiv.) 2-MeTHF, 30ºC 3) D-tartrate; 62%; (*D*-tartrate salt) 88% ee	Lindhagen et al.[158]
	Phospholipase-catalyzed transphosphatidylation	
#26	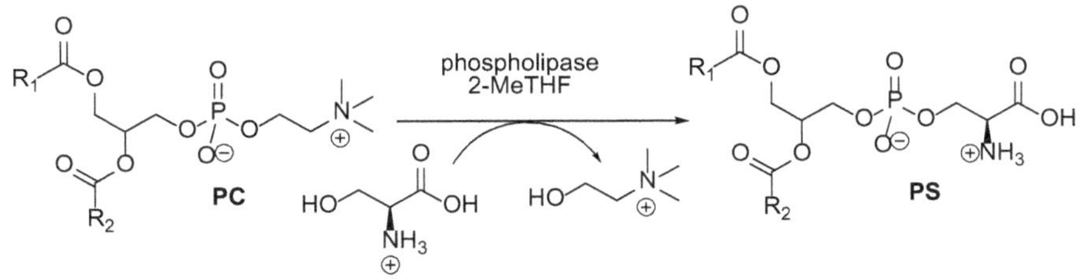 **PC** → (phospholipase, 2-MeTHF) **PS**	Duan and Hu[159]

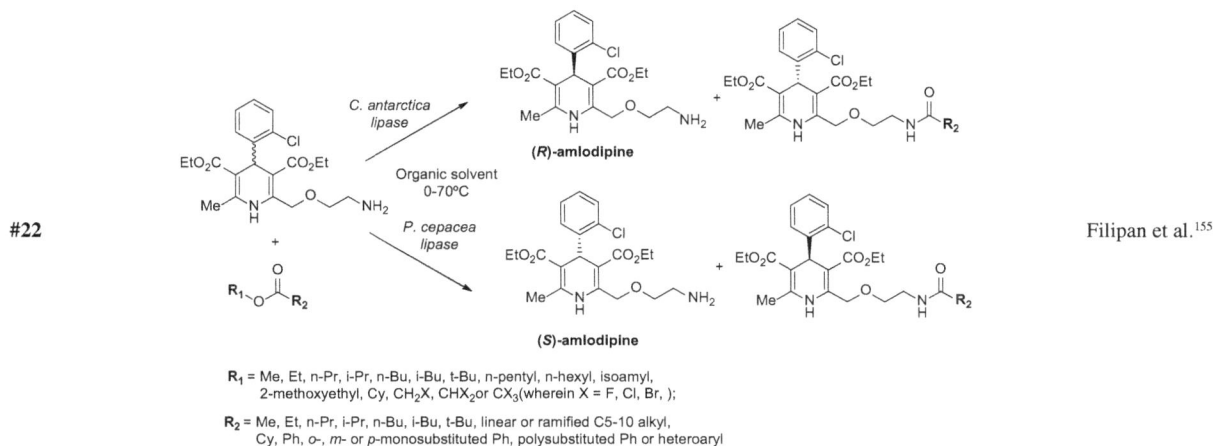

(*Continued*)

TABLE 6.1 (Continued)

Entry	Reaction Type Lipase-Catalyzed Acylation of Lineal Alcohols	Reference
	Lipase-catalyzed hydrolysis in biphasic media	
#27	tri-*O*-acetyl isatoribine → (Novozyme 435; phosphate buffer pH 6.3-6.5/ 2-MeTHF, 25-27°C, 16 h) → isatoribine pro-drug	Gallou et al.[160]
#28	R_1= Me, *t*-Bu, Bn; R_2= Me, *i*-Pr → (H_2O, lipase; 2-MeTHF, 28°C; conv. 40-53%, E = 2-93) → 38–55%, ee 15–94% + 11–50%, ee 17–96%	Torres et al.[161]
	Protease-catalyzed hydrolysis in biphasic media	
#29	licarbazepine methoxyacetate → (Protex 6L protease; 2-MeTHF/buffer pH 7.3, 30°C, conv. 53%) → 99.85–99.99% ee + 85% ee; → (chemical steps) → (*S*)-licarbazepine (eslicarbazepine)	Husain et al.[162]
	***O*-glycosylations**	
#30	andrographolide + glucose → (β-glucosidase, 2-MeTHF/water (90/10), 72 h, 40°C) → 68%	Chen et al.[163]
	Transglycosidations	
#31	R = H, OH; uridine + adenine → (NPs; phosphate buffer/co-solvent) → adenosine + uracil	Montilla Arevalo et al.[164]

Entry	Reaction Type — Lipase-Catalyzed Acylation of Lineal Alcohols	Reference
	Bioreductions	
#32	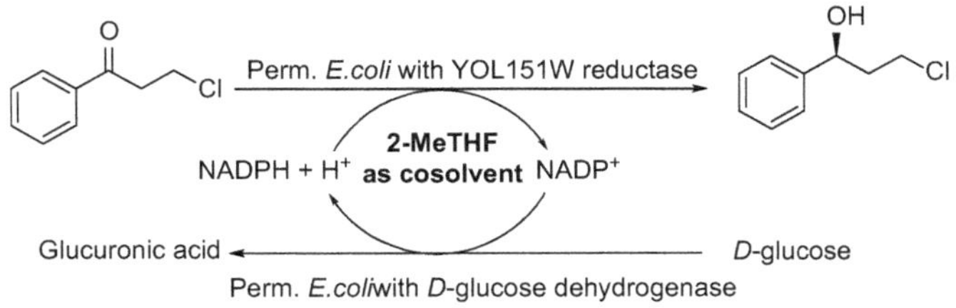 Perm. *E.coli* with YOL151W reductase; 2-MeTHF as cosolvent; NADPH + H^+ / $NADP^+$; Glucuronic acid ← D-glucose, Perm. *E.coli* with D-glucose dehydrogenase	Tian et al.[165]
#33	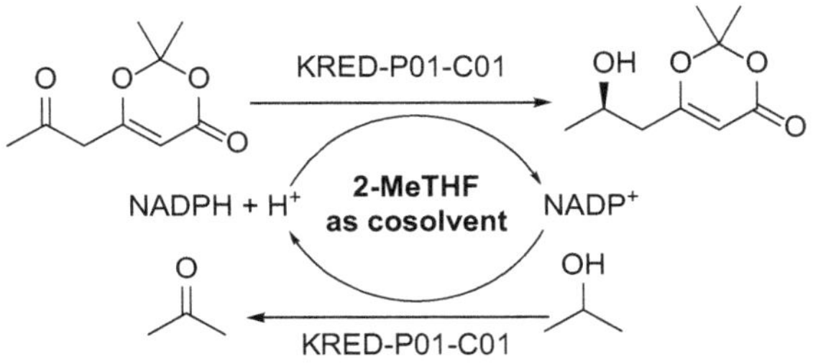 KRED-P01-C01; 2-MeTHF as cosolvent; NADPH + H^+ / $NADP^+$; KRED-P01-C01	Betori et al.[166]
#34	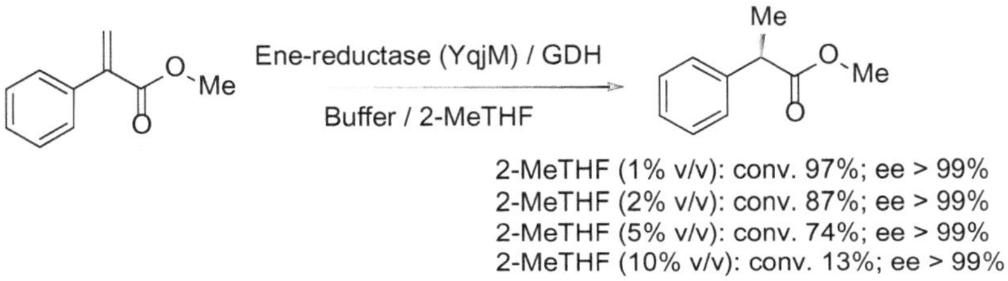 Ene-reductase (YqjM) / GDH; Buffer / 2-MeTHF 2-MeTHF (1% v/v): conv. 97%; ee > 99% 2-MeTHF (2% v/v): conv. 87%; ee > 99% 2-MeTHF (5% v/v): conv. 74%; ee > 99% 2-MeTHF (10% v/v): conv. 13%; ee > 99%	Pietruszka et al.[167]
	Carboligations	
#35	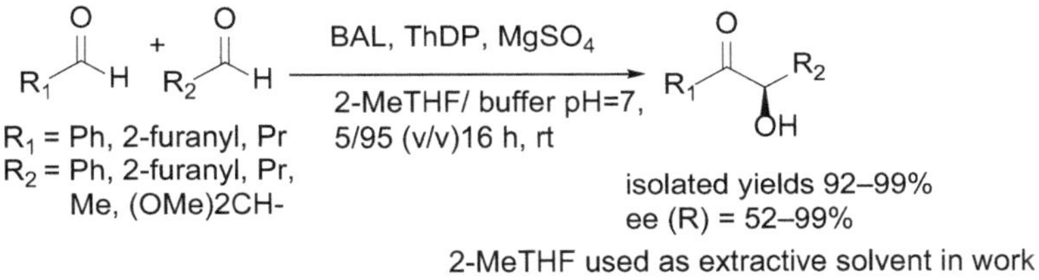 BAL, ThDP, $MgSO_4$; 2-MeTHF/ buffer pH=7, 5/95 (v/v)16 h, rt R_1 = Ph, 2-furanyl, Pr R_2 = Ph, 2-furanyl, Pr, Me, (OMe)2CH- isolated yields 92–99% ee (R) = 52–99% 2-MeTHF used as extractive solvent in work	Shanmuganathan et al.[168]
#36	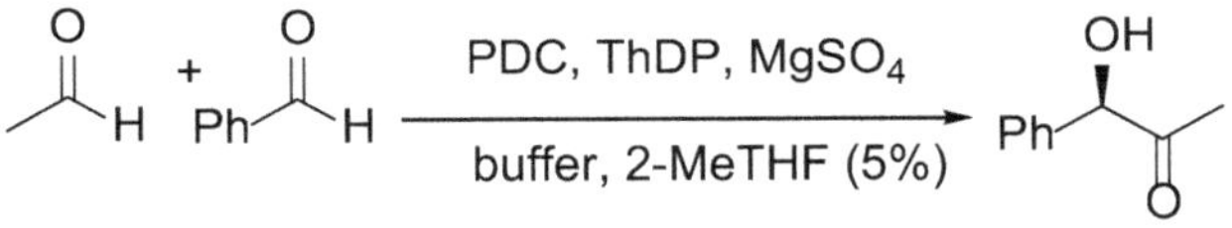 PDC, ThDP, $MgSO_4$; buffer, 2-MeTHF (5%)	Gerhards et al.[169]
	Biocatalytic cascades	
#37	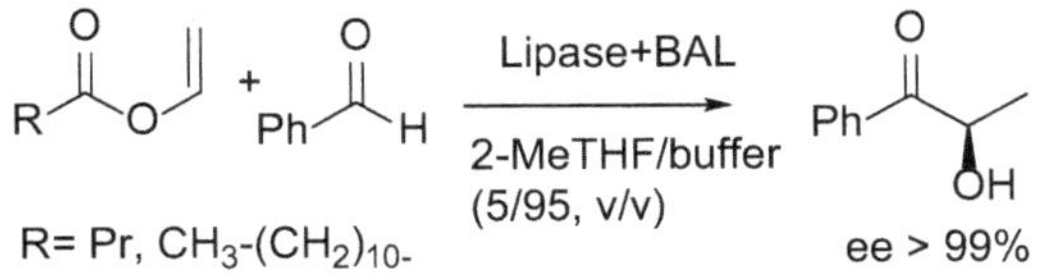 Lipase+BAL; 2-MeTHF/buffer (5/95, v/v) R= Pr, CH_3-$(CH_2)_{10}$- ee > 99%	Perez-Sanchez et al.[170]
#38	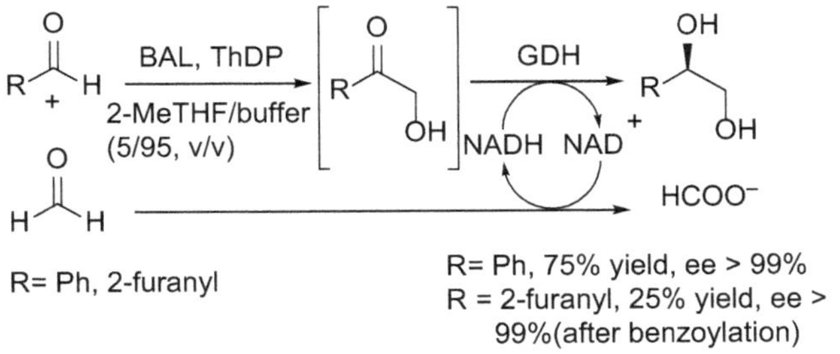 BAL, ThDP; 2-MeTHF/buffer (5/95, v/v); GDH; NADH / NAD+; $HCOO^-$ R= Ph, 2-furanyl R= Ph, 75% yield, ee > 99% R = 2-furanyl, 25% yield, ee > 99%(after benzoylation)	Shanmuganathan et al.[171]

(*Continued*)

TABLE 6.1 (Continued)

Entry	Reaction Type Lipase-Catalyzed Acylation of Lineal Alcohols	Reference
#39	2 **Ar**–OH —PpAOX (O_2 → H_2O_2; Catalase → H_2O + 1/2 O_2)→ 2 **Ar**–CHO —PfBAL→ **Ar**–CO–CH(OH)–**Ar** 13-96% yield, 92->99% ee **Ar** = Ph,*p*-OMe, *p*-OAC, *p*-Me,*p*-Cl, *m*-Me, 2-furyl	Schmidt et al.[172]
	Multicomponent reactions	
#40	NH_2 (**R₁**) + H–C=O (**R₂**) + (**R₃**, **R₄**) — dried yeast, 2-MeTHF, r.t. → **R₁** = H, Me, NO_2, Cl **R₂** = H, *p*-OMe, *m*-NO_2, *p*-Cl **R₃**, **R₄**= H, Me **R1=R2=R3= R4**=H, 75% (30h) **R1=R3=R4**=H; **R2**=*p*-Cl, 79% (48h) **R1**=Cl;**R2**=*m*-NO2; **R3=R4**=H, 70% (36h) **R1**=NO2; **R2**=*p*-OMe; **R3=R4**=Me, 72% (36h) **R1**=OMe; **R2**=H; **R3=R4**=H, 68% (24h) **R1**=Me; **R2**=*H*; **R3=R4**=H, 60% (28h)	Fang[173]
#41	H_2N–CO–NH_2 + CH₃COCH₂CO–**R** + H–C=O (**R₁**) — dried yeast, 2-MeTHF, r.t. → **R**= OEt, **R₁**=H,83% (24h) **R**= OEt, **R₁**=*p*-OMe, 56% (48h) **R**= OEt, **R₁**=3,4 di-OMe, 62% (30h) **R**= OEt,**R₁**=*p*-Cl, 66% (30h) **R**= OEt, **R₁**=*p*-NO_2, 79% (30h) **R**= Me, **R₁**=*o*-OH, 56% (30h)	Fang[174]

6.2.1 Lipase-Catalyzed Acylation of Alcohols in 2-MeTHF

As can be seen, most of the reported applications of 2-MeTHF in biotransformations are ***lipase-catalyzed acylation*** of alcohols, as these enzymes are highly stable and selective in organic solvents.[175–183] In this sense, Secundo and coworkers[132] have assessed the utility of 2-MeTHF in the transesterification of racemic sulcatol (**entry #01**), racemic α-cyclogeraniol (**entry #09**) or (±)-menthol (**entry #10**) with vinyl acetate, catalyzed by several commercial lipases. Similarly, Liu et al.[133] evaluated some 2-MeTHF and some other organic solvents for the resolution of rac-2-octanol via transesterification with vinyl acetate catalyzed by *Yarrowialipolytica* lipase (YLL) immobilized onto magnetic nanoparticles (**entry #02**). 2-MeTHF has also been shown to be an excellent solvent for the kinetic resolution of benzoins (Hoyos et al.,[134] Aires-Trapote et al.,[136] **entry #06**) as well as for their dynamic-kinetic resolution (Hoyos et al.,[134] **entry #03**; Nieguth et al.,[135] **entry #04**; Aires-Trapote et al.,[136] **entry #05**).

The lipase-catalyzed acylation of alcohols has also been focused in the ***preparation of APIs***. Thus, the preparation of *N*-Fmoc-doxorubicin-14-*O*-dicarboxylic acid mono-ester derivatives (intermediates in the synthesis of analogs of LHRH, luteinizing hormone-releasing hormone, to treat prostate cancer[184]) starting from *N*-Fmoc-doxorubicin and *bis*-acyl donor compounds (dicarboxylic acids or anhydrides) has been reported using lipases (CAL-B gave the best results) in different mixtures of organic solvents, as depicted in Table 6.1, **entry #07**.[137] Thus, the role of one of them is to facilitate azeotropic distillation (by removing water to drive the reaction to completion), while the other one is chosen to ensure lipase activity and a sufficient solubility of the starting materials. Herein, 2-MeTHF was used mixed with MEK, acetone, CPME, anisole, THF, MIBK, NMP, *t*AmylOH, *t*BuOH, cyclopentane, or Et2O, as well as in ternary mixtures. Overall, good yields (> 90%) were obtained, proceeding at a relatively high temperature for a biocatalytic procedure (90°C). In another relevant example, Bristol Myers Squibb has focused on chiral pyrimidinyl-piperidinyl-oxypyridone structures as tentative agonists of the G protein-coupled receptor 119 (GPR119, also named "glucose-dependent insulinotropic receptor"), a recent potential target for the development of oral antidiabetic drugs.[185] To that end, the scalable synthesis of enantiomers of *N*-substituted 3-hydroxypyrrolidin-2-ones has been recently reported, as resumed in Table 1, **entry #11**.[140] Lipase from *Pseudomonas cepacia* immobilized on polypropylene (PS30) catalyzed the enantioselective esterification of *rac*-1-(2-fluoro-4-iodophenyl)-3-hydroxypyrrolidin-2-one with succinic anhydride and 2-MeTHF at 4°C, leading to (*S*)-non-acylated derivative in high enantiomeric excess (ee) >99% and yield ~40%, after straightforward separation from the esterified (*R*)-derivative. Following the initial experiments, it was possible to scale-up to 50 g/L of substrate. After 8.5 h, immobilized enzyme was removed by simple filtration and the (*S*)-alcohol was isolated in ~40% yield and ee>99%. Remarkably, 2-MeTHF served both as a reaction medium for the biotransformation and as an extraction solvent to the desired compound from the reaction mixture, eliminating the use of chromatography.

A new concept recently proposed is the formation of **CO_2-expanded phases with 2-MeTHF**, leading to sustainable solvents with tailored properties.[138,139] In fact, the addition of CO_2 (up to 10 bar) to 2-MeTHF leads to solvents with altered properties (e.g., changes in the polarity and the hydrophobicity), by expanding the carbon dioxide in the organic solvent (the so-called CO_2-expanded liquid, CXLs). For biocatalysis, the set-up of CXL using 2-MeTHF leads to largely improved kinetic resolutions of bulky substrates, compounds that otherwise would display low or no activity in 2-MeTHF, as shown in Table 1, **entry #08**. It is believed that some flexibilization of the protein structure occurs, thus enabling the acceptance of extremely bulky substrates. This represents a promising proof-of-concept to open new biocatalytic alternatives.

Following the seminal work of Simeó et al.,[131] several examples of ***regioselective acylation of nucleosides and analogs*** (to increase the lipophilicity of the parent drug and to improve its pharmacokinetics[186,187]) using lipases on 2-MeTHF have been reported, e.g., the acylation of 8-choloroadenosine[141,142] (**entry # 13**), pyrimidine nucleosides[143] (**entry #14**) or cordycepin[144,145] (**entry #15**).

Modification of natural products may be an area of use for 2-MeTHF as well. Thus, polydatin (3,4,5-trihydroxystilbene-3-β-*D*-glucopyranoside) is one of the most common pharmacological constituents isolated from the root and rhizome of a traditional Chinese medicinal plant of *Polygonum cuspidatum*, and has been widely applied in anti-inflammatory, anti-oxidant and anti-angiogenesis chemotherapy.[188] The preparation of 6-*O*-acyl-polydatin prodrugs in 2-MeTHF has been described through lipase-catalyzed acylation (Table 1, **entry #16**) using different acyl donors.[146–148] The lipophilic 6”-O-sorboyl-polydatin exhibited an improved apoptosis-inducing capability compared to the parent drug. Similarly, the preparation of a lipophilic prodrug of gastrodin (one of the major active ingredients obtained from *Gastrodiaelata* Blume) has been reported by the enzymatic undecylenoylation in 2-MeTHF (Table 1, **entry #17**).[149] The highest catalytic activity and regioselectivity toward 7″-ester was reported using *P. cepacia* lipase and a cosolvent mixture of THF and 2-MeTHF (3/1, v/v), leading to a conversion >99%, with a regioselectivity of 93% for the major ester, and a good reaction rate (60.6 mM/h).

In the field of ***carbohydrate acylations***, Hu et al. described the use of a mixture of 2-MeTHF and t-butanol as the best reaction medium for the enzymatic synthesis of ascorbyl undecylenate (table 1, **entry #18**), a fat-soluble antioxidant, via transesterification of ascorbic acid (Vitamin C) with the corresponding vinyl ester, with high conversion (up to 95%) and total regioselectivity.[150,151] Alternatively, also mixtures of 2-methyl-2-butanol and 2-MeTHF may be feasible for the enzymatic synthesis of ascorbyl esters since the two tertiary alcohols have highly similar physicochemical properties. In these mixtures, 2-MeTHF played an important role in the enhancement of enzymatic acylation by lowering the apparent activation energy and increasing the affinity between enzyme and substrates, while the tertiary alcohols mainly aided the dissolution of the polar substrate. The approach was subject to a patent application[189] and different lipases were successfully assessed (from *C. antarctica*, *Thermomyces lanuginosus*, *Rhizomucor miehei*, *P. cepacia*, *C. rugosa*, *P. fluorescens* or *Penicillium expansum*). The same authors[151,190] reported the enzymatic synthesis of other ascorbyl fatty acid esters using directly vegetal oils as acyl donors for transesterification. Novozyme 435 resulted again in useful catalysts, and the best solvent was a binary mixture of 2-MeTHF and *tert*-butanol, leading to conversions of 70–73% after 24 h, with unsaturated fatty acid esters (oleate and linoleate, 80–90%) being the major products. The immobilized lipase kept the relative activity of 80% after reuse for six batches in the 2-MeTHF-containing system.

Likewise, Kanerva and coworkers[153] described the regioselective acylation of methyl α-*D*-galacto-, -gluco- and mannopyranosides with fluorinated β-lactams (catalyzed by lipases, to furnish 6-*O*-acylated glycopyranoside–β-amino acid conjugates (Table 1, **entry #20**). The use of β-lactams as acyl donor makes the process irreversible, rendering the (*S*)-6-*O*-acylated sugar and a mixture of (*R*)-lactam and (*S*)-β-amino acid (from background hydrolysis). In the initial screening, several immobilized lipases were tested, being *Burkholderia cepacia* lipase adsorbed on celite (lipase PS-D) the best option. The organic solvents tested were *tert*-amyl alcohol (*t*AmOH, water-soluble) and water-insoluble MTBE and 2-MeTHF. Although the starting methyl β-glycopyranosides were poorly soluble in these last two solvents, the presence of other reagents and products positively shifted the equilibrium. Conversion (53–60% after 48 h) and optical purity (98–99%) were similar for the three solvents. Remarkably, 2-MeTHF was also used by these authors in the chemical synthesis of racemic azetidinones via a modified Reformatsky addition.[153]

Chen et al.[152] have studied the enzymatic acylation of a lily polysaccharide (LP, one of the most active constituents of the bulbs of *Lilium lancifolium* Thunb, used to make traditional Chinese medicines and tonics) catalyzed by immobilized PSL-C (ceramic support) with vinyl acetate in organic solvents, Room-Temperature Ionic Liquids (RTILs) and RTILs-containing systems (Table 6.1, **entry #19**). In this case, the polysaccharide studied was composed of β-1,4-linked *D*-glucopyranose and *D*-mannopyranose units and the molar ratio of glucose to mannose is 2 to 1. The degree of substitution (DS, average number of acyl groups per sugar unit) of the modified LP was used to evaluate the extent of acylation and thus enzymatic activity. Best results were obtained in a mixture composed of 20% (v/v) 2-MeTHF and 1-butyl-3-methylimidazolium tetrafluoroborate ($[C_4MIm][BF_4]$). In fact, the RTIL improved the solubility of the starting LP, while the proper proportion of 2-MeTHF was crucial for lowering the viscosity of the reaction medium, without decreasing the enzymatic activity.

Lipase-catalyzed* N*-acylations (amidations) have been reported using with 2-MeTHF as well. Thus, van Pelt et al.[154] reported the use of *P. stutzeri* lipase for aminolysis reactions with bulky substrates using 2-MeTHF among other solvents, with excellent results; remarkably, the use of molecular sieves showed a significant effect on the aminolysis rate and amide yield, since it enabled the effective removal of the inhibiting by-product methanol from the reaction mixture. In another example, Filipan et al.[155] reported the kinetic resolution of racemic amlodipine

((*R*,*S*)-2-(2-amino-ethoxy-methyl)-3-ethoxy-carbonyl-4-(2-chlorophenyl)-5-methoxy-carbonyl-6-methyl-1,4-dihydropyridine), a representative of 1,4-dihydropyridine drugs used in the treatment of hypertension, angina pectoris and other vascular diseases, by *N*-acylation with carboxylic acid esters (Table 6.1, entry #22). Although 2-MeTHF was not the final choice for these acylations (toluene was used), the example was the first one in which 2-MeTHF was tested for amide synthesis. Analogous results were observed for the enzymatic resolution of several α-trifluoromethylated amines *via* kinetic resolution (KR) using Novozyme 435 and isopropyl acetate,[156] as shown in **entry #23**. 2-MeTHF led to moderate conversions and enantioselectivity, while best results were obtained with toluene, a solvent in which a DKR was developed by coupling the KR with racemization with palladium over Al_2O_3.

The ***formation of carbamates from amines by lipase-catalyzed acylation with carbonates*** is a very common strategy.[191] In this context, lipase from *P. cepacia* (PSC-II) was selected among some others for the kinetic resolution of racemic (3,4-dimethoxybicyclo [4.2.0]octa-1,3,5-trien-7-yl) methanamine *via* alkoxycarbonylation with diethyl carbonate, as shown in **entry #24**. 2-MeTHF was the best solvent to obtain an (*S*)-carbamate, subsequently converted into a (*S*) secondary amine, which was further submitted to reductive alkylation with benzazepine to furnish Ivabradine, an alternative drug for the treatment of stable angina pectoris in cases of intolerance or contraindications for β-blockers. It was possible to isolate enantiopure Ivabradine in a 30% overall yield.[157] In an analogous example, researchers at AstraZeneca have developed a scalable route to obtain (*R*)-allyl-(3-amino-2-(2-methylbenzyl)propyl) carbamate in four synthetic steps starting from 2-methylbenzyl chloride, in 32% overall yield (**entry #25**). The route was optimized for multi-kg production and the key-enzymatic step, namely the desymmetrization of 2-(2-methylbenzyl)propane-1,3-diamine using Amano PS-IM and diallyl carbonate, was carried out in 2-MeTHF, the best solvent after a solvent screening.[158]

2-MeTHF has also been used for enzyme-mediated ***trans-phosphatidylation of phosphatidylcholine (PC) with L-serine to synthesize phosphatidylserine (PS)***, a membrane phospholipid component with broad applications in the pharmaceutical and functional food (Table 1, entry **#26**).[192] After an enzymatic screening, the best results were obtained using PLD-1 (phospholipase D from *Streptomyces chromofuscus*) at 40°C, leading to a 90% yield of phosphatidylserine after 12 h without any by-product (phosphatidic acid).[193] This proof-of-concept paper triggered further research to assess the enzymatic synthesis of PS in other biosolvents such as terpenes (limonene and *p*-cymene,[103]) or CPME.[194]

6.2.2 Enzyme-Catalyzed Hydrolysis Using 2-MeTHF as (Co)solvent

2-MeTHF can be also used in ***lipase-catalyzed hydrolytic processes*** due to the low solubility of 2-MeTHF in water[65,66]; these systems could be biphasic systems or monophasic media in which 2-MeTHF behaves as cosolvent (< 5% v/v 2-MeTHF in water). For instance, isatoribine is a nucleoside analog potentially useful for the treatment of chronic hepatitis C and other viral infections.[195] Researchers from Novartis described a practical process for the synthesis of the isatoribine-prodrug, as shown in **entry #27**, in which regioselective enzymatic hydrolysis was the key step.[160] Initial hydrolysis was developed in acetone or *tert*-butanol and phosphate buffer. Under these conditions, completion was observed after 2 days, with substantial amounts of undesired mono and other diacetates. Furthermore, an overall low volume efficiency was observed (>50 L total organic solvent/kg product recovered), being not appropriate for large scale. The use of 2-MeTHF definitively proved to be highly beneficial, since the solubility of the final product was 50-fold higher than in MTBE, the extraction solvent used in the first-generation process. This led to a dramatic reduction of the solvent requirements, resulting in a direct upgrading in the reactor volume efficiency of ca. 30%. Moreover, the enzyme was stable in 2-MeTHF and negligible deactivation was observed even with relatively high substrate loadings (0.15 M) were set, leading to an overall 85% isolated yield after crystallization from MTBE/2-MeTHF. Finally, from a chemometric analysis, a promising >6-fold improvement in process mass intensity (PMI) was observed when 2-MeTHF was used as a cosolvent.

In another recent example, Rebolledo and coworkers[161] reported a chemoenzymatic approach for the synthesis of optically active 4-(3-acetoxyphenyl)-5-(alkoxycarbonyl)-6-methyl-3,4-dihydropyridin-2-ones (3,4-DHP-2-ones) and their hydroxyphenyl derivatives, the key step being a lipase-catalyzed hydrolysis reaction using 2-MeTHF as a cosolvent (**entry #28**). 3,4-DHP-2-ones, components of natural products,[196,197] are compounds possessing many therapeutic effects.[198–200] The first attempts to catalyze the acylation of the phenol moiety of these molecules with several lipases in 2-MeTHF were not satisfactory,[161] but the enzymatic hydrolysis of the phenolic esters in 2-MeTHF/water (99/1, V/V) led to good results by using lipase from *C. rugosa* (CRL), with good optical purity (ee = 94–99%) and yields.

Apart from lipases, ***other hydrolases*** are also compatible with 2-MeTHF as (co)solvent. For instance, in **entry #29**, a patent from Matrix Laboratories Limited[162] is illustrated. In this case, a protease is used to catalyze the hydrolysis of a racemic methoxy-acetate ester of licarbazepine to produce (*S*)-licarbazepine (eslicarbazapine), an active metabolite of oxcarbazepine (Trileptal®, Novartis), an anticonvulsant drug primarily used in the treatment of epilepsy.[201] After an enzymatic screening including several hydrolases (lipases and proteases), best results were obtained with Protex 6L protease in a mixture of 2-MeTHF/buffer to produce the (*R*)-alcohol (removed by chemical hemi-succinate formation also in 2-MeTHF) and the (*S*)-ester, which was subsequently hydrolyzed with NaOH in a mixture EtOH/2-MeTHF to finally afford eslicarbazepine (**entry #29**). Further studies reported also other biocatalytic procedures, such as the lipase-catalyzed acylation of licarbazepine (lipase from *C. rugosa*, vinyl benzoate and MTBE as solvent[202]) or the bio-reduction starting from oxcarbazepine with evolved keto reductases (aqueous medium with *iso*-propanol as auxiliary substrate[203]). Likewise, some glycosidases are also active 2-MeTHF/water systems for the glycosylation of molecules hardly soluble in water. Thus, β-glucosidase from almonds catalyzes the glycosylation of andrographolide (a diterpenoid lactone and a major constituent of *Andrographis paniculata* Nees, with potential antitumoral[204] or anti-inflammatory[205] activities), to render a glycoconjugate with enhanced water solubility, as shown in **entry #30**.[163] Also, nucleoside phosphorylases (NPs, EC 2.4.2.X) catalyze the reversible phosphorolysis/synthesis of nucleosides, namely a transglycosylation transferring the carbohydrate moiety from one heterocyclic base to another[206,207]). These enzymes are frequently used for synthesizing nucleosides and analogs, broadly applied as pharmaceuticals.[186,187,208] In many cases, it is required to use an organic cosolvent when either the substrates and/or the products bear low solubility in an aqueous medium. In this context, Montilla et al.[164] claimed the use of 2-MeTHF as a cosolvent in the transglycosylation of (desoxi) uridine with adenine to produce (desoxi) adenosine, using two enzymes from thermophilic archaea, a UPase from *Aeropyrumpernix* and a PNPase from *Sulfolobus solfataricus*, either isolated or expressed in *Escherichia coli* cells, at temperatures ranging from 60% to 100%, and 5–10% v/v organic cosolvent (**entry #31**). Similarly, Almendros et al.[209] reported on five putative encoding genes of *Thermus thermophilus* HB27, cloned and expressed in *E. coli*, two of them showing phosphorolytic activities against purine nucleosides and a third one showing phosphorolytic activity against pyrimidine nucleosides *in vitro*. Thus, nucleoside phosphorolysis assays were performed as phosphorolytic reactions on 2′-deoxyinosine, 2′-deoxyadenosine or 2′-deoxyuridine (30 min of incubation at 80°C, 50% V/V organic cosolvent), confirming better enzymatic stabilities in 2-MeTHF compared to DMSO or DMF.

6.2.3 Bioreductions Using 2-MeTHF as (Co)solvent

2-MeTHF has started to be considered a reference solvent in other biocatalytic processes as well, being frequently part of solvent screening programs. Moreover, industries are incorporating it when novel enzymatic processes are studied, for instance, for Baeyer–Villiger mono-oxygenases,[210] while novel alcohol dehydrogenases (ADHs) identified in the metagenomic studies[211] revealed high tolerance to 2-MeTHF.

The use of permeabilized whole-cells containing reductases for the stereoselective bioreduction of the prochiral ketone 3-chloro-1-phenyl-1-propanone with 2-MeTHF as cosolvent has been reported recently, as shown in **entry #32**.[165] This process was designed for the preparation of (*S*)-3-chloro-1-phenylpropanol, to afford a building block to furnish antidepressant drugs such as fluoxetine, tomoxetine or nisoxetine. Thus, two types of permeabilized recombinant cells were used simultaneously, namely *E. coli* containing the YOL151W reductase from *Saccharomyces cerevisiae* and another *E. coli* containing *D*-glucose dehydrogenase, for cofactor regeneration purposes (NADPH). The low solubility of the substrate was solved with the use of 2-MeTHF as a cosolvent (1–7% v/v), together with a surfactant (Triton X-100). Albeit substrate loadings were not very high (up to 60 mM), yields (98%) and enantioselectivities (> 99%) were excellent. Another recent work has focused on the enantioselective reduction of β-ketodioxinones (**entry #33**) by means of ketoreductases to afford β-hydroxydioxinones,[166] useful building blocks to produce a variety of natural products. In this case, different organic cosolvents were used (5–10% v/v), showing that the ketoreductase (a commercial enzyme provided by Codexis, KRED-P01-C01) was active when biosolvents like 2-MeTHF or CPME were used. In the case of 2-MeTHF, 74% yield and 98% ee were obtained).

Likewise, ene-reductases have been assessed with 2-MeTHF as well, for the synthesis of profen derivatives for the pharmaceutical industry, as shown in entry **#34**. In this case, ene-reductase YqjM from *Bacillus subtilis* (a homolog of the old yellow enzyme) was used, and low amounts of 2-MeTHF as a cosolvent (1–2% v/v) led to high conversions and excellent enantiomeric excess.[167] The enzyme was deleterious to higher amounts of cosolvent, and conversion decreased accordingly.

6.2.4 Use of Lyases and 2-MeTHF as (Co)solvent

2-MeTHF has been also used as a cosolvent (5% v/v) in aqueous-based biocatalytic processes, being an efficient alternative to DMSO and MTBE in benzaldehyde lyase (BAL) catalyzed reactions (**entry #35**), leading to quantitative yields and very high enantiomeric excesses of acyloins.[168] Similarly, the use of pyruvate decarboxylase (PDC) under the same reaction conditions (5% v/v of 2-MeTHF as (co)solvent) allowed the stereoselective synthesis of (*R*)-2-hydroxy-1-phenyl-propan-1-one ((*R*)-HPP, **entry #36**), a very useful chiral building block.[169]

6.2.5 Biocatalytic Cascades Using 2-MeTHF as (Co)solvent

2-MeTHF is very useful in different multistep enzymatic processes. The set-up of multistep reactions diminishes the wastes generated during downstream units, thus being nowadays a subject of intense research. In this area, Domínguez de María and coworkers[170] have evaluated the combination of lipases and lyases using 2-MeTHF as cosolvent, reporting excellent conversions and enantioselectivities for the preparation of (*R*)-HPP (**entry #37**). The same authors[171] reported the combination of BAL and glucose dehydrogenase (GDH) for the preparation of optically pure diols (**entry #38**). Following that line, in a recent example, Schmidt et al.[172] described a one-pot multistep enzymatic oxidation of aliphatic and benzylic alcohols to the corresponding aldehydes (catalyzed by a PpAOX from *Pichia pastoris* coupled to catalase) combined with a subsequent carboligation (mediated by benzaldehyde lyase from *P. fluorescens*) to furnish chiral α-hydroxy ketones (**entry #39**); in this system, 5% of 2-MeTHF resulted as the best option for solubilizing the substrates. Through this, the direct use of the reactive aldehyde intermediates is avoided.

On the other hand, biocatalytic multicomponent reactions catalyzed by whole cells have been recently assessed in 2-MeTHF, focusing on heterocycle synthesis. In that respect, a recent patent described a biocatalyzed three-component one-pot Hantzsch reaction for synthesizing different polyhydroacridines, using 2-MeTHF as the solvent.[173] Thus, by reacting aromatic aldehydes with aromatic amines and 1,3-cyclohexanedione derivatives at a molar ratio of 1:1:2 in 2-MeTHF, catalyzed by dry baker's yeast (5-12 wt.% of raw materials) at room temperature and pressure, differently substituted polyhydroacridines were obtained, with yields ranging from 60% to 79% after purification and recrystallization with ethanol (no ee values were reported) (**entry #40**). The process had been previously reported in phosphate buffer and glucose, under fermentative conditions,[212] but not in the pure organic solvent. Based on a previously reported yeast-catalyzed Biginelli synthesis in fermentative conditions,[213] the same author reported[174] a method for synthesizing 2,4-dihydropyrimidin-2-ones mixing urea, aromatic aldehydes and a 1,3-dicarbonylic compound with yields ranging from 56% to 83% (**entry #41**).

For both three-component one-pot syntheses, yields were similar to those reported in fermentative conditions. Albeit the processes conducted in organic solvents required longer reaction times, it may be expected that a more straightforward work-up procedure can be set, diminishing the waste formation associated with the extraction from water when fermentative processes are implemented.

6.2.6 Recent Examples of Biocatalysis in 2-MeTHF

In a very recent paper, Peris et al.[214] have reported the combination of multiple and consecutive multicatalytic steps in a single cascade combining organocatalytic supported ionic-liquid-like phases (SILLPs) with Novozym 435 for the preparation of optically pure chiral cyanohydrins (Scheme 6.2).

These cyanohydrins are very useful building blocks for the ulterior synthesis of a plethora of pharmaceuticals and agrochemicals.[215,216] The process depicted in Scheme 6.2 implies three consecutive steps and four catalytic reactions: an organocatalytic cyanosylilation of different aromatic aldehydes to furnish racemic *O*-sylilated α-aromatic acetonitriles (**step-1**), its transformation into the corresponding esters via two consecutive reactions (hydrolysis of the cyanosylil ether and subsequent acetylation, **step-2**) and, finally, enzymatic kinetic resolution of the racemic cyanohydrin ester by transesterification with propanol (**step-3**). The authors reported the optimization of each individual step under batch conditions, using SILLPs catalysts 1 and 2 shown in Scheme 6.2 for the initial steps (yields ranging from 95% to 99%); the enzymatic kinetic resolution was carried out using Novozym 435 and 2-MeTHF (yields from 52-57%, ee higher than 99% in all cases). The authors reported that the best SILLP catalyst for the second step was different from that used in the batch process; with this modification, they were able to obtain an excellent continuous process, leading to a noticeable space-time yield (STY) of 124 g g^{-1} cat h^{-1} L^{-1}, comparable to that obtained with hydroxynitrileliases and HCN[217] (a highly hazardous reagent), in a much greener manner.

Another recent example of lipase-catalyzed kinetic resolution using 2-MeTHF has been reported in a recent patent,[218] depicted in Scheme 6.3, for the preparation of an enantiopure precursor of Ozanimod (Zeposia®), a sphingosine-1-phosphate (S1P) receptor agonist acting as an immunomodulatory drug for the treatment of relapsing multiple sclerosis.[219]

In this sense, the kinetic resolution of racemic 1-amino-2,3-dihydro-1*H*-indene-4-carbonitrile was the key step for introducing the desired chirality. This was carried out by *N*-acylation with ethyl 2-methoxyacetate in 2-MeTHF, once again using commercial Novozym 435 as the catalyst, obtaining pure (*S*)-amine in an overall 20% yield in the reductive amination starting from the corresponding carbonylic compound.

Very recently, 2-MeTHF has become very useful in polymerizations to furnish polyesters. Thus, lipase-catalyzed preparation of adipate- and furan-2,5-dicarboxylate-based polyesters, using several 2-MeTHF and other neoteric solvents (2,2,5,5-tetramethyloxolane, 2,5-dimethyltetrahydrofuran and pinacolone) as substitutes for traditional solvents such as PhMe or THF, has been reported.[220] These polymers were prepared by enzymatic esterification of both diacids with 1,4-butanediol (BDO) or 1,8-octanediol (ODO), as shown in Scheme 6.4.

The solvent selection was based on their capability for dissolving both the monomers and the resulting polymer while retaining enzymatic activity. Additionally, the solvents should ideally be relatively volatile (70–139°C range, as set by the CHEM21 solvent-selection guide[128]), therefore following their easy removal from reaction media. In fact, solvents with higher boiling points would make distillation more energy-demanding. No significant differences were observed between the new

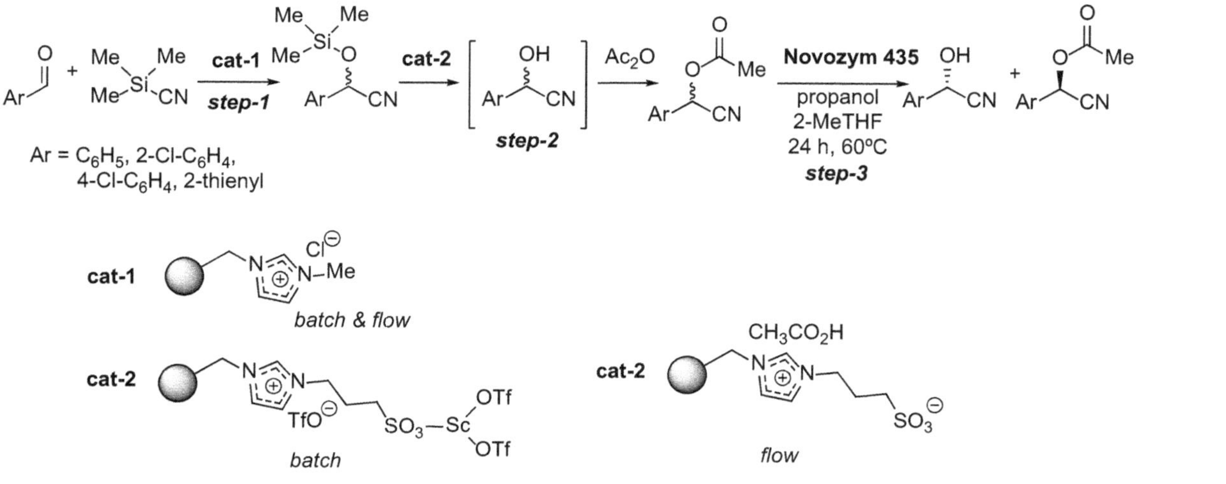

SCHEME 6.2 Chemoenzymatic cascade for the preparation of optically pure cyanohydrins.

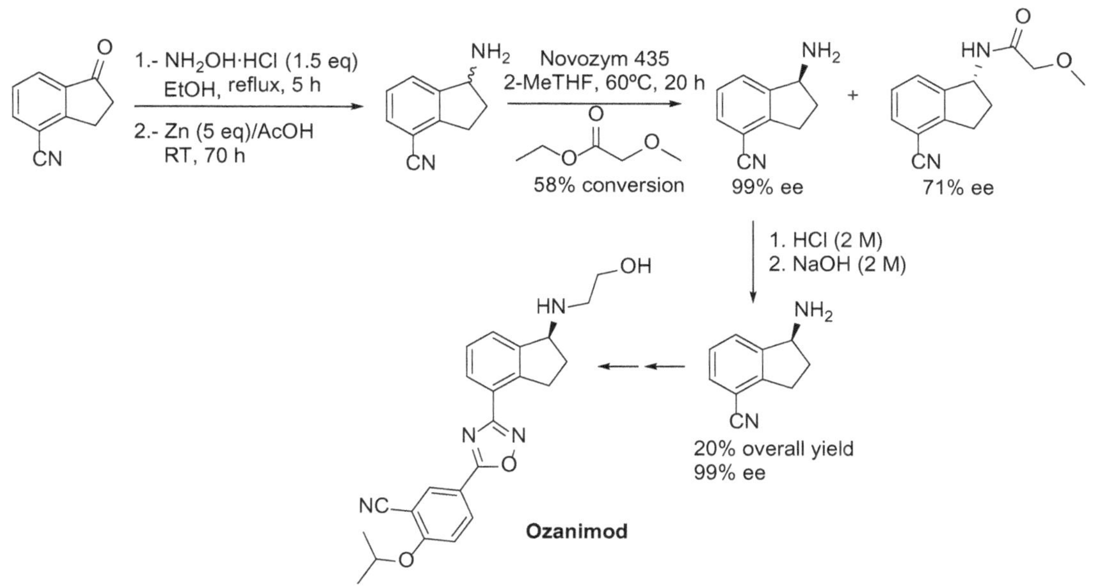

SCHEME 6.3 Chemoenzymatic synthesis of enantiopure cascade for the preparation of optically pure cyanohydrins.

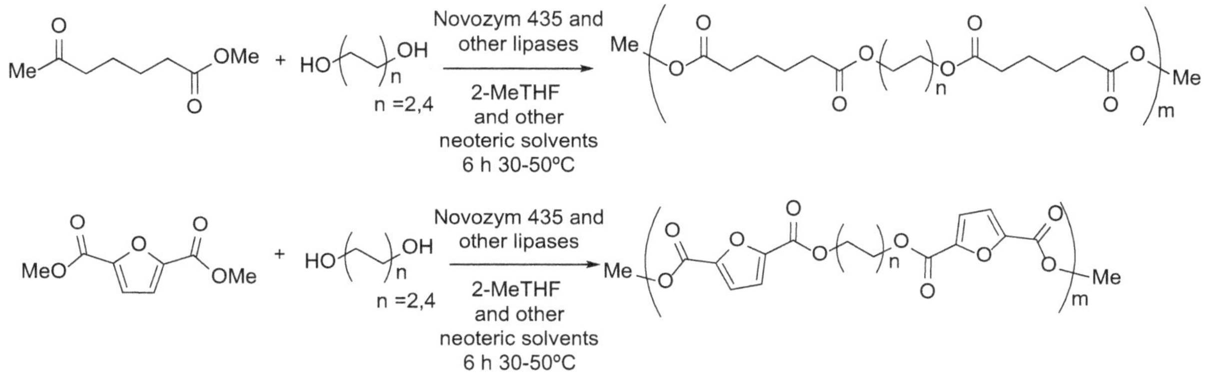

SCHEME 6.4 Enzymatic synthesis of polyesters in 2-MeTHF.

solvents in the reaction of dimethtyl adipate and diols catalyzed by Novozym 435 (best of the lipases tested) at reaction temperatures of 30°C and 50°C, in all cases being comparable or even better than conventional THF or PhMe, in terms of monomer conversions and molecular mass of the polymers. Synthesis of furan-derived polymers generally gave lower monomer conversions, although all the neoteric solvents allowed better conversions than THF of PhMe, with 2-MeTHF being one of the best alternatives.

In another example, Englezou et al.[221] have assessed the use of 2-MeTHF for ring-opening polymerizations (ROPs), comparing the lipase-catalyzed opening ofε-caprolactone (oxepan-2-one) with methyl-polyethylene glycol (mPEG) versus the ROP of lactide (lactone cyclic di-ester derived from lactic acid) with DBU (1,8-diazabicyclo[5.4.0]undec-7-ene), as shown in Scheme 6.5.

These linear amphiphilic block copolymers (polylactides or polycaprolactones) have been used for different biomedical applications,[222] after nanoprecipitation to generate nanoparticles, as they are biocompatible, mechanically strong and can be obtained starting from natural feedstocks. The use of 2-MeTHF as a solvent reduced the carbon footprint in the polymerization, leading to excellent results compared to classical solvents such as DBM or THF.

2,5-bis-(hydroxymethyl)furan (BHMF, Scheme 6.6) is an extremely attractive building block, easily obtainable by reduction of 5-(hydroxymethyl)furfural (HMF), an important and versatile bioderived platform chemical obtainable from renewable C5-C6 carbohydrate feedstocks[223] (see Scheme 6.1). In fact, BHMF is used as a building block for polymers, foams and crown ethers, and as an intermediate in the preparation of different pharmaceuticals [224]. The reduction of HMF to BHMF can be performed using different procedures; the employing of stoichiometric amounts of metal hydride salts such as $LiAlH_4$ had been the archetypical methodology,[225] although the generation of great amounts of waste and the high cost of metal hydrides prevent this methodology for being used at industrial scale. Thus, a great variety of catalytic pathways (conventional hydrogenation, hydrogen transfer procedures, photocatalytic and electrocatalytic reductions) have

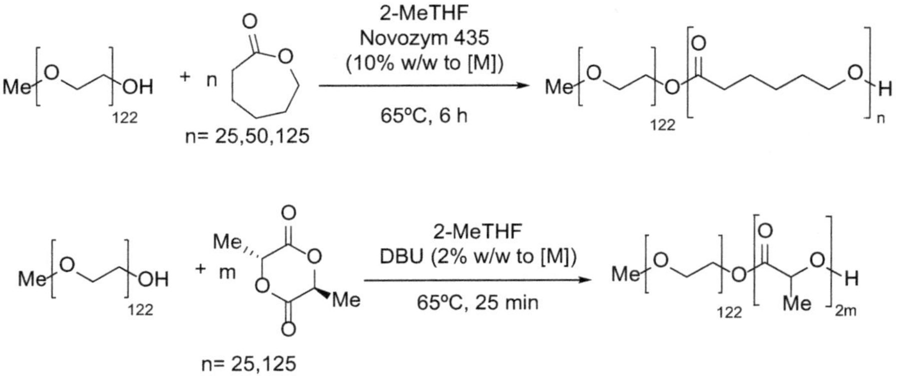

SCHEME 6.5 Enzymatic versus chemical synthesis of amphiphilic copolymers.

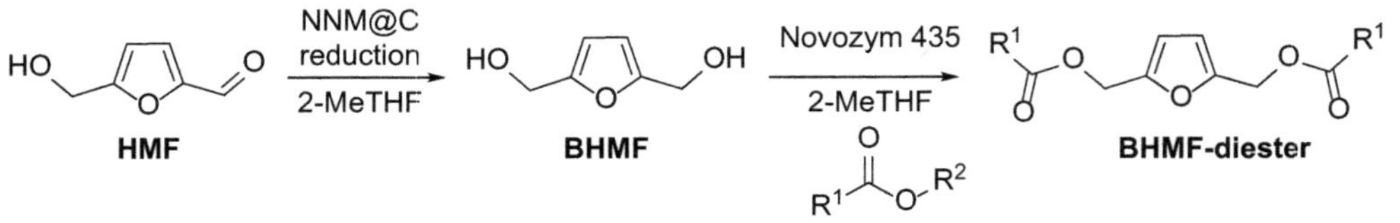

SCHEME 6.6 Chemoenzymatic synthesis of BHMF diesters.

been proposed[226–228]; additionally, also biocatalytic reductions,[229] mainly using whole cells,[230–232] have been reported for the transformation of HMF into BHMF. Very recently, Arias et al.[233] have described a chemoenzymatic protocol for the preparation of BHMF-derived diesters, through a first catalytic reduction using non-noble metal nanoparticles (NNM@C) and subsequent esterification of the diol using Novozym 435 (Scheme 6.6).

BHMF-diesters are bio-based plasticizers that can act as an alternative to petroleum-derived phthalates[229]; the use of 2-MeTHF as the solvent allowed the implementation of the cascade reaction depicted in Scheme 6.4, as it acts as an effective solvent both for the chemical and the enzymatic steps. In a batch mode reactor, using hexanoic acid as acyl donor and Novozym 435, esterification of BHMF led to 99% yield after 11 h reaction time. In the overall process starting from HMF, the chemical reduction using Co@C catalyst was carried out in 2-MeTHF at 110°C for 10 h; after that time, the catalyst was removed with a magnet, and subsequently, hexanoic acid, Novozym 435 and molecular sieves (to eliminate the water molecules produced upon esterification) were added, lowering the reactor temperature down to 35°C. Excellent results were reported (total yield 91%, total selectivity 96%) and similar performance was observed using butyric or octanoic acid, while acetic acid turned out not to be adequate for the esterification, due to its lower pKa. When conducting repetitive batches, a deactivation of both catalysts was observed, so that reaction time had to be increased; additionally, changing form esterification with carboxylic acids to transesterification with vinyl hexanoate led to much better results, as long as excellent yield (97%) and selectivity to BHMF dihexanoate (100%) were obtained in only 0.5 h, affording an overall diester yield of 89%. In a further step, the process was implemented in a continuous-flow reactor by coupling two fixed-bed reactors, one containing Co@C and the other the immobilized lipase, which afforded an overall yield of the desired diester close to 90% and remained stable for 60 h of operation.

In any case, the enzymatic esterification of BHMF with fatty acids (10:0, 12:0, 14:0, 16:0 or 18:0) in 2-MeTHF had been previously reported by Lăcătuş et al.[234] These authors pointed out that Novozym435 was the best catalyst, by comparing its performance with that from five other immobilized enzyme preparations commonly used for biofuel productions, also observing that the small water amount present in the enzymatic preparations was enough to allow the hydrolysis of diesters when incubated in dry 2-MeTHF. These authors optimized reaction variables such as molar ratio, temperature, reaction time, fatty acid alkane length or the amount of catalyst, to finally scale up to a preparative scale using capric, lauric, myristic, palmitic or stearic acid. This same group has recently reported the double esterification of BHMF with a fatty acid mixture (FAM) resulting from the hydrolysis of commercial sunflower oil.[235] Novozym 435 and 2-MeTHF were again selected for this esterification, although finally it was observed that conducting the reaction under solvent-free conditions (solid BHMF is scarcely soluble in the FAM, although as the reaction is progressing, the gradually formed diesters were acting as solvents) and removing the water under reduced pressure allowed the complete conversion of BHMF into esters at 60°C (Scheme 6.7).

In this sense, the kinetic resolution of racemic 1-amino-2,3-dihydro-1*H*-indene-4-carbonitrile was the key step for introducing the desired chirality. This was carried out by *N*-acylation with ethyl 2-methoxyacetate in 2-MeTHF, once again using commercial Novozym 435 as the catalyst, obtaining pure (*S*)-amine in an overall 20% yield in the reductive amination starting from the corresponding carbonylic compound.

Very recently, 2-MeTHF has become very useful in polymerizations to furnish polyesters. Thus, lipase-catalyzed preparation of adipate- and furan-2,5-dicarboxylate-based

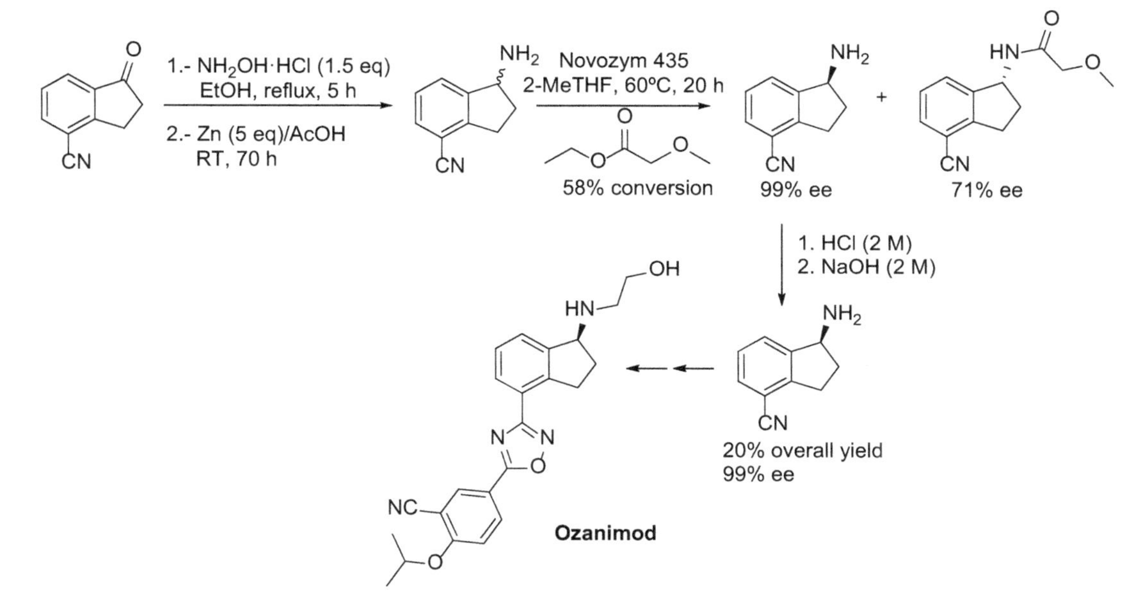

SCHEME 6.7 Chemoenzymatic synthesis of enantiopure cascade for the preparation of optically pure cyanohydrins.

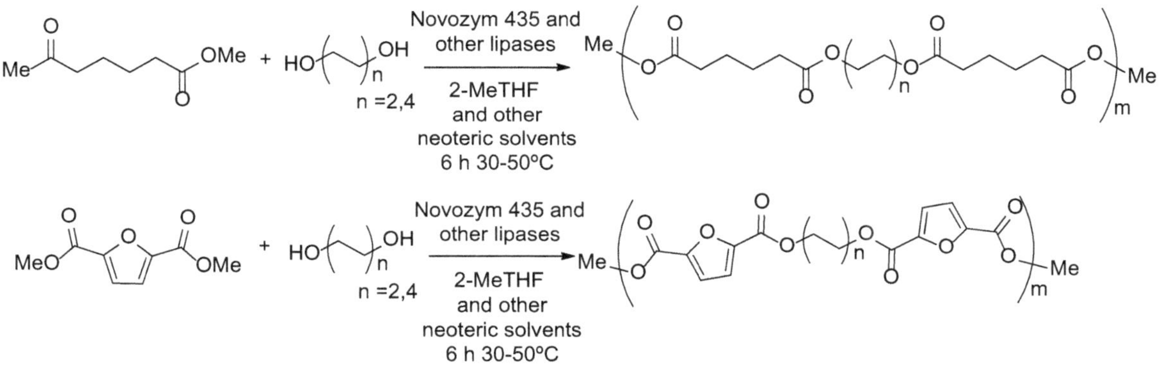

SCHEME 6.8 Enzymatic synthesis of polyesters in 2-MeTHF.

polyesters, using several 2-MeTHF and other neoteric solvents (2,2,5,5-tetramethyloxolane, 2,5-dimethyltetrahydrofuranand pinacolone) as substitutes for traditional solvents such as PhMe or THF, has been reported.[220] These polymers were prepared by enzymatic esterification of both diacids with 1,4-butanediol (BDO) or 1,8-octanediol (ODO), as shown in Scheme 6.8.

Solvent selection was based on their capability for dissolving both themonomers and the resulting polymer, while retaining enzymatic activity. Additionally, the solvents should ideally be relatively volatile (70–139°C range, as set by the CHEM21 solvent selection guide[128]), therefore llowing their easy removal from reaction media. In fact, solvents with higher boiling points would make distillation more energy-demanding. No significant differences were observed between the new solvents in the reaction of dimethtyl adipate and diols catalyzed by Novozym 435 (best of the lipases tested) at reaction temperatures of 30 and 50°C, in all cases being comparable or even better than conventional THF or PhMe, in terms of monomer conversions and molecular mass of the polymers. Synthesis of furan-derived polymers generally gave lower monomer conversions, although all the neoteric solvents allowed better conversions than THF of PhMe, with 2-MeTHF being one of the best alternatives.

In another example, Englezou et al.[221] have assessed the use of 2-MeTHF for ring opening polymerizations (ROPs), comparing the lipase-catalysed opening ofε-caprolactone (oxepan-2-one) with methyl-polyethylene glycol (mPEG) versus the ROP of lactide (lactone cyclic di-ester derived from lactic acid) with DBU (1,8-diazabicyclo[5.4.0]undec-7-ene), as shown in Scheme 6.9.

These linear amphiphilic block copolymers (polylactides or polycaprolactones) have been used for different biomedical applications,[222] after nanoprecipitation to generate nanoparticles, as they are biocompatible, mechanically strong and can be obtained starting from natural feedstocks. The use of 2-MeTHF as a solvent reduced the carbon footprint in the polymerization, leading to excellent results compared to classical solvents such as DBM or THF.

2,5-bis-(hydroxymethyl)furan (BHMF, Scheme 6.10) is an extremely attractive building block, easily obtainable by reduction of 5-(hydroxymethyl)furfural (HMF), an important and versatile bioderived platform chemical obtainable from renewable

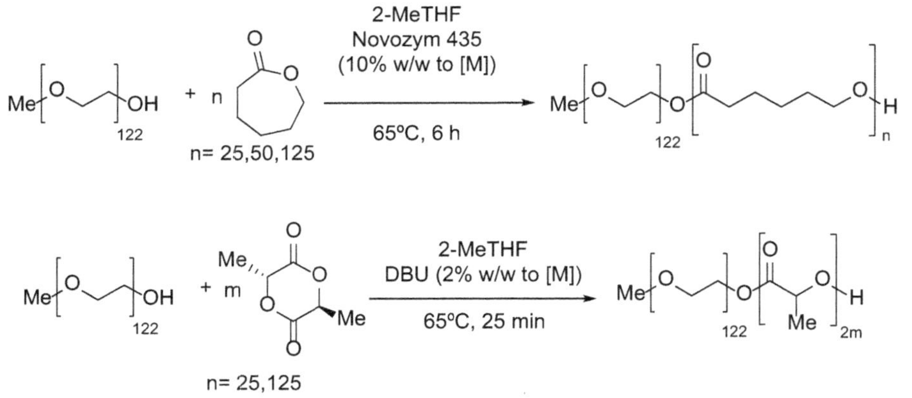

SCHEME 6.9 Enzymatic versus chemical synthesis of amphiphilic copolymers.

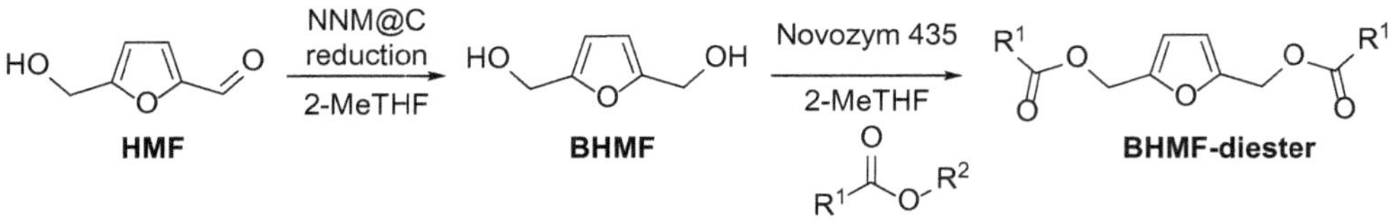

SCHEME 6.10 Chemoenzymatic synthesis of BHMF diesters.

C5-C6 carbohydrate feedstocks[223] (see Scheme 6.1). In fact, BHMF is used as building block for polymers, foams and crown ethers, and as an intermediate in the preparation of different pharmaceuticals 224. The reduction of HMF to BHMF can be performed using different procedures; the employ of stoichiometric amounts of metal hydride salts such as $LiAlH_4$ had been the archetypical methodology,[225] although the generation of great amounts of waste and the high cost of metal hydrides prevents this methodology for being used at industrial scale. Thus, a great variety of catalytic pathways (conventional hydrogenation, hydrogen transfer procedures, photocatalytic and electrocatalytic reductions) have been proposed[226–228]; additionally, also biocatalytic reductions,[229] mainly using whole cells,[230–232] have been reported for the transformation of HMF into BHMF. Very recently, Arias et al.[233] have described a chemoenzymatic protocol for the preparation of BHMF-derived diesters, through a first catalytic reduction using non-noble metal nanoparticles (NNM@C) and a subsequent esterification of the diol using Novozym 435 (Scheme 6.10).

BHMF-diesters are biobased plasticizers that can be act as an alternative to petroleum-derived phthalates[229]; the use of 2-MeTHF as solvent allowed the implementation of the cascade reaction depicted in Scheme 6.4, as it acts as an effective solvent both for the chemical and the enzymatic steps. In in a batch mode reactor, using hexanoic acid as acyl donor and Novozym 435, esterification of BHMF led to 99% yield after 11 h reaction time. In the overall process starting from HMF, the chemical reduction using Co@C catalyst was carried out in 2-MeTHF at 110°C for 10 h; after that time, the catalyst was removed with a magnet, and subsequently hexanoic acid, Novozym 435 and molecular sieves (to eliminate the water molecules produced upon esterification) were added, lowering reactor temperature down to 35°C. Excellent results were reported (total yield 91%, total selectivity 96%) and similar performance was observed using butyric or octanoic acid, while acetic acid turned out not to be adequate for the esterification, due to its lower pKa. When conducting repetitive batches, a deactivation of both catalysts was observed, so that reaction time had to be increased; additionally, changing form esterification with carboxylic acids to transesterification with vinyl hexanoate led to much better results, as long as excellent yield (97%) and selectivity to BHMF dihexanoate (100%) were obtained in only 0.5 h, affording an overall diester yield of 89%. In a further step, the process was implemented in a continuous-flow reactor by coupling two fixed-bed reactors, one containing Co@C and the other the immobilizedlipase, which afforded an overall yield of the desired diester of close to 90% and remained stable for 60 h of operation.

In any case, the enzymatic esterification of BHMF with fatty acids (10:0, 12:0, 14:0, 16:0 or 18:0) in 2-MeTHF had been previously reported by Lăcătuş et al.[234] These authors pointed out that Novozym435 was the best catalyst, by comparing its performance with that from five other immobilized enzyme preparations commonly used for biofuel productions, also observing that the small water amount present in the enzymatic preparations was enough to allow the hydrolysis of diesters when incubated in dry 2-MeTHF. These authors optimized reaction variables such as molar ratio, temperature, reaction time, fatty acid alkane length or amount of catalyst, to finally scale up to a preparative scale using capric, lauric, myristic, palmitic or stearic acid. This same group have recently reported the double esterification of BHMF with a fatty acid mixture (FAM) resulting from the hydrolysis of commercial sunflower oil.[235] Novozym 435 and 2-MeTHF were again selected for this esterification, although finally it was observed that conducting the reaction under solvent-free conditions (solid BHMF is scarcely soluble in the FAM, although as the reaction is progressing the gradually formed diesters were acting as solvents) and removing the water under reduced pressure allowed the complete conversion of BHMF into esters at 60°C.

6.3 Glycerol and Glycerol-Derived Solvents

Glycerol is undoubtedly a biogenic product broadly available, as the production of biodiesel starting from raw materials is increasingly growing[236,237] and the uses of glycerol are expanding.[238] Properties of glycerol as a biosolvent have been commented elsewhere in this book (see Chapter 12, Greener Synthesis of Potential Drugs, by Studzińska et al.), where some examples of its use in biotransformations can be found. As commented there, glycerol has been used as a solvent in redox whole-cells biocatalyzed processes to solve the problem caused by the low solubility of generally bulky prochiral ketones in aqueous reaction culture, thus leading to low yields. Thus, Wolfson and coworkers reported the reduction of prochiral β-keto esters and ketones by adding yeast powder or powder or immobilized yeast to pure glycerol medium to give high yield and high ee values.[239,240] Anyhow, glycerol is generally used as a cosolvent in aqueous media for bioreductions catalyzed by whole cells[241] or by isolated enzymes.[242–244] As also mentioned in Chapter 12 by Studzinska et al., glycerol can be used as a biogenic component of Natural Deep Eutectic Solvents (NDES), formed by a eutectic mixture of Lewis (or Brønsted) acids and bases. In this sense, the use of choline chloride/glycerol (ChCl/Gly) in biotransformations has increased in the last years, as already illustrated in Chapter 12. Particularly, the mixture ChCl/Gly at 1/2 (mol mol^{-1}), also known as glyceline, has been frequently used.[245]

As commented before, glycerol can be easily derivatized to furnish glycerol-based solvents (GBSs), which have received attention in the last years.[50,54] These solvents have been frequently used mixed with aqueous buffers; thus, a remarkable improvement in the synthesis of carbohydrates through glycosidase catalyzed processes was reported by Perez et al.,[246] showing how the presence of GBSs (ethereal type) compared to buffered media considerably decreased the hydrolytic activity of the β-galactosidase from *E. coli* in the synthesis of *N*-acetyl-*D*-allolactosamine *via* transglycosylation of p-nitrophenyl β-*D*-galactopyranoside as donor and *N*-acetylglucosamine as acceptor. Similar results were observed in the synthesis of *N*-acetyl-*D*-lactosamine by a β-galactosidase from *T. thermophilus*[247]; in fact, using GBSs as cosolvents, it was possible to reduce the amount of undesired self-condensation product derived from the donor [Gal-β(1→3)-Gal-β-pNP], leading to an increase in the synthesis of the target disaccharide. This fact was attributed to an alteration of the secondary and tertiary structure of the enzyme, causing a modification in the protein flexibility, which enabled the proper interaction of the substrate in the active site of the enzyme.[247] Analogously, Perez-Sanchez et al.[248,249] reported the effect of these GBSs in the activity of β-galactosidase from *B. circullans*, leading to *N*-acetyl-*D*-allolactosamine as the main product (quantitative conversion) versus *N*-acetyl-*D*-lactosamine, the major product when the same reaction is performed in buffered media. Some other examples of the use of ether-type GBSs as a cosolvent for enzymatic reactions have been reported so far.[250–252]

6.4 Gamma-Valerolactone (GVL)

γ-Valerolactone (GVL, see Scheme 6.1), a naturally occurring chemical found in fruits and generally used as a food additive, can be considered an archetypical sustainable liquid,[253] as it can be obtained from renewable sources,[69] and displays very useful properties: low melting (−31°C), high boiling (207°C) and flash (96°C) points, a definitive but acceptable smell for easy recognition of leaks and spills, nontoxicity and high water solubility, and therefore favoring its biodegradation; furthermore, its vapor pressure is 0.65 kPa at 25°C and it only increases to 3.5 kPa at 80°C. On the other hand, it does not suffer hydrolysis under neutral conditions and can be considered safe, as it did not form any measurable amount of peroxides in a glass flask kept under air for several weeks. Additionally, its extraction from reaction mixtures is very easy, as its high solubility in water allows the use of a certain amount of water for solubilizing and removing it; afterward, a simple distillation would separate GVL from water, allowing its reuse in another reaction cycle.[253]

Thus, GVL has been used as a green solvent in many cases (e.g., see the recent review of Vaccaro[254] and Gerardy et al.[223] and references cited therein); nevertheless, its use as a solvent for biotransformations is relatively scarce. Indeed, in different cases, GVL has been incorporated into the group of bio-based and green solvents tested for a particular biocatalyzed process, although its performance has not been the best. In this way, Zhu et al.[255] reported the use of recombinant *E. coli* cells harboring nitrile hydratase activity for transforming 2-amino-2, 3-dimethylbutyronitrile (ADBN) into 2-amino-2, 3-dimethylbutyramide (ADBA) in various green solvent/aqueous reaction systems, including GVL, but best results were obtained with methoxyperfluorobutane/H_2O (v/v, 10%) biphasic system. Iemhoff et al. included GVL in their assessment of the performance of Novozyme 435 in the esterification of racemic 2-phenylpropionic acid with EtOH in different bio-based solvents, although they selected p-cymene as the best option. Similarly, Paggiola et al. have recently investigated different bio-based solvents for the biocatalyze damidation reactions of various ester-amine combinations by *P. stutzeri* lipase[101]; once again, GVL was tested, although terpene-based solvents (terpinolene, p-cymene, *D*-limonene) were demonstrated to perform better.

Anyhow, there is an interesting reported example in which GVL was used for a biocatalytic process, more concretely for the phospholipase-mediated transphosphatidylation of phosphatidylcoline (PC) with *L*-serine (shown in Scheme 6.11) to furnish phosphatidylserine (PS). Among the phospholipases tested, the best results were obtained using the enzyme from *S. chromofuscus*, which allowed the preparation of PS, a compound with many applications in functional food and pharma industries, with 95% yield at 40°C.

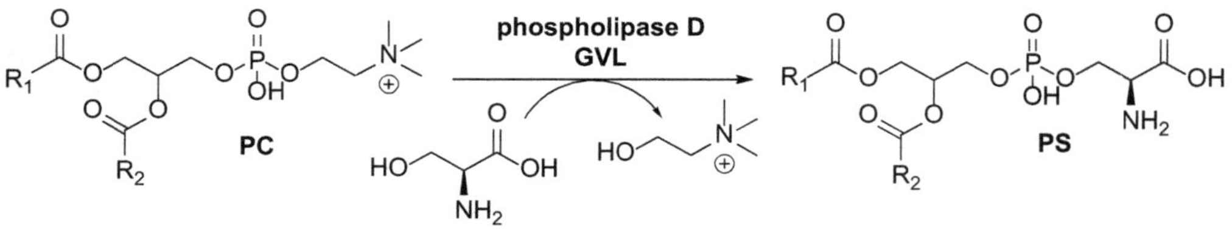

SCHEME 6.11 Phospholipase-mediated transphosphatidylation of phosphatidylcoline (PC) with *L*-serine.

6.5 Dihydrolevoglucosenone

6,8-Dioxabicyclo[3.2.1]octanone, also known as dihydrolevoglucosenone (or Cyrene™, name registered by Merck) has been synthesized from a number of different biomass starting materials under a variety of conditions, but most of these processes are initially from levoglucosenone (LGO, (1*S*,5*R*)-6,8-dioxabicyclo[3.2.1]oct-2-en-4-one), which in its turn is obtained from cellulose (see Scheme 6.1). Despite its structure being well known, the potential to use dihydrolevoglucosenone as a solvent has only been explored in the last years, as it can be considered a bio-available replacement for industrially relevant (but toxic) dipolar aprotic solvents, such as dimethylformamide (DMF), N-methyl-2-pyrrolidone (NMP), dimethylsulfoxide (DMSO) or dimethylacetamide (DMAc). In fact, dihydrolevoglucosenone is only barely eco-toxic and has dispersion parameters closest to DMSO, polarity closest to DMAc and hydrogen-bonding-like interactions similar toNMP. One of the key features of dihydrolevoglucosenone is its high miscibility with water, as it is in equilibrium with its ketal hydrate, in contrast to other ketones; additionally, the derivatization of dihydrolevoglucosenone with 1,2-diols allows the formation of novel solvent types (the so-called Cygnets solvents family). Dihydrolevoglucosenone has been used in many organic reactions, as recently reviewed, and shows also a great potential for liquid–liquid extractions. Regarding its use in biotransformations, it was included in some studies, such as the esterification of racemic 2-phenylpropionic acid with EtOH catalyzed by Novozyme 435 or the synthesis of β-sitosterol esters through the reaction of β-sitosterol and fatty acids using lipase from *C. rugosa*, but the results were not satisfactory. Nevertheless, very recently, Guajardo and Dominguez de María[67] have assessed its potential as a solvent in the regioselective esterification of glycerol with benzoic acid catalyzed by cross-linked aggregates of CALB (Scheme 6.12). The selection of alcohol and acid was based on their unpaired solubilities, leading to moderate-to-high conversions using conventional solvents; by adding a small amount of buffer (up to 2% v/v), a full conversion in the production of α-monobenzoate glycerol (α-MBG) was reported.

On the other hand, de Gonzalo has just published the use of Cyrene™ as the cosolvent in the bioreduction of α-ketoesters, using different commercial recombinant ADHs from Codexis, using either substrate- (*iso*propanol) or enzyme-coupled (glucose and glucose dehydrogenase, GDH) methodologies for cofactor recycling. When this solvent was employed in 2.5% v/v, it was feasible to furnish both the (*S*)- and the (*R*)-hydroxyesters (depending on the ADH used) with complete conversion. Interestingly, bioreductions catalyzed by the *iso*propanol-tolerant ADH KRED P2-D03 could be carried out using up to 30% v/v ofCyrene™ with only a small loss in the biocatalyst properties, whereas the substrate concentration can be increased up to 1.0 M, obtaining a higher conversion in the presence of this cosolvent compared to that reported in its absence. These promising results pave the way to increased use of this biosolvent in the near future.

6.6 Conclusion

As derived from all the data presented so far, a clear conclusion can be drawn: the potentiality of combining green biosolvents with biocatalysts for the sustainable production of high-valued compounds has not been fully exploited. Apart from 2-MeTHF, which has been thoroughly used and can be regarded as the archetypal biosolvent for biotransformations, there is still a long way ahead to consider the implementation

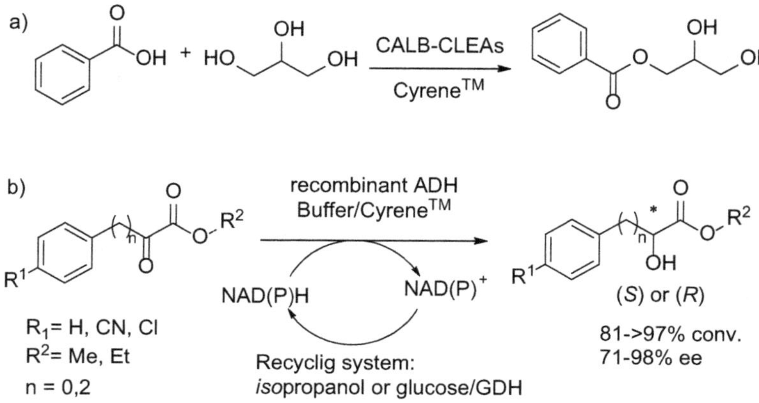

SCHEME 6.12 Recently reported biotransformations using dihydrolevoglucosenone (Cyrene™) as biosolvent.

of this green synergy as complete. We have presented different examples showing how it is possible to substitute classical and not environment-friendly organic solvents for more sustainable alternatives in enzyme-mediated processes, leading to excellent results; nevertheless, with the increasing demand for executing greener synthetic protocols, fostered by the increasing status of Circular Economy, we can foresee a vast increase in the implementation of biosolvents in the near future.

REFERENCES

1. Sheldon, R.A. Green solvents for sustainable organic synthesis: State of the art. *Green Chem.* 2005, *7*, 267–278.
2. Hackl, K.; Kunz, W. Some aspects of green solvents. *C. R. Chim.* 2018, *21*, 572–580.
3. Clark, J.H.; Farmer, T.J.; Hunt, A.J.; Sherwood, J. Opportunities for bio-based solvents created as petrochemical and fuel products transition towards renewable resources. *Int. J. Mol. Sci.* 2015, *16*, 17101–17159.
4. Clarke, C.J.; Tu, W.-C.; Levers, O.; Bröhl, A.; Hallett, J.P. Green and sustainable solvents in chemical processes. *Chem. Rev.* 2018, *118*, 747–800.
5. Constable, D.J.C.; Jimenez-Gonzalez, C.; Henderson, R.K. Perspective on solvent use in the pharmaceutical industry. *Org. Process Res. Dev.* 2007, *11*, 133–137.
6. Ratti, R. Industrial applications of green chemistry: Status, challenges and prospects. *SN Appl. Sci.* 2020, *2*, 7.
7. Sheldon, R.A. The greening of solvents: Towards sustainable organic synthesis. *Curr. Opin. Green Sustain. Chem.* 2019, *18*, 13–19.
8. Capello, C.; Fischer, U.; Hungerbuhler, K. What is a green solvent? A comprehensive framework for the environmental assessment of solvents. *Green Chem.* 2007, *9*, 927–934.
9. Jessop, P.G. Searching for green solvents. *Green Chem.* 2011, *13*, 1391–1398.
10. Cvjetko Bubalo, M.; Vidović, S.; Radojčić Redovniković, I.; Jokić, S. Green solvents for green technologies. *J. Chem. Technol. Biotechnol.* 2015, *90*, 1631–1639.
11. Calvo-Flores, F.G.; Monteagudo-Arrebola, M.J.; Dobado, J.A.; Isac-Garcia, J. Green and bio-based solvents. *Top. Curr. Chem.* 2018, *376*, 40.
12. Zhou, F.; Hearne, Z.; Li, C.J. Water—the greenest solvent overall. *Curr. Opin. Green Sustain. Chem.* 2019, *18*, 118–123.
13. Butler, R.N.; Coyne, A.G. Organic synthesis reactions on-water at the organic-liquid water interface. *Org. Biomol. Chem.* 2016, *14*, 9945–9960.
14. Hallett, J.P.; Welton, T. Room-temperature ionic liquids: Solvents for synthesis and catalysis. *Chemical Rev.* 2011, *111*, 3508–3576
15. Hulsbosch, J.; De Vos, D.E.; Binnemans, K.; Ameloot, R. Biobased ionic liquids: Solvents for a green processing industry? *ACS Sustain. Chem. Eng.* 2016, *4*, 2917–2931.
16. Lei, Z.G.; Chen, B.H.; Koo, Y.M.; MacFarlane, D.R. Introduction: Ionic liquids. *Chem. Rev.* 2017, *117*, 6633–6635.
17. Nasirpour, N.; Mohammadpourfard, M.; Heris, S.Z. Ionic liquids: Promising compounds for sustainable chemical processes and applications. *Chem. Eng. Res. Des.* 2020, *160*, 264–300.
18. Singh, S.K.; Savoy, A.W. Ionic liquids synthesis and applications: An overview. *J. Mol. Liq.* 2020, *297*, 23.
19. Knez, Z. Enzymatic reactions in subcritical and supercritical fluids. *J. Supercrit. Fluids* 2018, *134*, 133–140.
20. Knez, Z.; Markocic, E.; Leitgeb, M.; Primozic, M.; Hrncic, M.K.; Skerget, M. Industrial applications of supercritical fluids: A review. *Energy* 2014, *77*, 235–243.
21. Matsuda, T. Recent progress in biocatalysis using supercritical carbon dioxide. *J. Biosci. Bioeng.* 2013, *115*, 233–241.
22. Peach, J.; Eastoe, J. Supercritical carbon dioxide: A solvent like no other. *Beilstein J. Org. Chem.* 2014, *10*, 1878–1895.
23. Domínguez de María, P.; Guajardo, N.; Kara, S. Enzyme Catalysis: In DES, with DES, and in the Presence of DES. In *Deep Eutectic Solvents*, Ramón, D.J., Guillena, G., Eds. Wiley-VCH Verlag GmbH & Co. KGaA.: Weinheim, Germany, 2019; doi:10.1002/9783527818488. ch13, pp. 257–271.
24. Alonso, D.A.; Baeza, A.; Chinchilla, R.; Guillena, G.; Pastor, I.M.; Ramon, D.J. Deep eutectic solvents: The organic reaction medium of the century. *Eur. J. Org. Chem.* 2016, 612–632. doi:10.1002/ejoc.201501197.
25. Guajardo, N.; Muller, C.R.; Schrebler, R.; Carlesi, C.; de Maria, P.D. Deep eutectic solvents for organocatalysis, biotransformations, and multistep organocatalyst/enzyme combinations. *ChemCatChem* 2016, *8*, 1020–1027.
26. Juneidi, I.; Hayyan, M.; Hashim, M.A. Intensification of biotransformations using deep eutectic solvents: Overview and outlook. *Process Biochem.* 2018, *66*, 33–60.
27. Paiva, A.; Matias, A.A.; Duarte, A.R.C. How do we drive deep eutectic systems towards an industrial reality? *Curr. Opin. Green Sustain. Chem.* 2018, *11*, 81–85.
28. Tomé, L.I.N.; Baião, V.; da Silva, W.; Brett, C.M.A. Deep eutectic solvents for the production and application of new materials. *Appl. Mater. Today* 2018, *10*, 30–50.
29. Florindo, C.; Lima, F.; Ribeiro, B.D.; Marrucho, I.M. Deep eutectic solvents: overcoming 21st century challenges. *Curr. Opin. Green Sustain. Chem.* 2019, *18*, 31–36.
30. Marcus, Y. *Deep eutectic solvents*; Springer Nature Switzerland AG: Cham, Switzerland, 2019.
31. Patzold, M.; Siebenhaller, S.; Kara, S.; Liese, A.; Syldatk, C.; Holtmann, D. Deep Eutectic Solvents as efficient solvents in biocatalysis. *Trends Biotechnol.* 2019, *37*, 943–959.
32. Panic, M.; Bubalo, M.C.; Redovnikovic, I.R. Designing a biocatalytic process involving deep eutectic solvents. *J. Chem. Technol. Biotechnol.* 2020, 17. doi:10.1002/jctb.6545.
33. Płotka-Wasylka, J.; de la Guardia, M.; Andruch, V.; Vilková, M. Deep eutectic solvents vs ionic liquids: Similarities and differences. *Microchem. J.* 2020, *159*.
34. Tan, J.N.; Dou, Y.Q. Deep eutectic solvents for biocatalytic transformations: Focused lipase-catalyzed organic reactions. *Appl. Microbiol. Biotechnol.* 2020, *104*, 1481–1496.
35. Jerome, F.; Luque, R. *Bio-based solvents*; John Wiley & Sons: Hoboken, NJ, 2017.
36. Gu, Y.L.; Jerome, F. Bio-based solvents: An emerging generation of fluids for the design of eco-efficient processes in catalysis and organic chemistry. *Chem. Soc. Rev.* 2013, *42*, 9550–9570.
37. Lomba, L.; Zuriaga, E.; Giner, B. Solvents derived from biomass and their potential as green solvents. *Curr. Opin. Green Sustain. Chem.* 2019, *18*, 51–56.

38. de Jong, E.; Jungmeier, G. Biorefinery concepts in comparison to petrochemical refineries. In *Industrial Biorefineries and White Biotechnology*, Elsevier: 2015; doi:10.1016/B978-0-444-63453-5.00001-X. pp. 3–33.
39. Cherubini, F. The biorefinery concept: Using biomass instead of oil for producing energy and chemicals. *Energy Conv. Manag*. 2010, *51*, 1412–1421.
40. Vu, H.P.; Nguyen, L.N.; Vu, M.T.; Johir, M.A.H.; McLaughlan, R.; Nghiem, L.D. A comprehensive review on the framework to valorise lignocellulosic biomass as biorefinery feedstocks. *Sci. Total. Environ*. 2020, *743*, 140630.
41. Garlapati, V.K.; Chandel, A.K.; Kumar, S.P.J.; Sharma, S.; Sevda, S.; Ingle, A.P.; Pant, D. Circular economy aspects of lignin: Towards a lignocellulose biorefinery. *Renew. Sust. Energ. Rev*. 2020, *130*, 13.
42. Hassan, S.S.; Williams, G.A.; Jaiswal, A.K. Moving towards the second generation of lignocellulosic biorefineries in the EU: Drivers, challenges, and opportunities. *Renew. Sust. Energ. Rev*. 2019, *101*, 590–599.
43. Isikgor, F.H.; Becer, C.R. Lignocellulosic biomass: A sustainable platform for the production of bio-based chemicals and polymers. *Polym. Chem*. 2015, *6*, 4497–4559.
44. Gavahian, M.; Munekata, P.E.S.; Eş, I.; Lorenzo, J.M.; Mousavi Khaneghah, A.; Barba, F.J. Emerging techniques in bioethanol production: From distillation to waste valorization. *Green Chem*. 2019, *21*, 1171–1185.
45. Niphadkar, S.; Bagade, P.; Ahmed, S. Bioethanol production: insight into past, present and future perspectives. *Biofuels* 2018, *9*, 229–238.
46. Bušić, A.; Mardetko, N.; Kundas, S.; Morzak, G.; Belskaya, H.; Šantek, M.I.; Komes, D.; Novak, S.; Šantek, B. Bioethanol production from renewable raw materials and its separation and purification: A review. *Food Technol. Biotechnol*. 2018, *56*, 289–311.
47. Vohra, M.; Manwar, J.; Manmode, R.; Padgilwar, S.; Patil, S. Bioethanol production: Feedstock and current technologies. *J. Environ. Chem. Eng*. 2014, *2*, 573–584.
48. Gevorgyan, A.; Hopmann, K.H.; Bayer, A. Exploration of new biomass-derived solvents: Application to carboxylation reactions. *ChemSusChem* 2020, *13*, 2080–2088.
49. Len, C.; Luque, R. Continuous flow transformations of glycerol to valuable products: An overview. *Sust. Chem. Processes* 2014, *2*.
50. Garcia, J.I.; Garcia-Marin, H.; Pires, E. Glycerol based solvents: Synthesis, properties and applications. *Green Chem*. 2014, doi:10.1039/C3GC41857J.
51. Diaz-Alvarez, A.E.; Francos, J.; Lastra-Barreira, B.; Crochet, P.; Cadierno, V. Glycerol and derived solvents: New sustainable reaction media for organic synthesis. *Chem. Commun*. 2011, *47*, 6208–6227.
52. Gu, Y.L.; Jerome, F. Glycerol as a sustainable solvent for green chemistry. *Green Chem*. 2010, *12*, 1127–1138.
53. Garcia, J.I.; Garcia-Marin, H.; Mayoral, J.A.; Perez, P. Green solvents from glycerol. Synthesis and physico-chemical properties of alkyl glycerol ethers. *Green Chem*. 2010, *12*, 426–434.
54. Leal-Duaso, A.; Perez, P.; Mayoral, J.A.; Garcia, J.I.; Pires, E. Glycerol-derived solvents: Synthesis and properties of symmetric glyceryl diethers. *ACS Sustain. Chem. Eng*. 2019, *7*, 13004–13014.
55. Moity, L.; Benazzouz, A.; Molinier, V.; Nardello-Rataj, V.; Elmkaddem, M.K.; de Caro, P.; Thiebaud-Roux, S.; Gerbaud, V.; Marion, P.; Aubry, J.M. Glycerol acetals and ketals as bio-based solvents: positioning in Hansen and COSMO-RS spaces, volatility and stability towards hydrolysis and autoxidation. *Green Chem*. 2015, *17*, 1779–1792.
56. Ravichandiran, P.; Gu, Y. Low melting carbohydrate mixtures and aqueous carbohydrates – An effective green medium for organic synthesis. In *Carbohydrate Chemistry*, Rauter, A.P., Lindhorst, T.K., Queneau, Y., Eds. Royal Society of Chemistry: 2018; Vol. *43*, pp. 177–195.
57. Fischer, V.; Kunz, W. Properties of sugar-based low-melting mixtures. *Mol. Phys*. 2014, *112*, 1241–1245.
58. Paul, S.; Pradhan, K.; Das, A.R. Ethyl lactate as a green solvent: A promising bio-compatible media for organic synthesis. *Curr. Green Chem*. 2016, *3*, 111–118.
59. Dolzhenko, A.V. Ethyl lactate and its aqueous solutions as sustainable media for organic synthesis. *Sustain. Chem. Pharm*. 2020, *18*.
60. Diwan, F.; Shaikh, M.H.; Fatema, S.; Farooqui, M. Gluconic acid aqueous solution: A bio-compatible media for one-pot multicomponent synthesis of dihydropyrano [2,3-c] pyrazoles. *Org. Commun*. 2019, *12*, 188–201.
61. Yang, J.; Zhou, B.; Li, M.; Gu, Y. Gluconic acid aqueous solution: A task-specific bio-based solvent for ring-opening reactions of dihydropyrans. *Tetrahedron* 2013, *69*, 1057–1064.
62. Sheng, W.; Du, Y.; Tian, F.; Han, L.; Zhu, N. Gluconic acid aqueous solution as a green and sustainable solvent for the synthesis of bis (indolyl) methanes. *Chem. Bull*. 2012, *75*, 1026–1030.
63. Zhou, B.; Yang, J.; Li, M.; Gu, Y. Gluconic acid aqueous solution as a sustainable and recyclable promoting medium for organic reactions. *Green Chem*. 2011, *13*, 2204–2211.
64. Alcántara, A.R.; Dominguez de Maria, P. Recent advances on the use of 2-methyltetrahydrofuran (2-MeTHF) in biotransformations. *Curr. Green Chem*. 2018, *5*, 85–102.
65. Pace, V.; Holzer, W.; Hoyos, P.; Hernáiz, M.J.; Alcántara, A.R. 2-Methyltetrahydrofuran. In *Encyclopedia of Reagents for Organic Synthesis [Online]*, John Wiley & Sons, http://onlinelibrary.wiley.com/book/10.1002/047084289X [accessed date]. 2014; doi:10.1002/047084289X.rn01637
66. Pace, V.; Hoyos, P.; Castoldi, L.; Domínguez de María, P.; Alcántara, A.R. 2-Methyltetrahydrofuran (2-MeTHF): A biomass-derived solvent with broad application in organic chemistry. *ChemSusChem* 2012, *5*, 1369–1379.
67. Guajardo, N.; Domínguez de María, P. Assessing biocatalysis using dihydrolevoglucosenone (Cyrene™) as versatile bio-based (co)solvent. *Mol. Cat*. 2020, *485*.
68. Sherwood, J.; De Bruyn, M.; Constantinou, A.; Moity, L.; McElroy, C.R.; Farmer, T.J.; Duncan, T.; Raverty, W.; Hunt, A.J.; Clark, J.H. Dihydrolevoglucosenone (Cyrene) as a bio-based alternative for dipolar aprotic solvents. *Chem. Commun*. 2014, *50*, 9650–9652.
69. Alonso, D.M.; Wettstein, S.G.; Dumesic, J.A. Gamma-valerolactone, a sustainable platform molecule derived from lignocellulosic biomass. *Green Chem*. 2013, *15*, 584–595.
70. Wei, J.; Tang, X.; Sun, Y.; Zeng, X.; Lin, L. Applications of novel biomass-derived platform molecule γ-valerolactone. *Progr. Chem*. 2016, *28*, 1672–1681.

71. Vaccaro, L.; Santoro, S.; Curini, M.; Lanari, D. The emerging use of g-Valerolactone as a green solvent. *Chim. Oggi*. 2017, *35*, 46–48.
72. Al Musaimi, O.; El-Faham, A.; Basso, A.; de la Torre, B.G.; Albericio, F. γ-Valerolactone (GVL): An eco-friendly anchoring solvent for solid-phase peptide synthesis. *Tetrahedron Lett*. 2019, *60*.
73. Boissou, F.; Baranton, S.; Tarighi, M.; De Oliveira Vigier, K.; Coutanceau, C. The potency of γ-valerolactone as bio-sourced polar aprotic organic medium for the electrocarboxlation of furfural by CO_2. *J. Electroanal. Chem*. 2019, *848*.
74. Diwan, F.; Shaikh, M.H.; Shaikh, M.; Farooqui, M. γ-Valerolactone: Promising bio-compatible media for the synthesis of 2-arylbenzothiazole derivatives. *Org. Commun*. 2019, *12*, 1–13.
75. Rasool, M.A.; Vankelecom, I.F.J. Use of γ-valerolactone and glycerol derivatives as bio-based renewable solvents for membrane preparation. *Green Chem*. 2019, *21*, 1054–1064.
76. Shen, X.; Xia, D.; Xiang, Y.; Gao, J. γ-valerolactone (GVL) as a bio-based green solvent and ligand for iron-mediated AGET ATRP. *E-Polymers* 2019, *19*, 323–329.
77. Aricò, F. Isosorbide as biobased platform chemical: Recent advances. *Curr. Opin. Green Sustain. Chem*. 2020, *21*, 82–88.
78. Wilson, K.L.; Murray, J.; Sneddon, H.F.; Jamieson, C.; Watson, A.J.B. Dimethylisosorbide (DMI) as a bio-derived solvent for Pd-catalyzed cross-coupling reactions. *Synlett* 2018, *29*, 2293–2297.
79. Sambiagio, C.; Munday, R.H.; John Blacker, A.; Marsden, S.P.; McGowan, P.C. Green alternative solvents for the copper-catalysed arylation of phenols and amides. *RSC Adv*. 2016, *6*, 70025–70032.
80. Russo, F.; Galiano, F.; Pedace, F.; Aricò, F.; Figoli, A. Dimethyl isosorbide as a green solvent for sustainable ultrafiltration and microfiltration membrane preparation. *ACS Sustainable Chem. Eng*. 2020, *8*, 659–668.
81. McKinlay, J.B.; Vieille, C.; Zeikus, J.G. Prospects for a bio-based succinate industry. *Appl. Microbiol. Biotechnol*. 2007, *76*, 727–740.
82. Lopez-Garzon, C.S.; van der Wielen, L.A.M.; Straathof, A.J.J. Ester production from bio-based dicarboxylates via direct downstream catalysis: Succinate and 2,5-furandicarboxylate dimethyl esters. *RSC Adv*. 2016, *6*, 3823–3829.
83. Morales, M.; Ataman, M.; Badr, S.; Linster, S.; Kourlimpinis, I.; Papadokonstantakis, S.; Hatzimanikatis, V.; Hungerbühler, K. Sustainability assessment of succinic acid production technologies from biomass using metabolic engineering. *Energy Environ. Sci*. 2016, *9*, 2794–2805.
84. López-Garzón, C.S.; van der Wielen, L.A.M.; Straathof, A.J.J. Green upgrading of succinate using dimethyl carbonate for a better integration with fermentative production. *Chem. Eng. J*. 2014, *235*, 52–60.
85. Sangon, S.; Supanchaiyamat, N.; Sherwood, J.; McElroy, C.R.; Hunt, A.J. Direct comparison of safer or sustainable alternative dipolar aprotic solvents for use in carbon–carbon bond formation. *React. Chem. Eng*. 2020, *5*, 1798–1804.
86. Lammens, T.M.; Franssen, M.C.R.; Scott, E.L.; Sanders, J.P.M. Synthesis of biobased *N*-methylpyrrolidone by one-pot cyclization and methylation of γ-aminobutyric acid. *Green Chem*. 2010, *12*, 1430–1436.
87. Lopez, J.; Pletscher, S.; Aemissegger, A.; Bucher, C.; Gallou, F. N-Butylpyrrolidinone as alternative solvent for solid-phase peptide synthesis. *Org. Process Res. Dev*. 2018, *22*, 494–503.
88. Kumar, A.; Alhassan, M.; Lopez, J.; Albericio, F.; de la Torre, B.G. *N*-Butylpyrrolidinone for solid-phase peptide synthesis is environmentally friendlier and synthetically better than DMF. *ChemSusChem* 2020, doi:10.1002/cssc.202001647.
89. De La Torre, B.G.; Kumar, A.; Alhassan, M.; Bucher, C.; Albericio, F.; Lopez, J. Successful development of a method for the incorporation of Fmoc-Arg(Pbf)-OH in solid-phase peptide synthesis using *N*-butylpyrrolidinone (NBP) as solvent. *Green Chem*. 2020, *22*, 3162–3169.
90. Erny, M.; Lundqvist, M.; Rasmussen, J.H.; Ludemann-Hombourger, O.; Bihel, F.; Pawlas, J. Minimizing HCN in DIC/Oxyma-mediated amide bond-forming reactions. *Org. Process Res. Dev*. 2020, *24*, 1341–1349.
91. Sherwood, J.; Parker, H.L.; Moonen, K.; Farmer, T.J.; Hunt, A.J. N-Butylpyrrolidinone as a dipolar aprotic solvent for organic synthesis. *Green Chem*. 2016, *18*, 3990–3996.
92. De Schouwer, F.; Adriaansen, S.; Claes, L.; De Vos, D.E. Bio-based N-alkyl-2-pyrrolidones by Pd-catalyzed reductive N-alkylation and decarboxylation of glutamic acid. *Green Chem*. 2017, *19*, 4919–4929.
93. Löser, C.; Urit, T.; Bley, T. Perspectives for the biotechnological production of ethyl acetate by yeasts. *Appl. Microbiol. Biotechnol*. 2014, *98*, 5397–5415.
94. Zhang, S.; Guo, F.; Yan, W.; Dong, W.; Zhou, J.; Zhang, W.; Xin, F.; Jiang, M. Perspectives for the microbial production of ethyl acetate. *Appl. Microbiol. Biotechnol*. 2020, *104*, 7239–7245.
95. Tsolakis, N.; Bam, W.; Srai, J.S.; Kumar, M. Renewable chemical feedstock supply network design: The case of terpenes. *J. Clean. Prod*. 2019, *222*, 802–822.
96. Lamarche, M.; Dang, M.T.; Lefebvre, J.; Wuest, J.D.; Roorda, S. Limonene as a green solvent for depositing thin layers of molecular electronic materials with controlled interdiffusion. *ACS Sustainable Chem. Eng*. 2017, *5*, 5994–5998.
97. Pourreza, N.; Naghdi, T. D-Limonene as a green bio-solvent for dispersive liquid–liquid microextraction of β-cyclodextrin followed by spectrophotometric determination. *J. Ind. Eng. Chem*. 2017, *51*, 71–76.
98. Yadav, S.; Sharma, C.S. Novel and green processes for citrus peel extract: A natural solvent to source of carbon. *Polym. Bull*. 2018, *75*, 5133–5142.
99. El-Deen, A.K.; Shimizu, K. Application of D-limonene as a bio-based solvent in low density-dispersive liquid-liquid microextraction of acidic drugs from aqueous samples. *Anal. Sci*. 2019, *35*, 1385–1391.
100. Ma, J.; Chen, S.; Ye, C.; Li, M.; Liu, T.; Wang, X.; Song, Y. A green solvent for operating highly efficient low-power photon upconversion in air. *Phys. Chem. Chem. Phys*. 2019, *21*, 14516–14520.
101. Paggiola, G.; Derrien, N.; Moseley, J.D.; Green, A.; Flitsch, S.L.; Clark, J.H.; McElroy, C.R.; Hunt, A.J. Application of bio-based solvents for biocatalysed synthesis of amides with Pseudomonas stutzeri lipase (PSL). *Pure Appl. Chem*. 2020, *92*, 579–586.
102. Sadi, M.; Zeboudj, S.; Azri, Y.M.; Tou, I. D-Limonene as a green solvent to regenerate granular-activated carbon

saturated with phenol. *Sep. Sci. Technol.* 2020, *55*, 1776–1785.

103. Bi, Y.H.; Duan, Z.Q.; Du, W.Y.; Wang, Z.Y. Improved synthesis of phosphatidylserine using bio-based solvents, limonene and p-cymene. *Biotechnol. Lett.* 2015, *37*, 115–119.
104. de Oliveira Dias, A.; Gutiérrez, M.G.P.; Villarreal, J.A.A.; Carmo, R.L.L.; Oliveira, K.C.B.; Santos, A.G.; dos Santos, E.N.; Gusevskaya, E.V. Sustainable route to biomass-based amines: Rhodium catalyzed hydroaminomethylation in green solvents. *Appl. Catal. A Gen.* 2019, *574*, 97–104.
105. Labua, S.; Kenkhunthot, T. The extraction of lipid from microalgae found in brackish water by terpenes. *J. Curr. Sci. Tech.* 2020, *10*, 35–40.
106. Quinn, K.J.; Hu, Y.; Miller, P.J.; Walsh, R.T.; Caporello, M.A.; Maliszewski, M.L.; Markowski, J.H. Synthesis of the non-adjacent bis(tetrahydrofuran) core of squamostanin C by silicon-tethered, size-selective triple ring-closing metathesis. *Tetrahedron Lett.* 2019, *60*, 1773–1776.
107. Banach, A.; Ścianowski, J.; Uzarewicz-Baig, M.; Wojtczak, A. Terpenyl selenides: Synthesis and application in asymmetric epoxidation. *Eur. J. Org. Chem.* 2015, *2015*, 3477–3485.
108. Boariu, M.; Nica, L.M.; Marinescu, A.; Ganea, E.V.; Velea, O.; Pop, D.M.; Bretean, I.D.; Cirugeriu, L.E. Efficiency of eucalyptol as organic solvent in removal of gutta-percha from root canal fillings. *Rev. Chim.* 2015, *66*, 907–910.
109. Campos, J.F.; Berteina-Raboin, S. Eucalyptol as bio-based solvent for Migita–Kosugi–Stille coupling reaction on O,S,N-heterocycle. *Catal. Today* 2019, doi:10.1016/j.cattod.2019.11.004.
110. Campos, J.F.; Berteina-Raboin, S. Eucalyptol as a bio-based solvent for Buchwald-Hartwig reaction on O,S,N-heterocycles. *Catalysts* 2019, *9*.
111. Knothe, G.; Steidley, K.R. Fatty acid alkyl esters as solvents: Evaluation of the kauri-butanol value. comparison to hydrocarbons, dimethyl diesters, and other oxygenates. *Ind. Eng. Chem. Res.* 2011, *50*, 4177–4182.
112. da Roza Costa, M.B.; Nicolau, A.; Guzatto, R.; Angeloni, L.M.; Samios, D. Using biodiesel as a green solvent in the polymerization reactions: the attempt to separate the biodiesel from the polymer by thermal treatment. *Polym. Bull.* 2017, *74*, 2365–2378.
113. Kaur, R.; Khullar, P.; Gupta, A.; Ahluwalia, G.K.; Bakshi, M.S. Biodiesel as a non-aqueous medium for the synthesis of nanomaterials: Relevance to metallic particulate suspensions in biofuels and their removal. *Biofuels* 2019, doi:10.1080/17597269.2019.1594593.
114. Hu, J.; Du, Z.; Tang, Z.; Min, E. Study on the solvent power of a new green solvent: Biodiesel. *Ind. Eng. Chem. Res.* 2004, *43*, 7928–7931.
115. Srinivas, K.; Potts, T.M.; King, J.W. Characterization of solvent properties of methyl soyate by inverse gas chromatography and solubility parameters. *Green Chem.* 2009, *11*, 1581–1588.
116. Wildes, S. Methyl soyate: A new green alternative solvent. *Chem. Health Saf.* 2002, *9*, 24–26.
117. Clarke, C.J.; Tu, W.C.; Levers, O.; Brohl, A.; Hallett, J.P. Green and sustainable solvents in chemical processes. *Chemical Reviews* 2018, *118*, 747–800.
118. Monticelli, S.; Castoldi, L.; Murgia, I.; Senatore, R.; Mazzeo, E.; Wackerlig, J.; Urban, E.; Langer, T.; Pace, V. Recent advancements on the use of 2-methyltetrahydrofuran in organometallic chemistry. *Mon. Chem.* 2017, *148*, 37–48.
119. Rapinel, V.; Claux, O.; Abert-Vian, M.; McAlinden, C.; Bartier, M.; Patouillard, N.; Jacques, L.; Chemat, F. 2-Methyloxolane (2-MeOx) as sustainable lipophilic solvent to substitute hexane for green extraction of natural products. Properties, applications, and perspectives. *Molecules* 2020, *25*.
120. Slater, C.S.; Savelski, M.J.; Hitchcock, D.; Cavanagh, E.J. Environmental analysis of the life cycle emissions of 2-methyl tetrahydrofuran solvent manufactured from renewable resources. *J. Environ. Sci. Health Part A-Toxic/Hazard. Subst. Environ. Eng.* 2016, *51*, 487–494.
121. He, X.; Yu, D. Research trends in life cycle assessment research: A 20-year bibliometric analysis (1999–2018). *Environ. Impact Assess. Rev.* **2020**, *85*.
122. Khoo, H.H.; Wong, L.L.; Tan, J.; Isoni, V.; Sharratt, P. Synthesis of 2-methyl tetrahydrofuran from various lignocellulosic feedstocks: Sustainability assessment via LCA. *Resour. Conserv. Recycl.* 2015, *95*, 174–182.
123. Bluhm, K.; Seiler, T.B.; Anders, N.; Klankermayer, J.; Schaeffer, A.; Hollert, H. Acute embryo toxicity and teratogenicity of three potential biofuels also used as flavor or solvent. *Sci. Total Environ.* 2016, *566*, 786–795.
124. Bluhm, K.; Heger, S.; Redelstein, R.; Brendt, J.; Anders, N.; Mayer, P.; Schaeffer, A.; Hollert, H. Genotoxicity of three biofuel candidates compared to reference fuels. *Environ. Toxicol. Pharmacol.* 2018, *64*, 131–138.
125. Antonucci, V.; Coleman, J.; Ferry, J.B.; Johnson, N.; Mathe, M.; Scott, J.P.; Xu, J. Toxicological assessment of 2-methyltetrahydrofuran and cyclopentyl methyl ether in support of their use in pharmaceutical chemical process development. *Org. Process Res. Dev.* 2011, *15*, 939–941.
126. Parris, P.; Duncan, J.N.; Fleetwood, A.; Beierschmitt, W.P. Calculation of a permitted daily exposure value for the solvent 2-methyltetrahydrofuran. *Regul. Toxicol. Pharmacol.* 2017, *87*, 54–63.
127. Sicaire, A.G.; Vian, M.A.; Filly, A.; Li, Y.; Bily, A.; Chemat, F. 2-Methyltetrahydrofuran: Main properties, production processes, and application in extraction of natural products. In *Alternative Solvents for Natural Products Extraction*, Chemat, F., Vian, M.A., Eds. Springer-Verlag Berlin: Berlin, 2014; doi:10.1007/978-3-662-43628-8_12, pp. 253–268.
128. Prat, D.; Wells, A.; Hayler, J.; Sneddon, H.; McElroy, C.R.; Abou-Shehada, S.; Dunn, P.J. CHEM21 selection guide of classical- and less classical-solvents. *Green Chem.* 2015, doi:10.1039/C5GC01008J.
129. Nakamura, K.; Kinoshita, M.; Ohno, A. Effect of solvent on lipase-catalyzed transesterification in organic media. *Tetrahedron* 1994, *50*, 4681–4690.
130. Palmer, D.C.; Terradas, F. Regioselective enzymic deacylation of sucrose esters in anhydrous organic media. US5445951A, 1995.
131. Simeo, Y.; Sinisterra, J.V.; Alcantara, A.R. Regioselective enzymic acylation of pharmacologically interesting nucleosides in 2-methyltetrahydrofuran, a greener substitute for THF. *Green Chem.* 2009, *11*, 855–862.
132. Belafriekh, A.; Secundo, F.; Serra, S.; Djeghaba, Z. Enantioselective enzymatic resolution of racemic alcohols

by lipases in green organic solvents. *Tetrahedron Asymmetry* 2017, *28*, 473–478.
133. Liu, Y.; Guo, C.; Liu, C.Z. Development of a mixed solvent system for the efficient resolution of (R, S)-2-octanol catalyzed by magnetite-immobilized lipase. *J. Mol. Catal. B Enzym.* 2014, *101*, 23–27.
134. Hoyos, P.; Quezada, M.A.; Sinisterra, J.V.; Alcántara, A.R. Optimised dynamic kinetic resolution of benzoin by a chemoenzymatic approach in 2-MeTHF. *J. Mol. Catal. B Enzym.* 2011, *72*, 20–24.
135. Nieguth, R.; ten Dam, J.; Petrenz, A.; Ramanathan, A.; Hanefeld, U.; Ansorge-Schumacher, M.B. Combined heterogeneous bio- and chemo-catalysis for dynamic kinetic resolution of (rac)-benzoin. *RSC Adv.* 2014, *4*, 45495–45503.
136. Aires-Trapote, A.; Hoyos, P.; Alcantara, A.R.; Tamayo, A.; Rubio, J.; Rumbero, A.; Hernaiz, M.J. Covalent immobilization of *Pseudomonas stutzeri* lipase on a porous polymer: An efficient biocatalyst for a scalable production of enantiopure benzoin esters under sustainable conditions. *Org. Process Res. Develop.* 2015, *19*, 687–694.
137. Schuster, T.; Gerlach, M.; Schuch, F. Enzymatic process for the regioselective manufacturing of N-Fmoc-doxorubicin-14-O-dicarboxylic acid- mono esters. EP3045540A1, 2016.
138. Hoang, H.N.; Granero-Fernandez, E.; Yamada, S.; Mori, S.; Kagechika, H.; Medina-Gonzalez, Y.; Matsuda, T. Modulating biocatalytic activity toward sterically bulky substrates in CO_2-expanded biobased liquids by tuning the physicochemical properties. *ACS Sustain. Chem. Eng.* 2017, *5*, 11051–11059.
139. Hoang, H.N.; Nagashima, Y.; Mori, S.; Kagechika, H.; Matsuda, T. CO_2-expanded bio-based liquids as novel solvents for enantioselective biocatalysis. *Tetrahedron* 2017, *73*, 2984–2989.
140. Singh, A.; Falabella, J.; LaPorte, T.L.; Goswami, A. Enzymatic process for *N*-substituted (3*S*)- and (3*R*)-3-hydroxypyrrolidin-2-ones. *Org. Process Res. Dev.* 2015, *19*, 819–830.
141. Gao, W.-L.; Li, N.; Zong, M.-H. Highly regioselective synthesis of undecylenic acid esters of purine nucleosides catalyzed by *Candida antarctica* lipase B. *Biotechnol. Lett.* 2011, *33*, 2233–2240.
142. Gao, W.-L.; Liu, H.; Li, N.; Zong, M.-H. Regioselective enzymatic undecylenoylation of 8-chloroadenosine and its analogs with biomass-based 2-methyltetrahydrofuran as solvent. *Bioresour. Technol.* 2012, *118*, 82–88.
143. Gao, W.L.; Li, N.; Zong, M.H. Enzymatic regioselective acylation of nucleosides in biomass-derived 2-methyltetrahydrofuran: Kinetic study and enzyme substrate recognition. *J. Biotechnol.* 2013, *164*, 91–96.
144. Chen, Z.G.; Zhang, D.N.; Cao, L.; Han, Y.B. Highly efficient and regioselective acylation of pharmacologically interesting cordycepin catalyzed by lipase in the eco-friendly solvent 2-methyltetrahydrofuran. *Bioresour. Technol.* 2013, *133*, 82–86.
145. Zhang, D.N.; Guo, X.Y.; Yang, Q.H.Z.; Chen, Z.G.; Tao, L.J. An efficient enzymatic modification of cordycepin in ionic liquids under ultrasonic irradiation. *Ultrason. Sonochem.* 2014, *21*, 1682–1687.
146. Wang, Z.Y.; Bi, Y.H.; Yang, R.L.; Zhao, X.J.; Jiang, L.; Zhu, C.; Zhao, Y.P.; Jia, J.B. Enzymatic synthesis of Sorboyl-Polydatin prodrug in biomass-derived 2-Methyltetrahydrofuran and antiradical activity of the unsaturated acylated derivatives. *Biomed Res. Int.* 2016, doi:10.1155/2016/4357052.
147. Wang, Z.Y.; Du, W.Y.; Duan, Z.Q.; Yang, R.L.; Bi, Y.H.; Yuan, X.T.; Mao, Y.Y.; Zhao, Y.P.; Wu, J.; Jia, J.B. Efficient regioselective synthesis of the crotonyl polydatin prodrug by thermomyces lanuginosus lipase: A kinetics study in eco-friendly 2-Methyltetrahydrofuran. *Appl. Biochem. Biotech.* 2016, *179*, 1011–1022.
148. Wang, Z.Y.; Bi, Y.H.; Yang, R.L.; Zhao, X.J.; Jiang, L.; Ding, C.X.; Zheng, S.Y. Highly efficient enzymatic synthesis of novel polydatin prodrugs with potential anticancer activity. *Process Biochem. (Oxford, U. K.)* 2017, *52*, 209–213.
149. Yang, R.L.; Liu, X.M.; Chen, Z.Y.; Yang, C.Y.; Lin, Y.S.; Wang, S.Y. Highly efficient and enzymatic regioselective undecylenoylation of gastrodin in 2-methyltetrahydrofuran-containing systems. *PLoS One* 2014, *9*.
150. Hu, Y.D.; Qin, Y.Z.; Li, N.; Zong, M.H. Highly efficient enzymatic synthesis of an ascorbyl unstaturated fatty acid ester with ecofriendly biomass-derived 2-methyltetrahydrofuran as cosolvent. *Biotechnol. Prog.* 2014, *30*, 1005–1011.
151. Hu, Y.D.; Zong, M.H.; Li, N. Enzymatic synthesis and anti-oxidative activities of plant oil-based ascorbyl esters in 2-methyltetrahydrofuran-containing mixtures. *Biocatal. Biotransform.* 2016, *34*, 181–188.
152. Chen, Z.G.; Zhang, D.N.; Han, Y.B. Lipase-catalyzed acylation of lily polysaccharide in ionic liquid-containing systems. *Process Biochem.* 2013, *48*, 620–624.
153. Sundell, R.; Siirola, E.; Kanerva, L.T. Regio- and stereoselective lipase-catalysed acylation of methyl alpha-*D*-Glycopyranosides with fluorinated beta-lactams. *Eur. J. Org. Chem.* 2014, 6753–6760, doi:10.1002/ejoc.201402800.
154. van Pelt, S.; Teeuwen, R.L.M.; Janssen, M.H.A.; Sheldon, R.A.; Dunn, P.J.; Howard, R.M.; Kumar, R.; Martinez, I.; Wong, J.W. *Pseudomonas stutzeri* lipase: A useful biocatalyst for aminolysis reactions. *Green Chem.* 2011, *13*, 1791–1798.
155. Filipan, M.; Litvic, M.; Cepanec, I.; Vinkovic, V. Lipase-catalyzed resolution of (*R*,*S*)-2-(2-aminoethoxymethyl)-3-ethoxycarbonyl-4-(2-chlorophenyl)-5-methoxycarbonyl-6-methyl-1,4-dihydropyridine. HR2004000520A2, 2005.
156. Cheng, G.L.; Xia, B.; Wu, Q.; Lin, X.F. Chemoenzymatic dynamic kinetic resolution of alpha-trifluoromethylated amines: Influence of substitutions on the reversed stereoselectivity. *RSC Adv.* 2013, *3*, 9820–9828.
157. Pedragosa-Moreau, S.; Le Flohic, A.; Thienpondt, V.; Lefoulon, F.; Petit, A.M.; Rios-Lombardia, N.; Moris, F.; Gonzalez-Sabin, J. Exploiting the biocatalytic toolbox for the asymmetric synthesis of the heart-rate reducing agent ivabradine. *Adv. Synth. Catal.* 2017, *359*, 485–493.
158. Lindhagen, M.; Klingstedt, T.; Andersen, S.M.; Mulholland, K.R.; Tinkler, L.; McPheators, G.; Chubb, R. Development of a chemoenzymatic route to (*R*)-allyl-(3-amino-2-(2-methylbenzyl)propyl)carbamate. *Org. Process Res. Dev.* 2016, *20*, 65–69.
159. Duan, Z.Q.; Hu, F. Efficient synthesis of phosphatidylserine in 2-methyltetrahydrofuran. *J. Biotechnol.* 2013, *163*, 45–49.
160. Gallou, F.; Seeger-Weibel, M.; Chassagne, P. Development of a robust and sustainable process for nucleoside formation. *Org. Process Res. Dev.* 2013, *17*, 390–396.

161. Torres, S.Y.; Brieva, R.; Rebolledo, F. Chemoenzymatic synthesis of optically active phenolic 3,4-dihydropyridin-2-ones: a way to access enantioenriched 1,4-dihydropyridine and benzodiazepine derivatives. *Org. Biomol. Chem.* 2017, *15*, 5171–5181.
162. Husain, M.; Datta, D. Process for preparing (*S*)-(-)-10-acetoxy-10,11-dihydro-5*H*-dibenz[*b,f*]azepine-5-carboxamide and its esters thereof. WO2011045648A2, 2011.
163. Chen, Z.; Qiao, Y.; Huang, Y. Enzymatic method for preparing andrographolide glycoside derivative. CN105177090A, 2015.
164. Montilla Arevalo, R.; Deroncele Thomas, V.M.; Lopez Gomez, C.; Pascual Gilabert, M.; Estevez Company, C.; Castells Boliart, J. Nucleoside synthesis by recombinant thermophilic purine nucleoside phosphorylase and uridine phosphorylase. EP2338985A1, 2011.
165. Tian, Y.J.; Ma, X.Q.; Yang, M.; Wei, D.Z.; Su, E.Z. Synthesis of (*S*)-3-chloro-1-phenylpropanol by permeabilized recombinant *Escherichia coli* harboring *Saccharomyces cerevisiae* YOL151W reductase in 2-methyltetrahydrofuran cosolvent system. *Catal. Commun.* 2017, *97*, 56–59.
166. Betori, R.C.; Miller, E.R.; Scheidt, K.A. A biocatalytic route to highly enantioenriched β-hydroxydioxinones. *Adv. Synth. Catal.* 2017, *359*, 1131–1137.
167. Pietruszka, J.; Scholzel, M. Ene reductase-catalysed synthesis of (*R*)-profen derivatives. *Adv. Synth. Catal.* 2012, *354*, 751–756.
168. Shanmuganathan, S.; Natalia, D.; van den Wittenboer, A.; Kohlmann, C.; Greiner, L.; de Maria, P.D. Enzyme-catalyzed C-C bond formation using 2-methyltetrahydrofuran (2-MTHF) as (co)solvent: Efficient and bio-based alternative to DMSO and MTBE. *Green Chem.* 2010, *12*, 2240–2245.
169. Gerhards, T.; Mackfeld, U.; Bocola, M.; von Lieres, E.; Wiechert, W.; Pohl, M.; Rother, D. Influence of organic solvents on enzymatic asymmetric carboligations. *Adv. Synth. Catal.* 2012, *354*, 2805–2820.
170. Perez-Sanchez, M.; Dominguez de Maria, P. Lipase catalyzed in situ production of acetaldehyde: A controllable and mild strategy for multi-step reactions. *ChemCatChem* 2012, *4*, 617–619.
171. Shanmuganathan, S.; Natalia, D.; Greiner, L.; Dominguez de Maria, P. Oxidation-hydroxymethylation-reduction: A one-pot three-step biocatalytic synthesis of optically active α-aryl vicinal diols. *Green Chem.* 2012, *14*, 94–97.
172. Schmidt, S.; de Almeida, T.P.; Rother, D.; Hollmann, F. Towards environmentally acceptable synthesis of chiral alpha-hydroxy ketones via oxidase-lyase cascades. *Green Chem.* 2017, *19*, 1226–1229.
173. Fang, D. Biocatalytic synthesis method of polyhydroacridine derivatives. CN102443615A, 2012.
174. Fang, D. Method for biocatalytic synthesis of 3,4-dihydropyrimidin-2-one. CN102399836A, 2012.
175. Gotor-Fernandez, V.; Brieva, R.; Gotor, V. Lipases: Useful biocatalysts for the preparation of pharmaceuticals. *J. Mol. Catal. B Enzym.* 2006, *40*, 111–120.
176. Bornscheuer, U.T.; Flickinger, M.C. Lipases, synthesis of chiral compounds, aqueous and organic solvents. In *Encyclopedia of Industrial Biotechnology*, John Wiley & Sons: 2009; doi:10.1002/9780470054581.eib648.
177. de Miranda, A.S.; Miranda, L.S.M.; de Souza, R. Lipases: Valuable catalysts for dynamic kinetic resolutions. *Biotechnol. Adv.* 2015, *33*, 372–393.
178. Seddigi, Z.S.; Malik, M.S.; Ahmed, S.A.; Babalghith, A.O.; Kamal, A. Lipases in asymmetric transformations: Recent advances in classical kinetic resolution and lipase–metal combinations for dynamic processes. *Coord. Chem. Rev.* 2017, *348*, 54–70.
179. Chandra, P.; Enespa; Singh, R.; Arora, P.K. Microbial lipases and their industrial applications: A comprehensive review. *Microb. Cell Fact.* 2020, *19*, 169.
180. Priyanka, P.; Tan, Y.Q.; Kinsella, G.K.; Henehan, G.T.; Ryan, B.J. Solvent stable microbial lipases: Current understanding and biotechnological applications. *Biotechnol. Lett.* 2019, *41*, 203–220.
181. Filho, D.G.; Silva, A.G.; Guidini, C.Z. Lipases: Sources, immobilization methods, and industrial applications. *Appl. Microb. Biotech.* 2019, *103*, 7399–7423.
182. Daiha, K.D.; Angeli, R.; de Oliveira, S.D.; Almeida, R.V. Are lipases still important biocatalysts? A study of scientific publications and patents for technological forecasting. *PLoS One* 2015, *10*, 20.
183. Tarczykowska, A.; Sikora, A.; Marszall, M.P. Lipases – Valuable biocatalysts in kinetic resolution of racemates. *Mini-Rev. Org. Chem.* 2018, *15*, 374–381.
184. Schally, A.V.; Block, N.L.; Rick, F.G. Discovery of LHRH and development of LHRH analogs for prostate cancer treatment. *Prostate* 2017, *77*, 1036–1054.
185. Ritter, K.; Buning, C.; Halland, N.; Poverlein, C.; Schwink, L. G Protein-coupled receptor 119 (GPR119) agonists for the treatment of diabetes: Recent progress and prevailing challenges. *J. Med. Chem.* 2016, *59*, 3579–3592.
186. Condezo, L.A.; Jesús Fernández-Lucas; Carlos A. García-Burgos; Alcántara, A.R.; Sinisterra, J.V. Enzymatic synthesis of modified nucleosides. In *Biocatalysis in the Pharmaceutical and Biotechnology Industries*, Patel, R.N., Ed. Taylor & Francis Group: 2006; pp. 401–423.
187. Sinisterra, J.V.; Alcántara, A.R.; Almendros, M.; Hernáiz, M.J.; Sánchez-Montero, J.M.; Trelles, J. Enzyme-catalyzed synthesis of nonnatural or modified nucleosides. In *Encyclopedia of Industrial Biotechnology: Bioprocess, Bioseparation, and Cell Technology*, Flickinger, M.C., Ed. John Wiley & Sons: 2010; doi:10.1002/9780470054581.eib635.
188. Peng, W.; Qin, R.; Li, X.; Zhou, H. Botany, phytochemistry, pharmacology, and potential application of Polygonum cuspidatum Sieb.et Zucc.: A review. *J. Ethnopharmacol.* 2013, *148*, 729–745.
189. Li, N.; Zong, M.; Hu, Y. Enzymatic method for synthesis of ascorbic acid ester. CN103667384A, 2014.
190. Li, N.; Zong, M.; Hu, Y. In-situ enzymic preparation of an antioxidative oil. CN104140985A, 2014.
191. Gotor, V. Non-conventional hydrolase chemistry: Amide and carbamate bond formation catalyzed by lipases. *Bioorg. Med. Chem.* 1999, *7*, 2189–2197.
192. Scientific opinion on the substantiation of health claims related to phosphatidyl serine (ID 552, 711, 734, 1632, 1927) pursuant to Article 13(1) of Regulation (EC) No 1924/2006. *EFSA J.* **2010**, *8*, 1749.

193. Duan, Z.Q.; Hu, F. Efficient synthesis of phosphatidylserine in 2-methyltetrahydrofuran. *J. Biotechnol.* 2013, *163*, 45–49.
194. Qin, W.; Wu, C.J.; Song, W.; Chen, X.L.; Liu, J.; Luo, Q.L.; Liu, L.M. A novel high-yield process of phospholipase D-mediated phosphatidylserine production with cyclopentyl methyl ether. *Process Biochem.* 2018, *66*, 146–149.
195. Horscroft, N.J.; Pryde, D.C.; Bright, H. Antiviral applications of Toll-like receptor agonists. *J. Antimicrob. Chemother.* 2012, *67*, 789–801.
196. Yang, L.; Wang, D.X.; Zheng, Q.Y.; Pan, J.; Huang, Z.T.; Wang, M.X. Highly efficient and concise synthesis of both antipodes of SB204900, clausenamide, neoclausenamide, homoclausenamide and zeta-clausenamide. Implication of biosynthetic pathways of clausena alkaloids. *Org. Biomol. Chem.* 2009, *7*, 2628–2634.
197. Liu, D.; Yu, X.M.; Huang, L. Novel concise synthesis of (-)-clausenamide. *Chin. J. Chem.* 2013, *31*, 344–348.
198. Feng, Y.B.; LoGrasso, P.V.; Defert, O.; Li, R.S. Rho kinase (ROCK) inhibitors and their therapeutic potential. *J. Med. Chem.* 2016, *59*, 2269–2300.
199. Lopez-Tapia, F.; Walker, K.A.M.; Brotherton-Pleiss, C.; Caroon, J.; Nitzan, D.; Lowrie, L.; Gleason, S.; Zhao, S.H.; Berger, J.; Cockayne, D., et al. Novel series of dihydropyridinone P2X7 receptor antagonists. *J. Med. Chem.* 2015, *58*, 8413–8426.
200. Homan, K.T.; Larimore, K.M.; Elkins, J.M.; Szklarz, M.; Knapp, S.; Tesmer, J.J.G. Identification and structure-function analysis of subfamily selective G protein-coupled receptor kinase inhibitors. *ACS Chem. Biol.* 2015, *10*, 310–319.
201. Ambrosio, A.F.; Soares-Da-Silva, P.; Carvalho, C.M.; Carvalho, A.P. Mechanisms of action of carbamazepine and its derivatives, oxcarbazepine, BIA 2-093, and BIA 2-024. *Neurochem. Res.* 2002, *27*, 121–130.
202. El-Behairy, M.F.; Sundby, E. One-step lipase-catalysed preparation of eslicarbazepine. *Rsc. Adv.* 2016, *6*, 98730–98736.
203. Modukuru, N.K.; Sukumaran, J.; Collier, S.J.; Chan, A.S.; Gohel, A.; Huisman, G.W.; Keledjian, R.; Narayanaswamy, K.; Novick, S.J.; Palanivel, S.M., et al. Development of a practical, biocatalytic reduction for the manufacture of (*S*)-licarbazepine using an evolved ketoreductase. *Org. Process Res. Dev.* 2014, *18*, 810–815.
204. Kishore, V.; Yarla, N.S.; Bishayee, A.; Putta, S.; Malla, R.; Neelapu, N.R.R.; Challa, S.; Das, S.; Shiralgi, Y.; Hegde, G., et al. Multi-targeting andrographolide and its natural analogs as potential therapeutic agents. *Curr. Top. Med. Chem.* 2017, *17*, 845–857.
205. Duan, H.; Jing, M.; Li, Z.; Zhang, Z.; Wang, Y.; Xu, L.; Yu, P. Synthesis and biological evaluation of andrographolide derivatives as potential anti-inflammatory agents. *J. Pharm. Biomed. Sci.* 2017, *7*, 94–99.
206. Yehia, H.; Kamel, S.; Paulick, K.; Neubauer, P.; Wagner, A. Substrate spectra of nucleoside phosphorylases and their potential in the production of pharmaceutically active compounds. *Curr. Pharm. Design* 2017, *23*, 6913–6935.
207. Drenichev, M.S.; Alexeev, C.S.; Kurochkin, N.N.; Mikhailov, S.N. Use of nucleoside phosphorylases for the preparation of purine and pyrimidine 2-deoxynucleosides. *Adv. Synth. Catal.* 2018, *360*, 305–312.
208. Lucas, J.F. New trends in enzymatic synthesis of nucleic acid derivatives. *Curr. Pharm. Design* 2017, *23*, 6849–6850.
209. Almendros, M.; Berenguer, J.; Sinisterra, J.-V. Thermus thermophilus nucleoside phosphorylases active in the synthesis of nucleoside analogues. *Appl. Environ. Microbiol.* 2012, *78*, 3128–3135.
210. Goundry, W.R.F.; Adams, B.; Benson, H.; Demeritt, J.; McKown, S.; Mulholland, K.; Robertson, A.; Siedlecki, P.; Tomlin, P.; Vare, K. Development and scale-up of a biocatalytic process to form a chiral sulfoxide. *Org. Process Res. Dev.* 2017, *21*, 107–113.
211. Itoh, N.; Kariya, S.; Kurokawa, J. Efficient PCR-based amplification of diverse alcohol dehydrogenase genes from metagenomes for improving biocatalysis: Screening of gene-specific amplicons from metagenomes. *Appl. Environ. Microbiol.* 2014, *80*, 6280–6289.
212. Kumar, A.; Maurya, R.A. Bakers' yeast catalyzed synthesis of polyhydroquinoline derivatives via an unsymmetrical Hantzsch reaction. *Tetrahedron Lett.* 2007, *48*, 3887–3890.
213. Kumar, A.; Maurya, R.A. An efficient bakers' yeast catalyzed synthesis of 3,4-dihydropyrimidin-2-(1 H)-ones. *Tetrahedron Lett.* 2007, *48*, 4569–4571.
214. Peris, E.; Porcar, R.; Isabel Burguete, M.; Garcia-Verdugo, E.; Luis, S.V. Supported Ionic Liquid-Like Phases (SILLPs) as immobilised catalysts for the multistep and multicatalytic continuous flow synthesis of chiral cyanohydrins. *ChemCatChem* 2019, *11*, 1955–1962.
215. Wu, W.B.; Yu, J.S.; Zhou, J. Catalytic enantioselective cyanation: Recent advances and perspectives. *ACS Catal.* 2020, *10*, 7668–7690.
216. Zeng, X.P.; Sun, J.C.; Liu, C.; Ji, C.B.; Peng, Y.Y. Catalytic asymmetric cyanation reactions of aldehydes and ketones in total synthesis. *Adv. Synth. Catal.* 2019, *361*, 3281–3305.
217. Bracco, P.; Busch, H.; von Langermann, J.; Hanefeld, U. Enantioselective synthesis of cyanohydrins catalysed by hydroxynitrile lyases – A review. *Org. Biomol. Chem.* 2016, *14*, 6375–6389.
218. Loewe, J.; Uthoff, F.; Harms, C.; Groeger, H.; Donsbach, K. Enantioselective biocatalytic preparation of 4-cyano-substituted 1-aminoindane and ozanimod. WO2019197571A1, 2019.
219. Scott, F.L.; Clemons, B.; Brooks, J.; Brahmachary, E.; Powell, R.; Dedman, H.; Desale, H.G.; Timony, G.A.; Martinborough, E.; Rosen, H., et al. Ozanimod (RPC1063) is a potent sphingosine-1-phosphate receptor-1 (S1P(1)) and receptor-5 (S1P(5)) agonist with autoimmune disease-modifying activity. *Br. J. Pharmacol.* 2016, *173*, 1778–1792.
220. Pellis, A.; Byrne, F.P.; Sherwood, J.; Vastano, M.; Comerford, J.W.; Farmer, T.J. Safer bio-based solvents to replace toluene and tetrahydrofuran for the biocatalyzed synthesis of polyesters. *Green Chem.* 2019, *21*, 1686–1694.
221. Englezou, G.; Kortsen, K.; Pacheco, A.A.C.; Cavanagh, R.; Lentz, J.C.; Krumins, E.; Sanders-Velez, C.; Howdle, S.M.; Nedoma, A.J.; Taresco, V. 2-Methyltetrahydrofuran

(2-MeTHF) as a versatile green solvent for the synthesis of amphiphilic copolymers via ROP, FRP, and RAFT tandem polymerizations. *J. Polym. Sci.* 2020, *58*, 1571–1581.

222. Seyednejad, H.; Ghassemi, A.H.; van Nostrum, C.F.; Vermonden, T.; Hennink, W.E. Functional aliphatic polyesters for biomedical and pharmaceutical applications. *J. Control. Release* 2011, *152*, 168–176.

223. Gerardy, R.; Debecker, D.P.; Estager, J.; Luis, P.; Monbaliu, J.C.M. Continuous flow upgrading of selected C-2-C(6) platform chemicals derived from biomass. *Chem. Rev.* 2020, *120*, 7219–7347.

224. Liu, X.; Leong, D.C.Y.; Sun, Y.J. The production of valuable biopolymer precursors from fructose. *Green Chem.* 2020, *22*, 6531–6539.

225. De, S.; Kumar, T.; Bohre, A.; Singh, L.R.; Saha, B. Furan-based acetylating agent for the chemical modification of proteins. *Bioorg. Med. Chem.* 2015, *23*, 791–796.

226. Chen, S.; Wojcieszak, R.; Dumeignil, F.; Marceau, E.; Royer, S. How Catalysts and experimental conditions determine the selective hydroconversion of furfural and 5-hydroxymethylfurfural. *Chem. Rev.* 2018, *118*, 11023–11117.

227. Wei, J.N.; Wang, T.; Tang, P.F.; Tang, X.; Sun, Y.; Zeng, X.H.; Lin, L. Chemoselective hydrogenation of biomass-derived 5-hydroxymethylfurfural into furanyl diols. *Curr. Org. Chem.* 2019, *23*, 2155–2167.

228. Tang, X.; Wei, J.N.; Ding, N.; Sun, Y.; Zeng, X.H.; Hu, L.; Liu, S.J.; Lei, T.Z.; Lin, L. Chemoselective hydrogenation of biomass derived 5-hydroxymethylfurfural to diols: Key intermediates for sustainable chemicals, materials and fuels. *Renew. Sust. Energ. Rev.* 2017, *77*, 287–296.

229. Hu, L.; He, A.Y.; Liu, X.Y.; Xia, J.; Xu, J.X.; Zhou, S.Y.; Xu, J.M. Biocatalytic transformation of 5-hydroxymethylfurfural into high-value derivatives: Recent advances and future aspects. *ACS Sustain. Chem. Eng.* 2018, *6*, 15915–15935.

230. Xia, Z.H.; Zong, M.H.; Li, N. Catalytic synthesis of 2,5-bis(hydroxymethyl)furan from 5-hydroxymethylfurfual by recombinant *Saccharomyces cerevisiae*. *Enzyme Microb. Technol.* 2020, *134*, 5.

231. Xu, Z.H.; Cheng, A.D.; Xing, X.P.; Zong, M.H.; Bai, Y.P.; Li, N. Improved synthesis of 2,5-bis(hydroxymethyl)furan from 5-hydroxymethylfurfural using acclimatized whole cells entrapped in calcium alginate. *Bioresour. Technol.* 2018, *262*, 177–183.

232. Li, Y.M.; Zhang, X.Y.; Li, N.; Xu, P.; Lou, W.Y.; Zong, M.H. Biocatalytic reduction of HMF to 2,5-Bis(hydroxymethyl) furan by HMF-tolerant whole cells. *ChemSusChem* 2017, *10*, 372–378.

233. Arias, K.S.; Carceller, J.M.; Climent, M.J.; Corma, A.; Iborra, S. Chemoenzymatic synthesis of 5-hydroxymethylfurfural (HMF)-derived plasticizers by coupling HMF reduction with enzymatic esterification. *ChemSusChem* 2020, *13*, 1864–1875.

234. Lacatus, M.A.; Bencze, L.C.; Tosa, M.I.; Paizs, C.; Irimie, F.D. Eco-friendly enzymatic production of 2,5-Bis(hydroxymethyl)furan fatty acid diesters, potential biodiesel additives. *ACS Sustain. Chem. Eng.* 2018, *6*, 11353–11359.

235. Lacatus, M.A.; Dudu, A.J.; Bencze, L.C.; Katona, G.; Irimie, F.D.; Paizs, C.; Tosa, M.I. Solvent-free biocatalytic synthesis of 2,5-bis-(Hydroxymethyl)Furan fatty acid diesters from renewable resources. *ACS Sustain. Chem. Eng.* 2020, *8*, 1611–1617.

236. Bačić, M.; Ljubić, A.; Gojun, M.; Šalić, A.; Tušek, A.J.; Zelić, B. Continuous integrated process of biodiesel production and purification—The end of the conventional two-stage batch process? *Energies* 2021, *14*, 403.

237. Monteiro, M.R.; Kugelmeier, C.L.; Pinheiro, R.S.; Batalha, M.O.; Cesar, A.D. Glycerol from biodiesel production: Technological paths for sustainability. *Renew. Sust. Energ. Rev.* 2018, *88*, 109–122.

238. Ladero, M. New glycerol upgrading processes. *Catalysts* 2021, *11*, 103.

239. Wolfson, A.; Dlugy, C.; Tavor, D.; Blumenfeld, J.; Shotland, Y. Baker's yeast catalyzed asymmetric reduction in glycerol. *Tetrahedron-Asymmetry* 2006, *17*, 2043–2045.

240. Wolfson, A.; Dlugy, C.; Shotland, Y. Glycerol as a green solvent for high product yields and selectivities. *Environ. Chem. Lett.* 2007, *5*, 67–71.

241. Cheng, C.; Nian, Y.C. Enantioselective reduction of 4-phenyl-2-butanone to chiral 4-phenyl-2-butanol in glycerol modified Saccharomyces cerevisiae cell culture. *J. Mol. Catal. B-Enzym.* 2016, *123*, 141–146.

242. Fischer, T.; Pietruszka, J. Alcohol dehydrogenase-catalyzed synthesis of enantiomerically pure delta-lactones as versatile intermediates for natural product synthesis. *Adv. Synth. Catal.* 2012, *354*, 2521–2530.

243. Emmanuel, M.A.; Greenberg, N.R.; Oblinsky, D.G.; Hyster, T.K. Accessing non-natural reactivity by irradiating nicotinamide-dependent enzymes with light. *Nature* 2016, *540*, 414-+.

244. Burns, M.; Martinez, C.A.; Vanderplas, B.; Wisdom, R.; Yu, S.; Singer, R.A. A chemoenzymatic route to chiral intermediates used in the multikilogram synthesis of a gamma secretase inhibitor. *Org. Process Res. Dev.* 2017, *21*, 871–877.

245. Huang, L.; Bittner, J.P.; de Maria, P.D.; Jakobtorweihen, S.; Kara, S. Modeling alcohol dehydrogenase catalysis in deep eutectic solvent/water mixtures. *ChemBioChem* 2020, *21*, 811–817.

246. Perez, M.; Sinisterra, J.V.; Hernaiz, M.J. Hydrolases in green solvents. *Curr. Org. Chem.* 2010, *14*, 2366–2383.

247. Sandoval, M.; Ferreras, E.; Perez-Sanchez, M.; Berenguer, J.; Sinisterra, J.V.; Hernaiz, M.J. Screening of strains and recombinant enzymes from *Thermus thermophilus* for their use in disaccharide synthesis. *J. Mol. Catal. B-Enzym.* 2012, *74*, 162–169.

248. Perez-Sanchez, M.; Sandoval, M.; Hernaiz, M.J. Bio-solvents change regioselectivity in the synthesis of disaccharides using Biolacta beta-galactosidase. *Tetrahedron* 2012, *68*, 2141–2145.

249. Perez-Sanchez, M.; Sandoval, M.; Cortes-Cabrera, A.; Garcia-Marin, H.; Sinisterra, J.V.; Garcia, J.I.; Hernaiz, M.J. Solvents derived from glycerol modify classical regioselectivity in the enzymatic synthesis of disaccharides with Biolacta beta-galactosidase. *Green Chem.* 2011, *13*, 2810–2817.

250. Bayon, C.; Cortes, A.; Aires-Trapote, A.; Civera, C.; Hernaiz, M.J. Highly efficient and regioselective enzymatic

synthesis of beta-(1 -> 3) galactosides in biosolvents. *RSC Adv.* 2013, *3*, 12155–12163.
251. Bayon, C.; Moracci, M.; Hernaiz, M.J. A novel, efficient and sustainable strategy for the synthesis of alpha-glycoconjugates by combination of a alpha-galactosynthase and a green solvent. *RSC Adv.* 2015, *5*, 55313–55320.
252. Hoyos, P.; Bavaro, T.; Perona, A.; Rumbero, A.; Tengattini, S.; Terreni, M.; Hernaiz, M.J. Highly efficient and sustainable synthesis of neoglycoproteins using galactosidases. *ACS Sustain. Chem. Eng.* 2020, *8*, 6282–6292.
253. Horvath, I.T.; Mehdi, H.; Fabos, V.; Boda, L.; Mika, L.T. γ-Valerolactone – A sustainable liquid for energy and carbon-based chemicals. *Green Chem.* 2008, *10*, 238–242.
254. Vaccaro, L. Green shades in organic synthesis. *Eur. J. Org. Chem.* 2020, *2020*, 4273–4283.
255. Zhu, S.J.; Ma, X.Q.; Su, E.Z.; Wei, D.Z. Efficient hydration of 2-amino-2,3-dimethylbutyronitrile to 2-amino-2,3-dimethylbutyramide in a biphasic system via an easily prepared whole-cell biocatalyst. *Green Chem.* 2015, *17*, 3992–3999.

7

Palladium-Catalyzed Suzuki–Miyaura Cross-Coupling in Continuous Flows

Remi Nguyen, Virinder S. Parmar, and Christophe Len

CONTENTS

7.1 Introduction 119
7.2 Mechanism of Suzuki Cross-Coupling 120
7.3 Homogeneous Suzuki–Miyaura Cross-Coupling Reaction in Continuous Flows 121
7.4 Heterogeneous Suzuki–Miyaura Cross-Coupling Reaction in Continuous Flows 124
7.5 Concluding Remarks and Future Perspectives 130
References 131

7.1 Introduction

Among the major classes of reactions in organic chemistry, *C-C* bond formation *via* a cross-coupling reaction catalyzed by transition metals is undoubtedly the most important and has been exploited very widely in recent years. Palladium, undoubtedly the most widely used metal in organic synthesis, enables the generation of complex and functionalized organic molecules and its chemistry possesses different interesting facets such as heterogeneous and homogeneous catalysis under mild experimental conditions compatible with many functional groups.[1–5] Several palladium-catalyzed cross-coupling reactions viz. Heck,[6–11] Suzuki,[12–16] Sonogashira,[17–21] Stille,[22–25] Hiyama,[26] Negishi,[27] Kumada,[28] Murahashi[29] and Buchwald-Hartwig[30,31] have been developed over the years.

Due to the current impetus in promoting green chemistry for environment protection and sustainable development, both in academia and industry, chemists have recently established catalytic reactions based on renewable resources, atom economy, less hazardous chemical steps, safer (least toxic) solvents, auxiliaries and alternative technologies such as continuous flow, microwave irradiation, ultrasound irradiation, etc. In the context of green chemistry, the employment of catalysts and alternative media, as well as of different cross-coupling reactions such as Suzuki–Miyaura in batch reactors have been reported in aqueous media or in water as sole green safer solvent *via* conventional heating or microwave irradiation.[32–43] Continuous flow chemistry as an alternative technology offers significant processing advantages including improved thermal management, mixing control, wide applicability under a range of reaction conditions, scalability, energy efficiency, waste reduction, safety and the use of heterogeneous catalysts to name a few.[44–50] It is notable that microwave-assisted continuous flow reactions were developed to associate the two technologies[51,52] Two different types of reactors, i.e., micro and meso (or flow) reactors have been used; the two devices differ in their channel dimensions: from 10 to 300 μm for the microreactor (also called milli or mini) and from 300 μm to more than 5 mm for the mesoreactor. Several advantages and disadvantages are associated with the micro- and mesoreactors; the main advantages for the microreactor are the low material input, low waste output, excellent mass transfer properties and fast diffusive mixing, while the disadvantages are the low throughput, tendency to channel blockage and high pressure drop. In the case of solid handling due to confined conditions and an increase in the concentrations to have better productivity, the use of continuous sonication could prevent clogging.[53] For the mesoreactors, the advantages are the high throughput, low pressure drop and possibility to handle solids during heterogeneous catalysis. A few disadvantages for mesoreactors are poor mass transfer property, slower mixing, etc. Different studies have been reported describing the theory and practicalities of scaled-out micro- and mesoreactors but no practical examples of large-scale production have been reported. Several palladium-catalyzed cross-coupling reactions in continuous flow reactors are available in the literature at temperatures higher than 60°C,[54–66] however only a few studies have described micro- and mesoreactors for the *C–C* bond formation at temperatures lower than 60°C. In parallel with the synthesis of low molecular weight compounds, this technique has been applied by researchers in academic and industrial groups for the production of polymers as well.[67–70] This comprehensive review includes continuous flow selective palladium-catalyzed cross-coupling reactions accompanied by high energy efficiency at temperatures ranging between 0°C and 80°C.

DOI: 10.1201/9781003089162-7

7.2 Mechanism of Suzuki Cross-Coupling

The Suzuki–Miyaura cross-coupling reaction[12–16] is among the most versatile and frequently employed methods for the *C–C* bond formation, the most sought-after chemical methodology. It consists of the coupling of organoboron compounds (organoboranes, organoboronic acids, organoboronate esters and potassium trifluoroborate) with aryl, alkenyl and alkynyl halides, a large variety of boronic acids are commercially available. The complete Suzuki–Miyaura catalytic cycle involves oxidative addition, transmetallation and reductive elimination[13–15,71–73] steps. After the formation of the catalytic species Pd(0), generated *in situ* starting from palladium Pd(II) or directly from the Pd(0) derivatives, oxidative addition of the aryl halide ArX furnishes the palladium complex [ArPdXLn] which then gets converted to a nucleophilic palladium alkoxy complex [ArPdORLn]. This complex subsequently reacts with a neutral organoboron compound Ar'$B(OH)_2$ to afford the diaryl complex [ArPdAr'Ln] as an equilibrium mixture of the *cis–trans* products, followed by the reductive elimination of the *cis* form to give the biaryl derivative Ar–Ar' and Pd(0) (Scheme 7.1).[15]

The Suzuki–Miyaura cross-coupling reactions follow a heterogeneous catalysis pathway when the palladium catalysts on appropriate supports are employed.[74] Under these conditions, the palladium Pd(II) is first released from the surface of the solid support and the leached palladium then catalyzes the cross-coupling reaction through a (quasi)homogeneous mechanism (Scheme 7.2).[75–86]

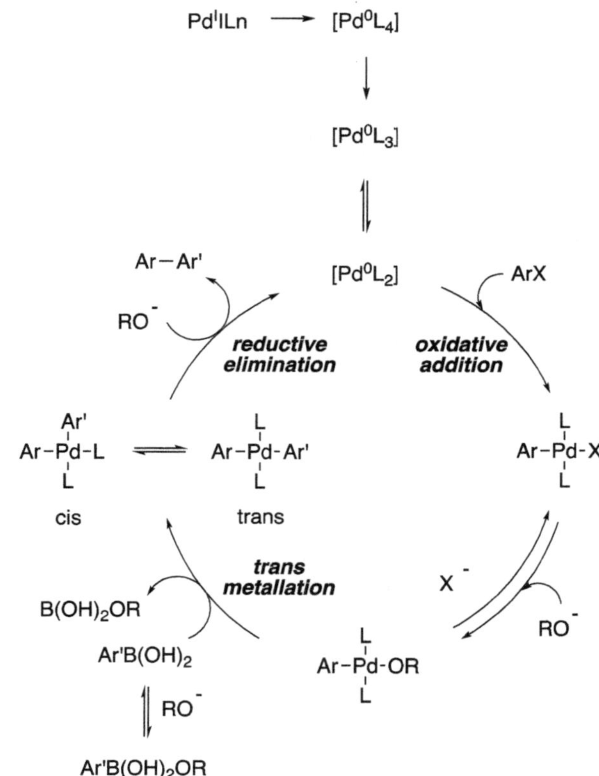

SCHEME 7.1 Mechanism of the homogeneous Suzuki–Miyaura coupling reaction.

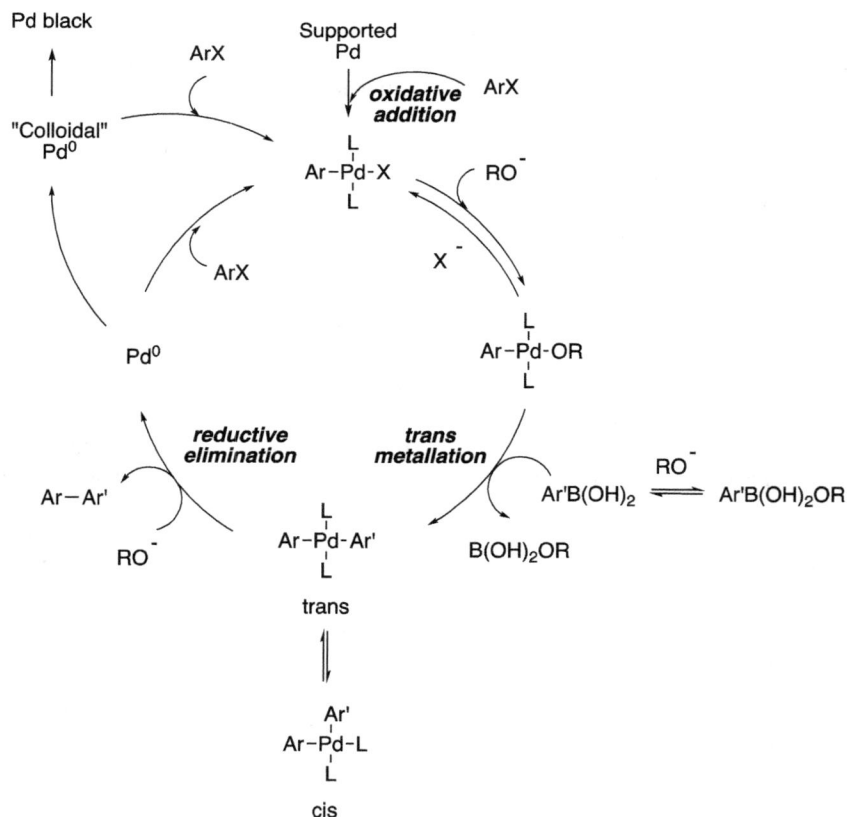

SCHEME 7.2 Mechanism of the heterogeneous Suzuki–Miyaura reaction on a solid support.

7.3 Homogeneous Suzuki–Miyaura Cross-Coupling Reaction in Continuous Flows

Buchwald reported an efficient synthesis of biaryls starting with the aryl halide substrates involving a successive lithiation/borylation/Suzuki–Miyaura cross-coupling sequence in three successive mesoreactors.[87] The aryl bromide undergoes the bromine–lithium exchange in the first step affording the corresponding aryl-lithium which reacts with the borate to form the boronate agent. One of the chief drawbacks of this nice concept was the formation of solids such as lithium triisopropylarylborate during the process; optimization of the nature of the solvent (THF and H_2O), the concentration of reagents and the use of acoustic irradiation have been reported to avoid the formation of such solids. The development of different reactors made from a perfluoroalkoxyalkane (PFA) tube having an inner diameter of 1 mm has been achieved.[87] The solutions of arylbromide in THF and that of *n*-butyllithium in hexane (1.6 M or 2.5 M) were injected simultaneously and got mixed at the T-shaped mixer; the mixture was delivered to the first reactor (reactor 1) at room temperature with a flow rate of 50–78 $\mu L \cdot min^{-1}$ and a varying residence time of 2–120 s. A solution of $B(OiPr)_3$ in THF (0.05 M) of appropriate concentration was injected with a flow rate of 1 $\mu L \cdot min^{-1}$ and mixed with the exiting stream of aryllithium derivative at a T-shape mixer; the mixed stream was introduced to the second reactor (reactor 2) at 60°C under acoustic irradiation with a residence time of 1 min. Aqueous KOH (0.87 M) and a solution of aryl halide (1.00 M) and XPhos precatalyst (a, 1 mol%) in THF were successively injected into the exiting stream with a flow rate of 100 $\mu L \cdot min^{-1}$ and 21–40 $\mu L \cdot min^{-1}$, respectively; the combined mixture was introduced to the third reactor (reactor 3) at 60°C under acoustic irradiation with a residence time of 10 min (Scheme 7.3). Reactors 2 and 3 were under constant sonication to avoid reactor clogging and ensure good mixing of the reagents during the formation of the borate and the Suzuki–Miyaura cross-coupling reaction.

Applications of the above methodology have been realized with a range of aryl halides (Figure 7.1), the limiting step of the process being the lithiation of aryl halides. Buchwald et al. have reported that the aryl bromides could be lithiated at room temperature, whatever the nature of the starting aryl bromide having different electronic and steric demands in *para, meta* and *ortho* positions, the aryllithium and then the corresponding lithium arylborate were obtained in good yields. For the third step, the Suzuki–Miyaura cross-coupling reaction with aryl bromides or chlorides with both electron-withdrawing and electron-donating substituents, afforded the target compounds in good yields. It was noteworthy that non-canonic heteroatomic halides such as quinoline, isoquinoline, pyrimidine and benzothiophene were good reagents for the continuous flow reactions.

It is noteworthy to note that the five-membered 2-heteroaromatic boronic acids are unstable at room temperature and consequently give low yields in the Suzuki–Miyaura cross-coupling reaction[88–93] in a continuous flow reactor. Consequently, Buchwald turned attention to the lithiation/borylation/Suzuki–Miyaura cross-coupling of heteroarenes such as thiophene and furan derivatives starting from furanic derivatives, selective deprotonation of the C-2 hydrogen at room temperature affording the corresponding lithium analog which reacted with borate to form the boronate agent. Then, conventional homogeneous Suzuki–Miyaura cross-coupling reaction furnished the target biaryl derivatives; after optimization of the first continuous flow process depicted in Scheme 7.3, the borylation was made at room temperature with a reduced time (6 s *vs* 60 s) and acoustic irradiation was not needed for this step in reactor 2 (Scheme 7.4).[87]

Application of this method extended the scope of the reaction[87]; starting from thiophene, 2-alkylthiophene and 2-alkylfuran, borylation in two steps was efficient, the coupling with different substituted aryls and heteroaromatic halides afforded

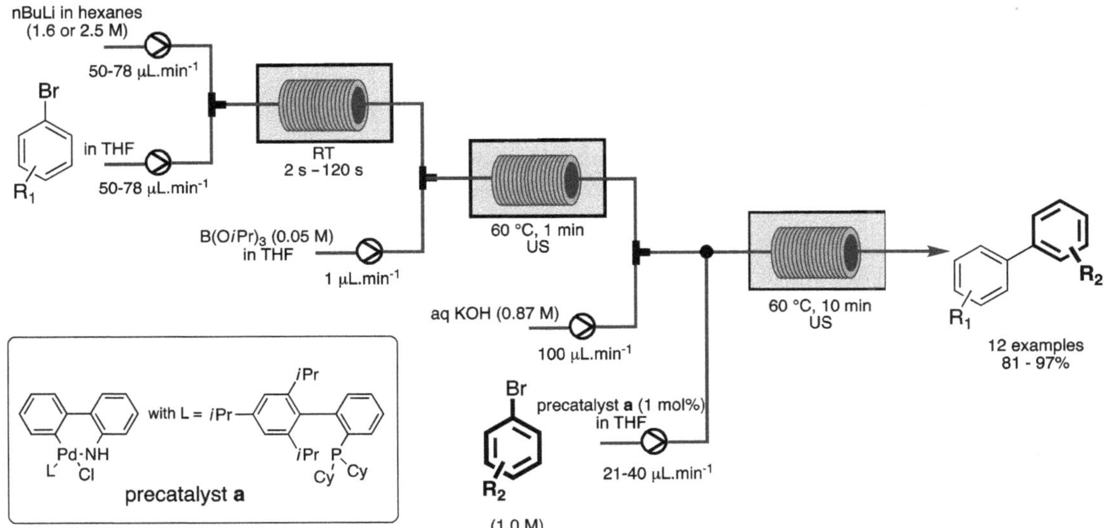

SCHEME 7.3 Lithiation/borylation/Suzuki–Miyaura cross-coupling sequence for the synthesis of biaryl derivatives in a microflow reactor.

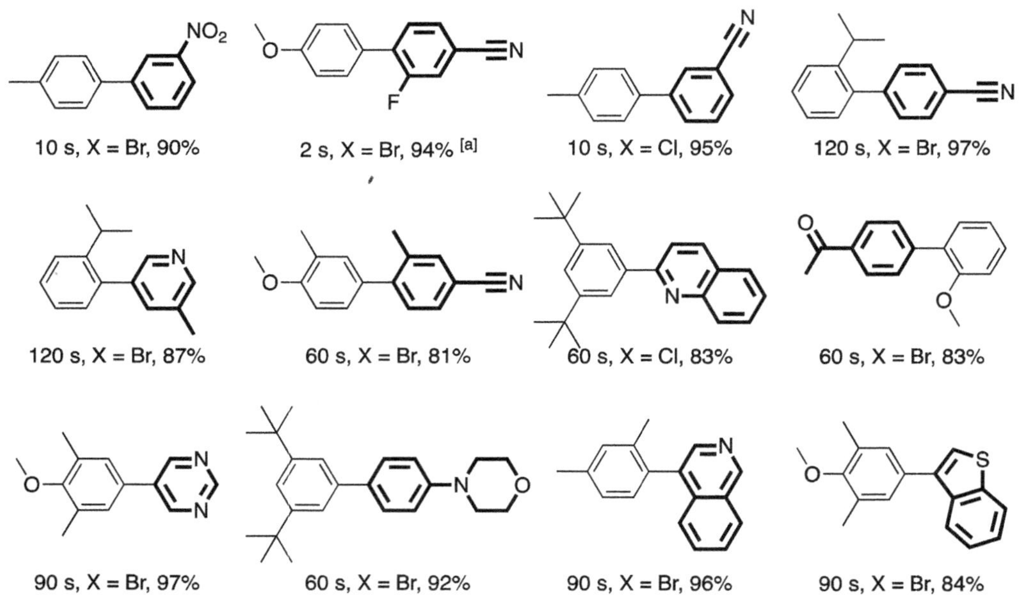

10 s, X = Br, 90% | 2 s, X = Br, 94% [a] | 10 s, X = Cl, 95% | 120 s, X = Br, 97%

120 s, X = Br, 87% | 60 s, X = Br, 81% | 60 s, X = Cl, 83% | 60 s, X = Br, 83%

90 s, X = Br, 97% | 60 s, X = Br, 92% | 90 s, X = Br, 96% | 90 s, X = Br, 84%

FIGURE 7.1 Substrate scope of continuous flow lithiation/borylation/Suzuki–Miyaura cross-coupling sequence starting from aryl halides.

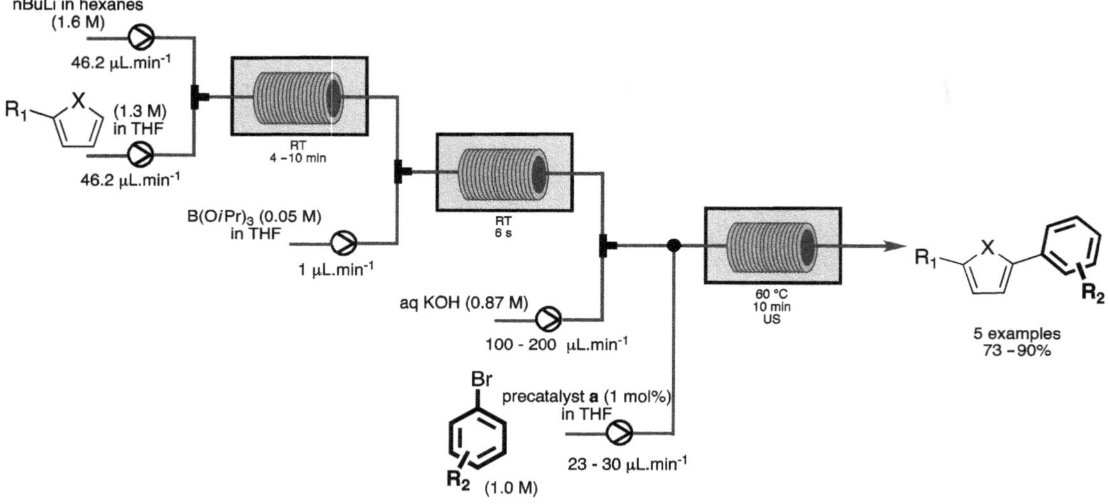

SCHEME 7.4 Lithiation/borylation/Suzuki–Miyaura cross-coupling sequence of heteroarenes with aryl halides in a continuous flow protocol at room temperature.

the target compounds in good yields (Figure 7.2). This novel process allows the use of low-cost heteroarenes instead of the more expensive and unstable 2-heteroaromatic boronic acids and 2-heteroaromatic bromides.

The synthetic potential of this methodology was demonstrated by the efficient synthesis of Diflunisal,[94,95] it was obtained in a multi-step sequence[87] in high yields. Starting from 4-bromoanisole, the lithiation/borylation followed by Suzuki–Miyaura cross-coupling with 1-bromo-2, 4-difluorobenzene permitted the synthesis of the key intermediate in the production of Diflunisal.

In order to develop an automated, droplet-flow microfluidic system for the Suzuki–Miyaura cross-coupling reaction, Buchwald and Jensen reported a systematic methodology providing key mechanistic insights.[96]

A three-step flow diazotization, iododediazotization and Suzuki–Miyaura cross-coupling reaction has been reported by Organ et al.[97]; starting from the arylamine, diazotation followed by the introduction of the iodide atom furnished the iodobenzene derivatives which was followed by the conventional Suzuki–Miyaura cross-coupling affording the biphenyl derivatives. Three reactors of different volumes were made with PFA capillary tube having an inner diameter of 1.52 mm, the residence time in the three reactors was adjustable by varying the length of the reactor tubing. The solutions of the aniline derivative in CH_3CN (0.42 M) and that of tBuONO in CH_3CN

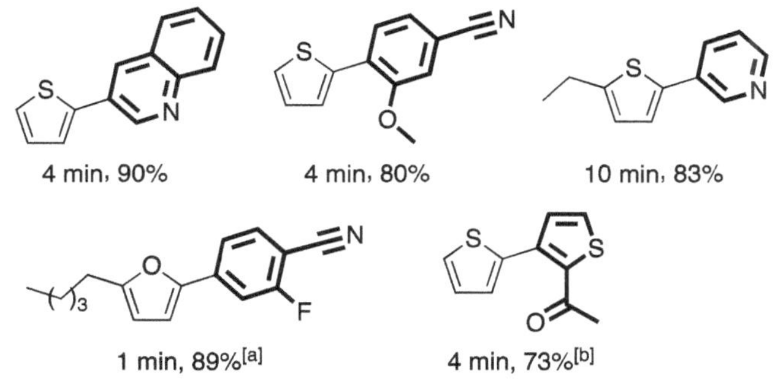

4 min, 90% 4 min, 80% 10 min, 83%

1 min, 89%[a] 4 min, 73%[b]

FIGURE 7.2 Substrate scope of continuous flow lithiation/borylation/Suzuki–Miyaura cross-coupling sequence starting from furan derivatives. (a) 0.44 M NaF aqueous solution was used instead of KOH. (b) 0.87 M KF aqueous solution was used instead of KOH.

(0.46 M) were injected simultaneously, followed by mixing with a T-mixer and injection of a solution of methanesulfonic acid in CH_3CN (0.46 M). The three solutions were run at a flow rate of 22 ΔL·min^{-1}. The mixed stream was introduced to the first reactor at room temperature with a residence time of 2.7 min, followed by the injection of a solution of nBuNI in CH_3CN (0.50 M) into the stream at the same flow rate. The combined mixture was introduced to the second reactor which was immersed in an ultrasonic bath; the residence time was 20 min at room temperature and the segmented effluent was temporarily collected in an intermediate reservoir. A solution of $[PdCl_2(PPh_3)_2]$ (5 mol%), CuI (10 mol%), *i*Pr_2NH (10 eq) in CH_3CN and a solution of boronic acid in MeOH (0.31 M) were injected simultaneously into the main stream and introduced to the third reactor at 60°C during 45 min (Scheme 7.5).[97] Application of this protocol with little variations to the production of several biphenyl compounds in satisfactory yields has been reported (Figure 7.3)[94] and could open new ways for industrial uses.[98,99]

Recently, highly reactive $Pd(OAc)_2$ was studied as homogeneous catalysts for the synthesis of biaryls having piperidylmethyl group using successive continuous lithiation, borylation and Suzuki–Miyaura coupling.[100] Using the same catalyst, sequential ortho-chemoselective Suzuki–Miyaura cross-coupling of 1,4-dibromo-2-nitrobenzene was described for the synthesis of precursors of biologically active natural product analogs containing p-terphenyl core.[101] A solution of 1,4-dibromo-2-nitrobenzene (0.125 M), 4-methoxyphenylboronic acid (0.187 M) and tetrabutyl ammonium acetate (0.375 M) in absolute ethanol and a solution $Pd(OAc)_2$ (7 mol%) in a mixture of absolute ethanol and THF (3:1, v/v) were injected simultaneously and got mixed at the T-shaped mixer; the mixture was delivered to the first reactor (reactor 1) at 25°C with a combined flow rate of 0.1 mL·min^{-1} (residence time of 140 min) (Scheme 7.6). Then, boronic acid (2 eq) was injected with a flow rate of 0.1 mL·min^{-1} and mixed with the exiting stream of bromo derivative at a T-shape mixer; the mixed stream was introduced to the second coil reactor (*reactor 2*) at 70°C with a residence time of 70 min. Even the combined residence times were long, the ortho-chemoselective permitted to produce unsymmetrically substituted *p*-terphenyl compounds in good overall yields (61–78%) (Figure 7.4).

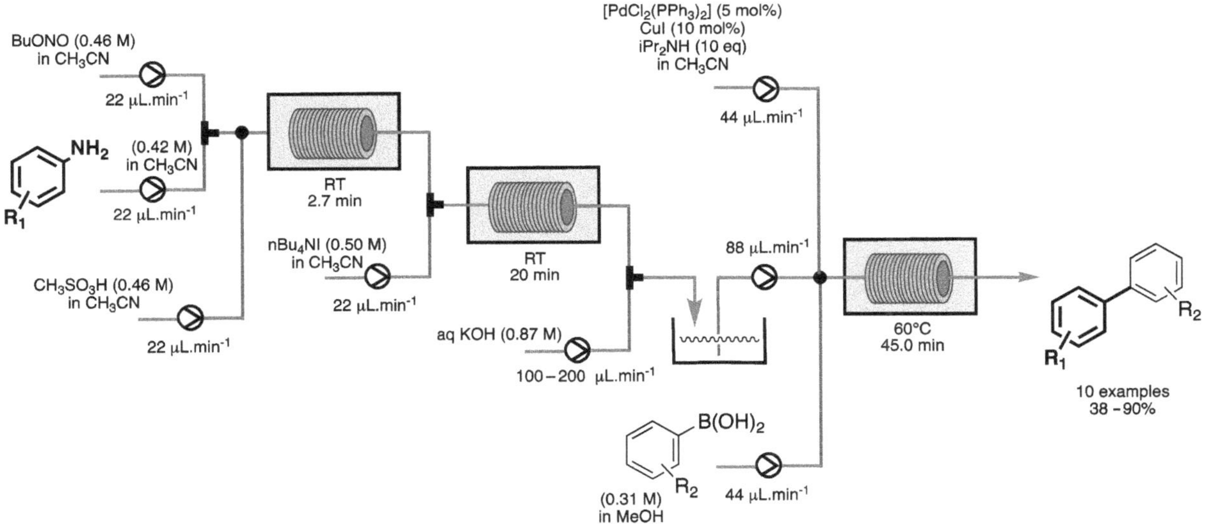

SCHEME 7.5 Diazotization/iododediazotization/Suzuki–Miyaura cross-coupling sequence of aniline derivatives with aryl halides in a continuous flow reactor.

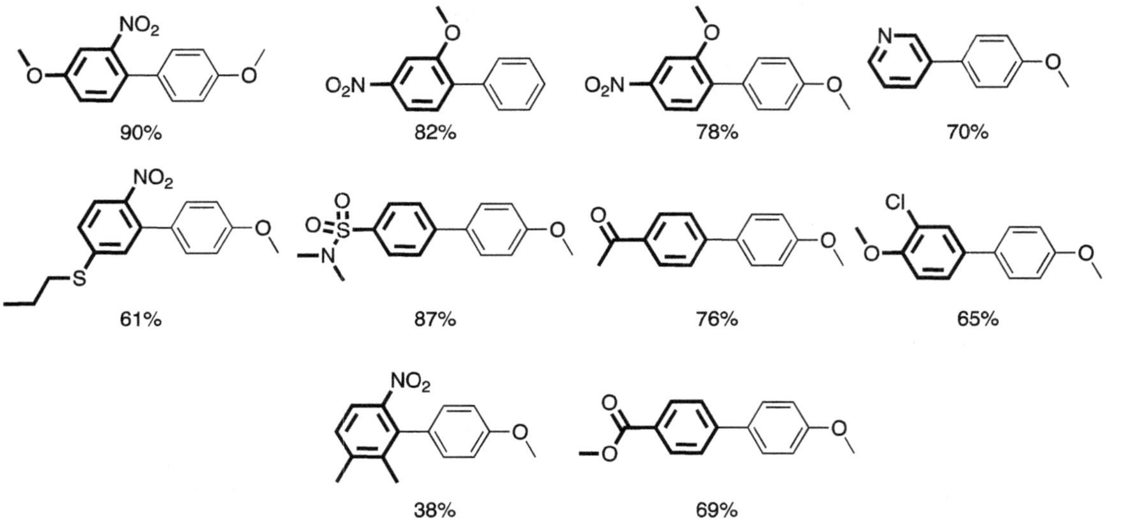

90% 82% 78% 70%

61% 87% 76% 65%

38% 69%

FIGURE 7.3 Substrate scope of continuous flow diazotization/iododediazotization/Suzuki–Miyaura cross-coupling sequence starting from aniline derivatives.

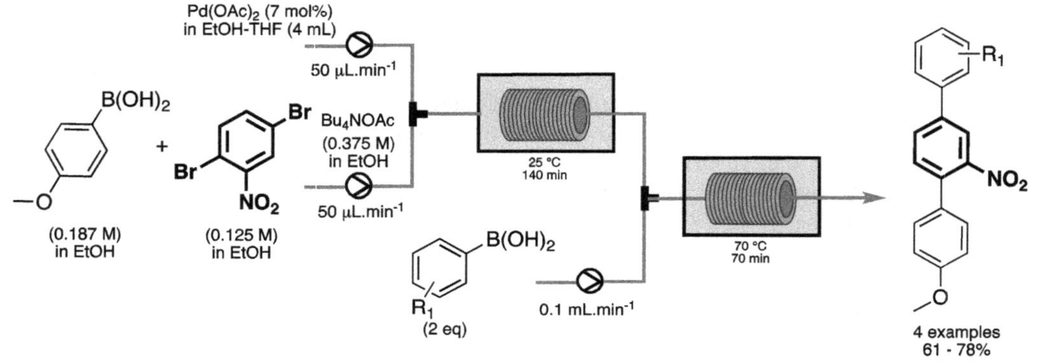

Pd(OAc)$_2$ (7 mol%) in EtOH-THF (4 mL)

50 μL.min^{-1}

B(OH)$_2$ + Br, Br, NO$_2$

Bu$_4$NOAc (0.375 M) in EtOH

(0.187 M) in EtOH (0.125 M) in EtOH

50 μL.min^{-1}

25 °C 140 min

B(OH)$_2$ R$_1$ (2 eq)

0.1 mL.min^{-1}

70 °C 70 min

R$_1$ NO$_2$

4 examples 61 - 78%

SCHEME 7.6 Ortho-chemoselective Suzuki–Miyaura cross-coupling sequence of 1,4-dibromo-2-nitrobenzene with aryl boronic acid in a continuous flow reactor.

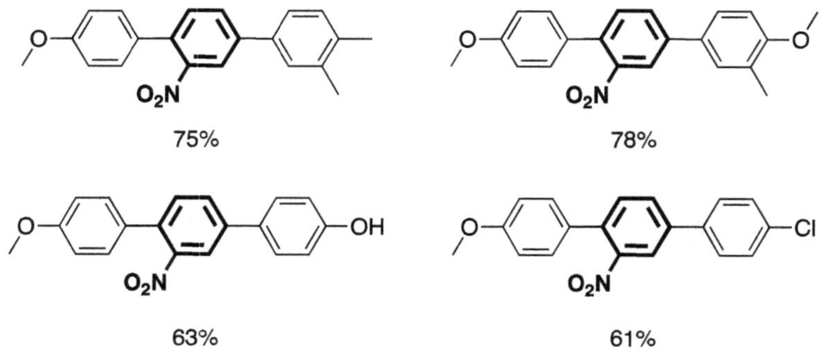

75% 78%

63% 61%

FIGURE 7.4 Substrate scope of ortho-chemoselective continuous flow Suzuki–Miyaura cross-coupling sequence starting from 1,4-dibromo-2-nitrobenzene.

7.4 Heterogeneous Suzuki–Miyaura Cross-Coupling Reaction in Continuous Flows

Suitable solid supports having Pd(II) species precursors to Pd(0) catalysts are currently available commercially but researchers prefer to design their own, home-made catalysts. Monguchi and Sajiki reported palladium on carbon-catalyzed Suzuki–Miyaura coupling reaction sequence in tandem using an efficient and continuous flow system (Scheme 7.7).[102] To investigate the scope of the reaction, a range of arylboronic acids and halogenobenzene derivatives were tested under mild conditions for 20 sec during a single-pass (Figure 7.5), the authors have reported the detection of little or no leaching (< 1 ppm).

Using the same apparatus H-Cube®, another group has recently reported the Suzuki–Miyaura cross-coupling in the

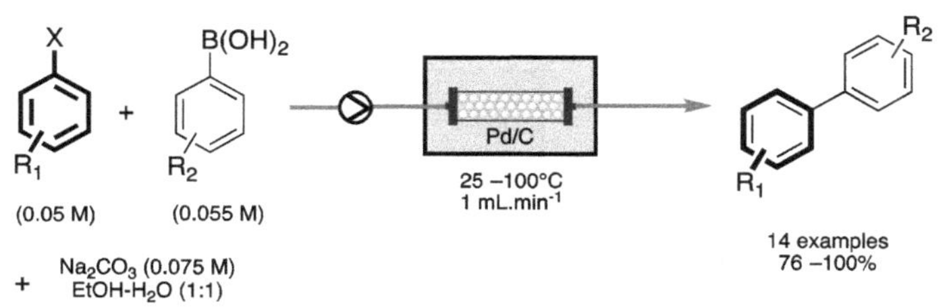

SCHEME 7.7 Suzuki–Miyaura cross-coupling in a continuous flow system employing H-Cube®.

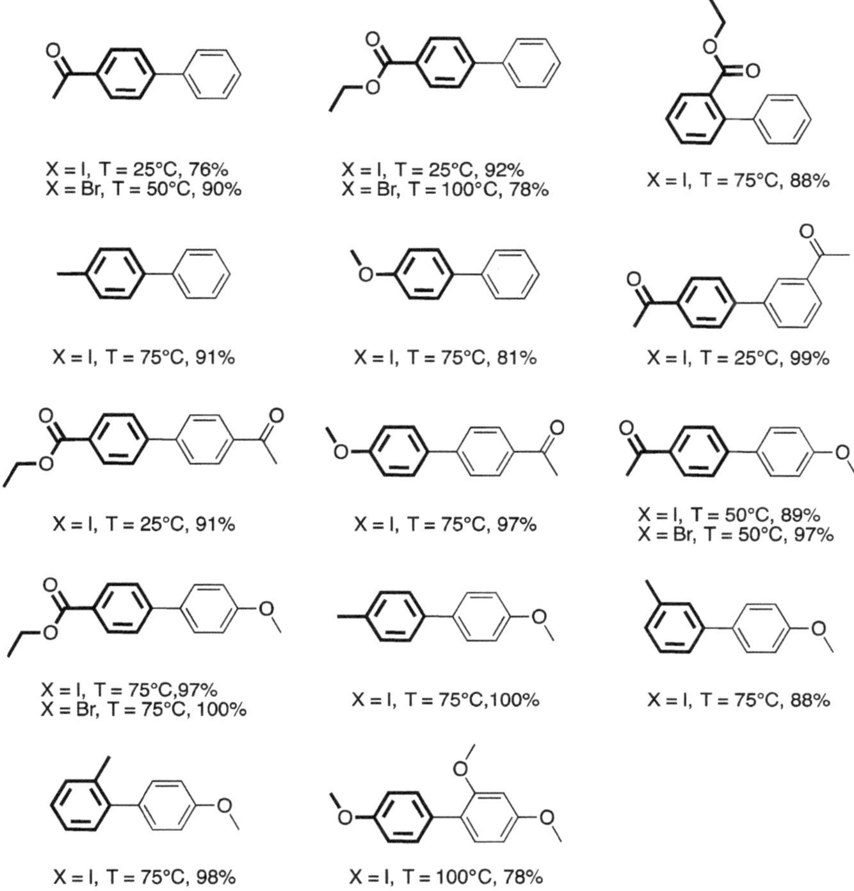

FIGURE 7.5 Substrate scope of continuous flow Suzuki–Miyaura cross-coupling sequence in a continuous flow system using H-Cube®.

presence of graphene-supported palladium nanoparticles.[103] A solution of 4-bromobenzaldehyde and phenylboronic acid dissolved in H_2O-EtOH-THF (1:1:1) mixture was injected at a flow rate of 0.2 mL·min^{-1} in less than 1 min at 135°C, the target biaryl compound was obtained in good yield with a conversion of 96% (Scheme 7.8).

In 2006, Canty et al. was the first group to develop a macroporous monolith support as a suitable substrate for anchoring a palladium complex for Suzuki–Miyaura cross-coupling continuous flow capillary microreactor.[104] Then, different groups reported the use of monolithic porous gel.[105–107] Nagaki et al. reported an efficient three-step flow sequence using Pd catalyst,[108] the aryllithium obtained from arylbromide reacted with $B(OMe)_3$. After the borylation reaction, the Suzuki–Miyaura cross-coupling reaction was carried out in the presence of immobilized Pd(0) on the polymer afforded the target 17 biaryl derivatives. In this efficient protocol, a solution of bromobenzene in THF (0.10 M) and a solution n-BuLi (0.6 M in hexane) were injected simultaneously at flow rates of 6.0 mL·min^{-1} and 1 mL·min^{-1}, respectively, in a micromixer (ID = 500 μm) and then in a reactor (ID = 1000 μm) during a residence time of 1.7 s. A solution of diluted $B(OMe)_3$ (0.12 M) in THF was injected at a flow rate of 6.0 mL·min^{-1} and the main stream was introduced into a micromixer (ID = 500 μm) and the second reactor (ID = 1000 μm) for a residence time of 2.0 s; the boronic acid solution was thus obtained was mixed with iodoaryl derivative (0.33 M) in methanol and the mixture was passed through the palladium catalyst at 100°C with a residence time of 4.7 min or at 120°C with a residence time of 9.4 min (Scheme 7.9).

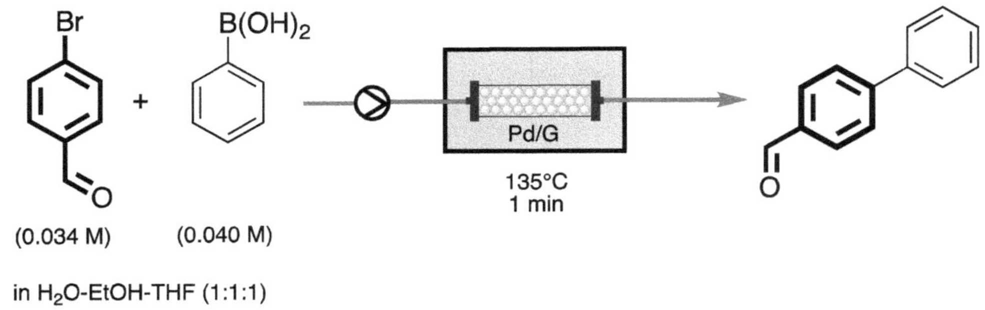

Br
B(OH)$_2$
+
Pd/G
135°C
1 min
(0.034 M) (0.040 M)
in H_2O-EtOH-THF (1:1:1)

SCHEME 7.8 Suzuki–Miyaura cross-coupling of aldehydes in a continuous flow system using H-Cube®.

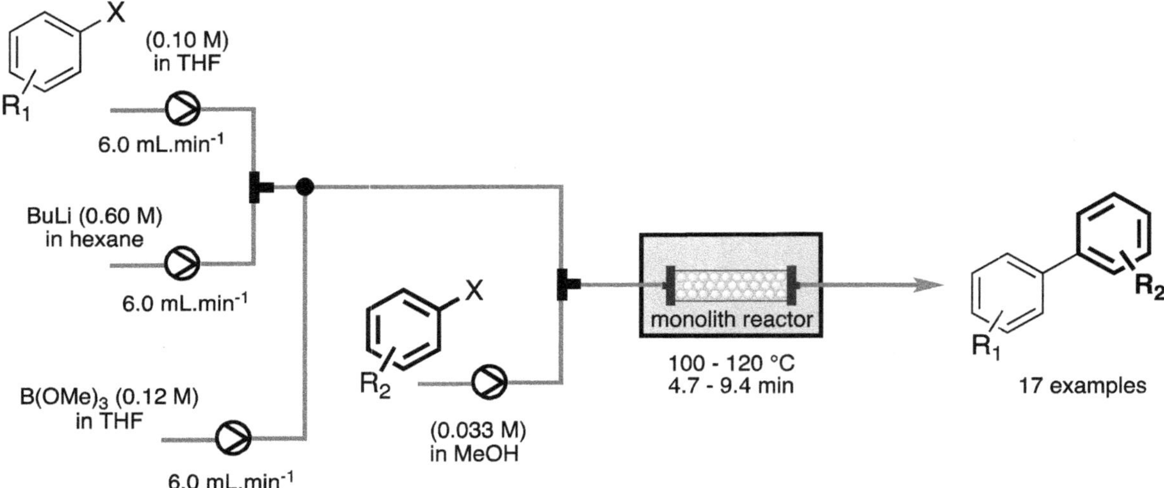

SCHEME 7.9 Halogen/lithium exchange/borylation/Suzuki–Miyaura cross-coupling sequence in tandem for the synthesis of biaryl derivatives in a microflow system.

Application of the above methodology was successfully applied to the cross-coupling of various functional aryl and heteroaryl iodides (Figure 7.6),[108] it has been seen that the cyano derivatives were well tolerated under these experimental conditions. Adapalene, the drug used for the treatment of acne was produced in 86% yield by applying this methodology.

A large-scale Suzuki–Miyaura cross-coupling reaction using solid-supported palladium (Pd^0) nano/microparticles and ultrasound irradiation (20 kHz) has recently been reported in continuous flows by Das et al.[109]

Using a "home-made" cartridges packed with spherical resin supported palladium catalysts (7% Pd/WA30) a continuous flow ligand-free Suzuki–Miyaura cross-coupling reaction of (hetero)aryl iodides, bromides and chlorides with (hetero)aryl boronic acids was developed [110] It is noticeable that his work permitted to create C–C bond starting from the iodo and bromo derivatives but chloro derivatives as well. In the case of aryl chlorides as substrates, the addition of TBAF in THF instead of the common inorganic base improved the reaction efficiency. A solution containing aryl chloride (0.125 M), aryl boronic acid (0.250 M) and TBAF (0.250 M) in THF was pumped into the 7% Pd/WA30-packed cartridge at 80°C with a flow rate of 0.05 mL·min^{-1} (Scheme 7.10). Under these conditions, it was possible to create a C–C bond starting from the corresponding chloride derivative although the reaction efficiency was affected by the electronic properties of the aromatics and the steric hindrances (Figure 7.7).

An impressive strategy based on the use of dendrimer-encapsulated Pd nanoparticles as a catalyst in flow reactors has been developed by the group of Verboom.[111,112] In contrast with the conventional heterogeneous Suzuki–Miyaura cross-coupling reaction using cartridge filled with solid catalysts, Verboom's method involved the anchoring of the Pd nanoparticles to the inner walls of the flow reactor. Aryl halides (10 mM) were mixed with boronic acid derivatives (15 mM) in ethanol at 80°C using n-Bu_4NOH (20 mM) as the base at 80°C and the solution was passed through the catalytic microreactor with a residence time of 13 min (Scheme 7.11). This strategy demonstrated the influence of dendrimers in the stabilization of the Pd NPs with low metal leaching[111,112] and allowed the formation of a large number of biaryl compounds having different types of substituents (Figure 7.8).

Another group has developed dendrimers for continuous flow Suzuki–Miyaura cross-coupling reaction[113] employing magnetic Fe_3O_4 fixation of dendron-functionalized iron oxide nanoparticles containing Pd nanoparticles. In this process, the non-covalent magnetic fixation of solid material inside the

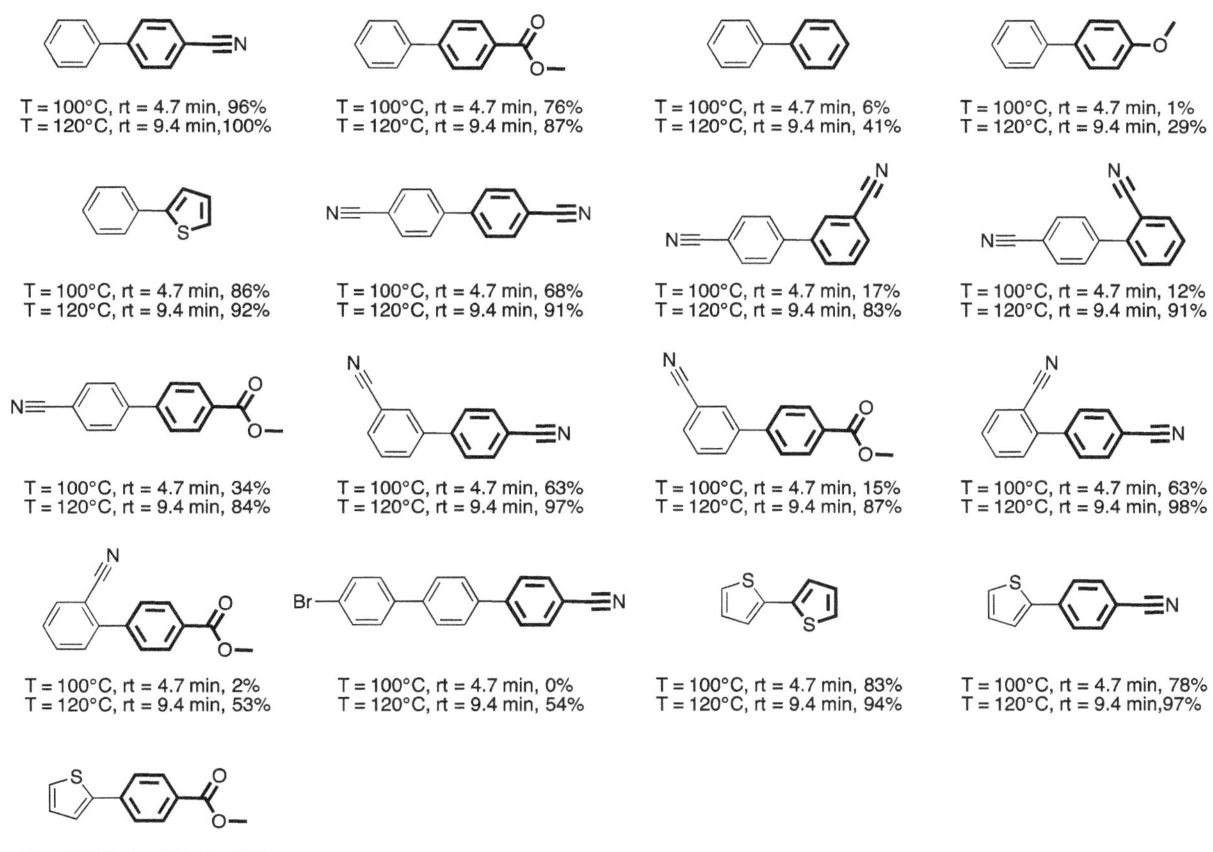

FIGURE 7.6 Substrate scope of continuous flow lithiation/borylation/Suzuki–Miyaura cross-coupling sequence in a continuous flow system using immobilized Pd on polymer monolith.

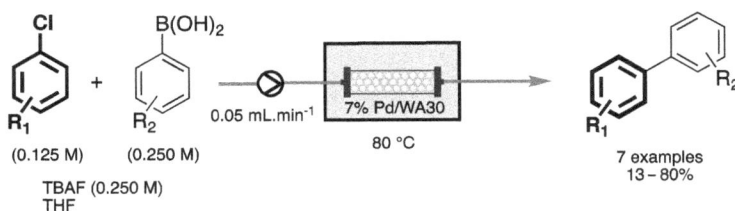

SCHEME 7.10 Suzuki–Miyaura cross-coupling of (hetero)aryl chlorides in a continuous flow system.

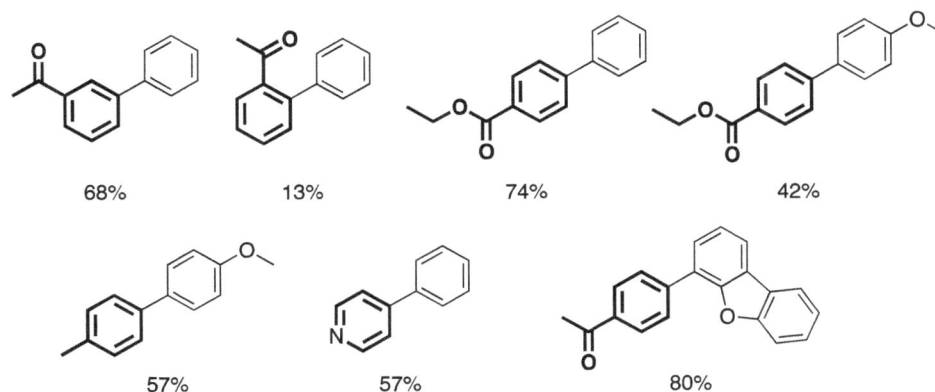

FIGURE 7.7 Scope of substrates of continuous flow Suzuki–Miyaura cross-coupling sequence using 7% Pd/WA30.

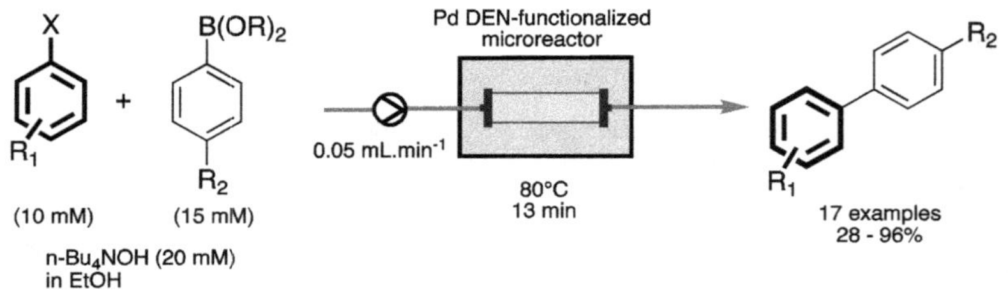

SCHEME 7.11 Suzuki–Miyaura cross-coupling in a flow system employing the dendrimer-encapsulated Pd nanoparticles.

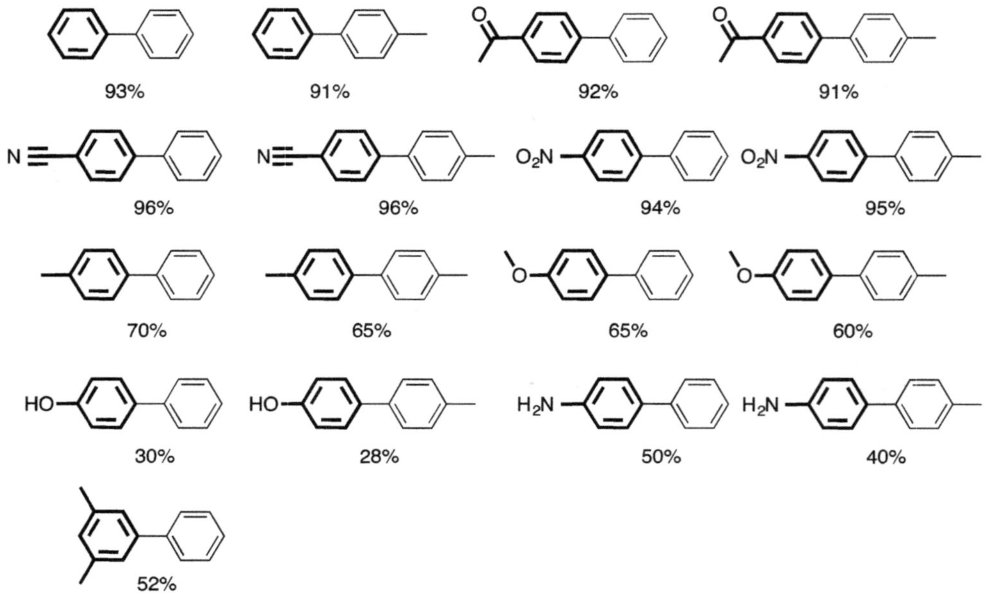

FIGURE 7.8 Substrate scope of continuous flow Suzuki–Miyaura cross-coupling sequence using Pd-encapsulated dendrimers microreactor.

glass reactor microstructures has been applied using external magnetic forces and the reversible immobilization of catalyst materials onto the walls of microchannels. Applications of this methodology have been extended successfully to produce a novel compound using 4-methoxy-1-bromobenzene and phenylboronic acid.

Another strategy reports the use of palladium nanoparticles immobilized in a polymer membrane for the Suzuki–Miyaura cross-coupling reaction.[114–116] The use of conventional silica and hierarchical bimodal porous silica gel as well as a supporting material for palladium catalysts was developed.[117] Alcazar et al. have reported an efficient cross-coupling reaction using a commercial heterogeneous silica-supported palladium catalyst and a mesoreactor,[118] these authors used a simple and efficient experimental set-up using a 6.6 mm (internal diameter) Omnifit column containing 1 g of heterogeneous catalyst and commercial boronic acids and aryl halides (Scheme 7.12). A solution of aryl halide in THF and a solution of boronic acid and base in water were pumped at 0.2 mL·min^{-1} speed with two independent pumps. The flow streams met at a T-shaped mixer and were subsequently passed through a column containing SiliaCat DPP-Pd as diphenylphosphine palladium (II) heterogeneous catalyst at 60°C with a residence time of 5 min; a biphasic solvent system such as THF-H_2O was used to ensure complete dissolution of any solid and avoiding any accompanied clogging.

Application of this strategy permitted the synthesis of several biaryl derivatives starting from halides/pseudohalides and (4-methoxyphenyl)boronic acid in excellent yields (Figure 7.9).[118] Whatever the leaving group on the benzene ring, the target biaryl derivatives were obtained in high yields; however, the use of aromatic ring bearing electron-donor groups such as 2,4-dimethoxy analogs gave lower yields (50%). It was notable that bromo- and chloropyridines provided good yields and the ester functionality was tolerated despite the use of the strong base KOH. Using this process, the authors claimed that the crude products are clean and free of phosphine ligand impurities avoiding the need for chromatographic purification; low leaching of palladium from the support and the stability of the catalyst after more than 30 cycles were additional features of this methodology. In order to further explore the scope of this useful reaction, Alcazar's group reported the use of bromobenzene and phenyltriflate as starting materials with different boronic acid derivatives [118], excellent yields were obtained even with commercial boronic acids or boronic esters.[119,120]

Recently, Kappe et al. have reported the synthesis of four commercial immobilized phosphine-based Pd catalysts.[121] One

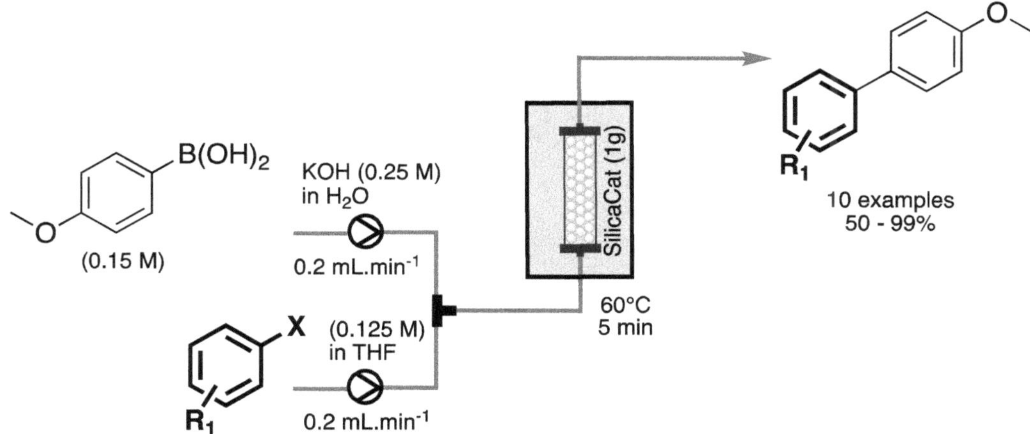

SCHEME 7.12 Continuous flow Suzuki–Miyaura cross-coupling sequence using supported SiliaCat DPPP-Pd catalyst.

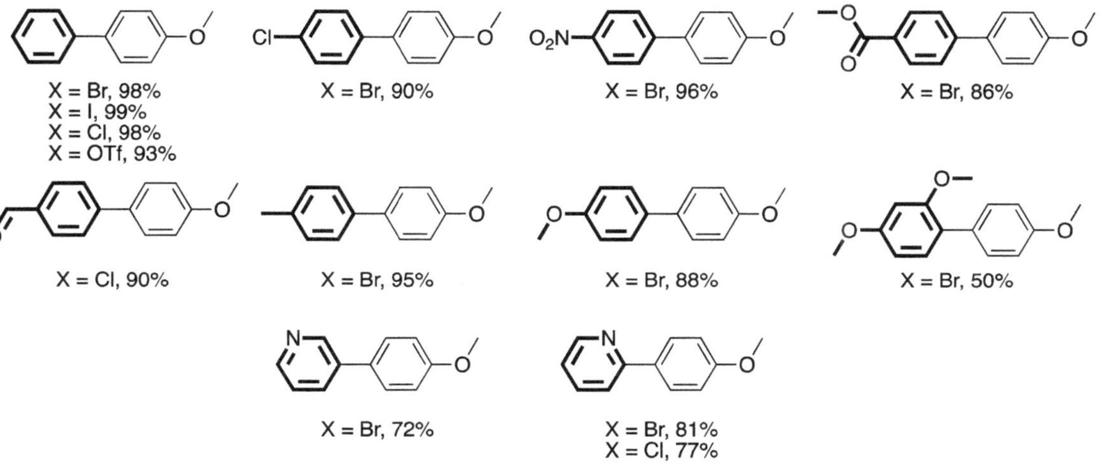

X = Br, 98%
X = I, 99%
X = Cl, 98%
X = OTf, 93%

X = Br, 90%

X = Br, 96%

X = Br, 86%

X = Cl, 90%

X = Br, 95%

X = Br, 88%

X = Br, 50%

X = Br, 72%

X = Br, 81%
X = Cl, 77%

FIGURE 7.9 Substrate scope of continuous flow Suzuki–Miyaura cross-coupling sequence with different aryl bromides and 4-methoxyphenylboronic acid on SiliaCat DPPP-Pd.

of them was SiliaCat DPP-Pd as diphenylphosphine palladium (II) heterogeneous catalyst,[122–124] the best process used two stock solutions. The first solution contained aryl halide (0.83 M) in THF and the second one contained phenylboronic acid (0.45 M) and K_2CO_3 (0.55 M) in a mixture of H_2O-EtOH (1:1). These two solutions were pumped separately at speeds of 0.055 mL·min^{-1} and 0.155 mL·min^{-1}, respectively, and mixed in a T-mixer and then introduced into the catalyst cartridge of the X-cube flow reactor at 80°C. Under these conditions, complete conversion of the starting materials was achieved in less than 20 min and almost quantitative yields of the biaryl target compound was obtained.

Metal–Organic Frameworks (MOFs)[125] and Covalent Organic Frameworks (COFs)[126] were described to assist the Suzuki–Miyaura cross-coupling reaction. COFs are different from MOFs since COFs do not contain metallic ions or heavy elements as part of their structures. $Pd(OAc)_2$@COF-300 was developed for the Suzuki–Miyaura cross-coupling reaction in continuous flows; the mixture of bromobenzene and phenylboronic acid in a solution of MeONa (2 M) in MeOH was injected into a glass column (Omnifit column with a volume of 6.3 mL) which was filled with glass beads (2 mm) and $Pd(OAc)_2$@COF-300 (100 mg). The residence time was 20 min and the temperature was maintained at 60°C; under these conditions, the maximal conversion was obtained between 20 and 40 min with a very high degree of selectivity.

An efficient approach has recently been reported for the production of furan-based biaryls[127] in which a mixture of the aryl halide, a boronic acid derivative and TBAF in methanol (0.37 M) was injected into an X-cube fitted with an FC1032 catalyst at a flow rate of 0.5 mL·min^{-1} at 120°C for 2 hours (Scheme 7.13). A series of furan derivatives were obtained in good yields (82–92%) using FC1032 catalyst as *t*-butyl based palladium polymer (Figure 7.10).[127]

An identical process has been developed by substituting the FC1032 catalyst with the $PdCl_2(PPh_3)_2$ DVB catalyst at a flow rate of 0.3 mL·min^{-1} at 120°C for 3 hours, it is well established that $PdCl_2(PPh_3)_2$ DVB is a more efficient catalyst than FC1032. It is significant to note that even by starting with the deactivated aryl bromides or aryl chlorides, $PdCl_2(PPh_3)_2$ DVB catalyst afforded the target furan-based biaryls in 83–92% yields (Figure 7.11).[127]

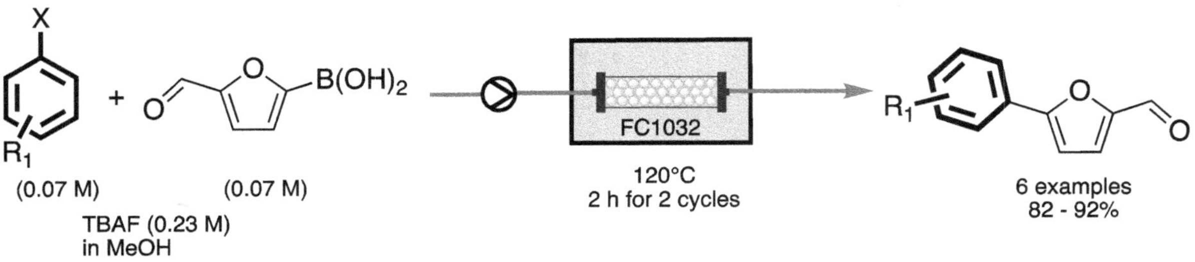

SCHEME 7.13 Suzuki–Miyaura cross-coupling in a continuous flow system X-Cube using the catalyst FC1032.

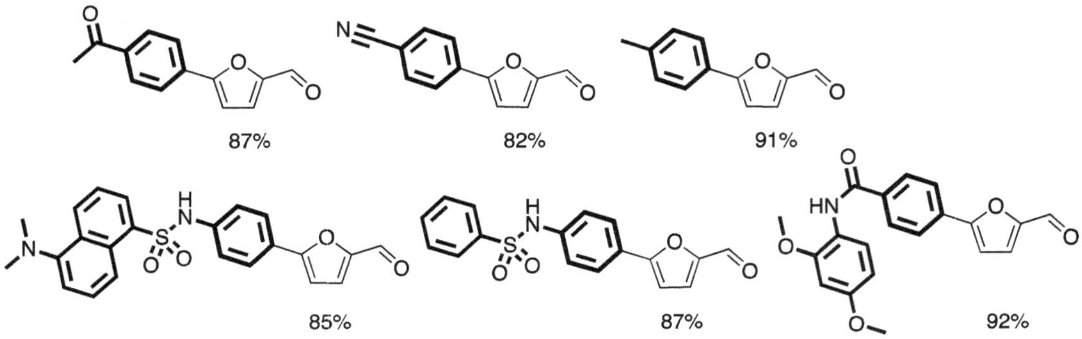

FIGURE 7.10 Substrate scope of the continuous flow Suzuki–Miyaura cross-coupling sequence using the catalyst FC1032.

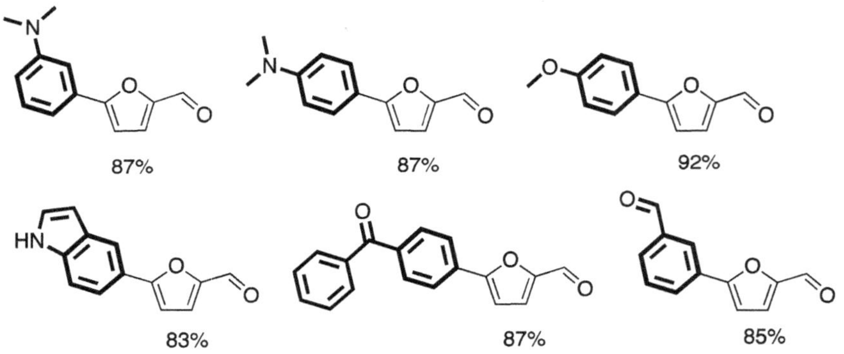

FIGURE 7.11 Substrate scope of continuous flow Suzuki–Miyaura cross-coupling sequence using the catalyst $PdCl_2(PPh_3)_2$ DVB.

7.5 Concluding Remarks and Future Perspectives

The main focus of this review has been the development of protocols for carrying out the Suzuki–Miyaura cross-coupling reactions in continuous flows. Homogeneous Suzuki–Miyaura cross-coupling reactions have been elaborated in two different elegant and modular strategies: (i) lithiation/borylation/homogeneous Suzuki–Miyaura sequence using a three-step triphasic flow system and (ii) diazotization/iododediazotization/homogeneous Suzuki–Miyaura sequence using a three-step triphasic flow system. Some groups have used Pd(0) as the active catalyst while others have preferred to start with Pd(II) as a precursor of Pd(0). Whatever the type of catalyst: homogeneous, heterogeneous, Pd(II) or Pd(0), the residence times were less than one hour and the Pd loading was low compared with the conversion, yield and selectivity. As mentioned by Kappe, "palladium which is leached from the support is most likely responsible for the catalysis, thus suggesting a (quasi)homogeneous mechanism", homogeneous metal catalyst/ligand system should probably be more efficient if the recycling of the catalyst could be improved. Future works in the field should focus on this aspect.

The chief parameters in these continuous flow reaction sequences are the concentrations, temperature, pressure, pumps, pipes, solvents and reactors, etc. Depending on these variations, the lifetime of the process could be longer or shorter and also the overall process could be greener and sustainable. Currently, no realistic study has been published on these crucial aspects. Other future work needs to focus on the nature of the materials, the designs of the reactor vis-à-vis the microfluidic system, the possibilities to work in high concentrations, etc. Chemists and chemical engineers, both in academia and industry need to pave the way to a more widespread implementation of continuous flow strategies for the production of industrially relevant products in the future. We strongly believe that the advantages of combining Suzuki–Miyaura cross-coupling reaction and flow processes presented in this review can stimulate further advances in the field from the younger generations of chemists and chemical engineers for the benefit of the chemical industry and humanity in the years to come.

REFERENCES

1. Yin, L., and Liebscher, J. Carbon–carbon coupling reactions catalyzed by heterogeneous palladium catalysts. *Chem. Rev.*, 2007, 107, 133–173.
2. Barnard, C. Palladium-catalyzed C-C coupling: then and now. *Platinium Metals Rev.*, 2008, 52, 38–45.
3. Johansson Seechurn, C.C.C., Kitching, M.O., Colacot, T.J., and Sniekus, V. Palladium-catalyzed cross-coupling: a historical contextual perspective to the 2010 Nobel prize. *Angew. Chem. Int. Ed.*, 2012, 51, 5062–5085.
4. Agrofoglio, L.A., Gillaizeau, I., and Saito, Y. Palladium-assisted routes to nucleosides. *Chem. Rev.*, 2003, 103, 1875–1916.
5. Polshettiwar, V., Len, C., and Fihri, A. Silica-supported palladium: sustainable catalysts for cross-coupling reactions. *Coord. Chem. Rev.*, 2009, 253, 2599–2626.
6. Heck, R.F. Acylation, methylation, and carboxyalkylation of olefins by group VIII metal derivatives. *J. Am. Chem. Soc.*, 1968, 90, 5518–5526.
7. Mizoroki, T., Mori, K., and Ozaki, A. Arylation of olefin with aryl iodide catalyzed by palladium. *Bull. Chem. Soc. Jpn.*, 1971, 44, 581.
8. Heck, R.F., and Nolley, J.P. Palladium-catalyzed vinylic hydrogen substitution reactions with aryls benzyl, and styryl halides. *J. Org. Chem.*, 1972, 37, 2320–2322.
9. Heck, R.F. Palladium-catalyzed reactions of organic halides with olefins. *Acc. Chem. Res.*, 1979, 12, 146–151.
10. Dieck, H.A., and Heck, R.F. Organophosphinepalladium complexes as catalysts for vinylic hydrogen substitution reactions. *J. Am. Chem. Soc.*, 1974, 96, 1133–1136.
11. Beletskaya, I.P., and Cheprakov, A.V. The Heck reaction as a sharpening stone of palladium catalysis. *Chem. Rev.*, 2000, 100, 3009–3066.
12. Miyaura, N., Yamada, K., and Suzuki, A. A new stereospecific cross-coupling by the palladium-catalyzed reaction of 1-alkenylboranes with 1-alkenyl or 1-alkynyl halides. *Tetrahedron Lett.*, 1979, 20, 3437–3440.
13. Miyaura, N., Yanagi, T., and Suzuki, A. The palladium-catalyzed cross-coupling reaction of phenylboronic acid with haloarenes in the presence of bases. *Synth. Commun.*, 1981, 11, 513–519.
14. Miyaura, N., and Suzuki, A. Palladium-catalyzed cross-coupling reactions of organoboron compounds. *Chem. Rev.*, 1995, 95, 2457–2483.
15. Suzuki, A. Recent advances in the cross-coupling reactions of organoboron derivatives with organic electrophiles. *J. Organomet. Chem.*, 1999, 576, 147–168.
16. Suzuki, A. Cross-coupling reactions of organoboranes: an easy way to construct C-C bonds. *Angew. Chem. Int. Ed.*, 2011, 50, 6722–6764.
17. Sonogashira, K., Tohda, Y., and Hagihara, N. A convenient synthesis of acetylenes: catalytic substitutions of acetylenic hydrogen with bromoalkenes, iodoarenes and brompyridines. *Tetrahedron Lett.*, 1975, 16, 4467–4470.
18. Paterson, I., Davies, R.D., and Marquez, R. Total synthesis of the *Callipeltoside aglycon*. *Angew. Chem. Int. Ed.*, 2001, 40, 603–607.
19. Toyota, M., Komori, C., and Ihara, M. A concise formal total synthesis of Mappicine and Nothapodytine B via an intramolecular hetero Diels-Alder reaction. *J. Org. Chem.*, 2000, 65, 7110–7113.
20. Nicolaou, K.C., and Dai, W.M. Chemistry and biology of the enediyne anticancer antibiotics. *Angew. Chem. Int. Ed.*, 1991, 30, 1387–1416.
21. Wu, R., Schumm, J.S., Pearson, D.L., and Tour, J.M. Convergent synthetic routes to orthogonally fused conjugated oligomers directed toward molecular scale electronic device applications. *J. Org. Chem.*, 1996, 61, 6906–6921.
22. Milstein, D., and Stille, J.K. A general, selective, and facile method for ketone synthesis from acid chlorides and organotin compounds catalyzed by palladium. *J. Am. Chem. Soc.*, 1978, 100, 3636–3638.
23. Milstein, D., and Stille, J.K. Palladium-catalyzed coupling of tetraorganotin compounds with aryl and benzyl halides. Synthetic utility and mechanism. *J. Am. Chem. Soc.*, 1979, 101, 4992–4998.
24. Stille, J.K. The palladium-catalyzed cross-coupling reactions of organotin reagents with organic electrophiles. *Angew. Chem. Int. Ed.*, 1986, 25, 508–524.
25. Espinet, P., and Echavarren, A.M. The mechanisms of Stille reaction. *Angew. Chem. Ed. Int.*, 2004, 43, 4704–4734.
26. Hiyama, T., and Hatanaka, Y. Palladium-catalyzed cross-coupling reaction of organometalloids through activation with fluoride ion. *Pure Appl. Chem.*, 1994, 66, 1471–1478.
27. King, A.O., Okukado, N., and Negishi, E. Highly general stereo-, regio-, and chemo-selective synthesis of terminal and internal conjugated enzymes by the Pd-catalyzed reaction of alkynylzinc reagents with alkenyl halides. *J. Chem. Soc., Chem. Commun.*, 1977, 683–684.
28. Tamao, K., Sumitani, K., and Kumada, M. Selective carbon-carbon bond formation by cross-coupling of Grignard reagents with organic halides. Catalysis by nickel-phosphine complexes. *J. Am. Chem. Soc.*, 1972, 94, 4374–4376.
29. Yamamura, M., Moritani, I., and Murahashi, S.I. The reaction of σ-vinylpalladium complexes with alkyllithiums. Stereospecific syntheses of olefins from vinyl halides and alkyllithiums. *J. Organomet. Chem.*, 1975, 91, C39–C42.
30. Hartwig, J.F. In *Handbook of organopalladium chemistry in organic synthesis*; Negishi, E., Ed.; Wiley-Interscience: New York, NY, 2003, p. 1051.
31. Jiang, L., and Buchwald, S.L. In *Metal-catalyzed cross-coupling reactions*; Meijere, A., Diederich, F., Ed.; Wiley-VCH, Weinheim, Germany, 2004, p. 699.
32. Polshettiwar, V., Decottignies, A., Len, C., and Fihri, A. Suzuki-Miyaura cross-coupling reactions in aqueous media: green and sustainable syntheses of biaryls. *ChemSusChem*, 2010, 5, 502–522.
33. Fihri, A., Luart, D., Len, C., Solhi, A., Chevrin, C., and Polshettiwar, V. Suzuki-Miyaura cross-coupling reactions

with low catalyst loading: a green and sustainable protocol in pure water. *Dalton Trans.*, 2011, 40, 3116–3121.

34. Sartori, G., Enderlin, G., Herve, G., and Len, C. Highly effective synthesis of C-5-substituted 2'-deoxyuridine using Suzuki-Miyaura cross-coupling in water. *Synthesis*, 2012, 44, 767–772.
35. Hassine, A., Sebti, S., Solhy, A., Zahouily, M., Len, C., Hedhili, M.N., and Fihri, A. Palladium supported on natural phosphate: catalyst for Suzuki coupling reactions in water. *Appl. Catal. A Gen.*, 2013, 450, 13–18.
36. Sartori, G., Enderlin, G., Herve, G., and Len, C. New efficient approach for the ligand-free Suzuki-Miyaura reaction of 5-iodo-2'-deoxyuridine in water. *Synthesis*, 2013, 45, 330–333.
37. Decottignies, A., Fihri, A., Azemar, G., Djedaini-Pilard, F., and Len, C. Ligandless Suzuki-Miyaura reaction in neat water with or without native β-cyclodextrin as additive. *Catal. Commun.*, 2013, 32, 101–107.
38. Gallagher-Duval, S., Herve, G., Sartori, G., Enderlin, G., and Len, C. Improved microwave-assisted ligand free Suzuki-Miyaura cross-coupling of 5-iodo-2'-deoxyuridine in pure water. *New J. Chem.*, 2013, 37, 1989–1995.
39. Enderlin, G., Sartori, G., Herve, G., and Len, C. Synthesis of 6-aryluridines via Suzuki-Miyaura cross-coupling reaction at room temperature under aerobic ligand-free conditions in neat water. *Tetrahedron Lett.*, 2013, 54, 3374–3377.
40. Herve, G., Sartori, G., Enderlin, G., Mackenzie, G., and Len, C. Palladium-catalyzed Suzuki reaction in aqueous solvents applied to unprotected nucleosides and nucleotides. *RSC Adv.*, 2014, 4, 18558–18594.
41. Herve, G., and Len, C. First ligand-free, microwave-assisted, Heck cross-coupling reaction in pure water on a nucleoside – Application to the synthesis of antiviral BVDU. *RSC Adv.*, 2014, 4, 46926–46929.
42. Lussier, T., Herve, G., Enderlin, G., and Len, C. Original access to 5-aryluracils from 5-iodo-2'-deoxyuridine via a microwave-assisted Suzuki-Miyaura cross-coupling/deglycosylation sequence in pure water. *RSC Adv.*, 2014, 4, 46218–46223.
43. Hassine, A., Bouhrara, M., Sebti, S., Solhy, A., Luart, D., Len, C., and Fihri, A. Natural phosphate-supported palladium: a highly efficient and recyclable catalyst for the Suzuki-Miyaura coupling under microwave irradiation. *Curr. Org. Chem.*, 2014, 18, 3141–3148.
44. Haswell, S.J., and Watts, P. Green chemistry: synthesis in micro reactors. *Green Chem.*, 2003, 5, 240–249.
45. Frost, C.G., and Mutton, L. Heterogeneous catalytic synthesis using microreactor technology. *Green Chem.*, 2010, 12, 1687–1703.
46. Wiles, C., and Watts, P. Continuous flow reactors: a perspective. *Green Chem.*, 2012, 14, 38–54.
47. Newman, S.G., and Jensen, K.F. The role of flow in green chemistry and engineering. *Green Chem.*, 2013, 15, 1456–1472.
48. Wiles, C., and Watts, P. Continuous process technology: a tool for sustainable production. *Green Chem.*, 2014, 16, 55–62.
49. Vaccaro, L., Lanari, D., Marrochi, A., and Strappaveccia, G. Flow approaches towards sustainability. *Green Chem.*, 2014, 16, 3680–3704.
50. Len, C., Bruniaux, S., Delbecq, F., and Parmar, V.S. Palladium-catalyzed Suzuki-Miyaura cross-coupling in continuous flow. *Catalysts*, 2017, 7, 146.
51. Daviot, L., Len, T., Li, C.S.K., and Len, C. Microwave-assisted homogeneous acid catalysis and chemoenzymatic synthesis of dialkylsuccinate in a flow reactor. *Catalysts*, 2019, 9, 272.
52. Monguchi, Y., Ichikawa, T., Yamada, T., Sawama, Y., and Sajiki, H. Continuous-flow Suzuki-Miyaura and Mizoroki-Heck reactions under microwave heating conditions. *Chem. Rec.*, 2019, 19, 3–14.
53. Falb, S., Tomaiuolo, G., Perazzo, A., Hodgson, P., Yaseneva, P., Zakrzewski, J., Guido, S., Lapkin, A., Woodward, R., and Meadows, R.E. A continuous process for Buchwald-Hartwig amination at micro-, lab-, and mesoscale using a novel reactor concept. *Org. Process Res. Dev.*, 2016, 20, 558–565.
54. Gemoets, H.P. L., Hessel, V., and Noel, T. Aerobic C-H olefination of indoles via a cross-dehydrogenative coupling in continuous flow. *Org. Lett.*, 2014, 16, 5800–5803.
55. Reynolds, W.R., Plucinski, P., and Frost, C.G. Robust and reusable supported palladium catalysts for cross-coupling reactions in flow. *Catal. Sci. Technol.*, 2014, 4, 948–954.
56. Bourne, S.L., O'Brien, M., Kasinathan, S., Koos, P., Tolstoy, P., Hu, D.X., Bates, R.W., Martin, B., Schenkel, B., and Ley, S.V. Flow chemistry syntheses of styrenes, unsymmetrical stilbenes and branched aldehydes. *ChemCatChem.*, 2013, 5, 159–172.
57. Peeva, L., da Silva Burgal, J., Vartak, S., and Livingston, A.G. Experimental strategies for increasing the catalyst turnover number in a continuous Heck coupling reaction. *J. Catal.*, 2013, 306, 190–201.
58. Sharma, S., Basavaraju, K.C., Singh, A.K., and Kim, D.P. Continuous recycling of homogeneous Pd/cu catalysts for cross-coupling reactions. *Org. Lett.*, 2014, 16, 3974–3977.
59. Peeva, L., Arbour, J., and Livingston, A. On the potential of organic solvent nanofiltration in continuous Heck coupling reactions. *Org. Process. Res. Dev.*, 2013, 17, 967–975.
60. Domier, R.C., Moore, J.N., Shaughnessy, K.H., and Hartman, R.L. Kinetic analysis of aqueous-phase Pd-catalyzed, Cu-free direct arylation of terminal alkynes using a hydrophilic ligand. *Org. Process. Res. Dev.*, 2013, 17, 1262–1271.
61. Tukacs, J.M., Jones, R.V., Darvas, F., Dibo, G., Lezsak, G., and Mika, L.T. Synthesis of γ-valerolactone using a continuous flow reactor. *RSC Adv.*, 2013, 3, 16283–16287.
62. Yang, G.R., Bae, G., Choe, J., Lee, S., and Song, K.H. Silica-supported palladium-catalyzed Hiyama cross-coupling reactions using continuous flow system. *Bull. Korean Chem. Soc.*, 2010, 31, 250–252.
63. Phan, N.T.S., Brown, D.H., and Styring, P. A facile method for catalyst immobilization on silica: nickel-catalyzed Kumada reactions in mini-continuous flow and batch reactors. *Green Chem.*, 2004, 6, 526–532.
64. Alonso, N., Zane Miller, L., M. de Munoz, J. Alcazar, and Tyler McQuade, D. Continuous synthesis of organozinc halides coupled to Negishi reactions. *Adv. Synth. Catal.*, 2014, 356, 3737–3741.

65. Tan, L.M., Sem, Z.Y., Chong, W.Y., Liu, X., Hendra, L., Kwan, W.L., and Ken Lee, C.L. Continuous flow Sonogashira C-C coupling using a heterogeneous palladium-copper dual reactor. *Org. Lett.*, 2013, 15, 65–67.
66. Zhang, H.H., Xing, C.H., Bouobda Tsemo, G., and Hu, Q.S. t-Bu_3P-coordinated 2-phenylaniline-based palladacycle complex as a precatalyst for the Suzuki cross-coupling polymerization of aryl dibromides with aryldiboronic acids. *ACS Macro Lett.*, 2013, 2, 10–13.
67. Schulte, N., Breuning, E., and Spreitzer, H. Method for the production of polymers, US 20080207851 A1 (dec 28, 2004)
68. Seyler, H., Jones, D.J., Holmes, A.B., and Wong, W.W.H. Continuous flow synthesis of conjugated polymers. *Chem. Commun.*, 2012, 48, 1598–1600.
69. Gao, M., Subbiah, J., Geraghty, P.B., Chen, M., Purushothaman, B., Chen, X., Qin, T., Vak, D., Scholes, F.H., Watkins, S.E., Skidmore, M., Wilson, G.J., Holmes, A.B., Jones, D.J., and Wong, W.W.H. Development of a high-performance donor-acceptor conjugated polymer: synergy in materials and device optimization. *Chem. Mater.*, 2016, 28, 3481–3487.
70. Mitchell, V.D., and Wong, W.W.H. In *Synthetic methods for conjugated polymer and carbon materials*; Leclerc, M., Morin, J.-F., Ed.; Wiley-VCH, Weinheim, Germany, 2017, p. 65.
71. Amatore, C., Jutand, A., and Le Duc, G. Kinetic data for the transmetallation/reductive elimination in palladium-catalyzed Suzuki-Miyaura reactions: unexpected triple role of hydroxide ions used as base. *Chem.–Eur. J.*, 2011, 17, 2492–2503.
72. Amatore, C., Jutand, A., and Le Duc, G. Mechanistic origin of antagonist effects of usual anionic bases (OH^-, CO_3^{2-}) as modulated by their countercations (Na^+, Cs^+, K^+) in palladium-catalyzed Suzuki-Miyaura reactions. *Chem.–Eur. J.*, 2012, 18, 6616–6625.
73. Carrow, B. P., and Hartwig, J. F. Distinguishing between pathways for transmetalation in Suzuki-Miyara reactions. *J. Am. Chem. Soc.*, 2011, 133, 2116–2119.
74. Cantillo, D., and Kappe, C.O. Immobilized transition metals as catalysts for cross-couplings in continuous flow – A critical assessment of the reaction mechanism and metal leaching. *ChemCatChem*, 2014, 6, 3286–3305.
75. Narayanan, R., and El-Sayed, M. A. Effect of catalysis on the stability of metallic nanoparticles: Suzuki reaction catalyzed by PVP-palladium nanoparticles. *J. Am. Chem. Soc.*, 2003, 125, 8340–8347.
76. Narayanan, R., and El-Sayed, M. A. Effect of catalytic activity on the metallic nanoparticle size distribution: electron-transfer reaction between $Fe(CN)_6$ and thiosulfate ions catalyzed by PVP-platinium nanoparticles. *J. Phys. Chem. B*, 2003, 107, 12416–12424.
77. De Vries, A.H.M., Mulders, J., Mommers, J.H.M., Henderickx, H.J.W., and De Vries, J.G. Homeopathic ligand-free palladium as a catalyst in the Heck reaction. A comparison with a palladacycle. *Org. Lett.*, 2003, 5, 3285–3288.
78. De Vries, J.G. A unifying mechanism for all high-temperature Heck reactions. The role of palladium colloids and anionic species. *Dalton Trans.*, 2006, 421–429.
79. Zhao, F., Bhanage, B.M., Shirai, M., and Arai, M. Heck reactions of iodobenzene and methyl acrylate with conventional supported palladium catalysts in the presence of organic and/and inorganic bases without ligands. *Chem. Eur. J.*, 2000, 6, 843–848.
80. Bhanage, B.M., Shirai, M., and Arai, M. Heterogeneous catalyst system for Heck reaction using supported ethylene glycol phase Pd/TPPTS catalyst with inorganic base. *J. Mol. Catal. A*, 1999, 145, 69–74.
81. Reetz, M.T., and Westermann, E. Phosphane-free palladium-catalyzed coupling reactions: the decisive role of Pd nanoparticles. *Angew. Chem. Int. Ed.*, 2000, 39, 165–168.
82. Reetz, M.T., Helbig, W., Quasier, S.A., Stimming, U., Breuer, N., and Vogel, R. Visualization of surfactants on nanostructured palladium clusters by a combination of STM and high-resolution TEM. *Science*, 1995, 267, 367–369.
83. Thathagar, M.B., Ten Elshof, J.E., Rothenberg, G. Pd nanoclusters in C-C coupling reactions: proof of leaching. *Angew. Chem. Int. Ed.*, 2006, 45, 2886–2890.
84. Gaikwad, A.V., Holuigue, A., Thathagar, M.B., Ten Elhof, J.E., and Rothenberg, G. Ion- and atom-leaching mechanisms from palladium nanoparticles in cross-coupling reactions. *Chem. Eur. J.*, 2007, 13, 6908–6913.
85. Ananikov, V.P., and Beletskaya, I.P. Toward the ideal catalyst: from atomic centers to a "cocktail" of catalysts. *Organometallics*, 2012, 31, 1595–1604.
86. Kashin, A.S., and Ananikov, V.P. Catalytic C-C and C-heteroatom bond formation reactions: in situ generated of preformed catalysts? Complicated mechanistic picture behing well-known experimental procedures. *J. Org. Chem.*, 2013, 78, 11117–11125.
87. Shu, W., Pellegatti, L., Oberli, M. A., and Buchwald, S. L. Continuous-flow synthesis of biaryls enabled by multistep solid-handling in a lithiation/borylation/Suzuki-Miyaura cross-coupling sequence. *Angew. Chem. Int. Ed.*, 2011, 50, 10665–10669.
88. Kabri, Y., Gellis, A., and Vanelle, P. Synthesis of original 2-substituted 4-arylquinazolines by microwave-irradiated Suzuki-Miyaura cross-coupling reactions. *Eur. J. Org. Chem.*, 2009, 4059–4066.
89. Gill, G. S., Grobelny, D. W., Chaplin, J.H., and Flynn, B. L. An efficient synthesis and substitution of 3-aroyl-2-bromobenzo[b]furans. *J. Org. Chem.*, 2008, 73, 1131–1134.
90. Organ, M., Calimsiz, S., Sayah, M., Hoi, K., and Lough, A. Pd-PEPPSI-IPent: an active, sterically demanding cross-coupling catalyst and its application in the synthesis of tetra-ortho-substituted biaryls. *Angew. Chem. Int. Ed.*, 2009, 48, 2383–2387.
91. Dang, T., and Chen, Y. One-pot oxidation and bromination of 3,4-diaryl-2,5-dihydrothiophenes using Br_2: synthesis and application of 3,4-diaryl-2,5-dibromothiophenes. *J. Org. Chem.*, 2007, 72, 6901–6904.
92. Maeda, H., Haketa, Y., and Nakanishi, T. Aryl-substituted C3-bridged oligopyrroles as anion receptors for formation of supramolecular organogels. *J. Am. Chem. Soc.*, 2007, 129, 13661–13674.
93. James, C. A., Coelho, A. L., Gevaert, M., Forgione, P., and Snieckus, V. Combined directed ortho and remote metalation-Suzuki cross-coupling strategies. Efficient

synthesis of heteroaryl-fused benzopyrannones from biaryl O-carbamates. *J. Org. Chem.*, 2009, 74, 4094–4103.

94. Giodiano, C., Coppi, L., and Minisci, F. US Patent 5312975, 1994.
95. Hannah, J., Ruyle, W.V., Jones, H., Matzuk, A.R., Kelly, K.W., Witzel, B.E., Holtz, W.J., Houser, R.A., Shen, T.Y., and Sarett, L.H. Novel analgesic-antiinflammatory salicylates. *J. Med. Chem.*, 1978, 21, 1093–1100.
96. Reizman, B.J., Wang, Y.M., Buchwald, S.L., and Jensen, K.F. Suzuki-Miyaura cross-coupling optimization enabled by automated feedback. *React. Chem. Eng.*, 2016, 1, 658–666.
97. Teci, M., Tilley, M., McGuire, M.A., and Organ, M.G. Using anilines as masked cross-coupling partners: design of a telescoped three-step flow diazotization, iododediazotization, cross-coupling process. *Chem. Eur. J.*, 2016, 22, 1–10.
98. Ormerod, D., Lefevre, N., Dorbec, M., Eyskens, I., Vloemans, P., Duyssens, K., Diez de la Torre, V., Kaval, N., Merkul, E., Sergeyev, S., and Maes, B.U.W. Potential of homogeneous Pd catalyst separation by ceramic membranes. Application to downstream and continuous flow processes. *Org. Process Res. Dev.*, 2016, 20, 911–920.
99. Cole, K.P., Campbell, B.M., Forst, M.B., McClary Groh, J., Hess, M., Johnson, M.D., Miller, R.D., Mitchell, D., Polster, C.S., Reizman, B.J., and Rosenmeyer, M. An automated intermittent flow approach to continuous Suzuki coupling. *Org. Process Res. Dev.*, 2016, 20, 820–830.
100. Takahashi, Y., Ashikari, Y., Takumi, M., Shimizu, Y., Jiang, Y., Higuma, R., Ishikawa, S., Sakaue, H., Shite, I., Maekawa, K., Aizawa, Y., Yamashita, H., Yonekura, Y., Colella, M., Luisi, R., Takegawa, T., Fujita, C., and Nagaki, A. Synthesis of biaryls having a piperidylmethyl group based on space integration of lithiation, borylation, and Suzuki-Miyaura coupling. *Eur. J. Org. Chem.*, 2020, 618–622.
101. Kazi, S.A., Campi, E.M., and Hearn, M.T.W. Flow reactor synthesis of unsymmetrically substituted p-terphenyls using sequentially selective Suzuki cross-coupling protocols. *Green Chem. Lett. Rev.*, 2019, 12, 377–388.
102. Hattori, T., Tsubone, A., Sawama, Y., Monguchi, Y., and Sajiki, H. Palladium on carbon-catalyzed Suzuki-Miyaura coupling reaction using an efficient and continuous flow system. *Catalysts*, 2015, 5, 18–25.
103. Brinkley, K.W., Burkholder, M., Siamaki, A.R., Belecki, K., and Gupton, B.F. The continuous synthesis and application of graphene supported palladium nanoparticles: a highly effective catalyst for Suzuki-Miyaura cross-coupling reactions. *Green Process Synth.*, 2015, 4, 241–246.
104. Bolton, K.F., Canty, A.J., Deverell, J.A., Guijt, R.M., Hilder, E.F., Rodemann, T., and Smith, J.A. Macroporous monolith supports for continuous flow capillary microreactors. *Tetrahedron Lett.*, 2006, 47, 9321–9324.
105. Matsumoto, H., Seto, H., Akiyoshi, T., Shibuya, M., Hoshino, Y., and Miura, Y. Macroporous gel with a permable reaction platform for catalytic flow synthesis. *ACS Omega*, 2017, 2, 8796–8802.
106. Nagaki, A., Hirose, K., Moriwaki, Y., Takumi, M., Takahashi, Y., Mitamura, K., Matsukawa, K., Ishizuka, N., and Yoshida, J.I. Suzuki-Miyaura coupling using monolithic Pd reactors and scaling-up by series connection of the reactors. *Catalysts*, 2019, 9, 300.
107. Ghobadi, S., Burkholder, M.B., Smith, S.E., Gupton, B.F., and Castano, C.E. Catalytically sustainable, palladium-decorated graphene oxide monoliths for synthesis in flow. *Chem. Eng. J.*, 2020, 381, 122598.
108. Nagaki, A., Hirose, K., Moriwaki, Y., Mitamura, K., Matsukawa, K., Ishizuka, N., and Yoshida, J. Integration of borylation of aryllithiums and Suzuki-Miyaura coupling using monolithic Pd catalyst. *Catal. Sci. Technol.*, 2016, 6, 4690–4694.
109. Shil, A. K., Guha, N.R., Sharma, D., and Das, P. A solid supported palladium (0) nano/microparticle catalyzed ultrasound induced continuous flow technique for large scale Suzuki reactions. *RSC Adv.*, 2013, 3, 13671–13676.
110. Yamada, T., Jiang, J., Ito, N., Park, K., Masuda, H., Furugen, C., Ishida, M., Otori, S., and Sajiki, H. Development of facile and simple processes for the heterogeneous Pd-catalyzed ligand-free continuous-flow Suzuki-Miyaura coupling. *Catalysts*, 2020, 10, 1209.
111. Ricciardi, R., Huskens, J., and Verboom, W. Dendrimer-encapsulated Pd nanoparticles as catalysts for C-C cross-couplings in flow microreactors. *Org. Biomol. Chem.*, 2015, 13, 4953–4959.
112. Ricciardi, R., Huskens, J., Holtkamp, M., Karst, U., and Verboom, W. Dendrimer-encapsulated palladium nanoparticles for continuous-flow Suzuki-Miyaura cross-coupling reactions. *ChemCatChem*, 2015, 7, 936–942.
113. Rehm, T.H., Bogdan, A., Hofmann, C., Lob, P., Shifrina, Z., Morgan, D.G., and Bronstein, L.M. Proof of concept: magnetic fixation of dendron-functionalized iron oxide nanoparticles containing Pd nanoparticles for continuous-flow Suzuki coupling reactions. *ACS Appl. Mat. Interfaces*, 2015, 7, 27254–27261.
114. Seto, H., Yoneda, T., Morii, T., Hoshino, Y., and Miura, Y. Membrane reactor immobilized with palladium-moaded polymer nanogel for continuous-flow Suzuki coupling reaction. *AIChe J.*, 2015, 61, 582–589.
115. Gu, Y., Favier, I., Pradel, C., Gin, D.L., Lahitte, J.F., Noble, R.D., Gomez, M., and Remigny, J.C. High catalytic efficiency of palladium nanoparticles immobilized in a polymer membrane containing poly(ionic liquid) in Suzuki-Miyaura cross-coupling reaction. *J. Membr. Sci.*, 2015, 492, 331–339.
116. Dai, Y., Formo, E., Li, H., Xue, J., and Xia, Y. Surface-functionalized electrospun titania nanofibers for the scavenging and recycling of precious metal ions. *ChemSusChem*, 2016, 9, 1–6.
117. Ashikari, Y., Maekawa, K., Takumi, M., Tomiyasu, N., Fujita, C., Matsuyama, K., Miyamoto, R., Bai, H., and Nagaki, A. Flow grams-per-hours production enabled by hierarchical bimodal porous silica gel supported palladium column reactor having low pressure drop. *Catal. Today*, 2020. DOI: 10.1016/j.cattod.2020.07.014
118. De Munoz, M., Alcazar, J., de la Hoz, A., and Diaz-Ortiz, A. Cross-coupling in flow using supported catalysts: mild, clean, efficient and sustainable Suzuki-Miyaura coupling in a single pass. *Adv. Synth. Catal.*, 2012, 354, 3456–3460.
119. Noel, T., Kuhn, S., Musacchio, A.J., Jensen, K.F., and Buchwald, S. L. Suzuki-Miyaura cross-coupling reactions in flow: multistep synthesis enabled by a microfluidic extraction. *Angew. Chem.*, 2011, 123, 6065–6068.

120. Noel, T., Kuhn, S., Musacchio, A.J., Jensen, K.F., and Buchwald, S. L. Suzuki-Miyaura cross-coupling reactions in flow: multistep synthesis enabled by a microfluidic extraction. *Angew. Chem. Int. Ed.*, 2011, 50, 5943–5946.
121. Greco, R., Goessler, W., Cantillo, D., and Kappe, C.O. Benchmarking immobilized di- and triaryphosphine palladium catalysts for continuous-flow cross-coupling reactions: efficiency, durability, and metal leaching studies. *ACS Catal.*, 2015, 5, 1303–1312.
122. Pandarus, V., Gingras, G., Beland, F., Ciriminna, R., and Pagliaro, M. Process intensification of the Suzuki-Miyaura reaction over sol-gel entrapped catalyst SiliaCat DPP-Pd under conditions of continuous flow. *Org. Process Res. Dev.*, 2014, 18, 1550–1555.
123. Pandarus, V., Gingras, G., Beland, F., Ciriminnia, R., and Pagliaro, M. Fast and clean borylation of aryl halides under flow using sol-gel entrapped SiliaCat DPP-Pd. *Org. Process Res. Dev.*, 2014, 18, 1556–1559.
124. Pandarus, V., Ciriminnia, R., Gingras, G., Beland, F., Drobod, M., Jina, O., and Pagliaro, M. Greening heterogeneous catalysis for fine chemicals. *Tetrahedron Lett.*, 2013, 54, 1129–1132.
125. Pascanu, V., Hansen, P.R., Bermejo-Gomez, A., Ayats, C., Platero-Prats, A.E., Johansson, M.J., Pericas, M.A., and Martin-Matute, B. Highly functionalized biaryls via Suzuki-Miyaura cross coupling catalyzed by Pd@MOF under batch and continuous flow regimes. *ChemSusChem*, 2015, 8, 123–130.
126. Goncalves, R.S.B., de Oliveira, A.B.V., Sindra, H.C., Archanjo, B.S., Mendoza, M.E., Carneiro, L.S.A., Buarque, C.D., and Esteves, P.M. Heterogeneous catalysis by covalent organic frameworks (COF): $Pd(OAc)_2$@COF-300 in cross-coupling reactions. *ChemCatChem*, 2016, 8, 743–750.
127. Trinh, T.N., Hizartzidis, L., Lin, A.J.S., Harman, D.G., McCluskey, A., and Gordon, C.P. An efficient continuous flow approach to furnish furan-based biaryls. *Org. Biomol. Chem.*, 2014, 12, 9562–9571.

8

Synthesis of Bioactive Heterocyclic Compounds

Athar Ata and Samina Naz

CONTENTS

8.1 Introduction 137
8.2 Microbial Reactions on Heterocyclic Natural Products 137
8.3 Synthesis of Heterocyclic Compounds 141
8.3.1 Quinoxaline Analogues 141
8.3.2 Dispiroheterocyclic Compounds 141
8.3.3 Synthesis of Heteroannulated 8-Nitroquinolines 143
8.3.4 Synthesis of Angular and Linear Fused Pyrazoloquinolines 143
8.4 Chemo-Enzymatic Synthesis of Bioactive Heterocyclic Natural Products 145
Acknowledgments 148
References 148

8.1 Introduction

Heteroatoms in cyclic organic compounds are an important fragment of a number of biologically active compounds. Recent estimates indicate that more than 85% of all biologically active chemical entities contain a heterocycle ring [1]. For instance, penicillin (**1**), a currently used antibiotic, contains β-lactam as an important bioactive part of this compound. Taxol (**2**), the most widely used anticancer pharmaceutical, contains oxetane ring in its structure [2]. This ring is considered a required pharmacophore for the expression of its anticancer activity. Similarly, other anticancer compounds, zanthosimuline (**3**), huajiasimuline (**4**) and pyrazolo [4,3-c]quinoline-3-one (**5**) contain heterocyclic rings incorporated into their structures [3–5]. Heterocycles play an important role in enhancing the physical and chemical properties of biologically active compounds such as solubility, lipophilicity, polarity, hydrogen bonding, etc. These properties make them excellent candidates for the drug-discovery process.

Presently, a number of research groups are involved in designing new synthetic methodologies for heterocyclic compounds using metal-catalyzed cross-coupling and hetero-coupling reactions, microwave-assisted synthesis and microbial reactions using whole-cell cultures of fungi [6–10]. A number of bioactive heterocyclic natural products have been reported in the literature. For example, topotecan (**6**) and irinotecan (**7**) are used to treat colon and ovarian cancer, respectively [10, 11]. Structures of bioactive heterocyclic compounds **1–7** are shown in Figure 8.1. Structural modifications of bioactive heterocyclic natural products not only help to study structure–activity relationships (SAR) but also aid in designing new analogues to overcome several issues of lead pharmaceuticals including solubility, lipophilicity, bioavailability, etc. Our research group is involved in discovering new health-related enzymes inhibiting natural products from medicinally important plants and studying their SAR [12–20]. We are also involved in designing new synthetic methodologies for the synthesis of heterocyclic compounds and structural modifications of bioactive heterocyclic natural products. To accomplish this task, we use green chemistry approaches such as microbial reactions, microwave-assisted synthesis, reusable catalysts and multicomponent reactions. Our recently obtained results in the area of heterocyclic chemistry are summarized as follows.

8.2 Microbial Reactions on Heterocyclic Natural Products

Microorganisms (bacteria and fungi) perform regio- and stereo-selective reactions on organic compounds. These microbial reactions include oxidation, reduction, Michael addition, aldol condensation umpolung-type and degradative reactions [21–26]. Microbial reactions produce new compounds that would be extremely difficult to synthesize chemically [27–29]. For instance, hydroxylation at inactivated methylene in terpenes and steroids is a very common microbial reaction [30]. This oxidation reaction activates oxidized carbon and adjacent carbons for further reactions to generate a library of compounds for drug discovery. Most interestingly, common microorganisms metabolize bioactive natural products similar to that shown by mammals. This similarity in the metabolic profile is due to the presence of an enzyme, cytochrome P-450 monooxygenase, in these organisms [31]. Microorganisms have been used successfully as *in vitro* models to mimic and predict the

DOI: 10.1201/9781003089162-8

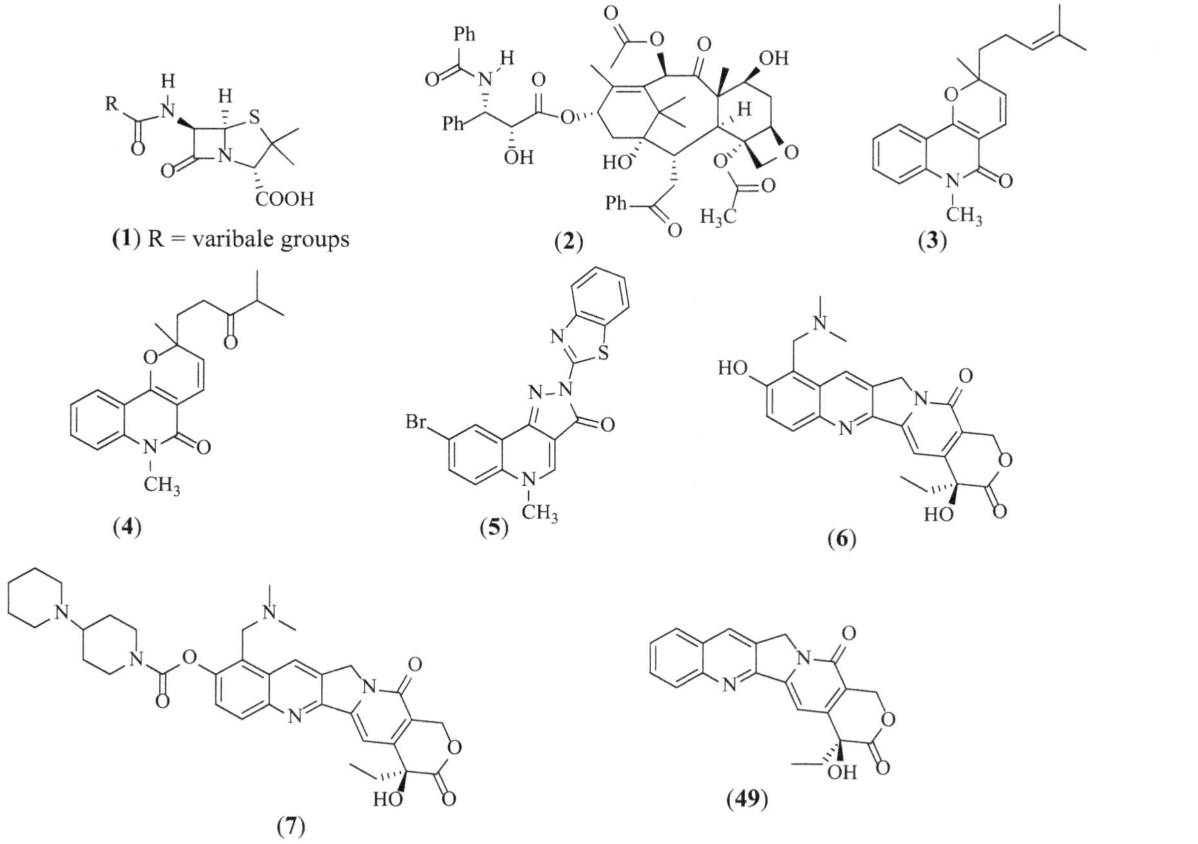

(1) R = varibale groups

(2)

(3)

(4)

(5)

(6)

(7)

(49)

FIGURE 8.1 Structures of compounds **1–7** and **49**.

metabolic fate of pharmaceutical agents in mammalian systems [32]. The mammalian system does not produce metabolites in large quantities and is difficult to identify, whereas microbial reactions can be used to produce metabolites in large quantities by using large-scale fermentation, and the metabolites can then be identified with the aid of spectroscopic methods. Metabolites obtained via microbial reactions have also shown similarity to those of human biotransformations [33]. Microbial reactions on moderate bioactive natural products produce metabolic products with improved bioactivity and with minimal toxicity [34].

α-Santonin (**8**), a sesquiterpene lactone, is found in several species of genus *Artemisia* [35]. It has been reported to treat nervous complaints. It also has anthelmintic activity [36]. This compound has also shown a number of other biological activities such as anti-inflammatory and antipyretic activities [37]. We performed microbial reactions on α-santonin (**8**) by incubating it with liquid cultures of *Rhizopus stolonifer* (ATCC 10404), *Cunninghamella bainieri* (ATCC 9244), *C. echinulata* (ATCC 9245) and *Mucor plumbeus* (ATCC 4740). These experiments showed that *R. stolonifer* metabolized compound **8** to afford 3,4-epoxy-α-santonin (**9**). Incubation of **8** with *C. bainieri*, *C. echinulata* and *M. plumbeus* biotransformed compound **8** to 3,4-dihydro-α-santonin (**10**) and 1,2-dihydro-α-santonin (**11**) [38]. These biotransformation results are summarized in Scheme 8.1. Compounds **9** and **10** showed moderate antibacterial activity against Gram-positive bacteria, *Staphylococcus aureus*, *Streptococcus agalactiae* with MIC values ≤8.0 and ≤32 μg/mL, respectively. Compound **11** was very weakly active in these bioassays with an MIC value of ≤128 μg/mL. The comparison of bioactivity data and structures of compounds **9–11** indicated that the presence of $\Delta^{1,2}$ double bond in these compounds is required for the expression of this bioactivity.

Sclareolide (**12**) has been isolated from a vascular plant, *Arnica angustifolia*. This compound has shown modest cytotoxcity against breast (MCF-7), colon (CKCO-1), lung (H-1299) and skin (HT-144) human cancer cell lines [39]. For this project, we screened five different fungi, namely *M. plumbeus* (ATCC 4740), *C. blakesleeana* (ATCC 9245), *C. echinulata* (ATCC 9244), *C. lunata* (ATCC 12017) and *Aspergillus niger* (ATCC 1004), for their potential to metabolize compound **12**. These efforts showed that *C. blakesleeana* is capable of performing ether forming, oxidation and reduction reactions on compound **12**. This fermentation experiment yielded three new compounds, namely, O^6-sclareolide (**13**), 3β,6α-dihydroxysclareolide (**14**) and 9-hydroxysclareolide (**15**), along with three previously reported compounds, 1β,3β-dihydroxysclareolide (**16**), 3-oxosclareolide (**17**) and 3β-hydroxysclareolide (**18**). Biotransformation experiments of compound **12** with *C. echinulata* yielded two new compounds, 5-hydroxysclareolide (**19**) and 7β-hydroxysclareolide (**20**). All of these microbial reactions are shown in Scheme 8.2. Compounds **13–20** exhibited acetylcholinesterase inhibitory activity (an enzyme

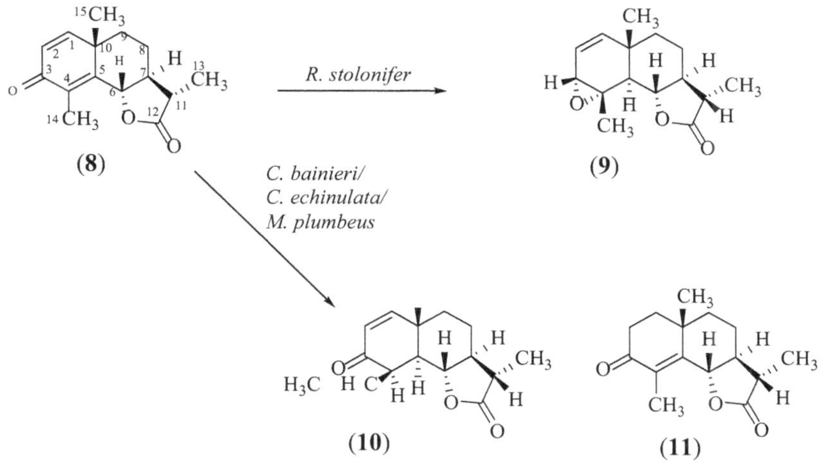

SCHEME 8.1 Biotransformation of α-santonin (**8**).

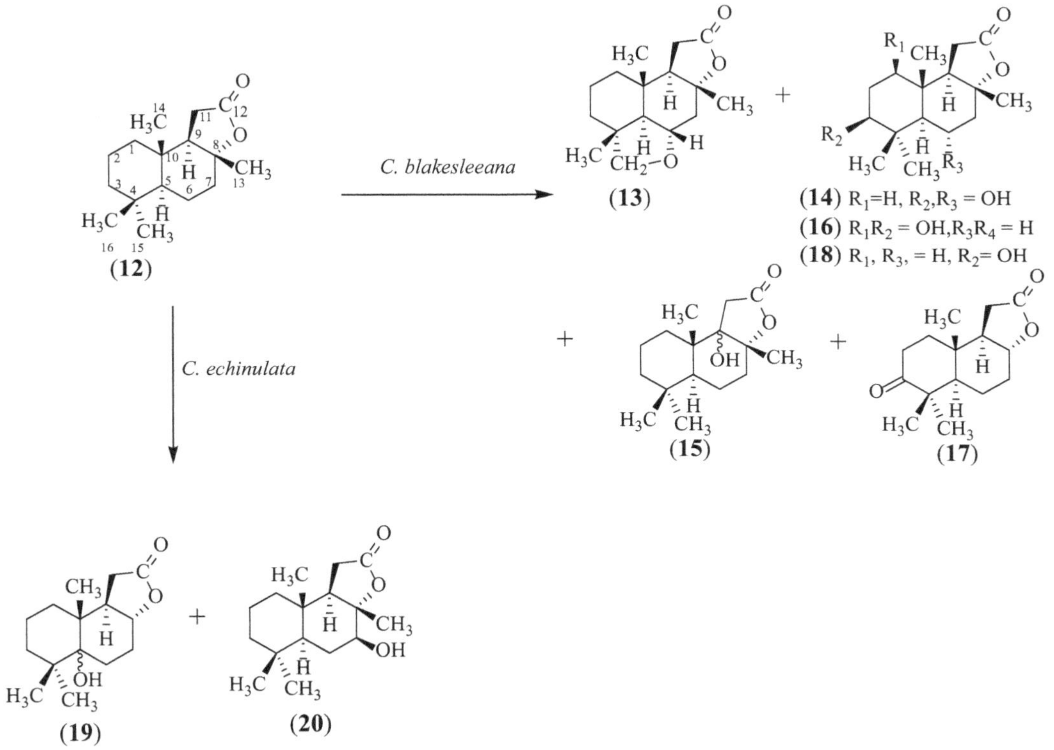

SCHEME 8.2 Biotransformation of sclareolide (**12**).

involved in the pathogenesis of Alzheimer's disease) with IC_{50} values (inhibition of enzyme activity by 50%) of 45, 82, 38, 98, 82. 100, 121 and 42 μM, respectively. Compound **12** was inactive in this bioassay. The moderate bioactivity of **13–20** might be due to the presence of hydroxyl group(s) at different carbon atoms. The bioactivity of compound **13** was hypothesized due to the presence of a tetrahydrofuran ring incorporated into its structure.

7α-Hydroxyfrullanolide (**21**), a sesquiterpenoid lactone, is a major constituent of *Sphaeranthus indicus*. This plant has been reported to treat wounds in folk medicines [40]. Our antimicrobial-directed phytochemical studies on the crude methanolic extract of this plant afforded compound **21**. Based on our phytochemical studies, we assumed that the wound-healing properties of this plant might be due to the presence of compound **21**. These studies validated the ethnomedicinal use of this plant by traditional healers. Compound **21** was significantly active in our antibacterial assay against Gram-positive bacteria. In order to study the SAR of this compound, we used a combination of chemical and microbial reactions to generate eight analogues of **21**. Incubation of compound **21** with liquid cultures of *C. echinulata* and *C. lunata* resulted in the

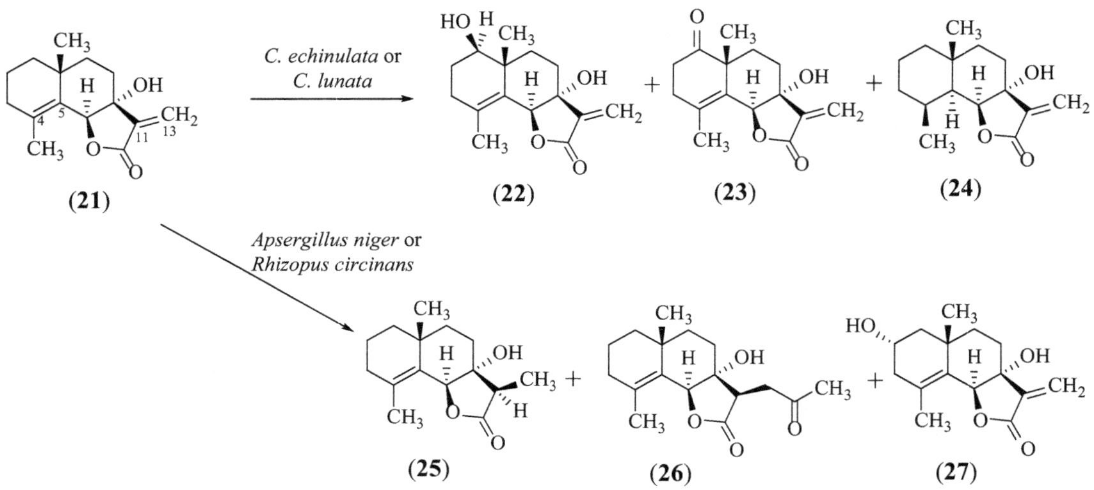

SCHEME 8.3 Biotransformation of 7α-hydroxyfrullanolide (**21**).

synthesis of three metabolites, namely, 1β,7α-dihydroxyfrullanolide (**22**), 1-oxo-7α-hydroxyfrullanolide (**23**) and 7α-hydroxy-4,5-dihydrofrullanolide (**24**). Microbial reactions using liquid cultures of *A. niger* and *R. circinans* afforded three metabolites, namely, 17α-hydroxy-11,13-dihydrofrullanolide (**25**), 13-acetyl-7α-hydroxyfrullanolide (**26**) and 2α,7α-dihydroxysphaerantholide, (**27**) [41]. During these microbial transformation experiments, we were unable to perform any reaction on endocyclic double ($\Delta^{4\text{-}5}$). We decided to incorporate 4,5-epoxy functionality by reacting compound **21** with *meta*-chloroperbenzoic acid. This reaction afforded two analogues, namely, 4α,5α-epoxy-7α-hydroxyfrullanolide (**28**) and 4β,5β-epoxy-7α-hydroxyfrullanolide (**29**) [41]. Microbial reactions on compound **21** are summarized in Scheme 8.3 and the synthesis of epoxy analogues of **21** is shown in Scheme 8.4. Compounds **21–29** exhibited antibacterial activity against *S. aureus*, *S. agalactiae*, *Escherichia coli* and *Pseudomonas aeruginosa* with MIC values in the range of ≤8–128 μg/mL. Compounds **22** and **23** were significantly active against *S. aureus* and *S. aglactiae* with MIC values of ≤8.0 μg/mL. These antibacterial activity data indicated that the higher potencies of **22** and **23** might be due to the presence of an oxygenated functionality at C-1, either in the form of hydroxyl or carbonyl functionality, in these compounds. However, the substitution of a hydroxyl group at C-2 in compound **27** did not play any role in bioactivity as **27** was weakly active with an MIC value of ≤128 μg/mL in this bioassay. These SAR studies also suggested that the presence of double bonds $\Delta^{4\text{-}5}$ and $\Delta^{11\text{-}13}$ in conjunction with a γ lactone moiety in compound **21** was a required pharmacophore for the expression of its antibacterial activity.

Strictosamide (**30**), an indole alkaloid, was identified as Glutathione *S*-Transferase (GST) inhibitor from a medicinally important plant, *Nauclea latifolia*. GST inhibitors have applications as adjuvants in overcoming the drug-resistance problems for anticancer and antiparasitic drugs during chemotherapy [42]. This compound was active in this bioassay with an IC_{50} value of 20.3 μM [43]. The potency of this compound was the same as that of the currently used standard, ethacrynic acid. Compound **30** was significantly active in our GST inhibitory assay and we made an effort to study its SAR by using microbial reactions.

Microbial reactions on compound **30** using fungi *C. blakesleeana* and *R. circinans* afforded three analogues, 10-hydroxystrictosamide (**31**), 10-β-glucosyloxyvincoside lactam (**32**) and 16,17-dihydro-10-β-glucosyloxyvincoside lactam (**33**). These compounds appeared to be formed in a sequence and this sequence was determined by carrying out time-dependent biotransformation experiments by incubating compounds **30** with the *R. circinans* culture. These experiments indicated that first of all, compound **31** was produced which further underwent glycosylation, followed by reduction of the $\Delta^{16\text{-}17}$ double bond to give compounds **32** and **33**, respectively. Compounds **31–33** exhibited anti-GST inhibition activity with IC_{50} values of 18.6, 12.3 and 46.6 μM, respectively. The bioactivity of **32** was approximately 2-fold higher compared to the parent compound (**30**) which might be due to the presence of a sugar moiety at C-10. This moiety might

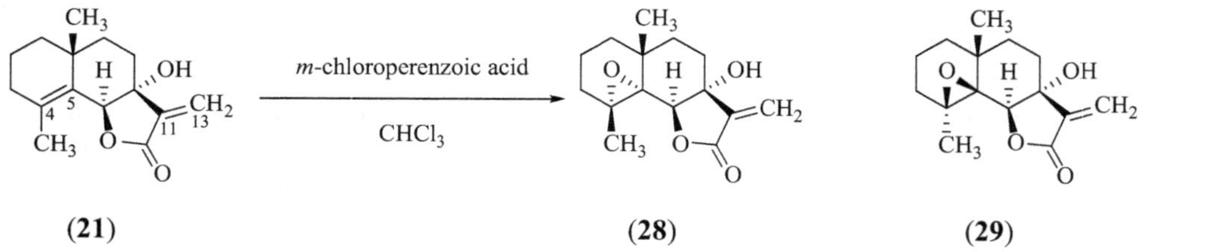

SCHEME 8.4 Synthesis of epoxy analogues of compound (**21**).

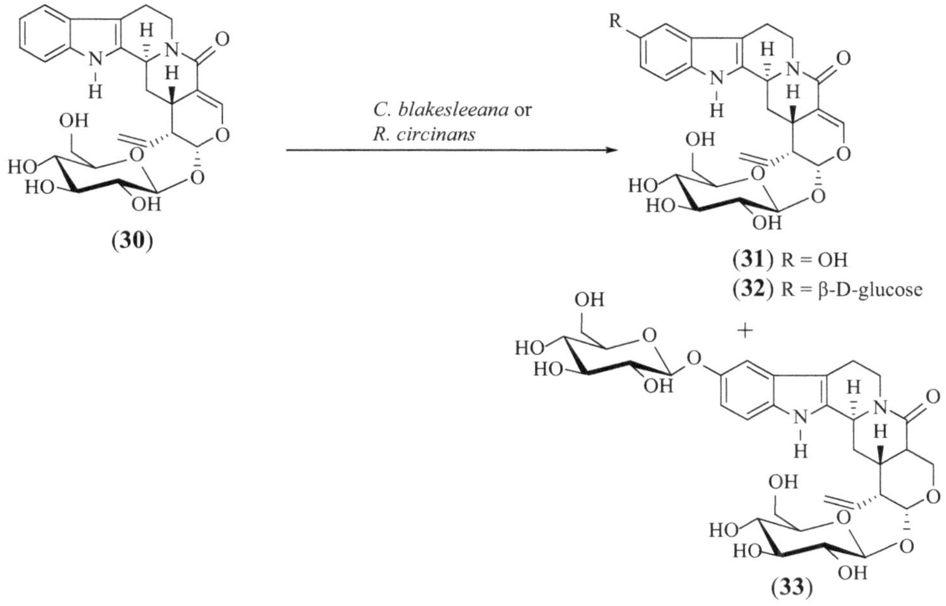

C. blakesleeana or R. circinans

(30) (31) R = OH (32) R = β-D-glucose + (33)

SCHEME 8.5 Microbial transformation of Strictosamide (**30**).

have increased its solubility in water for its better interactions with the GST enzyme. Microbial reactions of compound **30** are shown in Scheme 8.5.

8.3 Synthesis of Heterocyclic Compounds

8.3.1 Quinoxaline Analogues

These organic compounds containing benzopyrazine in their structures exhibit diverse biological activities including antimicrobial, antihypertensive, antitubercular, antidepressant, antimalarial, anti-inflammatory, anticonvulsant, anti-HIV, antidiabetic and anticancer [44–53].

We have synthesized a series of new N-(11H-Indeno[1,2-b] quinoxalin-11-ylidene)-benzohydrazide and evaluated them for their potential to inhibit the activity of α-glucosidase. This enzyme is involved in the digestion of carbohydrates by hydrolyzing the glycosidic bonds in carbohydrates to liberate free glucose and causing postprandial hyperglycemia. This results in type 2 diabetes mellitus that affects approximately over two billion people worldwide [54]. This health-related problem can be overcome by suppressing hyperglycemia that includes reduction of glucose absorption in the gut. The potent α-glucosidase inhibitors are reported to accomplish this task [55, 56]. The quinoxaline compound [1H-Indeno[1,2-b]quinoxalin-11-one] (**36**) was synthesized by refluxing an equimolar amount of *o*-phenylenediamine (**34**) and ninhydrin (**35**) using ethanol/acetic acid (1:1) as the solvent, as shown in Scheme 8.6.

The target compounds benzohydrazide analogues (**38**a–q) were prepared by reacting compound **36** with various benzohydroazides (**37**a–r) as shown in Scheme 8.7. Compounds **38**a–q exhibited anti-α-glucosidase activity with IC_{50} values of 62.51, 90.92, 53.42, 23.12, 112.53, 22.67, 25.91, 105.41, 99.85, 77.45, 37.72, 62.34, 700, 23.54, 29.47, 102.13 and 18.23 μM, respectively. These studies revealed that the presence of deactivating groups at *ortho*- and *para*-positions in compounds **38a–q** play a significant role in this bioactivity as **38a** was weakly active against α-glucosidase. Among these compounds, six analogues **38f**, **38d**, **38n**, **38g**, **38o** and **38k** were more potent than the currently used drug, acarbose (IC_{50} = 38.25 μM) [56].

8.3.2 Dispiroheterocyclic Compounds

Natural products containing spiropyrrolidinyl oxindole moiety have various biological applications including the inhibition of microtubule assembly, modulation of the function of the muscarinic serotonin receptor, a potent non-peptide inhibitor of the p53–MDM2 interaction and inhibitors of human NK-1 receptor [57–61]. Pyrrolothiazoles also find applications as hepatoprotective, antibiotics, antidiabetics and anticonvulsant agents [62–65]. The presence of 2,3-dihydro-4-quinolone ring system in alkaloids serves as important intermediates in organic synthesis and exhibits a wide range of pharmacological properties [66,

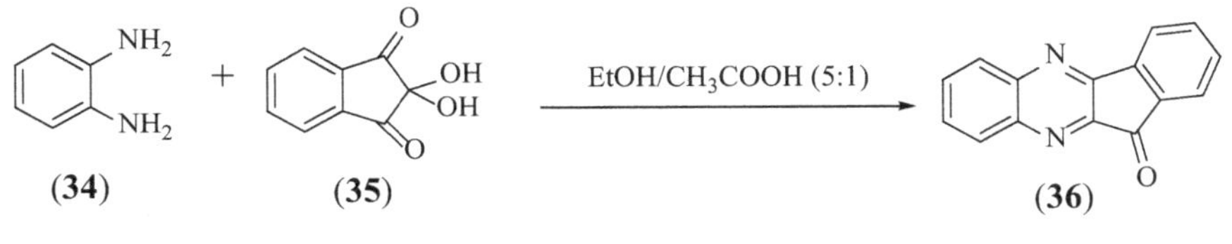

EtOH/CH_3COOH (5:1)

(34) (35) (36)

SCHEME 8.6 Synthesis of compound (**36**).

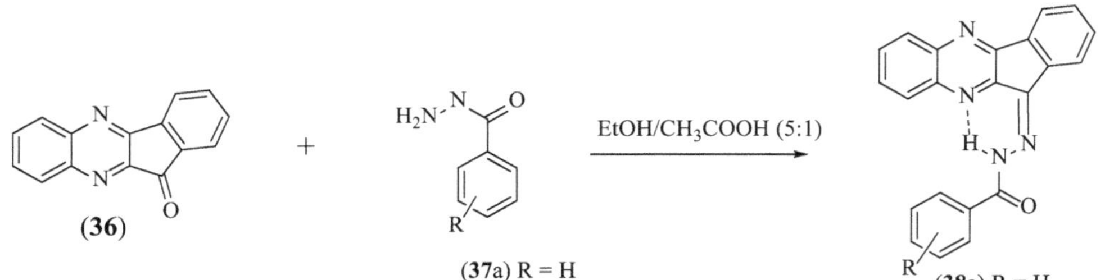

(36)

EtOH/CH_3COOH (5:1)

(37a) R = H
(37b) R = 3-F
(37c) R = 4-F
(37d) R = 2-Br
(37e) R = 3-Br
(37f) R = 4-Br
(37g) R = 2-Cl
(37h) R = 3-Cl
(37i) R = 4-Cl
(37j) R = 3-I
(37k) R = 4-I
(37l) R = 4-CH_3
(37m) R = 2-NH_2
(37n) R = 2-NO_2
(37o) R = 4-NO_2
(37p) R = 3-OH, 4-OCH_3
(37q) R = 4-OH

(38a) R = H
(38b) R = 3-F
(38c) R = 4-F
(38d) R = 2-Br
(38e) R = 3-Br
(38f) R = 4-Br
(38g) R = 2-Cl
(38h) R = 3-Cl
(38i) R = 4-Cl
(38j) R = 3-I
(38k) R = 4-I
(38l) R = 4-CH_3
(38m) R = 2-NH_2
(38n) R = 2-NO_2
(38o) R = 4-NO_2
(38p) R = 3-OH, 4-OCH_3
(38q) R = 4-OH

SCHEME 8.7 Synthesis of compounds (**38**a–r).

67]. Among these compounds are nitro-substituted 2,3-dihydro-4-quinolone derivatives that exhibit poly(ADP-ribosyl) transferase (PARP) inhibition and induce necrotic death of leukemic cells HL-6027 and used as important biosensors [68–70].

We have synthesized a series of rare classes of dispiroheterocycles containing spiropyrrolidine oxindole/spiro-thiapyrrolizidine oxindole and 2,3-dihydro-8-nitro-4-quinolone ring moieties. This dispiroheterocyclic containing 4-quinolone moiety was synthesized by 1,3 dipolar cycloaddition reaction of azomethine ylides with a newly prepared (E)-3-arylidene-2,3-dihydro-8-nitro-4-quinolone as dipolarophiles (**41**a–g). Compounds **41**a–g were prepared by stirring 2,3-dihydro-8-nitro-4-quinolone (**39**) with various aromatic aldehydes (**40**a–g) in the presence of pyrrolidine base at room temperature, as shown in Scheme 8.8 [71].

Compounds (**41**a–g) on performing 1,3 dipolar cycloaddition reaction with azomethine ylide (**44**) afforded novel dispiropyrrolidine oxindole derivatives (**45**a–g) with good yields in the range of 87–97%. For this reaction, we carried out a one-pot three-component reaction of isatin (**42**) with sarcosine (**43**) and compounds **41**a–g. This one-pot three-component reaction is shown in Scheme 8.9.

The successful application of this newly discovered cycloaddition reaction produced dispirothiapyrrolizidines **48**a–g. For this task, compounds **41**a–g was reacted with azomethine ylide (**47**), generated *in situ* by reacting isatin (**42**) with thiaproline (**46**), to afford the corresponding dispirothiapyrrolizidines **48**a–g with good yields in the range of 89–96%, as shown in Scheme 8.10. Compounds **45**a–g and **48**a–g were weakly active in our antibacterial assay.

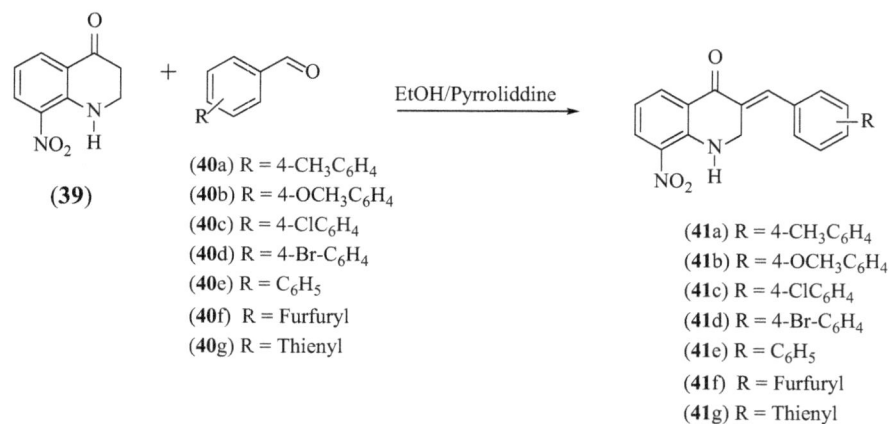

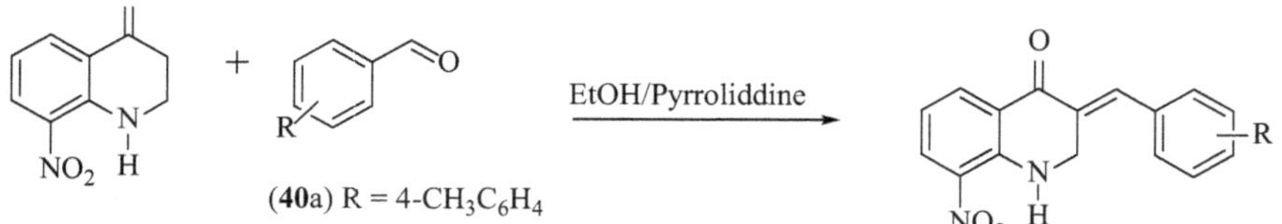

(39)

EtOH/Pyrroliddine

(40a) R = 4-$CH_3C_6H_4$
(40b) R = 4-$OCH_3C_6H_4$
(40c) R = 4-ClC_6H_4
(40d) R = 4-Br-C_6H_4
(40e) R = C_6H_5
(40f) R = Furfuryl
(40g) R = Thienyl

(41a) R = 4-$CH_3C_6H_4$
(41b) R = 4-$OCH_3C_6H_4$
(41c) R = 4-ClC_6H_4
(41d) R = 4-Br-C_6H_4
(41e) R = C_6H_5
(41f) R = Furfuryl
(41g) R = Thienyl

SCHEME 8.8 Synthesis of compounds (**41**a–g).

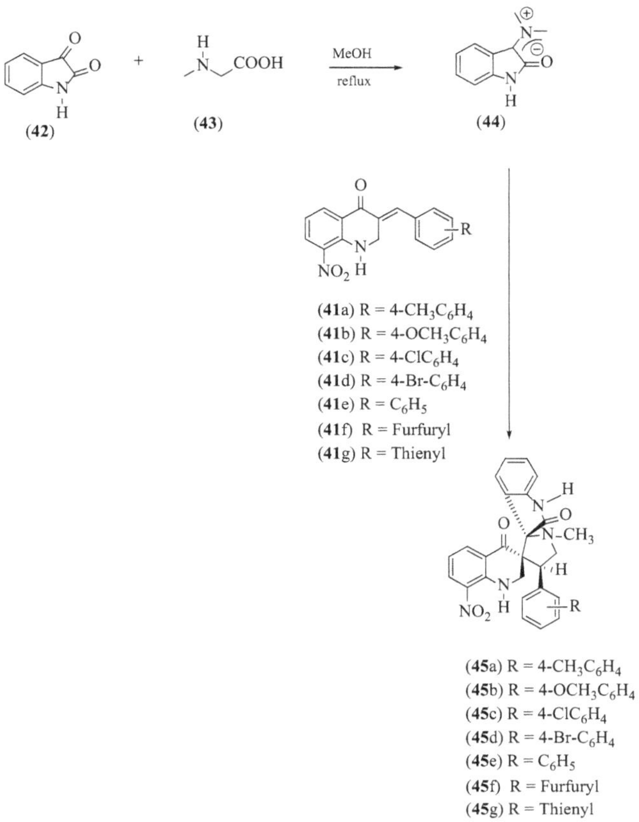

(42) + (43) $\xrightarrow[\text{reflux}]{\text{MeOH}}$ (44)

(41a) R = 4-$CH_3C_6H_4$
(41b) R = 4-$OCH_3C_6H_4$
(41c) R = 4-ClC_6H_4
(41d) R = 4-Br-C_6H_4
(41e) R = C_6H_5
(41f) R = Furfuryl
(41g) R = Thienyl

(45a) R = 4-$CH_3C_6H_4$
(45b) R = 4-$OCH_3C_6H_4$
(45c) R = 4-ClC_6H_4
(45d) R = 4-Br-C_6H_4
(45e) R = C_6H_5
(45f) R = Furfuryl
(45g) R = Thienyl

SCHEME 8.9 Synthesis of compounds **45**a–g.

8.3.3 Synthesis of Heteroannulated 8-Nitroquinolines

Heterocycles containing a nitrogen bridge are considered as a distinct class of anticancer drug candidates as these compounds are reported to induce cell apoptosis and many quinoline-containing compounds have been reported as antitumor agents [72, 73]. Examples include camptothecin (**49;** Figure 8.1), an anticancer alkaloid containing a quinoline moiety, which was isolated from *Camptotheca acuminate* [74]. It has also been reported in the literature that the efficacy of quinoline drugs can be improved by incorporating pyran or pyrazolo ring in quinoline moiety [75]. For instance, pyranoquinoline alkaloids and several pyrazoloquinoline synthetic analogues [zanthosimuline (**3**) and huajiaosimuline (**4**) exhibit anticancer activity [76]. Structures of compounds **3**, **4** and **49** are shown in Figure 8.1.

Based on the aforementioned importance of heteroannulated 8-nitroquinolines, we have synthesized nitroquinoline with pyran and pyrazolo moieties by adopting the green chemistry approach. For this task, we reacted 3-arrylidene-2,3-dihydro-8-nitro-4-quinolones (**41**a–i) with aromatic hydrazine derivatives (**50–51**) to afford compounds (**52**a–i) and (**53**a–i). These reactions were carried out in ethanol in a microwave reactor. Under these conditions, we were able to obtain compounds **52**a–i with good yields in the range of 46–87%. These reactions are summarized in Scheme 8.11. Though we used different solvents such as acetonitrile, dimethyl sulfoxide and chloroform, we observed that ethanol is a better solvent to get optimum yields of products as other solvents afforded very poor yields of products in the range of 10–20% [76].

To increase the number of new heterocyclic compounds in our library, we prepared fluoro analogues of 3-arrylidene-2,3-dihydro-8-nitro-4-quinolones (**54**a–b) using the same reaction conditions as those used for the synthesis of 3-arrylidene-2,3-dihydro-8-nitro-4-quinolones (**41**a–g). Compounds **54**a–b were reacted with hydrazine (**55**) under similar reaction conditions as those used for the synthesis of compounds of **52** and **53** to afford compounds **56**a–b. The synthesis of compounds **56**a–b is shown in Scheme 8.12. Both of these compounds were obtained with good yields. We also noted that the use of ethanol as a solvent provides better yields (73%) than other solvents. The use of microwave radiations as a catalyst has significantly reduced the reaction time [76].

For the expansion of the scope of this reaction, we reacted 3-arrylidene-2,3-dihydro-8-nitro-4-quinolones (**41**a–i) with malononitrile (**57**) using ethanol as the solvent and pyridine as a weak base. This reaction yielded compounds **58**a–i with moderate yields in the range of 46–66% along with compounds (**41**a–i). We assume that compounds **41**a–i resulted due to basic hydrolysis. This compound predominates when we use strong bases like pyrrolidine and Et_3N in various amounts [76]. The synthesis of compounds **58**a–i is shown in Scheme 8.13.

These aforementioned efforts resulted in the synthesis of several heteroannulated 8-nitroquinolines and we decided to evaluate all of these newly synthesized compounds for antimicrobial, anti-α-glucosidase and anticancer activities. All of these compounds exhibited moderate-to-strong antibacterial activity against Gram-positive (*S. aureus* and *S. agalactiae*) and Gram-negative bacteria (*E. coli and P. aeruginosa*) in our antimicrobial assay. The compound **58**a was significantly active compared to the rest of the compounds of this series. Based on these results, it was concluded that the electron-withdrawing group (F) at C-4 of the phenyl ring in **58**a is responsible for the enhanced antibacterial activity of this compound. All of these compounds were moderately active in our α-glucosidase inhibitory activity. In our anticancer assay, compounds **58**d and **58**i were significantly active against two cancer cell lines of breast adenocarcinoma (MCF-7) and lung adenocarcinoma (A549) with IC_{50} values of <20 and <23.6 μM, respectively. The enhanced bioactivity of these compounds might be due to the presence of the electron-withdrawing group and the thiophene group in **58**d and **58**i, respectively.

8.3.4 Synthesis of Angular and Linear Fused Pyrazoloquinolines

Azahetro-fused quinolones are an important group of heterocyclic compounds as they play an important role in medicinal chemistry research. These compounds exhibit various biological activities due to their structural and functional complexity [77, 78]. In this series, pyrazolo[3,4-b]quinoline and pyrazolo[4,3-c]quinoline are two important core structures of heterocycles. Pyrazolo[3,4-b]quinoline derivatives exhibit diverse biological activities including anticancer, antimycobacterial, antiviral, anti-inflammatory and antimalarial

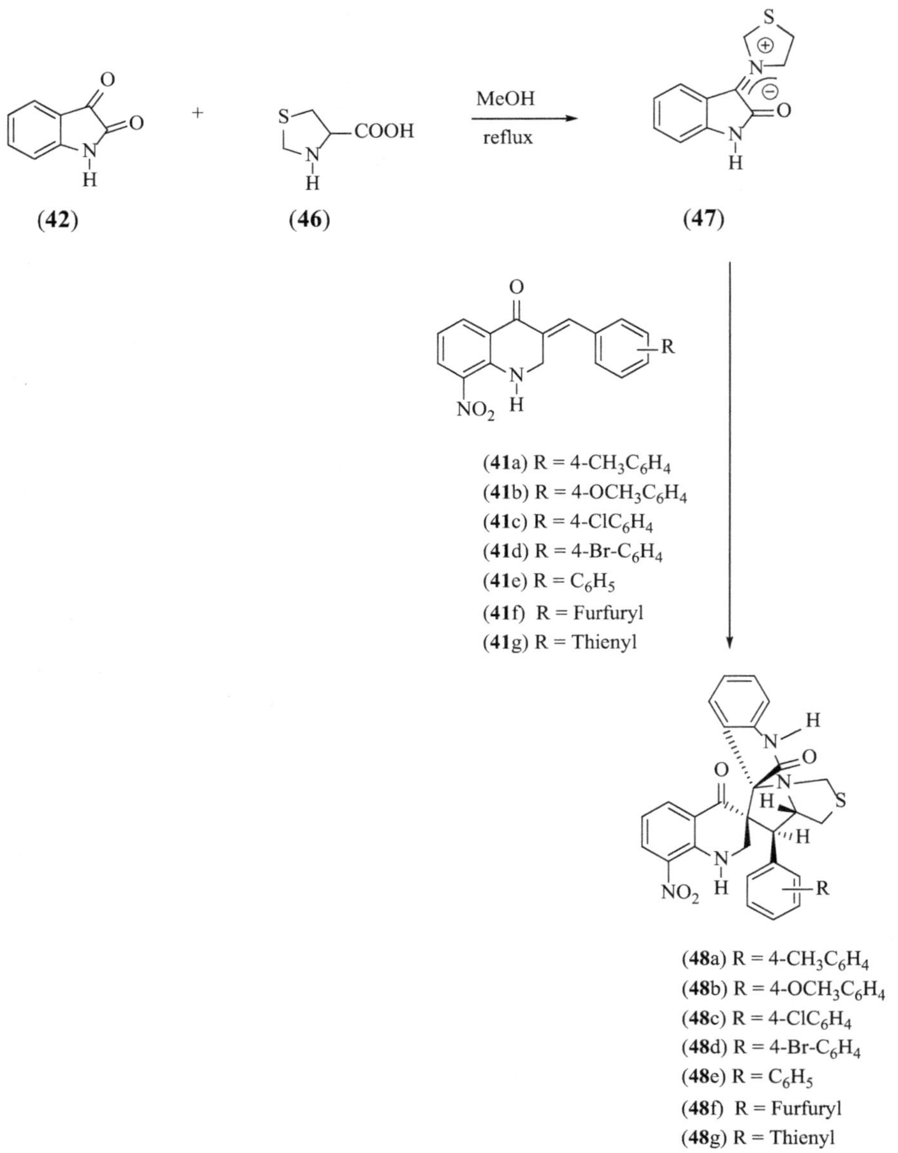

(42) (46) (47)

MeOH, reflux

(41a) R = 4-$CH_3C_6H_4$
(41b) R = 4-$OCH_3C_6H_4$
(41c) R = 4-ClC_6H_4
(41d) R = 4-Br-C_6H_4
(41e) R = C_6H_5
(41f) R = Furfuryl
(41g) R = Thienyl

(48a) R = 4-$CH_3C_6H_4$
(48b) R = 4-$OCH_3C_6H_4$
(48c) R = 4-ClC_6H_4
(48d) R = 4-Br-C_6H_4
(48e) R = C_6H_5
(48f) R = Furfuryl
(48g) R = Thienyl

SCHEME 8.10 Synthesis of compounds **48**a–g.

activities [79–82]. A number of reports describing the synthesis of linear pyrazolo[3,4-b]quinoline are present in the literature [80–82]. These syntheses have several drawbacks including poor yields due to multistep procedures, formation of by-products, long reaction times, harsh reaction conditions, high temperatures, etc. In this area, we have recently developed an efficient and selective synthetic strategy for the construction of highly substituted linear pyrazoloquinoline. We used enolizable 3-acyl quinolin-2-one **59**a–f for the reaction with aromatic hydrazine (**60**a–f) in the presence of Lewis acid $InCl_3$ to afford angular pyrazolo[4,3-c]quinolones (**61**a–p), as shown in Scheme 8.14.

These reactions proceed via nucleophilic substitution reaction to give aromatized heterocycles as shown in Scheme 8.15.

We observed that the presence of hydroxyl groups at C-2 and C-4 positions on reaction phenyl hydrazine (**60a–f**) predominantly afford an angular isomer. For linear isomer (**62**a–h), we noted that compounds having phenyl or methyl groups at the C-4 position enhance the nucleophilic attack at the C-2 position by phenyl hydrazine (**60**a–f) to give compounds (**61**a–h).

We used a green chemistry approach by adopting a conventional method and a microwave-assisted method one-pot synthesis to generate a series of pyrazolo[3,4-b]quinoline and pyrazolo[4,3-c]quinoline derivatives (**61**a–p and **62**a–f). During these experiments, we used various Lewis acid catalysts including $Yb(OTf)_3$, $Sc(OTf)_3$, $SnCl_4$, $AlCl_3$, $TiCl_4$, $ZnCl_2$, $FeCl_3$, BF_3-Et_2O and $InCl_3$. The reaction catalyzed by $InCl_3$ afforded pyrazolo[3,4-b]quinoline and pyrazolo[4,3-c] quinoline derivatives in 90–95% yield. The rest of the catalysts provided products in 60–75% yield. We also observed that the microwave-assisted method significantly reduced the reaction time (2–5 min) and performed these reactions in a regio-selective manner compared to the conventional

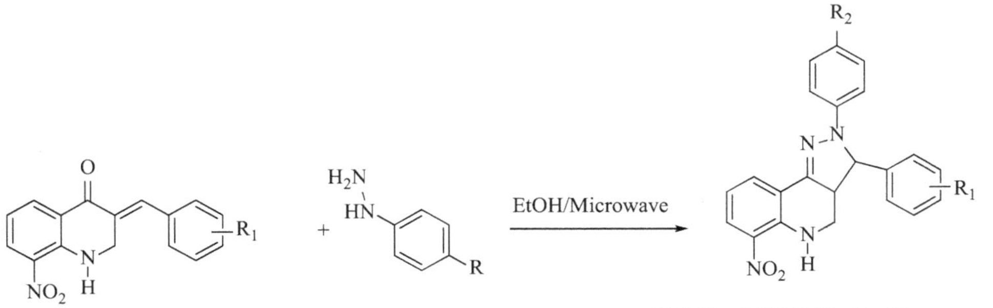

EtOH/Microwave

(**41**a) R = 4-$CH_3C_6H_4$
(**41**b) R = 4-$OCH_3C_6H_4$
(**41**c) R = 4-ClC_6H_4
(**41**d) R = 4-Br-C_6H_4
(**41**e) R = C_6H_5
(**41**f) R = Furfuryl
(**41**g) R = Thienyl
(**41**h) R = 2,4-Di$OCH_3C_6H_5$
(**41**i) R = 3,4-Di$OCH_3C_6H_5$

(**50**) R = H
(**51**)R = OCH_3

(**52**a) R = 4-$CH_3C_6H_4$, R_2=H
(**52**b) R = 4-$OCH_3C_6H_4$, R_2=H
(**52**c) R = 4-ClC_6H_4, R_2=H
(**52**d) R = 4-Br-C_6H_4, R_2=H
(**52**e) R = C_6H_5, R_2=H
(**52**f) R = Furfuryl, R_2=H
(**52**g) R = Thienyl, R_2=H
(**52**h) R = 2,4-Di$OCH_3C_6H_5$, R_2=H
(**52**i) R = 3,4-Di$OCH_3C_6H_5$, R_2=H
(**53**a) R = 4-$CH_3C_6H_4$, R_2=OCH_3
(**53**b) R = 4-$OCH_3C_6H_4$, R_2=OCH_3
(**53**c) R = 4-ClC_6H_4, R_2=OCH_3
(**53**d) R = 4-Br-C_6H_4, R_2=OCH_3
(**53**e) R = C_6H_5, R_2=OCH_3
(**53**f) R = Furfuryl, R_2=OCH_3
(**53**g) R = Thienyl, R_2=OCH_3
(**53**h) R = 2,4-Di$OCH_3C_6H_5$, R_2=OCH_3
(**53**i) R = 3,4-Di$OCH_3C_6H_5$, R_2=OCH_3

SCHEME 8.11 Synthesis of compounds **52**a–i and **53**a–i.

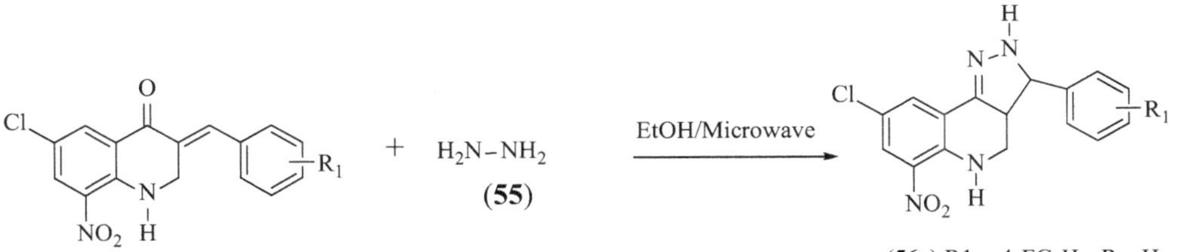

+ H_2N-NH_2 (**55**)

EtOH/Microwave

(**54**a)R = 4-FC_6H_4
(**54**b)R = 4-FC_6H_4

(**56**a) R1 = 4-FC_6H_4, R_2=H
(**56**b)R = 4-FC_6H_4, R_2=H

SCHEME 8.12 Synthesis of compounds **56**a–b.

method that requires 2–7 h for the completion of reactions (Scheme 8.16).

Compounds **61**a–p and **62**a–h were evaluated for α-glucosidase inhibitory activity. In this bioassay, compounds **61**a, **61**g, **61**h, **61**i, **61**k, **61**o and **61**p were found to be significantly active with IC_{50} values in the range of 57.5–187.5 μM. The bioactivity is decreeing in the order **61**a > **61**p > **61**i > **6**ih > **6**ik > **6**io > **6**ig. The higher potency of the fluorinated pyrazolo[3,4-b] quinoline **61**a and **61**i might be due to the presence of fluorine atom at the R_3 position. The bioactivity of compound **61h** was better than that of **61**k, which might be due to the electron-releasing group (Cl) at the R_2 position, to enhance binding of **61**h with the enzyme, whereas **61**k has electron-withdrawing group (NO_2) at R_2 position. The compound **61**p exhibits 2-fold stronger activity than compound **61**o. Compounds **61p** and **61o** contain bromine and methoxy groups, respectively, at the R_3 position. The presence of bromine at the R_3 position is responsible for the enhanced bioactivity of **61**p. During this bioassay, we also observed that compounds **61**a, **61**p and **61**i were potent than acarbose (IC_{50} = 115.8 μM), a currently used pharmaceutical in clinics to treat type 2 diabetes. But pyrazolo[4,3-c]quinoline analogs **62**a–h were found to be inactive in this bioassay.

8.4 Chemo-Enzymatic Synthesis of Bioactive Heterocyclic Natural Products

Our chemical studies on the crude methanolic extract of *Epilobium angustifolium* of Manitoban origin resulted in the isolation of two new potent α-glucosidase inhibitors **63** (IC_{50}

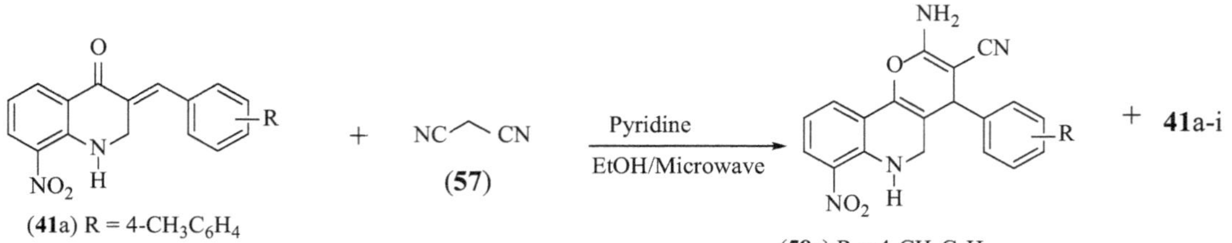

SCHEME 8.13 Synthesis of compounds **58**a–i.

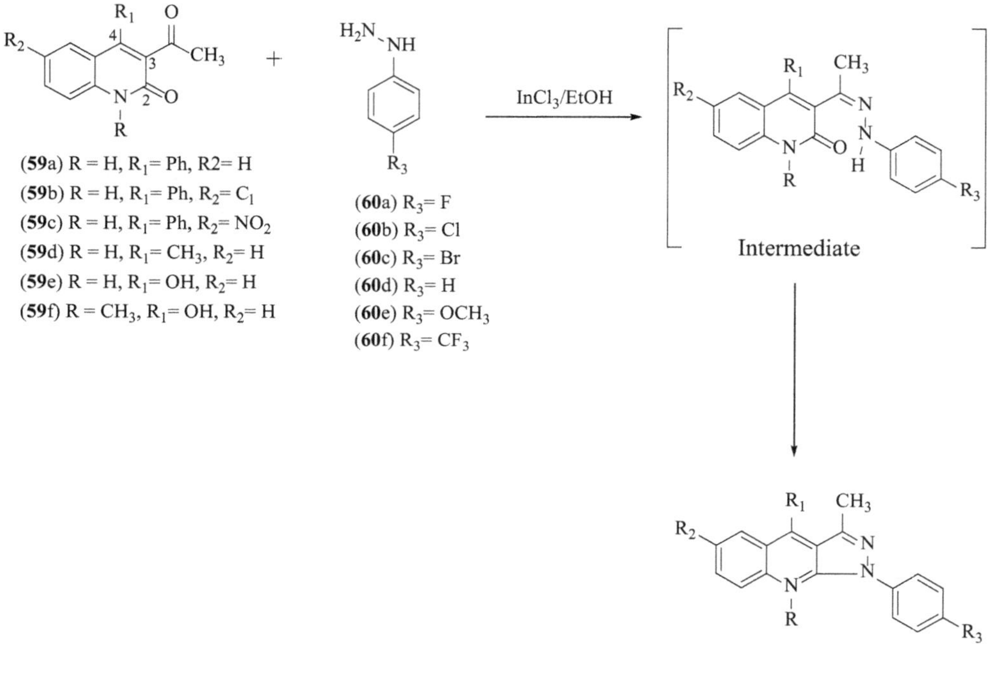

SCHEME 8.14 Synthesis of compounds **61**a–p.

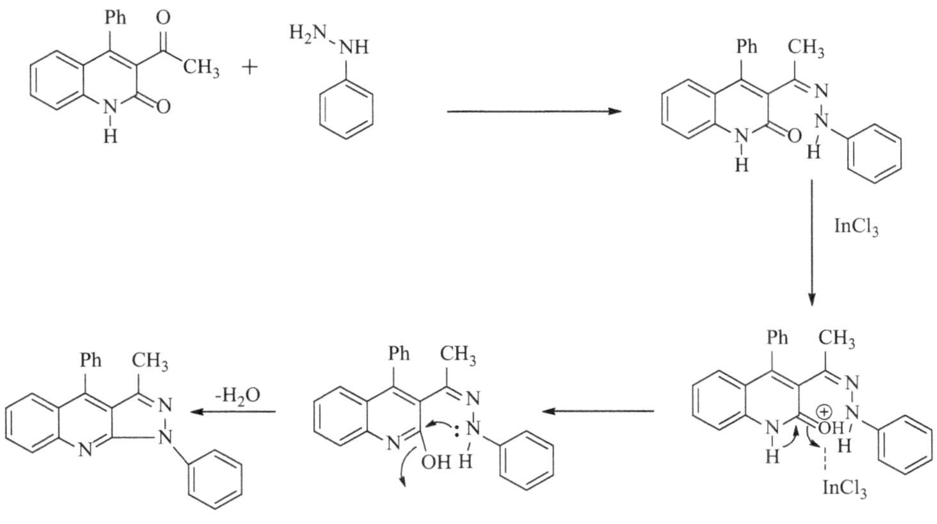

SCHEME 8.15 General mechanism for the formation of compounds (**61**a–p).

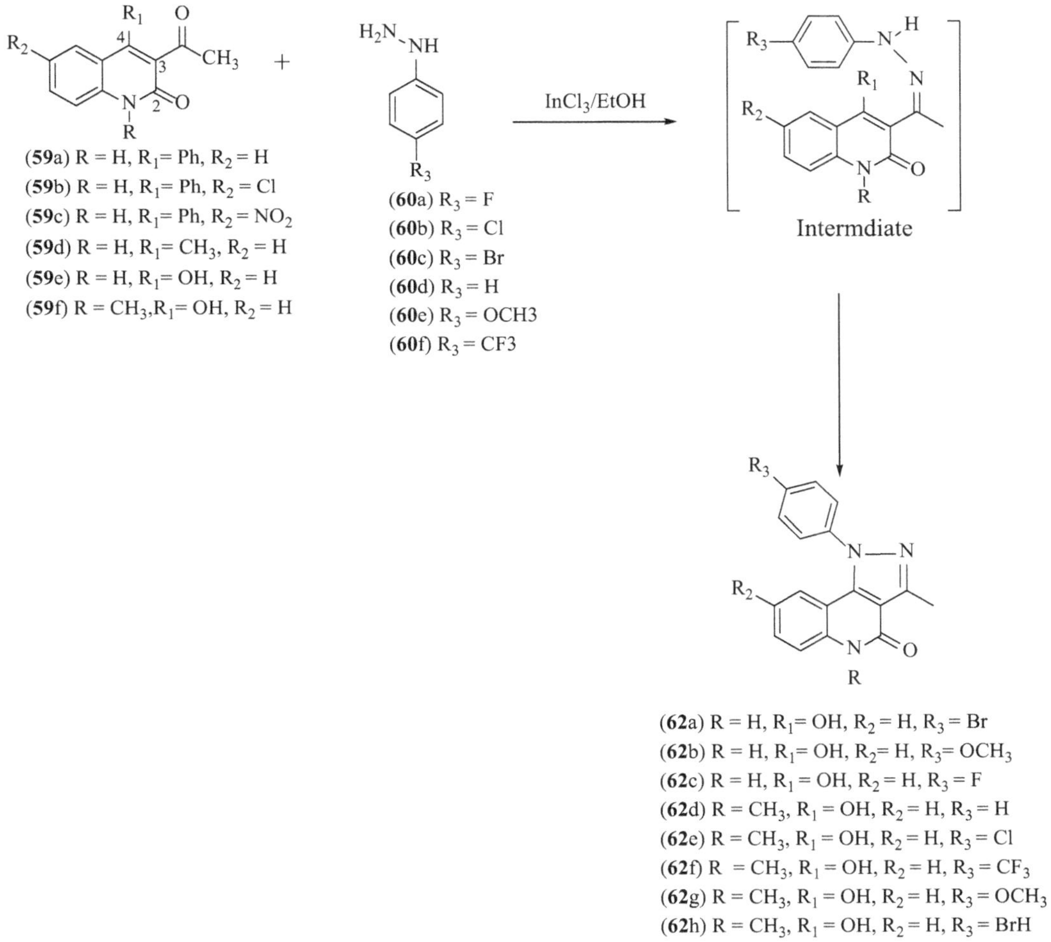

SCHEME 8.16 Synthesis of compounds **62**a–h.

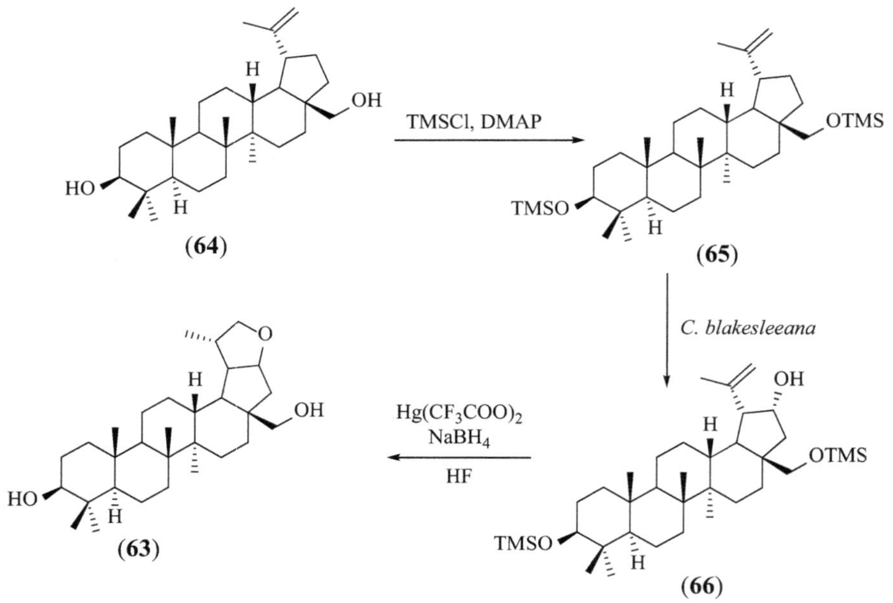

SCHEME 8.17 Chemo-enzymatic synthesis of compound **63**.

= 120 nM) [83]. The potent bioactivity of these compounds was due to the tetrahydrofram ring incorporated into their structures. This compound was a minor secondary metabolite of *E. angustifolium*. For its detailed *in vitro* and *in vivo* bioactivity studies, we have carried out chemo-enzymatic synthesis of compound **63** from commercially available betulin (**64**). Compound **64** on reaction with TMSCl and DMAP yielded compound **65**. Incubation of compound **65** with the whole-cell liquid culture of *C. blakesleeana* afforded 21-hydroxybetulin (**66**) which on reaction with $Hg(CF_3COO)_2$ and $NaBH_4$ followed by treatment with HF yielded compound **63**. This chemo-enzymatic synthesis of natural products using a combination of chemical and microbial reactions will be an important tool in natural-product-based medicinal chemistry research programs to scale up minor bioactive lead natural products from commercially available compounds. This green chemistry approach can help synthesize bioactive compounds that are difficult to synthesize using traditional synthetic chemistry to supply minor lead bioactive natural products for SAR studies or detailed *in vivo* and clinical testing. The chemo-enzymatic synthesis of **63** is shown in Scheme 8.17.

In conclusion, we have successfully applied microbial reactions in generating novel analogues of heterocyclic compounds. We have also discovered novel synthetic reactions in synthesizing new quinolone-type heterocyclic compounds. Some of these reactions have applications in organic synthesis. A few of these compounds have shown potent bioactivities that warrant further investigations to find out their biomedical applications. The moderately bioactive heterocycles are warranted for their SAR studies in order to improve their bioactivities. We have also developed a chemo-enzymatic approach to synthesize natural products analogues to study their SAR studies or detailed *in vitro* and *in vivo* studies.

ACKNOWLEDGMENTS

Athar Ata would like to thank all of his undergraduate and graduate students and collaborators who were involved in various aspects of synthetic chemistry described in this chapter, and their names are cited in the references. Funding for this research was provided by the Natural Sciences and Engineering Research Council of Canada.

REFERENCES

1. Gupta, G. K., Mittal, A., Kumar, V. 2014. *Letters in Organic Chemistry* 11: 273–286.
2. Rodrigues-Ferreira, S., Nehlig, A., Kacem, M., Nahmias, C. 2020. *Scientific Reports* 10: 13217.
3. Di Liberto, M. G., Caldo, A. J., Quiroga, A. D., Riveira, M. J., Derita, M. G. 2020. *ACS Omega* 5: 7481–7487.
4. Kumar, M., Sharma, K., Samarth, M., Kumar, A. 2010. *European Journal of Medicinal Chemistry* 45: 4467–4472.
5. Gaurav, A., Gautam, V. 2013. *Current Enzyme Inhibition* 9: 106–116.
6. Gopi, C., Krupamai, G., Dhanaraju, M. D. 2019. *Review Journal of Chemistry* 9: 255–289.
7. McAteer, C. H., Murugan, R., Rao, Y. V. S. 2017. *Advances in Heterocyclic Chemistry* 173–205.
8. Hooshmand, S. E., Halimehjani, A. Z. 2018. *Targets in Heterocyclic Systems* 22: 119–137.
9. Mishra, N. P., Mohapatra, S., Panda, P., Nayak, S. 2018. *Current Organic Chemistry* 22: 1959–1985.
10. Hosseini, H., Bayat, M. 2018. *Topics in Current Chemistry* 376: 1–67.
11. Sun, Y., Saha, L. K., Saha, S., Jo, U., Pommier, Y. 2020. *DNA Repair* 94: 102926.
12. Lam, C. W., Wakeman, A., James, A., Ata, A., Gengan, R. M., Ross, S. A. 2015. *Steroids* 95: 73–79.

13. Uvarani, C., Sankaran, M., Jaivel, N., Chandraprakash, K., Ata, A., Mohan, P. S. 2013. *Journal of Natural Products* 76: 993–1000.
14. Agomuoh, A. A., Ata, A., Udenigwe, C. C., Aluko, R. E., Irenus, I. 2013. *Chemistry & Biodiversity* 10: 401–410.
15. Mollataghi, A., Coudiere, E., Hadi, A. H. A., Mukhtar, M. R., Awang, K., Litaudon, M., Ata, A. 2012. *Fitoterapia* 83: 298–302.
16. Ata, A., Naz, S., Elias, E. M. 2011. *Pure and Applied Chemistry* 83: 1741–1749.
17. Ata, A., Tan, D. S., Matochko, W. L., Adesanwo, J. K., 2011. *Phytochemistry Letters* 4: 34–37.
18. Matochko, W. L., James, A., Lam, C. W., Kozera, D. J., Ata, A., Gengan, R. M. 2010. *Journal of Natural Products* 73: 1858–1862.
19. Uvarani, C., Arumugasamy, K., Chandraprakash, K., Sankaran, M., Ata, A., Mohan, P. S. 2015. *Chemistry & Biodiversity* 12: 358–370.
20. Arasakumar, T., Shyamsivappan, S., Gopalan, S., Ata, A., Mohan, P. S. 2019. *Synlett* 30: 63–68.
21. Batur, O. O., Kiran, I., Berger, R. G., Demirci, B. 2019. *Natural Volatiles & Essential Oils* 6: 8–15.
22. Song, C., Liu, J., Wang, H., Li, X., Liu, B., Zhang, M., Shan, X., Li, H., Gao, J., Qin, J. 2020. *Chemistry & Biodiversity* 17: e2000178.
23. Wang, Y., Xiang, L., Wang, Z., Li, J., Xu, J., He, X. 2020. *Bioorganic Chemistry* 101: 103870.
24. Shen, P., Zhang, J., Zhu, Y., Wang, W., Yu, B., Wang, W. 2020. *Bioorganic & Medicinal Chemistry* 28: 115465.
25. Parshikov, I. A., Netrusov, A. I., Sutherland, J. B. 2012. *Biotechnology Advances* 30: 1516–1523.
26. Parra, A., Rivas, F., Garcia-Granados, A., Martinez, A. 2009. *Mini-Reviews in Organic Chemistry* 6: 307–320.
27. Musharraf, S. G., Najeeb, A., Khan, S., Pervez, M., Ali, R. A., Choudhary, M. I. 2010. *Journal of Molecular Catalysis B: Enzymatic* 66: 156–160.
28. Ge, W., Li, N., Jiang, G. 2007. *Shengwu Jishu Tongbao* 2: 82–86.
29. Fodouop, C., Simeon, P., Kengni, A. D. M., Siddiqui, M., Fowa, A. B., Gatsing, D., Choudhary, M. I. 2020. *Steroids* 162: 108679.
30. Atta-ur-Rahman, Y. M., Farooq, A., Anjum, S., Choudhary, M. I. 1999. *Current Organic Chemistry* 3: 309–326.
31. Gnilka, R., Majchrzak, W., Wawrzenczyk, C. 2015. *Phytochemistry Letters* 13: 41–46.
32. Ahmad, M. S., Yousuf, S., Atia-Tul-Wahab, Jabeen, A., Atta-Ur-Rahman, Choudhary, M. I. 2017. *Steroids* 128: 75–84.
33. Arunrattiyakorn, P., Suwannasai, N., Aree, T., Kanokmedhakul, S., Ito, H., Kanzaki, H. 2014. *Journal of Molecular Catalysis B: Enzymatic* 102: 174–179.
34. Liu, D.-L., Liu, Y., Qiu, F., Gao, Y., Zhang, J.-Z. 2011. *Journal of Asian Natural Products Research* 13: 160–167.
35. Wang, J., Su, S., Zhang, S., Zhai, S., Sheng, R., Wu, W., Guo, R. 2019. *European Journal of Medicinal Chemistry* 175: 215–233.
36. Sharma, O. P., Pan, A., Hoti, S. L., Jadhav, A., Kannan, M., Mathur, P. P. 2012. *Medicinal Chemistry Research* 21: 2415–2427.
37. Kong, D.-X., Li, X.-J., Tang, G.-Y., Zhang, H.-Y. 2008. *ChemMedChem* 3: 233–236.
38. Ata, A., Nachtigall, J. A. 2004. *Zeitschrift fuer Naturforschung, C: Journal of Biosciences* 59: 209–214.
39. Ata, A., Conci, L. J., Betteridge, J., Orhan, I., Sener, B. 2007. *Chemical & Pharmaceutical Bulletin* 55: 118–123.
40. Pandey, P., Singh, D., Hasanain, M., Ashraf, R., Maheshwari, M., Choyal, K., Singh, A., Datta, D., Kumar, B., Sarkar, J. 2019. *Carcinogenesis* 40: 791–804.
41. Ata, A., Betteridge, J., Schaub, E., Kozera, D. J., Holloway, P., Samerasekera, R. 2009. *Chemistry & Biodiversity* 6: 1453–1462.
42. Iverson, C. D., Zahid, S., Li, Y., Shoqafi, A. H., Ata, A., Samarasekera, R. 2010. *Phytochemistry Letters* 3: 207–221.
43. Ata, A., Udenigwe, C. C., Matochko, W., Holloway, P., Eze, M. O., Uzoegwu, P. N. 2009. *Natural Product Communications* 4: 1185–1188.
44. Kurasawa, Y., Kim, H. O. 2002. *Journal of Heterocyclic Chemistry* 39: 551–570.
45. Monge, A., Palop, J. A., Urbasos, I., Fernández-Alvarez, E. 1989. *Journal of Heterocyclic Chemistry* 26: 1623–1626.
46. Vicente, E., Pérez-Silanes, S., Lima, L. M., Ancizu, S., Burguete, A., Solano, B., Villar, R., Aldana, I., Monge, A. 2009. *Bioorganic Medicinal Chemistry Letters* 17: 385–389.
47. Becker, I. 2008. *Journal of Heterocyclic Chemistry* 45: 1005–1022.
48. Guillon, J., Moreau, S., Mouray, E., Sinou, V., Forfar, I., Fabre, S. B., Desplat, V., Millet, P., Parzy, D., Jarry, C., Grellier, P. 2008. *Bioorganic Medicinal Chemistry* 16: 9133–9144.
49. Abouzid, K. A. M., Khalil, N. A., Ahmed, E. M., Abd El-Latif, H. A., El-Araby, M. E. 2010. *Medicinal Chemistry Research* 19: 629–642.
50. Wagle, S., Adhikari, A. V., Kumari, N. S. 2009. *European Journal of Medicinal Chemistry* 44: 1135–1143.
51. Kleim, J. P., Bender, R., Billhardt, U. M., Meichsner, C., Riess, G., Rösner, M., Winkler, I., Paessens, A. 1993. *Antimicrobial Agents Chemotherapy* 37: 1659–1664.
52. Kulkarni, N. V., Revankar, V. K., Kirasur, B. N., Hugar, M. H. 2012. *Medicinal Chemistry Research* 21: 663–671.
53. Amin, K. M., Ismail, M. M. F., Noaman, E., Soliman, D. H., Ammar, Y. A. 2006. *Bioorganic Medicinal Chemistry* 14: 6917–6923.
54. Ata, A. 2012. In *Studies in Natural Products Chemistry*, Atta-Ur-Rahman, Ed. Elsevier Science Publishers, Amsterdam, Vol. 38, pp. 225–245.
55. Atta-Ur-Rahman, Zareen, S., Choudhary, M.I., Akhtar, M.N., Khan, S.N. 2008. *Journal of Natural Products* 71: 910–913.
56. Khan, M. S., Munawar, M. A., Ashraf, M., Alam, U., Ata, A., Asiri, A. M., Kousar, S., Khan, M. A. 2014. *Bioorganic and Medicinal Chemistry* 22: 1195–1200.
57. Jossang, A., Jossang, P., Hadi, H. A., Sevenet, T., Bodo, B. 1991. *Journal of Organic Chemistry* 56: 6527–6530.
58. Cui, C. B., Kakeya, H., Osada, H. 1996. *Journal of Antibiotics* 49: 832–835.
59. Anderton, N., Cockrum, P. A., Colegate, S. M., Edgar, J. A., Flower, K., Vit, I., Willing, R. I. 1998. *Phytochemistry* 48: 437–439.
60. Usui, T., Kondoh, M., Cui, C.-B., Mayumi, T., Osada, H. 1998. *Biochemical Journal* 333: 543–548.

61. Kang, T. H., Matsumoto, K., Murakami, Y., Takayama, H., Kitajima, M., Aimi, N., Watanabe, H. 2002. *European Journal of Pharmacology* 444: 39–45.
62. Hasegawa, M., Nakayama, A., Yokohama, S., Hosokami, T., Kurebayashi, Y., Ikeda, T., Shimoto, Y., Ide, S., Honda, Y., Suzuki, N. 1995. *Chemical Pharmaceutical Bulletin* 43: 1125–1131.
63. Baldwin, J. E., Freeman, R. T., Lowe, C., Schofield, C. J., Lee, E. 1989. *Tetrahedron* 45: 4537–4550.
64. Aicher, T. D., Balkan, B., Bell, P. A., Brand, L. J., Cheon, S. H., Deems, R. O., Fell, J.B., Fillers, W. S., Fraser, J. D., Gao, J., Knorr, D. C., Kahle, G. G., Leone, C. L., Nadelsen, J., Simpson, R., Smith, H. C. 1998. *Journal of Medicinal Chemistry* 41: 4556–4566.
65. Trapani, G., Franco, M., Latrofa, A., Genchi, G., Brigiani, M., Mazzoccoli, M., Persichella, M., Serra, M., Biggio, G., Liso, G. 1994. *European Journal of Medicinal Chemistry* 29: 197–204.
66. Atwal, M. S., Bauer, L., Dixit, S. N., Gearien, J. E., Morris R.W. 1965. *Journal of Medicinal Chemistry* 8: 566–571.
67. Morrissey, I. Smith, J. T. 1995. *Journal of Medical Microbiology* 43: 4–15.
68. Jones, P., Kinzel, O., Koch, U., Ontoria, J. M., Pescatore, G., Scarpelli, R., Torrisi, C. 2009. International Patent WO 027,730.
69. Chandraprakash, K., Sankaran, M., Uvarani, C., Shankar, R., Ata, A., Dallemer, F. Mohan, P. S. 2013. *Tetrahedron Letters* 54: 3896–3901.
70. Fletcher, M. P., Diggle, S. P., Cámara, M., Williams P. 2007. *Nature Protocol* 2: 1254–1262.
71. Kumarasamy, C., Sankaran, M., Uvanrani, C., Shankar, R., Ata, A., Dallemer, F., Mohan, P. S. 2013. *Tetrahedron Letters* 54: 3896–3901.
72. Reddy, T. S., Reddy, V.G., Kulhari, H., Shukla, R., Kamal, A., Bansal, V. 2016. *European Journal of Medicinal Chemistry* 117: 157–166.
73. Yu, B., Wang, S. Q., Qi, P. P., Yang, D. X., Tang, K., Liu, H. M. 2016. *European Journal of Medicinal Chemistry* 124: 350–360.
74. Lee, H.-Y., Chang, C.-Y., Su, C.-J., Huang, H.-L., Mehndiratta, S., Chao, Y.-H., Hsu, C.-M., Kumar, S., Sung, T.-Y., Huang, Y.-Z., Li, Y.-H., Yang, C.-R., Liou, J.-P. 2016. *European Journal of Medicinal Chemistry* 122: 92–101.
75. Yamada, N., Kadowaki, S., Takahashi, K., Umezu, K. 1992. *Biochemical Pharmacology* 44: 1211–1213.
76. Arasakumar, T., Mathusalini, S., Gopalan, S., Shymsivappan, S., Ata, A., Mohan, P. S. 2017. *Bioorganic and Medicinal Chemistry Letters* 27: 1538–1546.
77. Kamel, M., Sherif, S., Issa, R. M., Abd-El-Hey, F. I. 1973. *Tetrahedron* 29: 221–225.
78. Moyer, M. P., Weber, F. H., Gross, J. L. 1993. *Journal of Medicinal Chemistry* 36: 940–940.
79. Karthikeyan, C., Lee, C., Moore, J., Mittal, R., Suswam, E. A., Abbott, K. L., Pondugula, S. R., Manne, U., Narayanan, N. K., Trivedi, P., Tiwari, A. K. 2015. *Biorganic and Medicinal Chemistry* 23: 602–611.
80. Parekh, N. M., Maheria, K. C. 2012. *Medicinal Chemistry Research* 21: 4168–4179.
81. Ouiroga, J., Diaz, Y., Bueno, J., Insuasty, B., Abonia, R., Ortiz, A., Nogueras, M., Cobo, J. 2014. *European Journal of Medicinal Chemistry* 74: 216–224.
82. Arasakumar, T., Mathusalini, S., Lakshmi, K., Mohan, P. S., Ata, A., Lin, C.-H. 2015. *Synthetic Communications* 46: 232–241.
83. Ata, A., Alhazmi, H. 2018. In *Asymmetric Synthesis of Drugs and Natural Products*. A. Nag Ed. CRC Press, Boca Raton, FL. pp. 191–204.

9

The Use of Small Particle Catalysts in Pursuit of Green and Sustainable Chemistry

John A. Glaser

CONTENTS

9.1 Introduction ... 151
9.2 Nanoscale Applications ... 151
9.3 Large-Scale Bulk Chemical Applications ... 152
9.3.1 Ammonia Synthesis ... 152
9.3.2 Hydrogen Peroxide Synthesis ... 154
9.4 Nanocatalyzed Organic Transformations ... 155
9.4.1 The Cross-Coupling Reaction ... 159
9.4.2 Recycling of Aqueous Micellar Solutions ... 163
9.4.3 Applications to Agricultural Products ... 165
9.4.4 Single-Atom Catalysts ... 166
9.4.5 Catalyst Recovery ... 168
9.5 Conclusions ... 169
Disclaimer ... 169
Copyright ... 169
References ... 169

9.1 Introduction

Synthetic chemistry ranges across a highly diverse field of applications.[1,2] The petrochemical industry provides a separation and manufacturing capability which takes a highly complex mixture of organic molecules found in petroleum crude and synthesizes an extensive spectrum of materials ranging from low molecular weight gases to macromolecules.[3] At a smaller scale, chemistry is used to synthesize products required for special applications such as pharmaceuticals, pesticides/fungicides, or fine chemicals. The process operations for these various chemical industry components have significant environmental footprints. Process emissions and significant waste streams become environmental hurdles that translate into economic burdens. The green chemistry perspective offers opportunities to reconsider how process chemistry and engineering can be utilized to mitigate these burdens. Twelve principles have been used to organize the green chemistry approach which identifies concerns such as safety, waste minimization, process efficiency, toxicity of materials, and energy consumption.[4–6]

As a pillar of green chemistry, the use of catalysis can be incorporated into the design of chemical products and processes to reduce or eliminate the use and generation of hazardous substances in a synthetic process.[7] A catalyst increases the rate of a reaction without modifying the overall standard Gibbs energy change in the reaction, Figure 9.1.[8]

The catalyst is not consumed in the reaction. The catalytic activity can be encountered in a single phase (homogeneous catalysis) or the reaction occurs at or near an interface between multiple phases (heterogeneous catalysis).

Each day, chemical processes generate significant waste streams of varying complexity. Chemical reactions conducted by using stoichiometric equivalents generally lead to the generation of large amounts of waste, detrimental energy and resource consumption, and undesired by-products. Currently, stoichiometric chemical processes are targeted for replacement by more efficient catalytic alternatives which can translate into a reduction of energy and resource consumption. Homogeneous or heterogeneous catalytic reactions are exploited to replace stoichiometric chemical processes in pursuit of green and sustainable chemistry objectives.[9]

9.2 Nanoscale Applications

The field of nanoscience has opened new chemical synthesis vistas offering opportunities employing nanoscale and smaller materials to catalyze chemical reactions with significant green and sustainable features.[10] The selection of nanomaterials in homogeneous and heterogeneous catalyst applications shapes the development of nanocatalytic applications. A notable property of nanoparticles is their large surface-to-volume ratio when

DOI: 10.1201/9781003089162-9

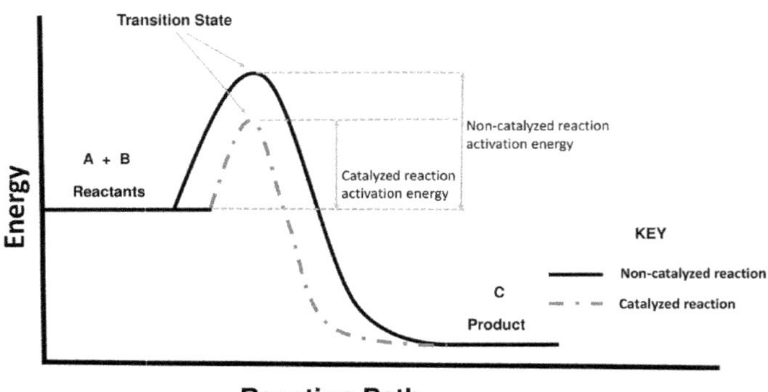

FIGURE 9.1 Exothermic catalytic reaction energy pathway showing changes to the rate of a reaction without modifying the overall standard Gibbs energy change.

contrasted with bulk materials.[11] Nanocatalyst catalytic activity rests on a series of properties, i.e., catalyst size, catalyst geometry, surface functionality, surface composition, aggregation attitude, physical environment, and chemical environment.[12] Catalyst designers are attracted by activity, selectivity, recoverability, and lifetime of these nanoscale materials.[13] Significant research and discovery have enabled the design and production of nanoscale particles having vastly different characteristics to meet the challenges of desired chemical conversion while adhering to objectives of green chemistry and sustainability (Figure 9.2).

The accessibility of the nanocatalyst atoms enhances the kinetics of more efficient chemical conversion of the reactants to products via the catalytic cycle when compared with conventional catalysts.[14,15] The understanding of catalytic structure and performance at the molecular level is required to transform novel nanomaterials into effective catalysts.[16] The features of high activity, safer reagents, waste minimization, energy efficiency, improved economy, and reduced global warming enhance the attractiveness of nanocatalysts.[17] The broad spectrum of nanocatalyst applications can be found as developing examples of new technology (Figure 9.3).

Organic synthesis, CO oxidation, water gas shift, selective oxidation, petroleum upgrade, NOX destruction, photocatalysis, and hydrogenation selectivity are massive topic research areas in which nanocatalysis is the subject of intensive investigation.[17]

9.3 Large-Scale Bulk Chemical Applications

Bulk chemical synthesis is the mainstay of the chemical industry, which ranges from separating valuable components of petroleum and other fossil-sourced materials to further manipulate separated materials to articles of commercial value. Biomass has become an emerging source of organic chemicals from a declared "green" source. Opportunities to build bulk chemicals from gaseous starting materials which are critically important to the production of food and other items deemed necessary for life are subjects of investigation.

9.3.1 Ammonia Synthesis

An indispensable agricultural feedstock and precursor to a vast array of chemicals having industrial and household utility, ammonia (NH_3) is an essential chemical intermediate to modern society.[18] Production of synthetic fertilizers for enhanced crop and nutrient production yield is a prominent use of NH_3. Current global food demands require the increase of NH_3 production to support the reasonably priced food needs

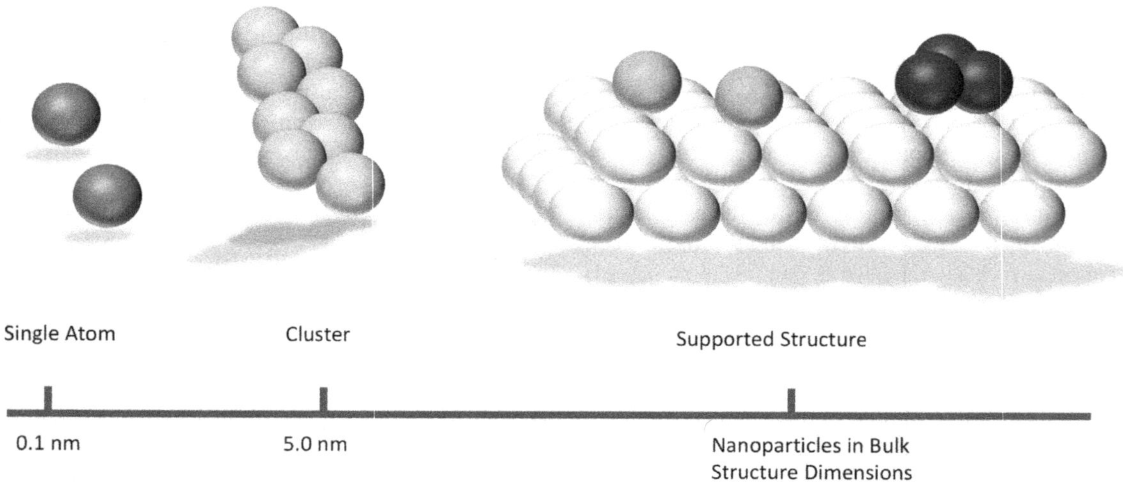

FIGURE 9.2 Physical states of nanoparticles.

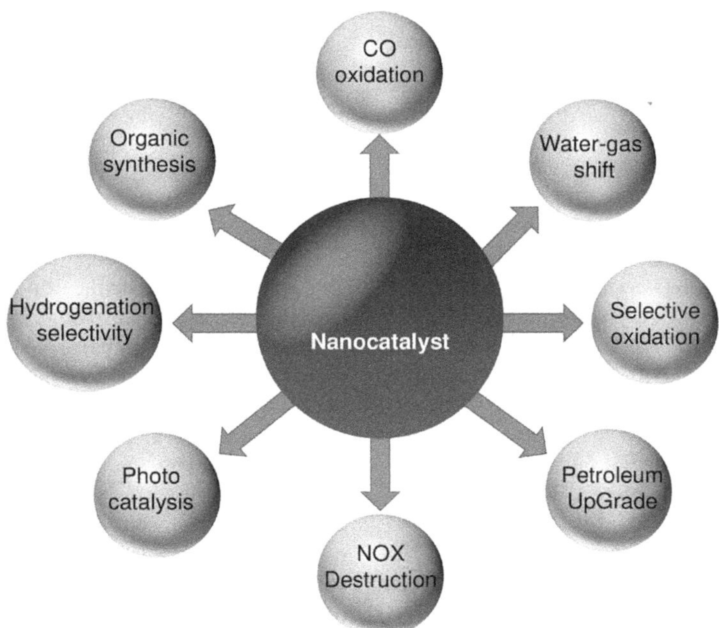

FIGURE 9.3 Nanocatalyst applications.

of a growing global population. The chemistry and economic affordability of NH_3 become pivotal to the desired production expansion. Processes leading to NH_3 have utilized four process conditions: thermocatalytic,[17,19] electrocatalytic,[20–22] photocatalytic,[23,24] and chemical looping processes.[25–27]

The current thermocatalytic Haber–Bosch (HB) synthesis process shortcomings can be easily identified in terms of high-energy consumption (27.4–31.8 GJ per metric ton NH_3), large greenhouse gas emissions (1.5–1.6 metric ton per metric ton NH_3), and overall process complexity. The HB process requires that a mixture of N_2 and H_2 pass over an iron catalyst promoted with K_2O and Al_2O_3 and at pressures of (20–40 MPa) and temperatures of 400–600°C. Ammonia yield is enhanced at lower temperatures and greater pressures since ammonia synthesis from N_2 and H_2 is an exothermic and entropy-decreasing reaction.[28] The HB process is not sustainable in its current form as it uses hydrogen sourced from steam methane reforming to break the strong triple bond of N_2. Use of the HB process consumes greater than 1% of the world's power production.[29]

A milder replacement technology for the HB process has been the subject of extensive research programs during the past century.[30–32] The formation of NH_3 from N_2 at mild conditions is one of the most challenging topics in catalysis (Figure 9.4).[33]

The iron catalyst BASF-S6-10 was recently developed for application in the HB process.[34] Having a surface area of 20 m^2/g, the catalyst is deployed as nanometer-sized particles containing reduced iron oxide in a framework of Al_2O_3 and CaO to stabilize the catalyst against agglomeration (Equation 9.1). Equation 9.1 BASF ammonia synthesis.

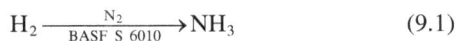

$$H_2 \xrightarrow[\text{BASF S 6010}]{N_2} NH_3 \tag{9.1}$$

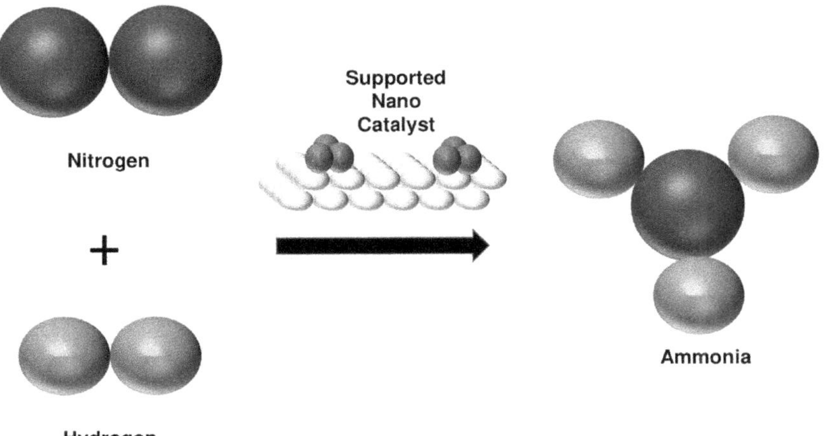

FIGURE 9.4 Ammonia synthesis over supported nanocatalyst.

The absence of Fe catalysts to support milder conditions after a century of discovery suggests progress may be enhanced by broadening the research scope to include alternative completely different catalysts.[35] Carbon-supported ruthenium (Ru)-based catalysts surfaced during the 20th century as second-generation catalyst candidates due to N_2 cleavage at lower temperatures than those required for a conventional Fe-based catalyst.[36] Recently introduced Ru-based catalysts have exhibited enhanced performance when compared with conventional iron-based catalysts. Continued optimization to approach theoretical minimum energy consumption is necessary. Nanocatalyst formulation is expected to increase surface area with enhanced structural activity through the increase of highly active site density. The control of catalyst particle size and shape in the nanometer range and stabilization of the particles by appropriate innovative preparation techniques are key factors for the development of commercial Ru nanocatalysts. Severe H_2 poisoning limits conventional Ru-based catalysts.[37] The cost of Ru catalysts limits their use.

An elaborate series of investigations has proceeded to the 3rd generation of catalysts involving advanced materials such as electrides, hydrides, nitrides, oxides, and oxy-hydrides-nitrides studied as support for NH_3 synthesis.[38–40] Researcher inventiveness has led to the development of multiple processes for the synthesis of NH_3 such as thermocatalytic, electrocatalytic, photocatalytic, and closed looping processes. In the thermocatalytic process, research has focused on iron, ruthenium, cobalt, molybdenum, and other transition metal materials.[41, 42] Gold, palladium, and other transition metals in the form of metal oxides, nitrides, and sulfides have been researched for their efficacy for NH_3 synthesis under electrocatalytic conditions.[43,44] Carbonaceous and biomimetic materials and modified semiconductors were found useful in photocatalytic NH_3 synthesis.[45] Closed looping NH_3 synthesis is based on metal nitrides, metal oxides, and alkali metal hydrides.[46,47]

The low-temperature, low-pressure synthesis of NH_3 is attractive for applications where a smaller-scale synthetic process provides availability that would significantly enable dedicated application technology.[48] Hydrogen has been identified as an energy carrier of the future.[49] Due to its low volumetric energy density as a gas at atmospheric conditions, hydrogen must be stored and transported effectively in a material having high gravimetric and volumetric hydrogen densities. Ammonia as a potential hydrogen carrier exhibits superior characteristics related to storage, transportation, and utilization.[50] As hydrogen assumes an expected role as part of the energy economy, the importance of ammonia for hydrogen storage may be realized.[51] These developments serve to enhance research leading to sustainable and economic synthetic routes to ammonia which may be aided by nanocatalysts.[52]

9.3.2 Hydrogen Peroxide Synthesis

One important key chemical in today's economy is hydrogen peroxide for use as an oxidant supporting sustainable development in today's global economy.[53] The increasing annual global consumption of hydrogen peroxide exceeds 5 million metric tons.[54] Recognized as a green oxidant, H_2O_2 is widely used in industrial and civilian applications where many materials are inefficient, and heavy metal stoichiometric oxidations are conducted.[55,56]

Current global commercial production of H_2O_2 utilizes a process based on a reduction/oxidation cycle involving a substituted anthraquinone.[57] This process employs hydrogen, anthraquinone, and air as raw materials for the hydrogen peroxide synthesis (Scheme 9.1).

Hydrogen is synthesized from steam by reforming methane, and oxygen is drawn from the atmosphere. A mixed solvent solution of 2-ethylanthraquinone is catalytically reduced to a mixture of the corresponding hydroquinone and tetra-hydroquinone with H_2 using palladium on alumina (Pd/Al_2O_3) catalyst at a temperature of 45°C. The reduction proceeds rapidly to consume all available hydrogen. Fixed bed catalyst reactors currently in use permit the necessary control of Pd in the catalyst which could catalytically decompose the newly formed H_2O_2. The cooled solution is non-catalytically oxidized through the introduction of air at low pressure. After a short residence time, a resultant solution is formed composed of H_2O_2 and regenerated anthraquinone is formed. Demineralized water is used to extract H_2O_2. Steam-heated distillation provides different commercial grades which are treated with stabilizers. This process has been criticized for energy consumption and as an unstainable production process based on the use of palladium as a critical material.[58] The Pd catalyst is largely conserved throughout the process and the waste streams consist of low-level contaminated water. Energy consumption of the commercial process is not as high as it is often claimed.[59] The commercial process has been staged at a large scale over the last century requiring

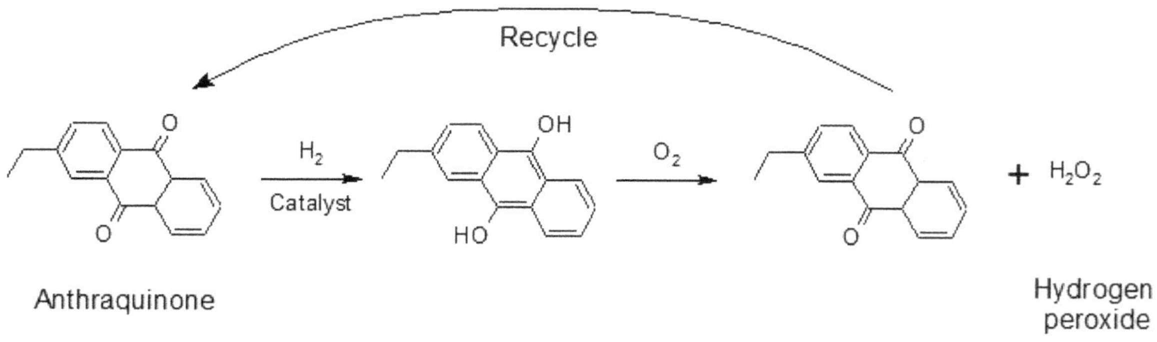

Recycle

Anthraquinone $\xrightarrow[\text{Catalyst}]{H_2}$ (hydroquinone) $\xrightarrow{O_2}$ (anthraquinone) + H_2O_2 (Hydrogen peroxide)

SCHEME 9.1 Anthraquinone-mediated hydrogen peroxide synthesis.

a large physical footprint and the plant is generally situated at central locations.[60]

Direct synthesis of hydrogen peroxide from hydrogen and oxygen is being actively studied as an alternative to the current manufacturing process (Equation 9.2).[61] Equation 9.2 Hydrogen peroxide synthesis from hydrogen and oxygen.

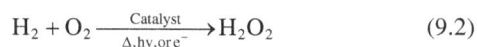

$$H_2 + O_2 \xrightarrow[\Delta, h\nu, or\, e^-]{Catalyst} H_2O_2 \tag{9.2}$$

The formation of hydrogen peroxide from the reaction of oxygen and hydrogen is esteemed as a "dream process" requiring a corresponding "dream catalyst".[62] Features of optimal catalyst design have been identified as the crucial features of metal selection, activity, and selectivity enhancement through metal promoters, support properties and role, suppression of side reactions via halide promoters, other promoters, particle morphology effects, catalyst morphology, and performance effects achieved through different synthetic method application, Figure 9.5.[63]

Gold nanoparticle-loaded metal oxides (Au/MOs) and Pd-nanocrystals-grafted-SiO_2 nanobeads encapsulated in a mesoporous silica shell are examples of active research searching for the nanocatalyst-based synthesis of H_2O_2.[64] Electrochemical synthesis of hydrogen peroxide via water oxidation continues to be a hot research topic with nanocatalyst requirements.[65]

Earth-abundant metals have become increasingly attractive as "green" replacements for rare noble metal catalysts.[66] The noble metals are in quite short supply so that process development utilizing them is fraught with economic issues and concerns of sustainability. Earth-abundant elements including Mn, Co, Ni, Cu, the early transition elements (Ti, V, Cr, Zr, Nb, and W), and derivative nanocomposites are under scrutiny as replacements for the rare noble catalysts currently in use.[67] Clearly, earth-abundant elements are recognized for their role in a wide spectrum of catalytic reactions and applications using nanoscience to afford an expanding venue for catalyst design and development.[68] Solutions to the various technical limitations of direct hydrogen peroxide synthesis can provide onsite synthesis capabilities that can be made adaptable to a process using the oxidative prowess of hydrogen peroxide.[69–71]

9.4 Nanocatalyzed Organic Transformations

Organic transformations conducted via nanocatalysis are a burgeoning area of research at academic and industrial application stages.[72] The interest in sustainable and economic organic synthetic technology is a significant driver for this development. The application of nanocatalysts to the discovery of catalytic organic transformations has become a crucial design component to advance the technology that is sensitive to environmental concerns.[73] Nanoparticles can agglomerate as clusters with an increase in particle size having lower surface area.[74] These conditions reflect a catalyst with fewer active sites on the accessible catalyst surface. Aggregation control can be pursued through the use of stabilizers, i.e., polymers or surfactants which shield nanoparticle surfaces thus controlling aggregation.[75] The electronic structure and catalytic activity of a nanoparticle can be altered by a surface-modifying process. Homogeneous reaction conditions are considered optimal to achieve maximal contact of reactive substrates. Homogeneous catalysis is defined as a catalytic process where reactive substrates and catalyst components form a single phase which contrasts with heterogeneous catalysis where multiple phases are employed.[76] Operationally, heterogeneous catalysis offers significant advantages for catalyst separation from products and unreacted starting substrates.[77,78] Strategies for catalyst separation in homogeneous reaction systems have offered means to recover the catalyst but not necessarily catalyst activity.[79] Catalyst recovery evaluation requires knowledge of the product yield, degree of conversion as a function of the catalytic cycle, the rate of product conversion, and turnover frequency along with the salient chemistry of the catalyst controlling recovery.[80] As a catalytic reaction proceeds, the catalyst may advance through different reactive forms involved in the chemical transformation which may correspond to energy maxima

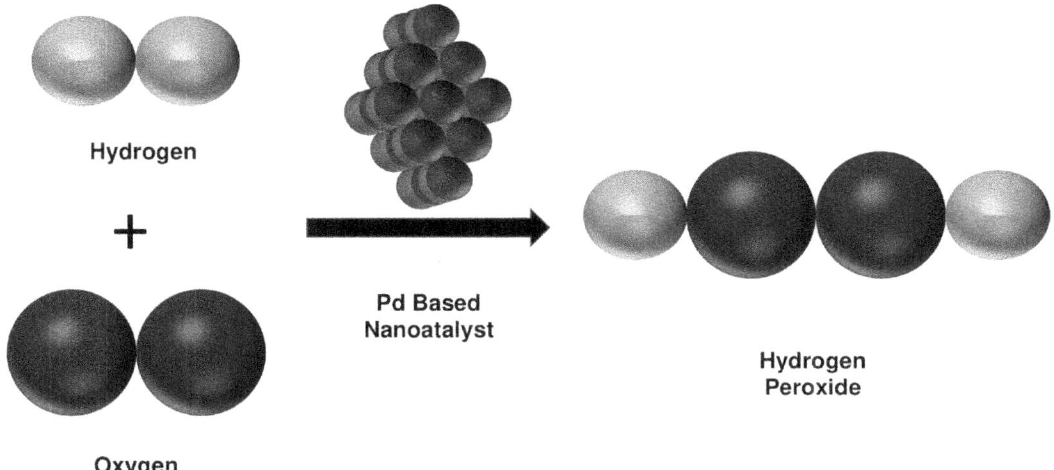

FIGURE 9.5 Hydrogen peroxide synthesis over Pd-based nanocatalyst.

or minima of rate-determining steps of the catalytic cycle.[81] Nano- and subnanoparticle catalysts have been developed with reliable catalyst synthesis and made durable using stabilization techniques to support a growing application research effort.[82]

Preparation techniques of nano- and subnanoparticles require special attention when synthesizing particles having different sizes by altering reaction conditions to optimize the dependence of catalytic process yield, process temperature on the particle size, particle surface characteristics, reaction media, and surface characteristics.[83] Spanning dimensions from subnano to nano, processes based on catalytic particles have been designed for a wide spectrum of applications ranging from oxidation, hydrogenation, CO insertion, and C-C coupling to condensation reactions.[84–88] Nanoparticle catalytic activities and selectivity are a function of the intrinsic size and shape-dependent properties of nanoparticles, which can be tweaked through judicious selection and modification of catalyst preparation conditions with any required function-defining materials used for support development and capping.[89,90]

Current chemistry is faced with the challenge of developing processes that meet the increasingly stringent requirements of sustainability, economic viability, and adherence to the principles of green chemistry and engineering.[91] Efficient, selective, and energy-conserving reactions are esteemed for the transformation of raw materials into valuable chemical products.[92] High selectivity can translate into waste reduction, required workup equipment at plant level reduction, and enabling of effective feedstock use.[93]

One of the controllable features of chemistry is chemoselectivity where a selective reactivity of a functional group in the presence of others can be elicited. An example of this amazing phenomenon can be found in the synthesis of anilines.[94] Chemoselective catalytic reduction of nitro aromatic compounds is the prominent means for conversion to synthesize functionalized anilines.[95] These amines are integral feedstock for the synthesis of dyes, pigments, pharmaceuticals, and agrochemicals. The hydrogenation conversion of 4-nitrostyrene portrayed in Scheme 9.2 depicts the products that can be formed by nanocatalyst selection.

Rather than a complete reduction of both functional groups as found in the platinum group metal catalysts, it is possible to select a clear reaction outcome of the nitro group reduction. Gold nanoparticle catalysts having dimensions of 3.5–4.0 nm deposited on TiO_2 and Fe_2O_3 catalyzed the selective hydrogenation of nitrostyrene.[96–98] Gold particle size and support materials were found to be critical to the selectivity of the catalytic conversion of nitrostyrenes to anilines. Selective reduction of a nitro group with other reducible functional groups present can be an important method of producing functionalized anilines for the synthesis of a variety of specific and fine chemicals.[99–101]

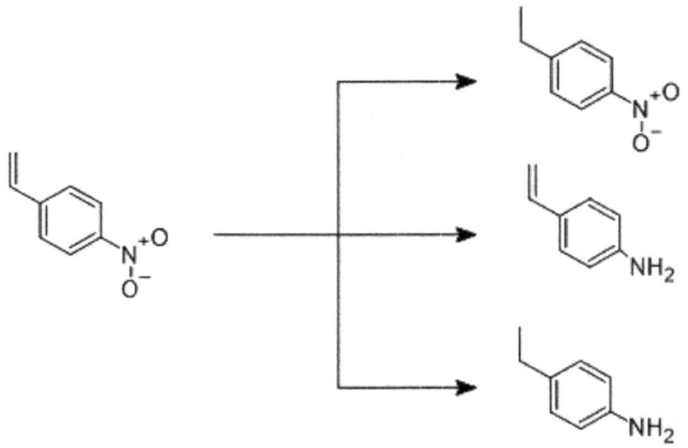

SCHEME 9.2 Nitrostyrene hydrogenation showing chemoselectivity.

The expanding catalog of organic transformations encompassed techniques and methodologies for nanocatalysts utilization to conduct oxidations and reductions of susceptible functionality using chemoselectivity, C–H activations, oxidative aminations, asymmetric hydrogenations, coupling reactions, and more complex scenarios such as click chemistry, domino and tandem reactions.[102] Accompanying these applications are the synthesis, characterization, and exploration of well-defined nanocatalysts which contribute to the desired chemical transformation. Silver nanoparticles have shown great utility in the synthetic area of oxidations and reductions.[103,104]

Hydrogen, a valuable chemical commodity for petroleum refining, ammonia synthesis, hydrogen peroxide production, and fine chemical/pharmaceutical industries, is in high demand.[105–107] It is also viewed as a non-polluting energetic fuel that could significantly contribute to the fuel requirements of non-polluting engines in the transportation sector and energy conversion technologies as part of the developing hydrogen economy.[108] Natural gas (methane) is the largest current source of hydrogen using steam reforming and the water gas shift reaction. Using waste plastics to generate hydrogen is a technical possibility that could help solve a major waste problem.[109–111] As an alternative feedstock the polyolefin, polypropylene has been treated with thermal catalytic reaction conditions to form H_2.[112] A nanocatalyst composed of iron deposited on a SiO_2 surface was shown to convert polypropylene to H_2 or carbon nanotubes depending on the temperature regime employed (Scheme 9.3).

Hydrogen formation could be favored in this catalytic system by temperature adjustment.[113,114]

The catalytic pyrolysis conversion of polyethylene (PE) requires the breaking of carbon–carbon long chains in preference to shorter chains.[115] An effective catalyst for this process was constructed as Pt nanoparticles on a $SrTiO_3$ nanocuboid

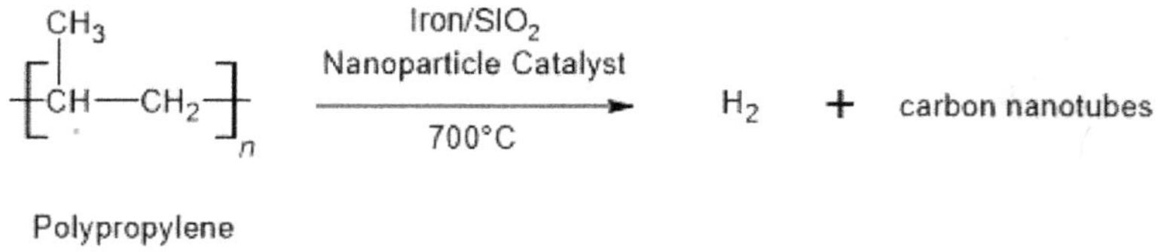

SCHEME 9.3 Nanocatalytic degradation of polypropylene.

support which converted polyethylene samples, having Mn of 8000–158,000 Da and Mw of 17,200–42,000 Da, into high-quality liquid products, such as lubricants and waxes, characterized by a narrow distribution of oligomeric chains (Scheme 9.4).

The waste plastic conversion was conducted at 170 psi H_2 and 300°C under solvent-free conditions.[116] Products with alkyl/aromatic chains can be synthesized using a variety of metal nanocatalysts.[117] Distances between catalytic sites where carbon–carbon bond cleavage occurs are important and these features can be designed into the catalyst. The polymeric chains are long and may adsorb onto multiple catalyst faces which may exert selective formation of specific chain lengths.

Sustainable alternative jet fuels derived from biomass have developed into an important technology to achieve high carbon efficiency and decarbonize the aviation industry.[118] Biomass feedstock such as corn stover, woody biomass, and cellulosic residues have been shown to be proficient sources of the remarkable chemical, 2,3-butanediol.[119] A nanocatalyst composed of copper supported on an H-ZSM-5 zeolite and under optimal reaction conditions of 250°C, 115 kPa, and 1.0 h^{-1} weight hourly space velocity, was recorded to show a conversion exceeding 97% and C_3+ olefin selectivity (Scheme 9.5).[120]

The sequence of steps converting the diol to fuel involves catalytic dehydration, oligomerization, and hydrogenation. The C_3–C_6 olefins formed in the initial step are oligomerized over Amberlyst-36 catalyst to form longer-chain hydrocarbons composed of greater than 70 wt% jet-range hydrocarbons having a composition of iso-olefins/iso-paraffins. Jet-range hydrocarbons synthesized by this process provide an overall carbon efficiency of 19–22% identifying it as a promising technology designed to meet the objective of sustainable alternative jet fuel production.[121]

The greatest mass of available renewable raw material is lignocellulosic biomass which has been converted to chemicals and fuels, assisting the global targets related to climate change. γ-Valerolactone (GVL) has recently been identified as a pivotal component of the building blocks available from renewable lignocellulosic biomass resources for energy, chemical, and material sector requirements.[122] Biomass-derived levulinic acid was converted to GVL using a ruthenium nanocatalyst deposited on Dowex 50-WX2-100 gel-type resin to chemoselectively hydrogenate the ketone carbonyl of levulinic acid, followed by acid-catalyzed lactonization (Scheme 9.6).[123,124]

The reaction was found to be intolerant to contamination in the levulinic acid feed that could be residual mineral acids and humins derived from the acid-catalyzed degradation of biomass. Catalyst deactivation caused by surface contamination with humins was the major obstacle. Clean feed presented no problem leading to longer catalyst life. Options of humin-resistant catalyst design have been pursued. The conversion reaction has been made compatible with a water milieu which can simplify operations and achieve a more environmentally

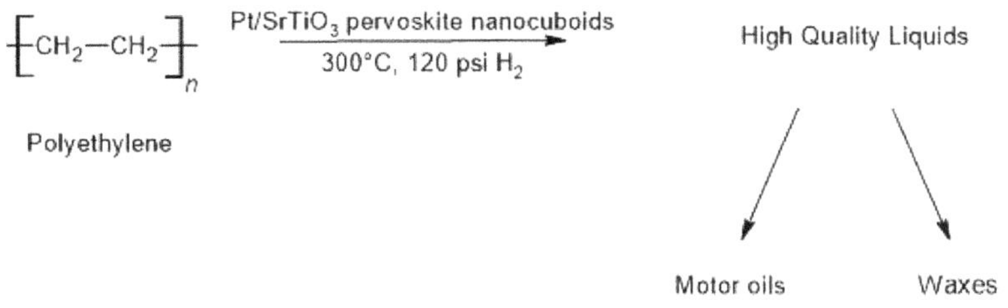

$-[CH_2-CH_2]_n-$ Polyethylene —Pt/SrTiO₃ pervoskite nanocuboids, 300°C, 120 psi H_2→ High Quality Liquids → Motor oils; Waxes

SCHEME 9.4 Nanocatalytic degradation of polyethylene.

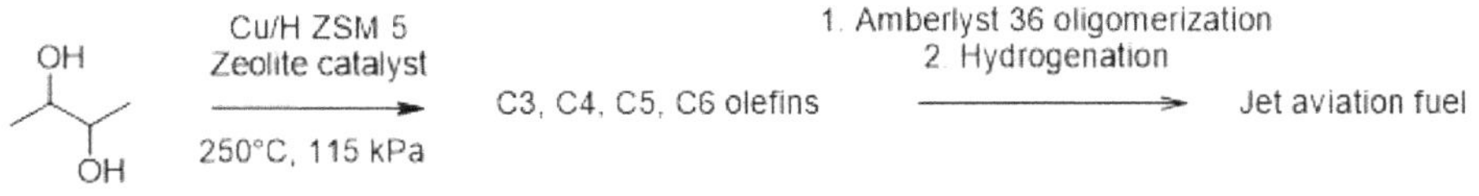

2,3-butanediol —Cu/H ZSM 5 Zeolite catalyst, 250°C, 115 kPa→ C3, C4, C5, C6 olefins —1. Amberlyst 36 oligomerization 2. Hydrogenation→ Jet aviation fuel

SCHEME 9.5 Product formation from 2,3-butanediol by nanocatalysts.

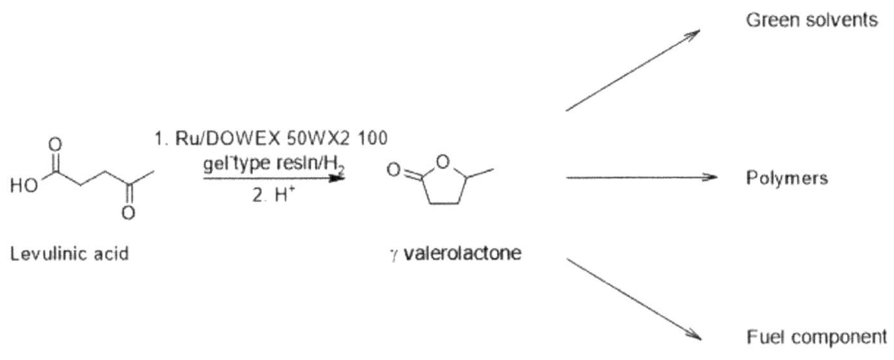

Levulinic acid —1. Ru/DOWEX 50WX2 100 gel type resin/H_2 2. H^+→ γ valerolactone → Green solvents; Polymers; Fuel component

SCHEME 9.6 Chemoselective reduction of Levulinic acid and formation of γ-valerolactone.

compatible footprint.[125–127] Development of a cost-effective process has become a real constraint of GVL large-scale production from lignocellulosic biomass. A new generation of biofuels may become available from this promising precursor chemical.

Heteroatoms constitute a common molecular fragment of many active medicinal ingredients and provide isosteric replacements for carbon in aliphatic/aromatic structures.[128] A variety of heterocyclic compounds can be synthesized by gold and silver nanoparticle-catalyzed reactions.[129] The diversity of heterocyclic systems that can be synthesized through NP-catalyzed reactions is amazingly broad. Green nanocatalysts used in the synthesis of various heterocycles show advantages such as short reaction time, high yield, inexpensive chemicals usage, easy workup procedure, and specific reaction conditions.[130,131] Acridines, coumarins, indoles, pyrazoles, quinoxalines, and 1,2,3 triazoles are but a few of the heterocyclic systems available through nanochemistry (Figure 9.6).

Some early perspectives suggested that nanocatalysts had not been applied to complex molecule synthesis. Many early applications focused on cross-couplings and oxidations/reductions. Opportunities using nanocatalysts in complex molecule synthesis were found in several areas.[132] Applications to large multiring systems have been instructive to understand the power of nanocatalyst transformations.[133] A pivotal application was utilized to complete the total synthesis of (±)-sorocenol B, a natural product exhibiting cytotoxic activity toward human cancer cell lines.[134] Key steps featured a silver nanoparticle (AgNP)-catalyzed Diels–Alder cycloaddition with a late-stage Pd(II)-catalyzed oxidative cyclization involved at a later stage (Scheme 9.7).[135]

The bicyclo[3.3.1] framework of sorocenol B was found to be accessible through the Pd(II)-catalyzed oxidative

Indoles 1,2,3-triazole Coumarins

Quinoxalines Benzofurans

Acridines Pyrazoles

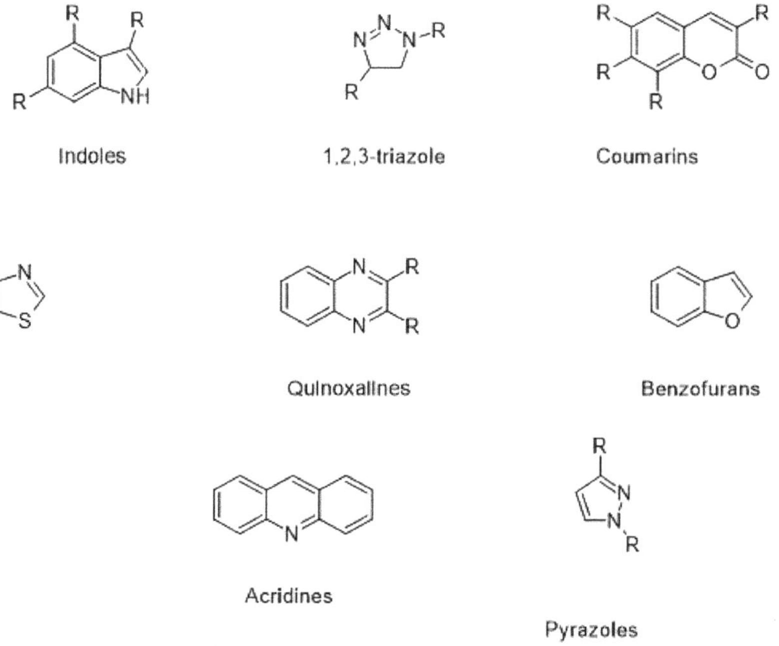

FIGURE 9.6 Heterocyclic systems available through the use of nanocatalysts.

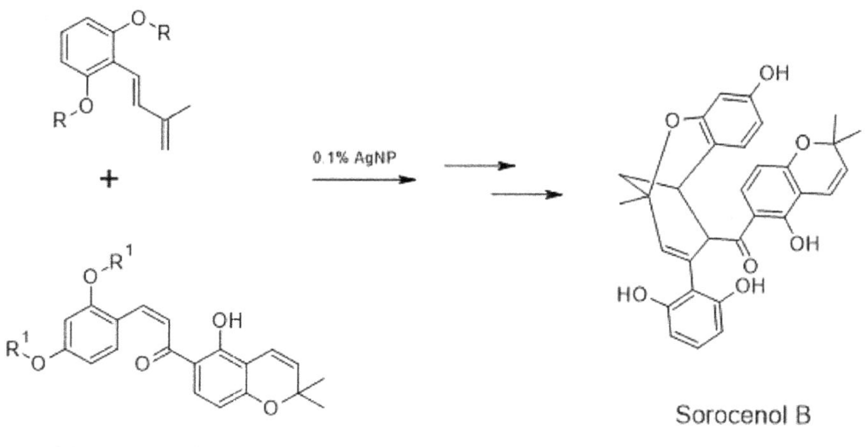

0.1% AgNP

Sorocenol B

SCHEME 9.7 Silver nanoparticle-catalyzed Diels–Alder cycloaddition.

cyclization of the endo cycloadduct. The synthetic direction for the sorocenol application was derived from the use of AgNPs as the catalyst for Diels–Alder cycloadditions of 2′-hydroxychalcones.[136] A similar application of nanoscale catalysts was utilized to synthesize sorbiterrin A, a member of a class of polyketides called sorbicillinoids having structural diversity and bioactivity. A bicyclic [3.3.1] ring system of the sorbiterrin A molecule was formed by the application of a distinct AgNP-catalyzed bridged aldol condensation (Scheme 9.8).[137]

Nanoparticle catalysis recently has exhibited extensive contributions to chemical synthesis applications. This advance is attributable to multidisciplinary collaborations between fruitful partnerships leading to remarkable catalytic activities and selectivities of nanocatalytic materials. The size and shape-dependent nanoparticle properties of intrinsic size and shape-dependence can be manipulated through the selection and modifications to preparation conditions, support materials, and capping agents. Nanocatalysts are beginning to offer features of catalyst recycling, continuous processing, and ease of separation, thereby offering green and cost-effective alternatives.

The iterative process prescribed for small-molecule drug discovery requires an appraisal and extension of existing knowledge to select new molecules as targets for synthesis and testing leading to molecules designed for improved treatment properties.[138] Clearly, the existing knowledge of a catalyst support's content and composition is important in a normative role for the construction of synthetic pathways.[139] The analysis of the chemical synthesis landscape for modern medicinal chemistry, discerned from the medicinal chemistry literature, identified a limited number of reactions dominating academic and industrial practice.[140] The economics of this information for the industry is quite important since the progression of a drug-discovery process can be expensive and time-consuming. With a process development duration constrained by the number of cycles of design, synthesis, and testing needed to improve the properties of successive generations of small molecules to match the various criteria required for the product candidate selection, it is necessary to conduct this investigation with optimal economic guidance.[141,142] The most common reactions found through the literature analysis were the Suzuki–Miyaura chemistry, amide bond formation, and a reduction of heterocyclic syntheses for the timeframes evaluated. Breakthrough synthetic transformations, such as ring-closing metathesis, C-H bond activation, or biocatalysis among other candidates, did not show comparable impact in the record of use. The cross-coupling reaction has become a workhorse for the pharmaceutical industry and despite its age, a series of modifications have simplified operations and reduced the reaction's environmental footprint.[143]

9.4.1 The Cross-Coupling Reaction

New carbon-to-carbon and carbon-to-heteroatom bonds formed using cross-coupling reactions are becoming a robust series of tools enabling the synthetic chemist to meet the continuing challenges of fine and pharmaceutical chemistry.[144] Cross-coupling reactions, discovered in the last half-century, are employed by industry and academia (Equation 9.3). Equation 9.3 Transition-metal-mediated cross-coupling reaction.

$$\underset{\text{electrophile}}{R\ X} + \underset{\text{nucleophile}}{R'\ M} \xrightarrow{\text{Catalyst}} R\ R' \qquad (9.3)$$

A selected series of cross-coupling reactions are portrayed (Scheme 9.9).

The named reactions of Heck, Stille, Suzuki, Sonogashira, Kumada, and Hiyama can be catalyzed using palladium catalysts and lead to different coupled products depending on the starting materials.[145] In each case, the scope of the reaction has been expanded to utilize a range of different organic structures having a wide spectrum of chemical functionalities, thereby expanding the synthetic utility of each cross-coupling reaction.

Despite the mature technology status, cross-coupling reactions continue to be improved by broadening the scope of application, efficient catalyst development, and expanding the reaction's practical utility. The importance of cross-coupling reactions has become recognized by the pharmaceutical/medicinal chemistry industry.[146] Examples of current pharmaceutical products synthesized using cross-coupling technology are shown in Scheme 9.10. The importance of this attention was derived from a general query of the breadth of the synthetic reaction technology currently available.[147]

Cross-coupling reactions have matured into a reliable technology that is used widely in both industry and academia.[148] Significant research related to the general use of these reactions has been pursued to search the scope of each of these reactions, ongoing design more efficient catalysts, and enhance the reactions' practicality.[149]

Recent advances to the Suzuki–Miyaura technology have included the development of improved ligands and nanoscale catalysts/precatalysts and a wider collection of electrophiles accessible for cross-coupling.[150] The mild reaction conditions, chemoselectivity, compatibility of diverse functional groups, predictable stereochemical results, and starting materials having non-toxic and air-stable property materials for the Suzuki–Miyaura reaction highlight its prominence as a premier cross-coupling technology at scales ranging from small to large.[151–153] A catalytic cycle depicting processes participating

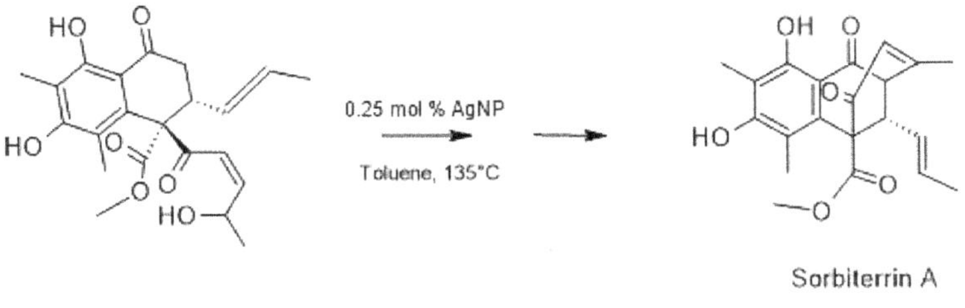

0.25 mol % AgNP

Toluene, 135°C

Sorbiterrin A

SCHEME 9.8 Nanoparticle-catalyzed bridged aldol condensation.

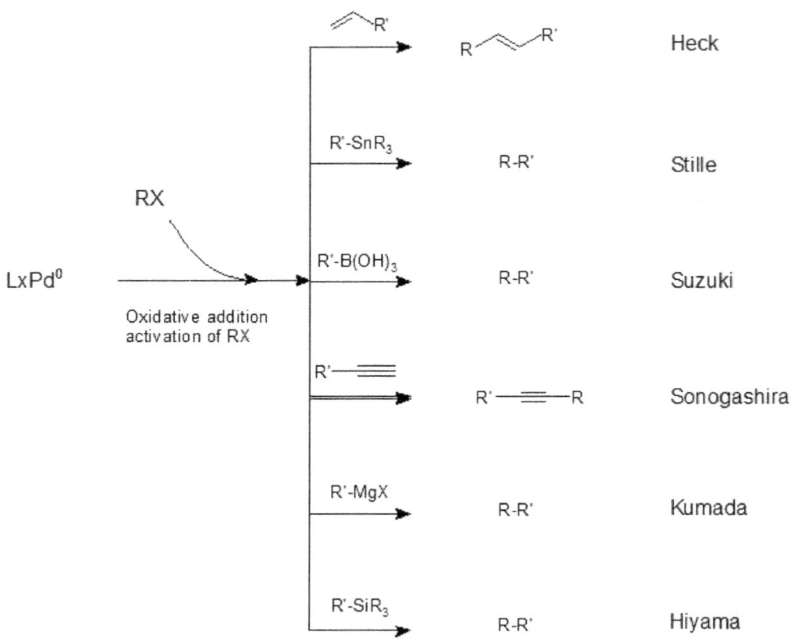

SCHEME 9.9 Selected series of cross-coupling reactions.

Losartan Etoricoxib Lapatinib

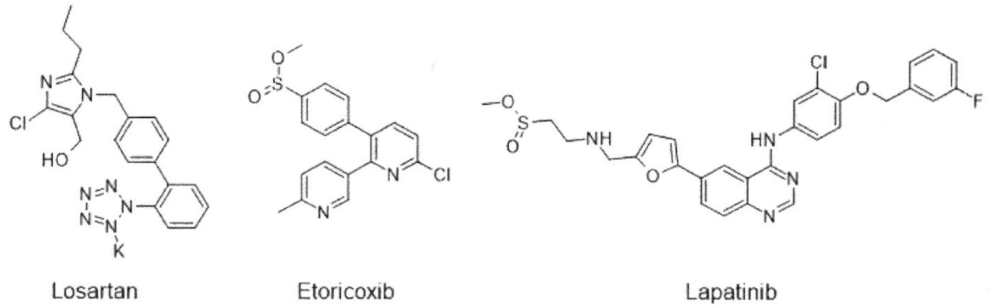

SCHEME 9.10 Current pharmaceutical products synthesized by cross-coupling technology.

in the catalyzed cross-coupling reactions of organometallic reagents between haloorganics and transition metal catalysts requires oxidative addition, transmetalation, and reductive elimination as basic processes (Scheme 9.11).[154] Oxidative addition and reductive elimination have been broadly investigated with detailed mechanisms for transmetalation recently elucidated.[155]

New carbon–carbon and carbon–heteroatom bonds arise from the reaction of organoboron molecules with halogenated organic substrates through the intermediacy of a metal catalyst.[156] Diverse starting materials having sp-, sp^2-, and sp^3-hybridized organoboron and halogenated organic reactants lead to a wide array of compatible component parts for the coupling reaction.[157] Reaction conditions continue to be optimized for the ligand system, solvents, and other reaction modifiers.

These cross-coupling reactions are now mature technology; however, there is still a significant amount of research in this area that aims to improve the scope of these reactions, develop more efficient catalysts, and make reactions more practical.[158,159] The importance of synthetic chemistry practices has changed considerably with the recognition of the importance of green and sustainable chemistry.[160] Reaction solvents have become one component of synthetic practice receiving significant scrutiny based on safety, economic, and waste treatment objectives. Solvent effects for palladium-catalyzed cross-couplings are poorly developed and require systematic evaluation for optimal reproducible results.[161,162]

Cross-coupling reactions are conducted in reaction mixtures containing solvents selected from a broad spectrum of candidates. For palladium-catalyzed cross-couplings, the solvent is recognized as being very important in most solution-based chemical transformations. The occurrence, rate, and selectivity of a reaction are significantly affected by the solvent selection, as well as reaction equilibria.[163] Separation effects such as partitioning or precipitation of reactants and products are based on solvent selection. Solvents can also be instrumental to the formation and removal of by-products or product separation. Catalyst stability or legend instability is functionally controlled by solvents as well as acid/base functionality changes and coordination. Beneficial effects to the catalyst lifetime can be accrued through optimal selection of solvent. Optimal process design benefits from solvent selection since a solvent can act as a heat sink to provide reaction temperature regulation and homogeneous reaction conditions to assist reaction kinetics and contact between reactants.

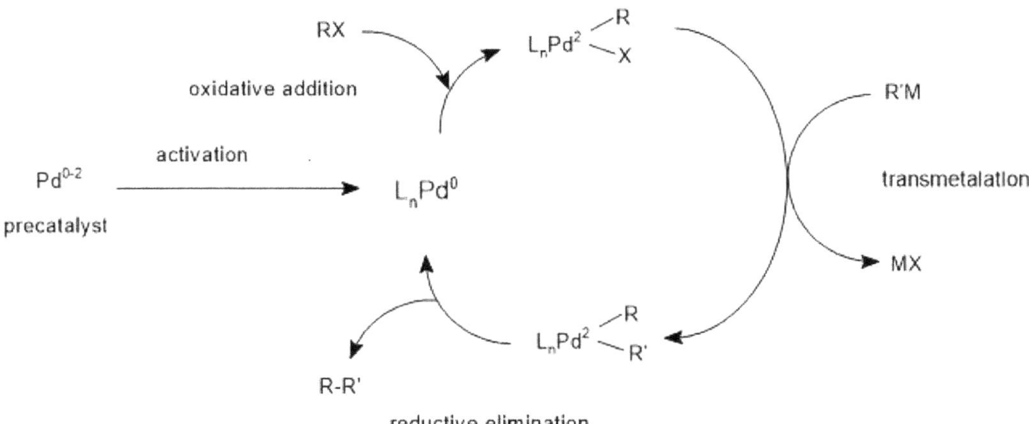

SCHEME 9.11 Catalytic cycle detailing the mechanistic components of the cross-coupling reaction.

An Astra Zeneca process team was tasked with the development of a large synthesis of savolitinib which was required in large quantities for the support of advanced clinical trials.[164] The complexity of the process design required a multidisciplinary contribution of information from analytical, physical, and synthetic organic chemistry and assisted by experimental design, process engineering, high-throughput experimentation, and solid-state chemistry. The isolated savolitinib product also exhibited different solid forms as anhydrous, hydrated, and solvated entities making isolation and purification a complex task. The team employed an established program of criteria, SELECT (Safety, Environmental, Legal, Economy, Control, Throughput) to guide the process design activities.[165] Catalyst and solvent medium compositions were evaluated using 96 experiment screens. The selected catalyst was the commercially available Pd-132, bis(di-tert-butyl(4-dimethylaminophenyl)phosphine without any size considerations and employed at 0.1 equiv. In the selected optimal conditions, a Suzuki–Miyaura cross-coupling was conducted as the last step (Scheme 9.12).

Removal of impurities including Pd was especially important to the success of this synthetic strategy. This effort exemplifies a successful large-scale strategy without the use of nanodimensioned catalysts. It also underscores the attractiveness of lower catalyst quantities to avoid Pd as an elemental impurity in the isolated product.

The solvent use in industrial R&D and academic facilities has been explored recently due to the economic burden associated with solvent use extending from purchasing to disposal where solvent recovery becomes more relevant.[166] The concern for operational safety has emphasized the importance of protective equipment and regulations are more restrictive toward the use of toxic and environmentally damaging substances. Solvent selection has been scrutinized with the development of tools designed to assist solvent replacement.[167,168] Solvent use becomes more restricted from considerations of good practice (health and safety), waste disposal, industrial emissions, and environmental protection. Tools have been developed to assist the proper selection of reactions solvents based on impacts of health, safety, waste, environment, and life cycle considerations.[169–175] Chemical legislation looms as a major controlling element to worldwide regulations targeting solvent use.

To measure the "greenness" of a chemical reaction or process, the Environmental Factor, or E Factor was developed to assess the amount of waste created as related to the mass of the isolated product (Equation 9.4). Equation 9.4 E Factor.

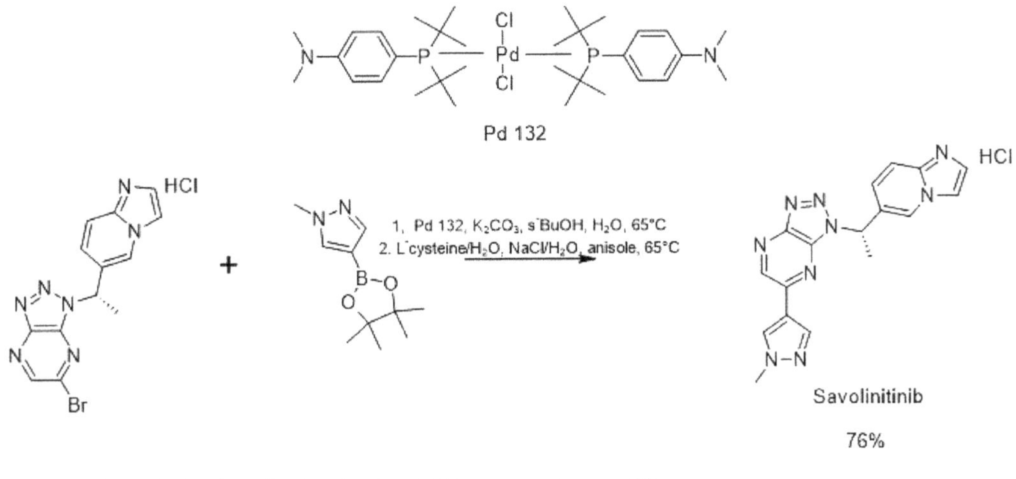

Pd 132

1. Pd 132, K_2CO_3, s-BuOH, H_2O, 65°C
2. L-cysteine/H_2O, NaCl/H_2O, anisole, 65°C

Savolinitinib

76%

SCHEME 9.12 Large-scale synthesis of savolitinib using Suzuki–Miyaura reaction conditions.

$$\text{E Factor} = \frac{\text{Kg of waste}}{\text{Kg of desired product}} \tag{9.4}$$

A small magnitude E Factor denotes a more environmentally acceptable process.[176,177] It is expected that the E Factors associated with specific reactions can be dramatically reduced by an order of magnitude for certain reaction types. When considering the features of a chemical reaction that are available for optimization in environmental terms, the reaction medium, energy input, safety, toxicity of chemical components and products, the reaction chemistry, and waste generation are obvious candidates to name a few. The reaction medium is by far the major feature translating into elevated E Factors. Reliance on reaction solvents implies corresponding workup steps which usually expand the waste mass and the economic burden of disposal. Organic solvents were chosen in support of organic chemical reactions for their ability to afford a manageable homogeneous milieu in which a reaction was expected to proceed. The use of certain chemically reactive reagents and systems also excluded certain solvents. The adage of "like dissolves like" reigned as an unbreakable maxim despite the unattractive risks accompanying organic solvents such as toxicity, flammability, explosivity, and disposal. Water had been consigned as an agent of wide incompatibility with most organic chemistry syntheses.[178,179]

Amphiphilic materials having a lipophilic portion that can act as the organic solvent when dispersed in water form spontaneously self-assembled micelles.[180–182] These surfactant structures can provide the basis for minimizing organic solvent use in organic synthesis.[183,184] The amounts of surfactant required to form a micellar array of nanocells have been found in the range of 10^{-3} to 10^{-4} M. In these micellar constructs, nanoreactors are reportedly formed which enhance the desired chemical transformation between catalysts and otherwise water-insoluble reactants.[185,186]

The amphiphilic properties permit access to two domains of remarkably different solvent compatibilities of water and organic phase that mediate reactions and exhibit self-assembly features resulting in the formation of micelles and vesicles. The behavior of these surfactant systems involves the micellar nanoparticles acting as the reaction solvent which is crucial to the reaction's outcome. The specificity of the surfactant nature is exceptionally important to a reaction's results which has spurred the invention of "designer" surfactants (Scheme 9.13).[187,188] Particle size and shape have been identified as key features of these synthetic tools.[189,190]

For the Suzuki–Miyaura cross-coupling reaction, micellar self-aggregating surfactants form nanoreactors in water that are characterized by reduced E Factors.[191,192] Simple coupling of single-ring aromatic reactants having various functional groups occurs easily in aqueous conditions to form substituted biphenyls (Scheme 9.14).[193,194]

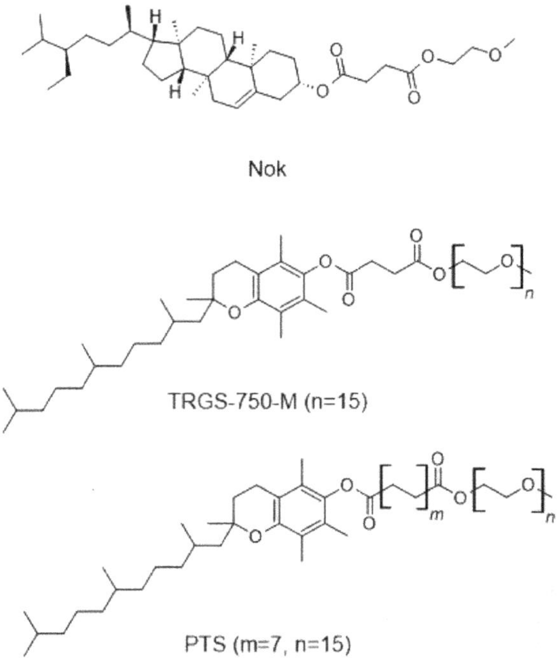

Nok

TRGS-750-M (n=15)

PTS (m=7, n=15)

SCHEME 9.13 Examples of designer surfactants used to improve cross-coupling synthesis yields and catalyst handling.

Several scales of operation were investigated and each exhibits significant E Factor reduction and the smaller-scale academic scale results of the metric research indicate that the E Factor reduction would be enhanced in larger-scale operations.[176]

Organic solvents used as traditional reaction media may require heating to initiate and/or drive reactions to completion but the aqueous micellar conditions enable the cross-coupling to proceed at room temperature. Side-product formation becomes an operational issue as certain solvents are heated, and impurities require purification schemes with attendant yield loss are required. Temperature control of the Suzuki–Miyaura reaction is a matter of general concern for the possibility of runaway reactions and explosive results when organic solvents are employed.[195] The surfactant-based reaction system at room temperature has produced clean and higher product yields. The surfactant-based reaction also enjoys the features of facile setup, simple reaction assembly, and significantly reduced reaction cleanup. Efficient stirring is required for the

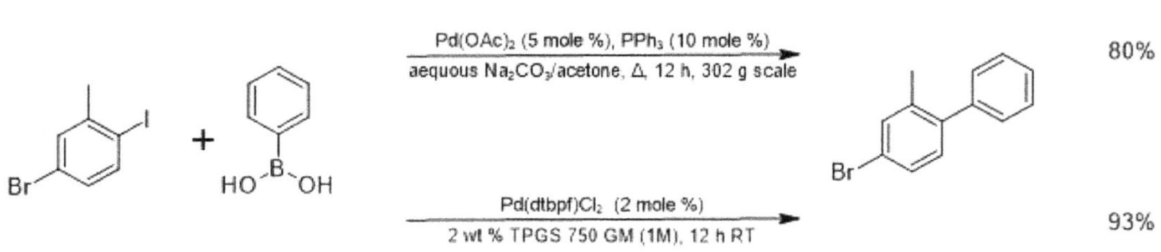

Pd(OAc)$_2$ (5 mole %), PPh$_3$ (10 mole %)
aequous Na$_2$CO$_3$/acetone, Δ, 12 h, 302 g scale

80%

Br + HO-B-OH → Br

Pd(dtbpf)Cl$_2$ (2 mole %)
2 wt % TPGS 750 GM (1M), 12 h RT

93%

SCHEME 9.14 Simple Suzuki–Miyaura cross-coupling E Factor analysis.

effective mixing of catalyst and reaction substrates to establish expected levels of conversion. The reaction in water is anticipated to have a heterogeneous composition. Upon reaction completion, a selected organic solvent is added to the reactor to afford in-reactor extraction and product separation.

9.4.2 Recycling of Aqueous Micellar Solutions

Aqueous reaction mixtures have been found to be tolerant to recycling which enhances E Factor reduction.[196] A simple biaryl coupling was employed to evaluate the effect of aqueous micellar solution recycling on the E Factor (Scheme 9.15).

After solvent extraction in the reactor with a minimum quantity of solvent assists product isolation. The aqueous phase contains the catalyst system (catalyst and surfactant) which is available for reuse. Surfactant degradation occurs with each cycle. After a series of reaction cycles of a biaryl coupling, the aqueous phase is contaminated with water-soluble reaction by-products and must be discarded. For the coupling under consideration, the conversion under aqueous and solvent conditions exhibit remarkable differences. The yields are quite dissimilar as well as the length of reaction time. The E Factors of the metric research are remarkably similar, but the non–aqueous conditions are not recycled.[176] The organic solvent used in reactor extraction can be partially recovered. A single recycle of a reaction enhances the E Factor reduction.[197,198]

Microemulsion droplets are formed with the addition of immiscible organic cosolvents (5–20%) to micelles in water that have a solvent-swollen oil-in-water composition. Micellar reactions are recognized to exhibit operational problems such as precipitation and reagent solubility which mars their reliability at a large-scale. Thermodynamically, microemulsion droplets are more stable, leading to the enhanced scale-up of similar chemical transformations. Application of these conditions to the Suzuki–Miyaura reaction with the addition of 10% toluene as a cosolvent to form a microemulsion resulted in shorter reaction times and increased yields for a reaction conducted open to the air (Scheme 9.16).[199]

Conventional micellar conditions of 2% surfactant produced low yields and partial loss of a protecting group for the same set of reactants. This variant of the cross-coupling reaction can offer milder, air-compatible reaction conditions, and higher yields. A significant modification of the cross-coupling reaction in water at room temperature is the ability to conduct compatible sequential reactions in the same reactor.

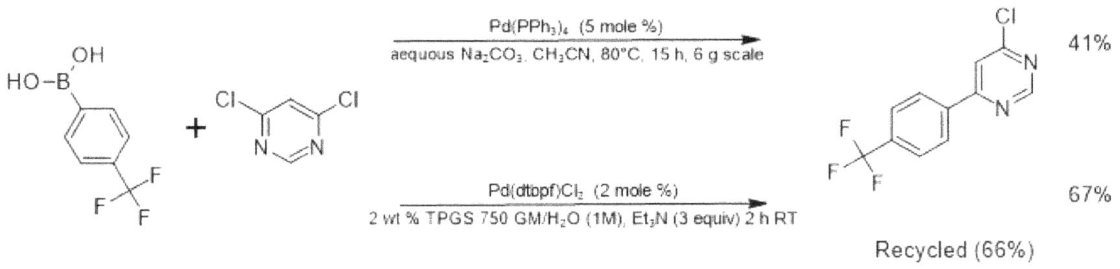

Pd(PPh$_3$)$_4$ (5 mole %)
aequous Na$_2$CO$_3$, CH$_3$CN, 80°C, 15 h, 6 g scale
41%

Pd(dtbpf)Cl$_2$ (2 mole %)
2 wt % TPGS 750 GM/H$_2$O (1M), Et$_3$N (3 equiv) 2 h RT
67%

Recycled (66%)

SCHEME 9.15 Suzuki–Miyaura cross-coupling medium recycling with E Factor analysis.

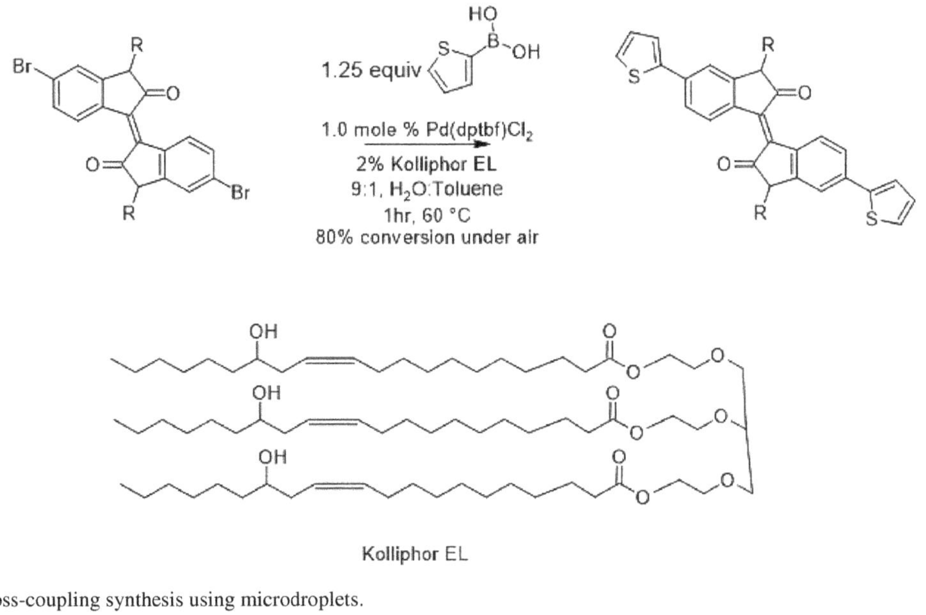

1.25 equiv
1.0 mole % Pd(dptbf)Cl$_2$
2% Kolliphor EL
9:1, H$_2$O:Toluene
1hr, 60 °C
80% conversion under air

Kolliphor EL

SCHEME 9.16 Cross-coupling synthesis using microdroplets.

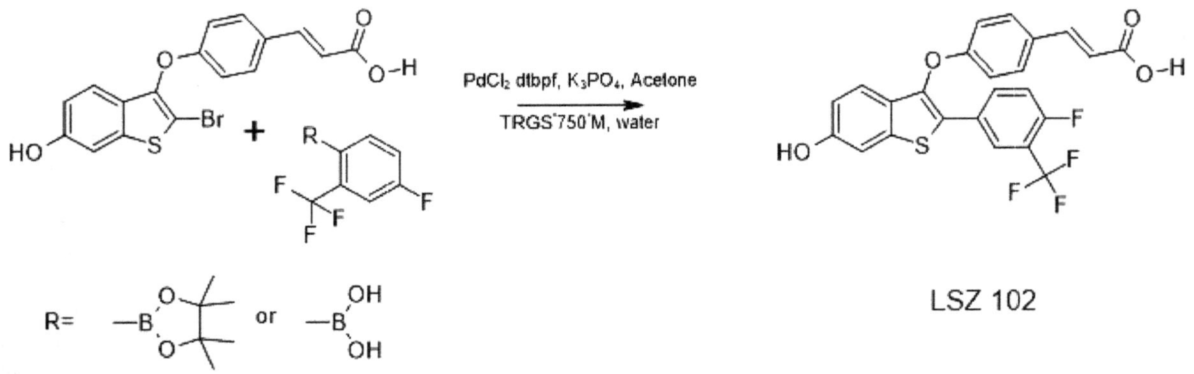

PdCl2 dtbpf, K_3PO_4, Acetone

TRGS'750'M, water

LSZ 102

SCHEME 9.17 Synthetic plan for drug candidate LSZ 102.

A Novartis research group designed a notable sequential set of reactions in the water milieu to accomplish the synthesis of the desired API through a cross-coupling step (Scheme 9.17).[200]

The clinical development candidate was positioned in Phase I/Ib trials designed to treat ERα positive breast cancer as a selective estrogen receptor-degrader (SERD) with potential antineoplastic activity. The multi-step synthesis yield was optimized to produce very satisfactory step yields. A kilogram scale protocol was developed for a Suzuki–Miyaura cross-coupling step leading to the synthesis of LSZ 102. Early synthetic activities using an aqueous/organic solvent medium were complicated by the formation of by-products including debrominated starting material requiring additional reaction workup. Reduction of the by-products became a major focus for synthetic design. Changes to the reaction catalyst and reaction composition continued to yield complex product compositions. The selection of the surfactant TPGS-750-M in water as a medium led to the improvement of product quality and selectivity. The change of reaction medium minimized typical impurities generated during the cross-coupling reaction.

The use of the surfactant TPGS-750-M as a medium with a catalytic amount of 1,1′-[bis(di-tert-butylphosphino)ferrocene] dichloropalladium(II) [PdCl2(dtbpf)] was found to provide the mild reaction conditions to minimize reactant degradation leading to greater product purity and conversion efficiency. Small quantities of water-miscible cosolvent (5–20 wt%) were used to assist the transport of the organic reactants into the surfactant micelles and the water phase to establish a concentration gradient which assists enhanced agitation and reaction kinetics in the reaction mixture.

The surfactant process was evaluated with the widely utilized Process Mass Intensity (PMI) (Equation 9.5) tool to estimate the reduction of environmental footprint and operating costs for the reaction.[201–203] Equation 9.5 Process Mass Intensity (PMI).

$$\text{Process Mass Intensity}\left(\text{PMI}\right) = \frac{\text{Mass of process or process step}}{\text{Mass of product}}$$

$$\text{PMI} = \frac{\begin{array}{c}\text{Mass}\left(\text{reactants}\right) + \text{Mass}\left(\text{reagents}\right) \\ +\text{Mass}\left(\text{catalyst}\right) + \text{Mass}\left(\text{solvents}\right)\end{array}}{\text{Mass}\left(\text{isolated product}\right)} \tag{9.5}$$

TABLE 9.1

PMI Analysis of Drug Candidate LSZ102 Synthesis Based on Solvents, Water, and Total Mass, Including Raw Materials

Metrics	Organic Solvent Process	Surfactant Process
PMI (overall)	116	68
PMI (solvents)	65	31
PMI (aqueous)	44	32

These evaluations were motivated by concerns for reaction and plant operation costs leading to optimizing the economic viability of the process. Assessment of the process PMI indicated a lowering of the PMI by 41% when compared with the organic solvent process including reagents and raw materials (Table 9.1).

The reduction of organic solvents was clearly responsible for the gain in the PMI contrast of the two processes. The minimization of solvent in the surfactant-based process assisted the aim to reduce the quantity of solvent further and enabled direct crystallization of the desired product from the reaction mixture. Catalyst recycling became possible with the use of a minimized amount of cosolvent required for extraction. It was found that the surfactants permitted the reaction mixture to be adjusted to high concentrations of reactants and the minimization of water required. Aqueous washing of the organic layers to remove salts was no longer required. These changes excluding water translated into a difference in the PMI for both processes of about 27% showing the improvements embodied in the surfactant process essentially attributed to wastewater reduction.

This analysis also revealed that the scaled-up version of the organic solvents process was about 33% more expensive than the surfactant process due to raw material and processing costs.

With water use reduction, plant operation costs were reduced by a streamlined workup and product-purification process.

Close attention to reaction stoichiometry assisted considerations of mass utilization, notably in the Pd catalyst reduction permitted by the surfactant-based reaction conditions for multiple couplings in the same reactor (Scheme 9.18).

A Novartis team established the synthesis of the target API in Scheme 9.18 at a large kg-scale in water employing several

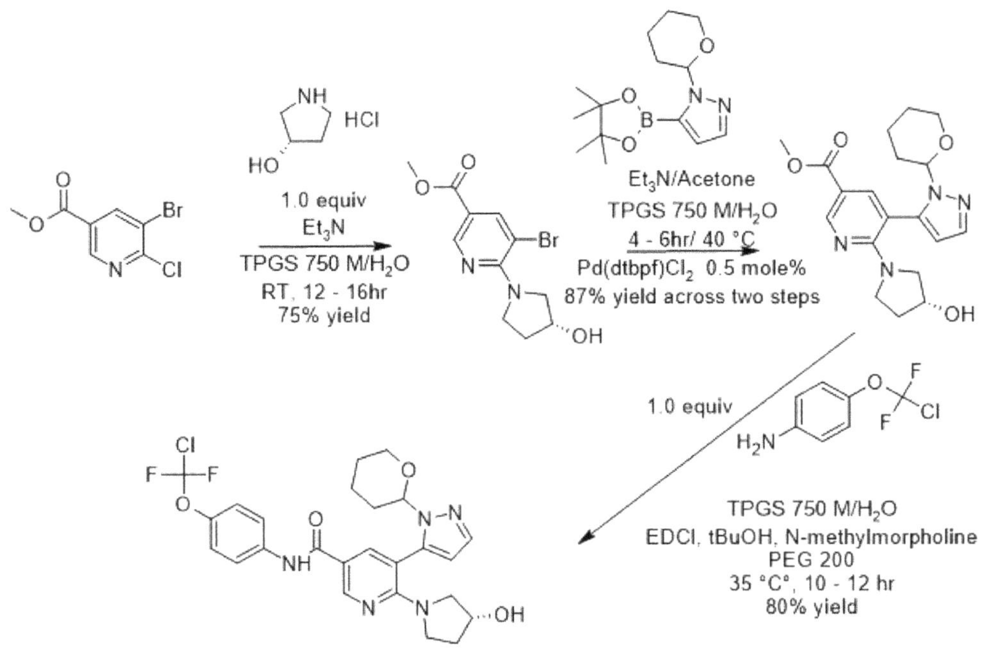

SCHEME 9.18 Multiple-step cross-coupling with recycling of catalyst.

complex synthetic steps.[204,205] Surfactant technology based on TPGS-750-M was found to be optimal with acetone and triethylamine as cosolvents and base. As a first attempt at a large-scale process used to synthesize multiple kilograms of the target API, the partitioning within the reactive system was found to be exceptionally important. For the coupling reaction, it was important to achieve successful synthesis to have the reactive coupling partners in the same compartment within the reaction system. This process was observed to function dependably and reproducibly when applied to the synthesis on a multiple kilogram scale, with the additional feature of being reasonably general on the scope when applied to a wide range of reactive components. The evaluation of environmental, cost, and productivity highlighted a 30% environmental footprint reduction, a productivity increase of approximately 70%, and a cost reduction of 30%.

9.4.3 Applications to Agricultural Products

One industrial sector using fine chemical manufacturing practice is agricultural chemical products including insecticides, herbicides, fungicides, and related chemicals.[206] These commodity fine chemicals are produced in multi-ton quantities. In excess of 4 million metric tons of agrochemical products are applied annually across the globe as reported by the Food and Agriculture Organization of the United Nations.[207]

The magnitude of the mass of matter synthesized by the industrial production of materials in support of agriculture offers a compelling scenario for the application of green chemistry principles to design the chemical manufacturing processes and products in a fine chemical industry that greatly exceeds the mass output of the pharmaceutical industry.[208] The reduction of environmental footprint and safety of the required chemical processes and products are essential assets important to the greening of agrochemistry.[209,210]

The fungicide, boscalid, has systemic and broad-spectrum properties with activity toward virtually all types of fungal disease including gray mold and powdery mildew.

As an active ingredient of BASF fungicides, boscalid has become a target for a more efficient synthesis based on Pd nanocatalysis in water. Earlier synthetic strategies were based on a multireactor sequence using organic solvents and Pd catalyst in unsustainable quantities to conduct a cross-coupling reaction. Utilizing recent research findings, surfactants enabled a cross-coupling reaction in a water-surfactant medium at room temperature was considered as a solution. A single reactor was employed to conduct the three-step reaction sequence with nanomicelles in water composing the reaction medium and an exceptionally low quantity of $Pd(OAc)_2$/SPhos catalyst of 700 ppm or 0.07 mol%. The initial step of the sequence was a Suzuki–Miyaura cross-coupling, followed by carbonyl iron powder reduction of the nitro group, and finally acylation with 2-chloronicotinyl chloride (Scheme 9.19).[211]

These three steps were employed to synthesize boscalid in an 83% yield. A high-yielding single reactor sequence demonstrates an environmentally responsible synthesis of the fungicide. The reaction sequence reflects the importance of the economic utility and waste avoidance accomplished by nanoparticle use.

The continued expansion of the global population requires improvements to maximize crop production. Weed control is a challenging process since weeds can compete with crops for essential nutrients very effectively. This competition can lead to decreased crop yield that may affect the availability and cost of crops to the consumer at several levels. Corteva Agriscience has selected classes of herbicides from the auxinic family that

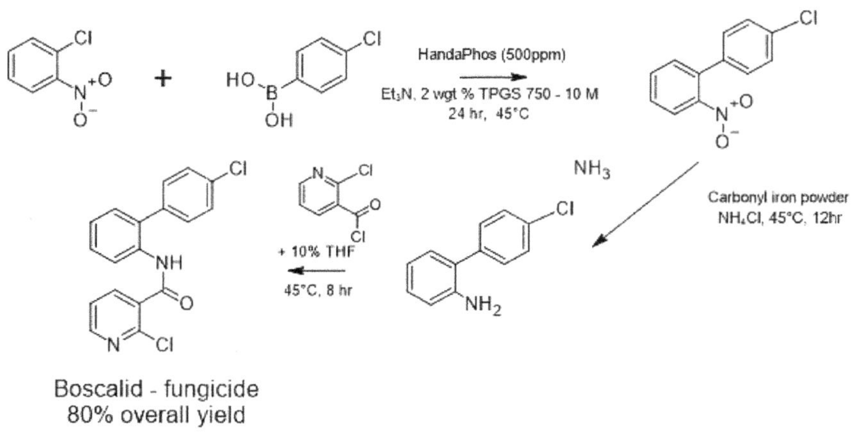

HandaPhos (500ppm)
Et_3N, 2 wgt % TPGS 750 - 10 M
24 hr, 45°C

NH_3

Carbonyl iron powder
NH_4Cl, 45°C, 12hr

+ 10% THF
45°C, 8 hr

Boscalid - fungicide
80% overall yield

SCHEME 9.19 Boscalid synthesis with catalyst recycling.

are recognized for effectiveness toward broadleaf weeds. The immense manufacturing volume of agrochemicals produced annually requires recognition and design of operations to understand the environmental footprint and process safety issues for optimal economic control. Corteva Agriscience's novel crop protection biaryl active ingredients, Arylex™ and Rinskor™, were targets of fine chemical synthesis since the herbicides were found to have desirable properties such as significant metabolite residual decrease in soil, use rate attenuation, and broader efficacy toward a range of weeds. The development of Rinskor™ was recognized by an ACS Presidential Green Chemistry Challenge Award, highlighting the agrochemical's favorable environmental and toxicity profile. Early versions of the synthesis employed organic solvents, microwave heating, and no catalyst recycling to provide the product at a reduced yield with a cross-coupling step. The quantity of Pd catalyst required for the early synthetic schemes was quite high (2–5 mol%, or 20,000–50,000 ppm). A green synthesis for each of the Arylex™ and Rinskor™ agrochemicals was conducted in a single reactor using water containing the surfactant, TPGS-750-M, (2 wt%) and a minimal organic cosolvent for the reaction medium with the Pd catalyst minimized to a few thousand ppm (10-fold reduction when compared with earlier synthetic conditions).[212] The optimized syntheses of each agrochemical require comparable but distinctly different reagents and conditions (Scheme 9.20).

This new synthetic technology sustainably employs nanomicelles as nanoreactors, leading to more environmentally attractive synthetic conversions. For Arylex™ recycling and scale-up reactions, the aqueous system could easily be recycled/reused several times with yields that are virtually the same. The calculated E factor was found to be 2.4, evaluating a measure of environmental acceptability and performance that is considerably less than historical estimates of process performance associated with the corresponding fine chemical industry. The application of aqueous micellar technology was effectively applied to the synthesis of active ingredients containing biaryl linkages via Suzuki–Miyaura coupling technology as sustainable and economically attractive.

9.4.4 Single-Atom Catalysts

Nanotechnology is a major force in the discovery of catalytic materials and processes due to the unique physicochemical and electronic properties arising from nanoscale-engineered materials.[213] Engineering below nanoscale and at the single-atom limit offers a synthesis of catalysts exhibiting high selectivity, enhanced efficiencies, and reduced costs.[214] A continuum of material composition ranging from molecules to solid-state circumstances involving small nanoparticles or nanoclusters to small clusters with identification based on molecular orbitals and larger nanoparticles characterized by energy band structures provides an extensive range of materials for use in catalyst design.[215]

Single-atom catalysts (SACs) are catalytic materials composed of the active metal species that are isolated single atoms attached to supports or incorporated as an alloy as a stabilizing property. SACs are candidates to become low-cost, efficient, and durable catalysts.[216] In the pursuit of the application of green synthesis to fine chemicals, SACs have been constructed for their distinct ability to catalyze a variety of organic transformations. SACs have been demonstrated to catalyze a wide assortment of chemical structure alterations important to the fine chemical industry such as selective hydrogenation of nitroarenes; reduction of ketones and other carbonyl group moieties; hydrogenation of alkenes and alkynes; selective oxidation of alcohols to carbonyl compounds; and cross-coupling reactions.[217,218] A wide landscape of industrially important reactions has been evaluated addressed through the inspection of SAC utility to highlight advantages of these catalysts (Figure 9.7).

A single-atom Pd catalyst on bimetallic oxides was applied to the Suzuki–Miyaura cross-coupling reaction with phosphine-free and open-air conditions at room temperature.[219] The catalyst is easily prepared by depositing Pd on bimetallic oxides (Pd-ZnO-ZrO_2). A significant synergetic effect of the bimetal oxide composition was observed. Application to a wide range of reactive substrates identified the catalyst's high activity and tolerance in the Suzuki–Miyaura reaction (Scheme 9.21).

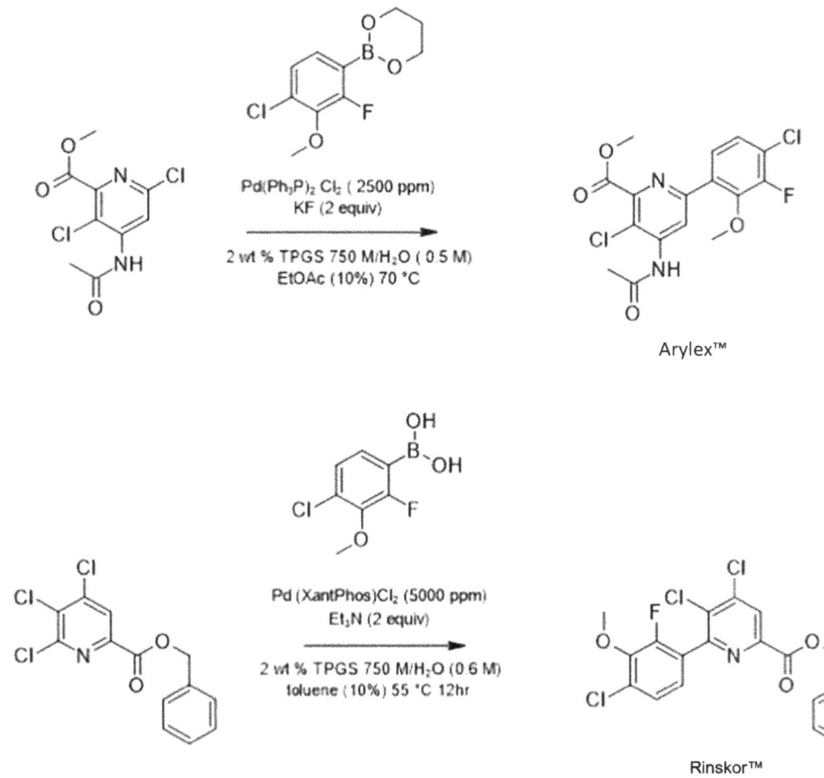

Pd(Ph3P)2 Cl2 (2500 ppm)
KF (2 equiv)
2 wt % TPGS 750 M/H2O (0.5 M)
EtOAc (10%) 70 °C
Arylex™
Pd (XantPhos)Cl2 (5000 ppm)
Et3N (2 equiv)
2 wt % TPGS 750 M/H2O (0.6 M)
toluene (10%) 55 °C 12hr
Rinskor™

SCHEME 9.20 Nanocatalytic syntheses of Arylex™ and Rinskor™ herbicides.

Nanocatalyst/Single Atom Catalyst
Efficiency
High Activity
Efficient Recovery
Low Preparation Cost
Sustainable Chemical Production

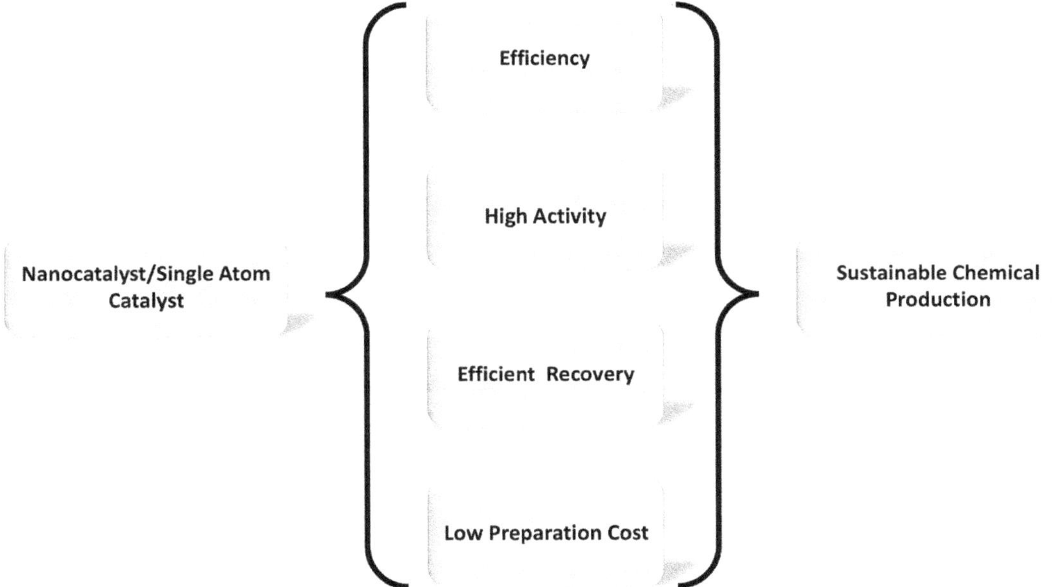

FIGURE 9.7 Comparison of nanocatalysts to single-atom catalysts as contributors to sustainable chemistry.

Physical characterization identified coordination of Pd single atoms with two oxygen atoms of the Pd-ZnO-ZrO_2 catalyst. The facile multi-gram scale preparation of the catalyst employs in situ co-precipitation in the isolation step. This new catalyst was found to be tolerant of room temperature, open-air conditions for the phosphine-free coupling reaction. The simple preparation of the catalyst and mild reaction conditions with high catalyst stability indicates that this catalyst may have potential real-world applications. The current reaction workup does not permit the easy recycling that has been developed for nanocatalysts in surfactant systems. Perhaps a marriage of the two techniques can be designed. Alternate approaches to this type of SAC catalysis used discrete cages of metal-organic polyhedra anchoring single Pd atom (MOP-BPY(Pd))

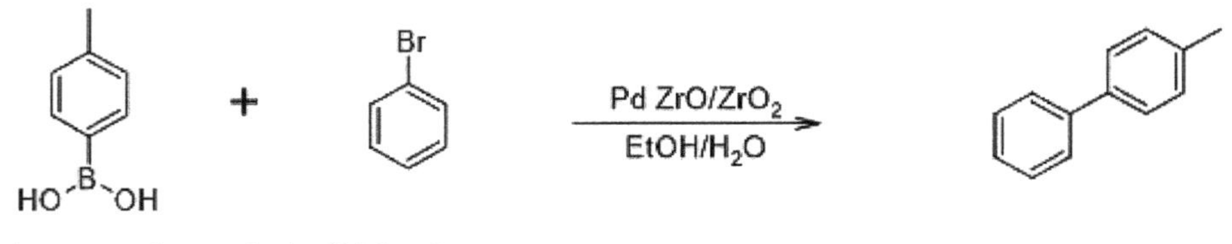

Br

HO B OH

Pd ZrO/ZrO_2

EtOH/H_2O

SCHEME 9.21 Single-atom catalyst synthesis of biphenyls.

and successfully performed a Suzuki–Miyaura cross-coupling reaction with various substrates in aqueous media with coupling efficiencies in excess of 90% for various substrates.[220]

Single-atom heterogeneous catalysts synthesized through the attachment of Pd atoms to exfoliated graphitic carbon nitride (Pd-ECN) form solid catalysts with the properties of high chemoselectivity and functional group tolerance comparable to current state-of-the-art homogeneous catalysts employed in Suzuki–Miyaura cross-couplings.[221] The structure of ECN which is composed of macroheterocyclic elements offers an adaptive coordination environment that facilitates each catalytic step and supports robust stability in flow systems. By structuring single atoms at the nanoscale in solid hosts for catalytic processes, SACs can now be designed to assist catalyst separation in their heterogenized states. These new traits and features are complementary to the nanocatalysts employed in designer surfactant systems which also permit separation with the added ability to recycle the catalyst and reaction medium multiple times.

The conversion of carbon dioxide into hydrocarbons and other organic chemicals is a major pursuit of many research efforts.[222,223] The ability to transform the greenhouse gas into commercially useful materials enables the utilization of a gas which is detrimental to the global atmosphere.[224]

The utilization of CO_2 as a resource can be observed through research efforts to convert the greenhouse gas to high-value chemical products via catalytic reactions.[225] Conversion to methanol, methane, formic acid, dimethyl carbonate, and other chemical materials illustrates how CO_2 can be transformed from an atmospheric pollutant to a chemical feedstock.[226] SACs have been recognized for their utility to accomplish these conversions due to their stability, catalytic activity, and atom efficiency. A catalyst based on Cu atoms embedded C2N monolayer (Cu/C2N), as an SAC for hydrogenation of CO_2 to formic acid has been evaluated using calculations to develop mechanisms that describe H_2 and CO_2 adsorption on the catalyst monolayer (Figure 9.8).[227]

The results highlight the high activity of the SAC Cu/C2N at room temperature conditions. This information suggests that the inexpensive Cu/C2N catalyst is a strong candidate for broad application to industrial CO_2 hydrogenation leading to value-added chemicals.

The efficiency of atom utilization is enhanced through downsizing catalyst size dimensions.[228] Complete metal atom utilization can be designed into SACs leading to superior catalytic activity when compared with nanoparticle catalysts. The design of SACs requires investigation and identification of distinctive properties of the catalysts and related structure-activity relationships to enable the optimal selection of metals, supports, and ligands necessary for catalyst composition and performance. Concerns for an understanding of single-atom active sites and the changes of reaction dynamics encountered during the reaction are critically important to the development of the corresponding reaction mechanisms which enable predictive directions for future research.[229,230]

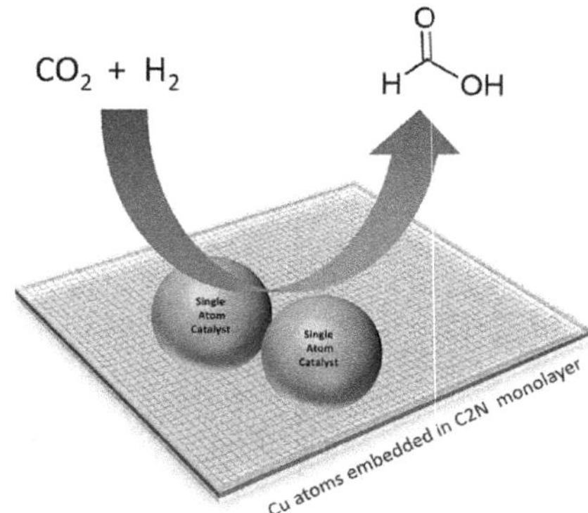

FIGURE 9.8 Single-atom-catalyst-mediated conversion of carbon dioxide to formic acid.

9.4.5 Catalyst Recovery

The use of rare and precious metals to conduct desired chemical transformation carries with it concerns for the judicious and proper use of these items in a sustainable manner so as to provide continued access to metals having exceptionally low concentrations in the planet's crust.[231,232] The sustainable approach to these economic constraints is to recover, recycle, and conserve these rare materials. The recovery of catalysts used in industrial applications has been extensively researched due to the amounts of precious metal catalysts involved.[78,80] This has been a major issue for items discarded as certain waste contains recoverable quantities of the metals of the platinum metal group. The electronics waste industry has been targeted for many metal-recovery ventures but this effort is constrained by the economics of operation.[233] The technical literature is quite rich in the disclosure of chemical and physical technologies exhibiting the ability for metal recovery using advanced metallurgical recovery technology.[234–237] Notable small-scale examples of recovery connected with palladium as a catalyst have been published.[238–241] Selected multilayer ceramic capacitors on printed circuit boards from electronic waste have been targeted as a potential source of palladium using chelating agents.[242]

As a primary process in hydrometallurgy, leaching transfers desired metals from a mineral or ore containing metals to an aqueous solution provides a means to concentrate the metals. The term lixiviant is used in hydrometallurgy to describe

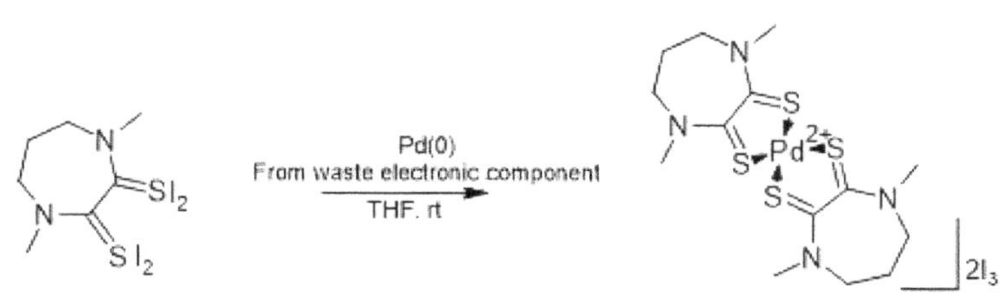

Pd(0)
From waste electronic component
THF, rt

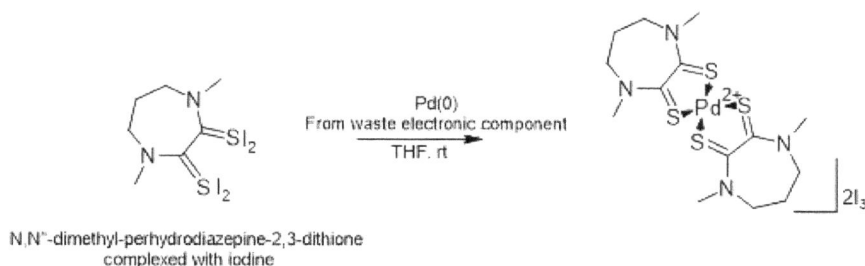

N,N'-dimethyl-perhydrodiazepine-2,3-dithione complexed with iodine

SCHEME 9.22 Pd catalyst recovery from electronic waste component using a chelating agent.

a liquid medium that can selectively extract the desired metal from an ore or metal mixture. It has been discovered that lixiviant compositions of dihalogen/S,S-ligands, N,N′-dimethylperhydrodiazepine-2,3-dithione (Me2dazdt) in the presence of iodine (I_2) for recovery of precious and noble metals from a non-ferrous metal fraction derived from shredded waste electronic equipment (Scheme 9.22).[242]

The process uses selective extraction of palladium into an organic phase by classical leaching techniques from a complex aqueous mixture of metal cations. The palladium is stabilized through the judicious selection of a surfactant during the back extraction forming an aqueous micellar solution that can be employed directly in a palladium-catalyzed synthetic reaction. The observed phase separation resulted in the clean distribution of components between the extraction phases. This process optimally offers palladium isolation and purification to form palladium precatalysts while controlling waste generation. The utility of a hydrometallurgical process, when combined with selected aqueous surfactants permits metal recovery from electronic waste and automobile catalytic converters, is remarkably straightforward, selective, and with minimal waste generation.

Clearly, the recovery of other rare metals in waste materials continues to be a targeted research area that can be quite rewarding. New chemistries and separation techniques offer abilities to accomplish the desired separation and concentration of the rare elements in waste materials. Optimally, these environmentally benign techniques, safe and minimizing waste economically, are waiting to be discovered.[243] Growing interest in the private sector and by major global governments seek to conserve rare raw materials through a reduction of their environmental burden and pursue closed-loop supply chains targeting manufacturing to be more sensitive to the value of rare element conservation.

9.5 Conclusions

The amazing realm of small dimension catalytic particles is broad with remarkable depth and continues to offer new discoveries to aid a myriad of inventions. Key elementary steps of reactant adsorption, intermediate diffusion, and product desorption proceed via electron transfer from the catalyst surface to the reactive species. The activation energy barrier of a catalytic reaction, determinant of reactivity and pathways selection, can be manipulated by the control of the catalyst surface distribution of valence electrons to favor product selectivity. The unique properties of nanoparticles have enabled their utilization in wide areas of application as catalysts and such is the case with the bulk synthesis of ammonia and hydrogen peroxide. The wide array of nanoscale synthetic transformation applications exhibits the versatility of synthetic catalytic activity that can be elicited through the judicious investigation of catalyst properties and systems. The utility of nanocatalysts to the fine chemical industry has been demonstrated along with systems engineering considerations to protect human health and safety. Attention given to understanding and optimizing the environmental footprint of these synthetic activities offers true claims to achieving sustainability. A significant increase in catalytic activity, selectivity, and stability can be gained through precise atomic-scale control of catalysts. The emerging utilization of single-atom catalytic systems offers new and expanding vistas designed to miniaturize many catalytic systems currently in use. The use of rare metal element catalysts continues with significant attention given to recovery and recycle components of a synthetic scheme. Recovery of rare metals from waste materials is clearly supportive of sustainability efforts but continued work is required to economize the metal recovery protocols at all levels. The general utility of small particle catalysts is clear to even the most casual observer with future discoveries and applications offering avenues to the accomplishment of synthetic chemical processes that are optimally designed for their environmental footprint and sustainability.

DISCLAIMER

This work is not a product of the U.S. Government or the U.S. Environmental Protection Agency. The author is not doing this work in any governmental capacity. The views expressed are his own and do not necessarily represent those of the United States or U.S. EPA.

COPYRIGHT

REFERENCES

1. Nicolaou KC. Organic synthesis: the art and science of replicating the molecules of living nature and creating others like them in the laboratory. *Proc Roy Soc A: Math, Phys Eng Sci*. 2014, 470, 20130690.

2. Ragan JA, Dreher SD. Excellence in Industrial Organic Synthesis 2019. The Past, Present, and Future. *J. Org. Chem.* 2019, 84, 4577–4579.
3. Matar S, Hatch LF, 2001. *Chemistry of Petrochemical Processes*, 2nd ed. Elsevier, p. 356.
4. Sheldon RA, Arends I, Hanefeld U. 2007, *Green Chemistry and Catalysis*, Wiley, p. 433.
5. Sheldon RA. Engineering a more sustainable world through catalysis and green chemistry. *J Roy Soc Interf.* 2016, 13, 20160087.
6. Jimenez-Gonzalez C, Constable DJC. 2011, *Green Chemistry and Engineering*, Wiley, p. 680.
7. Kharissova OV, Kharisov BI, Oliva González CM, Méndez YP, López I. 2019 Greener synthesis of chemical compounds and materials. *R. Soc. Open Sci.* 6: 191378.
8. A glossary of terms used in chemical kinetics, including reaction dynamics (IUPAC Recommendations 1996), *Pure Appl Chem.* 1996, 68, 149.
9. Copéret C, Chabanas M, Petroff Saint-Arroman R, Basset JM. Homogeneous and heterogeneous catalysis: bridging the gap through surface organometallic chemistry. *Angew Chem Int Ed.* 2003, 42, 156–81.
10. Prinsen P, Luque R. Introduction to Nanocatalysts. In: *Nanoparticle Design and Characterization for Catalytic Applications of Sustainable Chemistry*, eds Prinsen P, Luque R, Royal Society of Chemistry, pp. 1–36.
11. Singh SB, Tandon PK. Catalysis: a brief review on nano-catalyst. *J Energy Chem Eng.* 2014, 2, 106–15.
12. Yang F, Deng D, Pan X, Fu Q, Bao X. Understanding nano effects in catalysis. *Nat Sci Rev.* 2015, 2, 183–201.
13. Personick ML, Montemore MM, Kaxiras E, Madix RJ, Biener J, Friend CM. Catalyst design for enhanced sustainability through fundamental surface chemistry. *Phil. Trans. R. Soc. A.* 2016, 374, 20150077.
14. Cuenya BR. Synthesis and catalytic properties of metal nanoparticles: Size, shape, support, composition, and oxidation state effects. *Thin Solid Films.* 2010, 518, 3127–50.
15. Cuenya BR, Behafarid F. Nanocatalysis: size-and shape-dependent chemisorption and catalytic reactivity. *Surf Sci Rep.* 2015, 70, 135–87.
16. Ye R, Hurlburt TJ, Sabyrov K, Alayoglu S, Somorjai GA. Molecular catalysis science: Perspective on unifying the fields of catalysis. *Proc Nat Acad Sci US.* 2016, 113, 5159–66.
17. Liu H, Guan J, Mu X, Xu G, Wang X, Chen X. 2017. Nanaocatalysis, *Encyc Phys Org Chem.* 1st ed. Z. Wang ed. Wiley, pp. 1–75.
18. Marakatti VS, Gaigneaux EM. Recent Advances in Heterogeneous Catalysis for Ammonia Synthesis. *ChemCatChem.* 2020, 12, 1–21.
19. Nørskov J, Chen J, Miranda R, Fitzsimmons T, Stack R. Sustainable Ammonia Synthesis–Exploring the scientific challenges associated with discovering alternative, sustainable processes for ammonia production. US DOE Office of Science, Washington, DC; 2016, pp 33, doi:10.2172/1283146.
20. Singh AR, Rohr BA, Schwalbe JA, Cargnello M, Chan K, Jaramillo TF, Chorkendorff I, Nørskov JK. Electrochemical Ammonia Synthesis: The Selectivity Challenge. *ACS Catal.* 2017, 7, 706–709.
21. Wu T, Fan W, Zhang Y, Zhang F. Electrochemical Synthesis of Ammonia: Progress and Challenges. *Mater Today Phys.* 2020, 100310.
22. Qing G, Ghazfar R, Jackowski ST, Habibzadeh F, Ashtiani MM, Chen CP, Smith III MR, Hamann TW. Recent Advances and Challenges of Electrocatalytic N_2 Reduction to Ammonia. *Chem Rev.* 2020, 120, 12, 5437–5516.
23. Zhang S, Zhao Y, Shi R, Waterhouse GI, Zhang T. Photocatalytic ammonia synthesis: Recent progress and future. *Energy Chem.* 2019, 1, 100013.
24. Xue X, Chen R, Yan C, Zhao P, Hu Y, Zhang W, Yang S, Jin Z. Review on photocatalytic and electrocatalytic artificial nitrogen fixation for ammonia synthesis at mild conditions: Advances, challenges and perspectives. *Nano Res*, 2019, 1–21.
25. Gao W, Guo J, Wang P, Wang Q, Chang F, Pei Q, Zhang W, Liu L, Chen P. Production of ammonia via a chemical looping process based on metal imides as nitrogen carriers. *Nat Energy.* 2018, 3, 1067–75.
26. Swearer DF, Knowles NR, Everitt HO, Halas NJ. Light-driven chemical looping for ammonia synthesis. *ACS Energy Lett.* 2019, 4, 1505–12.
27. Sarafraz MM, Christo FC. Sustainable three-stage chemical looping ammonia production (3CLAP) process. *Energy Conv Manag.* 2021, 229, 113735.
28. Smith C, Hill AK, Torrente-Murciano L. Current and future role of Haber–Bosch ammonia in a carbon-free energy landscape. *Energy Environ Sci.* 2020,13, 331–44.
29. Demirhan CD, Tso WW, Powell JB, Pistikopoulos EN. Sustainable ammonia production through process synthesis and global optimization. *AIChE J.* 2019, 65, e16498.
30. Lan R, Irvine JT, Tao S. Synthesis of ammonia directly from air and water at ambient temperature and pressure. *Sci Rep.* 2013, 3, 1–7.
31. Lv XW, Weng CC, Yuan ZY. Ambient Ammonia Electrosynthesis: Current Status, Challenges and Perspective. *ChemSusChem.* 2020, 16, 3061–3073.
32. Barboun PM, Hicks JC. Unconventional Catalytic Approaches to Ammonia Synthesis. *Ann Rev Chem Biomol Eng.* 2020, 11, 503–21.
33. Pattabathula V, Richardson J. Introduction to ammonia production. *Chem. Eng. Prog.* 2016, 112, 69–75.
34. Kalidindi SB, Jagirdar BR. Nanocatalysis and prospects of green chemistry. *ChemSusChem.* 2012, 5, 65–75.
35. Marakatti VS, Gaigneaux EM. Recent Advances in Heterogeneous Catalysis for Ammonia Synthesis. *ChemCatChem.* 2020, 12, 1–21.
36. Saadatjou N, Jafari A, Sahebdelfar S. Ruthenium nanocatalysts for ammonia synthesis: a review. *Chem Eng Commun.* 2015, 202, 420–48.
37. Kim SY, Lee HW, Pai SJ, Han SS. Activity, selectivity, and durability of ruthenium nanoparticle catalysts for ammonia synthesis by reactive molecular dynamics simulation: The size effect. *ACS Appl Mater Interf.* 2018, 10, 26188–94.
38. Shi R, Zhang X, Waterhouse GI, Zhao Y, Zhang T. The Journey toward Low Temperature, Low Pressure Catalytic Nitrogen Fixation. *Adv Energy Mater.* 2020, 10, 2000659.
39. Ye TN, Lu Y, Kobayashi Y, Li J, Park SW, Sasase M, Kitano M, Hosono H. Efficient Ammonia Synthesis over Phase-Separated Nickel-Based Intermetallic Catalysts. *J Phys Chem C.* 2020, 124, 52, 28589–28595.

40. Schlögl R. Catalytic Synthesis of Ammonia—A "Never-Ending Story"? *Angew Chem Int Ed*. 2003, 42, 2004–8.
41. Humphreys J, Lan R, Tao S. Development and Recent Progress on Ammonia Synthesis Catalysts for Haber–Bosch Process. *Adv Energy Sustain Res*. 2021, 2, 2000043.
42. Tang Y, Kobayashi Y, Masuda N, Uchida Y, Okamoto H, Kageyama T, Hosokawa S, Loyer F, Mitsuhara K, Yamanaka K, Tamenori Y. Metal-Dependent Support Effects of Oxyhydride-Supported Ru, Fe, Co Catalysts for Ammonia Synthesis. *Adv Energy Mater*. 2018, 8, 1801772.
43. Qing G, Ghazfar R, Jackowski ST, Habibzadeh F, Ashtiani MM, Chen CP, Smith III MR, Hamann TW. Recent Advances and Challenges of Electrocatalytic N_2 Reduction to Ammonia. *Chem Rev*. 2020, 120, 5437–5516.
44. Wu T, Fan W, Zhang Y, Zhang F. Electrochemical Synthesis of Ammonia: Progress and Challenges. *Mater Today Phys*. 2020, 100310.
45. Xue X, Chen R, Yan C, Zhao P, Hu Y, Zhang W, Yang S, Jin Z. Review on photocatalytic and electrocatalytic artificial nitrogen fixation for ammonia synthesis at mild conditions: Advances, challenges and perspectives. *Nano Res*. 2019, 1–21.
46. Michalsky R, Avram AM, Peterson BA, Pfromm PH, Peterson AA. Chemical looping of metal nitride catalysts: low-pressure ammonia synthesis for energy storage. *Chem Sci*. 2015, 6, 3965–74.
47. Gao W, Guo J, Chen P. Hydrides, amides and imides mediated ammonia synthesis and decomposition. *Chinese J Chem*. 2019, 37, 442–51.
48. Wang Q, Guo J, Chen P. Recent progress towards mild-condition ammonia synthesis. *J Energy Chem*. 2019, 1, 6, 25–36.
49. Ball M, Weeda M. The hydrogen economy–vision or reality? *Intern J Hydrogen Energy*. 2015, 40, 7903–19.
50. Aziz M, Wijayanta AT, Nandiyanto AB. Ammonia as effective hydrogen storage: a review on production, storage and utilization. *Energies*. 2020, 13, 3062.
51. MacFarlane DR, Cherepanov PV, Choi J, Suryanto BH, Hodgetts RY, Bakker JM, Vallana FM, Simonov AN. A roadmap to the ammonia economy. *Joule*. 2020, 4, 1186–1205.
52. Muhammad ID, Awang M, Shaari KZ. Potentials of nano catalysts for energy reduction during synthesis of ammonia. *Int J Eng Innov Res*. 2014, 3, 904–8.
53. Ciriminna R, Albanese L, Meneguzzo F, Pagliaro M. Hydrogen peroxide: a key chemical for today's sustainable development. *ChemSusChem*. 2016, 9, 3374–81.
54. Goor G, Glenneberg J, Jacobi S, Dadabhoy J, Candido E. *Hydrogen Peroxide,* 2019 *Ullmann's Encyclopedia of Industrial Chemistry*, Wiley, https://doi.org/10.1002/14356007.a13_443.pub3.
55. Teong SP, Li X, Zhang Y. Hydrogen peroxide as an oxidant in biomass-to-chemical processes of industrial interest. *Green Chem*. 2019, 21, 5753–80.
56. Martin B, Sedelmeier J, Bouisseau A, Fernandez-Rodriguez P, Haber J, Kleinbeck F, Kamptmann S, Susanne F, Hoehn P, Lanz M, Pellegatti L. Toolbox study for application of hydrogen peroxide as a versatile, safe and industrially-relevant green oxidant in continuous flow mode. *Green Chem*. 2017, 19, 1439–48.
57. Campos-Martin JM, Blanco-Brieva G, Fierro JL. Hydrogen peroxide synthesis: an outlook beyond the anthraquinone process. *Angew Chem Int Ed*. 2006, 45, 6962–84.
58. Hargreaves JS, Chung YM, Ahn WS, Hisatomi T, Domen K, Kung MC, Kung HH. Minimizing energy demand and environmental impact for sustainable NH_3 and H_2O_2 production—A perspective on contributions from thermal, electro-, and photo-catalysis. *Appl Catal A*. 2020, 594, 117419.
59. Menegazzo F, Signoretto M, Ghedini E, Strukul G. Looking for the "Dream Catalyst" for Hydrogen peroxide production from hydrogen and oxygen. *Catalysts*. 2019, 9, 251.
60. Pegis ML, Wise CF, Martin DJ, Mayer JM. Oxygen reduction by homogeneous molecular catalysts and electrocatalysts. *Chem Rev*. 2018, 118, 2340–91.
61. Ranganathan S, Sieber V. Recent advances in the direct synthesis of hydrogen peroxide using chemical catalysis—A review. *Catalysts*. 2018, 8, 379.
62. Menegazzo F, Signoretto M, Ghedini E, Strukul G. Looking for the "Dream Catalyst" for Hydrogen peroxide production from hydrogen and oxygen. *Catalysts*. 2019, 9, 251.
63. Hutchings GJ, Lewis R. A review of recent advances in the direct synthesis of H_2O_2. *ChemCatChem*. 2019, 11, 298–308.
64. Yi Y, Wang L, Li G, Guo H. A review on research progress in the direct synthesis of hydrogen peroxide from hydrogen and oxygen: noble-metal catalytic method, fuel-cell method and plasma method. *Catal Sci Technol*. 2016, 6, 1593–610.
65. Seo MG, Kim HJ, Han SS, Lee KY. Direct synthesis of hydrogen peroxide from hydrogen and oxygen using tailored Pd nanocatalysts: a review of recent findings. *Catal Surveys Asia*. 2017, 21, 1–2.
66. Hunter BM, Gray HB, Muller AM. Earth-abundant heterogeneous water oxidation catalysts. *Chem Rev*. 2016, 116, 14120–36.
67. Bullock RM, Chen JG, Gagliardi L, Chirik PJ, Farha OK, Hendon CH, Jones CW, Keith JA, Klosin J, Minteer SD, Morris RH. Using nature's blueprint to expand catalysis with Earth-abundant metals. *Science*. 2020, 369, 6505, eabc3183.
68. Wang D, Astruc D. The recent development of efficient Earth-abundant transition-metal nanocatalysts. *Chem Soc Rev*. 2017, 6, 816–54.
69. García-Serna J, Moreno T, Biasi P, Cocero MJ, Mikkola JP, Salmi TO. Engineering in direct synthesis of hydrogen peroxide: targets, reactors and guidelines for operational conditions. *Green Chem*. 2014, 16, 2320–43.
70. Zhang X, Xia Y, Xia C, Wang H. Insights into Practical-Scale Electrochemical H_2O_2 Synthesis. *Trends Chem*. 2020, 2, 942–953.
71. Sun B, Zhu H, Liang W, Zhang X, Feng J, Xu W. A safe and clean way to produce H_2O_2 from H_2 and O_2 within the explosion limit range. *Internat J Hydrogen Energy*. 2019, 44, 19547–54.
72. Polshettiwar V, Asefa TN. *Nanocatalysis: Synthesis and applications*. Wiley 2013. pp 736.
73. Polshettiwar V, Varma RS. Green chemistry by nano-catalysis. *Green Chemistry*. 2010, 12(5), 743–54.
74. Yang F, Deng D, Pan X, Fu Q, Bao X. Understanding nano effects in catalysis. *Nation Sci Rev*. 2015, 2, 183–201.
75. Zhang W. Nanoparticle Aggregation: Principles and Modeling. In: Capco D., Chen Y. (eds) *Nanomaterial. Advances in Experimental Medicine and Biology*, vol. 811. Springer; 2014, pp. 19–43.

76. Wang H, Adeleye AS, Huang Y, Li F, Keller AA. Heteroaggregation of nanoparticles with biocolloids and geocolloids. *Adv Colloid Interf Sci.* 2015, 226, 24–36.
77. Copéret C, Chabanas M, Petroff Saint-Arroman R, Basset JM. Homogeneous and heterogeneous catalysis: bridging the gap through surface organometallic chemistry. *Angew Chem Int Ed.* 2003, 42, 156–81.
78. Cole-Hamilton DJ, Tooze RP, editors. *Catalyst separation, recovery and recycling: chemistry and process design.* Springer; 2006, pp 250.
79. Astruc D, Lu F, Aranzaes JR. Nanoparticles as recyclable catalysts: the frontier between homogeneous and heterogeneous catalysis. *Angew. Chem. Int. Ed.*, 2005, 44, 7852–7872.
80. Benaglia M, (ed). *Recoverable and recyclable catalysts.* Wiley; 2009, p. 471.
81. Gladysz JA. The Experimental Assay of Catalyst Recovery: General Concepts. In: *Recoverable and Recyclable Catalysts*, Edited by Maurizio Benaglia (ed), Wiley; 2009, pp. 1–14.
82. Somwanshi SB, Somvanshi SB, Kharat PB. Nanocatalyst: A Brief Review on Synthesis to Applications. *J Phys: Conf Series* 2020, 1644, 012046.
83. Verma A, Shukla M, Sinha I. 2019. Introductory Chapter: Salient Features of Nanocatalysis, In: *Nanocatalysts*, I Sinha and M Shukla (eds), IntechOpen; 2019. https://www.intechopen.com/books/nanocatalysts/introductory-chapter-salient-features-of-nanocatalysis.
84. Hemalatha K, Madhumitha G, Kajbafvala A, Anupama N, Sompalle R, Mohana Roopan S. Function of nanocatalyst in chemistry of organic compounds revolution: an overview. *J Nanomater.* 2013.
85. Chng LL, Erathodiyil N, Ying JY. Nanostructured catalysts for organic transformations. *Acc Chem Res.* 2013, 46, 1825–37.
86. Gawande MB. Sustainable nanocatalysts for organic synthetic transformations. *Org. Chem. Curr. Res.* 2014, 3, 100–37.
87. Khaturia S, Chahar M, Sachdeva H, Sangeeta. A Review: The Uses of Various Nanoparticles in Organic Synthesis. *J Nanomed Nanotech.* 2020, 10, 543.
88. Filice M, Palomo JM. Cascade reactions catalyzed by bionanostructures. *ACS Catal.* 2014, 4, 5, 1588–98.
89. Burda C, Chen X, Narayanan R, El-Sayed MA. Chemistry and properties of nanocrystals of different shapes. *Chem Rev.* 2005, 105, 1025–102.
90. Tao AR, Habas S, Yang P. Shape control of colloidal metal nanocrystals. *Small.* 2008, 4, 310–25.
91. Narayanan R. Synthesis of green nanocatalysts and industrially important green reactions, *Green Chem Let Rev.* 2012, 5, 707–725.
92. Beach ES, Cui Z, Anastas PT. Green Chemistry: A design framework for sustainability. *Energy Environ Sci.* 2009, 2, 1038–49.
93. Somorjai GA, Park JY. Molecular factors of catalytic selectivity. *Angew Chem Int Ed.* 2008, 47, 9212–28.
94. Corma A, Serna P. Chemoselective hydrogenation of nitro compounds with supported gold catalysts. *Science.* 2006, 313, 332–4.
95. Shimizu KI, Miyamoto Y, Kawasaki T, Tanji T, Tai Y, Satsuma A. Chemoselective hydrogenation of nitroaromatics by supported gold catalysts: mechanistic reasons of size-and support-dependent activity and selectivity. *J Phys Chem C.* 2009, 113, 7803–10.
96. Jagadeesh, R.V., Surkus, A.E., Junge, H. et al. Nanoscale Fe_2O_3-based catalysts for selective hydrogenation of nitroarenes to anilines. *Science.* 2013,342, 1073–1076.
97. Ciriminna R, Falletta E, Della Pina C, Teles JH, Pagliaro M. Industrial applications of gold catalysis. *Angew Chem Int Ed.* 2016, 55, 14210–7.
98. Westerhaus FA, Jagadeesh RV, Wienhöfer G, Pohl MM, Radnik J, Surkus AE, Rabeah J, Junge K, Junge H, Nielsen M, Brückner A. Heterogenized cobalt oxide catalysts for nitroarene reduction by pyrolysis of molecularly defined complexes. *Nat Chem.* 2013, 5, 537–43.
99. Zhang Y, Cui X, Shi F, Deng Y. Nano-gold catalysis in fine chemical synthesis. *Chem Rev.* 2012, 112, 2467–505.
100. Mallat T, Baiker A. Potential of gold nanoparticles for oxidation in fine chemical synthesis. *Ann Rev Chem Biomol Eng.* 2012, 3, 11–28.
101. Pareek V, Bhargava A, Gupta R, Jain N, Panwar J. Synthesis and applications of noble metal nanoparticles: a review. *Advan Sci Eng Med.* 2017, 9, 527–44.
102. Kharisov BI, Dias HR, Kharissova OV, Vázquez A. Ultrasmall particles in the catalysis. *J Nanopart Res.* 2014, 16, 2665.
103. Dong XY, Gao ZW, Yang KF, Zhang WQ, Xu LW. Nanosilver as a new generation of silver catalysts in organic transformations for efficient synthesis of fine chemicals. *Catal Sci Technol.* 2015, 5, 2554–74.
104. Bhosale A, Bhanage B. Silver nanoparticles: Synthesis, characterization and their application as a sustainable catalyst for organic transformations. *Curr Org Chem.* 2015, 19, 708–27.
105. Acar C, Dincer I. Review and evaluation of hydrogen production options for better environment. *J Clean Prod.* 2019, 218, 835–49.
106. Dincer I, Acar C. Review and evaluation of hydrogen production methods for better sustainability. *Internat J Hydrogen Energy.* 2015, 40, 11094–111.
107. Sahaym U, Norton MG. Advances in the application of nanotechnology in enabling a ‘hydrogen economy’. *J Mater Sci.* 2008, 43, 5395–429.
108. Fayaz H, Saidur R, Razali N, Anuar FS, Saleman AR, Islam MR. An overview of hydrogen as a vehicle fuel. *Renew Sustain Energy Rev.* 2012, 16, 5511–28.
109. Williams PT. Hydrogen and carbon nanotubes from pyrolysis-catalysis of waste plastics: A review. *Waste Biomass Valor.* 2020, 2, 1-
110. Mark LO, Cendejas MC, Hermans I. The Use of Heterogeneous Catalysis in the Chemical Valorization of Plastic Waste. *ChemSusChem.* 2020, 13, 5808–5836.
111. Vollmer I, Jenks MJ, Roelands MC, White RJ, van Harmelen T, de Wild P, van Der Laan GP, Meirer F, Keurentjes JT, Weckhuysen BM. Beyond mechanical recycling: Giving new life to plastic waste. *Angew Chem Int Ed.* 2020, 59, 15402–23.
112. Chung Y-H, Jou S. Carbon nanotubes from catalytic pyrolysis of polypropylene. *Mater Chem Phys.* 2005, 92, 256–259.
113. Yao D, Wang CH. 2020. Pyrolysis and in-line catalytic decomposition of polypropylene to carbon nanomaterials and hydrogen over Fe-and Ni-based catalysts. *Appl Energy*, 265, 114819.

114. Yao D, Li H, Dai Y, Wang CH. Impact of temperature on the activity of Fe-Ni catalysts for pyrolysis and decomposition processing of plastic waste. *Chem Eng J*. 2021, 408, 127268.
115. Kumagai S, Nakatani J, Saito Y, Fukushima Y, Yoshioka T. Latest Trends and Challenges in Feedstock Recycling of Polyolefinic Plastics. *J Japan Petrol Inst*. 2020, 63, 345–364.
116. Sánchez-Rivera KL, Huber GW. Catalytic Hydrogenolysis of Polyolefins into Alkanes. *ACS Cent. Sci*. 2021, 7, 17–19.
117. Celik G, Kennedy RM, Hackler RA, Ferrandon M, Tennakoon A, Patnaik S, LaPointe AM, Ammal SC, Heyden A, Perras FA, Pruski M. Upcycling single-use polyethylene into high-quality liquid products. *ACS Cent Sci*. 2019, 5, 1795–803
118. Zhang C, Hui X, Lin Y, Sung CJ. Recent development in studies of alternative jet fuel combustion: Progress, challenges, and opportunities. *Renew Sustain Energy Rev*. 2016, 54, 120–38.
119. Gutiérrez-Antonio C, Gómez-Castro FI, de Lira-Flores JA, Hernández S. A review on the production processes of renewable jet fuel. *Renew Sustain Energy Rev*. 2017, 79, 709–29.
120. Wang M, Dewil R, Maniatis K, Wheeldon J, Tan T, Baeyens J, Fang Y. Biomass-derived aviation fuels: Challenges and perspective. *Prog Energy Combust Sci*. 2019, 74, 31–49.
121. Adhikari SP, Zhang J, Guo Q, Unocic KA, Tao L, Li Z. A hybrid pathway to biojet fuel via 2, 3-butanediol. *Sustain Energy Fuels*. 2020, 4, 3904–14.
122. Yuan E, Ni P, Xie J, Jian P, Hou X. Highly Efficient Dehydrogenation of 2, 3-Butanediol Induced by Metal–Support Interface over Cu-SiO2 Catalysts. *ACS Sustain Chem Eng*. 2020, 8, 15716–31.
123. Tang X, Zeng X, Li Z, Hu L, Sun Y, Liu S, Lei T, Lin L. Production of γ-valerolactone from lignocellulosic biomass for sustainable fuels and chemicals supply. *Renew Sustain Energy Rev*. 2014, 40, 608–20.
124. Moreno-Marrodan C, Barbaro P. Energy efficient continuous production of γ-valerolactone by bifunctional metal/acid catalysis in one pot. *Green Chem*. 2014, 16.
125. Villa A, Schiavoni M, Chan-Thaw CE, Fulvio PF, Mayes RT, Dai S, More KL, Veith GM, Prati L. Acid-Functionalized Mesoporous Carbon: An Efficient Support for Ruthenium-Catalyzed γ-Valerolactone Production. *ChemSusChem*. 2015, 8, 2520–2528.
126. Tan J, Cui J, Deng T, Cui X, Ding G, Zhu Y, Li Y. Water-promoted hydrogenation of levulinic acid to γ-valerolactone on supported ruthenium catalyst. *ChemCatChem*. 2015, 7, 508–12.
127. Piskun AS, De Haan JE, Wilbers E, Van De Bovenkamp HH, Tang Z, Heeres HJ. Hydrogenation of levulinic acid to γ-valerolactone in water using millimeter sized supported Ru catalysts in a packed bed reactor. *ACS Sustain Chem Eng*. 2016, 4, 2939–50.
128. Jampilek J. Heterocycles in medicinal chemistry. *Molecules* 2019, 24, 3839
129. Kaur R, Bariwal J, Voskressensky LG, Van der Eycken EV. Gold and silver nanoparticle-catalyzed synthesis of heterocyclic compounds. *Chem Heterocyclic Comp*. 2018, 54, 241–8.
130. Yamane Y, Liu X, Hamasaki A, Ishida T, Haruta M, Yokoyama T, Tokunaga M. One-pot synthesis of indoles and aniline derivatives from nitroarenes under hydrogenation condition with supported gold nanoparticles. *Org Let*. 2009, 11, 5162–5.
131. Narayanan R. Synthesis of green nanocatalysts and industrially important green reactions. *Green Chem Let Rev*. 2012, 5, 707–25.
132. Liao H, Chou Y, Wang Y, Zhang H, Cheng T, Liu G. Multistep Organic Transformations over Base-Rhodium/Diamine-Bifunctionalized Mesostructured Silica Nanoparticles. *ChemCatChem*. 2017, 9, 3197–202.
133. Cong H, Porco Jr JA. Chemical synthesis of complex molecules using nanoparticle catalysis. *ACS Cat* 2012, 2, 65–70.
134. Cong H, Porco Jr JA. Total Synthesis of (±)-Sorocenol B Employing Nanoparticle Catalysis. *Org Let*. 2012, 14, 2516–9.
135. Chee CF, Lee YK, Buckle MJ, Abd RN. Synthesis of (±)-kuwanon V and (±)-dorsterone methyl ethers via Diels–Alder reaction. *Tet Let*. 2011, 52, 1797–9.
136. Cong H, Becker CF, Elliott SJ, Grinstaff MW, Porco Jr JA. Silver nanoparticle-catalyzed Diels– Alder cycloadditions of 2′-hydroxychalcones. *J Amer Chem Soc*. 2010, 132, 7514–8.
137. Qi C, Qin T, Suzuki D, Porco Jr JA. Total synthesis and stereochemical assignment of (±)-sorbiterrin A. *J Amer Chem Soc*. 2014, 136, 3374–7.
138. T. P. Kenakin, Chapter 11 The Drug Discovery Process. In: *A Pharmacology Primer*, 4th ed. 2014, Elsevier, pp. 281–320.
139. Brown DG, Bostrom J. Analysis of past and present synthetic methodologies on medicinal chemistry: where have all the new reactions gone? Miniperspective. *J Med Chem*. 2016, 59, 4443–58.
140. Roughley SD, Jordan AM. The medicinal chemist's toolbox: an analysis of reactions used in the pursuit of drug candidates. *J Med Chem*. 2011, 54, 3451–79.
141. Boström J, Brown DG, Young RJ, Keserü GM. Expanding the medicinal chemistry synthetic toolbox. *Nat Rev Drug Disc*. 2018, 10, 709–27.
142. Campos KR, Coleman PJ, Alvarez JC, Dreher SD, Garbaccio RM, Terrett NK, Tillyer RD, Truppo MD, Parmee ER. The importance of synthetic chemistry in the pharmaceutical industry. *Science*. 2019, 363, 6424.
143. Dombrowski AW, Gesmundo NJ, Aguirre AL, Sarris KA, Young JM, Bogdan AR, Martin MC, Gedeon S, Wang Y. Expanding the medicinal chemist toolbox: Comparing seven C (sp2)–C (sp3) cross-coupling methods by library synthesis. *ACS Med Chem Let*. 2020, 11, 597–604.
144. Shen HC. Selected Applications of Transition Metal-Catalyzed Carbon–Carbon Cross-Coupling Reactions in the Pharmaceutical Industry. In: Crawley ML, Trost BM (eds) *Applications of Transition Metal Catalysis in Drug Discovery and Development*. Wiley, New York, 2012.52, 25–43.
145. Nishihara Y, editor. *Applied cross-coupling reactions*. Heidelberg: Springer; 2013, pp 245.
146. Roughley SD, Jordan AM. The medicinal chemist's toolbox: an analysis of reactions used in the pursuit of drug candidates. *J Med Chem*. 2011, 54, 10, 3451–79.
147. Colacot TJ, editor. *New trends in cross-coupling: theory and applications*. Royal Society of Chemistry; 2014.
148. Lipshutz BH, Taft BR, Abela AR, Ghorai S, Krasovskiy A, Duplais C. Catalysis in the service of green chemistry:

Nobel prize-winning palladium-catalysed cross-couplings, run in water at room temperature: Heck, Suzuki-Miyaura and Negishi reactions carried out in the absence of organic solvents, enabled by micellar catalysis. *Platinum Metals Rev*. 2012, 56, 62.
149. Burke AJ, Marques CS. Chapt 1 Cross-Coupling Arylations: Precedents and Rapid Historical Review of the Field. In: *Catalytic Arylation Methods: From the Academic Lab to Industrial Processes*, AJ Burke and CS Marques editors, 2015 Wiley-VCH Verlag GmbH & Co. KGaA. pp. 1–94.
150. Lei A, Shi W, Liu C, Liu W, Zhang H, He C. *Oxidative cross-coupling reactions*. John Wiley & Sons. 2016, p. 230.
151. Miyaura, N.; Suzuki, A. Palladium-Catalyzed Cross-Coupling Reactions of Organoboron Compounds. *Chem. Rev*. 1995, 95, 2457–2483.
152. Suzuki, A. Recent advances in the cross-coupling reactions of organoboron derivatives with organic electrophiles, 1995–1998. *J. Organomet. Chem*. 1999, 576, 147–168.
153. Littke, A. F.; Fu, G. C. Palladium-catalyzed coupling reactions of aryl chlorides. *Angew Chem Int. Ed*. 2002, 41, 4176–4211.
154. Sun B, Ning L, Zeng HC. Confirmation of Suzuki–Miyaura cross-coupling reaction mechanism through synthetic architecture of nanocatalysts. *J Amer Chem Soc*. 2020, 142, 13823–32.
155. Zeng HC. Hierarchy Concepts in Design and Synthesis of Nanocatalysts. *ChemCatChem*. 2020, 12, 5303–11.
156. Biffis A, Centomo P, Del Zotto A, Zecca M. Pd metal catalysts for cross-couplings and related reactions in the 21st century: a critical review. *Chem Rev*. 2018, 118, 2249–95.
157. Trzeciak AM, Augustyniak AW. The role of palladium nanoparticles in catalytic C–C cross-coupling reactions. *Coord Chem Rev*. 2019, 384, 1–20.
158. Beletskaya IP, Alonso F, Tyurin V. The Suzuki-Miyaura reaction after the Nobel prize. *Coord Chem Rev*. 2019, 385, 137–73.
159. Lloyd-Jones, G. C.; Pagett, A. B. Suzuki–Miyaura Cross-Coupling, *Org. React*. 2019, 100, 9.
160. Li CJ, Trost BM. Green chemistry for chemical synthesis. *Proc Nat Acad Sci*. 2008, 105, 13197–202.
161. Sherwood J, Clark JH, Fairlamb IJ, Slattery JM. Solvent effects in palladium catalysed cross-coupling reactions. *Green Chem*. 2019, 21, 2164–213.
162. Reeves EK, Bauman OR, Mitchem GB, Neufeldt SR. Solvent Effects on the Selectivity of Palladium-Catalyzed Suzuki-Miyaura Couplings. *Israel J Chem*. 2020, 60, 406–9.
163. Mitrofanov I, Sansonetti S, Abildskov J, Sin G, Gani R. The solvent selection framework: Solvents for organic synthesis, separation processes and ionic liquids solvents. *Comp Aid Chem Eng*. 2012, 30, 762–766.
164. Adlington NK, Agnew LR, Campbell AD, Cox RJ, Dobson A, Barrat CF, Gall MA, Hicks W, Howell GP, Jawor-Baczynska A, Miller-Potucka L. Process Design and Optimization in the Pharmaceutical Industry: A Suzuki–Miyaura Procedure for the Synthesis of Savolitinib. *J Org Chem*. 2018, 84, 4735–47.
165. Butters M, Catterick D, Craig A, Curzons A, Dale D, Gillmore A, Green SP, Marziano I, Sherlock JP, White W. Critical Assessment of Pharmaceutical Processes -A Rationale for Changing the Synthetic Route. *Chem. Rev*. 2006, 106, 3002–3027.
166. Busacca CA, Fandrick DR, Song JJ, Senanayake CH. Transition metal catalysis in the pharmaceutical industry. In: Crawley, ML, Trost, BM (eds) *Applications of Transition Metal Catalysis in Drug Discovery and Development: An Industrial Perspective*. Wiley, New York, 2012. 52, 1–24.
167. Curzons AD, Constable DC, Cunningham VL. Solvent selection guide: a guide to the integration of environmental, health and safety criteria into the selection of solvents. *Clean Prod Proc*. 1999, 1, 82–90.
168. Reichardt C, Welton T. *Solvents and solvent effects in organic chemistry*. John Wiley & Sons, New York. 2011, p. 692.
169. Henderson RK, Jiménez-González C, Constable DJ, Alston SR, Inglis GG, Fisher G, Sherwood J, Binks SP, Curzons AD. Expanding GSK's solvent selection guide–embedding sustainability into solvent selection starting at medicinal chemistry. *Green Chem*. 2011, 13, 854–62.
170. Prat D, Pardigon O, Flemming HW, Letestu S, Ducandas V, Isnard P, Guntrum E, Senac T, Ruisseau S, Cruciani P, Hosek P. Sanofi's solvent selection guide: A step toward more sustainable processes. *Org Proc Res Dev*. 2013, 17, 1517–25.
171. Prat D, Hayler J, Wells A. A survey of solvent selection guides. *Green Chem*. 2014, 16, 4546–51.
172. Byrne FP, Jin S, Paggiola G, Petchey TH, Clark JH, Farmer TJ, Hunt AJ, McElroy CR, Sherwood J. Tools and techniques for solvent selection: green solvent selection guides. *Sustain Chem Proc*. 2016, 4, 1–24.
173. Piccione PM, Baumeister J, Salvesen T, Grosjean C, Flores Y, Groelly E, Murudi V, Shyadligeri A, Lobanova O, Lothschütz C., Solvent selection methods and tool. *Org Proc Res Dev*. 2019, 23, 998–1016.
174. Jimenez-Gonzalez C. Life cycle considerations of solvents. *Curr Opin Green Sustain Chem*. 2019, 18, 66–71.
175. Sheldon RA. The greening of solvents: Towards sustainable organic synthesis. *Curr Opin Green Sustain Chem*. 2019, 18, 13–9.
176. Lipshutz BH, Isley NA, Fennewald JC, Slack ED. On the Way Towards Greener Transition-Metal-Catalyzed Processes as Quantified by E Factors. *Angew Chem Int. Ed*. 2013, 52(42), 10952–8.
177. Sheldon RA. The E factor 25 years on: the rise of green chemistry and sustainability. *Green Chem*. 2017, 19, 18–43.
178. Cortes-Clerget M, Yu J, Kincaid JR, Walde P, Gallou F, Lipshutz BH. Water as the reaction medium in organic chemistry: from our worst enemy to our best friend. *Chem Sci*. 2021, 12, 4237–4266.
179. Kitanosono T, Masuda K, Xu P, Kobayashi S. Catalytic organic reactions in water toward sustainable society. *Chem Rev*. 2018, 118, 679–746.
180. Mishra M, Muthuprasanna P, Prabha KS, Rani PS, Satish IA, Ch IS, Arunachalam G, Shalini S. Basics and potential applications of surfactants-a review. *Int J Pharm Tech Res*. 2009, 1, 1354–1365.
181. Paprocki D, Madej A, Koszelewski D, Brodzka A, Ostaszewski R. Multicomponent reactions accelerated by aqueous micelles. *Front Chem*. 2018, 6, 502.
182. Cornils B, Herrmann WA, eds. *Aqueous-phase organometallic catalysis: concepts and applications*. John Wiley & Sons; 2004. p. 750.

183. Lipshutz BH. Synthetic chemistry in a water world. New rules ripe for discovery. *Cur Opin Green Sustain Chem.* 2018, 11, 1–8.
184. Sar P, Ghosh A, Scarso A, Saha B. Surfactant for better tomorrow: applied aspect of surfactant aggregates from laboratory to industry. *Res Chem Intermed.* 2019, 45, 6021–41.
185. Scarso A. Micellar nanoreactors. *Encyc Inorg Bioinorg Chem.* Wiley. 2011, 15, 1–16.
186. De Martino MT, Abdelmohsen LK, Rutjes FP, van Hest JC. Nanoreactors for green catalysis. *Beilstein J Org Chem.* 2018, 14, 716–33.
187. La Sorella G, Strukul G, Scarso A. Recent advances in catalysis in micellar media. *Green Chem.* 2015, 17, 644–83.
188. Lorenzetto T, Berton G, Fabris F, Scarso A. Recent designer surfactants for catalysis in water. *Catal Sci Technol.* 2020, 10, 4492–502.
189. Javadian S, Kakemam J. Intermicellar interaction in surfactant solutions; a review study. *J Mol Liquids.* 2017, 242, 15–28.
190. Svenson S. Controlling surfactant self-assembly. *Curr Opin Colloid Interf Sci.* 2004, 9, 201–12.
191. Lipshutz BH, Ghorai S. Transitioning organic synthesis from organic solvents to water. What's your E Factor? *Green Chem.* 2014, 16, 3660–79.
192. Andraos J. Relationships between step and cumulative PMI and E-factors: implications on estimating material efficiency with respect to charting synthesis optimization strategies. *Green Proc Synth.* 2019, 8, 324–36.
193. Ashcroft CP, Challenger S, Derrick AM, Storey R, Thomson NM. Asymmetric synthesis of an MMP-3 inhibitor incorporating a 2-alkyl succinate motif. *Org Proc Res Dev.* 2003, 7, 362–8.
194. Polshettiwar V, Decottignies A, Len C, Fihri A. Suzuki–Miyaura Cross-Coupling Reactions in Aqueous Media: Green and Sustainable Syntheses of Biaryls. *ChemSusChem.* 2010, 3, 502–22.
195. Wood-Black F, Blayney MB, Reid M, Montes I, Bayoumi AE, Sloan L, Rothbaum JO, Koudehi MF, Zibaseresht R, Bancroft L. Highlights: multilingual safety resources, Pd-catalyzed cross-coupling reactions, ethylene glycol purification, and more. *ACS Chem. Health Saf.* 2020, 27, 6, 313–315.
196. Yang Q, Babij NR, Good S. Potential safety hazards associated with Pd-catalyzed cross-coupling reactions. *Org Proc Res Dev.* 2019, 23, 2608–26.
197. Norman MH, Zhu J, Fotsch C, Bo Y, Chen N, Chakrabarti P, Doherty EM, Gavva NR, Nishimura N, Nixey T, Ognyanov VI. Novel vanilloid receptor-1 antagonists: 1. Conformationally restricted analogues of trans-cinnamides. *J Med Chem.* 2007, 50, 3497–514.
198. Gallou F, Isley NA, Ganic A, Onken U, Parmentier M. Surfactant technology applied toward an active pharmaceutical ingredient: more than a simple green chemistry advance. *Green Chem.* 2016, 18, 14–19.
199. Parmentier M, Gabriel CM, Guo P, Isley NA, Zhou J, Gallou F. Switching from organic solvents to water at an industrial scale. *Curr Opin Green Sustain Chem.* 2017, 7, 13–7.
200. Baenziger M, Baierl M, Devanathan K, Eswaran S, Fu P, Gschwend B, Haller M, Kasinathan G, Kovacic N, Langlois A, Li Y. Synthesis Development of the Selective Estrogen Receptor Degrader (SERD) LSZ102 from a Suzuki Coupling to a C–H Activation Strategy. *Org Proc Res Dev.* 2020, 24, 1405–19.
201. Kaldre D, Gallou F, Sparr C, Parmentier M. Interface-rich Aqueous Systems for Sustainable Chemical Synthesis. *CHIMIA Int J Chem.* 2019, 73, 714–9.
202. Kjell DP, Watson IA, Wolfe CN, Spitler JT. Complexity-based metric for process mass intensity in the pharmaceutical industry. *Org Proc Res Dev.* 2013, 17, 169–74.
203. Cespi D, Beach ES, Swarr TE, Passarini F, Vassura I, Dunn PJ, Anastas PT. Life cycle inventory improvement in the pharmaceutical sector: assessment of the sustainability combining PMI and LCA tools. *Green Chem.* 2015, 17, 3390–400.
204. Gallou F, Isley NA, Ganic A, Onken U, Parmentier M. Surfactant technology applied toward an active pharmaceutical ingredient: more than a simple green chemistry advance. *Green Chem.* 2016, 18, 14–9.
205. Gallou F. Sustainability as a Trigger for Innovation!. *Chimia.* 2020, 74, 538–48.
206. Tilman D, Balzer C, Hill J, Befort BL. Global food demand and the sustainable intensification of agriculture. *Proc Natl Acad Sci US.* 2011, 108, 20260–20264.
207. FAOSTAT Domain Pesticides Trade. Metadata, Release May 2020, fenixservices.fao.org › faostat › static.
208. Maienfisch P, Stevenson TM. Modern agribusiness-markets, companies, benefits and challenges. *ACS Symp Ser.* 2015, 1204, 1–13.
209. Devendar P, Qu RY, Kang WM, He B, Yang GF. Palladium-catalyzed cross-coupling reactions: a powerful tool for the synthesis of agrochemicals. *J Agric Food Chem.* 2018, 66, 8914–34.
210. Whiteker GT. Applications of the 12 Principles of Green Chemistry in the Crop Protection Industry. *Org Proc Res Dev.* 2019, 23, 2109–21.
211. Takale BS, Thakore RR, Mallarapu R, Gallou F, Lipshutz BH. A Sustainable 1-Pot, 3-Step Synthesis of Boscalid Using Part per Million Level Pd Catalysis in Water. *Org Proc Res Dev.* 2019, 24, 101–5.
212. Takale BS, Thakore RR, Irvine NM, Schuitman AD, Li X, Lipshutz BH. Sustainable and Cost-Effective Suzuki–Miyaura Couplings toward the Key Biaryl Subunits of Arylex and Rinskor Active. *Org Let.* 2020, 22, 4823–7.
213. Chen F, Jiang X, Zhang L, Lang R, Qiao B. Single-atom catalysis: Bridging the homo-and heterogeneous catalysis. *Chinese J Catal.* 2018, 39, 893–8.
214. Zhang H, Liu G, Shi L, Ye J. Single-atom catalysts: emerging multifunctional materials in heterogeneous catalysis. *Adv Energy Mater.* 2018, 8, 1701343.
215. Weon S, Huang D, Rigby K, Chu C, Wu X, Kim JH. Environmental materials beyond and below the nanoscale: single-atom catalysts. *ACS ES&T Eng.* 2020.
216. Parkinson GS. Single-Atom Catalysis: How Structure Influences Catalytic Performance. *Catal Let.* 2019, 9, 1137–46.
217. Zhang L, Ren Y, Liu W, Wang A, Zhang T. Single-atom catalyst: a rising star for green synthesis of fine chemicals. *Nat Sci Rev.* 2018, 5, 653–72.
218. Yan H, Su C, He J, Chen W. Single-atom catalysts and their applications in organic chemistry. *J Mater Chem A.* 2018, 6, 8793–814.

219. Ding G, Hao L, Xu H, Wang L, Chen J, Li T, Tu X, Zhang Q. Atomically dispersed palladium catalyses Suzuki–Miyaura reactions under phosphine-free conditions. *Comm Chem.* 2020, 3, 1–8.
220. Kim S, Jee S, Choi KM, Shin DS. Single-atom Pd catalyst anchored on Zr-based metal-organic polyhedra for Suzuki-Miyaura cross coupling reactions in aqueous media. *Nano Res.* 2020, 14, 486–92.
221. Chen Z, Vorobyeva E, Mitchell S, Fako E, Ortuño MA, López N, Collins SM, Midgley PA, Richard S, Vilé G, Pérez-Ramírez J. A heterogeneous single-atom palladium catalyst surpassing homogeneous systems for Suzuki coupling. *Nat Nanotech.* 2018, 13, 702–7.
222. Alper E, Orhan OY. CO_2 utilization: Developments in conversion processes. *Petrol.* 2017, 3, 109–26.
223. Gulzar A, Gulzar A, Ansari MB, He F, Gai S, Yang P. Carbon dioxide utilization: A paradigm shift with CO2 economy. *Chem Eng J Adv.* 2020, 100013.
224. Zhang Z, Pan SY, Li H, Cai J, Olabi AG, Anthony EJ, Manovic V. Recent advances in carbon dioxide utilization. *Renew Sustain Energy Rev.* 2020, 125, 109799.
225. Liu Q, Wu L, Jackstell R, Beller M. Using carbon dioxide as a building block in organic synthesis. *Nat Comm.* 2015, 6, 1–5.
226. Artz J, Müller TE, Thenert K, Kleinekorte J, Meys R, Sternberg A, Bardow A, Leitner W. 2018. Sustainable conversion of carbon dioxide: an integrated review of catalysis and life cycle assessment. *Chem Rev.* 2018,118, 434–504.
227. Ma J, Gong H, Zhang T, Yu H, Zhang R, Liu Z, Yang G, Sun H, Tang S, Qiu Y. Hydrogenation of CO2 to formic acid on the single atom catalysis Cu/C2N: A first principles study. *Appl Surf Sci.* 2019, 488, 1–9.
228. Zhang L, Doyle-Davis K, Sun X. Pt-Based electrocatalysts with high atom utilization efficiency: from nanostructures to single atoms. *Energy Environ Sci.* 2019, 12, 492–517.
229. Gawande MB, Fornasiero P, Zbořil R. Carbon-based single-atom catalysts for advanced applications. *ACS Catal.* 2020, 10, 2231–59.
230. De S, Dokania A, Ramirez A, Gascon J. Advances in the Design of Heterogeneous Catalysts and Thermocatalytic Processes for CO2 Utilization. *ACS Catal.* 2020, 10, 14147–85.
231. Erdmann L, Graedel TE. Criticality of non-fuel minerals: a review of major approaches and analyses. *Environ Sci Technol.* 2011, 45, 7620–30.
232. Graedel TE, Harper EM, Nassar NT, Nuss P, Reck BK. Criticality of metals and metalloids. *Proc Nat Acad Sci US.* 2015, 112, 4257–62.
233. Poli L. Endangered Element: A review on palladium. *McGill Green Chem. J.* 2010, 44, 56.
234. Gawande MB, Luque R, Zboril R. The rise of magnetically recyclable nanocatalysts. *ChemCatChem.* 2014, 6, 3312–3.
235. National Research Council US. 2012. The Role of the Chemical Sciences in Finding Alternatives to Critical Resources: A Workshop Summary. Washington, DC: The National Academies Press US. https://doi.org/10.17226/13366 p. 72.
236. Ashiq A, Kulkarni J, Vithanage M. Hydrometallurgical recovery of metals from E-waste. In: *Electronic waste management and treatment technology*, 2019. pp. 225–246. Butterworth-Heinemann.
237. Tu S, Yusuf S, Muehlfeld M, Bauman R, Vanchura B. The destiny of palladium: Development of efficient palladium analysis techniques in enhancing palladium recovery. *Org Proc Res Dev.* 2019, 23, 2175–80.
238. Lacanau V., Bonnete F., Wagner P., Contino-Pépin C., Schmitt M., et al. From Electronic Waste to Suzuki–Miyaura Cross-Coupling Reaction in Water: Direct Valuation of Recycled Palladium in Catalysis. *ChemSusChem*, 2020, 13, 5224–5230.
239. Jantan KA, Kwok CY, Chan KW, Marchiò L, White AJ, Deplano P, Serpe A, Wilton-Ely JD. From recovered metal waste to high-performance palladium catalysts. *Green Chem.* 2017, 19, 5846–53.
240. Serpe A, Artizzu F, Espa D, Rigoldi A, Mercuri ML, Deplano P. From trash to resource: a green approach to noble-metals dissolution and recovery. *Green Proc Synth.* 2014, 3, 141–6.
241. Serpe A, Bigoli F,Cabras MC, Fornasiero P, Graziani M, Mercuri ML, Montini T,Pilia L,Trogu EF, Deplano P. Pd dissolution through a mild and effective one-step reaction and its application for Pd-recovery from spent catalytic converters. *Chem Commun.* 2005, 38, 1040–1042.
242. Jantan KA, Chan KW, Melis L, White AJ, Marchio L, Deplano P, Serpe A, Wilton-Ely JD. From recovered palladium to molecular and nanoscale catalysts. *ACS Sustain Chem Eng.* 2019, 7, 12389–98.
243. Nelson JJ, Schelter EJ. Sustainable inorganic chemistry: metal separations for recycling, *Inorg Chem.* 2019, 58, 979–990.

10

Greener Organic Transformations by Plant-Derived Water Extract Ashes

Bipasa Halder and Ahindra Nag

CONTENTS

10.1 Introduction ... 177
10.2 Literature Survey ... 177
10.3 Organic Transformations by Plant-Derived Water Extract Ashes ... 179
10.4 Palladium-Mediated Cross-Coupling Reaction ... 179
10.5 Conclusion ... 182
References ... 187

10.1 Introduction

Along with, Suzuki–Miyaura cross-coupling reaction (SMCR) is known for synthesizing biaryl scaffolds in a wide range. This time, we focused on the generation of *in-situ* Pd nanoparticles for the formation of bi-aryl building blocks starting from aryl boronic acids and aryl halides, where the water extract of banana stem ash (WEBSA) was explored as co-solvent and base. The use of WEBSA bestowed this protocol with few added advantages such as short reaction time, eco-friendly medium, recovery of the catalyst with further use, and less rigorous reaction setup.

2,2′-aryl/alkyl-methylene-bis(3-hydroxy-2-cyclohexene-1-one),2,2′-aryl/alkyl-methylene-bis(3-hydroxy-5,5-dimethyl-2 cyclohexene-1-one) are widening up the interface between chemistry and pharmacology fields.[1] They have various biological and pharmaceutical activities such as antioxidant, antiviral, antibacterial, anti-inflammatory, tyrosinase inhibitors, lipoxygenase inhibitors, and xanthine oxidase inhibitotos.[2] They also show potent application in dermatological disorders along with hyperpigmentation and skin melanoma.[2c] Moreover, these compounds are important key intermediates for the synthesis of various heterocyclic compounds such as acridindione and xanthenedione derivatives.[3] Also, acridindione derivatives have been used for electron donors, electron acceptors, and in the photoinitiated polymerization of acrylates and methacrylates[4]; xanthenedione derivatives also have vast applications in biological and therapeutic properties.[2b,5] Besides, xanthenedione derivatives are also applied in laser technologies due to spectroscopic properties.[6] A glimpse of the structures of some biologically active tetraketone derivatives is summarized in Figure 10.1.

In addition, tetrahydrobenzo[*b*]pyrans are the other significant oxygen-containing cyclic 1,3-diketone-based heterocycles which become a zone of special flair to organic chemists and chemical biologists having a wide spectrum of biological activities. They belong to commonly occurring natural products family having antimicrobial, antimalarial, anticoagulant, anticancer, antibacterial, antianaphylactic, spasmolytic, diuretic, EAAT1 inhibitors, and calcium antagonistic properties and are of great biological and pharmacological importance.[7] Some structures of these compounds are displayed in Figure 10.2. Furthermore, they can be used as cognitive enhancers for the treatment of schizophrenia, Down's syndrome, Huntington's amyotrophic lateral sclerosis, AIDS-associated dementia, Alzheimer's disease, and Parkinson's disease.[7a] Besides, they have been applied in various fields like agrochemicals, laser dyes, cosmetics, pigments, optical brighteners, and fluorescence makers.[8]

10.2 Literature Survey

Literature report reveals that the most straightforward classical protocol for the synthesis of tetraketone derivatives has been achieved by the one-pot reaction between cyclic 1,3-diketones and various aryl aldehydes in the presence of a series of various catalysts such as nano-$ZnAl_2O_4$,[1] saccharine-based anion-functionalized ionic liquid [Bmim]Sac,[9] $CoFe_2O_4$,[10] piperidine,[11] tetraethyl ammonium bromide,[2b] diethylamine,[12] EDDA and $In(OTf)_3$,[13] ZnO,[14] silica-diphenic acid,[15] I_2,[16] 4 Å molecular sieves,[17] copper octoate,[18] $HClO_4$-SiO_2,[19] $CaCl_2$,[20] caffeinium hydrogen sulfate,[21] *L*-lysine,[22] taurine,[23] $ZrOCl_2$/$NaNH_2$,[24] *L*-histidine,[25] $SmCl_3$,[26] Pd(0) nps,[27] Cu(0) nps onto-silica,[28] Ni(0) nps-Mont,[29] immobilized Ni-Zn-Fe layered double hydroxide,[30] nano Fe/NaY zeolite,[31] $Yb(OTf)_3$-SiO_2 with aniline,[32] PVP-stabilized Ni nps,[33] Fe_3O_4@SiO_2-SO_3H,[34] Al-MCM-41,[35] SDS,[36] silica/HBF_4,[37] and baker's yeast.[38]

In addition, tetrahydrobenzo[*b*]pyran derivatives are synthesized from cyclic 1,3-diketone, aromatic aldehydes, and active methylene compounds (such as malononitrile or ethyl cyanoacetate) *via* one-pot three-component reactions (3-CRs) in

DOI: 10.1201/9781003089162-10

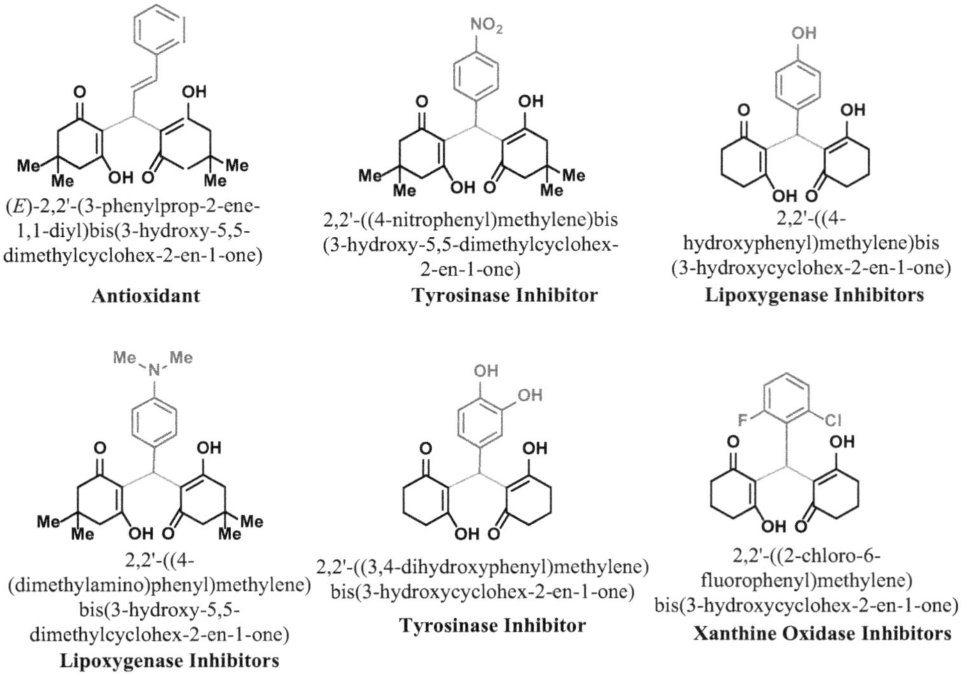

(*E*)-2,2'-(3-phenylprop-2-ene-1,1-diyl)bis(3-hydroxy-5,5-dimethylcyclohex-2-en-1-one)
Antioxidant

2,2'-((4-nitrophenyl)methylene)bis(3-hydroxy-5,5-dimethylcyclohex-2-en-1-one)
Tyrosinase Inhibitor

2,2'-((4-hydroxyphenyl)methylene)bis(3-hydroxycyclohex-2-en-1-one)
Lipoxygenase Inhibitors

2,2'-((4-(dimethylamino)phenyl)methylene)bis(3-hydroxy-5,5-dimethylcyclohex-2-en-1-one)
Lipoxygenase Inhibitors

2,2'-((3,4-dihydroxyphenyl)methylene)bis(3-hydroxycyclohex-2-en-1-one)
Tyrosinase Inhibitor

2,2'-((2-chloro-6-fluorophenyl)methylene)bis(3-hydroxycyclohex-2-en-1-one)
Xanthine Oxidase Inhibitors

FIGURE 10.1 Illustration of some biologically active bis-cyclohexenone derivatives.

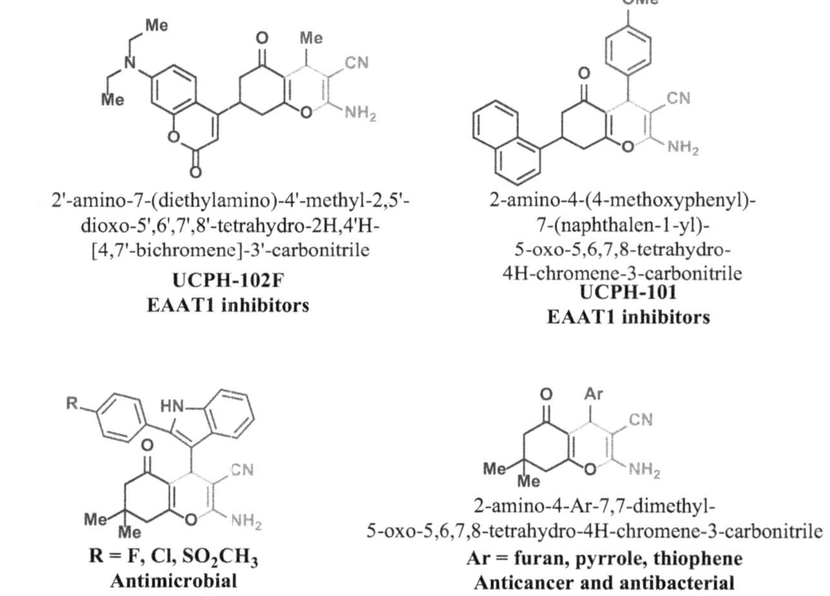

FIGURE 10.2 Examples of some biologically active tetrahydrobenzo[*b*]pyran derivatives.

the presence of various catalysts like $CoFe_2O_4$,[10] nano-SiO_2,[39] MPA,[40] Mn(III)-pentadentate Schiff base complex supported on MWCNTs,[41] nano-Fe_3O_4@TDI@TiO_2,[42] [Pyridin-SO_3H]Cl,[43] Fe_3O_4@SiO_2@TiO_2,[44] Na_2CO_3,[45] taurine,[23] triethanol ammonium acetate and triethanol ammonium formate,[46] I_2,[47] *N,N*-dimethylaminoethylbenzyldimethylammonium chloride,[48] 4-(dimethylamino)pyridine,[49] potassium phthalimide-*N*-oxyl,[50] MgO,[51] $NiFe_2O_4$ nps,[52] amberlite IRA-400 (OH^-),[53] [cmmim]Br and [cmmim][BF_4],[54] hexamethylenetetramine,[55] nano-zeolite clinoptilolite,[56] glutamic acid,[57] hexadecyldimethyl benzyl ammonium bromide,[58] *p*-dodecylbenzenesulfonic acid,[59] 2,2,2-trifluoroethanol,[60] urea,[61] {[HMIM]C$(CN)_3$},[62] triethanolamine,[63] and so on. Furthermore, a variety of greener technologies along with the use of electrolysis,[64] microwave heating,[65] grinding,[65a,66] ultrasound irradiation,[67] mechanochemical ball milling,[68] and visible-light-promoted system[69] are also disclosed as valuable alternative methods for the synthesis of tetraketone and tetrahydrobenzo[*b*]pyran derivatives in the literature. Apart from these, many of the reported approaches still suffer from several limitations such as longer reaction times, higher reaction temperatures, the use of the external acidic and basic catalysts, harsh reaction conditions, difficult work-up, and the requirement for special apparatus. To overcome of these limitations, a simple, new, efficient, effective, faster, and greener alternative route for the synthesis of tetraketone and tetrahydrobenzo[*b*]-pyran derivatives is still essential in terms of the synthetic methodology, and from the economic point of view.

10.3 Organic Transformations by Plant-Derived Water Extract Ashes

Recently, Nag et al. developed[70] a synthesis of tetraketone and tetrahydrobenzo[*b*]pyran derivatives by a greener and economically synthetic methodology using plant waste which has gained significant interest in recent years. The synthetic methodology of cyclic 1,3-diketone-based tetraketones from cyclic 1,3-diketones and arylaldehydes (Scheme 10.1) and tetrahydrobenzo[*b*]-pyran derivatives by multi-component reaction of cyclic 1,3-diketones, aryl aldehydes, and malononitrile (Scheme 10.2) at room temperature is a simple, new, efficient, fast, and high-yielding protocol.

A greener and feasible catalyst was prepared by water extract of tamarind seed ash (WETSA) and it acted as a basic medium (pH = 9.4). The hard tamarind seeds were separated and washed with distilled water to remove impurities, and then sun-dried for 4 days. Tamarind seeds were ground using mortar and pestle to make the tamarind seed powder. The tamarind seed ash was prepared by burning tamarind seed powder in a muffle furnace at 500°C for 2 h. Then, 1.0 g tamarind seed ash was suspended in 50 mL distilled water in 250 mL of a glass beaker and stirred for some time (30–40 min) at room temperature. Finally, the mixture was filtered and the filtrate was referred to as WETSA. The optimizing reaction condition for the one-pot synthesis of compounds is shown in Table 10.1.

With the optimized reaction conditions of diverse aryl aldehydes and cyclic 1,3-diketones, the synthesis of tetraketone derivatives is shown in Table 10.2.

With encouraging the results, then we extended to explore the scope of the reaction for the synthesis of tetrahydrobenzo[*b*]pyran derivatives (**5a-u**) in the presence of WETSA *via* one-pot reaction among aromatic aldehydes, cyclic 1,3-diketone, and malononitrile (Scheme 7.2), and the same reaction condition has been adopted in Table 10.1 (Entry 7). To synthesis of tetrahydrobenzo[*b*]pyran derivatives, aromatic aldehyde (**1**, 1 mmol), cyclic 1,3-diketone (**2**, 1 mmol), malononitrile (**4**, 1 mmol) were stirred in WETSA:EtOH (2:4) system. All the synthesized compounds are summarized in Table 10.3. From Table 10.3, we concluded that both aromatic aldehydes containing electron-donating as well as electron-withdrawing compounds show good-to-excellent yields. Moreover, all reactions proceeded smoothly, gave higher yields, and took lesser time in the case of 5,5-dimethylcyclohexane-1,3-dione (**2a**) compared to 1,3-cyclohexane dione (**2b**) (Table 10.3).

The plausible mechanistic path for the synthesis of tetraketone and tetrahydrobenzo[*b*]pyran derivatives is given in Scheme 10.3.

10.4 Palladium-Mediated Cross-Coupling Reaction

In the last decades, transition metal-mediated cross-coupling reactions have been synthesized under eco-friendly media, achieved enormous attention in green and sustainable chemistry.[71] The transition-metal-catalyzed cross-coupling reactions, namely, Suzuki–Miyaura,[72] Sonogashira,[73] Heck,[74] Negishi,[75] Stille,[76] Hiyama,[77] Kumada,[78] and Buchwald–Hartwig[79] reactions are very dynamic techniques for the formation of C–C and C–N bonds.[80] Transition-metal catalyzed cross-coupling reactions are presented in Figure 10.3.

Among these energetic conversions, the SMCR has been exploited as the most straightforward methodology which has increased interest both in academia and industry.[80,81] Biaryls are the essential building blocks in several natural products,[82] drug molecules,[81e,83] (Figure 10.4) chiral ligands, and catalysts[84] as well as in engineering materials such as polymers, liquid crystals, and molecular wires.[85] The biaryl core fragments drug

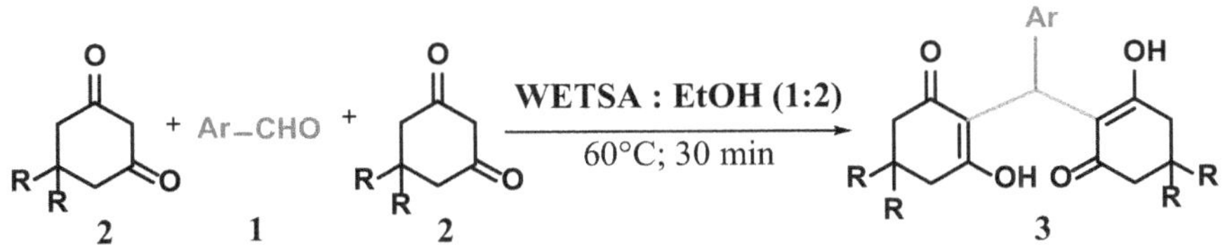

2 + Ar–CHO (1) + 2 → WETSA : EtOH (1:2), 60°C; 30 min → 3

2a, R = Me; 2b, R = H

SCHEME 10.1 Synthesis of tetraketone derivatives in WETSA:EtOH system.

2 + Ar–CHO (1) + NC–CH₂–CN (4) → WETSA : EtOH (1:2), 60°C; 45 - 60 min → 5

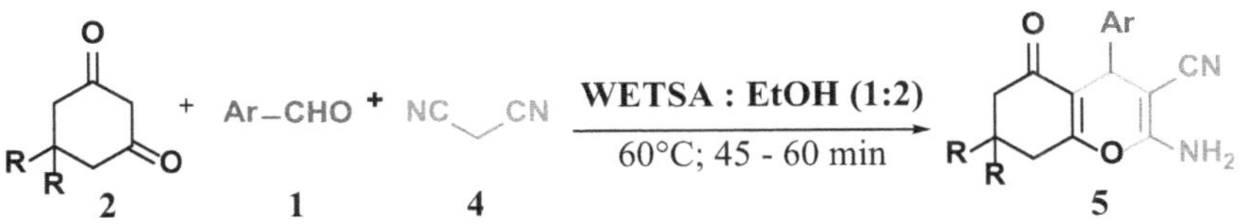

2a, R = Me; 2b, R = H

SCHEME 10.2 Synthesis of tetrahydrobenzo[*b*]pyran derivatives in WETSA:EtOH system.

TABLE 10.1
Optimization of the Reaction Condition for the Synthesis of Tetraketone **3a**[a]

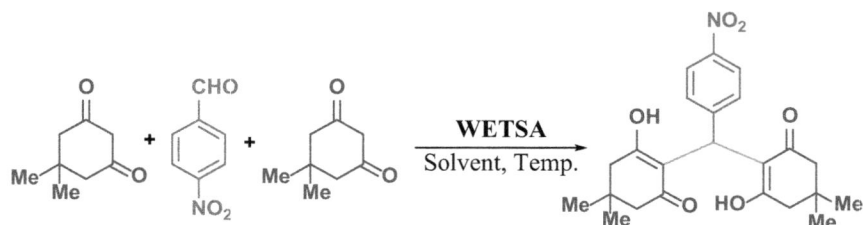

WETSA
Solvent, Temp.

Entry	WETSA-catalyst (mL)	Solvent (mL)	Temp. (°C)	Time (min)	Yield (%)[b]
1	-	H_2O (5)	r.t.	80	40
2	5	-	r.t.	60	66
3	2	DCM (4)	r.t.	60	55
4	2	Toluene (4)	r.t.	60	50
5	2	THF (4)	r.t.	60	48
6	2	MeOH (4)	r.t.	50	80
7	2	EtOH (4)	r.t.	50	93
8	2	EtOH (4)	60	30	95

[a] Reaction conditions: *p*-nitrobenzaldehyde (**1a**, 1 mmol), 5,5-dimethylcyclohexane-1,3-dione (**2a**, 2 mmol) under different conditions
[b] Isolated yields

TABLE 10.2
Substrate Scope for the Synthesis of Diverse Tetraketones Using WETSA[a,b]

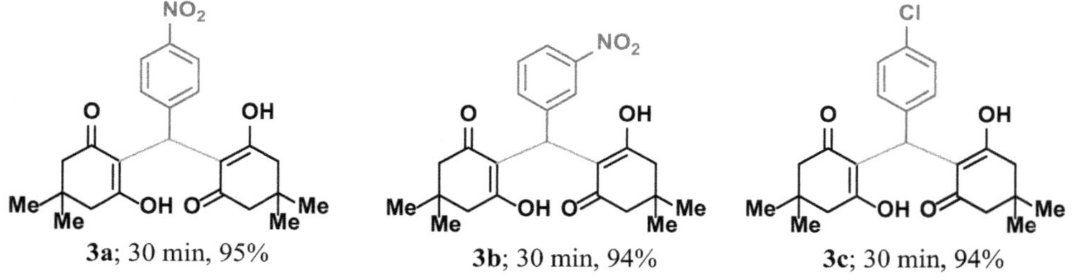

[a] Reaction conditions: aryl aldehyde (**1**, 1 mmol), cyclic 1,3-diketone (**2**, 1 mmol) in WETSA:ethanol (1:2, 6 mL) stirred at 60°C
[b] Isolated yields

molecules are the Bruton's tyrosine kinase (BTK) inhibitor ibrutinib,[86] 3-hydroxy-3-methylglutaryl coenzyme A (HMG-CoA) reductase inhibitor pitavastatin,[87] the Janus kinase (JAK) inhibitor baricitinib,[88] and the third epidermal growth factor receptor (EGFR) inhibitor osimertinib.[89]

The unique structural appearance, i.e., axial chirality, is the key feature of biaryls for the synthesis of chiral ligands, as, for example, BINOL, (*S*)-BINAP, (*R*)-SL-O103-1, (*R*)-BIPHEMP, and (*R*)-SLO106-1.[90] Therefore, biaryl motifs containing some examples of natural products, drugs, and ligands are highlighted in Figure 10.5. Biaryl skeleton systems have received great attention due to several biological and pharmacological properties like anti-inflammatory, antibiotic, antibacterial, anticancer, antifungal, antimicrobial, antitumor, antagonist, antiproliferative, antihypertension, antituberculosis, and analgesic.[81f,85f,91] Some examples of biologically active biaryl derivatives and electron-conducting materials are presented in Figure 10.5.

Palladium-catalyzed SMCR has been considered the most reliable and convenient method for the synthesis of biaryl derivatives which was first discovered by Suzuki and Miyaura in 1981 from aryl boranes and aryl halides.[92] The common mechanism of this reaction involves a catalytic cycle which includes three steps: oxidative addition, transmetalation, and reductive elimination.[72b]

Nag et al. on SMCR reported[unpublished data] a new, simple, and greener method for the synthesis of biaryl varieties **3a-o** with good to high yield from the reaction between varieties of aryl halides (**1a-g**) with several types of arylboronic acids (**2a-d**) using $Pd(OAc)_2$ catalyst in the presence of WEBSA and ethanol acts as a co-solvent and solvent, respectively (Scheme 10.4).

TABLE 10.3

Substrate Scope for the Synthesis of Diverse Tetrahydrobenzo[*b*]pyrans Using WETSA[a,b]

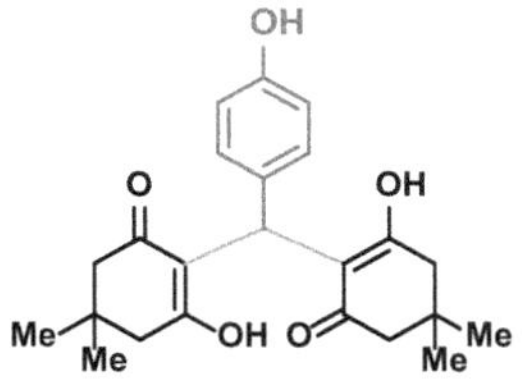

3d; 30 min, 94%

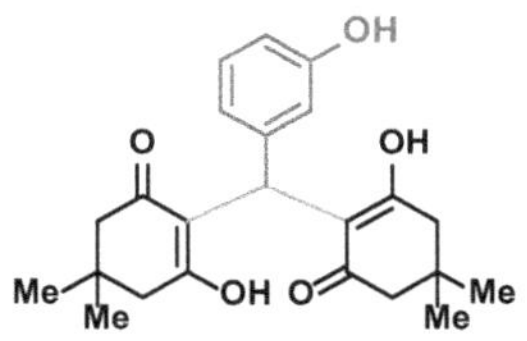

3e; 30 min, 93%

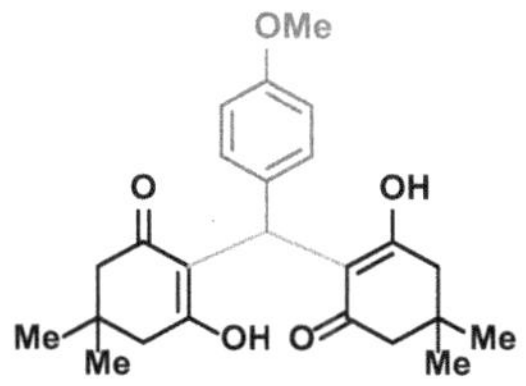

3f; 30 min, 93%

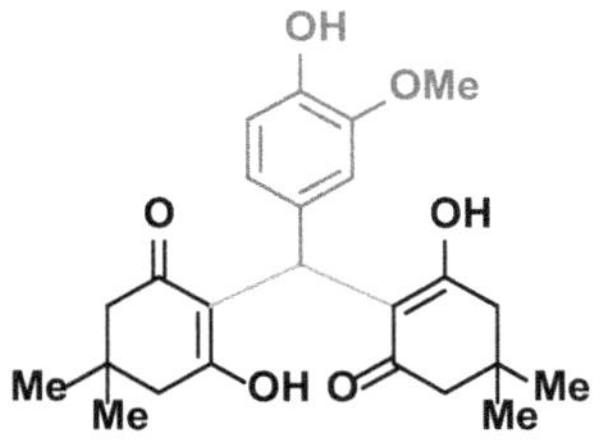

3g; 30 min, 90%

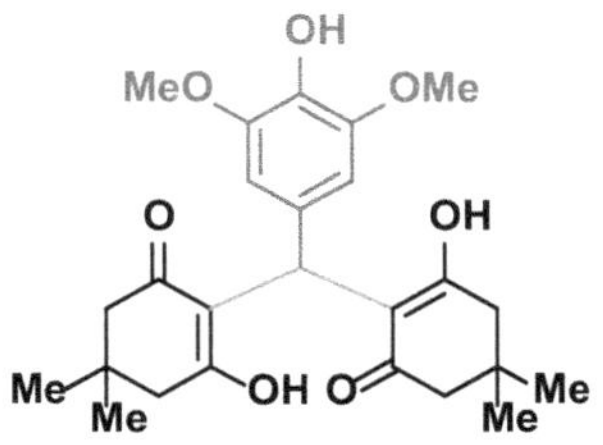

3h; 30 min, 90%

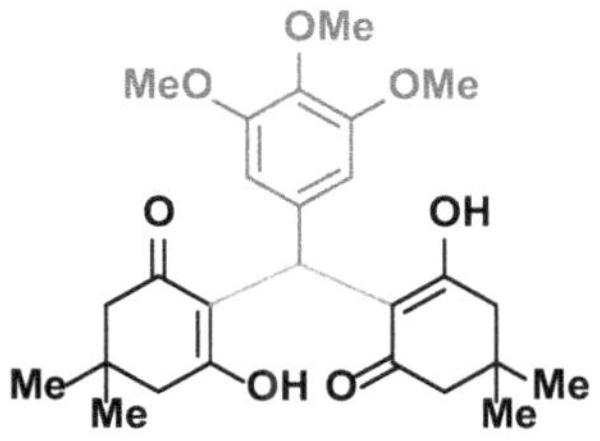

3i; 30 min, 88%

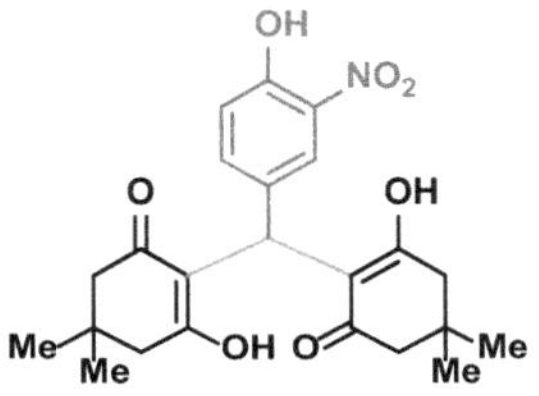

3j; 30 min, 85%

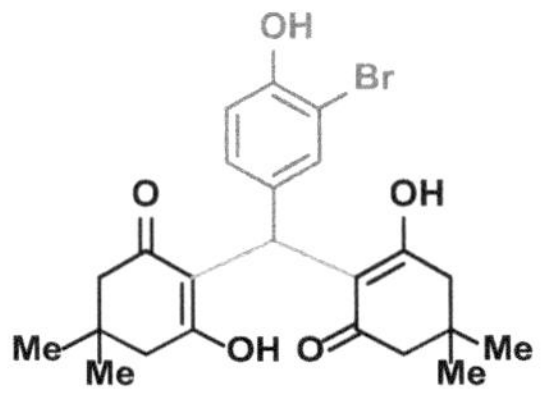

3k; 30 min, 82%

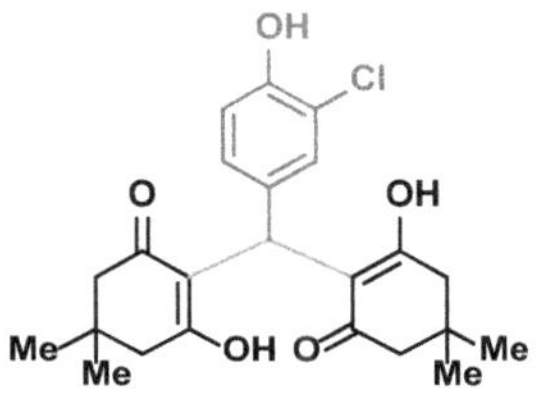

3l; 30 min, 82%

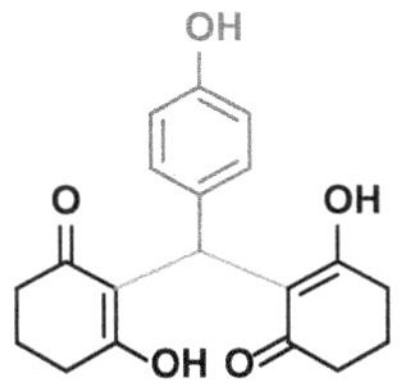

3m; 40 min, 88%

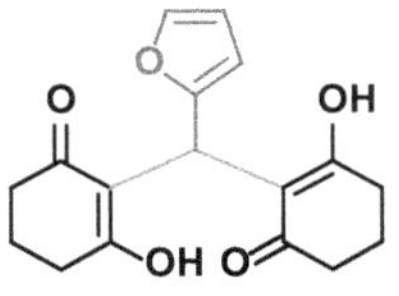

3n; 40 min, 85%

[a] Reaction conditions: aryl aldehyde (**1**, 1 mmol), cyclic 1,3-diketone (**2**, 1 mmol), malononitrile (**4**, 1 mmol) in WETSA:ethanol (1:2, 6 mL) stirred at 60°C

[b] Isolated yields.

The WEBSA acted as a basic medium (pH = 12.4) was prepared by simple filtration of banana stem ash suspended in distilled water. Literature reports revealed that many metal oxides are present in banana stem ash.[93]

The investigation was carried out by reacting aryl halides and arylboronic acids in WEBSA in the absence of any ligands. In order to explore the optimization reaction conditions for the synthesis of biaryl derivative, i.e., 4-methoxy-1,1′-biphenyl (**3c**), the reaction 1-bromo-4-methoxybenzene (**1c**) and phenylboronic acid (**2a**) was picked up as a model reaction at room temperature and the various conditions including solvents, catalyst loading, and time were examined. The results are summarized in Table 10.4. In the presence of WEBSA:MeOH (1:1; 8 mL), the reaction proceeded smoothly which gave 65% yield within 2 h (Table 10.4).

With the optimal conditions in hand, the scope and generality of the present approach, a variety of diverse aryl halides (**1a-g**) and arylboronic acids (**2a-d**) examined under this condition (Table 10.5) was further studied.

It was noticed that substituted aryl bromides gave good-to-excellent yield w.r.t aryl iodide. After completion of the reaction, all biaryl derivatives (**3a-o**) were purified by using column chromatography.

The probable mechanism for the formation of biaryl derivatives (**3-o**) is depicted in Scheme 10.5.[72b,94] From the mechanistic point of view, WEBSA acts as a base. At first, Pd(II) is reduced to Pd(0). Now, this Pd(0) system is ready for oxidative addition reaction in the presence of aryl halide (Ar–X). Due to oxidative addition Pd(0) is again converted to Pd(II) state. Then, the base converts this 'transmetallation-inactive' [Ar–Pd^{II}–X] intermediate to 'transmetallation-active' [Ar–Pd^{II}–B*]. Then, it reacts with arylboronic acid (Ar'B$(OH)_2$) and is followed by transmetallation generates [Ar–Pd^{II}–Ar']. At last by reductive elimination, biaryl derivative Ar–Ar' generates.

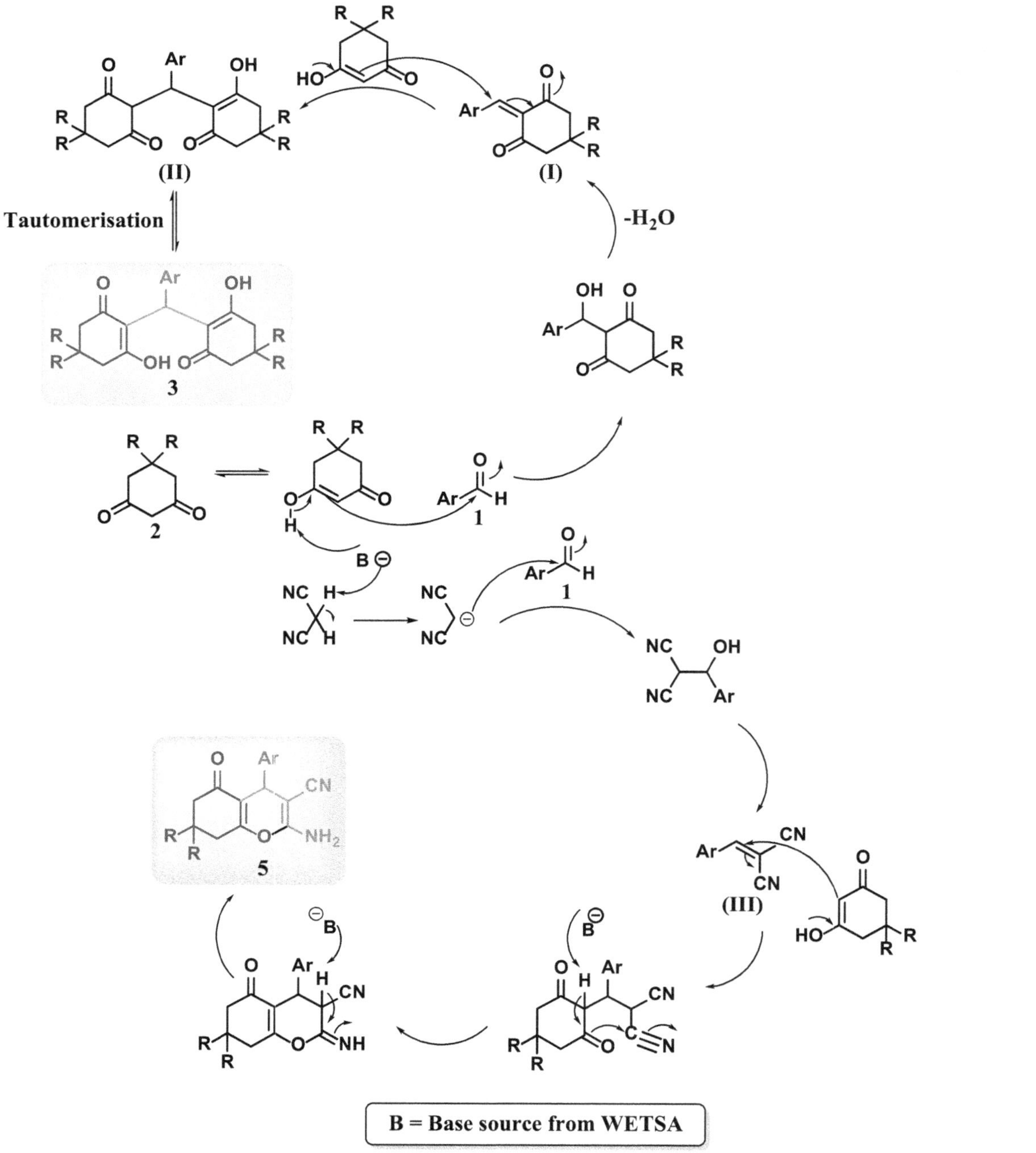

B = Base source from WETSA

SCHEME 10.3 A plausible mechanism with schematic representations for the synthesis of tetraketone and tetrahydrobenzo[*b*]pyran derivatives by WETSA.

10.5 Conclusion

This is a simple, novel, green, and efficient procedure for the synthesis of tetraketone and tetrahydrobenzo[*b*]pyran derivatives through Knoevenagel-Michael-type reaction in the presence of natural and eco-catalyst WETSA in ethanol at room temperature. This protocol gives many advantages such as short reaction time, high yield, easy separation, no column chromatography, clean reaction profile, reusable of the catalyst, and absence of corrosive reagents.

It is also a highly efficient, sustainable, and versatile protocol for the synthesis of a variety of biaryl derivatives using diverse aryl halides and arylboronic acids at room temperature in the presence of WEBSA as a natural feedstock extract, eco-catalyst, and renewable from agro-industrial waste bio-mass. The most attractive characteristics of this protocol are the operational simplicity, ease of catalyst preparation, clean reaction profiles, reuse of catalyst, broad substrate scopes, good yield of products, short reaction time, and environmentally benignity. From the environmental and economic point of view, this method greatly enhances the synthetic potency of biaryl derivatives and allows a new dimension in SMCRs.

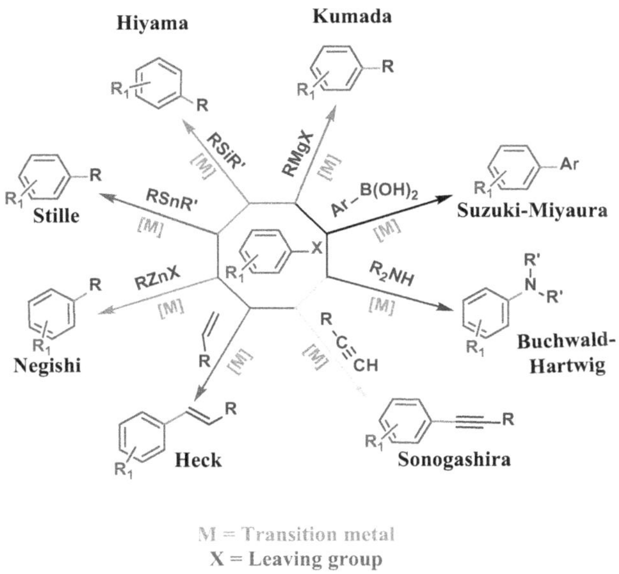

M = Transition metal
X = Leaving group

FIGURE 10.3 Representation of transition-metal-catalyzed cross-coupling reactions.

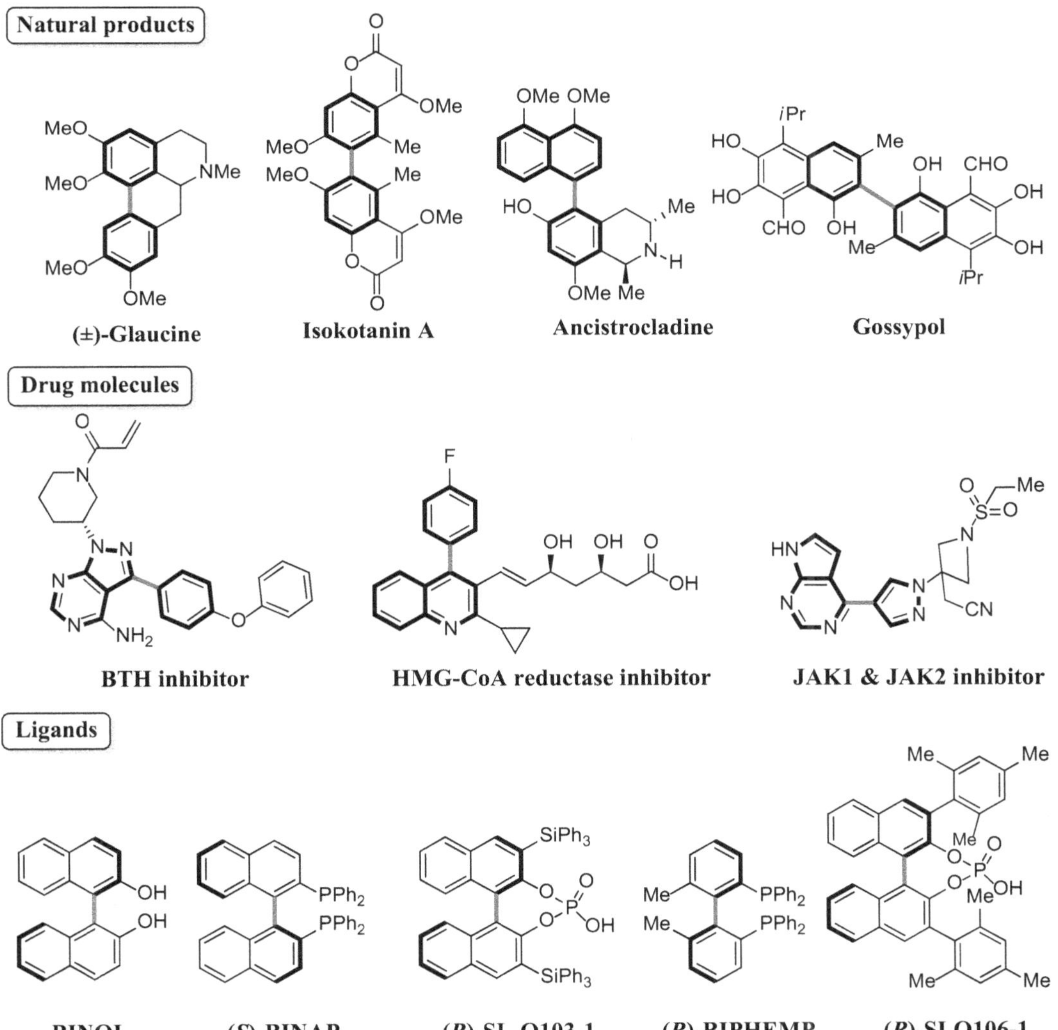

FIGURE 10.4 Some examples of natural products, drugs, and chiral ligands.

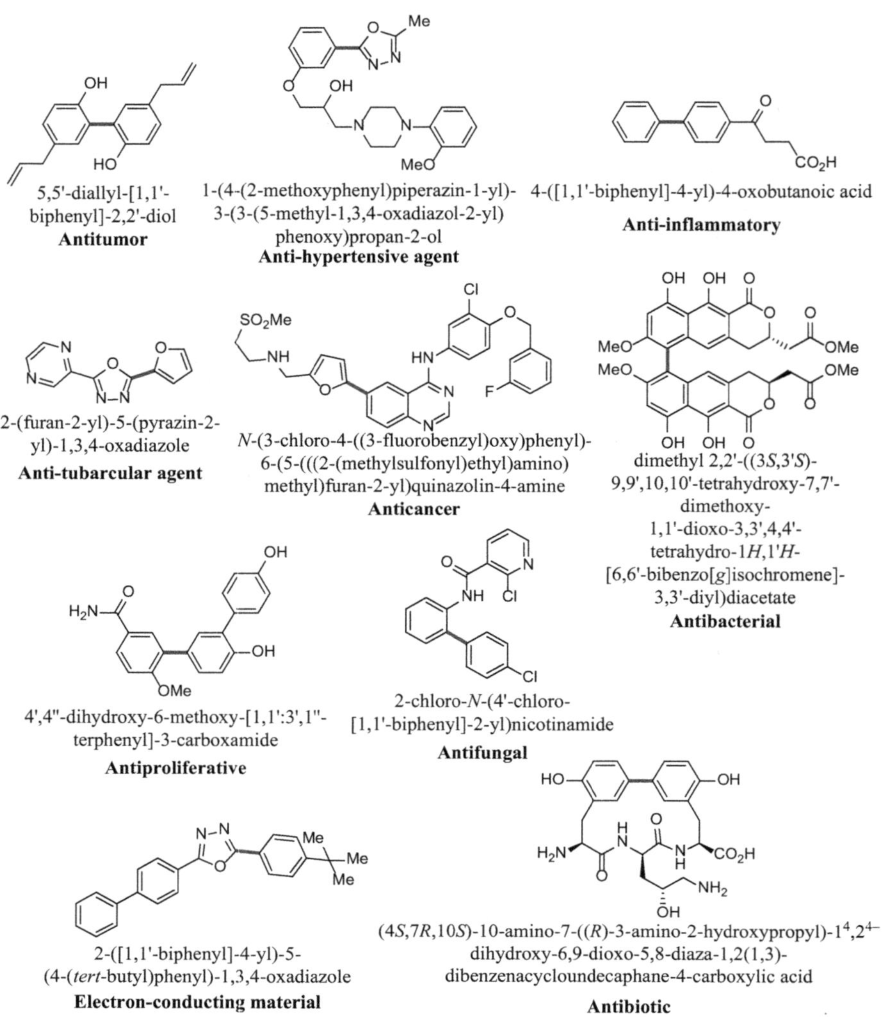

FIGURE 10.5 Illustration of some biaryl-based biologically active compounds and electron-conducting material.

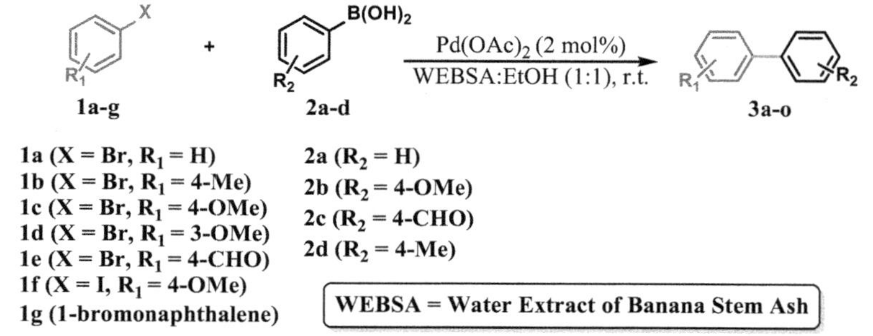

$Pd(OAc)_2$ (2 mol%)
WEBSA:EtOH (1:1), r.t.

1a-g + **2a-d** → **3a-o**

1a (X = Br, R_1 = H)
1b (X = Br, R_1 = 4-Me)
1c (X = Br, R_1 = 4-OMe)
1d (X = Br, R_1 = 3-OMe)
1e (X = Br, R_1 = 4-CHO)
1f (X = I, R_1 = 4-OMe)
1g (1-bromonaphthalene)

2a (R_2 = H)
2b (R_2 = 4-OMe)
2c (R_2 = 4-CHO)
2d (R_2 = 4-Me)

WEBSA = Water Extract of Banana Stem Ash

SCHEME 10.4 Suzuki–Miyaura coupling reaction in WEBSA:EtOH system.

TABLE 10.4

Screening of Catalyst Amount, WEBSA, Solvent and Time for the Synthesis of Compound (**3c**)[a]

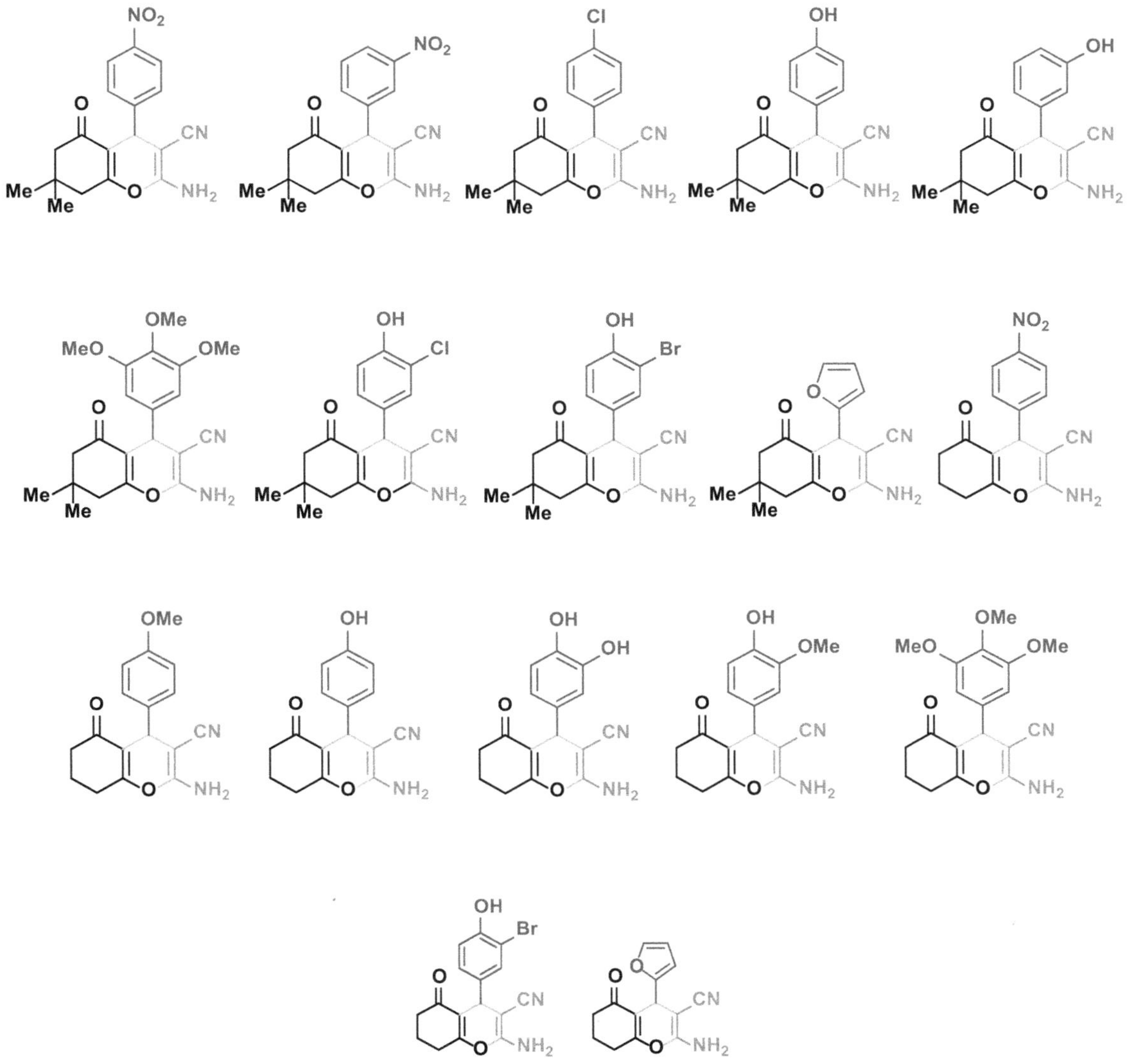

Entry	Pd Catalyst (mol%)	WEBSA (mL)	Solvent (mL)	Time (h)	Yield (%)[b]
1	$PdCl_2$ (1)	4	-	2	10
2	$PdCl_2$ (1)	4	-	2	10
3	$PdCl_2$ (1)	4	EtOH (4)	2	15
4	$Pd(PPh_3)_4$ (1)	4	-	2	-
5	$Pd(OAc)_2$ (1)	5	-	1	60
6	$Pd(OAc)_2$ (1)	4	MeOH (4)	2	65
7	$Pd(OAc)_2$ (2)	4	EtOH (4)	2	88
8	$Pd(OAc)_2$ (1)	5	EtOH (5)	3	80
9	$Pd(OAc)_2$ (0.5)	2	EtOH (2)	3	35
10	$Pd(OAc)_2$ (0.5)	2	EtOH (2)	4	35
11	$Pd(OAc)_2$(0.5)	2	EtOH (2)	3	30
12	$Pd(OAc)_2$(0.5)	WETSA (2)	EtOH (2)	3	18

[a] Reaction conditions: 1-bromo-4-methoxybenzene (**1c**, 1 mmol) and phenylboronic acid (**2a**, 1.2 mmol).

[b] Isolated yield.

TABLE 10.5
Substrate Scope for the Synthesis of Biaryl Derivatives[a]

Entry	R_1	X	R_2	Product	Time (h)	Yield (%)[b]
1	H	Br	H	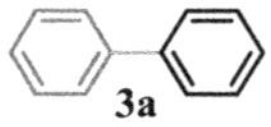 3a **3a**	2	90
2	4-Me	Br	H	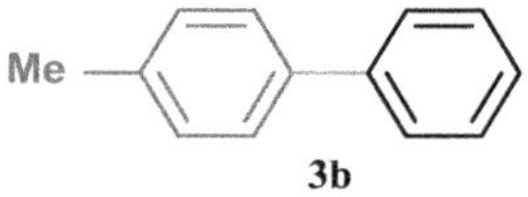 Me 3b **3b**	2	88
3	4-OMe	Br	H	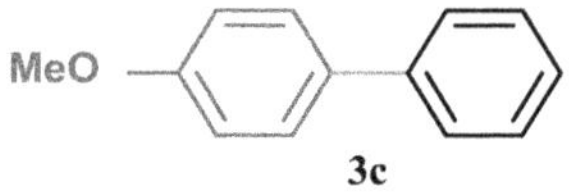 MeO 3c **3c**	2	88
4	4-Me	Br	4-OMe	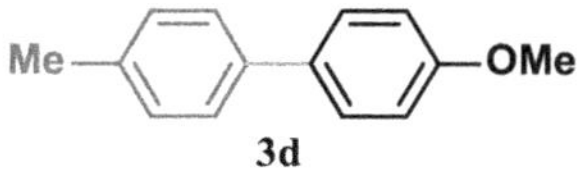 Me, OMe 3d **3d**	2	85
5	4-OMe	Br	4-OMe	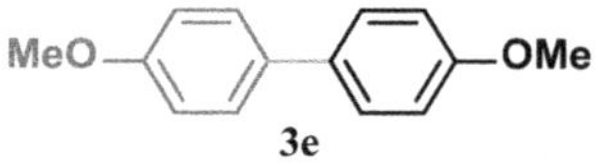 MeO, OMe 3e **3e**	2	85
6	4-OMe	I	H	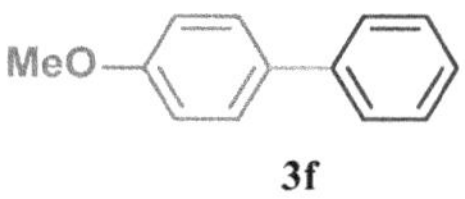 MeO 3f **3f**	3	82
7	3-OMe	Br	H	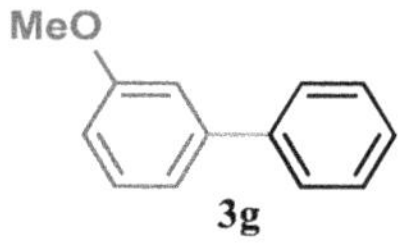 MeO 3g **3g**	2.5	82
8	3-OMe	Br	4-OMe	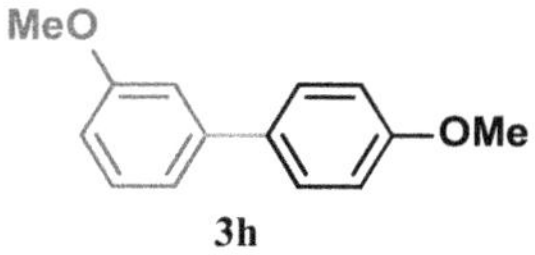 MeO, OMe 3h **3h**	2.5	80
9	4-CHO	Br	H	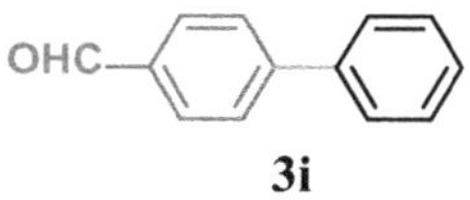 OHC 3i **3i**	3	80
10	4-CHO	Br	4-CHO	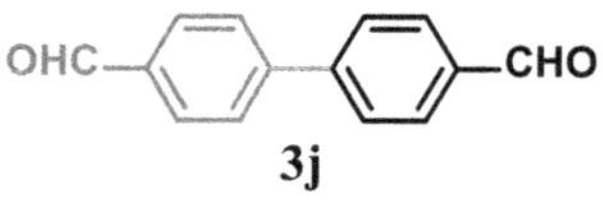 OHC, CHO 3j **3j**	3	80
11	4-OMe	Br	4-CHO	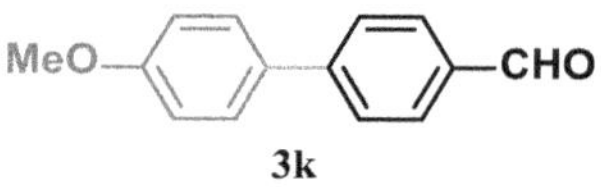 MeO, CHO 3k **3k**	3	78

Entry	R_1	X	R_2	Product	Time (h)	Yield (%)[b]
12	H	Br	4-CHO	3l	3	80
13	H	Br	4-OMe	3m	2	88
14	4-OMe	Br	4-Me	3n	2	84
15		Br	H	3o	3	75

[a] Reaction conditions: Aryl halide (1 mmol), arylboronic acid (1.2 mmol), $Pd(OAc)_2$ (2 mol%), WEBSA (4 mL), EtOH (4 mL) at room temperature.
[b] Isolated yield.

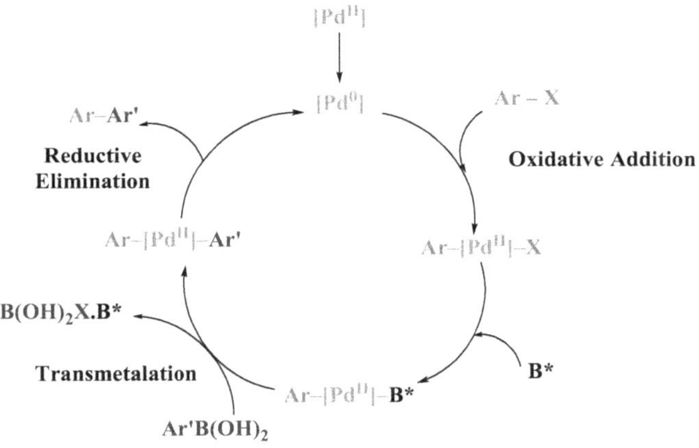

Ar – X = Aryl halide; $Ar'B(OH)_2$ = Arylboronic acid; B* = Base from WEBSA

SCHEME 10.5 A plausible mechanism with schematic representation for the synthesis of biaryl derivatives by WEBSA.

REFERENCES

1. Mandlimath, T. R.; Umamahesh, B.; Sathiyanarayanan, K. I. J. *Mol. Catal. A: Chem.* 2014, 391, 198–207.
2. (a) Vaid. R.; Gupta. M.; Kant. R.; Gupta, V. K. *J. Chem. Sci.* 2016, 128, 967–976. (b) Maharvi, G. M.; Ali, S.; Riaz, N.; Afza, N.; Malik, A.; Ashraf, M.; Iqbal, L.; Lateef, M. J. *Enzyme Inhib. Med. Chem.* 2008, 23, 62–69. (c) Khan, K. M.; Maharvi, G. M.; Khan, M. T. H.; Shaikh, A. J.; Perveen, S.; Begum, S.; Choudhary, M. I. *Bioorg. Med. Chem.* 2006, 14, 344–351. (d) Ali, S.; Maharvi, G. M.; Riaz, N.; Afza, N.; Malik, A.; Rehman, A. U.; Lateef, M.; Iqbal, L. *West Indian Med. J.* 2009, 58, 92–98. (e) Arora, S.; Joshi, G.; Kalra, S.; Wani, A. A.; Bharatam, P. V.; Kumar, P.; Kumar, R. *ACS Omega* 2019, 4, 4604–4614.
3. Agarwal, S.; Poddar, R.; Kidwai, M.; Nath, M. *ChemistrySelect* 2018, 3, 10909–10914.
4. (a) Timpe, H.-J.; Ulrich, S.; Fouassier, J.-P. *J. Photochem. Photobiol. A: Chem.* 1993, 73, 139–150. (b) Timpe, H.-J.; Ulrich, S.; Ali, S. *J. Photochem. Photobiol. A: Chem.* 1991, 61, 77–89. (c) Ulrich, S.; Timpe, H.-J.; Fouassier, J.-P.; Morlet-Savary, F. *J. Photochem. Photobiol. A: Chem.* 1993, 74, 165–170. (d) Timpe, H.-J.; Ulrich, S.; Decker, C.; Fouassier, J. P. *Macromolecules* 1993, 26, 4560–4566.
5. (a) Lambert, R. W.; Martin, J. A.; Merrett, J. H.; Parkes, E. B. K.; Thomas, G. *J. PCT Int Appl* WO9706178, 1997. (b) Carlin, G.; Djursäter, R.; Smedegård, G.; Gerdin, B. *Agents Actions* 1985, 16, 377–384.
6. Pohlers, G.; Scaiano, J. C.; Sinta, R. *Chem Mate.* 1997, 9, 3222–3230.

7. (a) Kiyani, H. Curr. Org. Synth. 2018, 15, 1043–1072. (b) Bonsignore, L.; Loy, G.; Secci, D.; Calignano, A. *Eur. J. Med. Chem.* 1993, 28, 517–520. (c) Kiyani, H.; Ghorbani, F. J. *Saudi Chem. Soc.* 2014, 18, 689–701. (d) Huynh, T. H. V.; Abrahamsen, B.; Madsen, K. K.; Gonzalez-Franquesa, A.; Jensen, A. A.; Bunch, L. *Bioorg. Med. Chem.* 2012, 12, 6831–6839.
8. (a) Abdel-Galil, F. M.; Riad, B. Y.; Sherif, S. M.; Elnagdi, M. H. *Chem. Lett.* 1982, 11, 1123–1126. (b) Hafez, E. A. A.; Elnagdi, M. H.; Elagamey, A. G. A.; El-Taweel, F. M. A. A. *Heterocycles* 1987, 26, 903–907. (c) Reynolds, G. A.; Drexhage, K. H. *Opt. Commun.* 1975, 13, 222–225. (d) Ellis, G. P. *In the Chemistry of Heterocyclic of Compounds. Chromenes, Harmones and Chromones*, John Wiley & Sons, New York, NY, 1977, chapter II, p. 1113. (e) Zollinger, H. *Color Chemistry*, Verlag Helvetica Chimica Acta: Zurikh and Wiley-VCH, Weinheim, 3rd edn., 2003. (f) Bissell, E. R.; Mitchell, A. R.; Smith, R. E. J. *Org. Chem.* 1980, 45, 2283–2287.
9. Sharma, H.; Srivastava, S. *New J. Chem.* 2019, 43, 12054–12058.
10. Rajput, J. K.; Kaur, G. *Catal. Sci. Technol.* 2014, 4, 142–151.
11. Khan, K. M.; Maharvi, G. M.; Khan, M. T. H.; Shaikh, A. J.; Perveen, S.; Begum, S.; Choudhary, M. I. *Bioorg. Med. Chem.* 2006, 14, 344–351.
12. Al-Majid, A. M.; Islam, M. S.; Barakat, A.; Al-Qahtani, N. J.; Yousuf, S.; Choudhary, M. I. *Arabian J. Chem.* 2017, 10, 185–193.
13. Jung, D. H.; Lee, Y. R.; Kim, S. H.; Lyoo, W. S. *Bull. Korean Chem. Soc.* 2009, 30, 1989–1995.
14. Maghsoodlou, M. T.; Habibi-Khorassani, S. M.; Shahkarami, Z.; Maleki, N.; Rostamizadeh, M. *Chin. Chem. Lett.* 2010, 21, 686–689.
15. Vaid, R.; Gupta, M.; Kant, R.; Gupta, V. K. *J. Chem. Sci.* 2016, 128, 967–976.
16. Kidwai, M.; Bansal, V.; Mothsra, P.; Saxena, S.; Somvanshi, R. K.; Dey, S.; Singh, T. P. J. *Mol. Catal. A: Chem.* 2007, 268, 76–81.
17. Magyar, Á.; Hell, Z. *Monatsh. Chem.* 2019, 150, 2021–2023.
18. Hekmatshoar, R.; Kargar, M.; Mostashari, A.; Hashemi, Z.; Goli, F.; Mousavizadeh, F. J. *Turkish Chem. Soc. Sect. Chem.* 2015, 2, 1–11.
19. Kantevari, S.; Bantu, R.; Nagarapu, L. J. *Mol. Catal. A: Chem.* 2007, 269, 53–57.
20. Ilangovan, A.; Muralidharan, S.; Sakthivel, P.; Malayappasamy, S.; Karuppusamy, S.; Kaushik, M. P. *Tetrahedron Lett.* 2013, 54, 491–494.
21. Agarwal, S.; Poddar, R.; Kidwai, M.; Nath, M. *ChemistrySelect* 2018, 3, 10909–10914.
22. Yan, Z.; Cuizhi, S.; Jun, L.; Zhicai, S. *Chin. J. Chem.* 2010, 28, 2255–2259.
23. Shirini, F.; Daneshvar, N. *RSC Adv.* 2016, 6, 110190–110205.
24. Heravi, M. R. P.; Piri, S. *J. Chem.* 2013, 1–5.
25. Zhang, Y.; Shang, Z. *Chin. J. Chem.* 2010, 28, 1184–1188.
26. Ilangovan, A.; Malayappasamy, S.; Muralidharan, S.; Maruthamuthu, S. *Chem. Cent. J.* 2011, 5, 81.
27. Saha, M.; Pal, A. K.; Nandi, S. *RSC Adv.* 2012, 2, 6397–6400.
28. Gupta, M.; Gupta, M. J. *Chem. Sci.* 2016, 128, 849–854.
29. Rahmani, S.; Zeynizadeh, B. *Res. Chem. Int.* 2019, 45, 1227–1248.
30. Gilanizadeh, M.; Zeynizadeh, B. *New J. Chem.* 2018, 42, 8553–8566.
31. Tajbakhsh, M.; Heidary, M.; Hossseinzadeh, R. *Res. Chem. Intermed.* 2015, 42, 1425–1439.
32. Rao, V. K.; Kumar, M. M.; Kumar, A. *Indian J. Chem.* 2011, 50B, 1128–1135.
33. Khurana, J.; Vij, K. *J. Chem. Sci.* 2012, 124, 907–912.
34. Nemati, F.; Heravi, M. M.; Rad, R. S. *Chin. J. Catal.* 2012, 33, 1825–1831.
35. Rastroshan, M.; Sayyahi, S.; Zare-Shahabadi, V.; Badri, R. J. *Iranian Chem. Res.* 2012, 5, 265–269.
36. Li-Bin, L.; Tong-Shou, J.; Li-Sha, H.; Meng, L.; Na, Q.; Tong-Shuang, L. *J. Chem.* 2006, 3, 117–121.
37. Ray, S.; Bhaumik, A.; Pramanik, M.; Butcher, R. J.; Yildirim, S. O.; Mukhopadhyay, C. *Catal. Commun.* 2014, 43, 173–178.
38. Ashtarian, J.; Heydari, R.; Maghsoodlou, M.-T.; Yazdani-Elah-Abadi, A. *Rev. Roum. Chim.* 2019, 64, 259–264.
39. Mollashahi, E.; Nikrafter, M. J. *Saudi Chem. Soc.* 2018, 22, 42–48.
40. Mashhadinezhad, M.; Mamaghani, M.; Rassa, M.; Shirini, F. *ChemistrySelect* 2019, 4, 4920–4932.
41. Rakhtshah, J.; Salehzadeh, S.; Zolfigol, M. A.; Baghery, S. *Appl. Organometal. Chem.* 2017, 31, e3690.
42. Tabrizian, E.; Amoozadeh, A. *Catal. Sci. Technol.* 2016, 6, 6267–6276.
43. Zolfigol, M. A.; Khazaei, A.; Moosavi-Zare, A. R.; Afsar, J.; Khakyzadeh, V.; Khaledian, O. *J. Chin. Chem. Soc.* 2015, 62, 398–403.
44. Khazaei, A.; Gholami, F.; Khakyzadeh, V.; Moosavi-Zare, A. R.; Afsar, J. *RSC Adv.* 2015, 5, 14305–14310.
45. Qareaghaj, O. H.; Mashkouri, S.; Naimi-Jamal, M. R.; Kaupp, G. *RSC Adv.* 2014, 4, 48191–48201.
46. Khazaei, A.; Nik, H. A. A.; Moosavi-Zare, A. R.; Afshar-Hezarkhani, H. Z. *Naturforsch.* 2018, 73, 707–712.
47. Bhosale, R. S.; Magar, C. V.; Solanke, K. S.; Mane, S. B.; Choudhary, S. S.; Pawar, R. P. *Synth. Commun.* 2007, 37, 4353–4357.
48. Chen, L.; Li, Y.-Q.; Huang, X.-J.; Zheng, W.-J. *Heteroatom Chem.* 2009, 20, 91–94.
49. Khan, A. T.; Lal, M.; Ali, S.; Khan, M. M. *Tetrahedron Lett.* 2011, 15, 5327–5332.
50. Dekamin, M. G.; Eslami, M.; Maleki, A. *Tetrahedron* 2013, 69, 1074–1085.
51. Seifi, M.; Sheibani, H. *Catal. Lett.* 2008, 126, 275–279.
52. Krishnan, K. K.; Dabholkar, V. V.; Gopinathan, A.; Jaiswar, R. J. *Chem. & Cheml. Sci.* 2018, 8, 66–74.
53. Khodaei, M. M.; Bahrami, K.; Farrokhi, A. *Synth. Commun.* 2010, 40, 1492–1499.
54. Moosavi-Zare, A. R.; Zolfigol, M. A.; Khaledian, O.; Khakyzadeh, V.; Farahani, M. D.; Kruger, H. G. *New J. Chem.* 2014, 38, 2342–2347.
55. Beheshtiha, S. Y.; Oskooie, H. A.; Pourebrahimi, F. S.; Zadsirjan, V. *Chem. Sci. Trans.* 2015, 4, 689–693.
56. Baghbanian, S. M.; Rezaei, N.; Tashakkorian, H. *Green Chem.* 2013, 15, 3446–3458.
57. Hatamjafari, F. *J. Chem. Health Risks* 2016, 6, 133–142.
58. Jin, T.-S.; Wang, A.-Q.; Shi, F.; Han, L.-S.; Liu, L.-B.; Li, T.-S. *Arkivoc* 2006, *xiv*, 78–86.
59. Sheikhhosseini, E.; Ghazanfari, D.; Nezamabadi, V. *Iran. J. Catal.* 2013, 3, 197–201.
60. Khaksar, S.; Rouhollahpour, A.; Talesh, S. M. *J. Fluorine Chem.* 2012, 141, 11–15.

61. Brahmachari, G.; Banerjee, B. *ACS Sustainable Chem. Eng.* 2014, 2, 411–422.
62. Zolfigol, M. A.; Bahrami-Nejad, N.; Afsharnadery, F.; Baghery, S. *J. Mol. Liq.* 2016, 221, 851–859.
63. Rahnamafa, R.; Moradi, L.; Khoobi, M. *Res. Chem. Intermed.* 2020, 46, 2109–2116.
64. (a) Kazemi-Red, R.; Azizian, J.; Kefayati, H. *J. Chin. Chem. Soc.* 2015, 62, 311–315. (b) Fotouhi, L.; Heravi, M. M.; Fatehi, A.; Bakhtiari, K. *Tetrahedron Lett.* 2007, 48, 5379–5381.
65. (a) Kumar, D.; Sandhu, J. S. *Synth. Commun.* 2010, 40, 510–517. (b) Ashry, E. S. H. E.; Awada, L. F.; Ibrahim, E. S. I.; Bdeewya, O. K. *Arkivoc* 2006, *ii*, 178–186. (c) El-Rahman, N. M. A.; El-Kateb, A. A.; Mady, M. F. *Synth. Commun.* 2007, 37, 3961–3970. (d) Pagore, V. P.; Tekale, S. U.; Jadhav, V. B.; Pawar, R. P. *Iran. J. Catal.* 2016, 6, 189–192. (e) Tu, S.-J.; Gao, Y.; Guo, C.; Shi, D.; Lu, Z. *Synth Commun.* 2002, 32, 2137–2141.
66. (a) Jin, T.-S.; Zhang, J.-S.; Wang, A.-Q.; Li, T.-S. *Synth. Commun.* 2005, 35, 2339–2345. (b) Jin, T.-S.; Wang, A.-Q.; Ma, H.; Zhang, J.-S.; Li, T.-S. *Indian J. Chem.* 2006, 45*B*, 470–474. (c) Guo, S.-B.; Wang, S.-X.; Li, J.-T. *Synth. Commun.* 2007, 37, 2111–2120. (d) Lian, X.-Z.; Huang, Y.; Li, Y.-Q.; Zheng, W.-J. *Monatsh. Chem.* 2008, 139, 129–131. (e) Gurumurthi, S.; Sundari, V.; Valliappan, R. *J. Chem.* 2009, 6, S466–S472. (f) Heydari, R.; Rahimi, R.; Kangani, M.; Yazdani-Elah-Abadi, A.; Lashkari, M. *Acta Chemica Iasi* 2017, 25, 163–178.
67. (a) Li, J.-T.; Li, Y.-W.; Song, Y.-L.; Chen, G.-F. *Ultrason. Sonochem.* 2012, 19, 1–4. (b) Esmaeilpour, M.; Javidi, J.; Dehghani, F.; Dodeji, F. N. *RSC Adv.* 2015, 5, 26625–26633. (c) Li, J.-T.; Xu, W.-Z.; Yang, L.-C.; Li, T.-S. *Synth. Commun.* 2004, 34, 4565–4571. (d) Azarifar, D.; Khatami, S.-M.; Zolfigol, M. A.; Nejat-Yami, R. *J. Iran. Chem. Soc.* 2014, 11, 1223–1230.
68. Dekamin, M. G.; Eslami, M. *Green Chem.* 2014, 16, 4914–4921.
69. Tiwari, J.; Saquib, M.; Singh, S.; Tufail, F.; Singh, M.; Singh, J. *J. Green Chem.* 2016, 18, 3221–3231.
70. Halder, B.; Maity, H. S.; Banerjee, F.; Kachave, A.; Nag, A. *Polycycl. Aromat. Compd.* 2020, doi:10.1080/10406638.2020.1858885.
71. (a) Hooshmand, S. S.; Heidari, B.; Sedghi, R.; Varma, R. S. *Green Chem.* 2019, 21, 381–405. (b) Sarmah, M.; Dewan, A.; Mondal, M.; Thakur, A. J.; Bora, U. *RSC Adv.* 2016, 6, 28981–28985. (c) Jana, R.; Pathak, T. P.; Sigman, M. S. *Chem. Rev.* 2011, 111, 1417–1492.
72. (a) Suzuki, A. *Pure Appl. Chem.* 1985, 57, 1749–1758. (b) Miyaura, N.; Suzuki, A. *Chem. Rev.* 1995, 95, 2457–2483.
73. Sonogashira, K. *Contemporary Organic Synthesis.* Pergamon Press, New York, 1991, vol. 3, p. 521.
74. Heck, R. F. *Org. React.* 2004, 27, 345–390.
75. (a) Negishi, E. *Acc. Chem. Res.* 1982, 15, 340–348. (b) Erdik, E. *Tetrahedron* 1992, 48, 9577–9648. (c) Negishi, E. I.; Liu, F. *Metal-catalyzed Cross-Coupling Reactions*, Wiley-VCH, Weinheim, Germany, 1998, pp. 1–47.
76. (a) Milstein, D.; Stille, J. K. *J. Am. Chem. Soc.* 1979, 101, 4981–4991. (b) Milstein, D.; Stille, J. K. *J. Am. Chem. Soc.* 1979, 101, 4992–4998. (c) Milstein, D.; Stille, J. K. *J. Org. Chem.* 1979, 44, 1613–1618.
77. Nakao, Y.; Hiyama, T. *Chem. Soc. Rev.* 2011, 40, 4893–4901.
78. Tamao, K.; Sumitani, K.; Kumada, M. *J. Am. Chem. Soc.* 1972, 94, 4374–4376.
79. (a) Yang, B. H.; Buchwald, S. L. *J. Organomet. Chem.* 1999, 576, 125–146. (b) Tewari, A.; Hein, M.; Zapf, A.; Beller, M. *Tetrahedron* 2005, 61, 9705–9709.
80. (a) Han, F.-S. *Chem. Soc. Rev.* 2013, 42, 5270–5298. (b) Len, C.; Bruniaux, S.; Delbecq, F.; Parmar, V. S. *Catalysts* 2017, 7, 146–168.
81. (a) Molander, G. A.; Trice, S. L. J.; Kennedy, S. M. *J. Org. Chem.* 2012, 77, 8678–8688. (b) Suzuki, A. *Angew. Chem. Int. Ed.* 2011, 50, 6722–6737. (c) Yet, L. *Privileged Structures in Drug Discovery: Medicinal Chemistry and Synthesis*, Wiley, 2018, pp. 83–135. (d) Liu, C.; Ji, C.-L.; Qin, Z.-X.; Hong, X.; Szostak, M. *iScience* 2019, 19, 749–759. (e) Brown, D. G.; Boström, J. *J. Med. Chem.* 2016, 59, 4443–4458. (f) González, J.; Dijk, L. V.; Goetzke, F. W.; Fletcher, S. P. *Nat. Protoc.* 2019, 14, 2972–2985.
82. (a) Kantham, S.; Chan, S.; McColl, G.; Miles, J. A.; Veliyath, S. K.; Deora, G. S.; Dighe, S. N.; Khabbazi, S.; Parat, M.-O.; Ross, B. P. *ACS Chem. Neurosci.* 2017, 8, 1901–1912. (b) Yadav, D. K.; Bharitkar, Y. P.; Hazra, A.; Pal, U.; Verma, S.; Jana, S.; Singh, U. P.; Maiti, N. C.; Mondal, N. B.; Swarnakar, S. J. *Nat. Prod.* 2017, 80, 1347–1353. (c) Bringmann, G.; Gulder, T.; Gulder, T. A. M.; Breuning, M. *Chem. Rev.* 2011, 111, 563–639. (d) Wencel-Delord, J.; Panossian, A.; Leroux, F. R.; Colobert, F. *Chem. Soc. Rev.* 2015, 44, 3418–3430. (e) Ashenhurst, J. A. *Chem. Soc. Rev.* 2010, 39, 540–548. (f) Bringmann, G.; Walter, R.; Weirich, R. *Angew. Chem. Int. Ed. Engl.* 1990, 29, 977–991. (g) Kozlowski, M. C.; Morgan, B. J.; Linton, E. C. *Chem. Soc. Rev.* 2009, 38, 3193–3207.
83. (a) Welsch, M. E.; Snyder, S. A.; Stockwell, B. R. *Curr. Opin. Chem. Biol.* 2010, 14, 347–361. (b) Khalid, S.; Zahid, M. A.; Ali, H.; Kim, Y. S.; Khan, S. *BMC Neurosci.* 2018, 19, 74–84. (c) Wu, P.; Nielsen, T. E.; Clausen, M. H. *Drug Discov. Today* 2016, 21, 5–10.
84. Kaye, S.; Fox, J. M.; Hicks, F. A.; Buchwald, S. L. *Adv. Synth. Catal.* 2001, 343, 789–794.
85. (a) Oriol, L.; Piñol, M.; Serrano, J. L.; Martínez, C.; Alcalá, R.; Cases, R.; Sánchez, C. *Polymer* 2001, 42, 2737–2744. (b) Schluter, A. D. *J. Polym. Sci. Part A: Polym. Chem.* 2001, 39, 1533–1556. (c) Kertesz, M.; Choi, C. H.; Yang, S. *Chem. Rev.* 2005, 105, 3448–3481. (d) Leclerc, N.; Sanaur, S.; Galmiche, L.; Mathevet, F.; Attias, A.-J.; Fave, J.-L.; Roussel, J.; Hapiot, P.; Lemaître, N.; Geffroy, B. *Chem. Mater.* 2005, 17, 502–513. (e) Yamamura, K.; Ono, S.; Tabushi, I. *Tetrahedron Lett.* 1988, 29, 1797–1798. (f) Simonetti, M.; Cannas D. M.; Larrosa, I. *Advances in Organometallic Chemistry*, 2017, 67, 299–399.
86. Zhan, M.; Deng, Y.; Zhao, L.; Yan, G.; Wang, F.; Tian, Y.; Zhang, L.; Jiang, H.; Chen, Y. *J. Med. Chem.* 2017, 60, 4023–4035.
87. Aberg, J. A.; Sponseller, C. A.; Ward, D. J.; Kryzhanovski, V. A.; Campbell, S. E.; Thompson, M. A. *Lancet HIV* 2017, 4, 284–294.
88. Flick, A. C.; Ding, H. X.; Leverett, C. A.; Kyne, R. E.; Liu, K. K.; Fink, S. J.; O'Donnell, C. J. *J. Med. Chem.* 2017, 60, 6480–6515.
89. (a) Johnson, C. N.; Erlanson, D. A.; Murray, C. W.; Rees, D. C. *J. Med. Chem.* 2017, 60, 89–99. (b) Li, Z. R.; Suo, F. Z.; Hu, B.; Guo, Y. J.; Fu, D. J.; Yu, B.; Zheng, Y. C.; Liu, H. M. *Bioorg. Chem.* 2019, 84, 164–169.

90. (a) Uemura, T.; Zhang, X. Y.; Matsumura, K.; Sayo, N.; Kumobayashi, H.; Ohta, T.; Nozaki, K.; Takaya, H. *J. Org. Chem.* 1996, 61, 5510–5516. (b) Patel, D. C.; Breitbach, Z. S.; Woods, R. M.; Lim, Y.; Wang, A.; Foss, F. W. Jr.; Armstrong, D. W. *J. Org. Chem.* 2016, 81, 1295–1299. (c) Tay, J. H.; Arguelles, A. J.; Nagorny, P. *Org. Lett.* 2015, 17, 3774–3777. (d) Ma, G.; Sibi, M. P. *Chem. Eur. J.* 2015, 21, 11644–11657.
91. (a) Cardullo, N.; Barresi, V.; Muccilli, V.; Spampinato, G.; D'Amico, M.; Condorelli, D. F.; Tringali, C. *Molecules* 2020, 25, 733–749. (b) Schäfer, P.; Palacin, T.; Sidera, M.; Fletcher, S. P. *Nat. Commun.* 2017, 8, 1–4. (c) Urquhart, A. S.; Hu, J.; Chooi, Y.-H.; Idnurm, A. *Fungal Biol. Biotechnol.* 2019, 6, 9–21. (d) Adamski-Werner, S. L.; Palaninathan, S. K.; Sacchettini, J. C.; Kelly, J. W. J. *Med. Chem.* 2004, 47, 355–374. (e) Ribeiro, P. R.; Ferraz, C. G.; Guedes, M. L. S.; Martins, D.; Cruz, F. G. *Fitoterapia* 2011, 82, 1237–1240. (f) Jain, Z. J.; Gide, P. S.; Kankate, R. S. *Arabian J. Chem.* 2017, 10, S2051–S2066. (g) Bhujabal, Y. B.; Vadagaonkar, K. S.; Kapdi, A. R. *Asian J. Org. Chem.* 2019, 8, 289–295.
92. Miyaura, S.; Yanagi, T.; Suzuki, A. *Synth. Commun.* 1981, 11, 513–519.
93. Ma, J. *Banana Pseudostem: Properties Nutritional Composition and Use as Food*, 2015, 1–229.
94. Malapit, C. A.; Bour, J. R.; Brigham, C. E.; Sanford, M. S. *Nature* 2018, 563, 100–104.

11

Application of Starch in the Synthesis of N-substituted Pyrroles by a Simple and Green Route

K. Arabpourian and Farahnaz K. Behbahani

CONTENTS

11.1 Introduction ... 191
11.2 Experimental ... 191
11.2.1 Synthesis of *N*-substituted Pyrroles Using Iron(III) Phosphate: General Procedure ... 191
11.2.2 Reusability of the Catalyst ... 192
11.2.3 Results and Discussion ... 192
11.3 Conclusions ... 192
References ... 193

11.1 Introduction

Starch, a glucose, is one of the most many homo polysaccharides manufactured by plants. Its native state is a granular form in plant cells that mostly exist in fruits, grains and roots/tubers.[1] This carbohydrate has been an unlimited matter of research for many decades. It is a cheap and easily available material with vast applications in the food and processing industry. The researchers are trying to improve its properties using a variety of protocols and thus enlarge its uses. In order to achieve chemical modifications, using organic acids has been recently paid the greatest attention, especially the use of starch in the food industry. Organic acids naturally exist in many plants and many of them are generally named as safe, making them ideal for the food industry. Ackar et al. reported several researches on starch modification such as esterification, etherification, cross-linking and dual modification with organic acids and their derivatives.[2] Other applications of starch also consist of the synthesis of starch-supported gold nanoparticles and their use in 4-nitrophenol reduction,[3] and starch-binding domain impacting catalysis in two *Lactobacillus* α-amylases.[4]

Also, the pyrrole nucleus exists in various biological activities as analgesic[5,6]CNS depressant,[7] antifungal,[8] antimycobacterial,[9,10] anticancer,[11] anticonvulsant[12,13] and anti-HIV[14] activities. The most general approach for the synthesis of *N*-substituted pyrroles is the Paal–Knorr reaction[15,16] in which 1,4-dicarbonyl compounds are resulted pyrroles using primary amines and 1,4-dicarbonyl compounds. Various methods for the synthesis of pyrrole derivatives have been developed such as $FePO_4$[17] and other references cited therein. Herein, we wish to present a mild and green route for the Paal–Knorr synthesis of *N*-substituted pyrroles using 1,4-dicarbonyl compounds and amines using starch as a green, readily available and reusable catalyst (Scheme 11.1).

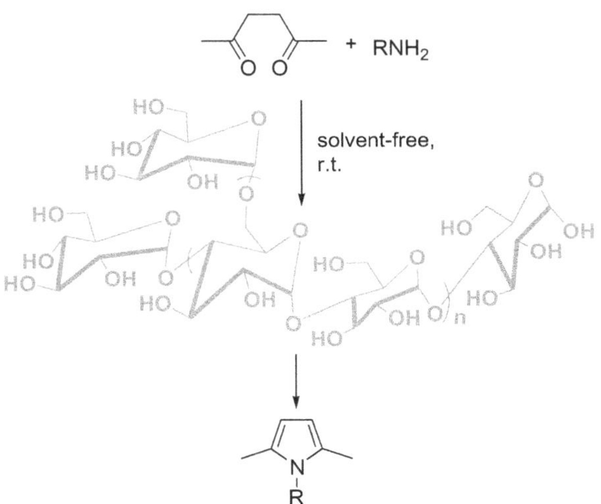

SCHEME 11.1 Starch-catalyzed preparation of *N*-substituted pyrroles.

11.2 Experimental

Melting points were measured using the capillary tube method with an electrothermal 9200 apparatus. IR spectra were recorded on a Bruker FT-IR spectrometer scanning between 4000 and 400 cm^{-1}. ^{1}HNMR spectra were obtained on a Bruker DRX-300MHz NMR instrument. GC/mass spectra were recorded on an Agilent 6890 GC Hp-5 capillary 30 m × 530 lm × 1.5 lm nominal operation at 70 eV.

11.2.1 Synthesis of *N*-substituted Pyrroles Using Iron(III) Phosphate: General Procedure

A mixture of amine (3 mmol), 2,5-hexanedione (3 mmol) and starch (wheat starch) (0.2 g) was stirred under solvent-free

DOI: 10.1201/9781003089162-11

condition and at room temperature. The progress of the reaction was monitored by TLC. After completion of the reaction, dichloromethane (10 mL) was added and the catalyst was filtered off. The organic layer was concentrated and underwent short-column chromatography (n-hexane/acetone). The evaporation of the solvent under the low pressure gave the products.

2-(2,5-dimethyl-1H-pyrrol-1-yl)-5-nitrobenzenamine (entry 11). Brown oil, IR (KBr, cm^{-1}): 3170, 3363 (NH_2), 3074 (CH, Aromatic), 2920 (CH, Aliphatic), 1623 (C=C, Aromatic), 1517 (N=O asymmetry), 1464, 1393 (N=O symmetry), 1309 (C-N). ^1H NMR spectrum, δ, ppm (J, Hz): 1.94 (6H, s, CH_3), 4.21 (2H, s, NH_2), 5.95 (2H, s, pyrrole), 7.25–8.15 (2H, m, Ar). MS (EI, 70 eV): m/z (%) = 231.25 $[M+1]^+$.

11.2.2 Reusability of the Catalyst

To evaluate of reusability of the catalyst, in the case of synthesis of 2,5-dimethyl-1H pyrrole, the removed catalyst was washed with CH_2Cl_2 (15 mL), dried at 50°C and was subjected to four runs (Table 11.1).

11.2.3 Results and Discussion

To optimize the catalyst amount in the Paal–Knorr type pyrrole synthesis, a mixture of aniline (4 mmol) and 2,5-hexadione (4 mmol) were mixed at ambient temperature under the solvent-free condition. The desired product was obtained after 480 min with a 20% yield. The yield of the product and the rate were increased in the presence of starch from 0.1 g to 0.3 g (Table 11.2). As shown in Table 11.2, the catalyst is an essential component for this reaction and the increase of the catalyst amount from 0.3 g to 0.3 g did not affect the yield or the rate.

TABLE 11.1

Reusability of Starch Catalyst in the Synthesis of 2,5-dimethyl-1-phenyl-1H-pyrrole

Run	Time (min)	Yield (%)
1	30	4
2	30	92
3	30	86
4	30	75

TABLE 11.2

Optimizing the Starch Amount in the Synthesis of 2,5-dimethyl-1-phenyl-1H-pyrrole

Entry	Starch (g)	Time (min)	Yield (%)
1	Free	480	20
2	0.1	64	85
3	0.2	30	94
4	0.3	30	94

Reaction condition: aniline (4 mmol), 2,5-hexadione (4 mmol) and starch under solvent-free condition and room temperature.

After optimizing the starch amount, the Paal–Knorr type pyrrole synthesis using arylamines was examined. To observe the effect of substituents, a variety of anilines with electron-donating (2-OH, 4-OMe and Me) and electron-withdrawing groups (Br, Cl and NO_2) were used (Table 10.3). These reactions were also carried out at room temperature without any solvent resulting in very good-to-high yields. It was also observed that, in spite of the longer reaction times, the product yields were mostly not as high as in Table 10.3. This was due to the decrease of aniline amino group basicity and nucleophilicity, which is caused by multiple factors. First, the electron-withdrawing nature of the phenyl ring decreases the electron density on the N. The direct attachment of the ring to the NH_2 group sterically hinders the attack on the N and also decreases reactivity. Based on the analysis of the yields, the substituents of the phenyl ring did not appear to have a significant impact on the reaction times. Finally, the electron-donating groups on phenyl ring increase the yield and reduce the reaction time relative to electron-withdrawing groups. Also, diamines were subjected to Paal–Knorr-type pyrrole synthesis (Table 11.3; entries 9, 10, 11 and 12). In these cases, *o*-phenylene diamine (Table 11.3; entries 10 and 11) reacted via one amino group due to the steric effect. Almost all compounds were characterized and compared with those of known samples in the literature.[18–21] In the case of Table 11.3, entry 11, we did not find any data to compare; so in our view, this compound is unknown and its physical and spectrum data are shown in Section 11.2.

The suggested mechanism for the synthesis of *N*-substituted pyrroles has been shown in Scheme 11.2. The proposed mechanism of starch-catalyzed transformation is shown in Scheme 10.2. At the first, the carbonyl group of diketone is activated by OH-groups of starch to obtain intermediate **1**. Then, an amine nucleophilic attack results in intermediate **2**. The desired product **4** is obtained after intramolecular nucleophilic attack of intermediate **2** after the dehydration of intermediate **3**.

To show the worthiness of this catalytic method compared with the reported protocols, we compared the results of the formation of 1-phenyl))-2,5-dimethyl-1H-pyrrol (Entry 1, Table 10.4) in the presence of a variety of catalysts. From the results given in Table 11.4, the advantages of this method are considerable, in terms of the use of the inexpensive catalysts that are very important in the chemical industry, especially when it is accompanied by easy separation and short reaction times unlike those in previously reported methods such as silica sulfuric acid[22] and $FePO_4$.[17]

11.3 Conclusions

An efficient, green and eco-friendly benign protocol for the synthesis of *N*-substituted dimethyl-pyrroles is presented. The reaction of 2,5-hexadione with a wide variety of amines provides high yields under mild conditions. The major advantages of this approach are: high atom economy, solvent-free reaction, reusability of the catalyst, short reaction time, no heating required and its waste-free nature as H_2O is the only byproduct.

TABLE 11.3

Synthesis of *N*-substituted Pyrroles Using Starch as a Catalyst

Entry	Amine	Product	Time (min)	Yield (%)	M. p (°C) [lit.]
1	Aniline		30	94	45–50[18]
2	4-Br-aniline		38	88	69–71[18]
3	4-OMe-aniline		22	95	56–58[18]
4	2-OH-aniline		40	80	92–96[18]
5	4-NO_2-aniline		200	72	127–130[18]
6	2,4-DiOMe-aniline		28	93	Oil[18]
7	2-Naphthyl amine		27	92	116–118[19]
8	4-Cl-aniline		40	91	Oil[18]
9	*m*-phenylene diamine		65	84	99–100[20]
10	*m*-phenylene diamine		68	90	70–73[21]
11	4-NO_2-1,2-phenylene diamine		110	65	Oil [new]
12	Ethylene diamine		10	97	118–120[18]

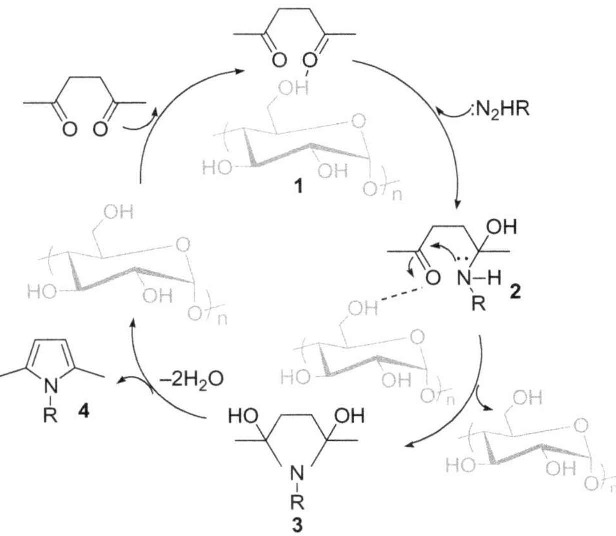

SCHEME 11.2 The mechanism of synthesis of *N*-substituted pyrroles.

TABLE 11.4

Comparison of Catalytic Activity in the Synthesis of 1-phenyl-2,5-dimethyl-1H-pyrrol Using Different Catalysts

Entry	Catalyst (mol% or g)	Time (min)	Temp. (°C)	Solvent	Yield (%)	Ref.
1	Silica sulfuric acid (10 mol%)	10	r.t.	Free	95	22
2	KSF (excess)	600	r.t.	-	95	19
3	$Bi(OTf)_3$ (5 mol%)	240	90		85	23
4	$Bi(NO_3)_3$ (100 mol%)	600	r.t.	-	96	19
5	$RuCl_3$ (5 mol%)	30	r.t.	-	94	19
6	Polystyrenesulfonate (18 % solution)	600	r.t.	EtOH/H_2O	96	19
7	$Sc(OTf)_3$ (10 mol%)	25	r.t.	free	93	23
8	$Y(OTf)_3$ (5 mol%)	30	r.t.	-	86	23
9	$FePO_4$ (10 mol%)	480	r.t.	-	99	17
10	Starch (0.2 g)	30	r.t.	-	94	This work

REFERENCES

1. Buléon, A., Colonna, P., Planchot, V., Ball, S., *Int. J. Biol. Macromol.*, 1998, vol. 23, pp. 85–112.
2. Ačkar, Đ., Babić, J., Jozinović, A., Miličević, B., Jokić, S., Miličević, R., Rajič, M., *Molecules*, 2015, vol. 20, pp. 19554–19570.
3. Chairam, S., Konkamdee, W., Parakhun, R., *J. Saudi Chem. Soc.*, 2017, vol. 216, pp. 656–663.
4. Rodrıguez-Sanoja, R., Ruiz, B., Guyot, J. P., Sanchez, S., *Appl. Environ. Microb.*, 2005, pp. 297–302.
5. Malinka, W., Dziuba, S.M., Rajtar, G., Rubaj, A., Kleinrokm Z., *IlFarmaco*, 1999, vol. 54, pp. 390–401.
6. Malinka, W., Kaczmarz, M., Redzicka, A., Filipek, B., Sapa, J., *IlFarmaco*, 2005, vol. 60, pp. 15–22.
7. Malinka, W., Dziuba, S. M., Rajtar, G., Rejdak, R., Rejdak, K., Kleinrok, Z., *Pharmazie*, 2000, vol. 55, pp. 9–16.
8. Seref, D., Ahmet, K.C., Nuri, K., *Eur. J. Med. Chem.*, 1999, vol. 34, pp. 275–278.

9. Delia, D., Lampis, G., Fioravanti, R., Biava, M., Porretta, C.G., Zanetti, S., Pompei, R., *Antimicrob. Agents Chemother.*, 1998, vol. 42, pp. 3035–3037.
10. Biava, M., Rossella, F., Giulio, C.P., Delia, D., Carlo, M., Pompei, R., *Bioorg. Med. Chem. Lett.*, 1999, vol. 9, pp. 2893–2896.
11. Miguel, F.B., Ascension, F., Mercedes, G., *Chem. Pharm. Bull.*, 1989, vol. 37, pp. 2710–2712.
12. Sorokina, I.K., Andreeva, N.I., Golovina, S.M., *Pharm. Chem. J.*, 1989, vol. 23, pp. 975–977.
13. Carson, J.R., Carmosin, R.J., Pitis, P.M., Vaught, J.L., Almond, H.R., Stables, J.P., Wolf, H.H., Swinyard, E.A., White, H.S., *J. Med. Chem.*, 1997, vol. 40, pp. 1578–1584.
14. Shibo, J., Hong, L., Shuwen, L., Qian, Z., Yuxian, H., Asim, K., *Antimicrob. Agents Chemother.*, 2004, vol. 48, pp. 4349–4359.
15. Paal, C., *Chem. Ber.*, 1884, vol. 17, pp. 2756–2767.
16. Knorr, L., *Chem. Ber.*, 1884, vol. 17, pp. 2863–2870.
17. Samadi, M., Behbahani, F.K., *J. Chil. Chem. Soc.*, 2015, vol. 60, pp. 2881–2884.
18. Patil, V., Sinaha, R., Masand, N., Jain, J., *Dig. J. Nanomater. Bios.*, 2009, vol. 4, pp. 471–477.
19. De Surya, K., *Catal. Lett.*, 2008, vol. 124, pp. 174–177.
20. Ghorbani-Vaghei, R., *S. Afr. J. Chem.*, 2009, vol. 62, pp. 33–38.
21. www.chemspider.com
22. Veisi, H., *Tetrahedron Lett.*, 2010, vol. 51, pp. 2109–2114.
23. Wu, H., Zheng, Z., Jin, C., Zhang, X., Su, W., Chen, J., *Tetrahedron Lett.*, 2006, vol. 47, pp. 5383–5387.

12

Greener Synthesis of Potential Drugs

Renata Studzińska, Renata Kołodziejska, and Daria Kupczyk

CONTENTS

12.1 Introduction ... 195
12.2 Greener Organic Synthesis of Potential Drugs ... 197
12.2.1 Quinoline and Quinoxaline Derivatives ... 197
12.2.1.1 Pyrrole Derivatives ... 198
12.2.1.2 Furan and Pyran Derivatives ... 201
12.2.1.3 Pyrazole Derivatives ... 202
12.2.1.4 Imidazole Derivatives ... 202
12.2.1.5 Thiazole Derivatives ... 203
12.2.1.6 Oxazole, Isoxazole, and Oxazines Derivatives ... 207
12.2.1.7 Other Potential Drugs ... 207
12.2.1.8 Substrates in the Synthesis of Potential Drugs ... 207
12.2.1.9 Biocatalysis ... 210
References ... 222

12.1 Introduction

The synthesis of active pharmaceutical ingredients (APIs) is a typical example of process chemistry, in which the need to build molecular complexity calls for multi-step chemistry with a large number of different solvents. This type of complex manufacturing generates a significant amount of waste [1]. Additionally, a work-up procedure after each reaction increases the solvent demand [2].

A survey by ACS Green Chemical Institute (GCI) showed that 56% of the materials used in pharmaceutical production were solvents, with water contributing a further 32% [3].

Solvents are used daily in many industrial processes as a reaction medium, in separation procedures, and as diluents [4]. They play an integral role in reaction mixtures. Intermolecular interactions with the solvent stabilize solutes, which can facilitate the desired equilibrium position, preferably adjust the kinetic profile of the reaction, as well as affect product selectivity [5].

Large quantities of organic solvents in relation to the reagents used are needed to carry out chemical reactions. In addition, many of the processes necessary to achieve sufficient product purity also require a large excess of solvents. The annual industrial-scale production of organic solvents has been estimated at almost 200 million tons [6].

On the one hand, the possibility of using a whole range of various organic solvents has led to significant progress in chemical synthesis, but on the other, the use of solvents causes a lot of environmental and health damage that has been observed since the 1940s [7].

The US Environmental Protection Agency (EPA) launched the "Alternative Synthesis Routines for Pollution Prevention" program, which has proposed green chemistry for new technological solutions where the main concept is to reduce the amount of toxic, unwanted waste that has a negative impact on the environment. The program includes an innovative approach in many fields of chemistry, those related to organic synthesis as well as in analytical chemistry by eliminating or reducing the amount of reagents and solvents [7].

Due to the fact that the tightening of laws and regulations increasingly limits the use of some solvents (including volatile organic compounds (VOCs) with low boiling points), and at the same time, the use of a solvent in many chemical processes is necessary, others are proposed as "green alternatives". With the advent of green chemistry (and especially research on ionic liquids), the number of chemicals considered as solvents has increased significantly, providing new opportunities in solvent applications [8].

"Green" solvents include, inter alia, water, ionic liquids (ILs), glycerol and deep solvents eutectic (DES), natural deep solvents eutectic (NADES), liquid polymers such as polyethylene glycol (PEG), and vegetable oils [2, 9].

Water possesses many advantages over conventional organic solvents. It has unique physicochemical properties, i.e. hydrogen bond formation, high heat capacity, large dielectric constant, and large temperature window, i.e. an area with an optimal temperature for the reaction environment. It is classified as a green solvent due to its low cost, ready availability, and nontoxic, nonpolluting, and nonflammable properties. Despite many advantages of water, it is not widely used as a

DOI: 10.1201/9781003089162-12

sole solvent for synthetic strategies, due to the fact that most organic compounds are not soluble in water [9].

Glycerol is a polyalcohol that has been utilized in many different fields such as the pharmaceutical and food industry, tobacco, and cellulose films. The sustainability and low cost of glycerol make it a good green solvent compared to other organic solvents which are hazardous, volatile, toxic, and harmful compounds. Despite the fact that glycerol is a solvent and selected for many reactions, there are some limits to its use. Glycerol is highly viscous and therefore it should be fluidified with a co-solvent or the reactions can be proceeded at temperatures higher than 60°C because then, its viscosity is much lower. Due to the presence of three hydroxyl groups which can be mentioned as acidic sites in the molecule, glycerol may be involved in the reaction. Glycerol has enough length and donor atom in which it can obtain complexes with metal catalysts resulting in unwanted side products and/or unreactivity of catalysis. Despite these limitations, glycerol is willingly used as a green solvent in organic synthesis also in the synthesis of biologically active compounds [9].

ILs are organic salts made up of various organic cations and inorganic or organic anions. The most common cations are 1,3-dialkylimidazolium, 1,1-dialkylpyrrolidinium, *N*-alkylpyridinium, or a quaternary ammonium salt (Scheme 12.1). Compared to standard organic solvents, ionic liquids have a low melting point so that most representatives of this class are liquid at room temperature.

By selecting different combinations of cations and anions and modifying the organic part (e.g. by introducing additional substituents), it is possible to design an ionic liquid with unique physical properties tailored to specific reaction requirements. ILs are a green alternative to conventional organic solvents, they have great potential due to their excellent chemical and thermal stability, high ionic conductivity and large electrochemical window, low vapor pressure, and the possibility of interaction with organic and inorganic substances. They are widely used, among others, in organic and inorganic synthesis, enzymatic and chemical catalysis, as well as biotransformation using whole cells of microorganisms [8–10]. A characteristic feature of ILs is the low melting point, in the range of 0–100°C. Most ionic liquids are liquid at room temperature (RILs) and remain as liquids within a broad temperature window below 300°C [8, 11, 12]. Compared to conventional organic solvents, ionic liquids are much more viscous. For commonly used ILs, the viscosity is in the range of 35–500 centipoise (cP) (for comparison: toluene – 0.6 cP, methanol – 0.5 cP, water – 0.9 cP). The density of ILs is in the range 1.1–1.6 g · cm^{-3} and is generally larger than the density of typical solvents or water [13–15]. Due to the diversity of ionic liquids in terms of their composition and properties, and the ability to "build" new ones, they are widely used in various organic syntheses.

Deep eutectic solvents (DESs) are now widely acknowledged as a new class of IL analogs because they share many characteristics and properties with ILs. The difference is that DESs are systems formed from a eutectic mixture of Lewis or Brønsted acids and bases which can contain a variety of anionic and/or cationic species while ILs are formed from systems composed primarily of one type of discrete anion and cation [16]. Different types of DESs have been described, and most contain hydrogen-bond acceptors (HBAs) and hydrogen-bond donors (HBDs) (Schemes 12.2). Alternatively, they may also contain metal salts or hydrated metal salts [2].

The melting points of DESs are lower than the individual components, and they are often prepared by mixing two solid reagents to form a liquid product (Scheme 12.3). Usually, they are obtained by the complexation of a quaternary ammonium salt with a metal salt or HBDs. The charge delocalization occurring through hydrogen bonding between, for example, a halide ion and the hydrogen-donor moiety is responsible for the decrease in the melting point of the mixture relative to the melting points of the individual components [2, 16].

DESs can be described by the general formula: Cat^+X^-zY, where Cat^+ is any ammonium, phosphonium, or sulfonium cation, and X is a Lewis base, generally, a halide anion. The complex anionic species are formed between X^- and either a Lewis or Brønsted acid Y (z is the number of Y molecules that interact with the anion) [16].

DES components often come from renewable sources (e.g. choline chloride (ChCl), urea, glycerol (Gly), lactic acid, carbohydrates, polyalcohols, amino acids, vitamins). Therefore, their biodegradability is extraordinarily high and their toxicity is non-existent or very low. In view of their minimal ecological footprint, cheapness of their constituents, tunability of their physico-chemical properties, and ease of preparation, DESs are successfully and progressively replacing often hazardous and VOCs in many fields of science inter alia as solvents for synthesis [17]. DESs composed entirely of plant metabolites (such as ammonium salts, sugars, and organic acids) are labeled natural deep eutectic solvents (NADES).

Many organic reactions are catalyzed by acids or bases and these components appear in DESs. Thus, a DES can be used not only as a reaction medium but also as a catalytic active species for some reactions, and, in some cases, can also represent part of the starting materials [18].

In addition to organic synthesis, in which the "green alternative" is the use of the solvents described above, biocatalysis, i.e. processes using isolated enzymes or whole microbial cells, fits perfectly in the global green chemistry trend. Biocatalysts are biodegradable, reusable, more efficient – thanks to unique features such as high selectivity – and they are widely used in industries such as food, pharmacology, medicine, and textiles. In addition, enzymes have unique properties that cannot

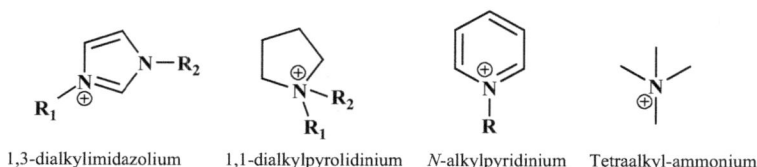

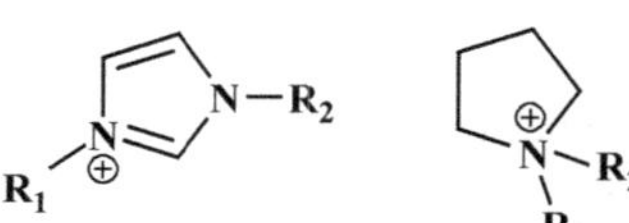

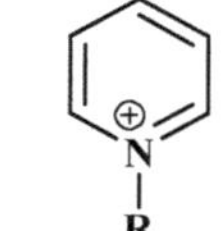

SCHEME 12.1 Most commonly used organic cations contained in ionic liquids.

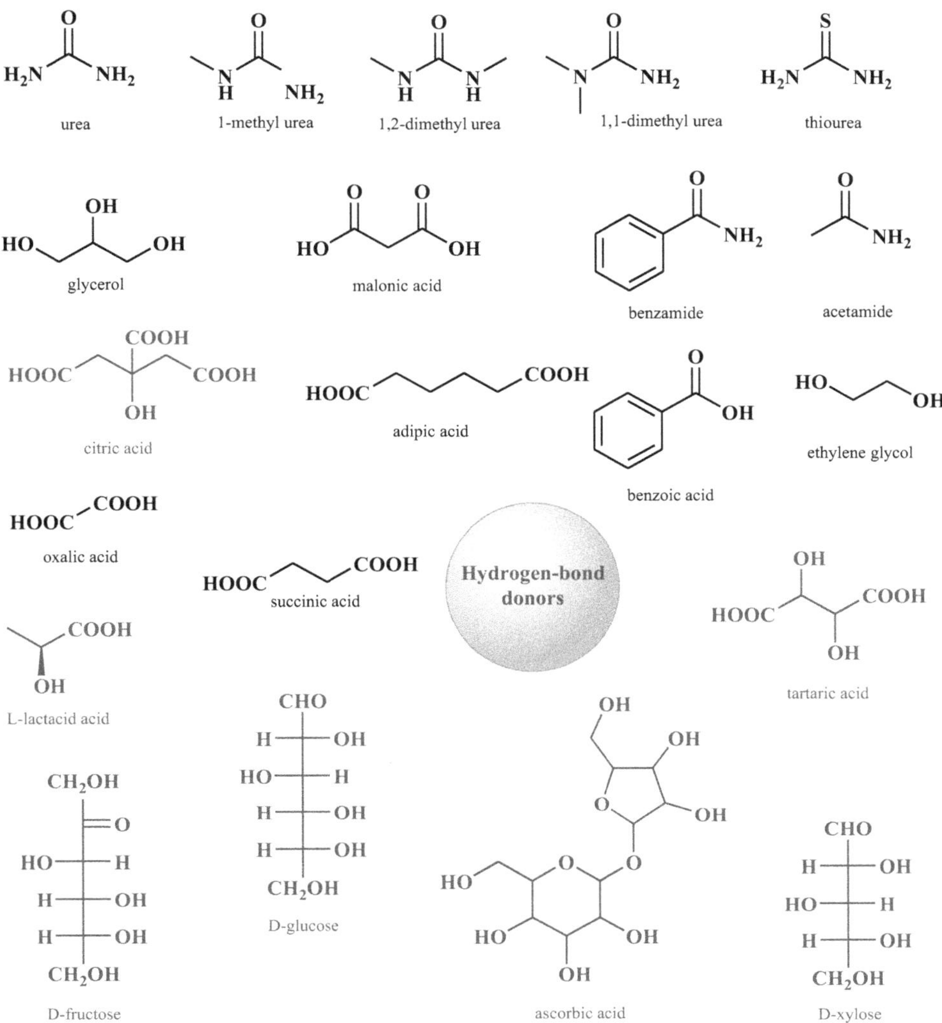

SCHEME 12.2 Examples of hydrogen-bond donors (NDES (natural deep eutectic solvents) ingredients are marked in blue).

always be replaced by synthetic chemical catalysts. High stereo selectivity is one of the most important and unique features of enzymes, especially in the drug-production process. Thanks to this property, enzymes can lower the price of chiral drugs. The process of biocatalysis is complemented by a "green" solvent, which, apart from being natural, non-toxic, cheap, easily available, biodegradable, biocompatible, non-volatile, non-flammable, chemically and thermally stable, reusable, and efficient, will also support the enzyme by increasing its activity, thermal stability and selectivity, the rate of reaction, and the degree of conversion [9, 19].

12.2 Greener Organic Synthesis of Potential Drugs

The pharmaceutical industry generates large amounts of waste, mainly of solvents; hence, measures are taken to reduce the amount of waste, including the use of the so-called "green" solvents. However, it should be noted that large amounts of organic solvents are used not only in the pharmaceutical industry, in the synthesis of drugs, but also in the synthesis of a huge amount of organic compounds, which, due to their structural similarity to drugs, can exhibit various biological activity with reduced toxicity, and which are synthesized in the search for new drugs. Therefore, in this chapter, examples of "green organic synthesis" of biologically active substances are given because they limit the use of organic solvents already at the stage of finding new drugs, but will also be important in reducing the amount of waste by the pharmaceutical industry after implementing a specific substance as a drug.

12.2.1 Quinoline and Quinoxaline Derivatives

Quinolines and their derivatives are ubiquitous in various natural products, and many of them have been widely recognized as important structural motifs to design tremendous synthetic drug candidates since they possess a broad spectrum of biological properties of being antimalarial [20], antibacterial [21], antipsychotic, anti-inflammatory [22], anticancer [23], anti-HIV [24], etc. [25]

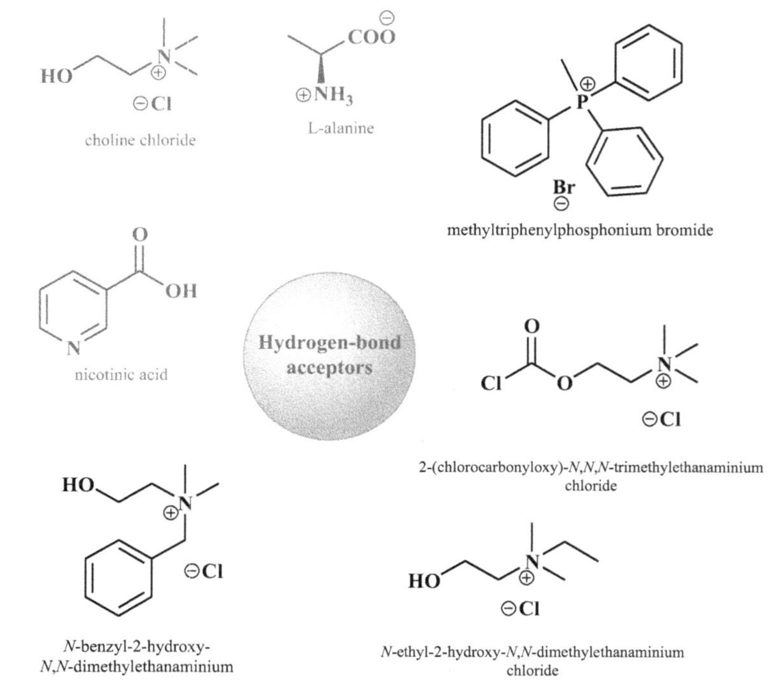

choline chloride

L-alanine

methyltriphenylphosphonium bromide

nicotinic acid

Hydrogen-bond acceptors

2-(chlorocarbonyloxy)-*N*,*N*,*N*-trimethylethanaminium chloride

N-benzyl-2-hydroxy-*N*,*N*-dimethylethanaminium

N-ethyl-2-hydroxy-*N*,*N*-dimethylethanaminium chloride

SCHEME 12.3 Examples of hydrogen-bond acceptors (NDES ingredients are marked in blue).

Quinoxaline derivatives also are well known in the pharmaceutical industry and have been shown to possess a broad spectrum of biological activities including antibacterial [26], antiviral [27], anti-inflammatory and analgesics [28], anticancer [29], and kinase inhibitory activities [30]. In addition, quinoxaline derivatives have been evaluated as anthelmintic agents [31]. That is why research is undertaken on the green synthesis of this group of compounds.

Haloquinolines and sulfonyl chlorides coupling reactions leading to 27 sulfonylated quinolines were carried out in water using Zn powder as the reductant. Aryl sulfonyl chlorides with a series of important functional groups (i.e. -Me, -OMe, -F, -Cl, -Br, -NO_2, -CN and -CF_3) on benzene ring with 2-chloroquinoline gave the desired 2-sulfonylquinolines with good-to-high yields (83–91%). A similar effect was observed for 4-chloroquinoline and sylchloride whereas the use of aliphatic sulfonyl chlorides such as benzylsulfochloride, *n*-propanesulfonyl chloride, and cyclopropylsulfonyl chloride allowed to obtain products with lower yields (46–72%) (Scheme 12.4) [32].

Various functionalized 2-sulfonylquinolines were synthesized from quinoline *N*-oxides and sulfonyl chlorides on water under base-free and open-air conditions. Conducting the reaction in the presence of Zn as the reducing agent, a total of 32 quinoline and isoquinoline derivatives were obtained with good yields between 61% and 91% [33].

Quinoxaline derivatives were also obtained in water under ultrasound irradiation. The reactions take place without a catalyst between suitable heterocyclic diketones and aniline derivatives. The use of microwave radiation reduced the reaction time from 10 min to 50 sec compared to conventional methods. Depending on the functional group used in the *ortho* aniline position, heterocyclic derivatives containing four fused rings with high yields (87–99%) were obtained [34].

The synthesis of quinoxalines from a variety of aromatic and aliphatic diketones and 1,2-diamines with cerium chloride was carried out in glycerol. The condensation took place faster when the reaction was carried out between aromatic diketones and *o*-phenylenediamines. Similarly, the reaction between aliphatic diketones and alicyclic diamine was comparatively slower with lesser yields. All the reactions were completed within 4–6 h at 75°C. The products were obtained with 75–95% yield [35].

12.2.1.1 Pyrrole Derivatives

Pyrrole derivatives are versatile pharmacophores possessing a variety of biological activities. In particular, pyrroles are found in many naturally occurring compounds such as heme, chlorophyll, and vitamin B_{12}. Therefore, many synthetic methods for the preparation of pyrrole derivatives have been reported in the literature inter alia in the field of green synthesis. Among them, the Paal–Knorr reaction remains one of the most attractive methods for the synthesis of pyrroles (Scheme 12.5). $Bi(OTf)_3$ immobilized in [BMIM]BF_4 was utilized as the catalytic system for the synthesis of pyrrole derivatives from 1,4-diketones. The reactions take place in an excess of ionic liquid, so it is not only a component of the catalytic system but also a solvent. Treatment of 1,4-diketones with aryl amines resulted in the formation of the corresponding pyrrole derivatives with 82–90% yield [36].

The combination of urea with choline chloride (as DES) are also effective solvents/catalysts for Paal-Knorr reactions to form either pyrrole. The reactions of simple amines with 2,5-hexanedione in 2 h afforded the pyrroles with near-quantitative yield. Attempting these reactions with less nucleophilic anilines was also successful, but only by applying longer reaction times (12 h). Again, the pyrrole products were obtained with near-quantitative yields. Only in the case of the very

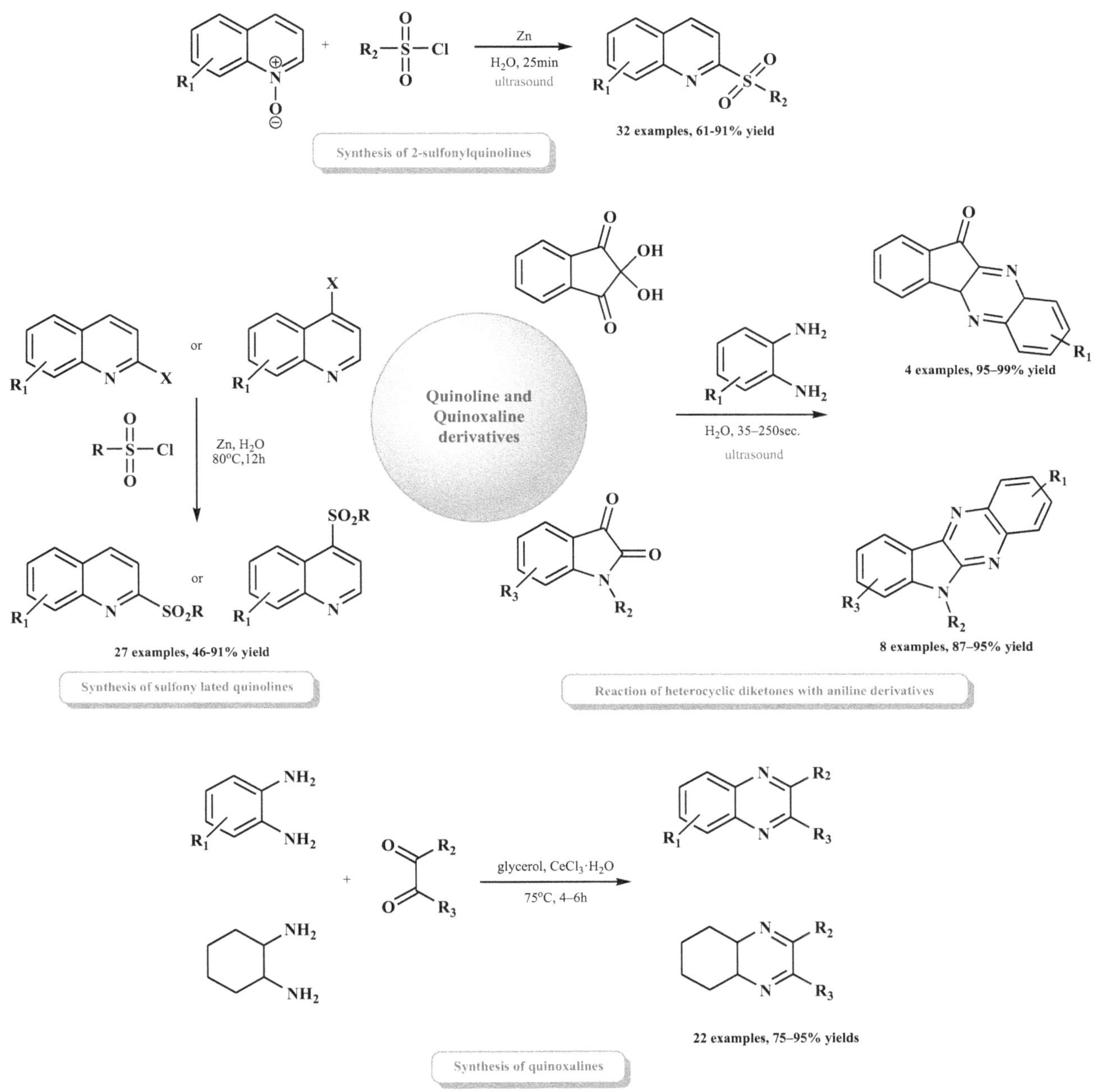

Zn, H_2O, 25min ultrasound

32 examples, 61-91% yield

Synthesis of 2-sulfonylquinolines

Zn, H_2O 80°C,12h

27 examples, 46-91% yield

Synthesis of sulfony lated quinolines

Quinoline and Quinoxaline derivatives

H_2O, 35–250sec. ultrasound

4 examples, 95–99% yield

8 examples, 87–95% yield

Reaction of heterocyclic diketones with aniline derivatives

glycerol, $CeCl_3 \cdot H_2O$ 75°C, 4–6h

22 examples, 75–95% yields

Synthesis of quinoxalines

SCHEME 12.4 Greener synthesis of quinoline and quinoxaline derivatives.

poorly nucleophilic 4-nitroaniline was a lower yield obtained. When dions with a phenyl group or groups were used, the reaction yields were also high, but with a longer time (24 h) [37].

Dipyrromethanes (or dipyrrilmethanes) are important building blocks for many of the structures of interest in the areas of porphyrins, material science, optics, and medicine. The reaction of ketones or aldehydes with pyrrole in water in the presence of hydrochloric acid, gave the corresponding dipyrromethane derivatives with good-to-excellent yields, after 30–45 min at 90°C [38, 39].

Isatin oximes constitute the central core of a wide variety of bioactive compounds and pharmaceutical molecules that exhibit potential biological activities. These compounds have been tested in many directions and show various biological activities. Isatin oximes have been proposed as agents against migraine and for coronary and ischemic heart disease, arteriospasm, cardiac arrhythmia, and hypertension. They also affect the central nervous system acting, among others, as anticonvulsant and antidepressant agents and exhibited high analgesic and anti-inflammatory activities. A clearly defined antileukemia activity was also discovered in the oxime derivatives of isatin [40]. Bezimidazole-isatin oximes inhibit RSV (respiratory syncytial virus) activity [41].

A green, metal-free synthesis of isatin oximes *via* radical coupling reactions of oxindoles with *t*-BuONO (*tert*-butyl nitrite) occurs in water. Reactions were carried out at room temperature. Using this method, 15 oximes of isatin were obtained which differed in substituents at the 5-position of the

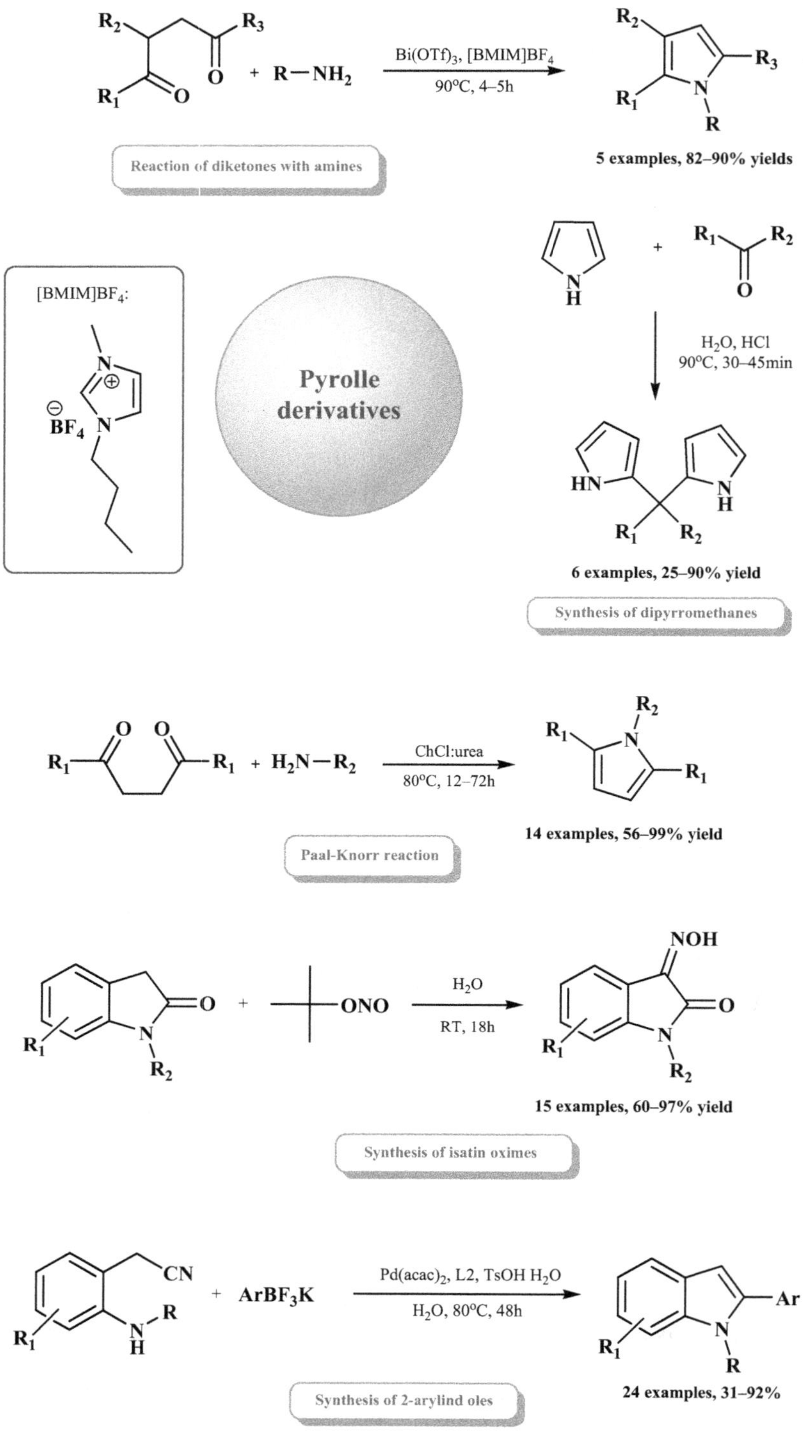

SCHEME 12.5 Greener synthesis of pyrrole derivatives.

oxindoles system and substituents at the nitrogen atom, with good yields (up to 97%). Unfortunately, the substrate *N*,2-diphenylacetamide was not suitable for the radical coupling reaction under the current conditions [42].

The synthesis of 2-arylindoles by palladium-catalyzed tandem addition/cyclization of 2-(2-aminoaryl)acetonitriles with potassium aryltrifluoroborates is also an example of the reaction in an aqueous medium. Pd(acac)$_2$ (Palladium(II) bis(acetylacetonate)) as a Pd sources and 4,4′-dimethyl-2,2′-bipyridine (L2) as a ligand were used. 24 various potassium aryltrifluoroborates differing substituents on the phenyl ring with 2-(2-aminophenyl)acetonitrile gave corresponding arylindoles with 31–92% yield. In the case of other 2-(2-aminoaryl)acetonitriles with bromo or iodo substituents in the phenyl ring, the reaction worked well and gave the corresponding products with 50–75% yield [43].

12.2.1.2 Furan and Pyran Derivatives

Pyrans are important core units in a number of natural products [44]. Compounds with a pyran ring system have many pharmacological properties and play important roles in biochemical processes.

They show, inter alia, antibacterial [45, 46] and antifungal activities [47]. In turn, some furan derivatives show anti-inflammatory [48], antiviral, antitumorigenic, psychotropic [49], as well as antibacterial and antifungal activities [47].

4*H*-pyran derivatives can be obtained in glycerol in the one-pot three-component reaction between aromatic or heteroaromatic aldehydes, carbonyl compounds possessing a reactive α-methylene group, and malononitrile or ethyl 2-cyanoacetate under catalyst-free conditions. All reactions are completed in short times, and 50 products are obtained with good-to-excellent yields (Scheme 12.6) [50].

Furan derivatives were obtained from 1,4-diketones using $Bi(OTf)_3$ immobilized in $[BMIM]BF_4$ as the catalyst. The reactions take place in an excess of the ionic liquid. In the case of

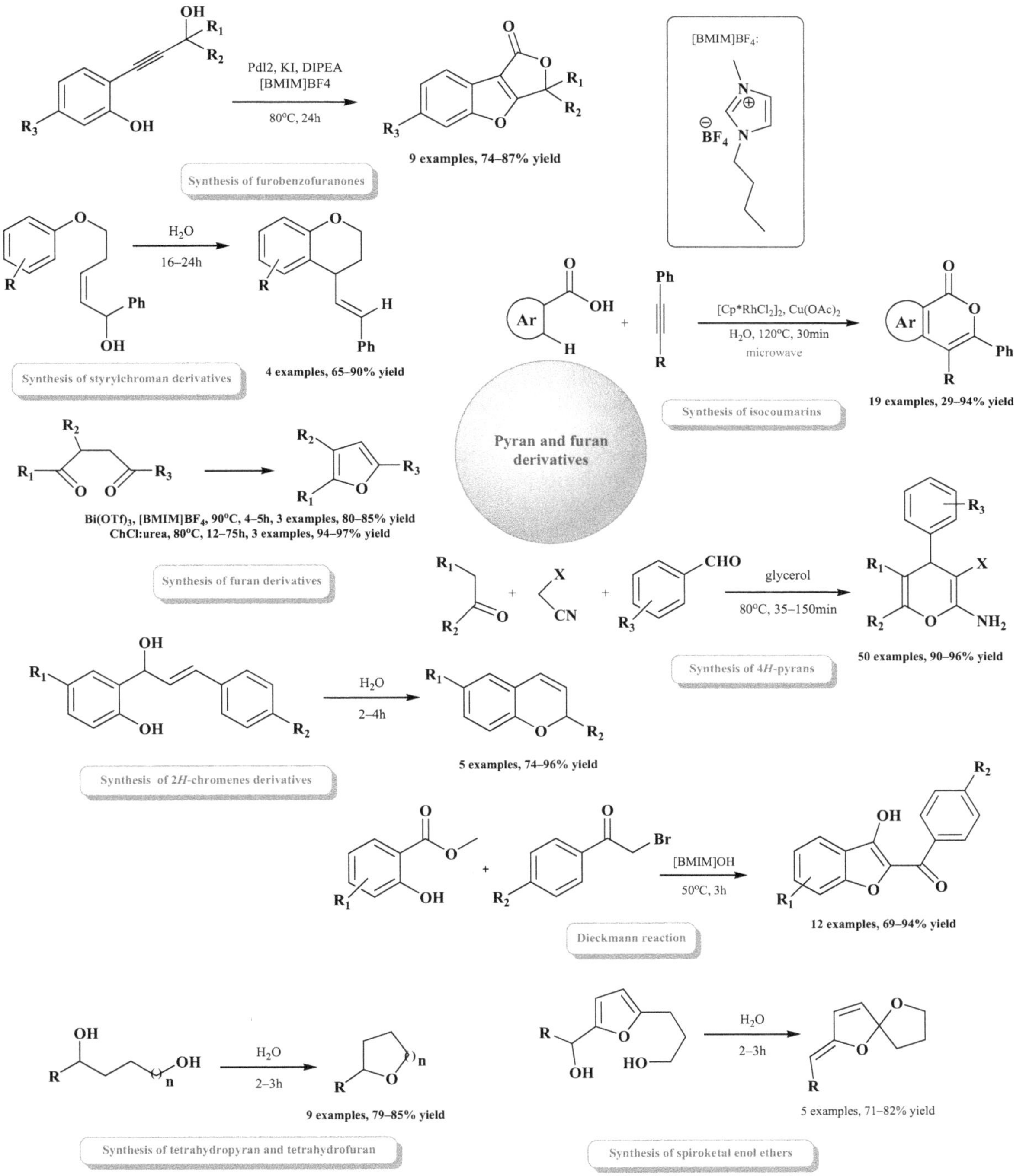

SCHEME 12.6 Greener synthesis of pyran and furan derivatives.

various substituted 1,4-diketones cyclizations, the corresponding trisubstituted furan derivatives were obtained with 80–85% yield [36]. Furan derivatives were also obtained with good yield in cyclization reactions of diones in the combination of urea with choline chloride (DES) as solvents/catalysts system [37].

Furo[3,4-*b*]benzofuran-1(3*H*)-ones were synthesized in reactions between 2-(3-hydroxy-1-yn-1-yl)phenols, CO, and oxygen in the presence of catalytic amounts of PdI_2 in conjunction with KI and diisopropylethylamine at 80°C for 24 h under 30 atm of a 1:4 mixture of CO–air. The process was not selective when carried out in classical organic non-nucleophilic solvents (such as MeCN or dimethoxymethane (DME)), leading to a mixture of the benzofurofuranone derivative and the benzofuran ensuing from simple cycloisomerization, whereas it turned out chemoselective toward the formation of the double cyclization compound in $[BMIM]BF_4$ as the reaction medium. Suitable furobenzofuranones were obtained with 74–87% yield [51].

Various 2-aroylbenzofuran-3-ols were synthesized by Dieckmann reaction of substituted methylsalicylates with 2-bromo-1-aroylethanones in the basic ionic liquid [BMIM]OH which acts as a base catalyst and as solvent. The products were obtained with much higher yields (69–94%) compared to the reactions carried out in other solvents with the same base [52].

Efficient cyclization reactions of π-activated alcohols can be carried out in water under refluxing conditions [53]. In the case of intramolecular etherification conducted in refluxing water without an additional catalyst, the phenoxy group of the substrate readily displaced the benzylic hydroxyl group to give 2*H*-chromene derivatives with 74–96% yield after 2–4 h. Intramolecular substitution reactions structurally similar to that of cinnamic alcohol diols did take place in water to give cyclic ethers with 70–79% yield. Under the same reaction conditions, the etherification of benzylic alcohols afforded spiro tetrahydrofurans and tetrahydropyrans with 79–85% yield. Spiroketal enol ether derivatives were synthesized by intramolecular cyclization of 2-furylcarbinols. Intramolecular nucleophilic substitution reactions of unsaturated alcohols by a phenyl ring carried out in water gave the styrylchroman derivatives with 65–90% yield.

The synthesis of isocoumarins *via* a water-mediated and Rh/Cu-catalyzed coupling reaction of benzoic acids with 1,2-diphenylethyne under microwave-assisted heating procedures was also described. Using $[Cp^*RhCl_2]_2$ as a catalyst and $Cu(OAc)_2{\cdot}H_2O$ as a co-catalyst, eight isocoumarins were obtained with different functional groups in the aromatic ring of carboxylic acids (halo-, methyl-, nitro- substituents) with 31–85% yield. Reactions with 1-(prop-1-ynyl)benzene and alkyne gave analogous products with 29–94% yield. Heteroarenes with carboxyl group and acrylic acid derivative were also used as substrates in this method; however, yields in these reactions were moderate (37–61%) [54].

12.2.1.3 Pyrazole Derivatives

Pyrazole derivatives are ubiquitous in organic molecules with valuable biological functions. In recent years, a variety of new pyrazole derivatives have been discovered with enriched bioactivities such as the reversal activity to fluconazole resistance [55], antimicrobial activity [56], anticancer activity [57], antiproliferative activity [58], ALK5 kinase inhibitory activity [59], and antioxidation activity [60]. Not surprisingly, the green synthesis of pyrazoles and their derivatives keeps receiving interest.

The cascade reactions between NH_2-functionalized enaminones and sulfonylhydrazines developed for the synthesis of fully substituted pyrazoles were carried out in water (Scheme 12.7). By making use of the hydrophilic primary amino group in the enaminones, the reactions proceed in the medium of pure water in the presence of molecular iodine as the catalyst, TBHP (*tert*-butyl hydroperoxide) as oxidant, and $NaHCO_3$ *via* cascade C–H sulfonylation and pyrazole annulation with 65–85% yield. The aryl sulfonyl hydrazines were well tolerated for the synthesis by affording the 20 products with good to excellent yields. The alkyl-based sulfonyl hydrazine, such as methyl sulfonyl hydrazine, however, was not successful [61].

Pyrazole derivatives such as pyrano[2,3-*c*]pyrazoles were received with high yields in multi-component reactions between aromatic aldehydes, malononitrile, hydrazine hydrate or phenyl hydrazine, and 1,3-dicarbonyl compounds using glycerol as both solvent and catalyst at ambient temperature [62]. Pyrano[2,3-*c*]pyrazoles were also formed in glycerol in two-step multicomponent reactions of arylhydrazine, β-oxo ester, alkene, and formaldehyde at 110°C [63]. In turn, arylselanylpyrazoles, were obtained from α-arylselanyl-1,3-diketones and arylhydrazines at 60°C under N_2 atmosphere with moderate-to-good yields [64].

12.2.1.4 Imidazole Derivatives

Compounds containing imidazole ring show various pharmacological activities, among others, antifungal, antiviral, antibacterial, anti-inflammatory, analgesic, antitubercular, antidepressant, anticancer, and antileishmanial activities [65].

2,4,5-triaryl imidazole derivatives were obtained in glycerol in reactions of benzil, benzaldehyde, and ammonium acetate, while in the case of synthesis of 1,2,4,5-tetraaryl imidazoles to the same components additionally the aromatic amine was added. The reactions were carried out at 90°C and were completed within 39–110 min for trisubstituted imidazoles and 2.15–4 h for tetrasubstituted imidazoles (Scheme 12.8). Whereas benzimidazoles were synthesized in catalyst-free condensation of *o*-phenylenediamine with several aromatic, heterocyclic, and aliphatic aldehydes in high yields [66].

The synthesis of novel benzimidazo-fused polyheterocycles *via* 6-*endo-dig* cyclization using readily accessible *o*-alkynylaldehyde with nucleophilic *o*-aryldiamine was carried out in water. As a result of the reaction, 36 condensed heterocyclic systems containing a furan, pyrrole, or thiophene ring were obtained with yields in the range of 68–92% [67].

Thiazolobenzimidazole derivatives have been synthesized in ionic liquids in the reaction of aromatic *o*-diamine with 2-mercaptoacetic acid and aldehyde. Reactions of this type were conducted in $[BMIM]BF_4$ and 1-methoxyethyl-3-methylimidazolium trifluoroacetate ([MOEMIM]TFA) and consequently, the products were obtained with slightly higher yields (80–94%) [68].

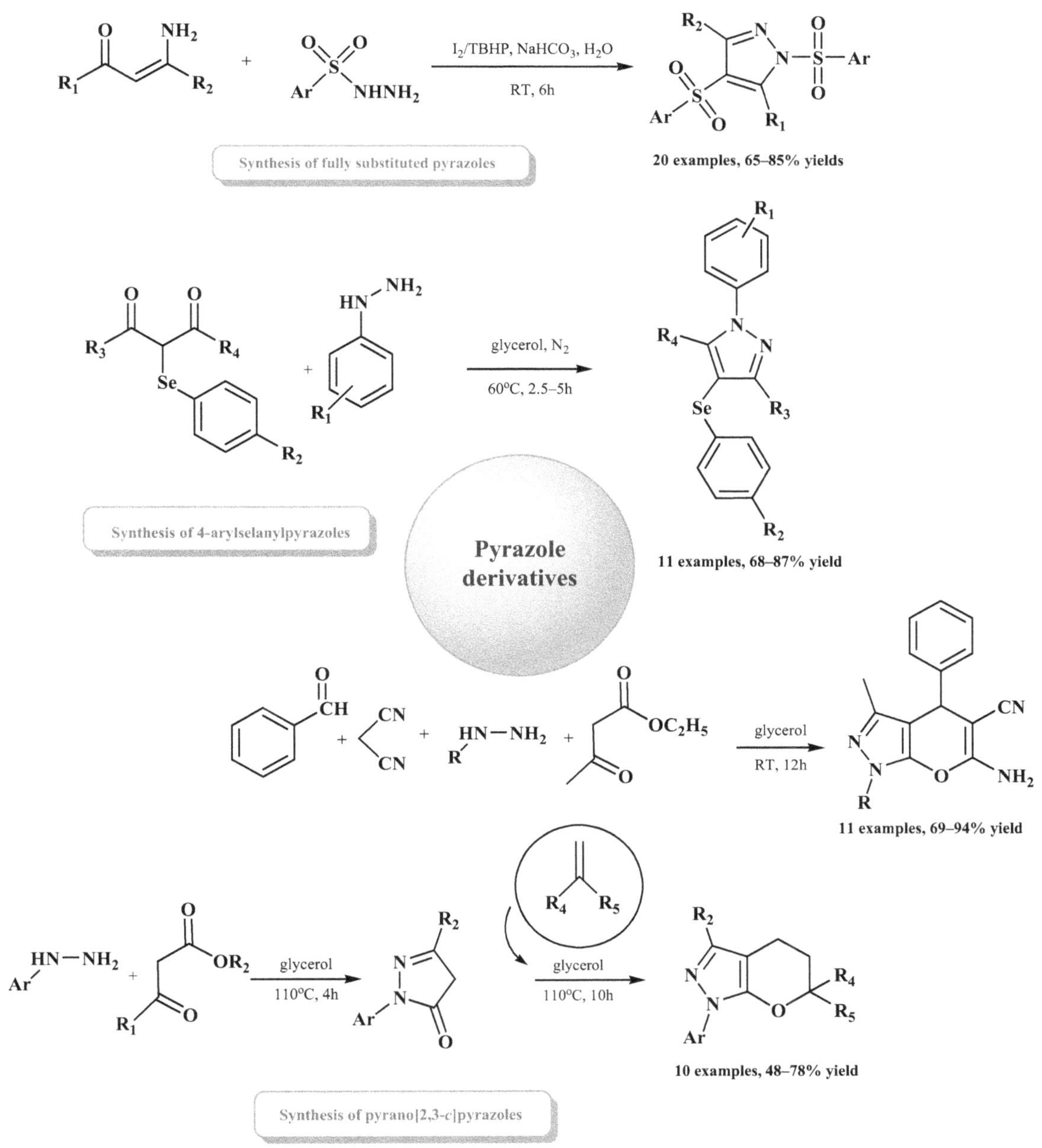

I₂/TBHP, NaHCO₃, H₂O

RT, 6h

Synthesis of fully substituted pyrazoles

20 examples, 65–85% yields

glycerol, N_2

60°C, 2.5–5h

Synthesis of 4-arylselanylpyrazoles

11 examples, 68–87% yield

Pyrazole derivatives

glycerol

RT, 12h

11 examples, 69–94% yield

glycerol

110°C, 4h

glycerol

110°C, 10h

10 examples, 48–78% yield

Synthesis of pyrano[2,3-c]pyrazoles

SCHEME 12.7 Greener synthesis of pyrazole derivatives.

The synthesis of 2-aminoimidazoles exploits the heterocyclodehydration process between α-chloroketones and guanidine derivatives. The combination of either glycerol or urea with choline chloride (ChCl) as DES proved to be effective for decreasing the reaction time to about 4–6 h in contrast to the 10–12 h usually required for the same reaction run in toxic and volatile organic solvents and under an argon atmosphere. Whereas the yields of the reactions were comparable or higher compared to reactions carried out in a conventional solvent (THF or ethanol) [69].

12.2.1.5 Thiazole Derivatives

Compounds in which a thiazole or hydrothiazole ring is a structural fragment play a significant role due to their biological activity. There are many drugs containing thiazole ring, including sulfathiazole (bacteriostatic effect), abafungin (antifungal activity), thiabendazole (anthelmintic effect), thiamine (vitamin B_1), or tiazofurin used in the treatment of cancer.

Dihydrothiazole derivatives show, among others, anticancer [70], antibacterial [71], and antiviral activities. [72] They were also tested for the selective inhibition of 11β-hydroxysteroid dehydrogenase [73–76]. Also noteworthy are the compounds in which the thiazole ring is condensed with other aromatic rings, especially containing heteroatoms. For example, thiazolopyrimidines play an important role in organisms due to their biological activity [77].

A one-pot synthesis of 2-arylbenzothiazoles occurs between aminothiophenols and aromatic aldehydes containing electron-withdrawing substituents on a phenyl ring, at the ambient temperature in glycerol under catalyst-free conditions with excellent yields (80–93%) (Scheme 12.9) [78].

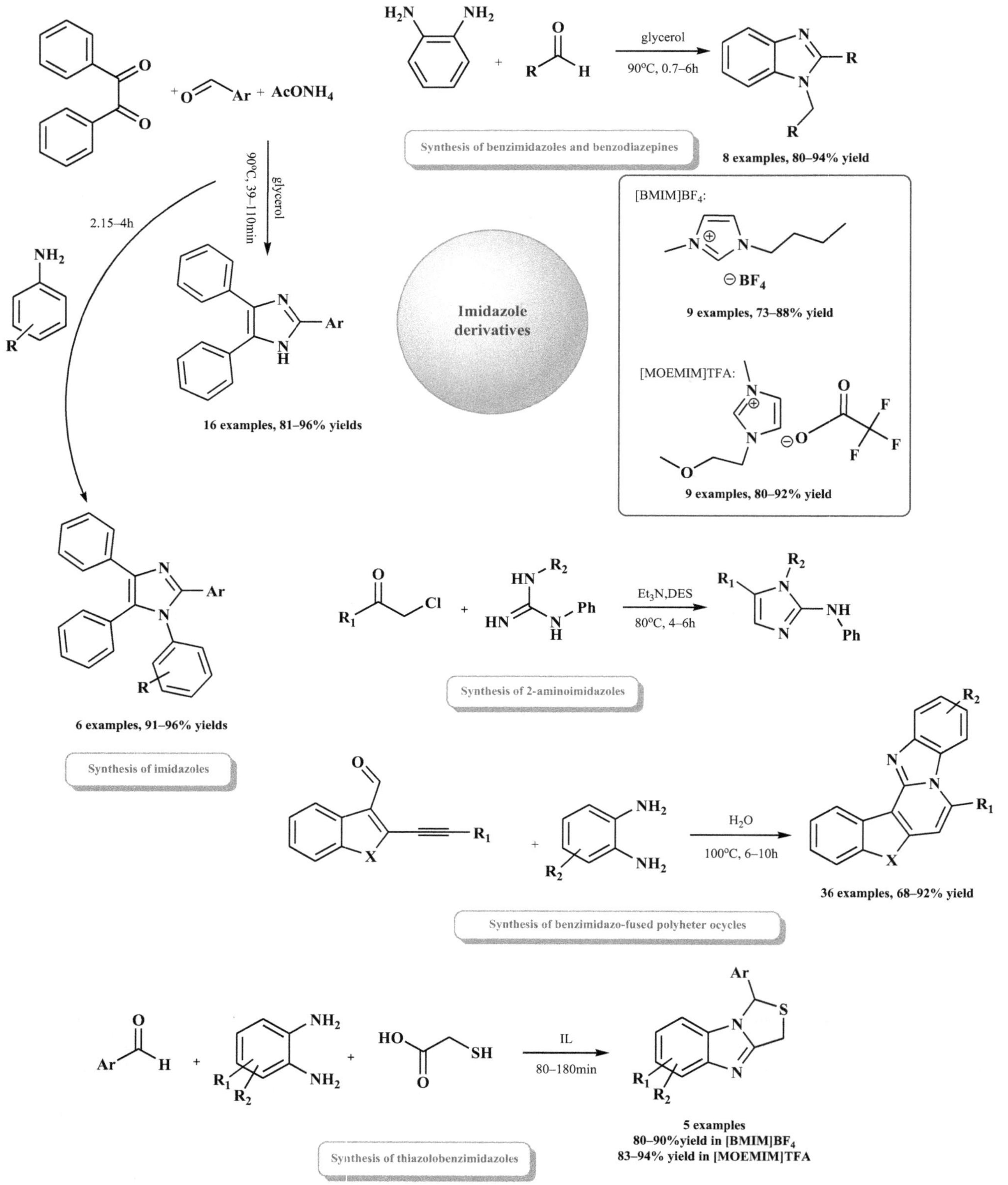

SCHEME 12.8 Greener synthesis of imidazole derivatives.

The synthesis of thiazole derivatives is an example of microwave-assisted reaction in glycerol as solvent. The substituted 2-cyanomethyl-4-phenylthiazoles were synthesized in the reaction of cyanoacetamide and 2-bromoacetophenones with substituents in the phenyl ring under focused microwave irradiation using glycerol as solvent. The method allows the synthesis of products with excellent yields up to 96% with short reaction times (few minutes) and the work-up is easy [79]. The synthesis of benzothiazoles *via* the condensation of 2-aminothiophenol with aldehydes under CEM-focused

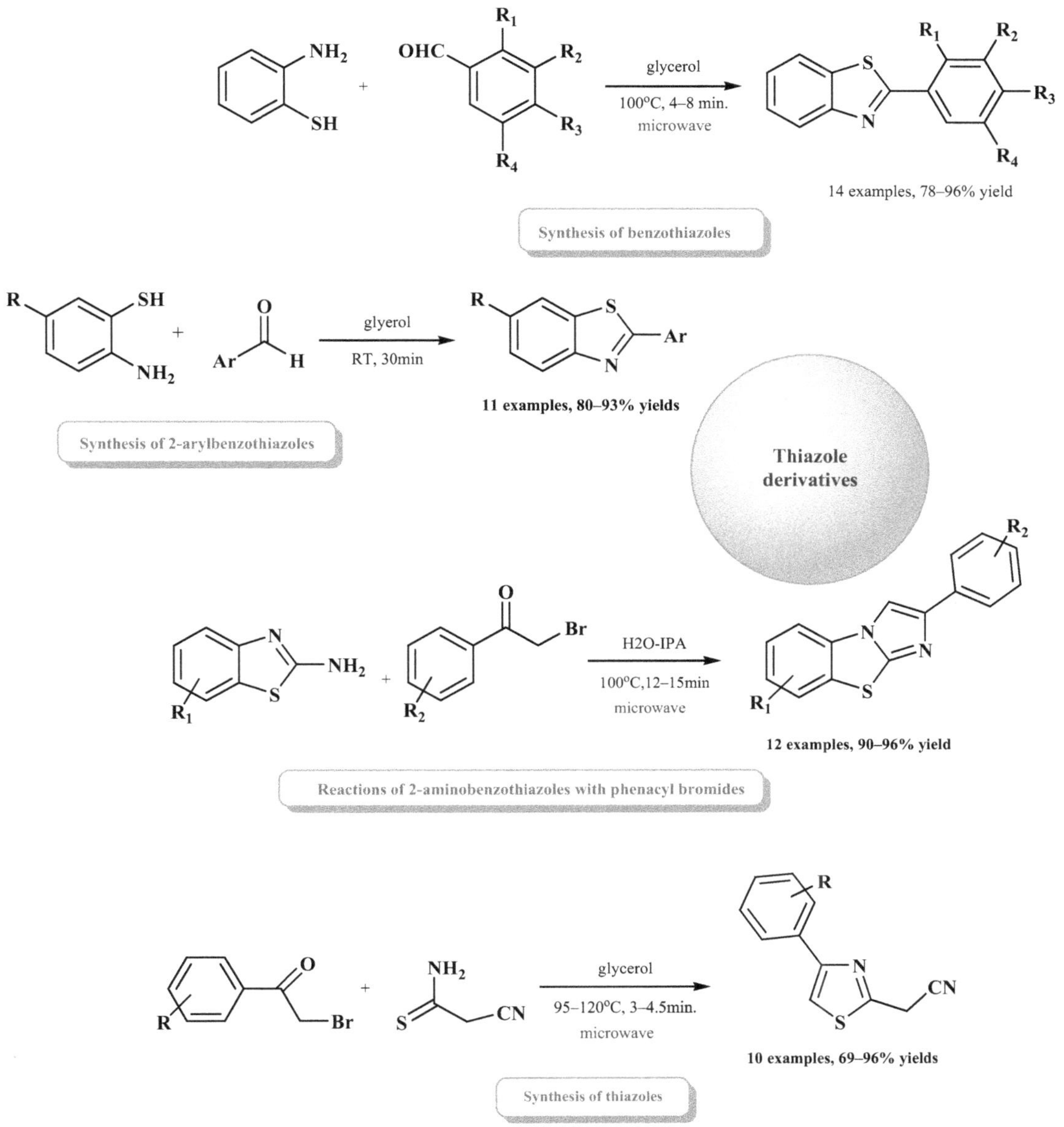

SCHEME 12.9 Greener synthesis of thiazole derivatives.

microwave-irradiation conditions with glycerol as a solvent without any catalyst was also carried out. The use of a variety of aromatic aldehydes bearing electron-withdrawing substituents, such as chloro, bromo, and nitro groups, afforded high yields of the products. In addition, a hydroxy group can be tolerated under these conditions, and good yields of the desired products were obtained. The reaction of heterocyclic aldehydes also provided the 2-heterocyclic-substituted benzothiazoles with good yields [80].

Heterocyclization of 2-aminobenzothiazole were carried out in microwave conditions. In an effort to solubilize organic compounds in water, isopropyl alcohol (IPA) as a polar protic organic co-solvent was used [81]. The reaction between substituted 2-aminobenzothiazoles and phenacyl bromides whether electron-withdrawing or electron-donating substituents were used, the corresponding benzo[*d*]imidazo[2,1-*b*]thiazole products with excellent yields (90–96%).

Ionic liquids are increasingly used as solvents for the synthesis of thiazole derivatives (Scheme 12.10). The synthesis of 2,4-diarylthiazoles from arylthioamides and α-bromoacetophenones in [BMIM]BF_4 was carried out under ultrasound irradiation at room temperature. These compounds were synthesized with excellent yields (91–98%) with a number of electron-deficient as well as electron-rich on both arylthioamides and α-bromoacetophenones. Although the differences in the transformation times are slight, generally, the electron-rich arylthioamides were superior to the electron-deficient ones in this regard. Similar results were also observed in the case of substituted α-bromoacetophenones. However, the longest reaction times are observed for the substrates bearing electron-withdrawing groups. When 1,3-benzenedithioamide was used in an analogous reaction, 1,3-dithiazolylbenzenes were obtained with 84–95% yield [82].

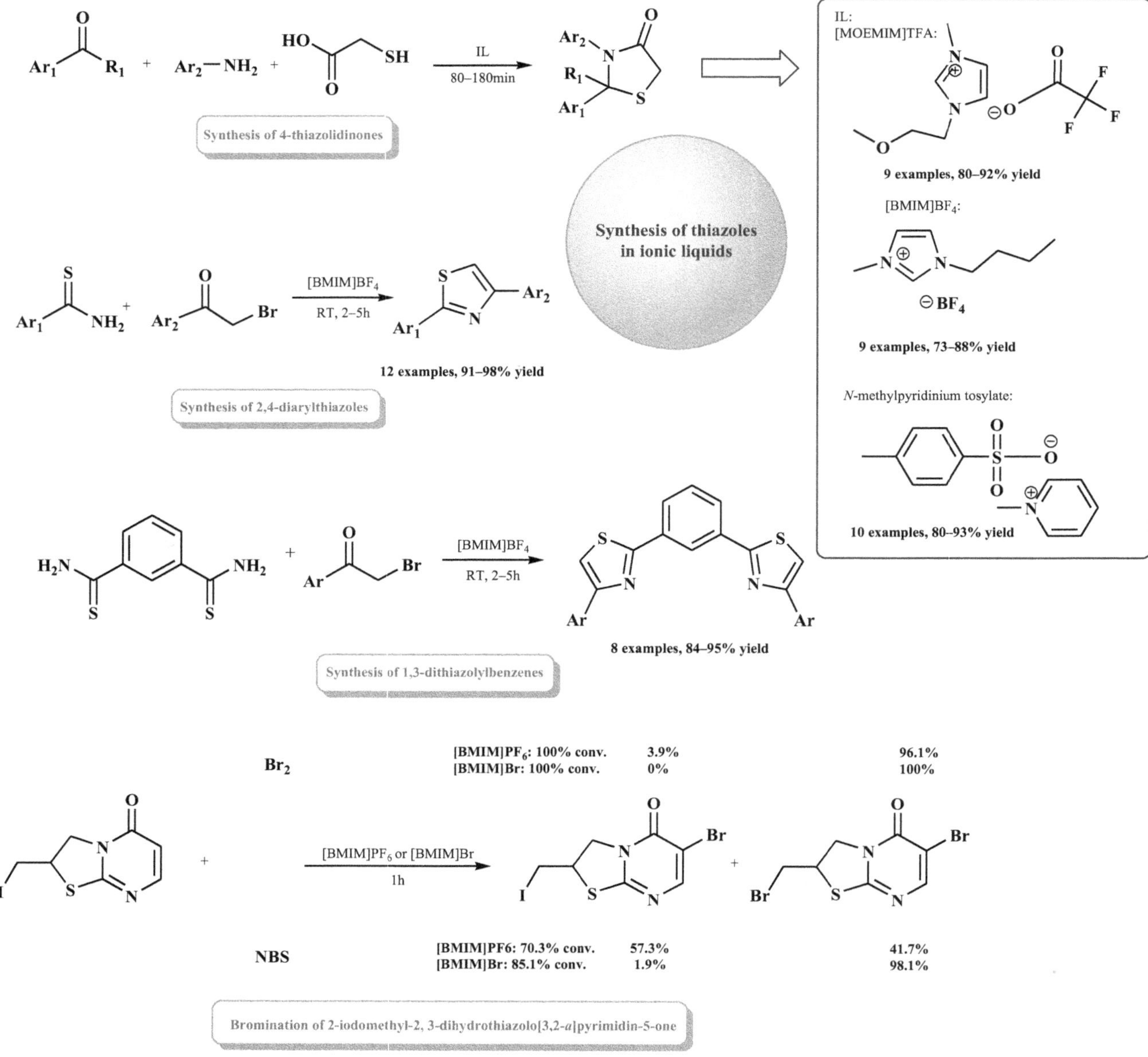

SCHEME 12.10 Greener synthesis of thiazole derivatives in ionic liquids.

The synthesis of thiazolidinones from hetero/aromatic amines, 2-mercaptoacetic acid, and carbonyl compounds was carried out in various ionic liquids. The use of 1-butyl-3-methylimidazolium tetrafluroborate [BMIM]BF_4 as a reaction medium allowed to obtain 2,3-disubstituted thiazolidinone derivatives with a yield of 73–88%. In the case of conducting the reaction in 1-methoxyethyl-3-methylimidazolium trifluoroacetate [MOEMIM]TFA gave the same products with higher yields. This may be attributed due to the ability of [MOEMIM]TFA to hydrogen bond with amines [68]. Other diaryl-substituted thiazolidinone derivatives have been synthesized in an analogous reaction in *N*-methylpyridinium tosylate. In this case, the reactions were carried out longer and at a higher temperature (3 h at 120°C) and the products were obtained with yields of 80–93% [83].

Ionic liquids as solvents have also been used in electrophilic and nucleophilic bromine substitution reactions in 2-iodomethyl-2,3-dihydrothiazolo[3,2-*a*]pyrimidin-5-one. The bromination reaction of 2-iodomethyl-2,3-dihydrothiazolo[3,2-*a*] pyrimidin-5-one using NBS (*N*-bromosuccinimide) and bromine, in two ionic liquids, [BMIM]PF_6 and [BMIM] Br, was conducted. The reaction led to two products: 6-bromo-2-(iodomethyl)-2,3-dihydrothiazolo-[3,2-*a*]pyrimidin-5-one and 6-bromo-2-(bromomethyl)-2,3-dihydrothiazolo[3,2-*a*] pyrimidin-5-one. Similar to organic solvents, in ionic liquid, the composition of the reaction mixture depended on the brominating agent used. Regardless of the solvent, in the bromination reaction with Br_2 the main reaction product was dibromo compound which was a result of the course of the two substitution reactions: electrophilic and nucleophilic. In [BMIM]

Br, the reaction occurred completely selectively. The use of the ionic liquid as a reaction medium increased the rate of reaction in comparison with organic solvents. Dibromo compound was predominantly formed also in [BMIM]Br using NBS as a brominating agent. In [BMIM]PF_6 dibromo compound was detected as 41.7% of the reaction mixture. In this case, the monobromo compound which was obtained as a result of electrophilic aromatic substitution was the main product of the bromination reaction (57.3%) [84].

12.2.1.6 Oxazole, Isoxazole, and Oxazines Derivatives

Five- and six-membered heterocycles containing nitrogen and oxygen atoms are also biological active compounds. Oxazoles and isoxazoles representatives show inter alia antimicrobial, anticancer, antitubercular, anti-inflammatory, and antidiabetic activity [85]. Oxazines derivatives additionally exhibit antioxidant and anti-osteoarthritis effects [86].

Cu-catalyzed synthesis of benzoxazoles from *o*-halobenzanilides in water was carried out using $Cu(OAc)_2$ as Cu sources, DPPAD (1-[2-(*N*-(3-diphenylphosphinopropyl))aminoethyl] pyrrolidine) as a ligand and triethyl amine as the base at 110°C (Scheme 12.11). Reactions were performed for 14 different bromobenzanilides andiodobenzanilides, differing in substituents in the phenyl ring, to obtain benzoxazole derivatives with 70–90% yield. Iodobenzanilides displayed good reactivity under these conditions, and benzoxazoles were obtained in higher yields than with bromobenzanilides [87].

The reactions between 2-aminobenzoxazoles and phenacyl bromides in (IPA: water system give *N*-alkylated 2-aminobenzo[*d*]oxazoles with 92–94% yield [81].

The synthesis of oxazole derivatives was carried out at room temperature by using a combination of ultrasound (US) and DES (ChCl:urea). Four novel derivatives of oxazole were synthesized in just 12–17 min by the use of ultrasound and DES as compared to conventional heating in DES that took 3–5 h. Ultrasonic irradiation also showed a significant improvement in reaction yield (82–90% vs 45–65%) [88].

Biologically active isoxazoline derivatives were also efficiently synthesized *via* cyclization reaction of a chalcone and hydroxylamine hydrochloride with good yields (up to 87%) and in small reaction times using butylmethylimidazolium bromide as the solvent and catalyst [89].

The one-pot chemoselective synthesis of 2*H*-benzo[*b*][1,4] oxazin-3(4*H*)-one derivatives was carried out in reaction of corresponding *o*-aminophenols with 2-bromoalkanoates using DBU (1,8-diazabicyclo[5.4.0]undec-7-ene) in the ionic liquid 1-methyl-3-octylimidazolium tetrafluoroborate [OMIM]BF_4. In all cases, the reactions were completed rapidly at room temperature and the products were easily separated by extraction with Et_2O. Both in the presence of electron-releasing and electron-withdrawing substituents in the *o*-aminophenol molecule, the products were obtained with high yields in short time periods [90].

12.2.1.7 Other Potential Drugs

Phthalimides were tested as anti-inflammatory [91], antibacterial, antioxidant, and hemolytic agents [92]. The synthesis of *N*-aryl phthalimide derivatives from phthalic anhydride and primary aromatic amines was carried out in green synthesis using glycerol as the solvent. The products containing an unsubstituted phenyl ring and with Me- or -OMe substituents in *m*-, *p*-, and/or *o*-position, and -Cl or NO_2 in *p*-position were obtained with good yields of up to 72% [93].

An interesting example of the use of glycerol as a solvent is the preparation of chitosan as a result of the deacetylation of chitin. Chitosan, as the only alkaline amino polysaccharide in nature, has been extensively applied as a promising and versatile biopolymer. Due to its physicochemical properties and biological functions, including nontoxicity, antimicrobial/antioxidant activities, biodegradability, and biocompatibility, it has been widely applied such as drug delivery, food and cosmetics preservation, beverage/juice purification, and metal chelation from wastewater. Traditional methods for chitin deacetylation require an aqueous solution with high alkali content as the reaction solvent, which generates huge environmental pressure for the treatment of vast alkaline wastewater. Liu et al. show that glycerol has a great potential to serve as an alternative green solvent for chitin deacetylation. However, the reaction conditions have impacts on the chitosan properties. The chitosan obtained by the traditional method had a much higher value of molecular weight. The chitin degradation occurred during the deacetylation reaction, which induces the reduction of chitosan molecular weight which indicates that the deacetylation and degradation of chitin proceeded concurrently during the reaction [94].

Benzodiazepines represent a significant class of biologically active nitrogen compounds and exhibit a number of important biological properties, such as anticonvulsant, antianxiety, anti-inflammatory, analgesic, hypnotic, antidepressive, antihistaminic, antiulcerative, antiallergic, and antipyretic. Their syntheses have received much attention in the field of medicinal and pharmaceutical chemistry. Benzodiazepines (Scheme 12.12) can be obtained by green synthesis starting from aromatic diamine and ketone. Reactions were carried out in glycerol. Benzodiazepine derivatives were obtained with 45–96% yield [95].

12.2.1.8 Substrates in the Synthesis of Potential Drugs

Many syntheses leading to the obtaining of potential drugs are multi-step reactions, in which the first step is to obtain a compound with a complex structure, containing specific functional groups. Therefore, the possibility of leading green syntheses of such compounds as amines, alcohols, nitriles, or carbonyl compounds play an important role in the formation of drugs as well as in the search for new biologically active compounds.

Phenacyl derivatives are important intermediates for the synthesis of valuable molecules applied to heterocyclic chemistry on account of the fact that many of them are subunits of natural products and pharmaceutical agents. They are substrates in synthesis i.e. furan, imidazole, thiazole, and oxazole derivatives [69, 81, 82, 88].

Nucleophilic substitution reactions in phenacyl bromide (α-bromo acetophenone) derivatives with azide, thiocyanate,

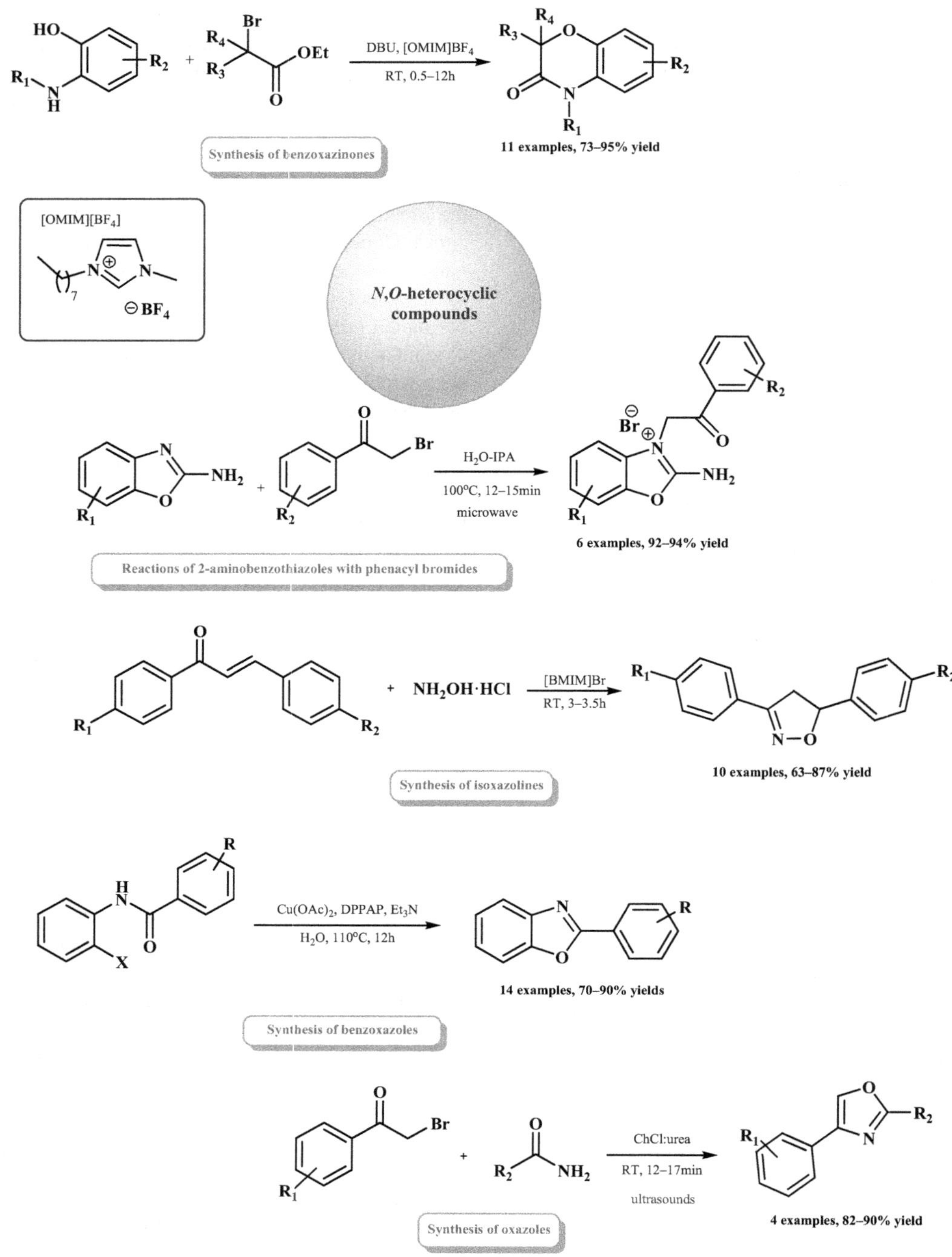

[OMIM][BF$_4$]

DBU, [OMIM]BF$_4$, RT, 0.5–12h

11 examples, 73–95% yield

Synthesis of benzoxazinones

N,O-heterocyclic compounds

H_2O-IPA, 100°C, 12–15min, microwave

6 examples, 92–94% yield

Reactions of 2-aminobenzothiazoles with phenacyl bromides

[BMIM]Br, RT, 3–3.5h

10 examples, 63–87% yield

Synthesis of isoxazolines

Cu(OAc)$_2$, DPPAP, Et$_3$N, H_2O, 110°C, 12h

14 examples, 70–90% yields

Synthesis of benzoxazoles

ChCl:urea, RT, 12–17min, ultrasounds

4 examples, 82–90% yield

Synthesis of oxazoles

SCHEME 12.11 Green synthesis of biological active five- and six-membered heterocycles with nitrogen and oxygen atoms.

cyanide, acetate, and iodine ions were carried out in water in the presence of nano-magnetite-supported organocatalyst PEG@ SiO_2@Fe_3O_4 (PEG bonded onto silica-coated ferrite and the nanoparticles). Fenacyl bromide and its three *para*-substituted derivatives were used for the reaction, resulting in 20 products with 75–94% yield (Scheme 12.13) [96].

Nitriles are also involved in the formation of biologically active compounds, including indoles [43], pyrans [50], pyranopyrazoles [62], or thiazoles [79]. The reaction of aromatic aldehydes with hydroxylamine hydrochloride, which results in aromatic nitriles is an example of a reaction using glycerol as the solvent. The reaction carried out at 90°C for benzaldehyde derivatives with electron-withdrawing substituents in the 2, 3, or 4 positions of the aromatic ring allowed to obtain 13 different nitriles in high yields [97].

Compounds containing hydroxyl and carbonyl groups also play an important role in the synthesis of potential drugs. They are building blocks, to form many heterocyclic systems.

NADES formed by D-glucose and racemic malic acid are suitable media to perform the enantioselective L-proline-catalyzed intermolecular aldol reaction between acetone and different aromatic aldehydes, creating simultaneously and

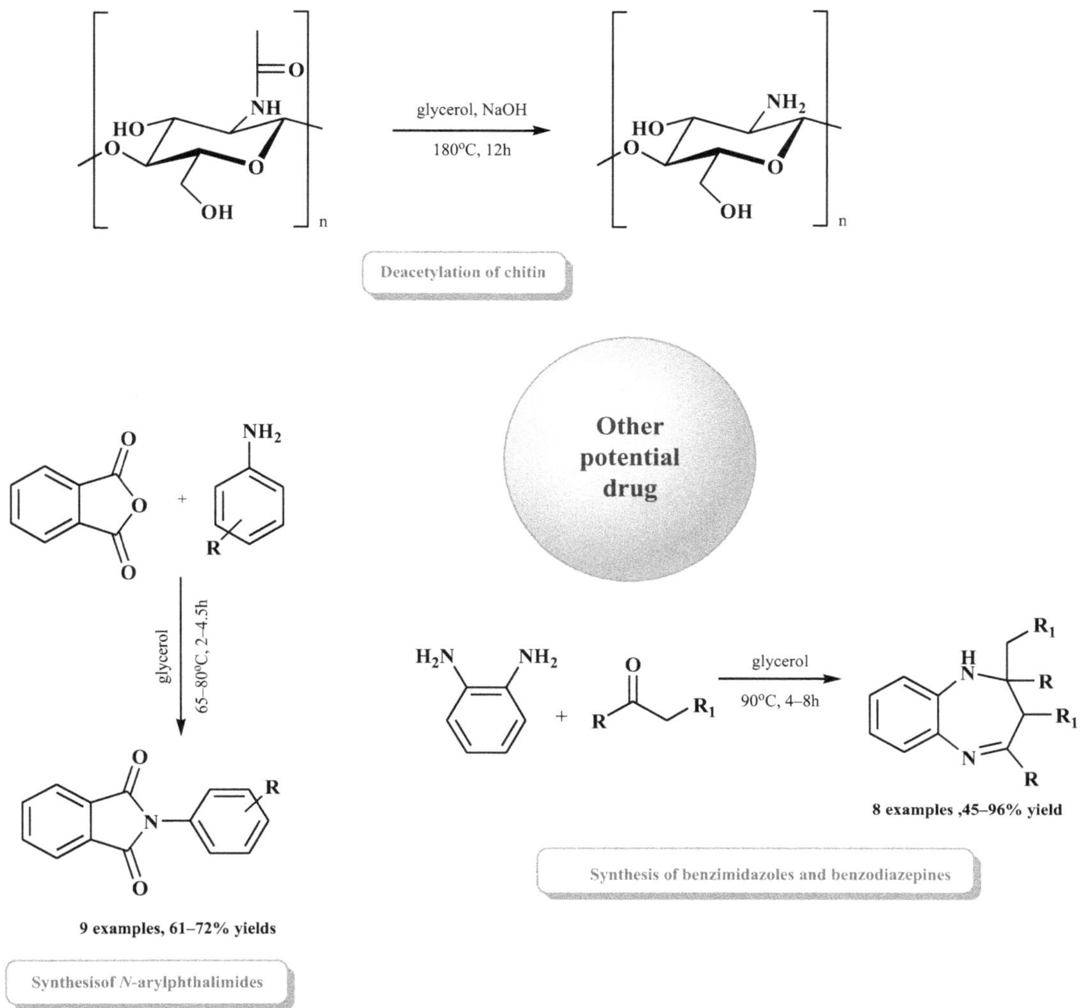

SCHEME 12.12 Greener synthesis of other potential drugs.

selectively a C–C bond and a new stereocenter. The products were obtained with good levels of enantioselectivities (60–78% ee) [98].

Noteworthy are reduction reactions of carbonyl compounds in which glycerol is both a solvent and a hydrogen source. The issue was described more extensively by Diaz-Alvarez et al.; therefore, no detailed data were provided in this study, but only general reaction patterns [99].

The oxidative hydroxylation of aryl boronic acids into phenols in the presence of a heterogeneous photocatalyst based on copper-doped g-C_3N_4 was carried out in water under room temperature in the irradiation of blue light (460 nm). Various phenols were received as products with high yields (76–99%) in a short reaction time [100].

Amines are an important group of compounds used in the synthesis of various biologically active heterocycles (see Schemes: A, B, E, F, G). One of the basic methods of amine synthesis is the reduction of nitro compounds. This reaction can also be performed as a green synthesis using glycerol as solvent and hydrogen source in the presence of magnetic ferrite nickel nanoparticles (Fe_3O_4-Ni) as catalyst [101]. Using this method, 19 different aryl amines were obtained with 84–94% yield (Scheme 12.14).

N-arylation of amines with aryl halides was also conducted in glycerol in the presence of KOH at 100°C within 15 h. Copper acetylacetonate ($Cu(acac)_2$) was used as the catalyst. 17 disubstituted amines were obtained with 70–98% yield [102].

N-acetylation of amines is an example of a reaction carried out in DESs for eight different aromatic amines, one alicyclic amine, and one hydrazide in two different DESs: ChCl:urea and ChCl:malonic acid. 1-Naphthylamine was acetylated with the highest yield (86%) and in the shortest time (15 min) in ChCl:urea DES. When aniline was acetylated in ChCl:urea DES, acetanilide was received with 91% yield in only 30 min. Of the substrates used, only when 2-nitro-4-(trifluoromethyl) aniline was used, no acetylated product was obtained in both DESs. Generally, ChCl:urea (1:2) was the better DES in terms of reaction time, temperature, yield, and purity of final compounds compared to ChCl:malonic acid system [103].

The ILs ($[CHCl][ZnCl_2]_2$) promoted direct nucleophilic substitution of alcohols with compounds containing amino groups

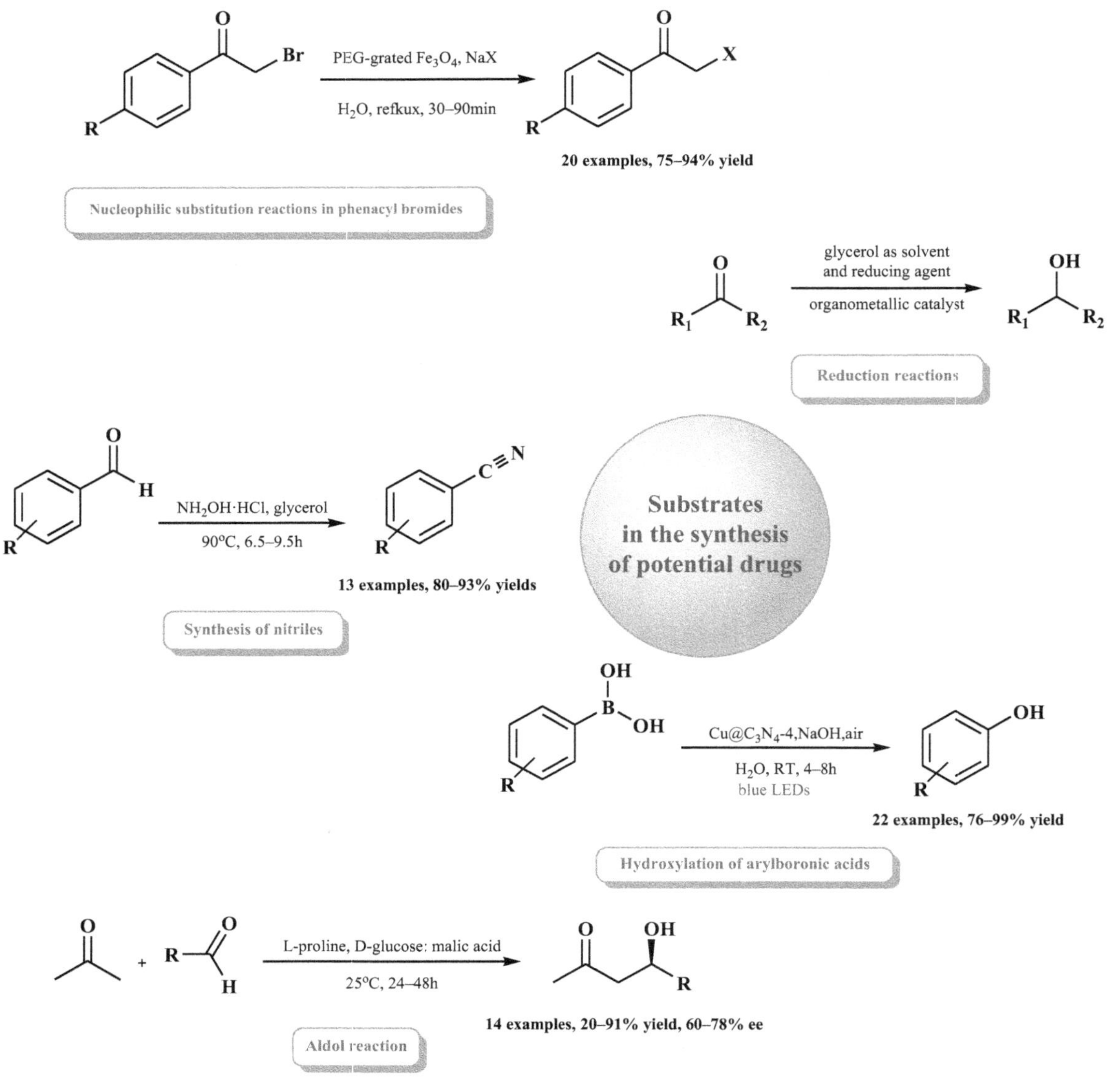

PEG-grated Fe_3O_4, NaX
H_2O, refkux, 30–90min
20 examples, 75–94% yield
Nucleophilic substitution reactions in phenacyl bromides

glycerol as solvent and reducing agent
organometallic catalyst
Reduction reactions

$NH_2OH{\cdot}HCl$, glycerol
90°C, 6.5–9.5h
13 examples, 80–93% yields
Synthesis of nitriles

Substrates in the synthesis of potential drugs

Cu@C_3N_4-4,NaOH,air
H_2O, RT, 4–8h
blue LEDs
22 examples, 76–99% yield
Hydroxylation of arylboronic acids

L-proline, D-glucose: malic acid
25°C, 24–48h
14 examples, 20–91% yield, 60–78% ee
Aldol reaction

SCHEME 12.13 Greener synthesis of substrates for potential drugs.

and played dual roles efficiently in these reactions as both catalysts and solvents, to give the products with good-to-excellent isolated yields (82–96%). It was suggested that the hydroxyl group on the choline cation was the major inducement for the formation of special microstructures which could provide adequate stability for the carbocation in the reaction systems and increase the reactivity and selectivity [104].

A range of α-aminophosphonates was synthesized in a three-component reaction in high yields using solvate ionic liquids (SILs) as the reaction media. The SILs used in the reaction are being equimolar mixtures of LiTFSI in tri- or tetra-glyme (referred to as $[G_3(Li)]$TFSI or $[G_4(Li)]$TFSI, respectively). The reactions were carried out within 5 min and precipitation of the products into water excluded the use of traditional work-up procedures, giving the products very high purity. Depending on the ionic liquid and amine used, the reaction yields ranged from 25% to 96% for aromatic monoamines and up to 64% for aromatic diamines [105].

12.2.1.9 Biocatalysis

In accordance with the concept of "Green Chemistry", such solutions in the synthesis of chemicals and pharmaceuticals should be promoted, which are based on the minimization or complete elimination of toxic reagents. In this context, biocatalysis in green solvents is the ideal solution. Biocatalysts are biodegradable, reusable, more efficient due to unique features such as high selectivity, enabling the production of chiral, optically pure drug-building blocks under mild conditions [9, 105].

The most commonly used solvents are called green, which successfully carry out biocatalytic processes, both with isolated enzymes as well as with whole cells of microorganisms are water, ILs, and NADES.

Some examples of the use of biocatalysts are presented below (mainly hydrolases and oxidoreductases) in the industrial synthesis of pure building blocks and biologically active compounds useful in the pharmaceutical industry.

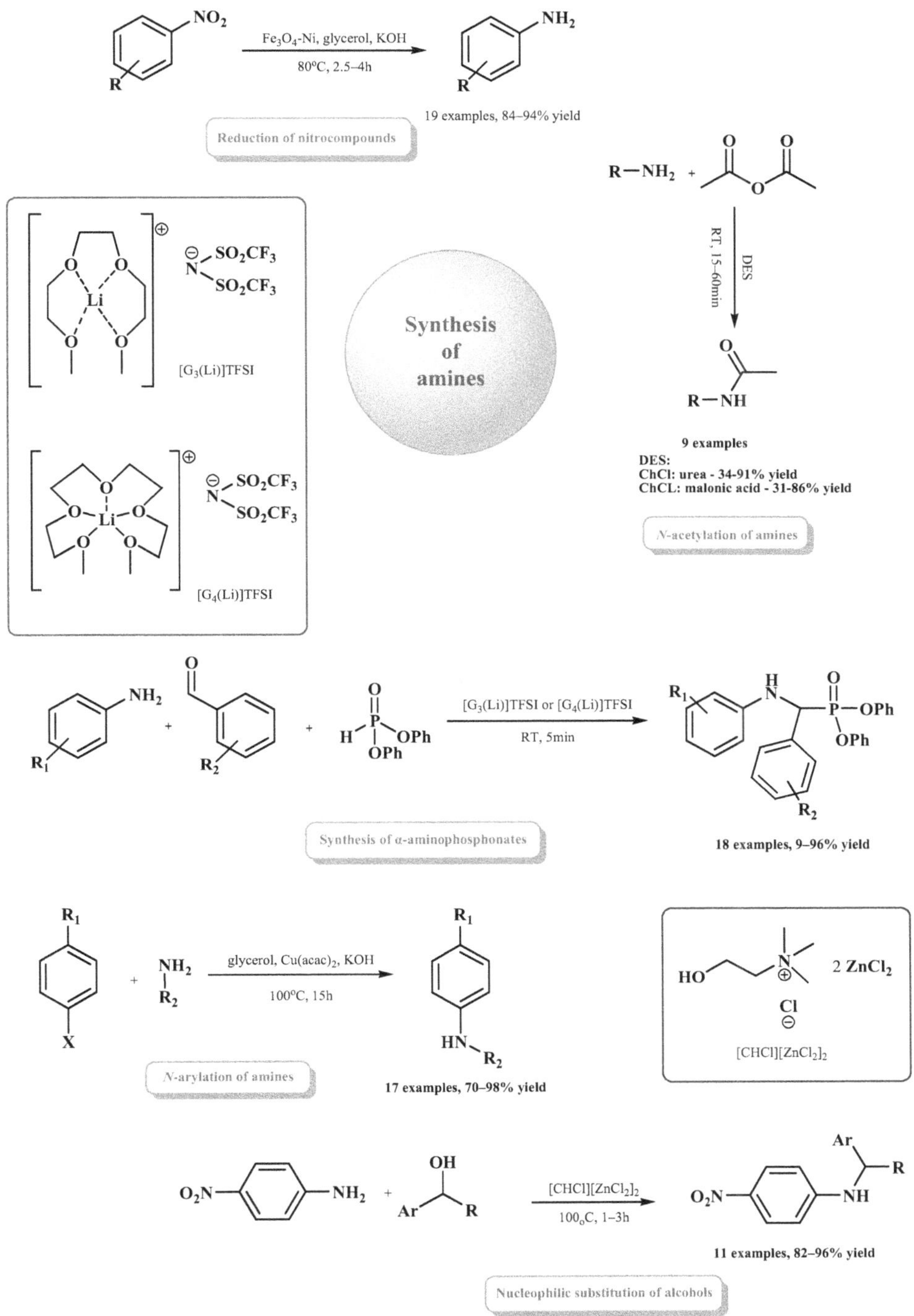

SCHEME 12.14 Greener synthesis and modifications of amines.

Hydrolases, more particularly lipases, present different advantages over other biocatalysts, as they require no cofactors for their catalytic behavior, and many of them are commercially available and easy to handle biocatalysts. Due to the availability, stability, and acceptability of a wide range of substrates, lipases are often used in bioorganic syntheses. They can catalyze numerous solvolytic reactions of a carboxyl group such as hydrolysis, transesterification (alcoholysis), esterification, acidolysis, and amino- or ammonolysis (amide synthesis). Moreover, lipases are characterized by high regio- and stereoselectivity. They are capable of carrying out reactions with one specific functional group on a substrate, kinetic resolution of racemic mixtures, and asymmetric biotransformation of prochiral compounds and *meso*-synthons [9, 106].

Per-*O*-acetylated thymidine is an interesting intermediate in the synthesis of different fungicidals, antitumors, and, especially, antiviral agents [107]. A green regioselective deprotection reaction in the presence of covalent immobilization of *Candida rugosa* lipase (CRL) was performed in water. The Ald-CRL preparation (CRL immobilized on Ald-Sepharose)

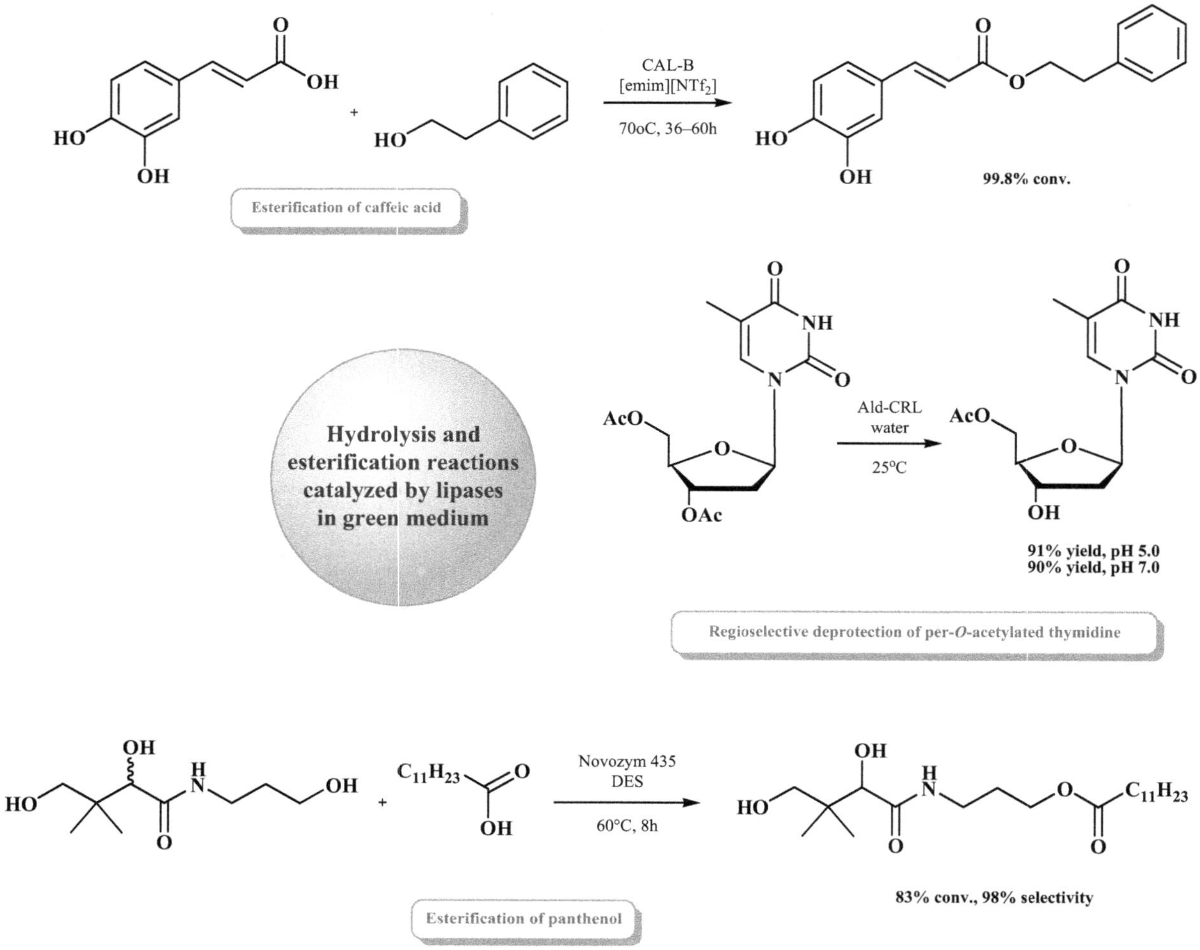

CAL-B
[emim][NTf$_2$]
70oC, 36–60h
99.8% conv.
Esterification of caffeic acid

Hydrolysis and esterification reactions catalyzed by lipases in green medium

Ald-CRL
water
25°C
91% yield, pH 5.0
90% yield, pH 7.0
Regioselective deprotection of per-*O*-acetylated thymidine

Novozym 435
DES
60°C, 8h
83% conv., 98% selectivity
Esterification of panthenol

SCHEME 12.15 Hydrolysis and esterification reactions catalyzed by lipases in green medium.

showed the highest activity (three times higher than the soluble enzyme) and regioselectivity in the monoacetylation at pH 5.0 and 7.0, producing the C-3 hydroxy monoacetylated thymidine with around 90% yield (Scheme 12.15) [108].

The esterification reaction catalyzed with lipase B from *Candida antarctica* (CAL-B) can be obtained acid phenethyl ester (CAPE) shows a wide range of biological activities i.e. antioxidant, antibacterial, anti-inflammatory, antitumor, and immunomodulatory activities. The esterification of caffeic acid (CA) with 2-phenylethanol was carried out in the ionic liquid of bis[(trifluoromethyl) sulfonyl]imide 1-ethyl-3-methylimidazolium ([EMIM]NTf$_2$). Under optimized conditions (60 h, 73.7°C, the molar ratio of alcohol to caffeic acid (27.1:1), and the weight ratio of enzyme to CA (17.8:1)), the product can be obtained with the maximum degree of conversion to 99.8% (Scheme 12.15) [109].

An example of the use of the green NDES solvent in the esterification reaction catalyzed by Novozym 435 (lipase B from *C. antarctica*) is the synthesis of panthenyl monoacyl esters (PMEs), which are used in the production of pharmacological and cosmetic preparations [110]. Panthenol ((2,4-dihydroxy-*N*-(3-hydroxypropyl)-3,3-dimethylbutanamide called pro-vitamin of B$_5$) was esterified with higher fatty acids such as capric, lauric, myristic, palmitic, oleic, and linoleic acids. Mixtures of panthenol with free fatty acids were seen to act as DESs that were excellent reaction media for the biocatalytic synthesis of PMEs (i.e. up to 83% conversion and 98% selectivity in the case of panthenyl monolaurate) – Scheme 12.15 [111].

The lipases can be successfully employed in the reaction to create a new C–S bond and C–C (Scheme 12.16).

An example would be thiolysis and hydrotiolation reactions in which C–S bond formation occurs in the molecule. PPL, lipase from porcine pancreas, in conjunction with water on reaction with different thiophenols and styrene oxides, undergo thiolysis with C–S bond formation without the use of any catalysts toward the formation of β-hydroxysulfides with 86–94% yield with high regioselectivity at room temperature. Furthermore, PPL also facilitates thiophenols to undergo hydrothiolation with styrenes or phenylacetylenes in sole water and thus forming linear thioethers or vinyl sulfides respectively *via* C–S bond formation with 81–96% yield (Scheme 12.16). Compounds containing a C–S bond belong to one of the most useful and versatile synthons in organic synthesis, important building blocks for various natural products and pharmaceuticals [112].

The synthesis of bis(indolyl)methanes catalyzed by lipase TLIM (lipase from *Thermomyces lanuginosus* immobilized

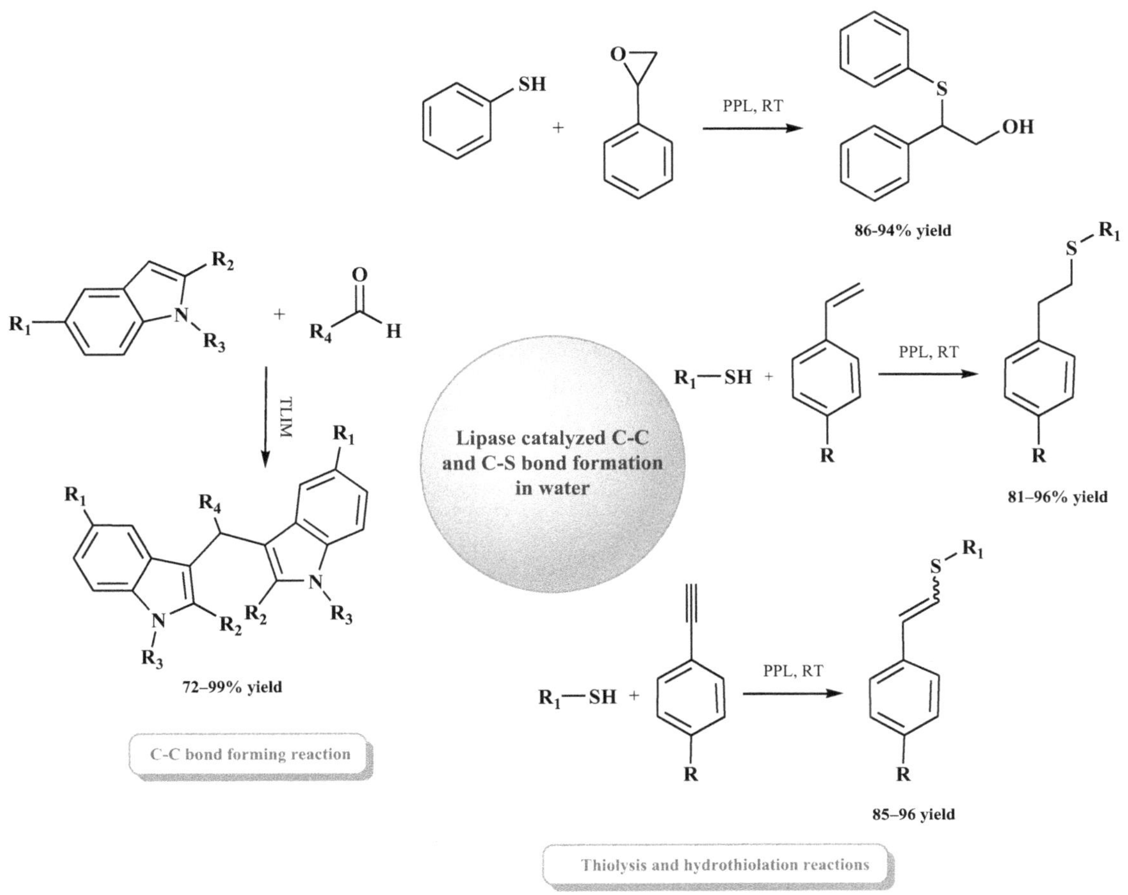

SCHEME 12.16 Lipase-catalyzed C–C and C–S bond formation in green solvent.

on particle silica gel) through the reactions of indole with aromatic and aliphatic aldehydes in pure water is an example of C–C bond-forming reactions. Reactions proceed with high yields (72–99%) – Scheme 12.16 [113]. Bis(indolyl) methanes make great contributions in the pharmacological industry. They may act as agonists of the immunostimulatory orphan G protein-coupled receptor GPR84, as antagonists of *Leishmania donovani* promastigotes as well as axenic amastigotes [114], and also as probes for monitoring the interactions of related molecules with transport and response-mediating proteins [115]. In addition, bis(indolyl) methanes play an important role in the treatment of prostate cancer [116], colon cancer [117], pancreatic cancer [118], and breast cancer [119].

Amides are ubiquitous in nature, including many important biological compounds as well as many drugs [106]. CALB was the catalyst in ammoniolysis reactions using the ionic liquids as reaction media. The reaction of ethyl octanoate with ammonia in the ionic liquid [BMIM]BF_4 gave octanamide with a yield of 40% (Scheme 12.17) [120]. Another enzyme belonging to the class of hydrolases used in biocatalysis in the synthesis of peptides is penicillin G (PGA) acylase. It is used in the pharmaceutical industry for the production of semi-synthetic β-lactam antibiotics such as cephalosporins and penicillins, for example, cephalexin, cefadroxil, cefazolin, ampicillin, amoxicillin [121]. Amoxicillin, prepared by the reaction catalyzed by penicillin G acylase in ionic liquid [BMIM]PF_6 mixed with water in the ratio 75:25 v/v (Scheme 12.17). In this system, an increase in selectivity by as much as 400% was observed compared to the reaction carried out in a completely aqueous medium. Amoxicillin has a broad spectrum of activity, high solubility, a high rate of absorption, and is stable under acid conditions, allowing the oral administration of this drug, which resists the gastric pH [122].

Lipases are commonly the biocatalyst of choice for the synthesis of chiral compounds, through the kinetic resolution (KR) of racemic mixtures or the enantioselective enzymatic desymmetrization (EED) of prochiral and *meso* compounds [106, 123].

The kinetic resolution reaction catalyzed by various commercial lipases (including CALB and *Rhizopus delemar* lipase) makes it possible to obtain (*S*)-1-(2,6-dichloro-3-fluorophenyl) ethanol. High enantioselective enzymatic hydrolysis was observed in phosphate buffer solution (pH 7.2) and provided the (*S*)-ester and (*R*)-alcohol with ee ranging from 80% to 97% (Scheme 12.18) [123]. (*S*)-1-(2,6-dichloro-3-fluorophenyl)ethanol is an intermediate in the production of the potent anticancer compound Crizotinib [124].

Another KR enzymatic reaction catalyzed by Novozym 435 allows receiving the precursor γ-aminobutyric acid derivative (GABA) – pregabalin (3-aminomethyl-5-methylhexanoic acid), which has anticonvulsant, anxiolytic-like, and analgesic

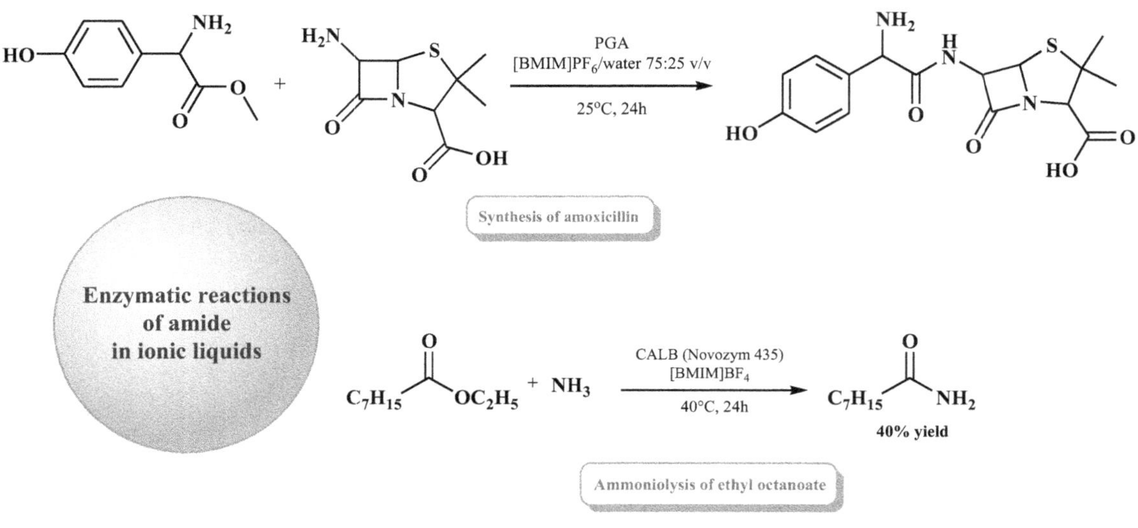

SCHEME 12.17 Enzymatic reactions of amide in green solvent.

properties [125]. Enzymatic hydrolysis of ethyl 2-(nitromethyl)alkanoates performed in a buffer solution at pH 7.4, at room temperature affording (*S*)-γ-nitro acids with moderate conversion (18–24%) and ee ranging between 65% and 94% (Scheme 12.18). Reduction of the nitro group, carried out with hydrogen, in the presence of Ra-Ni as a catalyst, under atmospheric pressure, gave 5-methyl-3-aminomethylhexanoic acid allowed to obtain therapeutically useful compound (*S*)-(+)-pregabalin [126].

The immobilized lipase from *P. cepacia* (PS-C, Amano) was used for the kinetic resolution of different substituted 1,2-diols in [BMIM]PF_6.1,2-Diols are important precursors and intermediates for a variety of synthetic applications and this functionality is found in a number of pharmaceuticals or their intermediates [127, 128]. In the lipase-mediated transesterification in ionic liquid medium, the acetylation process initially takes place regioselectively on the primary hydroxyl group of the 1,2-diols to afford their (*R*)-monoacetates in 1–2 h. Later, the formation of diacetate products takes place in an enantioselective manner. Various substituted 1,2-diols were converted enantioselectively into (*S*)-diacetates with high enantioselectivity (>99%) (Scheme 12.18) [127].

CALB was employed in the desymmetrization of the prochiral diethyl 3-[3′,4′ dichlorophenyl]glutarate a chiral intermediate for the synthesis of antagonists of tachykinins receptors NK1 and NK2, compounds with potential activity in the treatment of asthma, arthritis, and migraine [129, 130]. This enzyme provided a high selectivity of the reaction resulting in the ester with the (*S*)-configuration with ee >99% and an average isolated yield of 80% (Scheme 12.19). The reaction was carried out in a phosphate buffer solution [131].

On the other hand, the *R* monoester of 3-phenylglutarate was obtained in an enzymatic reaction in DES. The use of choline chloride ChCl:urea/phosphate buffer 50% (v/v) as a reaction medium increased the selectivity of Novozym 435 by 16% (ee = 88%) with respect to the one in 100% phosphate buffer (ee = 76%).Optically pure (99% ee) product was obtained under optimized conditions for high substrate concentrations (well above the solubility limit, 27-fold) in ChCl:urea/phosphate buffer 50% (v/v) at 30°C (Scheme 12.19) [132].

Prochiral pyrimidine acyclonucleoside propane-1,3-diol derivatives have also been subjected to enantioselective enzymatic desymmetrization in the ionic liquid using different acyl donors. Enantiomerically pure propane-1,3-diol derivatives constitute the interesting family of substrates since, in most cases, they are precursors in the synthesis of biologically active compounds. For example, 2-substituted-propane-1,3-diols are the key intermediates in synthesis of immunosuppressants [133], amino acids derivatives [134], antifungal agents [135] and antitumor antibiotics [136, 137].

The prochiral 1-((1,3-dihydroxypropan-2-yl)oxy) methyl)-5-methylpyrimidine-2,4(1*H*,3*H*)-dione were used as starting materials in the enzymatic transesterification. This compound is known as a micromole uridine phosphorylase inhibitor from *Escherichia coli* and compounds with a similar structure are reported to inhibit *in vitro* growth of cancer cells [138]. In the case of using lipase B from *C. antarctica* (CALB) as the catalyst and [BMIM]PF_6 as the solvent, an increase in yield (41–45%) and enantioselectivity (57–61% ee) of the monoacyl derivative was observed compared to pyridine (14–33% yield and 14–22% ee) when vinyl acetate or vinyl benzoate was used as an acyl donor [139]. When using Lipase Amano PS from *Burkholderia cepacia* (BCL) in [BMIM]PF_6, comparable ee values (84%) were obtained, but much lower yield compared to conventional organic solvent (TBME) (52% in TBME and 18% in [BMIM]PF_6) [140].

Whereas, the acylation of 1-((1,3-dihydroxypropan-2-yloxy) methyl)-5,6,7,8-tetrahydroquinazoline-2,4(1*H*,3*H*)-dione in [BMIM]PF_6 in the presence of BCL was more efficient, and the product was obtained with a higher enantiomeric excess (95% ee at 50°C) if vinyl benzoate was used as acyl donor and the same ee (99% ee) with vinyl butyrate compared to mixtures ionic liquid/TBME (Scheme 12.19) [141].

The diacetate of *meso*-2-(2-propynyl)cyclohexane-1,2,3-triol was asymmetrized by hydrolysis with *C. antarctica* lipase in pH 7 phosphate buffer to give the corresponding 2-substituted

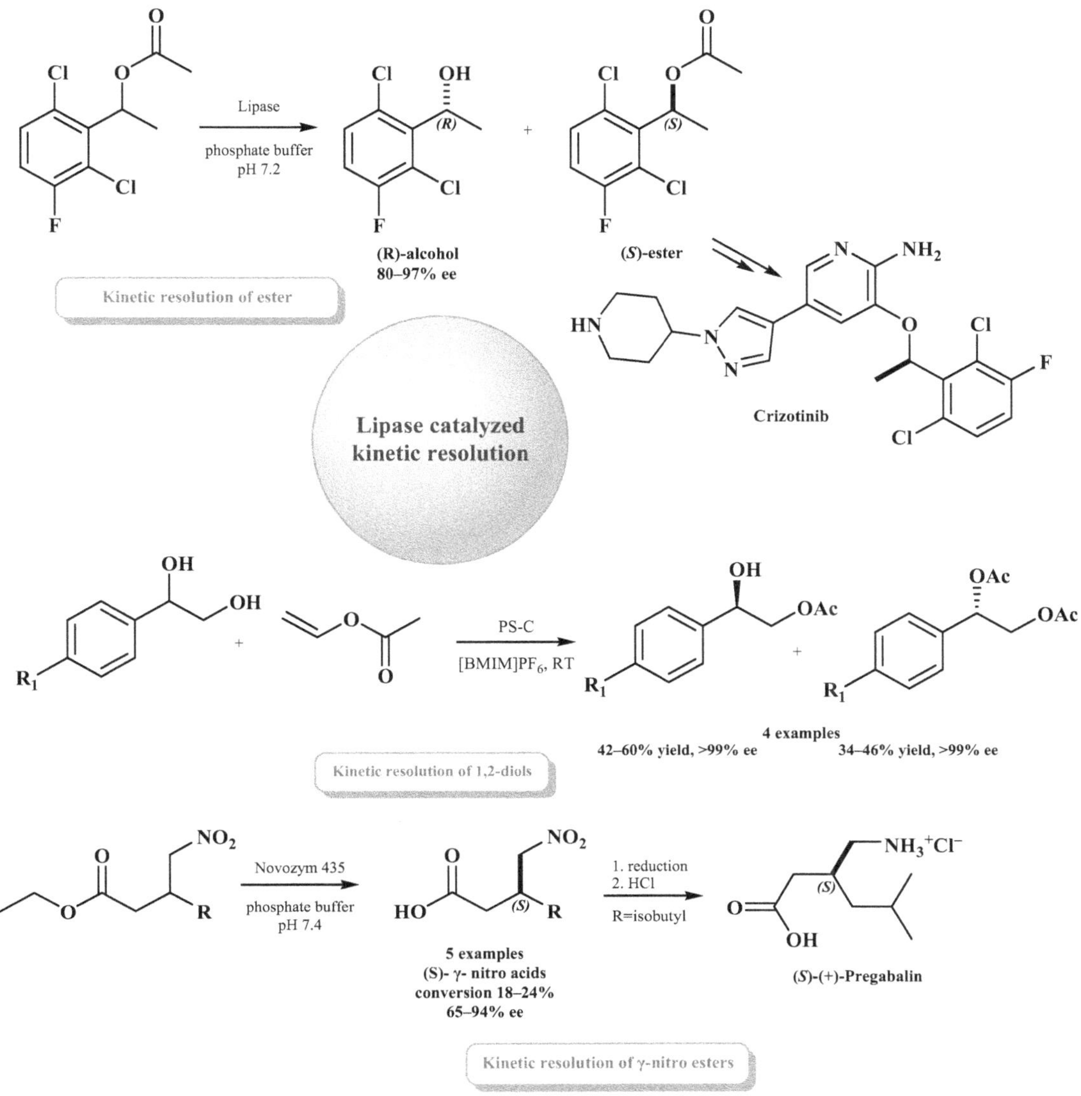

SCHEME 12.18 Lipase-catalyzed kinetic resolution in green solvent.

monoesters with good yield (80%) and excellent ee (>99%) – Scheme 12.20. (1*S*,2*R*,3*R*)-2,3-dihydroxy-2-vinylcyclohexyl acetate was used as the starting material for the total synthesis of aquayamycin [142].

Catalytic properties of esterase also found an application in the hydrolysis reaction of *meso* compounds. One of the intermediates in the synthesis of oseltamivir phosphate (Tamiflu), a prodrug used in the prevention and treatment of influenza infections, was obtained *via* an enzymatic desymmetrization of a *meso* 1,3-cyclohexanedicarboxylic acid diester. Pig liver esterase (PLE) provided a high selectivity of the reaction resulting in the product with the (*S*)-configuration with enantiomeric excess of 96–98% ee and with nearly quantitative yield. This reaction was performed in a TRIS buffer with pH 8.0 at 35°C (Scheme 12.20) [143].

Meso 1,3-imidazolin-2-one derivative diacetates proved to be excellent substrates for PLE (Scheme 12.20). The product of enzymatic hydrolysis in buffer solution served as a starting material for the synthesis of the vitamin (+)-biotin [144].

Biocatalytic reduction reactions of the carbonyl group are widely described in the literature, both those leading to alcohols and reductive amination reactions, which when using keto acids as substrates, lead to enantiomerically pure amino acids. The reduction reactions take place with the use of dehydrogenases, which when used in the reaction of prochiral substrates, often allow obtaining a chiral product with a high enantiomeric excess [125].

The application of isolated enzymes from the class of oxidoreductases and whole cells of organisms in enantioselective enzymatic desymmetrization of prochiral ketones leads to a broad spectrum of chiral alcohols and hydroxyl esters used as intermediates in the syntheses of many pharmaceuticals and compounds presenting a potential biological activity [125, 145, 146].

The whole-cell-mediated asymmetric reduction of α-, β-, and γ-chloroalkyl arylketones to optically active chlorohydrins was carried out in aqueous media. Among the various whole cells screened (baker's yeast, *Kluyveromyces*

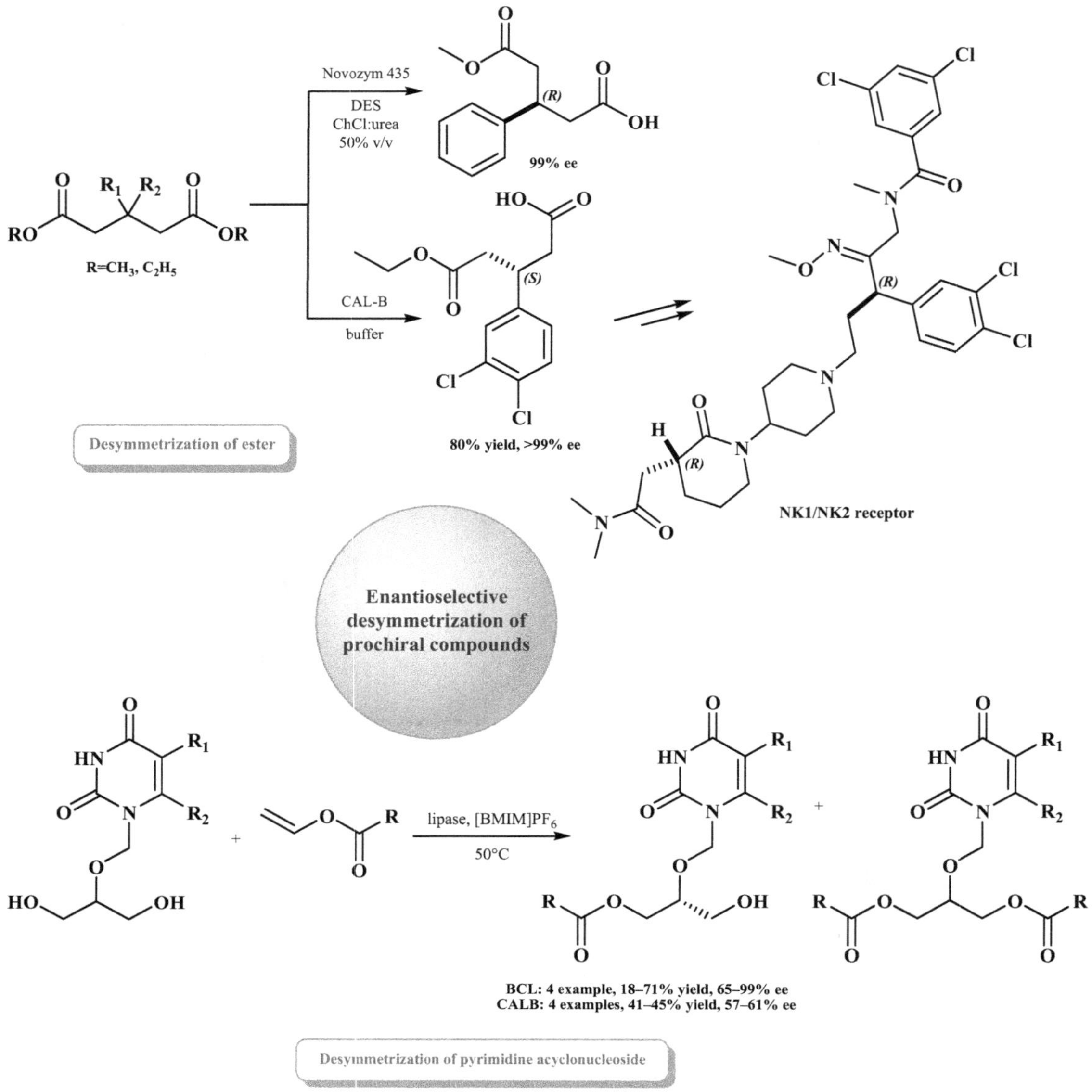

SCHEME 12.19 Enantioselective desymmetrization of prochiral compounds in green solvent.

marxianus CBS 6556, *Saccharomyces cerevisiae* CBS 7336, *Lactobacillus reuteri* DSM 20016), baker's yeast was the one providing the best yields and the highest enantiomeric ratios in the bioreduction of the ketones. In the case of β-chloroketones, the conversion to (*S*)-chlorohydrines were 12–85% with the best enatiomeric ratio 95:5 (90% ee). Among the γ-chloroketones, only γ-chloropropyl phenylketone was biotransformed to (*S*)-chlorohydrine in 49% with 90% ee. Baker's yeast successfully reduced α-chloroacetophenone and α-chloro-*p*-chloroacetophenone, providing the expected (*R*)-chlorohydrins with 53% and 64% yields, respectively, and with up to 80% ee. On the other hand, the *anti*-Prelog stereo-preference of *L. reuteri* DSM 20016 furnished (*S*)-2-chloro-1-phenylethanol with a 28% yield but with a higher stereoselectivity (92% ee) in comparison with baker's yeast. Thus, baker's yeast and *L. reuteri* DSM 20016 behave as two complementary whole-cell biocatalysts for the synthesis of optically active 2-chloro-1-arylethanols because of their ADHs opposite stereopreference, though with their own substrate specificity (Scheme 12.21) [147].

Chlorohydrin (1*S*,2*R*)-[3-chloro-2-hydroxy-1-(phenylmethyl)propyl]carbamic 1,1-dimethylethyl ester was prepared in the diastereoselective reduction of an appropriate α-chloroketone using microbial cultures among which three strains of *Rhodococcus sp.* gave >90% yield with a diastereomeric purity of >98% and an ee of 99.4% (Scheme 12.21). This compound is the key chiral intermediate required for the total synthesis of the HIV protease inhibitor, atazanavir [148, 149].

The water–glycerol combination leads to a homogenous system and avoids mass transfer limitations, which is common in biphasic systems. Furthermore, glycerol solubilizes several organic compounds but is immiscible with hydrophobic solvents, which makes possible, easy product recovery by extraction. Due to these properties, the study of the effect of the glycerol as a co-solvent in the bioreduction of

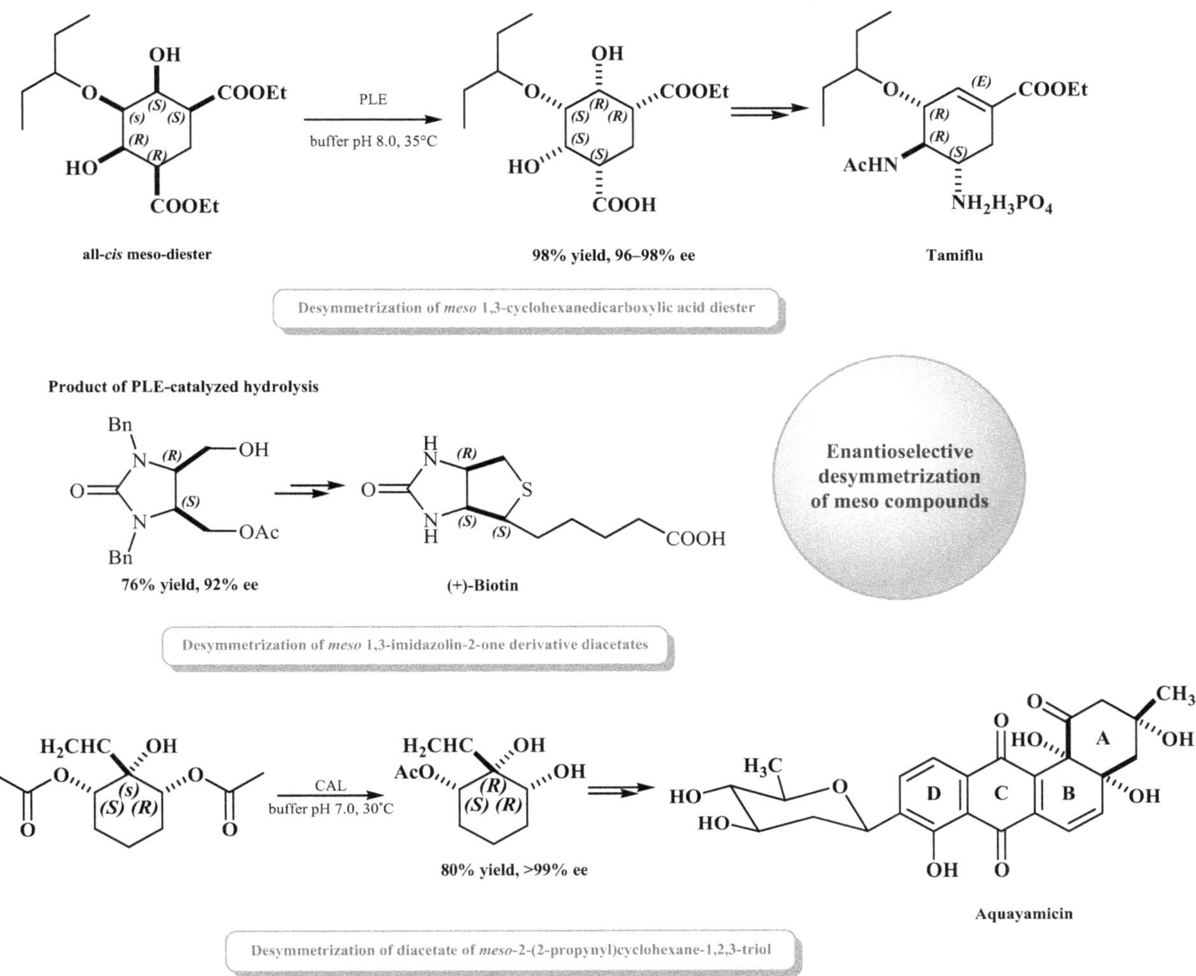

SCHEME 12.20 Enantioselective desymmetrization of *meso* compounds in green solvent.

haloacetophenones mediated by whole cells of *Aspergillus terreus* SSP 1498 and *Rhyzopus oryzae* CCT 4964 were carried out. For most haloacetophenones, the bioreduction provided the desired chiral alcohols in high enantiomeric excess (up to >99%) and conversion up to >99% (Scheme 12.21) [150].

The reduction of carbonyl compounds in an ionic liquid [EMIM]BF_4/water system, by immobilized *Geotrichumcandidum* cells containing an alcohol dehydrogenase on water-absorbing polymer was proceeded. Acetophenone derivatives, benzyl acetone, 2-hexanone, β-keto ester, and fluorinated epoxy ketone were used as substrates, and it was found that all of them were reduced with moderate-to-excellent yields (23–96%), and the enantioselectivity was excellent for all of the substrates tested ((*S*)-enantiomers were obtained with ee >99%) – Scheme 12.21 [151].

The microbiological reduction of α- and β-oxoesters and prochiral benzofuran and pyrimidine derivatives were subjected to selective bioreduction by Boni Protect fungicide containing live cells of *Aureobasidium pullulans* in aqueous media (Scheme 12.22). The reactions were carried out in phosphate buffer solution in the presence of glucose or sucrose as the energy source at 30°C. Reduction of α- and β-oxoesters by *A. pullulans* leads to 100% conversion to the corresponding (*S*)-hydroxyester with high ee (77–99%). Only in the case of *p*-chlorophenyl oxoester, the formation of the *R*-enantiomer was observed. The microbiological reduction of this oxoester by *S. cerevisiae* gave the same enantiomers with lower yield and ee sometimes in reverse configuration [152].

Whole-cell biocatalysis with the use of baker's yeast was conducted in different mixtures of water with DES (ChCl:glycerol). Enantioselective keto ester reduction at 10 and 50 vol.% water in the DES was observed for long reaction times (>200 h), which suggests that the whole cells remain stable in these solvents. Baker's yeast might maintain its integrity, and intracellular oxidoreductases still seem to be active after prolonged times. By changing the proportion of the DES added, a complete inversion of enantioselectivity is observed, from approximately 95% enantiomeric excess (ee) (*S*) in pure water to approximately 95% ee (*R*) in the pure DES (Scheme 12.22). Presumably, some (*S*)-oxidoreductases present in baker's yeast are inhibited by DESs [153].

The reaction in glycerol was carried out for prochiral β-keto esters such as methyl and ethyl acetoacetate with free baker's

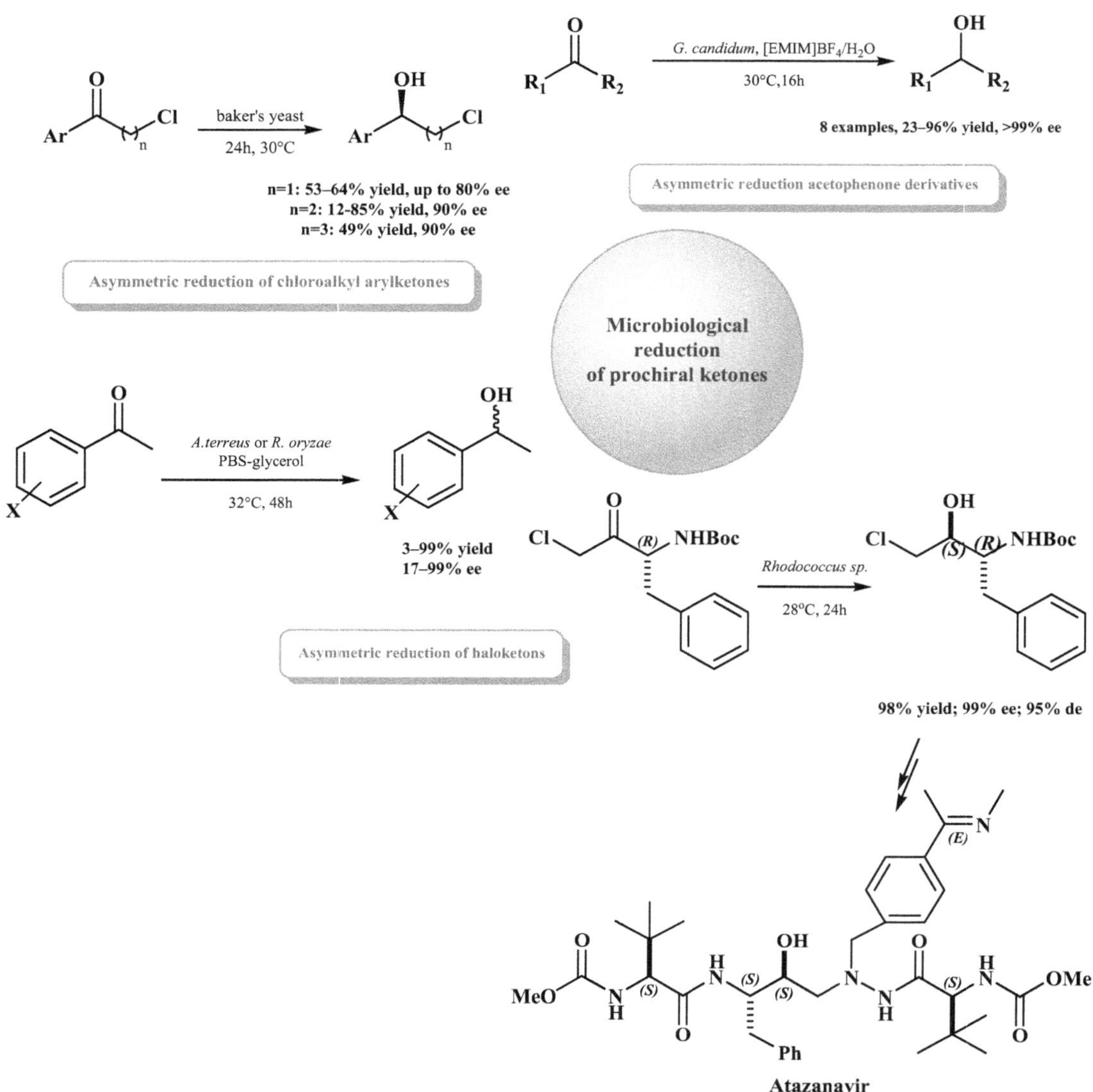

SCHEME 12.21 Microbial reduction of prochiral compounds in green solvent.

yeast (FBY) and immobilized baker's yeast (IBY). IBY always yielded higher activity than the corresponding reaction with FBY, while the enantioselectivity of the (*S*)-products were very high (95–99% ee) with both catalysts (Scheme 12.22) [154].

The selective reduction of α-chloro-β-ketoester, depending on enantio-complementary baker's yeast enzymes used, led to (*R*)- or (*S*)-enantiomers at high ee and de. In the presence of YDL124w reductase, syn-(2*S*,3*R*)-hydroxyester is prepared, whereas a short-chain YGL039w dehydrogenase gave a mixture of two diastereomers, *syn*-(2*R*,3*S*)- and *anti*-(2*S*,3*S*)-alcohol, in 9:1 ratio (Scheme 12.22). Both products are the precursors of the *N*-benzoyl phenylisoserine taxol side chain [155]. Taxol has emerged as the drug of choice for treating certain types of ovarian and breast cancers because it blocks microtubule disassembly [156].

Enzymatic reductive amination of α-keto acids using amino acid dehydrogenases is considered to be one of the most useful methods for the preparation of chiral unnatural amino acids because the enzymes have good stability, broad substrate specificity, and very high enantioselectivity [157].

The non-proteinogenic amino acid, which is a key intermediate required for the synthesis of drugs (e.g. Saxagliptin – a dipeptidyl peptidase IV inhibitor under development for the treatment of type 2 diabetes mellitus), can also be obtained by reductive amination in water. 2-(3-hydroxy-1-adamantyl)-2-oxoethanoic acid was converted to (*S*)-3-hydroxyadamantylglycine by reductive amination using a phenylalanine dehydrogenase from *Thermoactinomyces intermedius* (expressed in a modified form in *Pichia pastoris* or *E. coli*) with 95–100% yield (Scheme 12.22). NAD (nicotinamide adenine dinucleotide) produced during the reaction was recycled to NADH using formate dehydrogenase [158].

Oxygen- and nitrogen-containing heterocycles are ubiquitous in natural products and biologically active compounds, and are also very common in many pharmaceuticals [125].

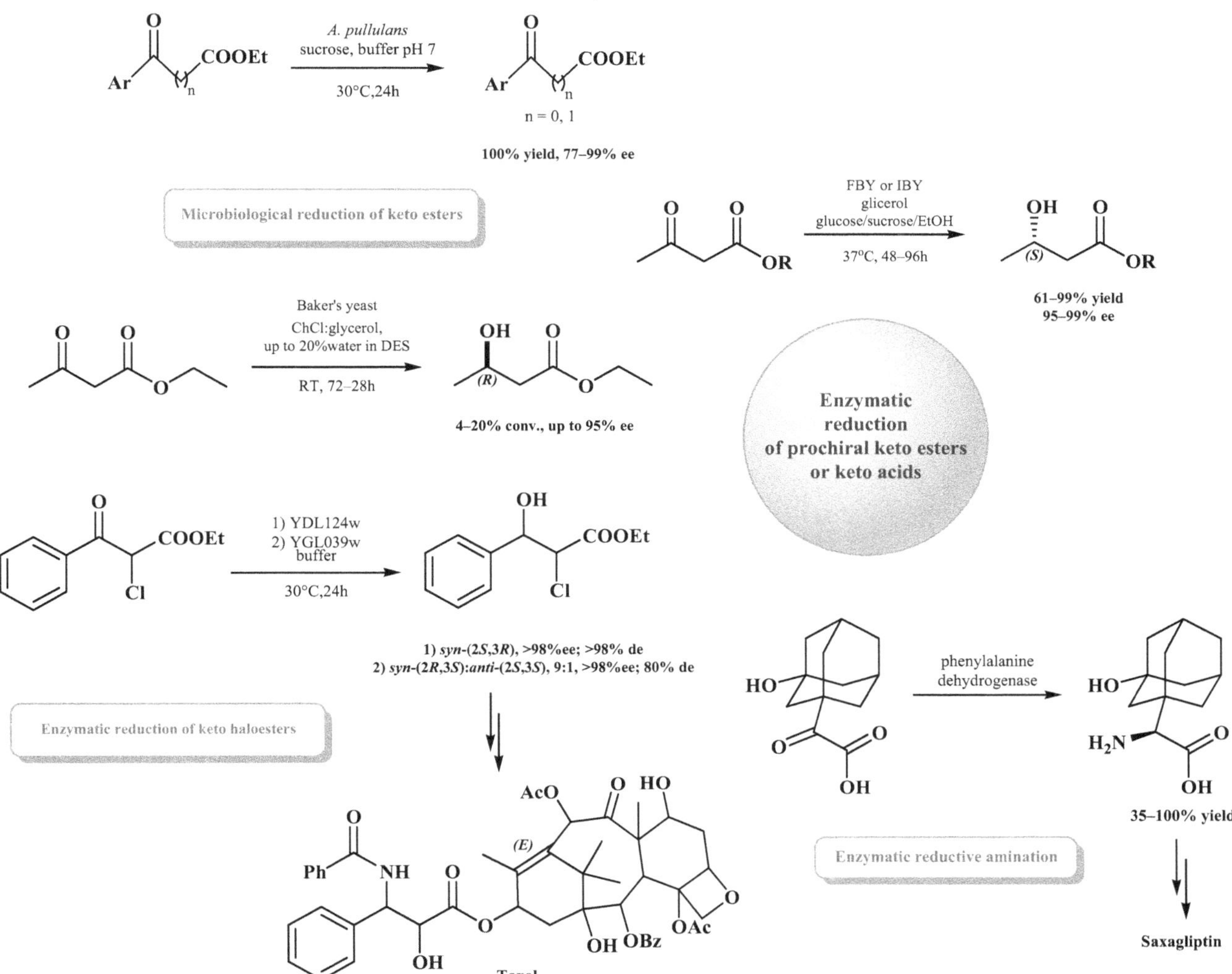

SCHEME 12.22 Enzymatic reduction of prochiral keto esters or keto acids in green solvent.

An enantioselective microbial reduction of 6-oxobuspirone to either (*R*)- or (*S*)-6-hydroxybuspirone was developed. The use of *Rhizopus stolonifer* SC 13898, *R. stolonifer* SC 16199, *Neurospora crassa* SC 13816, *Mucor racemosus* SC 16198, and *Pseudomonas putida* SC 13817 led to the (*S*)-hydroxybuspirone in >95% ee, while the yeast strains *Hansenula polymorpha* SC 13845 and *Candida maltose* SC 16112 gave the (*R*)-enantiomer in >60% reaction yield and >97% ee (Scheme 12.23) [159]. Buspirone is a drug used for the treatment of anxiety and depression that is thought to produce its effects by binding to the serotonin 5HT1A receptor [160].

Benzofuran is considered a very important oxygen-containing heterocyclic compound due to its diverse biological profile. Many of the clinically approved medicines are synthetic and naturally occurring substituted benzofuran derivatives, which display various biological activities including antitumor, antiarrhytmic, antidepressant, antihyperglycemic, antimicrobial, antibiotic, antiparasitic [161].

The stereoselective bioreduction of prochiral ketones of benzofuran despite similar chemical structure, of the ketones occurred with varying efficiency and selectivity for each of them. Secondary alcohols of the *S* and *R* configuration were obtained with enantioselectivity up to 99%. The unsymmetrical methyl ketones were biotransformed with the highest selectivity. *A. pullulans* microorganism is less effective in the reduction of unsymmetrical halomethyl ketones. The presence of a heteroatom in the alkyl group significantly decreases the selectivity of the process [162]. In contrast, reductions of benzofuran derivatives carried out in the presence of *Pythium oligandrum* were characterized by a high degree of conversion (73–98%) and led to enantiomerically pure products (Scheme 12.23) [163].

The microbe-catalyzed reduction of prochiral bulky–bulky pyrimidine base ketones to corresponding chiral alcohols by Boni Protect provides enantiomerically pure products (96–99% ee) with *R* configuration (Scheme 12.23) [164].

The synthesis of (*S*)-E-2-{3-[3-[2-(7-chloro-2-quilolinyl) ethenyl]phenyl]-3-hydroxypropyl}-benzoic acid methyl ester is an interesting example of the microbial reduction application in pharmacology (Scheme 12.23). This compound is the key intermediate in the synthesis of an antiasthma drug

(Montelukast) and was prepared by the reduction of the corresponding ketoester with *Microbacterium campoquemadoensis* (MB5614) or *Mucor hiemalis* [165].

Vitamin D_3 (VD_3, cholecalciferol) is generally known for its regulatory function in calcium and phosphorous homeostasis, but it is also recognized as important for extraskeletal functions such as immune function, cancer prevention, and hypertension prevention [166].

Vitamin D_3 converted into its biologically active form by consecutively acting cytochrome P450 monooxygenases (CYPs) by hydroxylation in the liver yields 25-hydroxy-VD_3 (25OHVD_3, calcidiol), which is converted into calcitriol ($25(OH)_2VD_3$) in the kidney. The regioselective, ferricyanide-dependent hydroxylation of VD_3 and proVD_3 (7-dehydrocholesterol) into the corresponding tertiary alcohols catalyzed by steroid C25 dehydrogenase (C25DH) occurs in water with greater than 99% yield (Scheme 12.24) [167].

Among the oxidoreductases, reductases are also used in the synthesis of biologically active compounds. The enantioselective reduction of α-methylene nitrile derivatives catalyzed by baker's yeast which contain ene-reductases (BY; *S. cerevisiae*) in most cases leads to the corresponding (*R*)-2-arylpropanenitriles with high conversion values. A biphasic ionic liquid–water system 1:10 with $[BMIM]PF_6$ was used

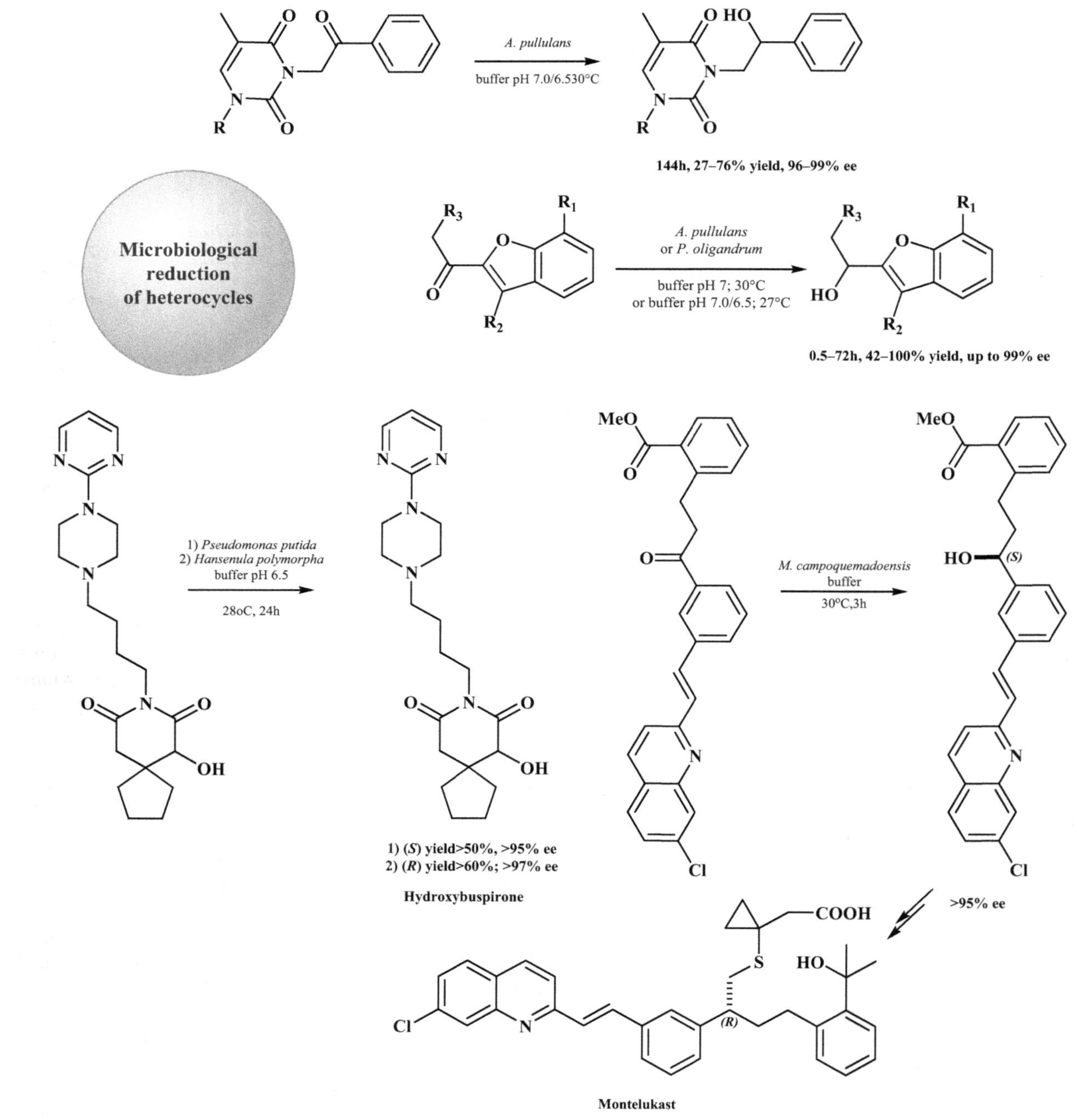

Microbiological reduction of heterocycles

A. pullulans, buffer pH 7.0/6.530°C

144h, 27–76% yield, 96–99% ee

A. pullulans or *P. oligandrum*, buffer pH 7; 30°C or buffer pH 7.0/6.5; 27°C

0.5–72h, 42–100% yield, up to 99% ee

1) *Pseudomonas putida* 2) *Hansenula polymorpha*, buffer pH 6.5, 28oC, 24h

1) (*S*) yield>50%, >95% ee
2) (*R*) yield>60%; >97% ee

Hydroxybuspirone

M. campoquemadoensis, buffer, 30°C,3h

>95% ee

Montelukast

SCHEME 12.23 Microbiological reduction of heterocycles in green solvent.

as the solvent. Depending on the aromatic substituent type, a conversion of 8–100% and high ee values (up to 99%) were observed (Scheme 12.24) [168].

Green asymmetric synthesis can also be carried out in the presence of lyases; they are involved in forming the C–C bond and the bond-cleavage process. Benzaldehyde was reacted with hydrogen cyanide in a two-phase solvent system aqueous buffer and ionic liquids [EMIM]BF_4, [PMIM]BF_4, and [BMIM]BF_4 in the presence of the hydroxynitrile lyases from *Prunus amygdalus* (*Pa*HNL) and *Hevea brasiliensis* (*Hb*NHL). Excellent conversion (86–98%) to cyanohydrin (mandelonitrile) and high ee values (74–97%) were obtained in most cases. Only the use of *Hb*HNL in [BMIM]BF_4 gave the product a slight enantiomeric excess (24% ee) (Scheme 12.24) [169].

The nitrile functional group is a pharmacophore in many biologically active compounds and a versatile precursor of other functionalities such as carboxylic acids, amides, aldehydes, ketones, and amines [170].

Transferases are enzymes that catalyze the transfer of a group of atoms, such as amine, carboxyl, carbonyl, methyl, acyl, glycosyl, and phosphoryl from a donor substrate to an acceptor compound. For example, the acyl transferase from *Mycobacterium smegmatis* (MsAcT) catalyzes transesterification reactions in aqueous media. In the case of the reaction of aliphatic alcohols with ethyl acetate, primary alcohols were almost completely esterified (conversion 90–98%). The secondary alcohols were converted; however, reactions were slow (conversion 7–21% in 120–300 min.) and MsAcT displayed enantioselectivity for the (*S*)-enantiomer of them.

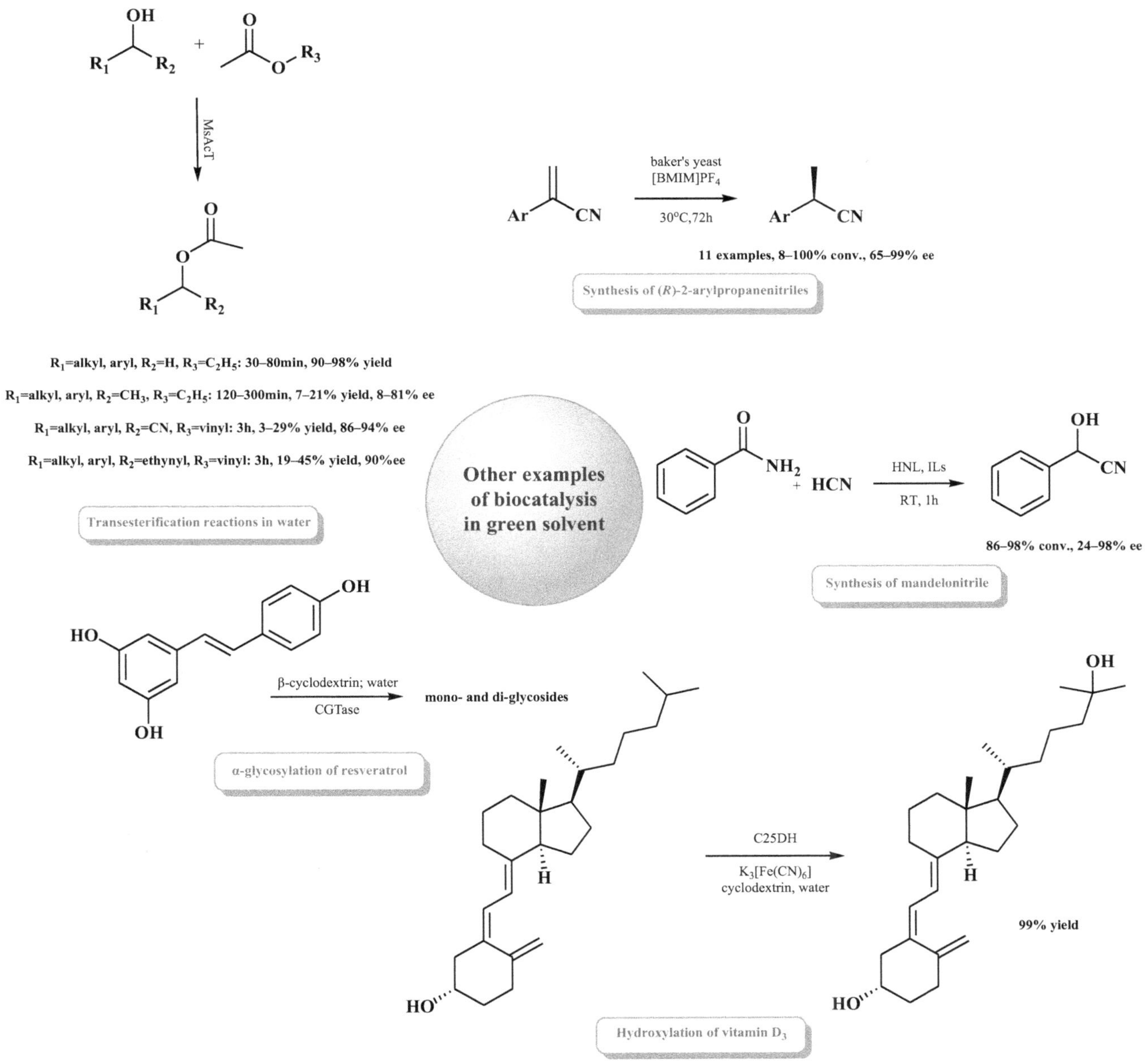

SCHEME 12.24 Other examples of biocatalysis in green solvent.

MsAcT-catalyzed kinetic resolution of cyanohydrins and secondary alkynols by transesterification with vinyl acetate as acyl donors gave better yields and ee compared to secondary alkanols (Scheme 12.24) [171].

A single organic solvent-free enzymatic α-glycosylation of resveratrol occurs directly from the β-cyclodextrin-resveratrol complex in water using β-cyclodextrin as glycoside-donor and cyclodextringlucanotransferase (CGTase) as the enzyme. The reaction performed obtained 35% molar yield. Two mono-glycosides: 3-*O*-α-D-glucosyl-resveratrol and 4'-*O*-α-D-glucosyl-resveratrol and two di-glycosides: 3-*O*-α-D-maltosyl-resveratrol and 4'-*O*-α-D-maltosyl-resveratrol were obtained (Scheme 12.24). Stilbenes, especially *trans*-resveratrol (3,5,4′-trihydroxystilbene), have a great interest in pharmacy to their antioxidant properties [172].

REFERENCES

1. S. Abou-Shehada, J. H. Clark, G. Paggiola, J. Sherwood, Tunable solvents: shades of green, *Chem. Eng. Process.*, 2016, 99, 88–96.
2. C. J. Clarke, W.-Ch. Tu, O. Levers, A. Brohl, J. P. Hallett, Green and sustainable solvents in chemical processes, *Chem. Rev.*, 2018, 118, 747–800.
3. C. Jimenez-Gonzalez, C. S. Ponder, Q. B. Broxterman, J. B. Manley, Using the right green yardstick: why process mass intensity is used in the pharmaceutical industry to drive more sustainable processes, *Org. Process Res. Dev.*, 2011, 15, 912–917.
4. A. Wolfson, Ch. Dlugy, Y. Shotland, Glycerol as a green solvent for high product yields and selectivities, *Environ. Chem. Lett.*, 2007, 5, 67–71.
5. C. Reichardt, T. Welton, *Solvents and Solvent Effects in Organic Chemistry*, 4th ed., Wiley-VCH, Weinheim, 2010.
6. J. H. Clark, T. J. Farmer, A. J. Hunt, J. Sherwood, Opportunities for bio-based solvents created as petrochemical and fuel products transition towards renewable resources, *Int. J. Mol. Sci.*, 2015, 16, 17101–17159.
7. B. A. de Marco, B. S. Rechelo, E. G. Tótoli, A. C. Kogawa, H. R. N. Salgado, Evolution of green chemistry and its multidimensional impacts: a review, *Saudi Pharm. J.*, 2019, 27, 1–8.
8. M. Moniruzzaman, K. Nakashima, N. Kamiyaa, M. Goto, Recent advances of enzymatic reactions in ionic liquids, *Biochem. Eng. J.*, 2010, 48, 295–314.
9. N. Menges, The role of green solvents and catalysts at the future of drug design and of synthesis. Green Chemistry.s.l.: H. E.-D. M. Saleh and M. Kolle, *IntechOpen*, 2018, 5, 73–100.
10. L. Fan, H. Li, Q. Chen, Applications and mechanisms of ionic liquids in whole-cell biotransformation, *Int. J. Mol. Sci.*, 2014, 15, 12196–12216.
11. G. Quijano, A. Couvert, A. Amrane, Ionic liquids: applications and future trends in bioreactor technology, *Bioresour. Technol.*, 2010, 101, 8923–8930.
12. S. Zhang, N. Sun, X. He, X. Lu, X. Zhang, Physical properties of ionic liquids: database and evaluation, *J. Phys. Chem. Ref. Data*, 2006, 35, 1475–1517.
13. H. Olivier-Bourbigou, L. Magna, D. Morvan, Ionic liquids and catalysis: recent progress from knowledge to applications, *App. Catal. A-Gen.*, 2010, 373, 1–56.
14. J. Wilkes, Properties of ionic liquid solvents for catalysis, *J. Mol. Catal. A Chem.*, 2004, 214, 11–17.
15. H. Zhao, Review: current studies on some physical properties of ionic liquids, *Phys. Chem. Liq.*, 2003, 41, 545–557.
16. E. L. Smith, A. P. Abbott, K. S. Ryder, Deep eutectic solvents (DESs) and their applications, *Chem. Rev.*, 2014, 114, 11060–11082.
17. F. M. Perna, P. Vitale, V. Capriati, Deep eutectic solvents and their applications as green solvents, *Curr. Opin. Green Sustain. Chem.*, 2020, 21, 27–33.
18. D. A. Alonso, A. Baeza, R. Chinchilla, G. Guillena, I. M. Pastor, D. J. Ramon, Deep eutectic solvents: the organic reaction medium of the century, *Eur. J. Org. Chem.*, 2016, 612–632.
19. J.-N. Tan, Y. Dou, Deep eutectic solvents for biocatalytic transformations: focused lipase-catalyzed organic reactions, *Appl. Microbiol. Biotechnol.*, 2020, 104, 1481–1496.
20. M. Jain, S. I. Khan, B. L. Tekwani, M. R. Jacob, S. Singh, P. P. Singh, R. Jain, Synthesis, antimalarial, antileishmanial, and antimicrobial activities of some 8-quinolinamine analogues, *Bioorg. Med. Chem.*, 2005, 13, 4458–4466.
21. B. Vaitilingam, A. Nayyar, P. B. Palde, V. Monga, R. Jain, S. Kaurb, P. P. Singh, Synthesis and antimycobacterial activities of ring-substituted quinolinecarboxylic acid/ester analogues. Part 1, *Bioorg. Med. Chem.*, 2004, 12, 4179–4188.
22. G. Roma, M. Di Braccio, G. Grossi, F. Mattioli, M. Ghia, 1,8-Naphthyridines IV. 9- Substituted *N,N*-dialkyl-5-(alkylamino or cycloalkylamino) [1,2,4]triazolo[4,3-*a*][1,8] naphthyridine-6-carboxamides, new compounds with anti-aggressive and potent anti-inflammatory activities, *Eur. J. Med. Chem.*, 2000, 35, 1021–1035.
23. S. T. Hazeldine, L. Polin, J. Kushner, K. White, N. M. Bouregeois, B. Crantz, E. Palomino, T. H. Corbett, J. P. Horwitz, II. Synthesis and biological evaluation of some bioisosteres and congeners of the antitumor agent, 2-{4-[(7-chloro-2-quinoxalinyl)oxy]phenoxy}propionic Acid (XK469), *J. Med. Chem.* 2002, 45, 3130–3137.
24. S. Chen, R. Chen, M. He, R. Pang, Z. Tan, M. Yang, Design, synthesis, and biological evaluation of novel quinoline derivatives as HIV-1 Tat–TAR interaction inhibitors, *Bioorg. Med. Chem.*, 2009, 17, 1948–1956.
25. W. Wei, J. Wen, D. Yang, X. Sun, J. You, Y. Suo, H. Wang, Iron-catalyzed three-component tandem process: a novel and convenient synthetic route to quinoline-2,4-dicarboxylates from arylamines, glyoxylic esters, and a-ketoesters, *Tetrahedron*, 2013, 69, 10747–10751.
26. A. Jaso, B. Zarranz, I. Aldana, A. Monge, Synthesis of new quinoxaline-2-carboxylate 1,4-dioxide derivatives as anti-*Mycobacterium tuberculosis* agents, *J. Med. Chem.* 2005, 48, 2019–2025.
27. H. An, F. Rong, J. Wu, C. Harris, S. Chow, Quinoxaline derivatives having Antiviral activity, US Patent, 7,189,724 B2, 2007.
28. A. A. Abu-Hashem, M. A. Gouda, F. A. Badria, Synthesis of some new pyrimido[2'1',2,3]thiazolo[4,5-*b*]quinoxaline derivatives as anti-inflammatory and analgesic agents, *Eur. J. Med. Chem.*, 2010, 45, 1976–1981.
29. M. N. Noolvi, H. M. Patel, V. Bhardwaj, A. Chauhan, Synthesis and in vitro antitumor activity of substituted quinazoline and quinoxaline derivatives: search for anticancer agent, *Eur. J. Med. Chem.*, 2011, 46, 2327–2346.

30. V. Desplat, A. Geneste, M.-A. Begorre, S. Fabre, S. Brajot, S. Massip, D. Thiolat, D. Mossalayi, Ch. Jarry, J. Guillon, Synthesis of New Pyrrolo[1,2-*a*]quinoxaline Derivatives as Potential Inhibitors of Akt Kinase, *J. Enzyme Inhib. Med. Chem.*, 2008, 23, 648–658.
31. M. H. Fisher, A. Lusi, J. R. Egerton, Anthelmintic Dihydroquinoxalino[2,3-*b*]quinoxalines, *J. Pharm. Sci.*, 1977, 66, 1349–1352.
32. P. Bao, L. Wang, Q. Liua, D. Yang, H. Wang, X. Zhao, H. Yue, W. Wei, Direct coupling of haloquinolines and sulfonyl chlorides leading to sulfonylated quinolines in water, *Tetrahedron Lett.*, 2019, 60, 214–218.
33. L.-Y. Xie, Y.-J. Li, J. Qu, Y. Duan, J. Hu, K.-J. Liu, Z. Cao, W.-M. He, Base-free, ultrasound accelerated one-pot synthesis of 2-sulfonylquinolines on water, *Green Chem.*, 2017, 19, 5642–5646.
34. A. Mishra, S. Singh, M. A. Quraishi, V. Srivastava, A catalyst-free expeditious green synthesis of quinoxaline, oxazine, thiazine, and dioxin derivatives in water under ultrasound irradiation, *Org. Prep. Proced. Int.*, 2019, 51, 345–356.
35. A. V. Narsaiah, J. K. Kumar, Glycerin and $CeCl_3{\cdot}7H_2O$: a new and efficient recyclable reaction medium for the synthesis of quinoxalines, *Synth. Commun.*, 2012, 42, 883–892.
36. J. S. Yadav, B. V. S. Reddy, B. Eeshwaraiah, M. K. Gupta, $Bi(OTf)_3$/[bmim]BF_4 as novel and reusable catalytic system for the synthesis of furan, pyrrole and thiophene derivatives, *Tetrahedron Lett.*, 2004, 45, 5873–5876.
37. S. Handy, K. Lavender, Organic synthesis in deep eutectic solvents: Paal-Knorr reactions, *Tetrahedron Lett.*, 2013, 54, 4377–4379.
38. A. J. F. N. Sobral, N. G. C. L. Rebanda, M. da Silva, S. H. Lampreia, M. R. Silva, A. M. Beja, J. A. Paixao, A. M. d'A. Rocha Gonsalves, One-step synthesis of pipyrromethanes in water, *Tetrahedron Lett.*, 2003, 44, 3971–3973.
39. A. J. F. N. Sobral, Synthesis of *meso*-diethyl-2,2'-dipyrromethane in water, *J. Chem. Educ.*, 2006, 83(11), 1665–1666.
40. E. Abele, R. Abele, O. Dzenitis, E. Lukevics, Indole and isatin oximes: synthesis, reactions, and biological activity, *Chem. Heterocycl. Compd.*, 2003, 39, 3–35.
41. N. Sin, B. L. Venables, K. D. Combrink, H. B. Gulgeze, K.-L. Yu, R. L. Civiello, J. Thuring, X. A. Wanga, Z. Yang, L. Zadjura, A. Marino, K. F. Kadow, Ch. W. Cianci, J. Clarke, E. V. Genovesi, I. Medina, L. Lamb, M. Krystal, N. A. Meanwell, Respiratory syncytial virus fusion inhibitors. Part 7: structure–activity relationships associated with a series of isatin oximes that demonstrate antiviral activity in vivo, *Bioorg. Med. Chem. Lett.*, 2009, 19, 4857–4862.
42. W.-T. Wei, W.-M. Zhu, W.-W. Ying, Y. Wu, Y.-L. Huang, H. Liang, Metal-free synthesis of isatin oximes *via* radical coupling reactions of oxindoles with *t*-BuONO in water, *Org. Biomol. Chem.*, 2017, 15, 5254–5257.
43. S. Yu, K. Hu, J. Gong, L. Qi, J. Zhu, Y. Zhang, T. Cheng, J. Chen, Palladium-catalyzed tandem addition/cyclization in aqueous medium: synthesis of 2-arylindoles, *Org. Biomol. Chem.*, 2017, 15, 4300–4307.
44. T. C. McKee, R. W. Fuller, C. D. Covington, J. H. Cardellina, R. J. Gulakowski, B. L. Krepps, J. B. McMahon, M. R. Boyd, New pyranocoumarins isolated from *Calophyllum lanigerum* and *Calophyllum Teysmannii*, *J. Nat. Prod.* 1996, 59, 754–758.
45. G. H. Sayed, M. E. Azab, K. E. Anwer, Conventional and microwave-assisted synthesis and biological activity study of novel heterocycles containing pyran moiety, *J. Heterocyclic Chem.*, 2019, 56, 2121–2133.
46. E. Sari, H. Aslan, Ş. Dadi, A. Öktemer, E. Logoglu, Biological activity studies of some synthesized novel furan and pyran derivatives, *J. Sci.*, 2017, 30, 49–55.
47. A. H. Bedair, H. A. Emam, N. A. El-Hady, K. A. R. Ahmed, A. M. El-Agrody, Synthesis and antimicrobial activities of novel naphtho[2,1-*b*]pyran, pyrano[2,3-*d*]pyrimidine and pyrano[3,2-*e*][1,2,4]triazolo[2,3-*c*]-pyrimidine derivatives, *Il Farmaco*, 2001, 56, 965–973.
48. G. Zeni, D. S. Ludtke, C. W. Nogueira, R. B. Panatieri, A. L. Braga, C. C. Silveira, H. A. Stefani, J. B. T. Rocha, New acetylenic furan derivatives: synthesis and anti-inflammatory activity, *Tetrahedron Lett.*, 2001, 42, 8927–8930.
49. E. Lukevits, L. Demicheva, Biological activity of furan derivatives (review), *Chem. Heterocycl. Compd.*, 1993, 29, 243–267.
50. H. R. Safaei, M. Shekouhy, S. Rahmanpur, A. Shirinfeshan, Glycerol as a biodegradable and reusable promoting medium for the catalyst-free one-pot three component synthesis of 4*H*-pyrans, *Green Chem.*, 2012, 14, 1696–1704.
51. R. Mancuso, R. Miliè, A. P. Piccionello, D. Olivieri, N. D. Ca, C. Carfagna, B. Gabriele, Catalytic carbonylative double cyclization of 2-(3-hydroxy-1-yn-1-yl)phenols in ionic liquids leading to furobenzofuranone derivatives, *J. Org. Chem.*, 2019, 84, 7303–7311.
52. S. G. Patil, V. V. Bhadke, R. R. Bagul, Synthesis of 2-aroylbenzofuran-3-ols using basic ionic liquid [bmIm] OH, *J. Chem. Pharm. Res.*, 2012, 4, 2832–2835.
53. F.-Z. Zhang, Y. Tian, G.-X. Li, J. Qu, Intramolecular etherification and polyene cyclization of π-activated alcohols promoted by hot water, *J. Org. Chem.*, 2015, 80(2), 1107–1115.
54. Q. Li, Y. Yan, X. Wang, B. Gong, X. Tang, J. J. Shi, H. E. Xu, W. Yi, Water as a green solvent for efficient synthesis of isocoumarins through microwave-accelerated and Rh/Cu-catalyzed C–H/O–H bond functionalization, *RSC Adv.*, 2013, 3, 23402–23408.
55. C.-Y. Cui, J. Liu, H.-B. Zheng, X.-Y. Jin, X.-Y. Zhao, W.-Q. Chang, B. Sun, H.-X. Lou, Diversity-oriented synthesis of pyrazoles derivatives from flavones and isoflavones leads to the discovery of promising reversal agents of fluconazole resistance in *Candida albicans*, *Bioorg. Med. Chem. Lett.*, 2018, 28, 1545–1549.
56. D. V. Sowmya, G. Lakshmi Teja, A. Padmaja, V. Kamala Prasad, V. Padmavathi, Green approach for the synthesis of thiophenyl pyrazoles and isoxazoles by adopting 1,3-dipolar cycloaddition methodology and their antimicrobial activity, *Eur. J. Med. Chem.*, 2018, 143, 891–898.
57. M. Ch. Sau, Y. Rajesh, M. Mandal, M. Bhattacharjee, Copper catalyzed regioselective *N*-alkynylation of pyrazoles and evaluation of the anticancer activity of ethynyl-pyrazoles, *Chemistry Select.*, 2018, 3, 3511–3515.
58. J. Li, H. Huo, R. Guo, B. Liu, L. Li, W. Dan, X. Xiao, J. Zhang, B. Shi, Facile and efficient access to Androsten-17-(1′,3′,4′)-pyrazoles and Androst-17b-(1′,3′,4′)-pyrazoles via Vilsmeier reagents, and their antiproliferative activity evaluation in vitro, *Eur. J. Med. Chem.*, 2017, 130, 1–14.

59. E. Rezníckova, L. Tenora, P. Pospísilova, J. Galeta, R. Jorda, K. Berka, P. Majer, M. Potacek, V. Krystof, ALK5 kinase inhibitory activity and synthesis of 2,3,4-substituted 5,5-dimethyl-5,6-dihydro-4H-pyrrolo[1,2-*b*]pyrazoles, *Eur. J. Med. Chem.*, 2017, 127, 632–642.
60. S. Durgamma, A. Muralikrishna, V. Padmavathi, A. Padmaja, Synthesis and antioxidant activity of amido-linked benzoxazolyl/benzothiazolyl/benzimidazolyl-pyrroles and pyrazoles, *Med. Chem. Res.*, 2013, 23, 2916–2929.
61. Y. Guo, G. Wang, L. Wei, J.-P. Wan, C-H sulfonylation and pyrazole annulation for fully substituted pyrazole synthesis in water using hydrophilic enaminones, *J. Org. Chem.*, 2019, 84(5), 2984–2990.
62. A. A. Hamid, M. Abd-Elmonem, A. M. Hayallah, F. A. A. Elsoud, K. U. Sadek, Glycerol: a promising benign solvent for catalyst free one-pot multi-component synthesis of pyrano[2,3-*c*]pyrazoles and tetrahydro-benzo[*b*]pyrans at ambient temperature, *Chemistry Select.*, 2017, 2, 10689–10693.
63. J.-N. Tan, M. Li, Y. Gu, Multicomponent reactions of 1,3-disubstituted 5-pyrazolones and formaldehyde in environmentally benign solvent systems and their variations with more fundamental substrates, *Green Chem.*, 2010, 12, 908–914.
64. J. E. R. Nascimento, D. H. de Oliveira, P. B. Abib, D. Alves, G. Perin, R. G. Jacob, Synthesis of 4-arylselanylpyrazoles through cyclocondensation reaction using glycerol as solvent, *J. Braz. Chem. Soc.*, 2015, 26, 1533–1541.
65. K. Shalini, P. K. Sharma, N. Kumar, Imidazole and its biological activities: a review, *Der. Chem. Sin.*, 2010, 1, 36–47.
66. F. Nemati, M. M. Hosseini, H. Kiani, Glycerol as a green solvent for efficient, one-pot and catalyst free synthesis of 2,4,5-triaryl and 1,2,4,5-tetraaryl imidazole derivatives, *J. Saudi Chem. Soc.*, 2016, 20, 503–508.
67. P. K. Mishra, A. K. Verma, Metal-free regioselective tandem synthesis of diversely substituted benzimidazo-fused polyheterocycles in aqueous medium, *Green Chem.*, 2016, 18, 6367–6372.
68. A. K. Yadav, M. Kumar, T. Yadav, R. Jain, An ionic liquid mediated one-pot synthesis of substituted thiazolidinones and benzimidazoles, *Tetrahedron Lett.*, 2009, 50, 5031–5034.
69. M. Capua, S. Perrone, F. Maria Perna, P. Vitale, L. Troisi, A. Salomone, V. Capriati, An expeditious and greener synthesis of 2-aminoimidazoles in deep eutectic solvents, *Molecules*, 2016, 21, 924–935.
70. S. Chen, L. Chen, N. Le, C. Zahao, A. Sidduri, J. Ping Lou, C. Michoud, L. Portland, N. Jackson, J. Liu, F. Konzelman, F. Chi, C. Tovar, Q. Xiang, Y. Chen, Y. Wen, L. T. Vassiley, Synthesis and activity of quinolinyl-methylyne-thiazolinones as potent and selective cyclin dependet kinase 1 inhibitors, *Bioorg. Med. Chem. Lett.*, 2007, 17, 2134–2138.
71. V. V. Dabholkar, S. D. Shah, V. M. Dave, Lacatums of thiazol-4-ones, *Der Pharma Chem.*, 2015, 7, 163–166.
72. G. H. Al-Ansary, M. A. H. Ismail, D. A. Abou El Ella, S. Eid, K. A. M. Abouzid, Molecular design and synthesis of HCV inhibitors based on thiazolone scaffold, *Eur. J. Med. Chem.*, 2013, 68, 12–32.
73. J. Scott, F. W. Goldberg, A. V. Turnbull, Medicinal chemistry of inhibitors of 11beta-hydroxysteroid dehydrogenase type 1 (11beta-HSD1), *J. Med. Chem.*, 2014, 57, 4466–4486.
74. D. J. Jean, C. Yuan, E. A. Bercot, R. Cupples, M. Chen, J. Fretland, C. Hale, R. W. Hungate, R. Komorowski, M. Veniant, M. Wang, X. Zhang, C. Fotsch, 2-(S)-Phenethylaminothiazolones as potent, orally efficacious inhibitors of 11β-hydroxysteriod dehydrogenase type 1, *J. Med. Chem.*, 2007, 50, 429–432.
75. R. Studzińska, R. Kołodziejska, D. Kupczyk, W. Płaziński, T. Kosmalski, A novel derivatives of thiazol-4(5*H*)-one and their activity in the inhibition of 11β-hydroxysteroid dehydrogenase type 1, *Bioorg. Chem.*, 2018, 79, 115–121.
76. R. Studzińska, R. Kołodziejska, W. Płaziński, D. Kupczyk, T. Kosmalski, K. Jasieniecka, B. Modzelewska-Banachiewicz, Synthesis of the *N*-methyl derivatives of 2-aminothiazol-4(5*H*)–one and their interactions with 11βHSD1 – molecular modeling and *in vitro* studies, *Chem. Biodivers.*, 2019, 16, e1900065, 1–10.
77. R. Studzińska, D. Kupczyk, A. Płazińska, R. Kołodziejska, T. Kosmalski, B. Modzelewska-Banachiewicz, Thiazolo[3,2-*a*]pyrimidin-5-one derivatives as a novel class of 11β-hydroxysteroid dehydrogenase inhibitors, *Bioorg. Chem.*, 2018, 81, 21–26.
78. K. U. Sadek, R. A. Mekheimer, A. M. A. Hameed, F. Elnahas, M. H. Elnagdi, Green and highly efficient synthesis of 2-arylbenzothiazoles using glycerol without catalyst at ambient temperature, *Molecules*, 2012, 17, 6011–6019.
79. T. Deligeorgiev, S. Kaloyanova, N. Lesev, R. Alajarín, J. J. Vaquero, J. Álvarez-Builla, An environmentally benign synthesis of 2-cyanomethyl-4-phenylthiazoles under focused microwave irradiation, *Green Sustain. Chem.*, 2011, 1, 170–175.
80. X.-Z. Zhang, W.-J. Zhou, M. Yang, J.-X. Wang, L. Baic, Microwave-assisted synthesis of benzothiazole derivatives using glicerol as green solvent, *J. Chem. Res.*, 2012, 489–491.
81. N. Mukku, B. Maiti, On water catalyst-free synthesis of benzo[d]imidazo[2,1-*b*] thiazoles and novel *N*-alkylated 2-aminobenzo[d]oxazoles under microwave irradiation, *RSC Adv.*, 2020, 10, 770–778.
82. J. Noei, A. R. Khosropour, Ultrasound-promoted a green protocol for the synthesis of 2,4-diarylthiazoles under ambient temperature in [bmim]BF_4, *Ultrason. Sonochem.*, 2009, 16, 711–717.
83. D. Lingampalle, D. Jawale, R. Waghmare, R. Mane, Ionic liquid–mediated, one-pot synthesis for 4-thiazolidinones, *Synth. Commun.*, 2010, 40, 2397–2401.
84. R. Studzińska, R. Kołodziejska, T. Kosmalski, B. Modzelewska-Banachiewicz, Regioselective bromination of 2-idomethyl-2,3-dihydrothiazolo[3,2-a]pyrimidin-5-one, *Heterocycles*, 2016, 92, 2271–2277.
85. S. Kakkar, B. Narasimhan, A comprehensive review on biological activities of oxazole derivatives, *BMC Chem.*, 2019, 13:16
86. D. S. Zinad, A. Mahal, R. K. Mohapatra, A. K. Sarangi, M. R. F. Pratama, Medicinal chemistry of oxazines as promising agents in drug discovery, *Chem. Biol. Drug. Des.*, 2020, 95, 16–47.
87. Y. Tang, M. Li, H. Gao, G. Rao, Z. Mao, Efficient Cu-catalyzed intramolecular *O*-arylation for synthesis of benzoxazoles in water, *RSC Adv.*, 2020, 10, 14317–14321.
88. B. S. Singh, H. R. Lobo, D. V. Pinjari, K. J. Jarag, A. B. Pandit, G. S. Shankarling, Ultrasound and deep eutectic

solvent (DES): a novel blend of techniques for rapid and energy efficient synthesis of oxazoles, *Ultrason. Sonochem.*, 2013, 20, 287–293.

89. J. Safaei-Ghomi, M. A. Ghasemzadeh, Synthesis of some 3,5-diaryl-2-isoxazoline derivatives in ionic liquids media, *J. Serb. Chem. Soc.*, 2012, 77, 733–739.
90. A. Sharifi, M. Barazandeh, M. S. Abaee, M. Mirzaei, [Omim][BF_4], a green and recyclable ionic liquid medium for the one-pot chemoselective synthesis of benzoxazinones, *Tetrahedron Lett.*, 2010, 51, 1852–1855.
91. D.-H. Bach, J.-Y. Liu, W. K. Kim, J.-Y. Hong, S. H. Park, D. Kim, S.-N. Qin, T.-T.-T. Luu, H. J. Park, Y.-N. Xu, S. K. Lee, Synthesis and biological activity of new phthalimides as potential anti-inflammatory agents, *Bioorg. Med. Chem.*, 2017, 25, 3396–3405.
92. P. S. Nayab, M. Irfan, M. Abid, M. Pulaganti, Ch. Nagaraju, S. K. Chitta, Rahisuddin, Experimental and molecular docking investigation on DNA interaction of N-substituted phthalimides: antibacterial, antioxidant and hemolytic activities, *Luminescence*, 2016, 1–11.
93. H. R. Lobo, B. S. Singh, G. S. Shankarling, Deep eutectic solvents and glycerol: a simple, environmentally benign and efficient catalyst/reaction media for synthesis of *N*-aryl phthalimide derivatives, *Green Chem. Lett. Rev.*, 2012, 5, 487–533.
94. C. Liu, G. Wang, W. Sui, L. An, Ch. Si, Preparation and characterization of chitosan by a novel deacetylation approach using glycerol as green reaction solvent, *ACS Sustain. Chem. Eng.*, 2017, 5, 4690–4698.
95. C. S. Radatz, R. B. Silva, G. Perin, E. J. Lenardão, R. G. Jacob, D. Alves, Catalyst-free synthesis of benzodiazepines and benzimidazoles using glycerol as recyclable solvent, *Tetrahedron Lett.*, 2011, 52, 4132–4136.
96. A. Amini, S. Sayyahi, S. J. Saghanezhad, N. Taheri, Integration of aqueous biphasic with magnetically recyclable systems: polyethylene glycol–grafted Fe_3O_4 nanoparticles catalyzed phenacyl synthesis in water, *Catal. Commun.*, 2016, 78, 11–16.
97. A. P. Ingale, S. M. Patil, S. V. Shinde, Catalyst-free, efficient and one pot protocol for synthesis of nitriles from aldehydes using glycerol as green solvent, *Tetrahedron Lett.*, 2017, 58, 4845–4848.
98. D. Gonzalez-Martinez, V. Gotor, V. Gotor-Fernandez, Application of deep eutectic solvents in promiscuous lipase-catalysed aldol reactions, *Eur. J. Org. Chem.*, 2016, 1513–1519.
99. A. E. Díaz-Álvarez, V. Cadierno, Glycerol: a promising green solvent and reducing agent for metal-catalyzed transfer hydrogenation reactions and nanoparticles formation, *Appl. Sci.*, 2013, 3, 55–69.
100. M. H. Muhammad, X.-L. Chen, Y. Liu, T. Shi, Y. Peng, L. Qu, B. Yu, Recyclable Cu@C_3N_4-catalyzed hydroxylation of aryl boronic acids in water under visible-light: synthesis of phenols under ambient condition and room temperature, *ACS Sustain. Chem. Eng.*, 2020, 8(7), 2682–2687.
101. M. B. Gawande, A. K. Rathi, P. S. Branco, I. D. Nogueira, A. Velhinho, J. J. Shrikhande, U. U. Indulkar, R. V. Jayaram, C. Amjad, A. Ghumman, N. Bundaleski, O. M. N. D. Teodoro, Regio- and Chemoselective Reduction of Nitroarenes and Carbonyl Compounds over Recyclable Magnetic Ferrite_Nickel Nanoparticles (Fe_3O_4-Ni) by Using Glycerol as a Hydrogen Source, *Chem. Eur. J.*, 2012, 18, 12628–12632.
102. P. K. Khatri, S. L. Jain, Glycerol ingrained copper: an efficient recyclable catalyst for the *N*-arylation of amines with aryl halides, *Tetrahedron Lett.*, 2013, 54, 2740–2743.
103. A. Amić, M. Molnar, An improved and efficient *N*-acetylation of amines using choline chloride based deep eutectic solvents, *Org. Prep. Proced. Int.*, 2017, 49, 249–257.
104. A. Zhu, L. Li, J. Wang, K. Zhuo, Direct nucleophilic substitution reaction of alcohols mediated by a zinc-based ionic liquid, *Green Chem.*, 2011, 13, 1244–1250.
105. D. J. Eyckens, L. C. Henderson, Synthesis of α-aminophosphonates using solvate ionic liquids, *RSC Adv.*, 2017, 7, 27900–27904.
106. D. M. Solano, P. Hoyos, M. J. Hernáiz, A. R. Alcántara, J. M. Sánchez-Montero, Industrial biotransformations in the synthesis of building blocks leading to enantiopure drugs, *Bioresource Technol.*, 2012, 115, 196–207.
107. L. P. Jordheim, D. Durantel, F. Zoulim, C. Dumontet, Advances in the development of nucleoside and nucleotide analogues for cancer and viral diseases, *Nat. Rev. Drug Discov.*, 2013, 12, 447–464.
108. C. W. Rivero, J. M. Palomo, Covalent immobilization of *Candida rugosa* lipase at alkaline pH and their application in the regioselective deprotection of per-*O*-acetylated thymidine, *Catalysts*, 2016, 6, 115–125.
109. S. H. Ha, T. V. Anh, Y.-M. Koo, Optimization of lipase-catalyzed synthesis of caffeic acid phenethyl ester in ionic liquids by response surface methodology, *Bioprocess Biosyst. Eng.*, 2013, 36, 79–807.
110. D. L. Bissett, Common cosmeceuticals, *Clin. Dermatol.*, 2009, 27, 435–445.
111. P. Lozano, E. Alvarez, S. Nieto, R. Villa, J. Bernal, A. Donaire, Biocatalytic synthesis of panthenyl monoacyl esters in ionic liquids and deep eutectic solvents, *Green Chem.*, 2019, 21(12), 3353–3361.
112. Saima, A. G. Lavekar, S. Kumar, S. K. Rastogi, A. K. Sinha, Biocatalysis for C–S bond formation: porcine pancreatic lipase (PPL) catalysed thiolysis/hydrothiolation reactions in sole water, *Synth. Commun.*, 2019, 49, 2029–2043.
113. Y. Fu, Z. Lu, K. Fang, X. He, H. Xu, Y. Hu, Enzymatic approach to cascade synthesis of bis(indolyl)methanes in pure water, *RSC Adv.*, 2020, 10, 10848–10853.
114. S. B. Bharate, J. B. Bharate, S. I. Khan, B. L. Tekwani, M. R. Jacob, R. Mudududdla, R. R. Yadav, B. Singh, P. R. Sharma, S. Maity, B. Singh, I. A. Khan, R. A. Vishwakarma, Discovery of 3,3'-diindolylmethanes as potent antileishmanial agents, *Eur. J. Med. Chem.*, 2013, 63, 435–443.
115. K. K.-W. Lo, K. H.-K. Tsang, W.-K. Hui, N. Zhu, Luminescent rhenium(i) diimine indole conjugates – photophysical, electrochemical and protein-binding properties, *Chem. Commun.*, 2003, 21, 2704–2705.
116. K. Abdelbaqi, N. Lack, E. T. Guns, L. Kotha, S. Safe, J. T. Sanderson, Antiandrogenic and growth inhibitory effects of ring-substituted analogs of 3,3′-diindolylmethane (Ring-DIMs) in hormone-responsive LNCaP human prostate cancer cells, *Prostate*, 2011, 71(13), 1401–1412.
117. A. Lerner, M. Grafi-Cohen, T. Napso, N. Azzam, F. Fares, The indolic diet-derivative, 3,3'-diindolylmethane, induced apoptosis in human colon cancer cells through

upregulation of NDRG1, *J. Biomed. Biotechnol.*, 2012, 2012, 256178–256183.

118. K. Yoon, S.-O. Lee, S.-D. Cho, K. Kim, S. Khan, S. Safe, Activation of nuclear TR3 (NR4A1) by a diindolylmethane analog induces apoptosis and proapoptotic genes in pancreatic cancer cells and tumors, *Carcinogenesis*, 2011, 32(6), 836–842.
119. Y. S. Kim, J. A. Milner, Targets for indole-3-carbinol in cancer prevention, *J. Nutr. Biochem.*, 2005, 16(2), 65–73.
120. R. Madeira Lau, F. van Rantwijk, K. R. Seddon, R. A. Sheldon, Lipase-catalyzed reactions in ionic liquids, *Org. Lett.*, 2000, 2(26), 4189–4191.
121. R. P. Elander, Industrial production of b-lactam antibiotics, *Appl. Microbiol. Biotechnol.*, 2003, 61, 385–392.
122. S. C. Pereira, R. Bussamara, G. Marin, R. L. C. Giordano, J. Dupont, R. de Campos Giordano, Enzymatic synthesis of amoxicillin by penicillin G acylase in the presence of ionic liquids, *Green Chem.*, 2012, 14, 3146–3156.
123. A. C. L. de Melo Carvalho, T. de Sousa Fonseca, M. C. de Mattos, M. da C. F. de Oliveira, T. L. G. de Lemos, F. Molinari, D. Romano, I. Serra, Recent advances in lipase-mediated preparation of pharmaceuticals and their intermediates, *Int. J. Mol. Sci.*, 2015, 16, 29682–29716.
124. J. J. Cui, M. Tran-Dubé, H. Shen, M. Nambu, P.-P. Kung, M. Pairish, L. Jia, J. Meng, L. Funk, I. Botrous, M. McTigue, N. Grodsky, K. Ryan, E. Padrique, G. Alton, S. Timofeevski, S. Yamazaki, Q. Li, H. Zou, J. Christensen, B. Mroczkowski, S. Bender, R. S. Kania, M. P. Edwards, Structure based drug design of crizotinib (PF-02341066), a potent and selective dual inhibitor of mesenchymal-epithelial transition factor (c-MET) kinase and anaplastic lymphoma kinase (ALK), *J. Med. Chem.*, 2011, 54, 6342–6363.
125. R. N. Patel, Biocatalysis: synthesis of key intermediates for development of pharmaceuticals, *ACS Catal.*, 2011, 1, 1056–1074.
126. F. Felluga, G. Pitacco, E. Valentin, C. D. Venneri, A facile chemoenzymatic approach to chiral non-racemic β-alkyl-γ-amino acids and 2-alkylsuccinic acids. A concise synthesis of (*S*)-(+)-Pregabalin, *Tetrahedron: Asymm.*, 2008, 19, 945–955.
127. A. Kamal, G. Chouhan, Chemoenzymatic synthesis of enantiomerically pure 1,2-diols employing immobilized lipase in the ionic liquid [bmim]PF_6, *Tetrahedron Lett.*, 2004, 45, 8801–8805.
128. D. Bianchi, A. Bosetti, P. Cesti, P. Golini, Enzymatic resolution of 1,2-diols:preparation of optically pure dropropizine, *Tetrahedron Lett.*, 1992, 33, 3231–3234.
129. P. C. Ting, J. F. Lee, J. C. Anthes, N.-Y. Shih, J. J. Piwinski, Synthesis and NK1/NK2 receptor activity of substituted-4(*Z*)-(methoxyimino)pentyl-1-piperazines, *Bioorg. Med. Chem. Lett.*, 2000, 10 (20), 2333–2335.
130. G. A. Reichard, Z. T. Ball, R. Aslanian, J. C. Anthes, N.-Y. Shih, J. J. Piwinski, The design and synthesis of novel NK1/NK2 dual antagonists, *Bioorg. Med. Chem. Lett.*, 2000, 10(20), 2329–2332.
131. M. J. Homann, R. Vail, B. Morgan, V. Sabesan, C. Levy, D. R. Dodds, A. Zaks, Enzymatic hydrolysis of a prochiral 3-substituted glutarate ester, an intermediate in the synthesis of an NK1/NK2 dual antagonist, *Adv. Synth. Catal.*, 2001, 343, 744–749.
132. Y. Fredes, L. Chamorro, Z. Cabrera, Increased selectivity of Novozym 435 in the asymmetric hydrolysis of a substrate with high hydrophobicity through the use of deep eutectic solvents and high substrate concentrations, *Molecules*, 2019, 24(4), 792–801.
133. T. Tsuji, Y. Iio, T. Takemoto, T. Nishi, Enzymatic desymmetrization of 2-amino-2-methyl-1,3-propanediol: asymmetric synthesis of (*S*)-*N*-Boc-*N,O*-isopropylidene-α-methylserinal and (4*R*)-methyl-4-[2-(thiophen-2-yl)ethyl] oxazolidin-2-one, *Tetrahedron: Asymm.*, 2005, 16, 3139–3142.
134. J. Y. Choi, R. F. Borch, Highly efficient synthesis of enantiomerically enriched 2-hydroxymethylaziridines by enzymatic desymmetrization, *Org. Lett.*, 2007, 18, 215–218.
135. B. Morgan, D. R. Dodds, A. Zaks, D. R. Andrews, R. Klesse, Enzymatic desymmetrization of prochiral 2-substituted-1,3-propanediols: a practical chemoenzymatic synthesis of a key precursor of SCH51048, a broad-spectrum orally active antifungal agent, *J. Org. Chem.*, 1997, 62, 7736–7743.
136. I. M. Fellows, D. E. Kaelin, S. F. Martin, Application of ring-closing metathesis to the formal total synthesis of (+)–FR900482, *J. Am. Chem. Soc.*, 2000, 122, 10781–10787.
137. D. B. Kastrinsky, D. L. Boger, Effective asymmetric synthesis of 1,2,9,9a-tetrahydrocyclopropa[*c*]benzo[*e*]indol-4-one (CBI), *J. Org. Chem.*, 2004, 69, 2284–2289.
138. A. Drabikowska, L. Lissowska, M. Dramiński, A. Zgit-Wróblewska, D. Shugar, Acyclonucleoside analogues consisting of 5-and 5,6-substituted uracils and different acyclic chains: inhibitory properties vs purified *E. coli* uridine phosphorylase, *Z. Naturforsch*, 1987, 42, 288–296.
139. R. Kołodziejska, A. Karczmarska-Wódzka, A. Wolan, M. Dramiński, *Candida antarctica* lipase B catalyzed enantioselective acylation of pyrimidine acyclonucleoside, *Biocatal, Biotransformation*, 2012, 30, 426–430.
140. R. Kołodziejska, M. Gorecki, J. Frelek, M. Dramiński, Enantioselective enzymatic desymmetrization of the prochiral pyrimidine acyclonucleoside, *Tetrahedron: Asymm.*, 2012, 23, 683–689.
141. R. Kołodziejska, M. Kwit, R. Studzińska, M. Jelecki, Enantio- and diastereoselective acylation of prochiral hydroxyl groupof pyrimidine acyclonucleosides, *J. Mol. Catal. B Enzym.*, 2016, 133, 98–106.
142. T. Matsumoto, T. Konegawa, H. Yamaguchi, T. Nakamura, T. Sugai, K. Suzuki, Lipase-catalyzed asymmetrization of diacetate of *meso*-2-(2-propynyl)cyclohexane-1,2,3-triol toward the total synthesis of Aquayamycin, *Synlett*, 2001, 10, 1650–1652.
143. U. Zutter, H. Iding, P. Spurr, B. Wirz, New, efficient synthesis of oseltamivir phosphate (Tamiflu) via enzymatic desymmetrization of a *meso*-1,3-cyclohexanedicarboxylic acid diester, *J. Org. Chem.*, 2008, 73, 4895–4902.
144. Y. F. Wang, C. J. Sih, Bifunctional chiral synthon via biochemical methods. Chiral precursors to (+)-biotin and (-)-A-Factor, *Tetrahedron Lett.*, 1984, 25, 4999–5002.
145. D. M. Tsachen, L. M. Fuentes, J. E. Lynch, W. L. Laswell, R. P. Volante, I. Shinkai, The stereoselective preparation of β-hydroxy esters using a yeast reduction in an organic solvent, *Tetrahedron Lett.*, 1997, 8, 1049–1054.
146. K. Mori, Synthesis of optically active pheromones, *Tetrahedron*, 1989, 45, 3233–3298.
147. P. Vitale, A. Digeo, F. M. Perna, G. Agrimi, A. Salomone, A. Scilimati, C. Cardellicchio, V. Capriati, Stereoselective chemoenzymatic synthesis of optically active aryl-substituted oxygen-containing heterocycles, *Catalysts*, 2017, 7, 37–50.

148. R. N. Patel, L. Chu, R. Mueller, Diastereoselective microbial reduction of (*S*)-[3-chloro-2-oxo-1-(phenylmethyl)propyl] carbamic acid, 1,1-dimethylethyl ester, *Tetrahedron: Asymm.*, 2003, 14, 3105–3109.
149. G. Bold, A. Faessler, H.-G. Capraro, R. Cozens, T. Klimkait, J. Lazdins, J. Mestan, B. Poncioni, J. Roesel, D. Stover, M. Tintelnot-Blomley, F. Acemoglu, W. Beck, E. Boss, M. Eschbach, T. Huerlimann, E. Masso, S. Roussel, K. Ucci-Stoll, D. Wyss, M. Lang, New aza-dipeptide analogues as potent and orally absorbed HIV-1 protease inhibitors: candidates for clinical development, *J. Med. Chem.*, 1998, 41(8), 3387–3401.
150. L. H. Andrade, L. Piovan, M. D. Pasquini, Improving the enantioselective bioreduction of aromatic ketones mediated by *Aspergillus terreus* and *Rhizopus oryzae*: the role of glycerol as a co-solvent, *Tetrahedron: Asymm.*, 2009, 20, 1521–1525.
151. T. Matsuda, Y. Yamagishi, S. Koguchi, N. Iwai, T. Kitazume, An effective method to use ionic liquids as reaction media for asymmetric reduction by *Geotrichum candidum*, *Tetrahedron Lett.*, 2006, 47, 4619–4622.
152. R. Kołodziejska, R. Studzińska, M. Kwit, M. Jelecki, A. Tafelska-Kaczmarek, Microbiological bio-reduction of prochiral carbonyl compounds by antimycotic agent Boni Protect, *Catal. Commun.*, 2017, 101, 81–84.
153. Z. Maugeri, P. Dominguez de Maria, Whole-cell biocatalysis in deep-eutectic-solvents/aqueous mixtures, *ChemCatChem*, 2014, 6, 1535–1537.
154. A. Wolfson, Ch. Dlugy, D. Tavor, J. Blumenfeld, Y. Shotland, Baker's yeast catalyzed asymmetric reduction in glicerol, *Tetrahedron: Asymm.*, 2006, 17, 2043–2045.
155. B. D. Feske, I. A. Kaluzna, J. D. Stewart, Enantiodivergent, biocatalytic routes to both Taxol side chain antipodes, *J. Org. Chem.*, 2005, 70, 9654–9657.
156. M. E. Wall, M. C. Wani, In Taxane anticancer agents: basic science and current status; G. Georg, I., Ed.; *ACS Symposium Series* 583; American Chemical Society: Washington, DC, 1995, 18–30.
157. R. N. Patel, Biocatalysis for synthesis of pharmaceuticals, *Bioorg. Med. Chem.*, 2018, 26, 1252–1274.
158. R. L. Hanson, S. L. Goldberg, D. B. Brzozowski, T. P. Tully, D. Cazzulino, W. L. Parker, O. K. Lyngberg, T. C. Vu, M. K. Wong, R. N. Patela, Preparation of an amino acid intermediate for the dipeptidyl peptidase IV inhibitor, Saxagliptin, using a modified phenylalanine dehydrogenase, *Adv. Synth. Catal.*, 2007, 349, 1369–1378.
159. R. Patel, L. Chu, V. Nanduri, J. Li, A. Kotnis, W. Parker, M. Liu, R. Mueller, Enantioselective microbial reduction of 6-oxo-8-[4-[4-(2-pyrimidinyl)-1-piperazinyl]butyl]-8-azaspiro[4.5]decane-7,9-dione, *Tetrahedron: Asymm.*, 2005, 16, 2778–2783.
160. J. P. Yevich, J. S. New, W. G. Lobeck, P. Dextraze, E. Bernstein, D. P. Taylor, F. D. Yocca, M. S. Eison, D. L. Temple, Jr., Synthesis and biological characterization of α-(4-fluorophenyl)-4-(5-fluoro-2-pyrimidiny1)-l-piperazinebutanol and analogues as potential atypical antipsychotic agents, *J. Med. Chem.*, 1992, 35, 4516–4525.
161. H. K. Shamsuzzaman, Bioactive benzofuran derivatives: a review, *Eur. J. Med. Chem.*, 2015, 97, 483–504.
162. R. Kołodziejska, R. Studzińska, A. Tafelska-Kaczmarek, H. Pawluk, B. Stasiak, M. Kwit, A. Woźniak, Effect of chemical structure of benzofuran derivatives and reaction conditions on enantioselective properties of *Aureobasidium pullulans* microorganism contained in Boni Protect antifungal agent, *Chirality*, 2020, 32, 407–415.
163. R. Kołodziejska, R. Studzińska, A. Tafelska-Kaczmarek, H. Pawluk, M. Kwit, B. Stasiak, A. Woźniak, The application of safe for humans and the environment Polyversum antifungal agent containing living cells of *Pythium oligandrum* for biotransformation of prochiral ketones, *Bioorg. Chem.*, 2019, 92, 1–7.
164. R. Kołodziejska, R. Studzińska, H. Pawluk, A. Karczmarska-Wódzka, A. Woźniak, Enantioselective bioreduction of prochiral pyrimidine base derivatives by Boni Protect fungicide containing live cells of *Aureobasidium pullulans*, *Catalysts*, 2018, 8, 1–9.
165. A. Shafee, H. Motamedi, A. King, Purifcation, characterization and immobilization of an NADPH-dependent enzyme involved in the chiral specific reduction of the keto ester M, an intermediate in the synthesis of an anti-asthma drug, Montelukast, from *Microbacterium campoquemadoensis* (MB5614), *Appl. Microbiol. Biotechnol.*, 1998, 49, 709–717.
166. N. Khazai, E. S. Judd, V. Tangpricha, Calcium and vitamin D: skeletal and extraskeletal health, *Curr. Rheumatol. Rep.*, 2008, 10, 110–117.
167. M. Warnke, T. Jung, J. Dermer, K. Hipp, N. Jehmlich, M. von Bergen, S. Ferlaino, A. Fries, M. Mîller, M. Boll, 25-Hydroxyvitamin D_3 synthesis by enzymatic steroid side-chain hydroxylation with water, *Angew. Chem.*, 2016, 128, 1913–1916.
168. E. Brenna, M. Crotti, F. G. Gatti, A. Manfredi, D. Monti, F. Parmeggiani, S. Santangelo, D. Zampieri, Enantioselective synthesis of (*R*)-2-arylpropanenitriles catalysed by ene-reductases in aqueous media and in biphasic ionic liquid–water systems, *ChemCatChem.*, 2014, 6, 2425–2431.
169. R. P. Gaisberger, M. H. Fechter, H. Griengl, The first hydroxynitrile lyase catalysed cyanohydrin formation in ionic liquids, *Tetrahedron Asymm.*, 2004, 15, 2959–2963.
170. F. F. Fleming, L. Yao, P. C. Ravikumar, L. Funk, B. C. Shook, Nitrile-containing pharmaceuticals: efficacious roles of the nitrile pharmacophore, *J. Med. Chem.*, 2010, 53, 7902–7917.
171. N. de Leeuw, G. Torrelo, C. Bisterfeld, V. Resch, L. Mestrom, E. Straulino, L. van der Weel, U. Hanefeld, Ester synthesis in water: *Mycobacterium smegmatis* acyl transferase for kinetic resolutions, *Adv. Synth. Catal.*, 2018, 360, 242–249.
172. T. Marie, G. Willig, A. R. S. Teixeira, E. G. Barboza, A. Kotland, A. Gratia, E. Courot, J. Hubert, J.-H. Renault, F. Allais, Enzymatic synthesis of resveratrol α-glycosides from β-cyclodextrin-resveratrol complex in water, *ACS Sustainable Chem. Eng.* 2018, 6, 5370–5380

13

Selected Green Efforts to Utilization of Carbohydrates

Michela I. Simone

CONTENTS

13.1 Introduction ... 229
13.2 Research in Uses of Carbohydrates Falls into Two Principal Categories ... 229
13.3 Chemical and Enzymatic Syntheses of Medicinally Valuable Intermediates and Drugs from Carbohydrates ... 229
13.4 Carbohydrates as Biomass for Renewable Energy Production via HTC ... 232
13.5 Future Outlook ... 233
References ... 233

13.1 Introduction

Imperative to all areas of modern chemical synthesis is the introduction of green chemistry principles in order to prevent waste, maximize incorporation of all materials in the final products, generate substances possessing little or no toxicity to human health and the environment, preserve efficacy of function, reduce the use of auxiliary substances, imbue energy efficiency into the chemical process, use of renewable feedstocks, reduce derivatives, use catalysts, carry out real-time analysis for pollution prevention and minimize the potential for chemical accidents.[1,2] Ultimately, these principles lay out a path for sustainability, which is of overarching importance in medicinal chemistry as the modern pharmaceutical industry is faced with new challenges, such as the expansion into and the production of molecularly complex[3] and high Fsp^3 index drugs that expand the traditional chemical space occupied by medicinal chemistry.[4–9]A number of fine-chemical companies are currently seizing on the opportunity to provide sp^3-rich fragments.[10,11]

Monosaccharides and carbohydrates are ideal starting materials for the synthesis of high Fsp^3 index drugs, possessing already a number of highly desirable properties, including several consecutive stereocenters, with functional groups (usually hydroxyls) that can be selectively manipulated, highest Fsp^3 index (1 for monosaccharide ring structures), with all D-stereo chemical configurations naturally available (albeit some of them as rare sugars), generally in large quantities and relatively cheaply.

13.2 Research in Uses of Carbohydrates Falls into Two Principal Categories

1. In medicinal chemistry, where chemical and enzymatic syntheses of medicinally valuable intermediates and drugs from carbohydrates expand the chemical space traditionally occupied by medicinal chemistry and provide the much-needed access to high Fsp^3 index drugs. Another emerging area utilizes carbohydrates as renewable media to green production of stable nanoparticles where the carbohydrates act as reducing and stabilizing agents[12–14] for further potential biomedical and pharmaceutical applications.
2. In renewable energy, where carbohydrates are used as biomass for renewable energy production. This is a promising field where processes such as hydrothermal carbonization (HTC) in water represent an important potential green pathway to sustainable fuels.

13.3 Chemical and Enzymatic Syntheses of Medicinally Valuable Intermediates and Drugs from Carbohydrates

In recent years, many syntheses have been developed from monosaccharides to drugs and glycans.[15] The link between glycan structure and functions is becoming as important as the correlation between protein structure and functions.[16] One such glycan is heparin and its analogues, which are prescribed as anticoagulants[17] and potentially for axon regeneration and extension.[18] Of these, Fondaparinux is an example (Figure 13.1).

Among the most successful drugs based on sugar scaffolds are probably the antivirals zanamivir (Relenza) and oseltamivir (Tamiflu), which are competitive neuraminidase ligands (Figure 13.1). They are effective against both type A and type B influenzas.[19] They work through blocking the enzyme-binding site and preventing the release of the new virions from the host cell. The antiviral activity of zanamivir in vitro is dependent on the assay and the viral strain utilized. The IC_{50} and IC_{90} range from 0.005 to 16.0 μM and 0.05 to >100 μM, respectively (1 μM = 0.33 μg/mL).[20] More recently, multiple molecules of zanamivir were also covalently conjugated to a flexible poly-L-glutamine (PGN) linker.[21,22] This conjugation strategy

DOI: 10.1201/9781003089162-13

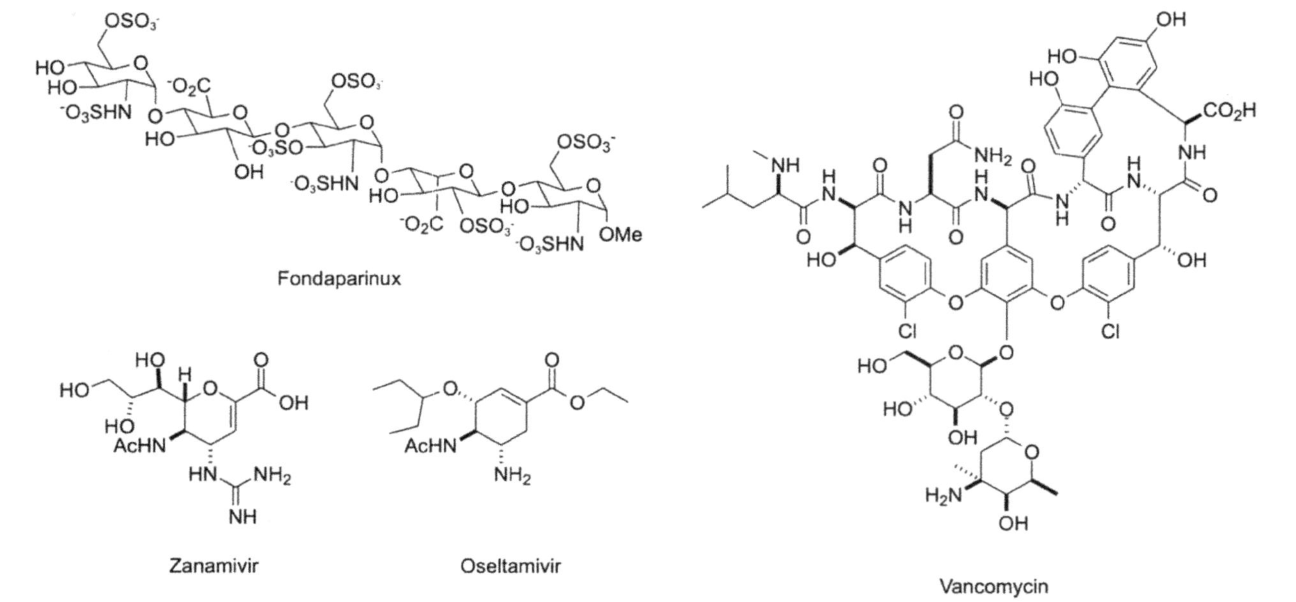

Fondaparinux

Zanamivir

Oseltamivir

Vancomycin

FIGURE 13.1 Selected drugs and glycan mimetics.

allows the enhancement of the anti-influenza virus activity by 1,000- to 10,000-fold more potent than monomeric zanamivir in plaque-reduction assays and, importantly, is effective even against zanamivir- and oseltamivir-resistant influenza viruses. Conjugated zanamivir possesses the same mode of action as zanamivir itself, and additionally also synergistically inhibits early stages of influenza virus infection, thus contributing to the markedly increased antiviral potency, which is thought to derive from interference with the intracellular trafficking of the endocytosed viruses and the subsequent virus–endosome fusion. This strategy confers a unique mechanism of antiviral action potentially useful for minimizing drug resistance.

Several synthetic strategies are available for the synthesis of zanamivir[23–29] with some of them utilizing the expensive sialic acid as starting material. More recently, a synthetic pathway from inexpensive D-glucono-δ-lactone was devised, which provided the total synthesis of zanamivir in 14 steps with a 12% overall yield.[30]

Oseltamivir (Tamiflu, Figure 13.1), approved by the FDA in 1999, has been one of the mainstay neuraminidase inhibitors against seasonal influenza and stockpiled also in view of pandemic influenza. Though a number of recent studies have questioned the risk–benefit ratio of this drug.[31,32] More than 70 Tamiflu synthetic procedures have been developed in the past 20 years in order to achieve a significantly efficient, safe, cost-effective and environmentally benign synthetic procedure.[33] (-)-shikimic acid was used as starting material and remains the current industrial synthetic starting material. The original synthesis of oseltamivir carboxylate proceeded in 15% overall yield over 14 synthetic steps despite using protecting group chemistry.[34] The Roche approach led to a 12-step route starting from (-)-shikimic acid with an overall yield of ~35%,[35] which went hand in hand with improved sourcing and purification of (-)-shikimic acid. A shorter (8 steps) approach toward Tamiflu with an improved 47% overall yield was developed by Shi et al.[36] The overall yield improved slightly (35–47%) and the number of transformations was considerably reduced relative to the Roche industrial approach (12 steps to eight steps).[35,36] This approach represents a model of atom economy since no protecting group manipulations were needed.

In the first step of bacterial and viral infections, adhesion to the cell surface occurs via recognition of the host glycans projecting out of adhesins and lectins. Microbes use a plethora of proteins to interact with host carbohydrates conjugated to surface proteins. Hence, glycomimetics able to block this interaction can be effective antibacterial agents (Figure 13.2).[37,38] An example is represented by the interaction of *Escherichia coli* strains with host cells. This interaction impinges on three fibrial lectins and was most effectively disrupted by multivalent glycons linked via hydrophobic aglycons. The first generation of ω-alkyl (methyl to octyl) α-D-mannopyranosides

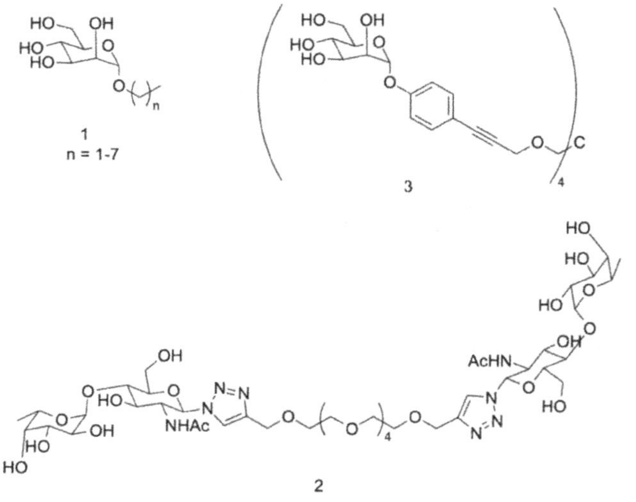

1
n = 1-7

3

2

FIGURE 13.2 Selected examples of glycomimetics with antibacterial properties.

1 (13.2) revealed decreasing K_d values with increasing alkyl chain length (5 nM for n = 7). The inclusion of multivalent glycons further decreased the K_d (0.45 nM for **3**).[39–41]

Ligands for the lectin PA-IIL from *Pseudomonas aeruginosa* were targeted with multivalent compounds, where linear dimer **2** was the most effective (K_d90 nM).[42] The synthetic development of such multivalent ensembles usually proceeds via click chemistry (e.g., azide alkyne cycloaddition), which allows high efficiency and selectivity, wide scope and high atom economy.[42,43]

In the infection area, there is a continued need for new-class antibacterial agents in order to provide effective therapeutics against infections, in particular, caused by extensively drug-resistant and pan-drug-resistant Gram-negative bacteria. Currently, the antibacterial agents in clinical development are predominantly derivatives of well-established antibiotic classes.[44]

Examples of antibacterials containing carbohydrate moieties include glycopeptides vancomycin, teicoplanin and semi-synthetic telavancin. Dalbavancin and oritavancin are structurally related to vancomycin and were developed to improve vancomycin's duration of action and tolerability.

They are used for treating multi-resistant *Staphylococcus aureus* infections and enterococcal infections which are resistant to beta-lactams and other antibiotics. The inhibition of peptidoglycan synthesis via the inhibition of transglycosylation is common to glycopeptides (such as vancomycin) and lipoglycopeptides (such as oritavancin). This occurs through the binding to D-alanyl-D-alanine stem termini of the C55-lipid transporter in Gram-positive bacteria. Additional inhibition of transpeptidation, the other essential enzymatic step in peptidoglycan polymerization, by oritavancin provides an extra binding mode from vancomycin and contributes to oritavancin's activity versus vancomycin-resistant organisms.[45]

The hydrophobic 4'-chlorobiphenylmethyl group also interacts and disrupts the cell membrane, resulting in depolarization, permeabilization and concentration-dependent, rapid cell death. This mechanism is shared with telavancin.[46]

Syntheses of such significantly larger and complex molecules present great synthetic challenges. Three total syntheses of the vancomycin aglycon[47–49] and two total syntheses of vancomycin[50,51] have been made available, with many reviews covering the synthetic approaches to glycopeptide antibiotics utilized over the years.[52–54]

One of the latest synthetic efforts in the area provides vancomycin in 19 steps, a significantly improved 5% overall yield and atroposelectivity ratios compared to previous works. This includes a protecting group-free two-step enzymatic glycosylation of the vancomycin aglycon, paving the way for large-scale synthetic preparation of analogues that directly address the underlying mechanism of resistance to vancomycin.[51,55]

A class of antitumor carbohydrate-based vaccines is being investigated that are immunotherapy that trains and activates the immune system in the human body to eliminate cancer cells. Cancer cells tend to express particular proteins or anomalous glycosylation patterns on their surface called tumor-associated carbohydrate antigens (TACAs).[56] Examples of these include Globo H, GM2 and STn (Figure 13.3).

An immunotherapy vaccine available for prostate cancer, approved by the FDA in 2010, is Provenge® which works on the premise that 95% of prostate cancer cells display Prostatic acid phosphatase.[57] TACAs fall into four main categories: (1) The Globo series including Globo H, SSEA4 and SSEA3. These are glycolipids overexpressed in breast, prostate, lung, ovary and colon cancer cells; (2) the gangliosides, e.g., GD2, GD3, GM2, GM3 and fucosyl GM1 which overexpress on melanoma, neuro- blastoma, sarcoma and B-cell lymphoma; (3) the blood group TACAs, e.g., Lewis^x, Lewis^y, sialyl Lewis^x and sialyl Lewis^a and are overexpressed on breast, prostate, lung colon and ovary cancer cells; (4) the glycoprotein including Thomsennouveau (Tn), Thomsen–Friendreich (TF) and sialyl-Tn (STn) which attach at the serine/threonine on the mucin and overexpress in epithelial cancer cells (breast, ovary and prostate).[58–65]

The administration of TACA antibodies confirmed that patients with GM2 antibodies against the melanoma differentiation antigen GM2 produced IgM antigens in most patients, had improved prognosis and demonstrated clinical benefit in terms of disease-free intervals from GM2 antibody induction.

Similarly, preclinical data indicate that ch14.18, a monoclonal antibody against the tumor-associated disialoganglioside GD2, had activity against neuroblastoma and that this activity was enhanced when ch14.18 was combined with granulocyte–macrophage colony-stimulating factor or interleukin-2.[61,62]

Covalent conjugation of TACAs to immunogenic carrier proteins is carried out to induce a T-cell mediated immune response. The main and potential issues related to this class of vaccines targeting TACAs are varying degrees of immune response were detected for various carrier proteins, the low

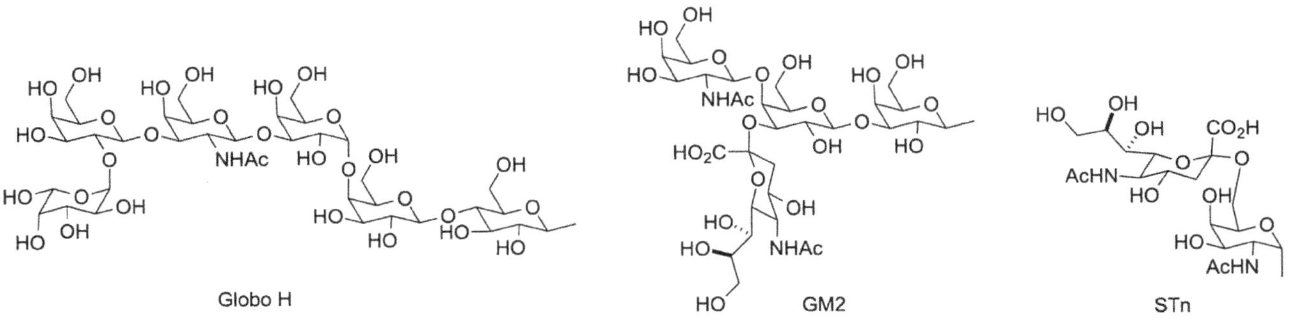

Globo H GM2 STn

FIGURE 13.3 Selected TACAs from the Globo series (Globo H), the gangliosides (GM2) and the mucin-attached glycans (STn).

immunogenity elicited by the vaccine, the challenging and laborious synthetic protocols[66,67] to the production of TACAs and their analogues (including low yields and reproducibility). Furthermore, the molecular tether linking TACAs to the carrier proteins also plays a role in eliciting the immune response (this can, in turn, produce non-specific immune responses that can suppress carbohydrate-specific antibody production), as well as the number and types of TACAs attached to the tether.

In recent years, notable greener syntheses to high-value intermediates and glycosylated drugs have also been provided from monosaccharide starting materials by a variety of methodologies,[68–73] including automated glycan assembly[74,75] and enzymatic methodologies. The latter includes the process of Izumoring[76] that allows the production of rare sugars from starch, whey or hemicellulosic waste with D-glucose obtained from starch, whey and hemicellulose, D-galactose from whey and D-xylose from hemicellulose. All monosaccharides (tetroses to hexoses) were produced via this methodology using the four enzymes D-tagatose 3-epimerase, aldose isomerase, aldose reductase and oxidoreductase, or whole cells as biocatalysts.

13.4 Carbohydrates as Biomass for Renewable Energy Production via HTC

The chemical, mechanistic and thermodynamic complexities of the Maillard reaction render the study of HTC processes very challenging, with recent efforts attempting to further a general mechanistic rationalization,[77] though research is still significantly away from a process that is both sustainable and reliable.

Some considerations about reaction conditions follow. Hydrothermal technologies effect chemical and physical transformations of biomass at high temperatures (200–600°C) and high pressures (5–40 MPa) in hot compressed water (HCW)[78] (either subcritical or supercritical[79]) as reaction medium.[80] Several definitions are in use for subcritical water.[81–88] A focus on reactions occurring in subcritical water is of relevance here.

Parameters that influence product outcomes of HTC processes include temperature, residence time,[81,89] starting materials, catalysts,[90–93] pH, etc.

Temperature is crucial in the formation of hydrothermal carbons (HCs) from glucose, with a minimum temperature of 160°C to effect the formation of solid residues.[94,95] At 180°C–200°C, the HCs formed are rich in carbonyl functionalities[96] and interconnected furans.[96–106]

Increasing the temperature affords an increase in the degree of aromatization to structures akin to those formed during pyrolysis.[100,107–109] Temperature is also critical in determining preferred reaction mechanisms,[81,110] with ionic processes prevailing at lower temperatures (subcritical HCW) and radical formation at higher temperatures (supercritical HCW).[106,111,112] Radical reaction mechanisms usually lead to a highly diverse, random product palette,[113–116] and finally to gas formation.[116–121]

By heating water at high pressures, a phase transition to steam is prevented. This avoids a large enthalpic penalty,[122–124] but opens the door to a set of new challenges, which include unknown or largely uncharacterized reaction pathways and kinetics, inadequate catalysts and solid management protocols, a need for novel materials to withstand the high temperature, high-pressure and corrosive environments of hydrothermal media.[122]

The physical properties of water change significantly at HTC temperatures and pressures.[81] In particular, its density,[125,126] viscosity[81,127] and surface tension decrease, while its diffusivity increases and solvation properties change. These lead to an enhancement of mass transfer and ultimately to an increase in observed reaction rates.[128]

Furthermore, water becomes highly corrosive at HTC temperatures and pressures, with detection of corrosion toward stainless steel after 40–50 uses of the reaction vessel for temperatures up to 350°C.[129–136]

At 30 MPa, in the 25–450°C temperature range, the solvation properties of water change dramatically from being a polar, highly hydrogen-bonded solvent to behaving more like a polar/non-polar organic solvent (depending on the temperature).[79,137–139] The solubilities of organic compounds[140,141] and gases[122,142,143] increase by several orders of magnitude, with lignin and cellulose becoming soluble.[144,145] Whereas, the solubility of inorganic ionic compounds, dependent on the dielectric constant, decreases as the structures of electrolyte solutions change, due to rearrangement of hydration shells (dehydration effect) and ion-pairing occurs.[146–151] The dielectric constant of water decreases from 80 (at r.t.) to 2 (just below 300°C),[137,152–154] due to a significantly diminished and changed hydrogen bond network.[134,138,155–158] It is crucial to investigate the phase diagrams of multicomponent systems to steer toward suitable temperature/pressure combinations for HTC. This is done keeping in mind that the addition of an electrolyte to water appreciably shifts the critical point of water, and that now all multicomponent phase diagrams are available. In those cases, models obtained from computer simulations are advantageous.[87] Critical point parameters of solvents are 303.9 K, 7.375 MPa for CO_2, 405.3 K, 11.350 MPa for NH_3 and 646.9 K, 22.060 MPa for H_2O. Multicomponent phase diagrams for H_2O/CO_2[155] and H_2O/NH_3.[159]

As the ion product pK_w bottoms out at 11 at 350°C, water becomes a greater source of protons and hydroxide ions to then rise again by 5 orders of magnitude at 500°C.[153,160,161]

In subcritical water, the following are generally enhanced: acidic and basic catalysis involving charged species,[162] reactions involving polar transition states (e.g., nucleophilic substitutions and eliminations)[88,163,164] and ionic reactions involving water as a reactant (e.g., hydrolyses and dehydrations).[88,122,165–170] With the note that minimal variations in temperature, greatly affect the density and hence can significantly influence reaction rates and outcomes.[138,171,172]

Studies of selected organic reactions in sub-and supercritical fluids[173–176] relevant to HTC conditions include methanol oxidation to water and carbon dioxide,[177–179] methane oxidation,[179–182] eliminations,[164] dehydration of alcohols to alkenes,[164,173,183,184] probably fastest in subcritical water[162] which proceeds by E2 mechanism with high selectivity[183] and is followed by rearrangements. Also relevant are ether transformations,[82,162,185–187] acetic acid oxidation,[188–195] phenol oxidation,[196–202] reduction of aldehydes and ketones, Diels–Alder reactions, alkylations, oxidation of benzylic position, ester hydrolysis, transesterification, studies on polycyclic aromatic hydrocarbon stabilities and decarboxylations.

Thermodynamics calculations for the alcohol/alkene dehydration/hydration show that, for a 1 M solution of alcohol at 400°C and 34.6 MPa, the ethanol/ethylene equilibrium lies in favor of the alkene by 26/74 mol%, whereas the n-propanol/propylene system equilibrates at 3/97 mol%.[187] In the case of acid-catalyzed dehydration of propanol at 34.5 MPa and 375°C, the sole formation of water and propene was seen, with no *n*-propanol left, but some 2-propanol, and consistent with an E2 mechanism.[184] HTC of extended polyol systems (whether polymeric or monomeric) such as cellulose, hemicellulose, sucrose and glucose, is dominated by dehydration reactions, where primary alcohols dehydrate via E2, S_N2 and Ad_E3 pathways. Tertiary alcohols dehydrate via E1 and Ad_E2 mechanisms. Secondary alcohols seem to proceed via both E2 and E1 mechanisms. Ethers seem to form via substitution.

13.5 Future Outlook

It is clear that improved processes for the utilization of carbohydrates and their analogues, both as fine chemicals in medicinal chemistry and as biomass in renewable energy production, are crucial to the sustainable production of high Fsp^3 index drugs and new fuels. Increasing structural complexity, via saturation and introduction of more than one chiral center, correlates with success as leads transition from discovery, through clinical testing, to drugs. Similarly, a better mechanistic understanding of the Maillard reaction and the factors affecting its equilibria would provide more efficient pathways to biomass recycling. Both areas would benefit from mimicking the exquisite pathways devised by Nature.

REFERENCES

1. Anastas, P. T.; Warner, J. C., *Green Chemistry: Theory and Practice*. Oxford University Press: New York, 1998.
2. https://www.acs.org/content/acs/en/greenchemistry/principles/12-principles-of-green-chemistry.html (accessed 25 August 2020).
3. Méndez-Lucio, O.; Medina-Franco, J. L., The many roles of molecular complexity in drug discovery. *Drug Discov. Today* 2017, *22*, 120–126.
4. Lovering, F.; Bikker, J.; Humblet, C., Escape from flatland: Increasing saturation as an approach to improving clinical success. *J. Med. Chem.* 2009, *52*, 6752–6756.
5. Hanby, A. R.; Troelsen, N. S.; Osberger, T. J.; Kidd, S. L.; Mortensen, K. T.; Spring, D. R., Fsp3-rich and diverse fragments inspired by natural products as a collection to enhance fragment-based drug discovery. *Chem. Commun.* 2020, *56*, 2280–2283.
6. Kombo, D. C.; Tallapragada, K.; Jain, R.; Chewning, J.; Mazuro, A. A.; Speake, J. D.; Hause, T. A.; Toler, S., 3D molecular descriptors important for clinical success. *J. Chem. Inf. Model.* 2013, *53*, 327–342.
7. Vincetti, P.; Kaptein, S. J. F.; Costantino, G.; Neyts, J.; Radi, M., Scaffold morphing approach to expand the toolbox of broad-spectrum antivirals blocking dengue/Zika replication. *ACS Med. Chem. Lett.* 2019, *10*, 558–563.
8. Lenci, E.; Innocenti, R.; Menchi, G.; Trabocchi, A., Diversity-Oriented Synthesis and Chemoinformatic Analysis of the Molecular Diversity of sp3-Rich Morpholine Peptidomimetics. *Front. Chem.* 2018, https://doi.org/10.3389/fchem.2018.00522.
9. Meanwell, N. A., Improving Drug Design: An Update on Recent Applications of Efficiency Metrics, Strategies for Replacing Problematic Elements, and Compounds in Nontraditional Drug Space. *Chem. Res. Toxicol.* 2016, *29*, 564–616.
10. https://enamine.net/fragments/fragment-collection/sp3-rich-fragments (accessed 25 August 2020).
11. https://www.chembridge.com/screening_libraries/targeted_libraries/spirocycle-library/index.php; (accessed 25 August 2020).
12. Ali, K.; Ahmed, B.; Dwivedi, S.; Saquib, Q.; Al-Khedhairy, A. A.; Musarrat, J., Microwave Accelerated Green Synthesis of Stable Silver Nanoparticles with Eucalyptus globulus Leaf Extract and Their Antibacterial and Antibiofilm Activity on Clinical Isolates. *PLoS ONE* 2015, *10*, e0131178.
13. Shervani, Z.; Yamamoto, Y., Carbohydrate-directed synthesis of silver and gold nanoparticles: Effect of the structure of carbohydrates and reducing agents on the size and morphology of the composites. *Carbohydr. Res.* 2011, *346*, 651–658.
14. Vigneshwaran, N.; Nachane, R. P.; Balasubramanya, R. H.; Varadarajan, P. V., A novel one-pot 'green' synthesis of stable silver nanoparticles using soluble starch. *Carbohydr. Res.* 2006, *341*, 2012–2018.
15. Valverde, P.; Ardá, A.; Reichardt, N.-C.; Jiménez-Barbero, J.; Gimeno, A., Glycans in drug discovery. *Med. Chem. Commun.* 2019, *10*, 1678–1691.
16. Varki, A.; Lowe, J. B., *Essentials of Glycobiology. Chapter 6 Biological Roles of Glycans*. 2nd ed.; Cold Spring Harbor (NY): Cold Spring Harbor Laboratory Press, 2009.
17. Linhardt, R. J., 2003 Claude S. Hudson Award Address in Carbohydrate Chemistry. Heparin: Structure and Activity. *J. Med. Chem.* **2003**, *46*, 2551–2564.
18. Ohtake, Y.; Li, S. X., Molecular mechanisms of scar-sourced axon growth inhibitors. *Brain Res.* 2015, *1619*, 22–35.
19. Eiland, L. S.; Eiland, E. H., Zanamivir for the prevention of influenza in adults and children age 5 years and older. *Ther. Clin. Risk Manag.* 2007, *3*, 461–465
20. Kimberlin, D. W., *Principles and Practice of Pediatric Infectious Disease. Part IV. Chapter 295 - Antiviral Agents*. 3rd ed.; 2008.
21. Weight, A. K.; Belser, J. A.; Tumpey, T. M.; Chen, J.; Klibanov, A. M., Zanamivir conjugated to poly-L-glutamine is much more active against influenza viruses in mice and ferrets than the drug itself. *Pharm. Res.* 2014, *31*, 466–474.
22. Lee, C. M.; Weight, A. K.; Haldar, J.; Wang, L.; Klibanov, A. M.; Chen, J., Polymer-attached zanamivir inhibits synergistically both early and late stages of influenza virus infection. *Proc. Natl. Acad. Sci. U. S. A.* 2012, *109*, 20385–20390.
23. Itzstein,M.V.;Wu,W.-Y.;Jin,B.,Thesynthesisof2,3-didehydro-2,4-dideoxy-4-guanidinyl-N-acetylneuraminic acid: A potent

influenza virus sialidase inhibitor. *Carbohydr. Res.* 1994, *259*, 301–305.

24. Scheigetz, J.; Zamboni, R.; Bernstein, M. A.; Roy, B., A Synthesis of 4-α-Guanidino-2-Deoxy-2,3-Didehydro N-Acetylneuraminic Acid. *Org. Prep. Proced. Int.* 1995, *27*, 637–644.
25. Chandler, M.; Bamford, M. J.; Conroy, R.; Lamont, B.; Patel, B.; Patel, V. K.; Steeples, I. P.; Storer, R.; Weir, N. G.; Wright, M.; Williamson, C. J., Synthesis of the potent influenza neuraminidase inhibitor 4-guanidino Neu5Ac2en. X-Ray molecular structure of 5-acetamido-4-amino-2,6-anhydro-3,4,5-trideoxy-D-erythro-L-gluco-nononic acid. *J. Chem. Soc., Perkin Trans.* 1995, *1*, 1173–1180.
26. Zhu, X.-B.; Wang, M.; Wang, S.; Z.-J. Yao, Concise synthesis of zanamivir and its C4-thiocarbamido derivatives utilizing a [3+2]-cycloadduct derived from d-glucono-δ-lactone. *Tetrahedron* 2012, *68*, 2041–2044.
27. Magano, J., Synthetic Approaches to the Neuraminidase Inhibitors Zanamivir (Relenza) and Oseltamivir Phosphate (Tamiflu) for the Treatment of Influenza. *Chem. Rev.* 2009, *109*, 4398–4438.
28. Tian, J.; Zhong, J.; Li, Y.; Ma, D., Organocatalytic and scalable synthesis of the anti-influenza drugs zanamivir, laninamivir, and CS-8958. *Angew. Chem., Int. Ed.* 2014, *53*, 13885–13888.
29. Nitabaru, T.; Kumagai, N.; Shibasaki, M., Catalytic Asymmetric antiselective nitroaldol reaction En Route to Zanamivir. *Angew. Chem., Int. Ed.* 2012, *51*, 1644–1647.
30. Lin, L.-Z.; Fang, J.-M., Total Synthesis of Anti-Influenza Agents Zanamivir and Zanaphosphor via Asymmetric Aza-Henry Reaction. *Org. Lett.* 2016, *18*, 4400–4403
31. Gupta, Y. K.; Meenu, M.; Mohan, P., The Tamiflu fiasco and lessons learnt. *Indian J. Pharmacol.* 2015, *47*, 11–16.
32. Jefferson, T.; Jones, M.; Doshi, P.; Spencer, E. A.; Onakpoya, I.; Heneghan, C. J., Oseltamivir for influenza in adults and children: Systematic review of clinical study reports and summary of regulatory comments. *BMJ* 2014, *348*, g2545.
33. Sagandira, C. R.; Mathe, F. M.; Guyo, U.; Watts, P., The evolution of Tamiflu synthesis, 20 years on: Advent of enabling technologies the last piece of the puzzle? *Tetrahedron* 2020, *76*, 131440
34. Kim, C. U.; Lew, W.; Williams, M. A.; Liu, L. Z.; Swaminathan, S.; Bischofberger, N.; Chen, M. S.; Mendel, D. B.; Tai, C. Y.; Laver, W. G.; Stevens, R. C., Influenza Neuraminidase Inhibitors Possessing a Novel Hydrophobic Interaction in the Enzyme Active Site: Design, Synthesis, and Structural Analysis of Carbocyclic Sialic Acid Analogues with Potent Anti-Influenza Activity. *J. Am. Chem. Soc.* 1997, *119*, 681–690.
35. Federspiel, M.; Fischer, R.; Hennig, M.; Mair, H.-J.; Oberhauser, T.; Rimmler, G.; Albiez, T.; Bruhin, J.; Estermann, H.; Gandert, C.; Gockel, V.; Gotzo, S.; Hoffmann, U.; Huber, G.; Janatsch, G.; Lauper, S.; Rockel-Stabler, O.; Trussardi, R.; Zwahlen, A. G., Industrial synthesis of the key precursor in the synthesis of the anti-influenza drug oseltamivir phosphate (Ro 64–0796/002, GS-4104–02): Ethyl (3R,4S,5S)–4,5-epoxy-3-(1-ethyl-propoxy)-cyclohex-1-ene-1-carboxylate. *Org. Process Res. Dev.* 1999, *3*, 266–274.
36. Nie, L.; Shi, X.; Ko, K. H.; Lu, W., A short and practical synthesis of oseltamivir phosphate (Tamiflu) from (-)-shikimic acid. *J. Org. Chem.* 2009, *74*, 3970–3973.
37. Unione, L.; Gimeno, A.; Valverde, P.; Calloni, I.; Coelho, H.; Mirabella, S.; Poveda, A.; Ardá, A.; Jimenez-Barbero, J., Glycans in Infectious Diseases. A Molecular Recognition Perspective. *Curr. Med. Chem.* 2017, *24*, 4057–4080.
38. Imberty, A.; Chabre, Y. M.; Roy, R., Glycomimetics and glycodendrimers as high affinity microbial anti-adhesins. *Chem. Eur. J.* 2008, *14*, 7490–7499.
39. Touaibia, M.; Wellens, A.; Shiao, T. C.; Wang, Q.; Sirois, S.; Bouckaert, J.; Roy, R., Mannosylated G(0) dendrimers with nanomolar affinities to *Escherichia coli* FimH. *ChemMedChem* 2007, 1190–1201.
40. Bouckaert, J.; Berglund, J.; Schembri, M.; Gents, E. D.; Cools, L.; Wuhrer, M.; Hung, C.-S.; Pinkner, J.; Slättegard, R.; Savialov, A.; Choudhury, D.; Langermann, S.; Hultgren, S. J.; Wyns, L.; Klemm, P.; Oscarson, S.; Knight, S. D.; Greve, H. D., Receptor binding studies disclose a novel class of high-affinity inhibitors of the *Escherichia coli* FimH adhesin. *Mol. Microbiol.* 2005, *55*, 441–445.
41. Stoker, A. W., RPTPs in axons, synapses and neurology. *Semin. Cell Dev. Biol.* 2015, *37*, 90–97.
42. Marotte, K.; Préville, C.; Sabin, C.; Moumé-Pymbock, M.; Imberty, A.; Roy, R., Synthesis and binding properties of divalent and trivalent clusters of the Lewis a disaccharide moiety to Pseudomonas aeruginosa lectin PA-IIL. *Org. Biomol. Chem.* 2007, *5*, 2953–2961.
43. https://www.organic-chemistry.org/namedreactions/click-chemistry.shtm (accessed February 2021).
44. Theuretzbacher, U.; Bush, K.; Harbarth, S.; Paul, M.; Rex, J. H.; Tacconelli, E.; Thwaites, G. E., Critical analysis of antibacterial agents in clinical development. *Nat. Rev. Microbiol.* 2020, *18*, 286–298.
45. Monteiro, J. M.; Covas, G.; Rausch, D.; Filipe, S. R.; Schneider, T.; Sahl, H.-G.; Pinho, M. G., The pentaglycine bridges of Staphylococcus aureus peptidoglycan are essential for cell integrity. *Nat. Sci. Reports* 2019, *9*, 5010.
46. Brade, K. D.; Rybak, J. M.; Rybak, M. J., Oritavancin: A New Lipoglycopeptide Antibiotic in the Treatment of Gram-Positive Infections. *Infect. Dis. Ther.* 2016, *5*, 1–15.
47. Boger, D. L.; Miyazaki, S.; Kim, S. H.; Wu, J. H.; Castle, S. L.; Loiseleur, O.; Jin, Q., Total synthesis of the vancomycin aglycon. *J. Am. Chem. Soc.* 1999, *121*, 10004–10011.
48. Nicolaou, K. C.; Takayanagi, M.; Jain, N. F.; Natarajan, S.; Koumbis, A. E.; Bando, T.; Ramanjulu, J. M., Total synthesis of vancomycin aglycon, part 3: Final stages. *Angew. Chem., Int. Ed.* 1998, *37*, 2717–2719.
49. Evans, D. A.; Wood, M. R.; Trotter, B. W.; Richardson, T. I.; Barrow, J. C.; Katz, J. L., Total syntheses of vancomycin and eremomycin aglycons. *Angew. Chem., Int. Ed.* 1998, *37*, 2700–2704.
50. Nicolaou, K. C.; Mitchell, H. J.; Jain, N. F.; Winssinger, N.; Hughes, R.; Bando, T., Total synthesis of vancomycin. *Angew. Chem., Int. Ed.* 1998, *38*, 240–244.
51. Nakayama, A.; Okano, A.; Feng, Y.; Collins, J. C.; Collins, K. C.; Walsh, C. T.; Boger, D. L., Enzymatic Glycosylation of Vancomycin Aglycon: Completion of a Total Synthesis of Vancomycin and N- and C-Terminus Substituent Effects of the Aglycon Substrate. *Org. Lett.* 2014, *16*, 3572–3575.
52. Walsh, C. T., Vancomycin resistance: Decoding the molecular logic. *Science* 1993, *261*, 308–310.
53. Walsh, C., Deconstructing vancomycin. *Science* 1999, *284*, 442–443.

54. James, R. C.; Pierce, J. G.; Okano, A.; Xie, J.; Boger, D. L., Redesign of glycopeptide antibiotics: Back to the future. *ACS Chem. Biol.* 2012, *7*, 797–804.
55. Moore, M. J.; Qu, S.; Tan, C.; Cai, Y.; Mogi, Y.; Keith, D. J.; Boger, D. L., Next-Generation Total Synthesis of Vancomycin. *J. Am. Chem. Soc.* 2020, *142*, 16039–16050
56. Mettu, R.; Chen, C.-Y.; Wu, C.-Y., Synthetic carbohydrate-based vaccines: Challenges and opportunities. *J. Biomed. Sci.* 2020, *27*, 1–22.
57. Kong, H. Y.; Byun, J., Emerging Roles of Human Prostatic Acid Phosphatase. *Biomol. Ther.* 2013, *21*, 10–20.
58. Hakomori, S., Antigen structure and genetic basis of histo-blood groups A, B and O: Their changes associated with human cancer. *Biochim. Biophys. Acta.* 1999, *1473*, 247–266.
59. Glinsky, G. V.; Ivanova, A. B.; Welsh, J.; McClelland, M., The role of blood group antigens in malignant progression, apoptosis resistance, and metastatic behavior. *Transfus. Med. Rev.* 2000, *14*, 326–350.
60. Hattrup, C. L.; Gendler, S. J., Structure and function of the cell surface (tethered) mucins. *Annu. Rev. Physiol.* 2008, *70*, 431–457.
61. Chang, W. W.; Lee, C. H.; Lee, P.; Lin, J.; Hsu, C. W.; Hung, J. T.; Lin, J. J.; Yu, J. C.; Shao, L. E.; Yu, J.; Wong, C.-H.; Yu, A. L., Expression of Globo H and SSEA3 in breast cancer stem cells and the involvement of fucosyl transferases 1 and 2 in Globo H synthesis. *Proc. Natl. Acad. Sci. U. S. A.* 2008, *105*, 11667–11672.
62. Heimburg-Molinaro, J.; Lum, M.; Vijay, G.; Jain, M.; Almogren, A.; Rittenhouse-Olson, K., Cancer vaccines and carbohydrate epitopes. *Vaccine* 2011, *29*, 8802–8826.
63. Lou, Y. W.; Wang, P. Y.; Yeh, S. C.; Chuang, P. K.; Li, S. T.; Wu, C. Y.; Khoo, K. H.; Hsiao, M.; Hsu, T. L.; Wong, C. H., Stage-specific embryonic antigen-4 as a potential therapeutic target in glioblastoma multiforme and other cancers. *Proc. Natl. Acad. Sci. U. S. A.* 2014, *111*, 2482–2487.
64. Livingston, P. O.; Zhang, S.; Lloyd, K. O., Carbohydrate vaccines that induce antibodies against cancer. 1. Rationale. *Cancer Immunol. Immunother.* 1997, *45*, 1–9.
65. Hamilton, W. B.; Helling, F.; Lloyd, K. O.; Livingston, P. O., Ganglioside expression on human malignant melanoma assessed by quantitative immune thin-layer chromatography. *Internat. J. Cancer* 1993, *53*, 566–573.
66. Zhu, J.; Warren, J. D.; Danishefsky, S. J., Synthetic Carbohydrate-Based Anticancer Vaccines: The Memorial Sloan-Kettering Experience. *Exp. Rev. Vaccines* 2009, *8*, 1399–1413.
67. Reily, C.; Stewart, T. J.; Renfrow, M. B.; Novak, J., Glycosylation in health and disease. *Nat. Rev. Nephrol.* 2019, *15*, 346–366.
68. Hotchkiss, D.; Soengas, R.; Simone, M.; Ameijde, J. v.; Hunter, S.; Cowley, A. R.; Fleet, G. W. J., Kiliani on ketoses: Branched carbohydrate building blocks from D-fructose and L-sorbose. *Tetrahedron Lett.* 2004, *45*, 9461–9464.
69. Soengas, R.; Izumori, K.; Simone, M.; Watkin, D. J.; Skytte, U. P.; Soetaert, W.; Fleet, G. W. J., Kiliani reactions on ketoses: Branched carbohydrate building blocks from D-tagatose and D-psicose. *Tetrahedron Lett.* 2005, *46*, 5755–5759.
70. Booth, K. V.; Cruz, F. P. D.; Hotchkiss, D. J.; Jenkinson, S. F.; Jones, N. A.; Weymouth-Wilson, A. C.; Clarkson, R.; Heinz, T.; Fleet, G. W. J., Carbon-branched carbohydrate chirons: Practical access to both enantiomers of 2-C-methyl-ribono-1,4-lactone and 2-C-methyl-arabinonolactone. *Tetrahedron: Asymm.* 2008, *19*, 2417–2424.
71. Filice, M.; Palomo, J. M., Monosaccharide derivatives as central scaffolds in the synthesis of glycosylated drugs. *RSC Adv.* 2012, *2*, 1729–1742.
72. Simone, M. I.; Soengas, R.; Newton, C. R.; Watkin, D. J.; Fleet, G. W. J., Branched tetrahydrofuran α,α-disubstituted δ-sugar amino acid scaffolds from branched sugar lactones: A new family of foldamers? *Tetrahedron Lett.* 2005, *46*, 5761–5765.
73. Zhang, Y.; Wang, F., Carbohydrate drugs: Current status and development prospect. *Drug Disc. Ther.* 2015, *9*, 79–87
74. Joseph, A. A.; Pardo-Vargas, A.; Seeberger, P. H., Total synthesis of polysaccharides by automated glycan assembly. *J. Am. Chem. Soc.* 2020, *142*, 8561–8564.
75. Panza, M.; Pistorio, S. G.; Stine, K. J.; Demchenko, A. V., Automated chemical oligosaccharide synthesis: Novel approach to traditional challenges. *Chem. Rev.* 2018, *118*, 8105–8150
76. Granström, T. B.; Takata, G.; Tokuda, M.; Izumori, K., Izumoring: A novel and complete strategy for bioproduction of rare sugars. *J. Biosci Bioeng.* 2004, *97*, 89–94.
77. Latham, K. G.; Simone, M. I.; Dose, W. M.; Allen, J. A.; Donne, S. W., Synchrotron based NEXAFS study on nitrogen doped hydrothermal carbon: Insights into surface functionalities and formation mechanisms. *Carbon* 2017, *114*, 566–578.
78. Kruse, A.; Dinjus, E., Hot compressed water as reaction medium and reactant: Properties and synthesis reactions. *J. Supercrit. Fluids* 2007, *39*, 362–380.
79. Watanabe, M., Chemical reactions of C(1) compounds in near-critical and supercritical water. *Chem. Rev.* 2004, *104*, 5803–5821.
80. Yakaboylu, O.; Harinck, J.; Smit, K. G.; Jong, W. D., Supercritical water gasification of biomass: A literature and technology overview. *Energies* 2015, *8*, 859–894.
81. M. Möller; P. Nilges; F. Harnisch; Schröder, U., Subcritical water as reaction environment: Fundamentals of hydrothermal biomass transformation. *ChemSusChem* 2011, *4*, 566–579.
82. J. Krammer; Vogel, H., Hydrolysis of esters in subcritical and supercritical water. *J. Supercrit. Fluids* 2000, *16*, 189–206.
83. Y. Yu; X. Lou; Wu, H., Some recent advances in hydrolysis of biomass in hot-compressed water and its comparisons with other hydrolysis methods. *Energy Fuels* 2008, *22*, 46–60.
84. Z. Srokol; A.-G. Bouche; A. van Estrik; R. C. Strik; T. Maschmeyer; Peters, J. A., Hydrothermal upgrading of biomass to biofuel; studies on some monosaccharide model compounds. *Carbohydr. Res.* 2004, *339*, 1717–1726.
85. D. Bröll; C. Kaul; A. Krämer; P. Krammer; T. Richter; M. Jung; H. Vogel; Zehner, P., *Angew. Chem.* 1999, *111*, 3180–3196.
86. D. Bröll; C. Kaul; A. Krämer; P. Krammer; T. Richter; M. Jung; H. Vogel; Zehner, P., Chemistry in supercritical water. *Angew. Chem. Int. Ed.* 1999, *38*, 2998–3014.
87. A. A. Galkin, V. V. Lunin, Subcritical and supercritical water: A universal medium for chemical reactions. *Russ. Chem. Rev.* 2005, *74*, 21–35.

88. Akiya, N.; Savage, P. E., Roles of water for chemical reactions in high-temperature water. *Chem. Rev.* 2002, *102*, 2725–2750.
89. Titirici, M.-M., *Sustainable Carbon Materials from Hydrothermal Processes*. Wiley Online Library: 2013.
90. J. Zhang, B. Hou, A. Wang, Z. L. Li, H. Wang, T. Zhang, Kinetic study of Retro-Aldol condensation of glucose to glycolaldehyde with ammonium metatungstate as the catalyst. *AIChE J.* 2014, *60*, 3804–3813.
91. F. S. Asghari; Yoshida, H., Dehydration of fructose to 5-hydroxymethylfurfural in sub-critical water over heterogeneous zirconium phosphate catalysts. *Carbohydr. Res.* 2006, *341*, 2379–2387.
92. M. Bicker, S. Endres, L. Ott, H. Vogel, Catalytical conversion of carbohydrates in subcritical water: A new chemical process for lactic acid production. *J. Mol. Catal. A Chem.* 2005, *239*, 151–157.
93. J. Jow, G. L. Rorrer, M. C. Hawley, Dehydration of D-Fructose to Levulinic Acid over LZY Zeolite Catalyst. *Biomass* 1987, *14*, 185–194.
94. M. Sevilla; A. B. Fuertes, Chemical and structural properties of carbonaceous products obtained by hydrothermal carbonization of saccharides. *Chem. Eur. J.* 2009, *15*, 4195–4203.
95. Fagerson, I. S., Thermal degradation of carbohydrates; a review. *J. Agric. Food Chem.* 1969, *17*, 747–750.
96. C. Falco; N. Baccile; Titirici, M.-M., Morphological and structural differences between glucose, cellulose and lignocellulosic biomass derived hydrothermal carbons. *Green Chem.* 2011, *13*, 3273–3281.
97. K. Latham; E. A. Locke, Molecular Structures Driving Pseudocapacitance in Hydrothermal Nanostructured Carbons. *in press* 2015.
98. N. Baccile, E. A. Appel, Structural characterization of hydrothermal carbon spheres by advanced solid-state MAS 13C NMR investigations. *J. Phys. Chem. C* 2009, *113*, 9644–9654.
99. K. Aydincak, Synthesis and characterization of carbonaceous materials from saccharides (glucose and lactose) and two waste biomasses by hydrothermal carbonization. *Ind. Eng. Chem. Res.* 2012, *51*, 9145–9152.
100. A. Chuntanapum; Matsumura, Y., Formation of Tarry Material from 5-HMF in Subcritical and Supercritical Water. *Ind. Eng. Chem. Res.* 2009, *48*, 9837–9846.
101. A. Chuntanapum; T. L.-K. Yong; S. Miyake; Y. Matsumura, Behavior of 5-HMF in subcritical and supercritical water. *Ind. Eng. Chem. Res.* 2008, *47*, 2956–2962.
102. A. Chuntanapum, Y. Matsumura, Char formation mechanism in supercritical water gasification process: A study of model compounds. *Ind. Eng. Chem. Res.* 2010, *49*, 4055–4062.
103. M. M. Titirici; A. Thomas; S. H. Yu; J. O. Müller; M. Antonietti, Direct synthesis of mesoporous carbons with bicontinuous pore morphology from crude plant material by hydrothermal carbonization. *Chem. Mat.* 2007, *19*, 4205–4212.
104. Kuster, B. F. M., 5-Hydroxymethylfurfural (HMF). A review focusing on its manufacture. *Starch/Staerke* 1990, *42*, 314–321.
105. J. N. Chheda; G. W. Huber; J. A. Dumesic, Liquid-phase catalytic processing of biomass-derived oxygenated hydrocarbons to fuels and chemicals. *Angew. Chem. Int. Ed.* 2007, *48*, 7164–7183.
106. D. Knezevic; W. P. M. V. Swaaij; S. R. A. Kersten, Hydrothermal conversion of biomass: I, glucose conversion in hot compressed water. *Ind. Eng. Chem. Res.* 2009, *48*, 4731–4743.
107. M. M. Titirici; R. J. White; C. Falco; M. Sevilla, Black perspectives for a green future: Hydrothermal carbons for environment protection and energy storage. *Energy Environ. Sci.* 2012, *5*, 6796-.
108. T. M. Aida; K. Tajima; M. Watanabe; Y. Saito; K. Kuroda; T. Nonaka; H. Hattori; R. L. Smith Jr; K. Arai, Reactions of D-fructose in water at temperatures up to 400°C and pressures up to 100 MPa. *J. Supercrit. Fluids* 2007, *42*, 110–119.
109. G. C. A. Luijkx; F. van Rantwijk; H. V. Bekkum, Hydrothermal formation of 1,2,4-benzenetriol from 5-hydroxymethyl-2-furaldehyde and D-fructose. *Carbohydr. Res.* 1993, *242*, 131–139.
110. M. J. Antal Jr; A. Brittain; C. DeAlmeida; S. Ramayya; J. C. Roy, In *Heterolysis and homolysis in supercritical water*, Supercritical Fluids: Chemical and Engineering Principles and Applications (ACS Symposium Series 329), (ACS Symposium Series 329): 1987; pp. 77–86.
111. M. Watanabe; T. Sato; H. Inomata; R. L. Smith; K. Arai; A. Kruse; E. Dinjus, Chemical Reactions of C1 Compounds in Near-Critical and Supercritical Water. *Chem. Rev.* 2004, *104*, 5803–5821.
112. Y. Matsumura; S. Yanachi; T. Yoshida, Glucose Decomposition Kinetics in Water at 25 MPa in the Temperature Range of 448–673 K. *Ind. Eng. Chem. Res.* 2006, *45*, 1875–1879.
113. S. Yaman, Pyrolysis of biomass to produce fuels and chemical feedstocks. *Energy Convers. Manage.* 2004, *45*, 651–671.
114. A. V. Bridgwater; D. Meier; D. Radlein, An overview of fast pyrolysis of biomass. *Org. Geochem.* 1999, *30*, 1479–1493.
115. H. Pakdel; C. Roy, Hydrocarbon content of liquid products and tar from pyrolysis and gasification of wood. *Energy Fuels* 1991, *5*, 427–436.
116. P. T. Williams; J. Onwudili, Composition of Products from the Supercritical Water Gasification of Glucose: A Model Biomass Compound. *Ind. Eng. Chem. Res.* 2005, *44*, 8739–8749.
117. J. H. Marsman; J. Wildschut; F. Mahfud; H. J. Heeres, Identification of components in fast pyrolysis oil and upgraded products by comprehensive two-dimensional gas chromatography and flame ionisation detection. *J. Chromatogr. A* 2007, *1150*, 21–27.
118. P. Basu; V. Mettanant, Biomass Gasification in SupercriticalWater – A Review. *Int. J. Chem. React. Eng.* 2009, *7*, R3.
119. T. Yoshida; Y. Oshima; Y. Matsumura, Gasi cation of biomass model compounds and real biomass in supercritical water. *Biomass Bioenergy* 2004, *26*, 71–78.
120. T. Yoshida; Y. Matsumura, Gasification of Cellulose, Xylan, and Lignin Mixtures in Supercritical Water. *Ind. Eng. Chem. Res.* 2001, *40*, 5469–5474.

121. I.-G. Lee; M.-S. Kim; Ihm, S.-K., Gasification of Glucose in Supercritical Water. *Ind. Eng. Chem. Res.* 2002, *41*, 1182–1188.
122. A. A. Peterson; F. Vogel; R. P. Lachance; M. Fröling; M. J. Antal; Tester, J. W., Thermochemical biofuel production in hydrothermal media: A review of sub- and supercritical water technologies. *Energy Environ. Sci.* 2008, *1*, 32–65.
123. A. Kruse, Supercritical water gasification. *Biofuel Bioprod. Biorefin.* 2008, *2*, 415–437.
124. D. C. Elliott, Catalytic hydrothermal gasification of biomass. *Biofuel Bioprod. Biorefin.* 2008, *2*, 254–265.
125. H. R. Patrick, K. Griffith, C. L. Liotta, C. A. Eckert, Near-Critical Water: A Benign Medium for Catalytic Reactions. *Ind. Eng. Chem. Res.* 2001, *40*, 6063–6067.
126. W. Wagner, A. Pruß, The IAPWS Formulation 1995 for the thermodynamic properties of ordinary water substance for general and scientific use. *J. Phys. Chem. Ref. Data* 2002, *31*, 387–535.
127. Marcus, Y., On transport properties of hot liquid and supercritical water and their relationship to the hydrogen bonding. *Fluid Phase Equilib.* 1999, *164*, 131–142.
128. Atkins, P. W., *Physikalische Chemie*. Wiley VCH: Weinheim, 1996.
129. D. B. Mitton, E. H. Han, S. H. Zhang, K. E. Hautanen, R. M. Latanision, Degradation in Supercritical Water Oxidation Systems. *ACS Symp. Ser.* 1997, *670*, 242–254.
130. D. Castello, A. Kruse, L. Fiori, Biomass gasification in supercritical and subcritical water: The effect of the reactor material. *Chem. Eng. J.* 2013, *228*, 535–544.
131. C. Friedrich, P. Kritzer, N. Boukis, G. Franz, E. Dinjus, The corrosion of tantalum in oxidizing sub- and supercritical aqueous solutions of HCl, H_2SO_4 and H_3PO_4. *J. Mater. Sci.* 1999, *34*, 3137–3143.
132. V. Casal, H. Schmidt, SUWOX – a facility for the destruction of chlorinated hydrocarbons. *J. Supercrit. Fluids* 1998, *13*, 269–276.
133. P. Kritzer, E. Dinjus, Factors controlling corrosion in high-temperature aqueous solutions: A contribution to the dissociation and solubility data influencing corrosion properties. *J. Supercrit. Fluids* 1999, *15*, 205–227.
134. A. G. Kalinichev, S. V. Churakov, Size and topology of molecular clusters in supercritical water: A molecular dynamics simulation. *Chem. Phys. Lett.* 1999, *302*, 411–417.
135. L. B. Kriksunov, D. D. MacDonald, Corrosion in supercritical water oxidation systems: A phenomenological analysis. *J. Electrochem. Soc.* 1995, *142*, 4069–4073.
136. D. B. Mitton, J.-H. Yoon, J. A. Cline, H.-S. Kim, N. Eliaz, R. M. Latanision, Corrosion studies in supercritical water oxidation systems. *ACS Symp. Ser.* 1995, *608*, 327–337.
137. P. Kritzer, E. Dinjus, An assessment of supercritical water oxidation (SCWO): Existing problems, possible solutions and new reactor concepts. *Chem. Eng. J.* 2001, *83*, 207–214.
138. R. van Eldik, C. Hubbard, *Chemistry under Extreme or Non-Classical Conditions*. Wiley: New York, 1997.
139. R. Biswas, Ion solvation dynamics in supercritical water. *Chem. Phys. Lett.* 1998, *290*, 223–228.
140. H. Weingärtner; Franck, E. U., *Angew. Chem.* 2005, *117*, 2730–2752.
141. H. Weingärtner; E. U. Franck, Supercritical water as a solvent. *Angew. Chem. Int. Ed.* 2005, *44*, 2672–2692.
142. P. Khuwijitjaru; S. Adachi; R. Matsuno, Solubility of saturated fatty acids in water at elevated temperatures. *Biosci. Biotechnol. Biochem.* 2002, *66*, 1723–1726.
143. J. F. Connolly, Solubility of hydrocarbons in water near the critical solution temperatures. *J. Chem. Eng. Data* 1966, *11*, 13–16.
144. S. Kumar, R. B. Gupta, Hydrolysis of microcrystalline cellulose in subcritical and supercritical water in a continuous flow reactor. *Ind. Eng. Chem. Res.* 2008, *47*, 9321–9329.
145. E. Kamio, S. Takahashi, H. Noda, C. Fukuhara, T. Okamura, Liquefaction kinetics of cellulose treated by hot compressed water under variable temperature conditions. *J. Mater. Sci.* 2008, *43*, 2179–2188.
146. S. H. Lee, P. T. Cummings, J. M. Simonson, R. E. Mesmer, Molecular dynamics simulations of the limiting conductance of NaCl in supercritical water. *Chem. Phys. Lett.* 1998, *293*, 289–294.
147. S. L. Wallen, J. L. Fulton, The ion pairing and hydration structure of Ni 2+ in supercritical water at 425°C determined by x-ray absorption fine structure and molecular dynamics studies. *J. Chem. Phys.* 1998, *108*, 4039–4046.
148. Y. Ikushima; N. Saito; Arai, M., Raman spectral studies of aqueous zinc nitrate solution at high temperatures and at a high pressure of 30 MPa. *J. Phys. Chem. B* 1998, *102*, 3029–3035.
149. M. Kubo, R. M. Levy, P. J. Rossky, N. Matubayasi, M. Nakahara, Chloride ion hydration and diffusion in supercritical water using a polarizable water model. *J. Phys. Chem. B* 2002, *106*, 3979–3986.
150. Y. Ikushima, N. Saito, R. Arai, Raman spectral studies of aqueous strontium nitrate solution up to 450°C and 40 MPa. *Bull. Chem. Soc. Jpn.* 1998, *71*, 1763–1769.
151. A. A. Chialvo, J. M. Simonson, R. E. Mesmer, Raman spectral studies of aqueous zinc nitrate solution at high temperatures and at a high pressure of 30 MPa. *J. Phys. Chem. B* 1998, *102*, 3029–3035.
152. Archer, D.; Wang, P., The dielectric constant of water and Debye–Hückel limiting law slopes. *J. Phys. Chem. Ref. Data* 1990, *19*, 371–411.
153. S. N. V. K. Aki, J. Feng, J. E. Chateauneuf, J. F. Brennecke, Generation of xanthenium and 9-phenylxanthenium carbocations in subcritical water and reactivity with amylamine. *J. Phys. Chem. A* 2001, *105*, 8046–8052.
154. X. Amashukeli, C. C. Pelletier, J. P. Kirby, F. J. Grunthaner, Subcritical water extraction of amino acids from Atacama Desert soils. *J. Geophys. Res.* 2007, *112*, G04S16.
155. Franck, E. U., *Ber. Bunsen-Ges. Phys. Chem.* 1984, *88*, 820-.
156. Marcus, Y., 1999, *Fluid Phase Equil.* (164).
157. C. H. Uffindell, A. I. Kolesnikov, J.-C. Li, J. Mayers, Inelastic Neutron scattering study of water in the sub and supercritical region. *Physica B* 2000, *276–278*, 444–445.
158. H. Touba, G. Ali Mansoorir, Structure and property prediction of sub- and supercritical water. *Fluid Phase Equilib.* 1998, *150–151*, 459–468.
159. Cynn, H. C.; Boone, S.; Koumvakalis, A.; Nicol, M.; Stevenson, D. J., Phase diagram for ammonia-water mixtures at high pressures: Implications for icy satellites. *Proceed. 19th Lunar Planetary Sci. Conference* 1989, 433–441.

160. Bandura, A.; S. Lvov, The ionization constant of water over wide ranges of temperature and density. *J. Phys. Chem. Ref. Data* 2006, *35*, 15–30.
161. H. R. Patrick, K. Griffith, C. L. Liotta, C. A. Eckert, *J. Phys. Chem. Ref. Data. Ion product of water substance, 0–1000°C, 1–10,000 bars New International Formulation and its background.* 1981; Vol. 10.
162. M. J. Antal Jr; A. Brittain; C. DeAlmedia; S. Ramayya; Roy, J. C., Heterolysis and homolysis in supercritical water, in supercritical fluids: Chemical and engineering principles and applications. *ACS Symp. Ser.* 1987, *329*, 77–86.
163. X. Xu; C. D. Almeida; Antal, J. M. J., Mechanism and kinetics of the acid-catalyzed dehydration of ethanol in supercritical water. *J. Supercrit. Fluids* 1990, *3*, 228–232.
164. B. Kuhlmann; E. M. Arnett; Siskin, M., Classic organic reactions in pure superheated water. *J. Org. Chem.* 1994, *59*, 3098–3101.
165. M. Sasaki; M. Furukawa; K. Minami; T. Adschiri; Arai, K., Kinetics and mechanism of cellobiose hydrolysis and Retro-Aldol condensation in subcritical and supercritical water. *Ind. Eng. Chem. Res.* 2002, *41*, 6642–6649.
166. B. M. Kabyemela; M. Takigawa; T. Adschiri; R. M. Malaluan; Arai, K., Mechanism and kinetics of cellobiose decomposition in sub- and supercritical water. *Ind. Eng. Chem. Res.* 1998, *37*, 357–361.
167. S. E. Hunter, P. E. Savage, Recent advances in acid- and base-catalyzed organic synthesis in high-temperature liquid water. *Chem. Eng. Sci.* 2004, *59*, 4903–4909.
168. C. M. Comisar, S. E. Hunter, A. Walton, P. E. Savage, Effect of pH on ether, ester, and carbonate hydrolysis in high-temperature water. *Ind. Eng. Chem. Res.* 2008, *47*, 577–584.
169. G. W. Huber; S. Iborra; Corma, A., Synthesis and transportation fuels from biomass: Chemistry, catalysts, and engineering. *Chem. Rev.* 2006, *52*, 4044–4098.
170. F. Behrendt; Y. Neubauer; M. Oevermann; B. Wilmes; Zobel, N., Direct Liquefaction of Biomass. *Chem. Eng. Technol.* 2008, *31*, 667–677.
171. C. A. Eckert, K. Chandler, Tuning fluid solvents for chemical reactions. *J. Supercrit. Fluids* 1998, *13*, 187–195.
172. Schmidt, E., *Properties of water and steam in SI-units. KJ, bar. 0–800°C. 0–1000 bar.* Springer: Berlin, 1969.
173. Savage, P. E., Organic chemical reactions in supercritical water. *Chem. Rev.* 1999, *99*, 603–621.
174. M. Siskin, A. R. Katritzky, A review of the reactivity of organic compounds with oxygen-containing functionality in superheated water. *J. Anal. Appl. Pyrolysis* 2000, *54*, 193–214.
175. M. Siskin, A. R. Katritzky, Reactivity of organic compounds in superheated water: General background. *Chem. Rev.* 2001, *101*, 825–835.
176. A. R. Katritzky, D. A. Nichols, M. Siskin, R. Murugan, M. Balasubramanian, Reactions in high-temperature aqueous media. *Chem. Rev.* 2001, *101*, 837–892.
177. E. E. Brock, P. E. Savage, Detailed chemical kinetics model for supercritical water oxidation of C1 compounds and HP. *AIChE J.* 1995, *41*, 1875–1888.
178. P. Dutournie, C. Aymonier, A. Gratias, F. Cansell, Determination of hydrothermal oxidation reaction heats by experimental and simulation investigations. *Ind. Eng. Chem. Res.* 2001, *40*, 114–118.
179. E. E. Brock, P. E. Savage, Detailed chemical kinetics model for supercritical water oxidation of C1 compounds and H2. *AIChE J.* 1995, *41*, 1874–1888.
180. P. E. Savage, J. Rovira, N. Stylski, C. J. Martino, Oxidation kinetics for methane/methanol mixtures in supercritical water. *J. Supercrit. Fluids* 2000, *17*, 155–170.
181. R. R. Steeper, S. F. Rice, I. M. Kennedy, J. D. Aiken, Kinetic measurements of methane oxidation in supercritical water. *J. Phys. Chem.* 1996, *100*, 184–189.
182. P. E. Savage, J. Yu, N. Stylski, E. E. Brock, Kinetics and mechanism of methane oxidation in supercritical water. *J. Supercrit. Fluids* 1998, *12*, 141–153.
183. N. Akiya; Savage, P. E., Kinetics and Mechanism of Cyclohexanol Dehydration in High-Temperature Water. *Ind. Eng. Chem. Res.* 2001, *40*, 1822–1831.
184. Narayan, R.; M. J. Antal Jr, Influence of pressure on the acid catalyzed rate constant for 1-propanol dehydration in supercritical water. *J. Am. Chem. Soc.* 1990, *112*, 1927–1931.
185. J. M. L. Penningera; R. J. A. Kersten; Baur, H. C. L., Reactions of diphenylether in supercritical water—mechanism and kinetics. *J. Supercrit. Fluids* 1999, *16*, 119–132.
186. M. M. Elbaccouch; Elliott, J. R., High-pressure vapor-liquid equilibrium for dimethyl ether + 2-propanol and dimethyl ether + 2-propanol + water. *J. Chem. Eng. Data* 2001, *46*, 675–678.
187. D. Gao, T. Kobayashi, S. Adachi, Kinetics of sucrose hydrolysis in a subcritical water-ethanol mixture. *J. Appl. Glycosci.* 2014, *61*, 9–13.
188. Lee, D. S., Heterogeneous oxidation kinetics of acetic acid in supercritical water. *Environ. Sci. Technol.* 1996, *30*, 3487–3492.
189. P. Dell'Orco; B. Foy; E. Wilmanns; L. Le; J. Ely; K. Patterson; Buelow, S., Hydrothermal oxidation of organic compounds by nitrate and nitrite. *ACS Symp. Ser. 608* 1995, 179–196.
190. K. C. Chang; L. X. Li; Gloyna, E. F., Supercritical water oxidation of acetic acid by potassium permanganate. *J. Hazard. Mater.* 1993, *33*, 51–62.
191. P. E. Savage; Smith, M. A., Kinetics of acetic acid oxidation in supercritical water. *Environ. Sci. Technol.* 1995, *29*, 216–221.
192. D.-S. Lee, E. F. Gloyna, L. Li, Efficiency of H_2O_2 and O_2 in supercritical water oxidation of 2,4-dichlorophenol and acetic acid. *J. Supercrit. Fluids* 1990, *3*, 249–255.
193. P. Dutournie, Determination of hydrothermal oxidation reaction heats by experimental and simulation investigations. *Ind. Eng. Chem. Res.* 2011, *40*, 114–118.
194. J. C. Meyer, P. A. Marrone, J. W. Tester, Acetic acid oxidation and hydrolysis in supercritical water. *AIChE J.* 1995, *41*, 2108–2121.
195. M. Krajc, The role of catalyst in supercritical water oxidation of acetic acid. *J. Appl. Catal., B: Environ.* 1997, *13*, 93–103.
196. J. R. Portela, E. Nebot, E. Martinez de la Ossa, Kinetic comparison between subcritical and supercritical

water oxidation of phenol. *Chem. Eng. J.* 2001, *81*, 287–299.
197. X. Zhang, P. E. Savage, Fast catalytic oxidation of phenol in supercritical water. *Catal. Today* 1998, *40*, 333–342.
198. S. Gopalan, P. E. Savage, Reaction mechanism for phenol oxidation in supercritical water. *J. Phys. Chem.* 1994, *98*, 12646–12652.
199. P. E. Savage; Gopalan, S., Phenol oxidation in supercritical water. *ACS Symp. Ser.* 1995, *608*, 217–231.
200. A. Santos, F. Garsia Ochoa, Overall rate of aqueous-phase catalytic oxidation of phenol: pH and catalyst loading influences. *Catal. Today* 1999, *48*, 109–117.
201. Y. Oshima, M. Toda, T. Chommanad, S. Koda, Phenol oxidation kinetics in supercritical water. *J. Supercrit. Fluids* 1998, *13*, 241–246.
202. J. Vicente, R. Rosal, M. Diaz, Noncatalytic oxidation of phenol in aqueous solutions. *Ind. Eng. Chem. Res.* 2002, *41*, 46–51.

14

Greener Synthesis of Natural Products

Renata Kołodziejska, Renata Studzińska, Hanna Pawluk, and Alina Woźniak

CONTENTS

14.1 Introduction ... 241
14.2 Enzymes in Asymmetric Synthesis ... 243
14.3 Enzymatic Synthesis of Chiral Bioflavonoids ... 247
14.4 Enzymatic Synthesis of Chiral Alkaloids ... 252
14.5 Enzymatic Synthesis and Biotransformation of Terpenoids ... 262
References ... 282

14.1 Introduction

In recent years, more and more emphasis has been placed on protecting the earth against human devastation. In 1992, after assessing the impact of manufacturing processes of active pharmaceutical ingredients (APIs) on the degree of environmental pollution, it was found that the chemical and pharmaceutical industry generates too high an Environmental Factor (E factor) (kg waste/kg product) higher than 100, indicating the huge amount of residues.[1] Anastas and Warner first developed the 12 principles of "Green Chemistry" in order to promote green and sustainable chemical manufacturing processes.[2] These principles are presented in Figure 14.1.[3]

Biocatalysis (biotransformation) has emerged as one of the methods perfectly fitting in the global "Green Chemistry" trend, which is based on the use of isolated enzymes from biological preparations (microorganisms, plants, and animals) and by microorganisms and plants in the forms of growing or resting cells. The enzymatic catalyst is derived from renewable resources and is biocompatible, biodegradable, and essentially non-hazardous. By the application of biocatalysis, toxic reagents, fewer by-products, and production waste are avoided because they are executed under mild conditions and achieve high yields with excellent chemo-, regio-, and stereoselectivity.[4,5]

Applying enzymatic synthesis for chiral biologically active molecules or their precursors with defined chirality is particularly important. The demand for optically pure biologically active compounds is very valid for the pharmaceutical industry; often enantiomers could have dramatically different biological activities and toxicity. They are distinguished by biological systems and may have different pharmacokinetic properties (absorption, distribution, biotransformation, and excretion) and pharmacologic effects.[6,7] An infamous example of disregarding the optical purity of the drug is the administration of thalidomide in the form of a racemic mixture in the mid-1950s as a sedative, used in pregnant women. Only one enantiomer had a therapeutic effect, the (*S*)-isomer is a potent teratogen, acting on fetal DNA.

Microbes and enzymes due to their high enantioselectivity are extremely desirable in pharmacy, in the synthesis of chiral drugs; they offer benefits such as minimizing side reactions, not

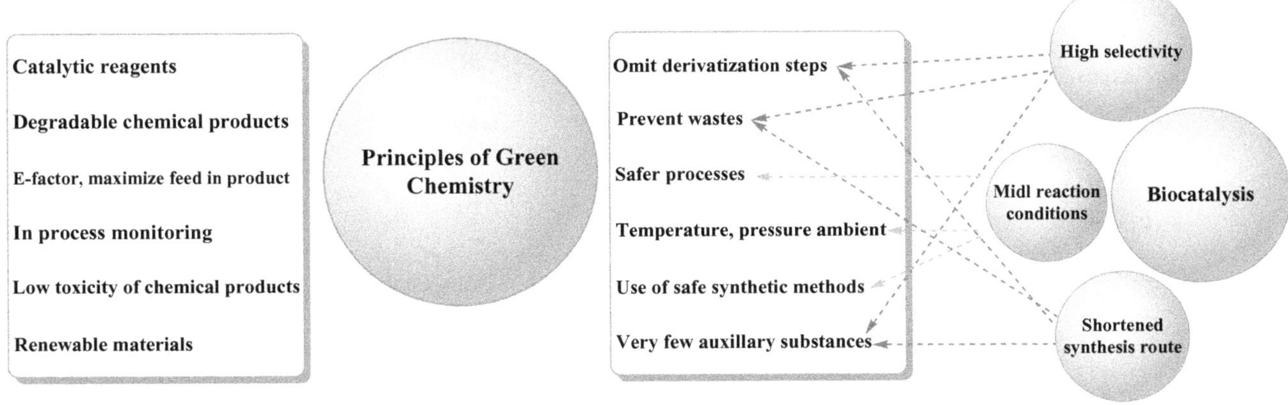

FIGURE 14.1 Principles of "Green Chemistry".

DOI: 10.1201/9781003089162-14

requiring protection and deprotection steps, and thus enabling the shorter synthesis. Stereoselective biotransformation enables obtaining optically pure biologically active compounds in the asymmetric synthesis (desymmetrization) of achiral compounds (prochiral and *meso* compounds) and in the enzymatic kinetic resolution (EKR) of racemic mixtures.[8–10]

According to the definition, desymmetrization refers to compounds with elements of symmetry. The achiral molecule is transformed into a chiral molecule in such a manner that the formation of stereoisomeric products occurs in unequal amounts. Chirality is generated in prochiral compounds with sp^3 or sp^2 hybridization, or in *meso* synthons and centrosymmetric compounds.

As a result of the transformation, prochiral compounds lose elements of symmetry. The prochiral center may be an sp^3 hybridized carbon atom containing two of the same ligands or an sp^2 hybridized carbon atom in an unsymmetrically substituted unsaturated compound (carbonyl compounds, alkenes).[11]

In the sp^3 prochiral molecule of $R_1R_2CX_2$-type, identical X substituents become enantiotopic or diastereotopic under the influence of a chiral agent (the diastereotopic one, if apart from the prochiral group a chirality center in a molecule appears). In order to determine the configuration of the prochiral substituents, one of them is first designated as X_a and the other as X_b. The molecule should be oriented in such a way that the substituent X_a, for example, is located as far away from the observer as possible. For other groups or atoms, priority should be assigned according to the Cahn–Ingold–Prelog system (CIP). If we move from the most important substituent to the least important clockwise (e.g., $R_1 > R_2 > X_b$), the prochiral substituent X_a is designated as the pro-*R* one (Figure 14.2). In contrast, the X_b ligand is described as pro-*S*. A stereogenic center is formed when one of the identical pro-*R* or pro-*S* ligands is replaced by another group or atom. In most cases, the substitution of the pro-*R* will produce an *R*-configuration molecule, whereas substitution of a pro-*S* ligand may lead to an optical *S*-isomer.[8,12–14]

The prochiral trigonal carbon atom of $R_1R_2C{=}X$ molecule has two enantiofacial sides *re* and *si*. In order to differentiate both faces, the priority of substituents R_1, R_2, and X directly related to the prochiral atom should be determined according to the CIP rules. If the priority of substituents decreases clockwise then the face is designed as *re*, while if it decreases counter-clockwise then the face is designed as *si*, as shown in Figure 14.3.[12–14]

Similarly to the prochiral compounds, possessing tetrahedral carbon atoms, breaking the symmetry in *meso* and centrosymmetric substrates leads to the optically active products (Figure 14.4).[15,16]

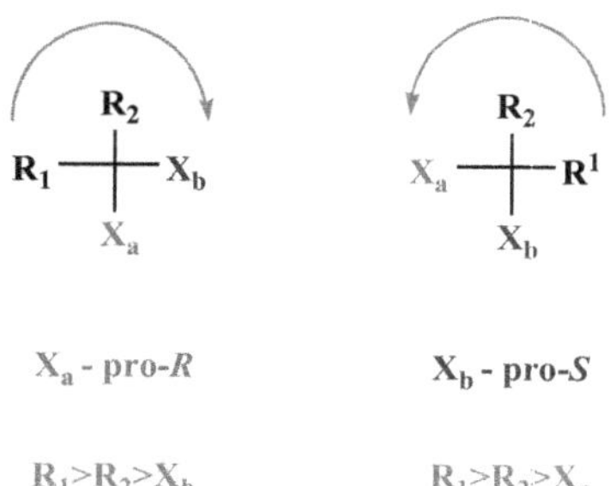

FIGURE 14.2 Prochiral tetragonal carbon atom.

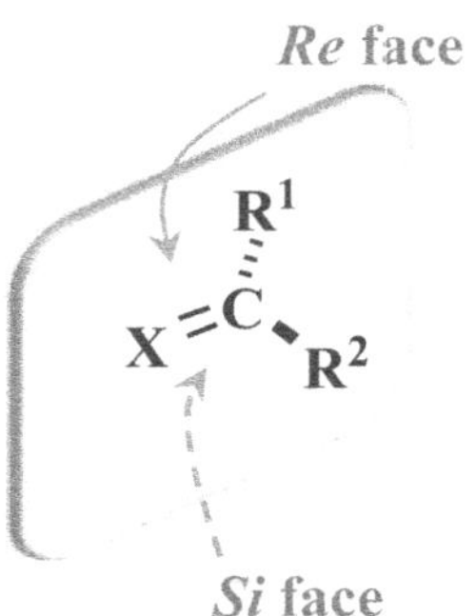

X>R¹>R²

FIGURE 14.3 Prochiral trigonal carbon atom.

In KR, only one of the enantiomers of the racemic mixture undergoes transformation and an optical isomer with a specific configuration is formed.[12–14,16]

The kinetic resolution is based on the difference in the reaction rate of enantiomers with the enzyme catalyst, resulting in an enantioenriched sample of the less reactive stereoisomer. The maximum yield of KR is 50%, provided the enzyme exhibits high selectivity. In order to improve the performance of biotransformation, a method of dynamic kinetic resolution (DKR) was developed in which the enzymatic process combines with a simultaneous *in situ* racemization of the less reactive enantiomer, thus, producing optically active products in up to quantitative yield (Figure 14.5). For DKR to be effective, certain requirements have to be fulfilled:

- The enzyme should display high specificity for one of the enantiomers *R* or *S* ($k_R >> k_S$ or $k_S >> k_R$),
- The substrate must racemize at least as fast as the subsequent enzymatic reaction ($k_{rac(Sub)} \geq k_R$ or k_S),
- Racemization of the product should not occur under the reaction conditions or should be minimal.

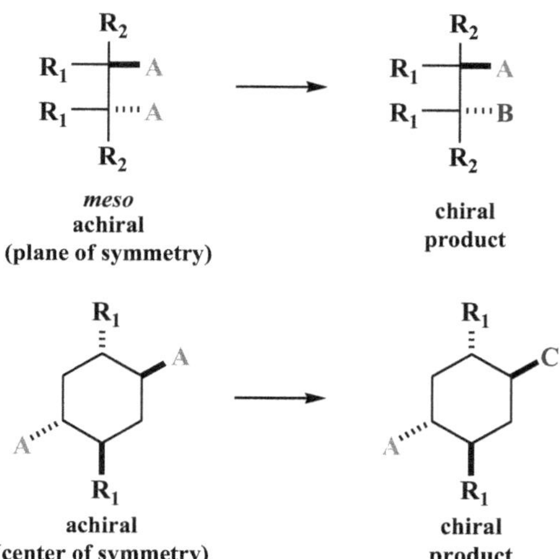

FIGURE 14.4 *Meso* and centrosymmetric substrates.

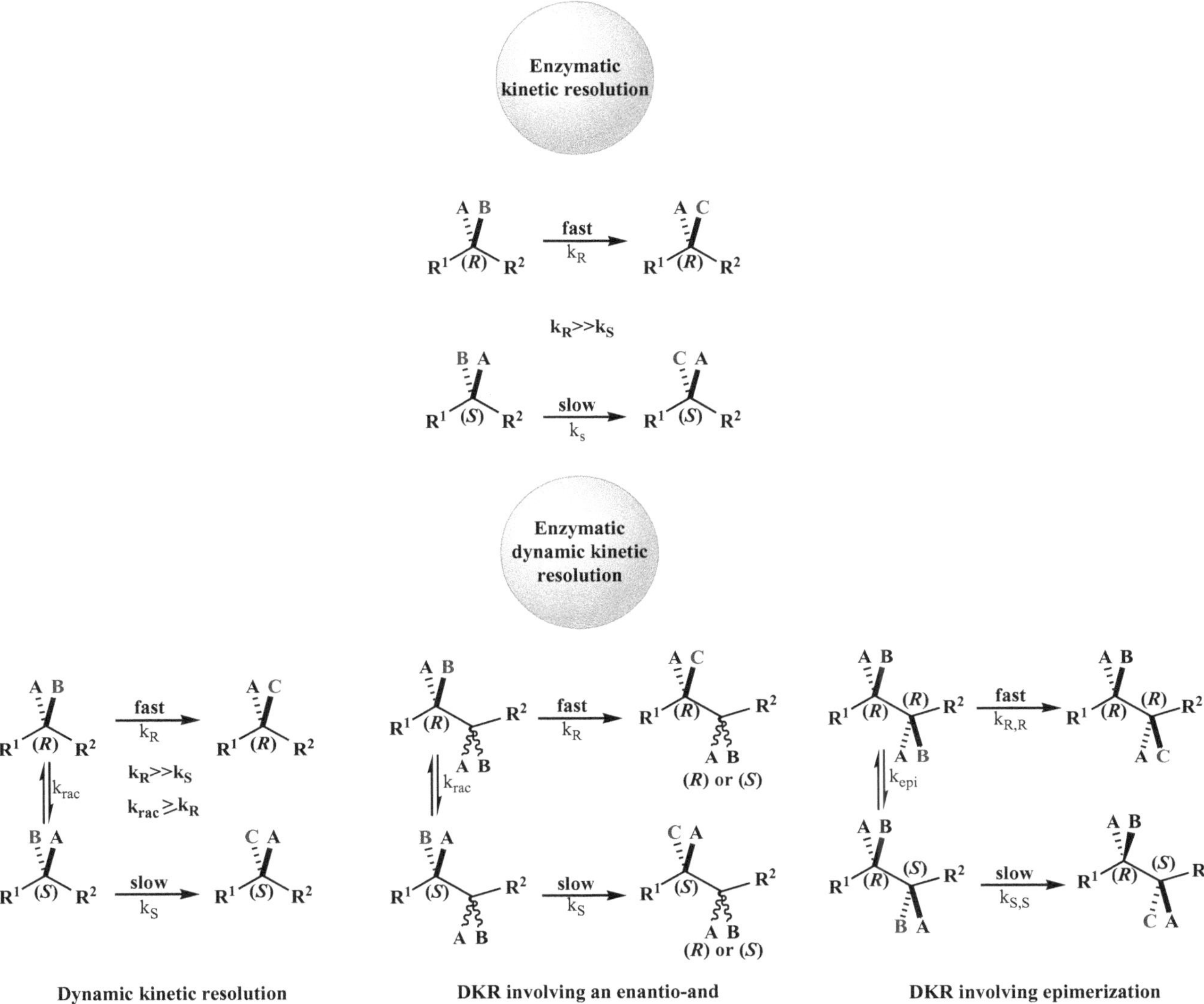

FIGURE 14.5 Kinetic resolution methods.

14.2 Enzymes in Asymmetric Synthesis

Chiral discrimination of enantiomers and prochiral, *meso* compounds is possible due to the stereoselective property of enzymes, one of the enantiomers, prochiral, *meso* compounds fits into the active center better than its mirror image. The enzymes are divided into six main classes; each enzyme corresponds to an EC number, consisting of a series of four consecutive digits, separated by dots (Figure 14.6). The EC classification system is derived from the biochemical function of enzymes in living systems.[17]

Of all types of enzymes used in biosynthesis, undoubtedly, hydrolases belong to the most-explored group of biocatalysts. They are characterized by high availability, broad substrate specificity and do not require the use of cofactors that would have to be recycled. Hydrolytic enzymes include proteases, esterases, and lipases that hydrolyze amide and ester linkages. The mechanism of enzymatic hydrolysis catalyzed by hydrolases is similar to that observed in the chemical hydrolysis with the use of a base. A nucleophilic group from the enzyme active side attacks the carbon atom of the substrate carbonyl group forming an acyl-enzyme intermediate. This nucleophilic group initiating the biotransformation can be either the serine hydroxyl group (e.g., pig liver esterase, subtilisin, and the majority of lipases), an aspartic acid carboxyl group (e.g., pepsin), or a cysteine tiol functionality (e.g., papain). The product is formed by the tetrahedral intermediate deacylation and the enzyme is regenerated. The resulting product depends on the nucleophile used.[8,16]

Hydrolases have also been applied in catalysis in acylation reactions, Baeyer–Villiger addition, carbon–carbon bond formations, including aldol reactions, Knoevenagel condensations, Michael additions, and Henry reactions.[8,16,18–22] Mainly, lipases are used in asymmetric synthesis; they effectively produce enantiomerically pure compounds from prochiral substrates and in EKR. Some examples of their applications in asymmetric synthesis are shown in Figure 14.7.[23–26]

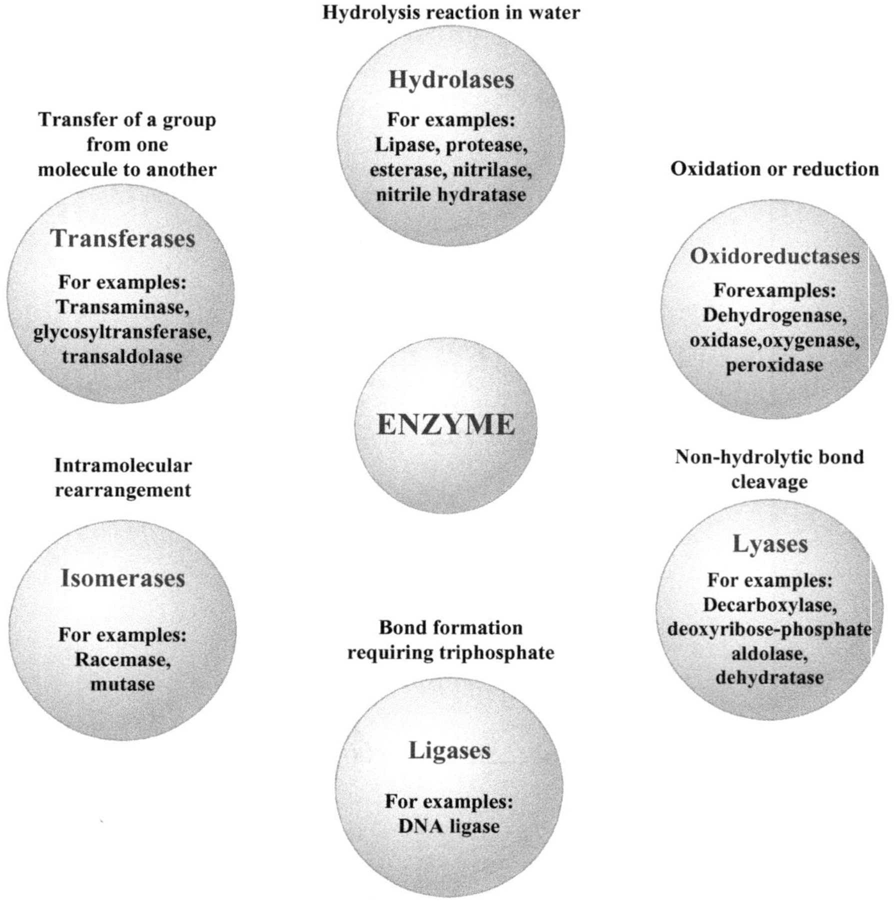

Hydrolysis reaction in water

Hydrolases
For examples:
Lipase, protease,
esterase, nitrilase,
nitrile hydratase

Transfer of a group
from one
molecule to another

Transferases
For examples:
Transaminase,
glycosyltransferase,
transaldolase

Oxidation or reduction

Oxidoreductases
Forexamples:
Dehydrogenase,
oxidase,oxygenase,
peroxidase

ENZYME

Intramolecular
rearrangement

Isomerases
For examples:
Racemase,
mutase

Non-hydrolytic bond
cleavage

Lyases
For examples:
Decarboxylase,
deoxyribose-phosphate
aldolase,
dehydratase

Bond formation
requiring triphosphate

Ligases
For examples:
DNA ligase

FIGURE 14.6 Classification of enzymes.

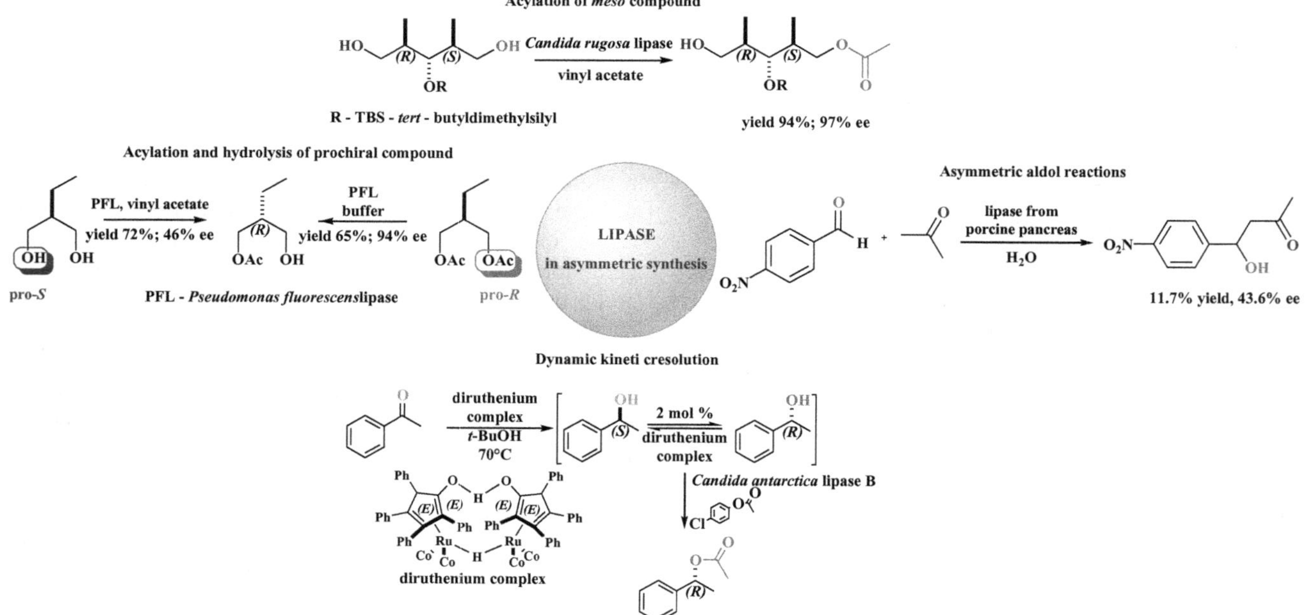

FIGURE 14.7 Application of lipases in asymmetric synthesis.

The next group of enzymes willingly used in asymmetric synthesis is oxidoreductases. Oxidoreductases include, among others: oxidases, hydroperoxidases, peroxidases, catalases, oxygenases, hydroxylases, dehydrogenases, and reductases. Of the oxidoreductases, dehydrogenases are particularly explored, which oxidize primary and secondary alcohols and reduce the carbonyl double bond to form an alcohol moiety. Oxygenases to carry out the reaction need oxygen as a co-substrate, catalyze the oxidation reactions of C-H and C=C binding, belong to the second important subgroup of oxidoreductases used in asymmetric synthesis. They are divided into dioxygenases, which incorporate two oxygen atoms into the substrate, and monooxygenases catalyze the incorporation of one oxygen atom into the hydroxylated substrate. In contrast, oxidases, enzymes that catalyze the transfer of hydrogen to oxygen, resulting in the formation of water or hydrogen peroxide, and reductases, which reduce olefins to alkanes, are used to a small extent in biosynthesis.

The reduction of a carbonyl compound by means of alcohol dehydrogenase (ADH) takes place when the transfer of one of the hydrogen atoms from the dihydropyridine ring of the cofactor (in most cases, it is nicotinamide adenine dinucleotide (NADH) or its phosphorylated derivative (NADPH)) to the particular face of the carbonyl group occurs. The pro-(*S*)- or pro-(*R*)-hydride (H_S and H_R, respectively) of the cofactor (NAD(P)H) attacks the *si* or *re* face of an sp^2 hybridized carbon atom of the carbonyl group to form (*R*)- or (*S*)-alcohols. The reaction range comprises reduction of, for example, ketones, α- and β-keto esters, and α-keto acids by isolated enzymes or whole cells of microorganisms, plants, or animals – Figure 14.8.[27–31]

Monooxygenases (MO) catalyze different reactions, such as hydroxylation, epoxidation, oxidative deamination, or *N*- and *S*-oxidation. Asymmetric synthesis mainly utilizes the transformation of prochiral alkanes in the hydroxylation process, prochiral alkenes in the epoxidation reaction, and the Baeyer–Villiger oxidation reaction (Figure 14.9).[32–34]

Of the dioxygenases, toluene (TDO), naphthalene (NDO), and biphenyl (BPDO) dioxygenase are explored in asymmetric synthesis. These enzymes are found mainly in prokaryotic microorganisms. An example of the application of this type of enzyme may be the selective oxidation of dialkyl or alkyl-aryl sulfides to the corresponding sulfoxides or the use of their natural function of the oxidation of arenes to arene *cis*-dihydroxydiols (Figure 14.10).[35,36]

Oxidases oxidize many organic compounds such as alcohols, aldehydes, monosaccharides, amino acids, amines, phenols, steroids, thiol, and nitro compounds by transferring electrons from the substrate to molecular oxygen. In asymmetric synthesis, they have been used, among others, for α-hydroxylation of fatty acids and oxidation of amines to appropriate imines or imine ions (Figure 14.11).[37,38]

Reductases, on the other hand, reduce activated double bonds in alkenes with an electron-withdrawing substituent, i.e., unsaturated nitro-compounds, aldehydes, ketones, esters, and carboxylic acids to the corresponding saturated molecules (Figure 14.12).[39–41] The reduction of alkenes is catalyzed by both the whole cells (microorganisms, plant cells) as well as isolated enzymes so-called enolate reductases. The addition of a hydrogen molecule to the carbon–carbon double bond catalyzed by enoate reductases proceeds in a *trans*-fashion with

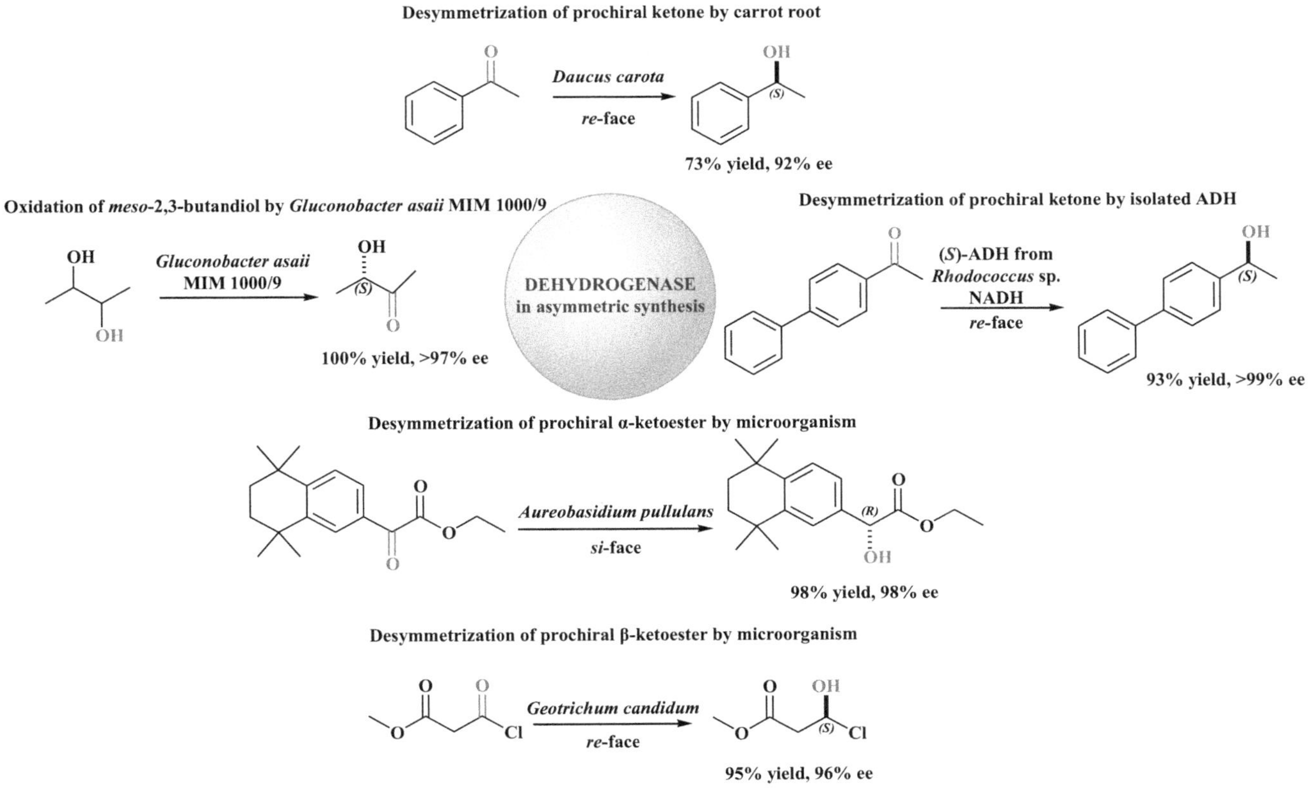

Desymmetrization of prochiral ketone by carrot root

Daucus carota

re-face

73% yield, 92% ee

Oxidation of *meso*-2,3-butandiol by *Gluconobacter asaii* MIM 1000/9

Gluconobacter asaii MIM 1000/9

100% yield, >97% ee

DEHYDROGENASE in asymmetric synthesis

Desymmetrization of prochiral ketone by isolated ADH

(*S*)-ADH from *Rhodococcus* sp. NADH

re-face

93% yield, >99% ee

Desymmetrization of prochiral α-ketoester by microorganism

Aureobasidium pullulans

si-face

98% yield, 98% ee

Desymmetrization of prochiral β-ketoester by microorganism

Geotrichum candidum

re-face

95% yield, 96% ee

FIGURE 14.8 Application of dehydrogenases in asymmetric synthesis.

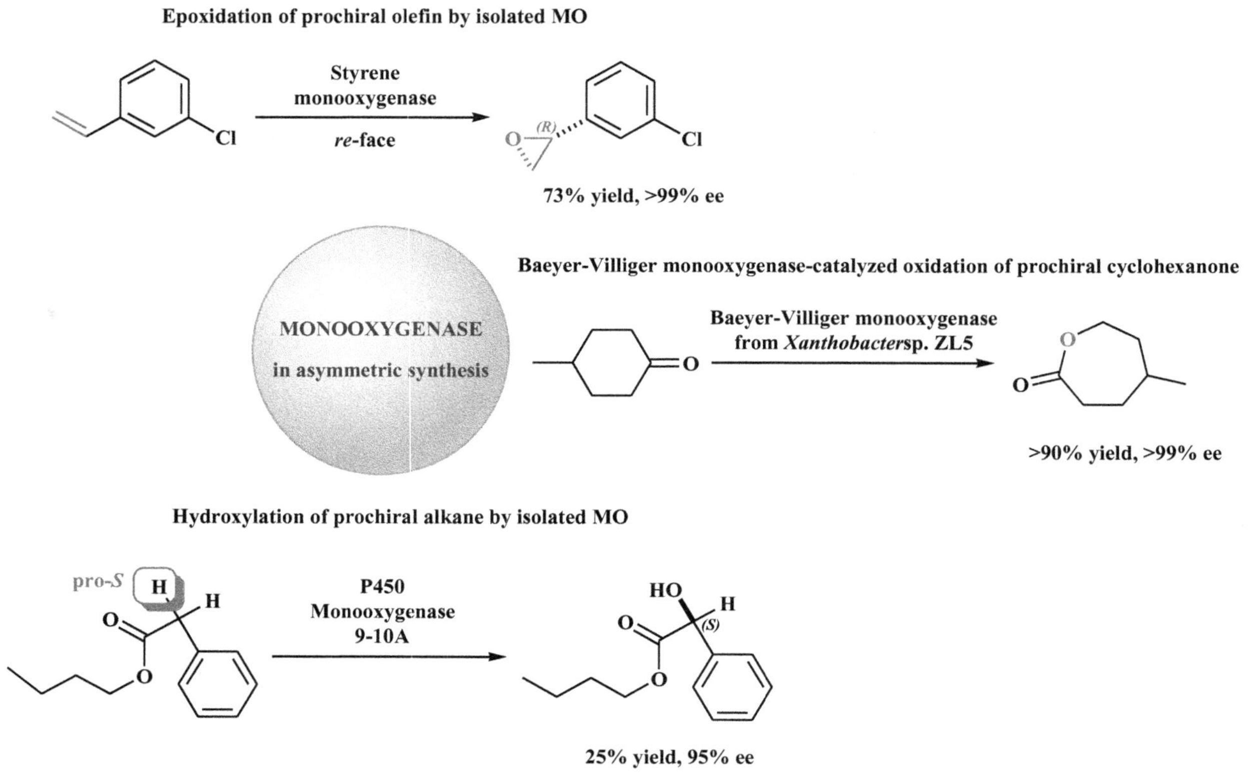

Epoxidation of prochiral olefin by isolated MO
Styrene monooxygenase
re-face
Cl
(R)
O
Cl
73% yield, >99% ee
MONOOXYGENASE in asymmetric synthesis
Baeyer-Villiger monooxygenase-catalyzed oxidation of prochiral cyclohexanone
Baeyer-Villiger monooxygenase from Xanthobactersp. ZL5
O
>90% yield, >99% ee
Hydroxylation of prochiral alkane by isolated MO
pro-S
H
P450 Monooxygenase 9-10A
HO
(S)
25% yield, 95% ee

FIGURE 14.9 Application of monooxygenases in asymmetric synthesis.

Bio-oxidation of sulfides by recombinant E.coli JM109 (pDTG141 expressing anaphthalene dioxygenase (NDO)gene from Pseudomonas sp. NCIB 9816-4
Recombinant E. coli JM109 (pDTG141)
(S)
98% yield, 98% ee
Dihydroxylation of dioxole catalysed by recombinant E. coli cells expressing a naphthalene dioxygenase from Pseudomonas putida G7
DIOXYGENASE in asymmetric synthesis
Recombinant E. coli
OH
(R)
(S)
64% yield, >98% ee

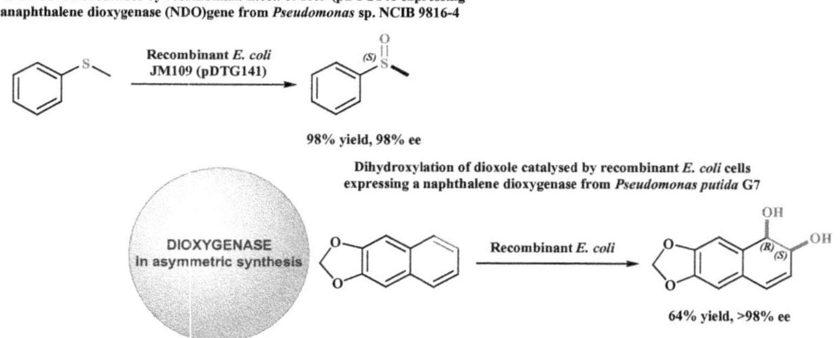

FIGURE 14.10 Application of dioxygenases in asymmetric synthesis.

α-hydroxylation of carboxylic acid by the α-oxidase of peas (Pisum satiVum)
OH
α-oxidase (Pisum sativum)
O2
(R)
100% yield, >99% ee
OXIDASE in asymmetric synthesis
Oxidation of amine by monoamine oxidase from Aspergillus niger
monoamine oxidase from Aspergillus niger
N
NH3 BH3
(R)
H
>97% ee

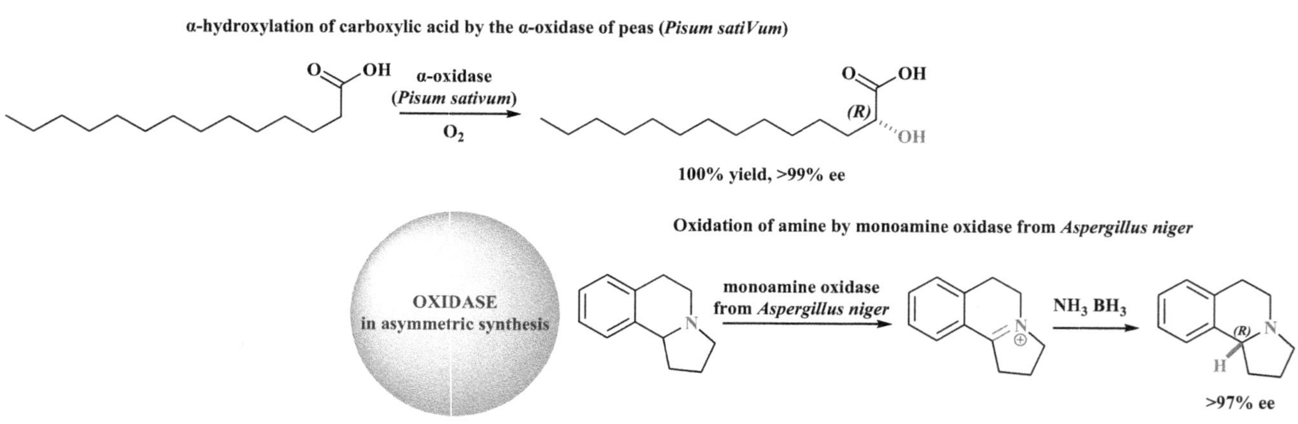

FIGURE 14.11 Application of oxidases in asymmetric synthesis.

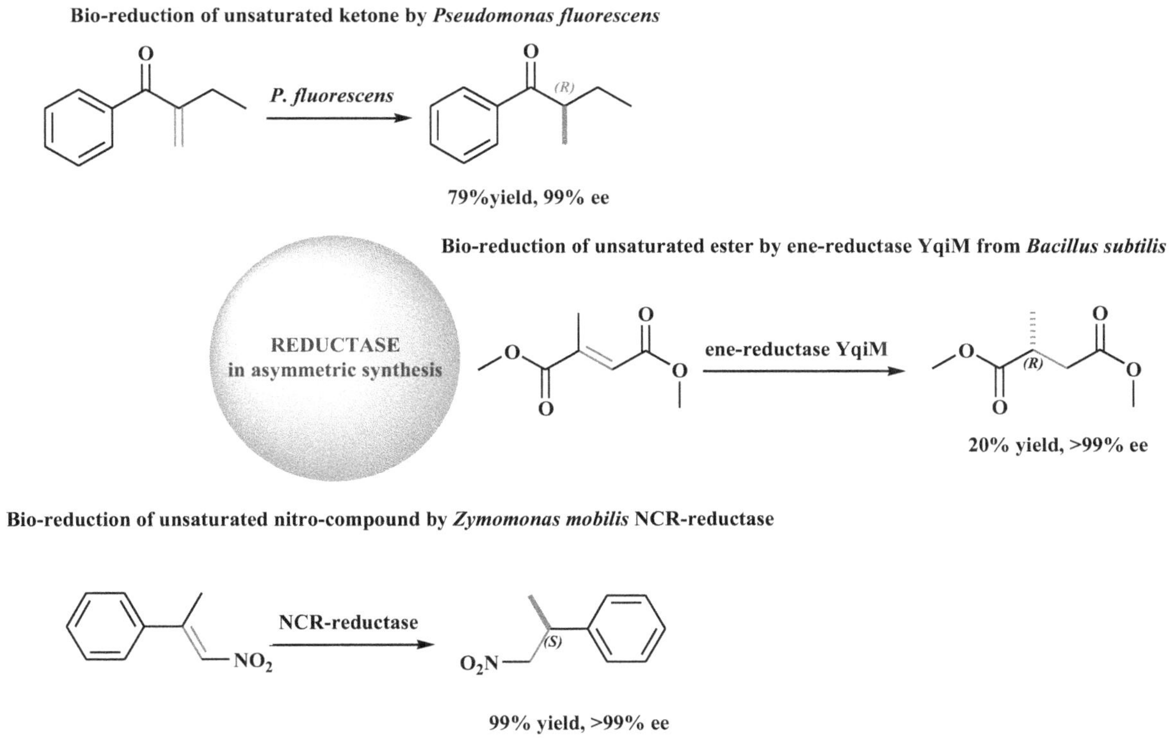

Bio-reduction of unsaturated ketone by *Pseudomonas fluorescens*

P. fluorescens

79%yield, 99% ee

Bio-reduction of unsaturated ester by ene-reductase YqiM from *Bacillus subtilis*

REDUCTASE in asymmetric synthesis

ene-reductase YqiM

20% yield, >99% ee

Bio-reduction of unsaturated nitro-compound by *Zymomonas mobilis* NCR-reductase

NCR-reductase

99% yield, >99% ee

FIGURE 14.12 Application of reductases in asymmetric synthesis.

absolute stereospecificity. A hydride (derived from a reducted flavin cofactor) is stereoselectively transferred onto the β carbon, while a proton (derived from the solvent) is transferred onto the α carbon from the opposite side of the double bond.[39–41]

Transaminases (TAs) classified as transferases, participate in the metabolism of amino acids, transform the carbonyl function into an amino moiety. The transaminase-catalyzed reaction requires pyridoxal phosphate (PLP) as a cofactor. Of the two types of PLP-dependent TAs classified according to the type of substrate they convert, the use of α-TAs, exclusively converting α-amino and α-keto acids, is more limited, while ω-TAs can accept substrates with a distal carboxylate group. Depending on the type of transaminase, both keto acids and ketones are tolerated as substrates, thus leading to enantiomerically enriched amino acids and amines, respectively (Figure 14.13).[42–44]

Asymmetric synthesis can also be carried out in the presence of lyases, they are involved in the forming C–C bond and bond cleavage process other than hydrolysis and redox reactions. Lyases catalyze (usually reversible) the addition of nucleophiles to the C=O double bond, in a manner mechanistically most often categorized as aldol and Claisen condensation or acyloin reaction. Aldolases, the most commonly used lyases, can be divided into four groups, according to the structure of the donor substrate accepted by the enzyme. The most useful families are the dihydroxyacetone phosphate (DHAP) or dephosphorylated ketone (DHA)-dependent aldolase, pyruvate, or phosphoenolpyruvate-dependent aldolase and aldolase which utilize acetaldehyde or glycine as the donor (Figure 14.14).[45–50]

Due to the subject of this chapter, enzymatic kinetic separation will be discussed on examples related to the biotransformation of chiral natural compounds. The term "natural product" defines a wide range of naturally occurring organic compounds, biologically active with a complex structure, possessing multiple functional groups. Natural compounds are most often primary or secondary metabolites of various medicinal plants; these are so-called phytochemicals. Secondary metabolites synthesized by plants successfully used in natural medicine belong to compounds with high therapeutic efficacy. Natural compounds include alkaloids, terpenes, flavonoids, saponins, steroids, glycosides, and tannins. Biocatalysis is successfully used in the chemistry of natural products. In view of the highly selective property of protein catalysts, many natural products can be obtained that are rare or synthesized in low yields by traditional chemical approaches. Biocatalysis may serve as useful models for plant, microbial, and mammalian metabolic systems by affording metabolites of natural products similar to those produced by these complex organisms.[51] The synthesis of chiral natural compounds using enzymes as catalysts will be discussed in the example of flavonoids, alkaloids, and terpenes.

14.3 Enzymatic Synthesis of Chiral Bioflavonoids

The group of natural biologically active compounds includes bioflavonoids; that are found primarily in fruits and vegetables and are responsible for their intense coloration. Bioflavonoids are plant metabolites that play a protective role against stress, and insect and mammalian consumption, a signaling function and a key function in plant growth and development. Flavonoids are based on the backbone of 2-phenylchromane

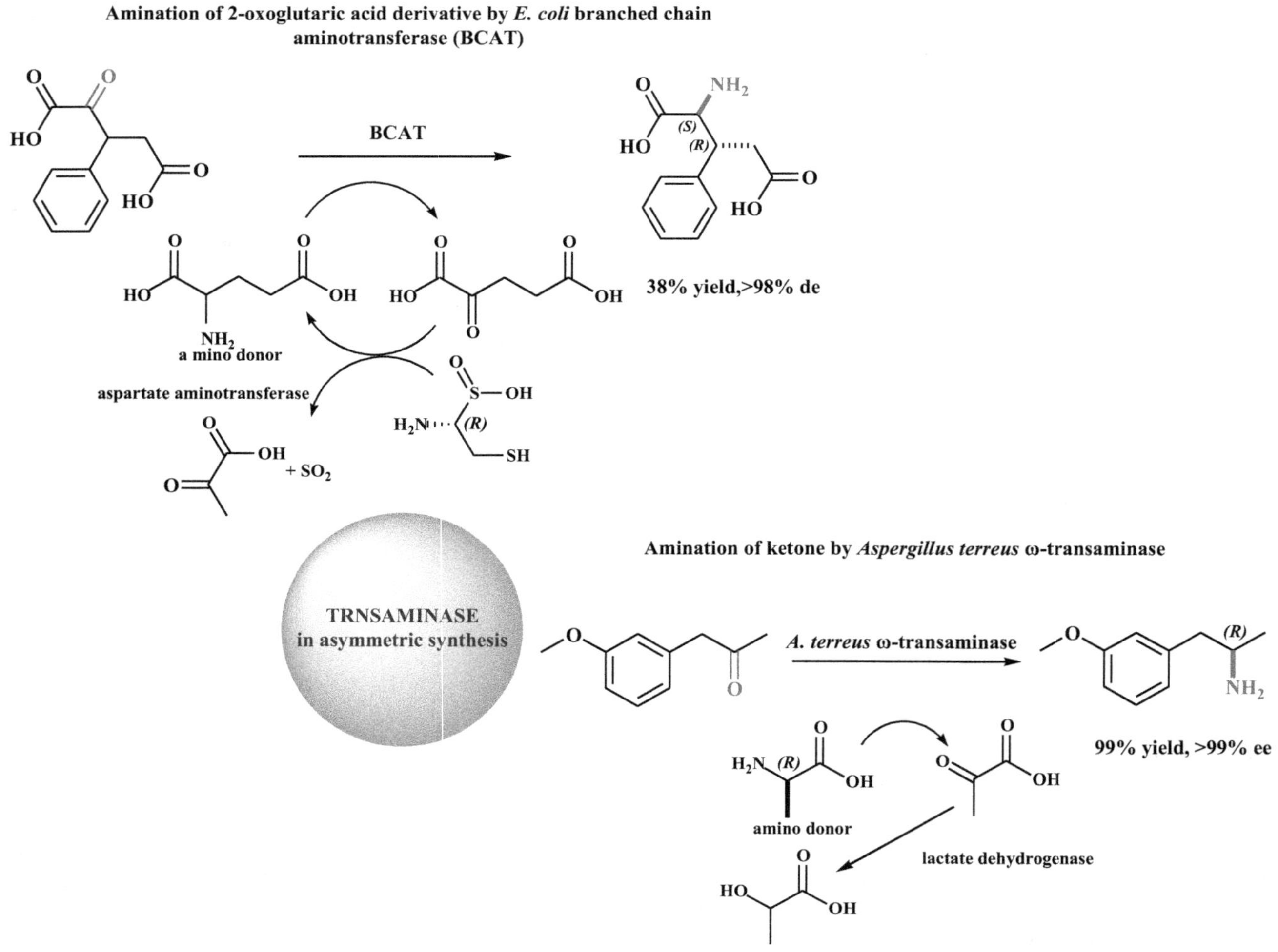

Amination of 2-oxoglutaric acid derivative by *E. coli* branched chain aminotransferase (BCAT)

BCAT

a mino donor

aspartate aminotransferase

$+ SO_2$

38% yield,>98% de

TRNSAMINASE
in asymmetric synthesis

Amination of ketone by *Aspergillus terreus* ω-transaminase

A. terreus ω-transaminase

99% yield, >99% ee

amino donor

lactate dehydrogenase

FIGURE 14.13 Application of transaminases in asymmetric synthesis.

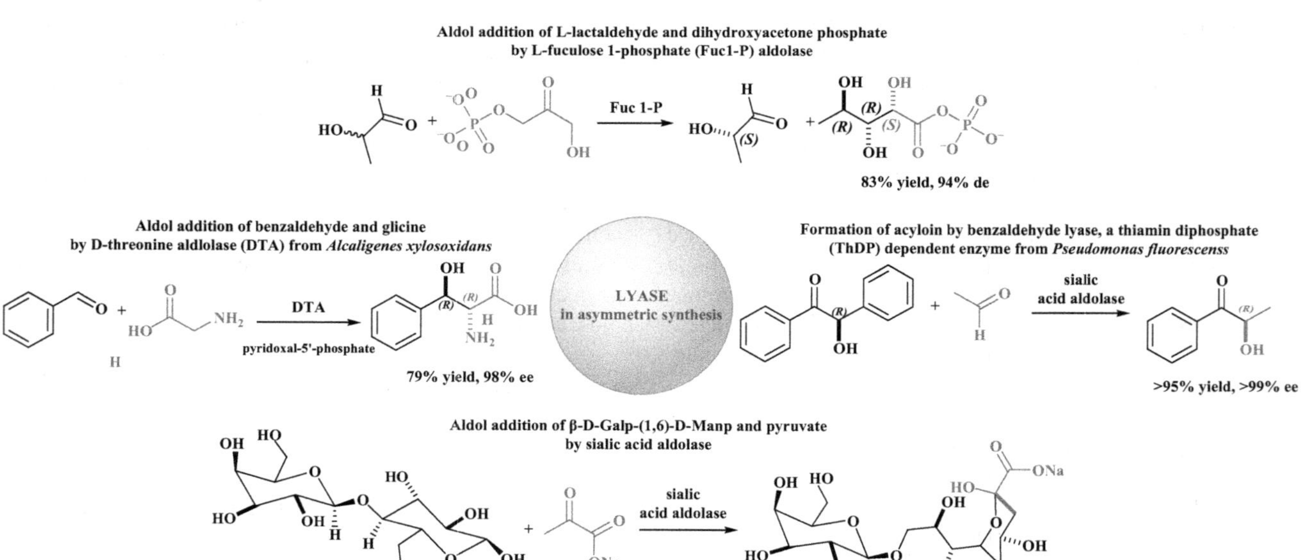

FIGURE 14.14 Application of lyases in asymmetric synthesis.

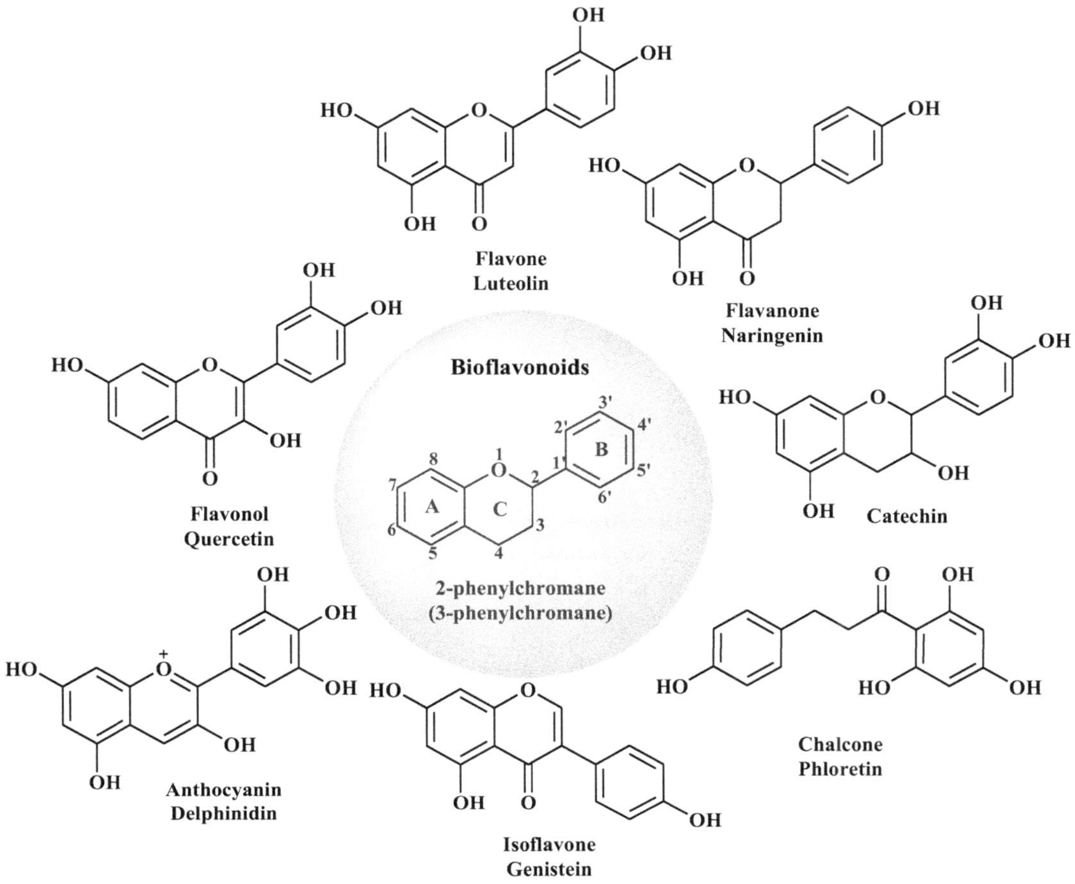

FIGURE 14.15 The molecular structure of flavonoids.

or 3-phenylchromane. The molecular structure of each group of flavonoids is depicted in Figure 14.15.[52,53] The privileged structure of bioflavonoids enables interaction with many receptors in the human body, which undoubtedly translates into their wide range of biological activity. These natural molecules have been shown anticancer, antitumor, antibacterial, antimicrobial, antioxidant, estrogenic, and antiestrogenic properties.[54,55]

Microbiological biotransformation strategies for the production of flavonoids have attracted considerable interest because they allow obtaining new biologically active flavonoids. The process of microbial biotransformation mainly concerns hydroxylation, dehydroxylation, *O*-methylation, *O*-demethylation, glycosylation, deglycosylation, dehydrogenation, hydrogenation, C ring cleavage of the benzo-γ-pyrone system, cyclization, and carbonyl reduction.[56] A few of them relate to asymmetrical biotransformations, which aim to obtain optically pure bioflavonoids. Some examples of the use of microorganisms in the desymmetrization of prochiral compounds are presented below.

Biotransformation of bavachinin, one of the flavanones, separated from *Psoralea corylifolia*, was carried out in the presence of three fungi species, *Aspergillus flavus* ATCC 30899, *Cunninghamella elegans* CICC 40250, and *Penicillium raistrickii* ATCC 10490.[57] The dried fruits from *P. corylifolia* containing this natural compound are used in traditional medicine Chinese for the treatment of asthma, diarrhea, vitiligo, alopecia areata, enuresis, pollakiuria, waist and knee psychroalgia, kidney weak, bone fracture, osteomalacia, and osteoporosis.[58] Two biocatalyst systems, *A. flavus* ATCC 30899 and *C. elegans* CICC 40250 fungal cells, in a hydroxylation reaction, modified the bavachinin A-ring in the sp^2 hybridized prochiral center in the C-2″= 3″ position to give the optically pure (*S*)-6-((*R*)-2,3-dihydroxy-3-methylbutyl)-2-(4-hydroxyphenyl)-7-methoxychromen-4-one. In contrast, *P. raistrickii* ATCC 10490 cell cultures possessed the ability to reducte at the prochiral center C-4 of the substrate C-ring, resulting in (2*S*,4*R*)-2-(4-hydroxyphenyl)-7-methoxy-6-(3-methylbut-2-en-1-yl)chromen-4-ol (Figure 14.16). These compounds were tested for antitumor activity, (2*S*,4*R*)-2-(4-hydroxyphenyl)-7-methoxy-6-(3-methylbut-2-en-1-yl)chromen-4-ol showed a strong inhibitory effect on the human breast cancer cell line (MCF-7) and slightly lower inhibition activities against Hep G2, HeLa, Hep-2, and A549 cells lines.[57]

Similarly, in the microbiological hydroxylation reaction of cannflavin A, methylated isoprenoid flavone isolated from cannabis, chiral products resulting from desymmetrization of the prochiral center with sp^2 hybridization were obtained. The incubation of cannflavin A with *Mucor ramannianus* (ATCC 9628) yielded 6″*S*,7″-dihydroxycannflavin A, 6″*S*,7″-dihydroxycannflavin A 7-sulfate, and 6″*S*,7″-dihydroxycannflavin A 4′-*O*-α-L-rhamnopyranoside (Figure 14.17).[59]

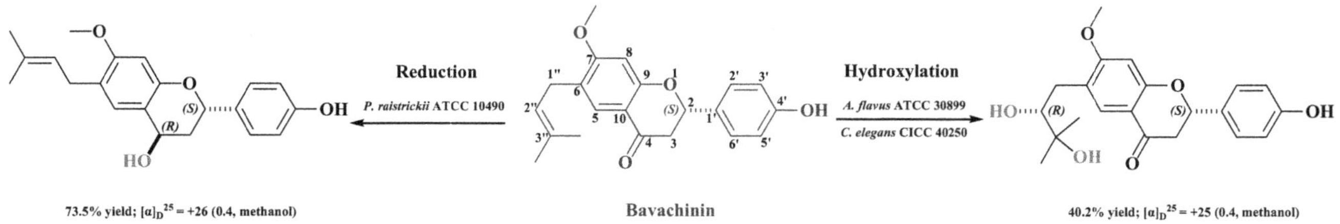

FIGURE 14.16 Biotransformation of bavachinin.

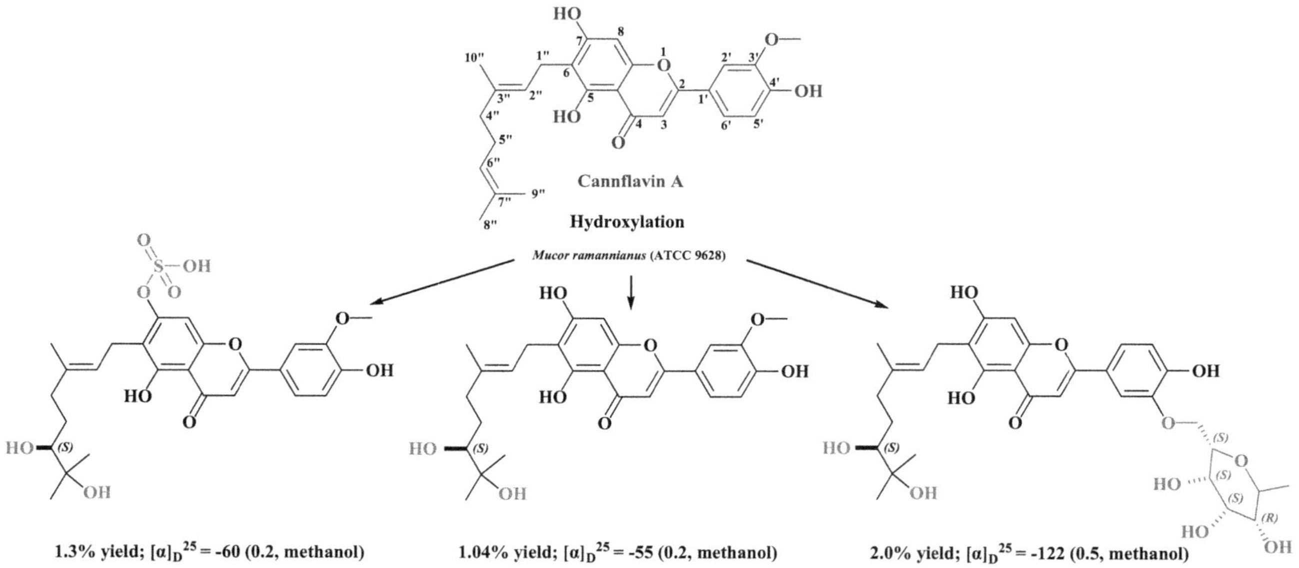

FIGURE 14.17 Biotransformation of cannflavin A.

Optically active bioflavonides are also obtained in a microbial asymmetric cyclization reaction from prochiral substrates. The biotransformation was subjected to xanthohumol, a chalcone-type prenylflavanoid, which is a constituent of beer and some dietary supplements. Its ability to inhibit CYP enzymes[60] and to induce quinone reductase has suggested that xanthohumol may have cancer-chemopreventive properties.[61] Xanthohumol has also been patented as a drug for osteoporosis treatment.[62] The microbiological transformation of xanthohumol using the culture broth of *Pichia membranifaciens* afforded three metabolites, including one chiral (2*S*)-2″-(2‴-hydroxyisopropyl) dihydrofurano[2″,3″:7,8]-4′-hydroxy-5-methoxyflavanone.[63] [2″,3″:7,8]-bicyclic ring system was created as a result of a nucleophilic attack of the carbon atom in the C-2 position by the C-8′ phenolic group (first an epoxidation of the prenyl group, then an intramolecular attack of the neighboring hydroxy group) while the stereogenic center in the C-2 position is created by forming the C ring (cyclization-type chalcone–flavanone). The biotransformation by *Cunninghamella echinulata* NRRL 3655 gave (2*S*)-8-[4″-hydroxy-3″-methyl-(2″-Z)-butenyl]-4′,7-dihydroxy-5-methoxyfavanone, and (2*S*)-8-[5″-hydroxy-3″-methyl-(2″-E)-butenyl]-4′,7-dihydroxy-5-methoxyfavanone. As in the[64] presence of *P. membranifaciens*, the chiral center in these metabolites is in C-2 position after cyclization. The biotransformation of xanthohumol by *Aspergillus ochraceus* was also examined, in which two chiral products were identified: (2*S*,2″*S*)-2″-(2‴-hydroxyisopropyl)-dihydrofurano[4″,5″:7,8]-4′-hydroxy-5-methoxyflavanone and (2*S*,2″*R*)-2″-(2‴-hydroxyisopropyl)-dihydrofurano[4″,5″:7,8]-4′-hydroxy-5-methoxyflavanone (Figure 14.18). Similar to *P. membranifaciens*, these metabolites were formed by oxidation of the prenyl side chain, followed by cyclization to a furan ring and cyclization-type chalcone–flavanone.[65]

Kinetic resolution of racemic is, next to asymmetric synthesis, a very good tool for achieving enantiomeric purity of optical isomers; it can be successfully used in the preparation of enantiomerically enriched bioflavonoids. Both flavone enantiomers were obtained by selective EKR in a hydrolysis or acylation reaction (Figure 14.19). For example, the hydrolysis of racemic flavanol acetate catalyzed by lipase PS (lipase from *Burkholderia cepacia*, previously classified as *Pseudomonas cepacia*), carried out at an ambient temperature, leads to a mixture of flavanol acetate and flavanol with an enantiomeric purity of 95% ee and 93% ee, respectively.[66] However, in reaction with the AY lipase (lipase from *Pseudomonas fluorescens*) in phosphate buffer (pH 6.8) at room temperature, the kinetic resolution of flavanol acetate gives the acetate and flavanol, with an optical purity of 96% ee and 88% ee.[67] While the kinetic resolution of flavone oxime acetate in the presence of PPL lipase (porcine pancreas lipase) allows obtaining a mixture of flavone oxime and its acetate with about 80% enantiomeric excess.[68] The treatment of racemic flavanol with PS lipase and vinyl acetate gave a mixture of *cis*-flavanol acetate and *cis*-flavanol that can be converted to (2*R*)-flavone or (2*S*)-flavone in one or two steps.[69]

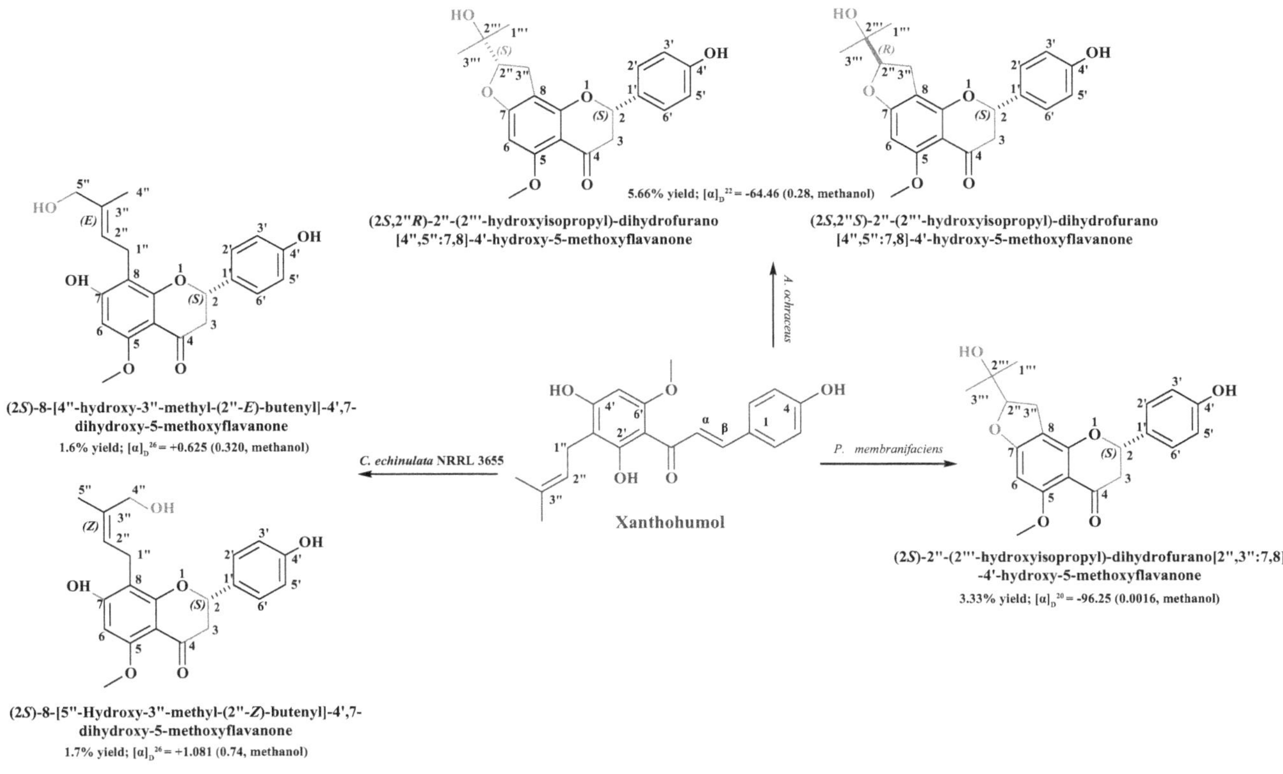

(2*S*,2″*R*)-2″-(2‴-hydroxyisopropyl)-dihydrofurano [4″,5″:7,8]-4′-hydroxy-5-methoxyflavanone

5.66% yield; $[\alpha]_D^{22}$ = -64.46 (0.28, methanol)

(2*S*,2″*S*)-2″-(2‴-hydroxyisopropyl)-dihydrofurano [4″,5″:7,8]-4′-hydroxy-5-methoxyflavanone

A. ochraceus

(2*S*)-8-[4″-hydroxy-3″-methyl-(2″-*E*)-butenyl]-4′,7-dihydroxy-5-methoxyflavanone

1.6% yield; $[\alpha]_D^{26}$ = +0.625 (0.320, methanol)

C. echinulata NRRL 3655

Xanthohumol

P. membranifaciens

(2*S*)-2″-(2‴-hydroxyisopropyl)-dihydrofurano[2″,3″:7,8] -4′-hydroxy-5-methoxyflavanone

3.33% yield; $[\alpha]_D^{20}$ = -96.25 (0.0016, methanol)

(2*S*)-8-[5″-Hydroxy-3″-methyl-(2″-*Z*)-butenyl]-4′,7-dihydroxy-5-methoxyflavanone

1.7% yield; $[\alpha]_D^{26}$ = +1.081 (0.74, methanol)

FIGURE 14.18 Biotransformation of xanthohumol.

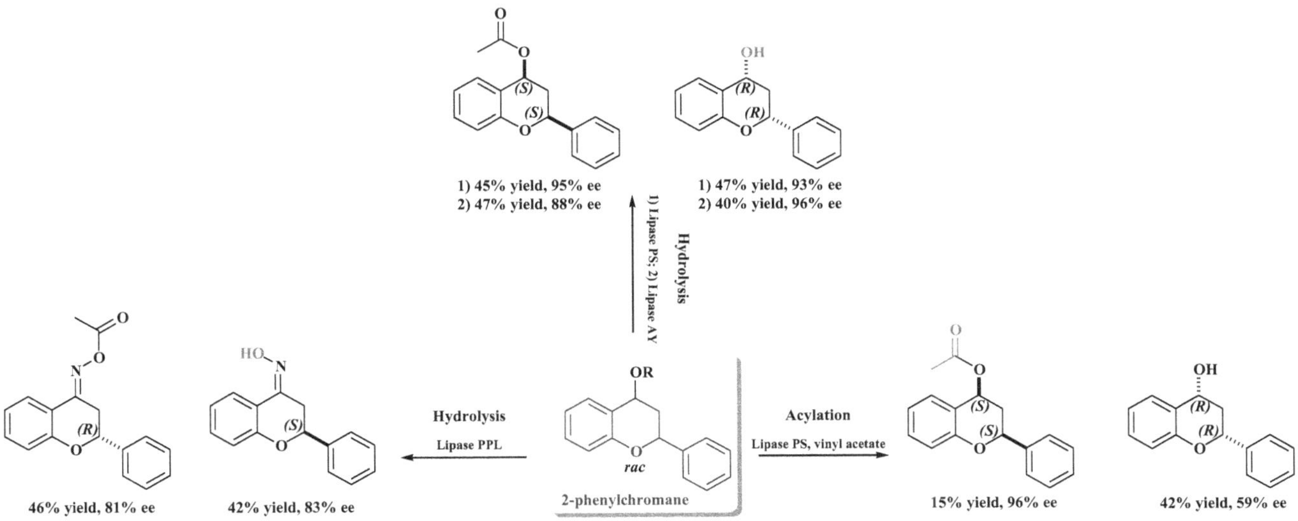

1) 45% yield, 95% ee
2) 47% yield, 88% ee

1) 47% yield, 93% ee
2) 40% yield, 96% ee

Hydrolysis
1) Lipase PS; 2) Lipase AY

Hydrolysis
Lipase PPL

Acylation
Lipase PS, vinyl acetate

rac
2-phenylchromane

46% yield, 81% ee

42% yield, 83% ee

15% yield, 96% ee

42% yield, 59% ee

FIGURE 14.19 Kinetic resolution of racemic flavanol derivatives.

Enantiomerically enriched 2-methylchroman-4-ones in the combined Mitsunobu reaction and EKR were obtained (Figure 14.20). The inversion of homoallyl alcohol with phenol in the Mitsunobu reaction leads to the alkyl phenyl ether which after oxidation to carboxylic acid has been subjected to enzymatic transesterification to form a mixture of butyl ester and acid. Enzymatic hydrolysis of butyl ester and subsequent by Friedel–Crafts acylation generated optically pure chromanone antipodes.[70]

Optically pure bioflavonoids can be obtained by kinetic resolution by microorganisms. Biotransformation in the presence of *Aspergillus niger* MB, 6-acetoxy-, 6-propionoxy-, and 6 butyryloxyflavanone allows obtaining enantiomerically pure (-)-(*S*)-6,4’-dihydroxyflavanone (Figure 14.21). In the first step of biotransformation, the hydrolysis of the ester linkage produces racemic 6-hydroxyflavanone, which in the next stage through selective hydroxylation at the C-4’ position on the B ring and at the C-2 position on the A ring

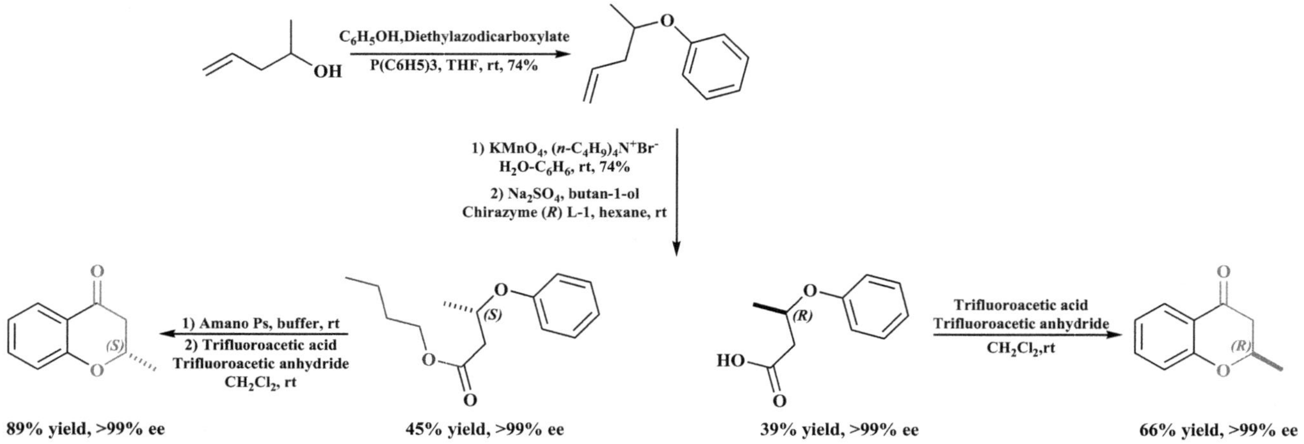

C_6H_5OH,Diethylazodicarboxylate
P(C6H5)3, THF, rt, 74%

1) $KMnO_4$, $(n\text{-}C_4H_9)_4N^+Br^-$
$H_2O\text{-}C_6H_6$, rt, 74%
2) Na_2SO_4, butan-1-ol
Chirazyme (*R*) L-1, hexane, rt

1) Amano Ps, buffer, rt
2) Trifluoroacetic acid
Trifluoroacetic anhydride
CH_2Cl_2, rt

Trifluoroacetic acid
Trifluoroacetic anhydride
CH_2Cl_2,rt

89% yield, >99% ee
45% yield, >99% ee
39% yield, >99% ee
66% yield, >99% ee

FIGURE 14.20 Synthesis of enantiomerically enriched 2-methylchroman-4-ones.

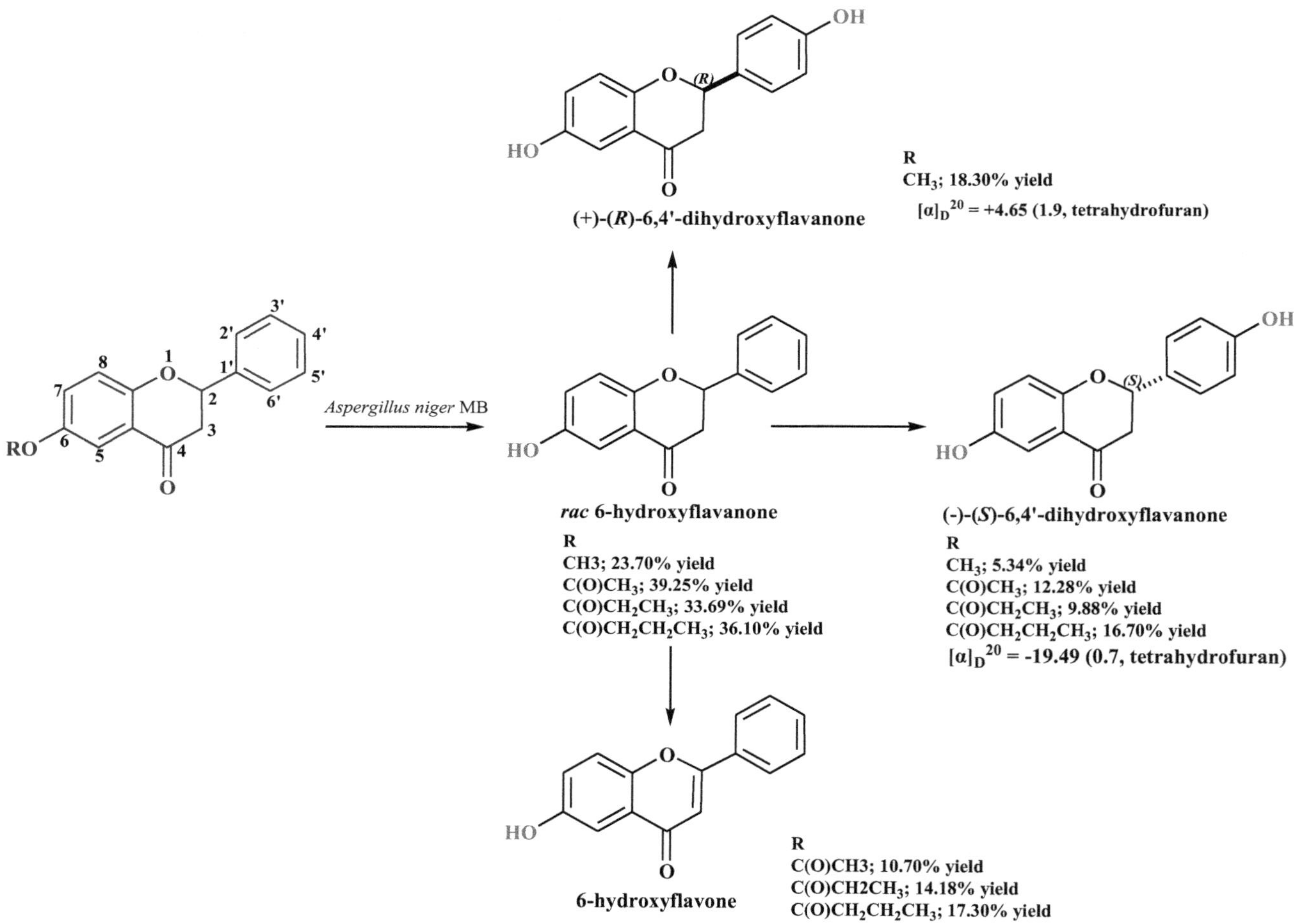

FIGURE 14.21 Biotransformation of flavanone derivatives.

gives (-)-(*S*)-6,4'-dihydroxyflavanone and 6-hydroxyflavone. 6-hydroxyflavone is formed by dehydration leading to a double bond between C-2 and C-3. The biotransformation of 6-methoxyflavanone is analogical with the difference that in the second step, enantioselective hydroxylation leads to the opposite enantiomer: (+)-(*R*)-6,4-dihydroxyflavanone.[71]

14.4 Enzymatic Synthesis of Chiral Alkaloids

The next group of biotransformed compounds in the enzymatic desymmetrization and kinetic resolution are alkaloids. According to the definition, "An alkaloid is a cyclic organic compound containing nitrogen in a negative oxidation state

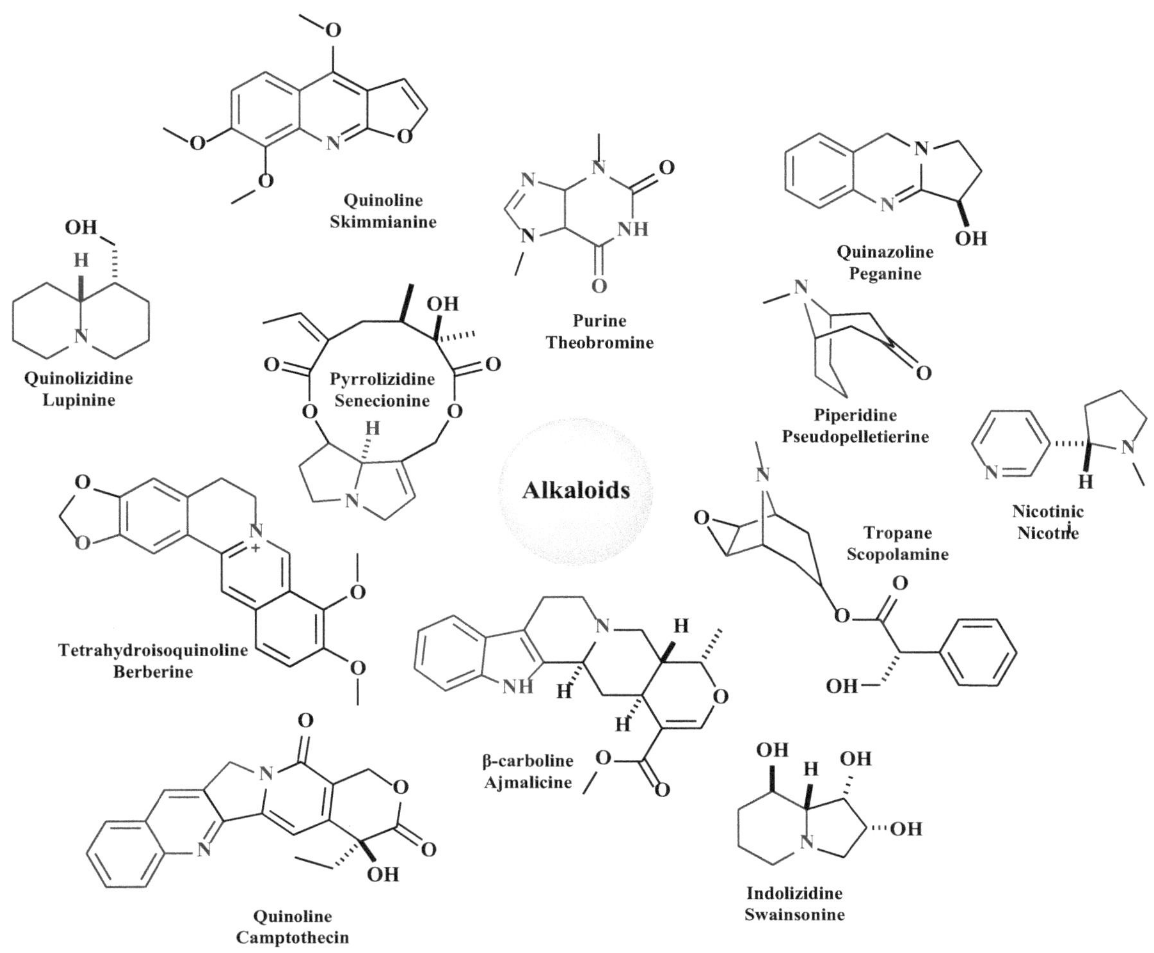

FIGURE 14.22 The molecular structure of alkaloids.

which is of limited distribution among living organisms".[72,73] Alkaloids are based on the backbone of β-carbline, indolizidine, nicotinic, piperidine, pyrrolizidine, purine, tetrahydroisoquinoline, tropane, quinoline, quinazoline, and quinolizidine (Figure 14.22). The natural functions of plant alkaloids are mainly associated with the interaction of the plant with other organisms – for instance, protection against herbivores or pathogens or the attraction of pollinating insects, seed dispensing animals, or root nodule bacteria. Many alkaloids exert a strong pharmacological effect that has been exploited for thousands of years for both therapeutic and recreational purposes.[74]

There are three main strategies for the enzymatic synthesis of alkaloids:

- Enzymatic synthesis of chiral building blocks, transformed into target compounds using chemical strategies,
- EKR and desymmetrization of alkaloids,
- Chemoenzymatic synthesis of alkaloids via biocatalytic C–N and/or C–C bond formation in the asymmetric key step.

In each strategy, the target compound is a naturally occurring, optically pure alkaloid. The most common enzymatic method is biocatalysis in the synthesis of chiral building blocks. The following are some examples of the use of protein catalysts in the enzymatic process primarily of *meso* and prochiral precursor desymmetrization.

One of the important building blocks in the synthesis of alkaloids is *meso* cyclohex-4-ene 1,2-diacetate or its diol (Figure 14.23). The hydrolysis of enantiotopic ester groups of diacetate or acylation of hydroxyl groups of diol, in a PLE (esterase from porcine live)/PPL (lipase from porcine pancreas) catalyzed reaction, leads to a (1*S*,2*R*)-monoester ((1*S*,2*R*)-cyclohexenedimethanol monoacetate) with 96% or >99% enantiomeric excess, respectively. This monoester is a precursor in the multi-stage synthesis of indole alkaloids, for example, (–)-alloyohimbane, (–)-akagerine, and polycyclic (+)-scholarisine A.[74–77] The opposite enantiomer obtained in stereoselective monohydrolysis with PPL was used to get (+)-meroquinene, a product of cinchonine degradation and a key intermediate in the synthesis of cinchonine alkaloids and indole alkaloid (–)-antirhine.[78,79]

The hydrolytic desymmetrization of *meso* diacetates of cycloheptene derivatives leads to monoacetate (1*S*,6*R*) with 97% ee using lipase PS (Figure 14.24). Several 4-hydroxypiperidine derivatives from this building block were obtained, such as *cis*-4-hydroxy-2-pipecolic acid, hydroxylated quinolizidine, and piperidine alkaloid (-)-halosaline.[80,81]

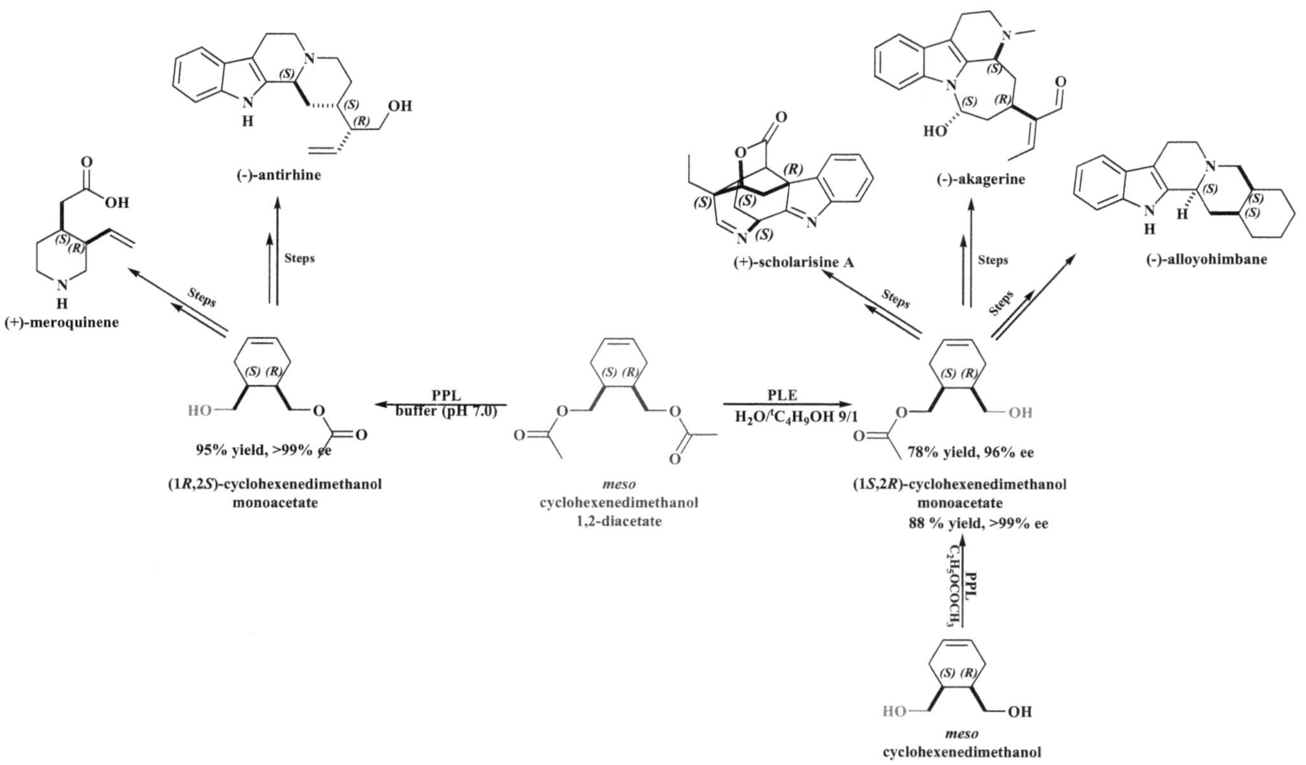

FIGURE 14.23 Kinetic resolution of *meso* cyclohex-4-ene 1,2-diacetate or its diol.

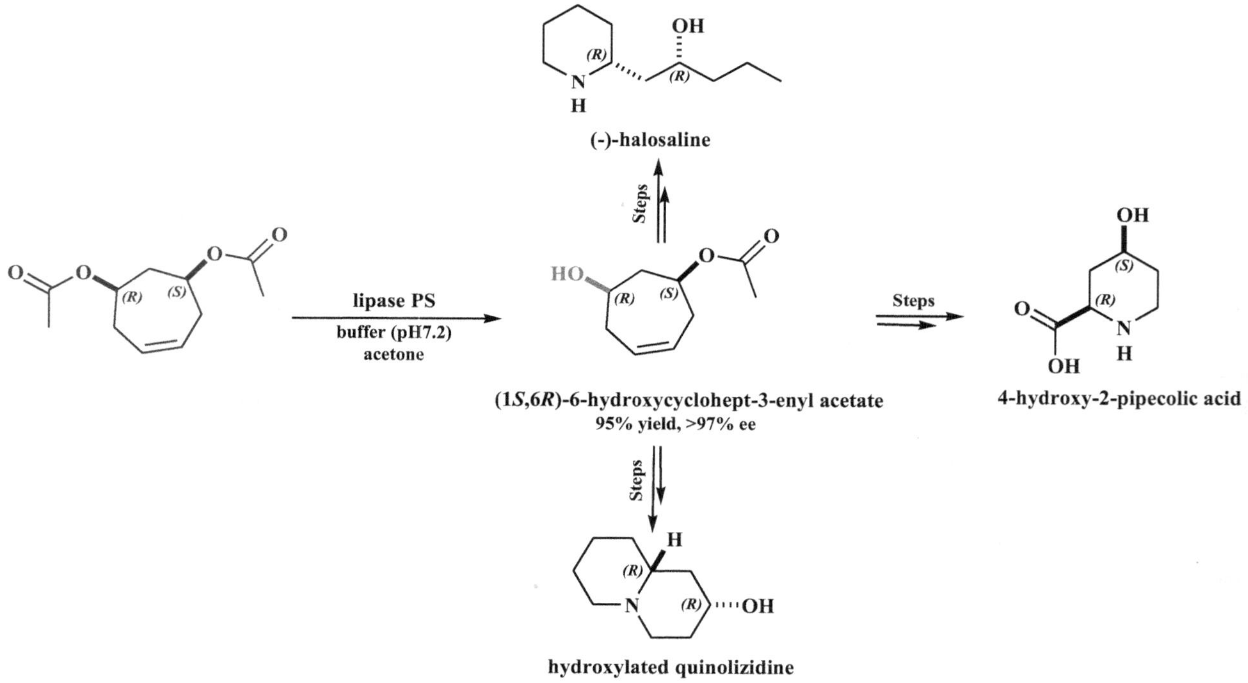

FIGURE 14.24 Desymmetrization of *meso* diacetates of cycloheptene derivatives.

Alkaloids are also obtained from piperidine building blocks in the enzymatic desymmetrization of the respective achiral *meso* compounds. The stereoselective transesterification of *N*-protected *meso*-9-azabicyclo[3.3.1]nonanediol, as well as the stereoselective hydrolysis of the corresponding diacetate, leads to an efficient synthesis of both enantiomers of the alkaloid synthon 3-piperidinol, such as (+)-dihydropinidine and (-)-cassine, in high yield and good enantiomeric excess (ee = 90% for *R*, 80% for *S*).[82–84]

The (2*R*,6*S*)-monoacetates, precursor (+)-dihydropinidine, and the dendrobate frog alkaloids (+)-hydroxypiperidine 241D and (-)-indolizidine 167B, were obtained with high yield and excellent > 99% ee selectivity in the hydrolytic desymmetrization of *meso cis*-2,6- and *cis,cis*-2,4,6-substituted piperidines with lipase from *A. niger*.[85,86] The transesterification of the corresponding diol in the presence of lipase from *Candida antarctica* gave an opposite enantiomer with a high enantiomeric excess of 95%.[87,88]

The structural motif containing 3,5-*cis*-disubstituted piperidines occurs in many natural products, for example, in indole alkaloids of the ibogan- and tacaman-type, or in (–)-sparteine, which has found wide application as chiral ligands in asymmetric synthesis. Stereoselective monoacetylation of the analogous piperidine-3,5-dimethanols, catalyzed by a lipase from *P. fluorescens*, turned out to be efficient and provided (3*S*,5*R*)-monoacetates with excellent optical purity (ee> 98%) (Figure 14.25). The opposite enantiomers can be obtained using the catalytic properties of the same enzyme from corresponding diacetates.[89] The chiral building blocks thus obtained served as the basis for the preparation of several compounds featuring *cis*-fused piperidine rings, for example, the 3,7-diazabicyclo[3.3.1]nonane derivative, alkaloid (–)-cytisine, analog (+)-sparteine.[90–92]

Oxidoreductases also deserve attention, in addition to hydrolases, in enzymatic synthesis of chiral building blocks. Toluene dioxygenase (TDO), belonging to the class of oxidoreductases, has been used in the preparation of arene *cis*-dihydrodiols versatile chiral building blocks offering ample possibilities for further functionalization. The systematic name of this enzyme class is toluene, NADH: oxygen oxidoreductase (1,2-hydroxylating), in the presence of oxygen and the NADH cofactor, convert toluene to (1*S*,2*R*)-3-methylcyclohexa-3,5-diene-1,2-diol.

The microbial oxidation of chlorobenzene with *Pseudomonas putida* enabled the transformation of chlorobenzene to (1*S*,2*S*)-3-chlorocyclohexa-3,5-diene-1,2-diol; this compound served as a chiral building block in the synthesis of iminocyclitol alkaloids such as (+)-kifunensine, (+)-mannojirimycin, (+)-1-deoxygalactonojirimycin, and (-)-1-deoxymaninejirimycin.[93–96] Generally, various halogenodiols obtained in asymmetric biotransformation with TDO have been widely used as building blocks in the synthesis of natural products in recent years. For example, in the production of the cytotoxic alkaloids of *Amaryllidaceae pancratistatin*, narciclasine, lycoricidine, and their analogues, morphinan alkaloids such as morphine, codeine, and lycorenine-type alkaloids: (+)-clividine and (-)-narseronine (Figure 14.26).[97–103]

An interesting example of using the catalytic properties of TDO and lipase is the chemo-enzymatic synthesis of (-)-7-deoxypancratistatin (Figure 14.27). In the first step of the TDO-catalyzed oxidation reaction, *para*-dihalobenzene was transformed into the corresponding diol with 20% ee, in subsequent stages in chemical synthesis of optically impure conduramine A derivative was obtained, which became the subject of the EKR of PPL-lipase-catalyzed. This step enriched the post-reaction mixture with a product in 98% ee, which was transformed into a natural alkaloid in the following stages.[104]

Asymmetric biotransformation is a less commonly used method to obtain enantiomerically and diasteromerically enriched alkaloids, differentiation of optical isomers by kinetic resolution of racemates is more common. The synthesis of (-)-lobeline, a bioactive alkaloid also known as Indian tobacco, is an example of asymmetric biotransformation. It was synthesized via the stereoselective desymmetrization of

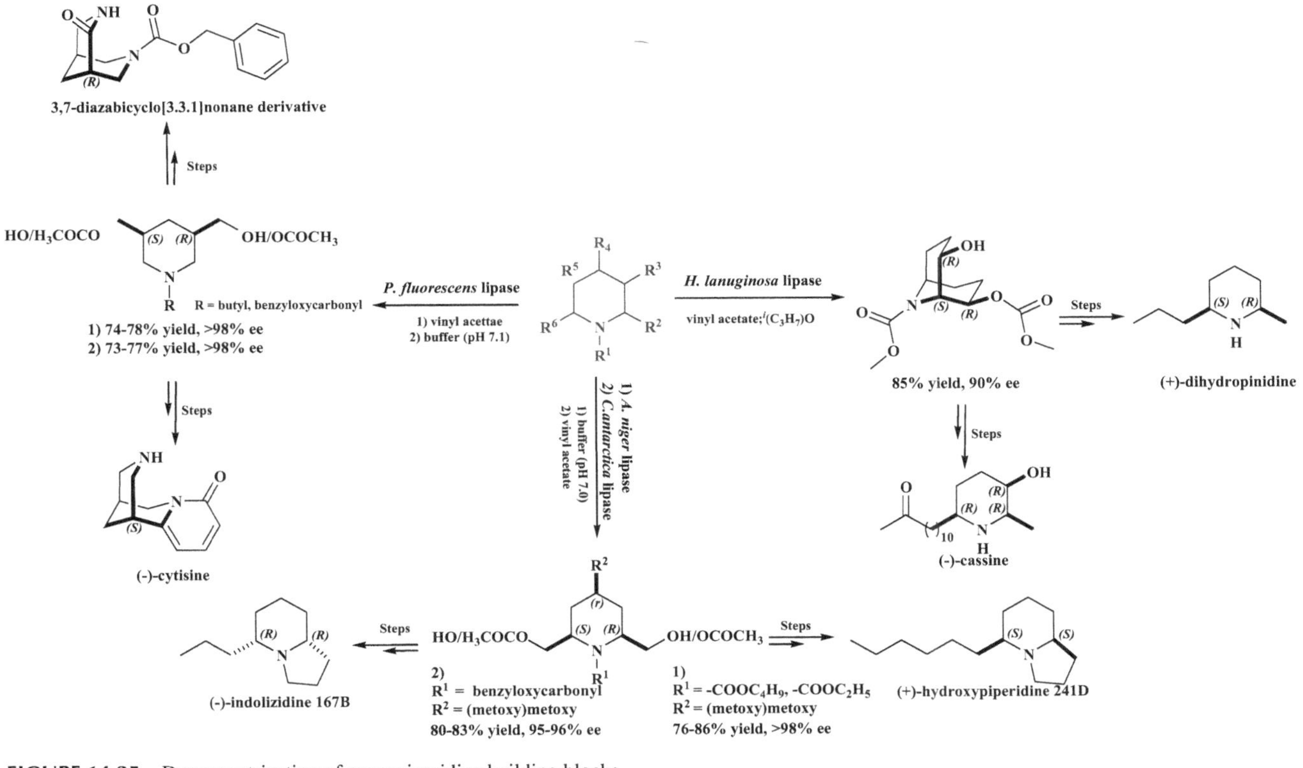

FIGURE 14.25 Desymmetrization of *meso* piperidine building blocks.

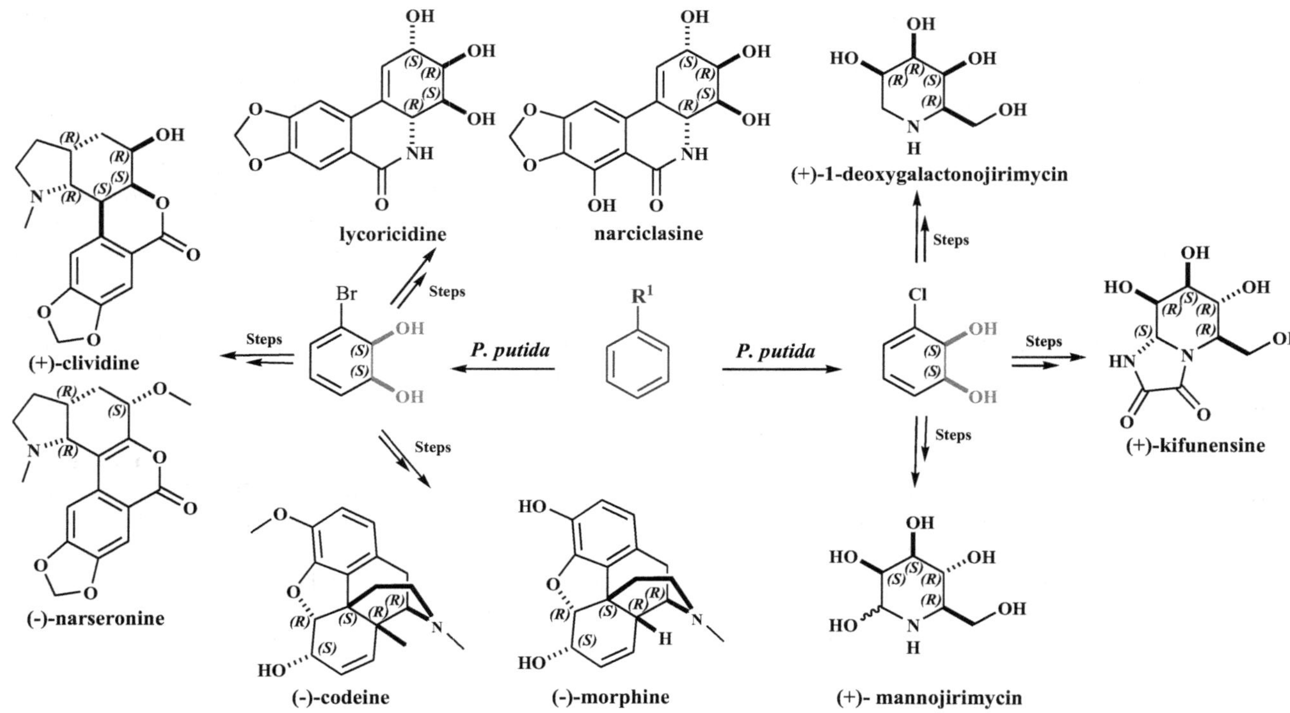

FIGURE 14.26 Synthesis of alkaloids from halogenodiols obtained in asymmetric biotransformation with toluene dioxygenase.

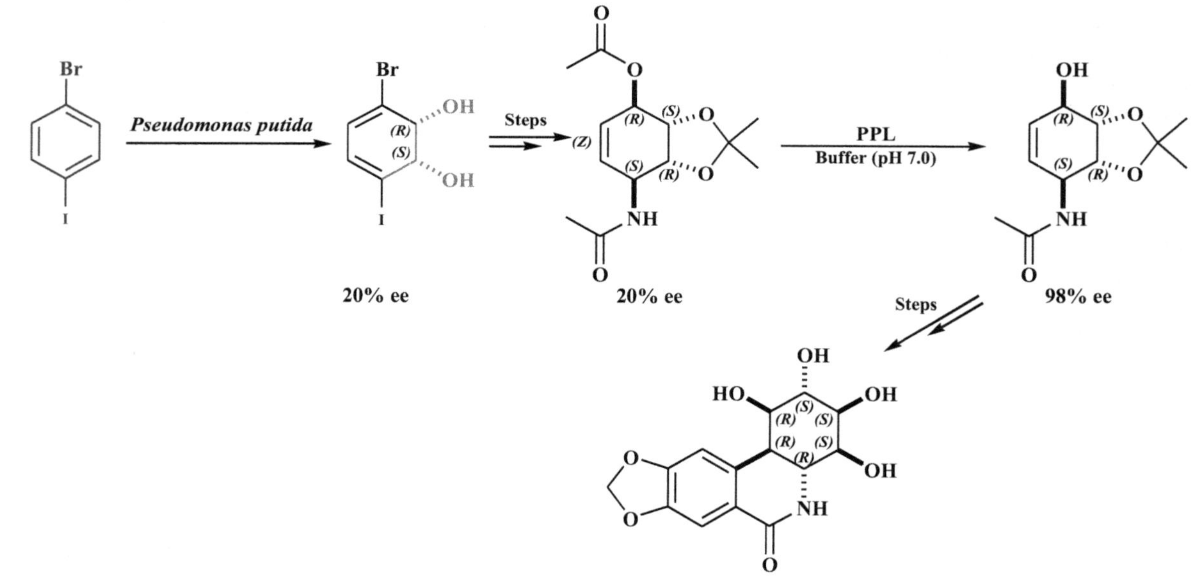

FIGURE 14.27 Chemo-enzymatic synthesis of (-)-7-deoxypancratistatin.

*meso*lobelanidine catalyzed by CAL-B. This enzyme allowed the preparation of (*S*)-monoester of high optical purity (Figure 14.28).[105]

One of the alkaloids separated in the EKR is mappicine and 9-methoxy derivative (Figure 14.29), which possesses strong selective activity against the herpes viruses HSV-1 and HSV-2 and human cytomegalovirus (HCMV). These compounds are produced by *Nothapodytesfoetida* with a small yield. The synthesis of (*S*)-and (*R*)-mappicines and their analogues were carried out as a result of the transformation of camptothecins, the more abundant alkaloid. The kinetic resolution of chiral esters provides (*S*)-mappicine and (*R*)-mappicine with 97% ee in hydrolysis by Baker's yeast or Amano PS lipase, respectively. Using the same catalysts allows (*S*)- and (*R*)-9-methoxymappicine to be obtained from 95% ee with yeast and 87% ee with lipase Amano Ps.[106] In the presence of other commercially available QLM lipase (lipase B from *C. antarctica*), another alkaloid – (*R*)-(2)-desprenylcarbochostatin A was

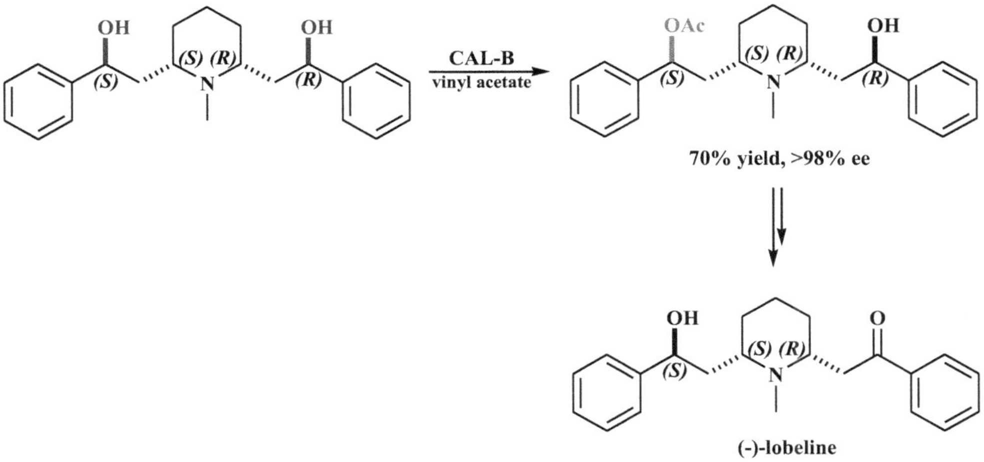

OH (S) (S) (R) N OH (R)

CAL-B
vinyl acetate

OAc (S) (S) (R) N OH (R)

70% yield, >98% ee

OH (S) (S) (R) N O

(-)-lobeline

FIGURE 14.28 Desymmetrization of *meso*lobelanidine.

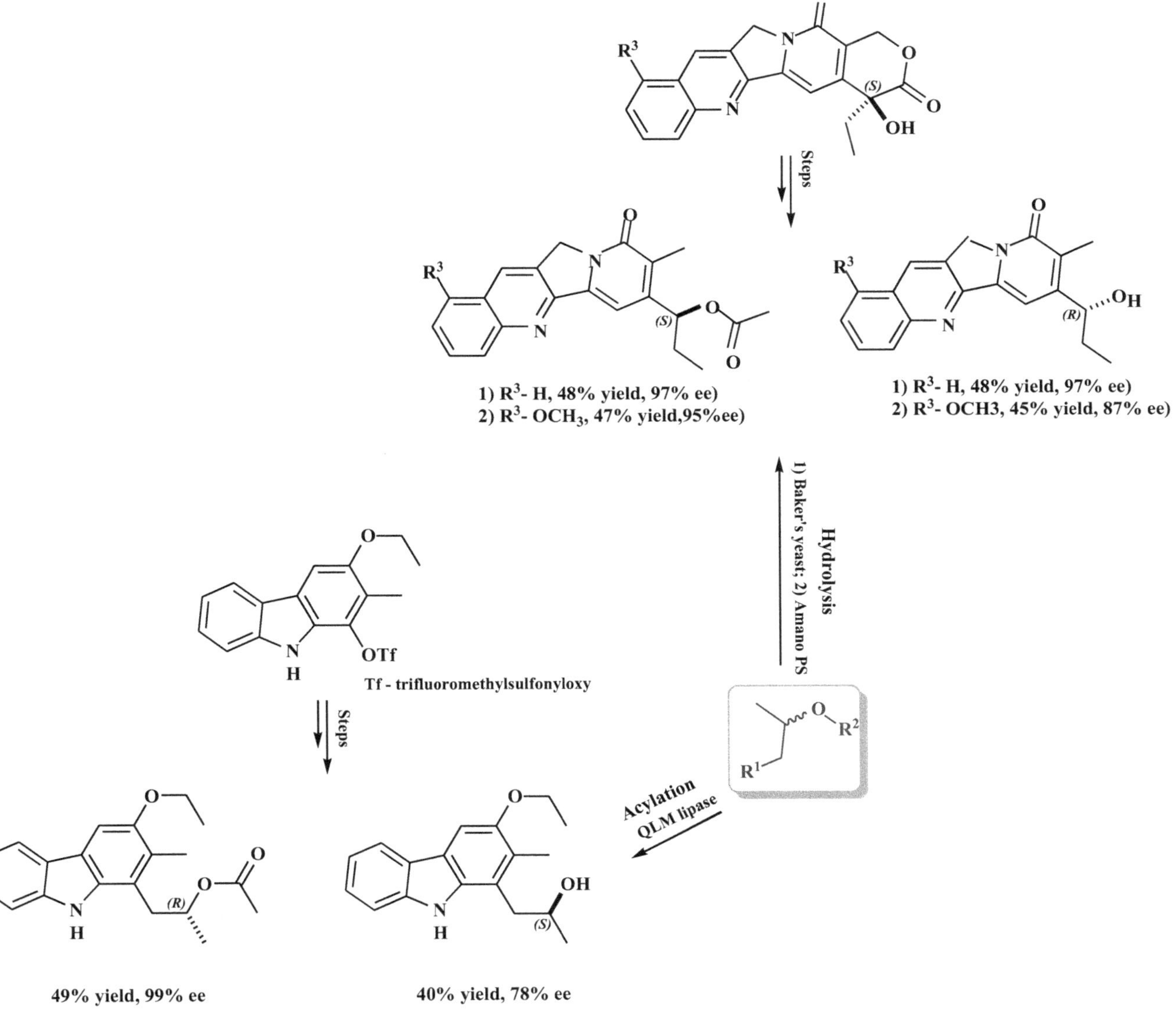

FIGURE 14.29 Kinetic resolution of mappicine and 9-methoxy derivative.

obtained by chemo-enzymatic reaction. (*R*)-(-)-6-desprenyl-carquinostatin A is a derivative of carquinostatin A, the carbazole-3,4-quinone alkaloid, was isolated from *Streptomyces exfoliates* 2419-SVT2. Carquinostatin A was shown to be a potent neuronal cell-protecting substance that exhibits a free radical scavenging activity.[107] An asymmetric synthesis of the core carbazole structure, 6-desprenyl-carquinostatin toward a total synthesis of carquinostatin A was carried out. In one step, as a result of a selective acylation reaction, lipase enabled the kinetic resolution of the racemate to provide the (*R*)-acetate of 6-desprenyl-carquinostatin and the (*S*)-alcohol in 99% and 78% ee, respectively.[108]

Vasicinone is one of the pyrrolo[2,1-*b*]quinazoline alkaloids isolated from the aerial parts of *Adhatoda vasica* was synthesized in a chemo-enzymatic process by an azidoreductive cyclization strategy employing TMSCl-NaI and bakers' yeast. The optical purity of the product was ensured in the lipase-catalyzed kinetic resolution reaction. Acetyl vasicinone was enzymatically hydrolyzed employing lipase PS "Amano" to its (*R*)-alcohol and (*S*)-acetate in 98% ee. In contrast, racemic vasicinone was resolved by transesterification; (*R*)-acetate was obtained in >99% ee employing lipase PS in THF (Figure 14.30).[109] Most of the quinazoline alkaloids are known to have a broad spectrum of pharmacological activity, particularly, bronchodilatory activity. *A. vasica*, an evergreen subherbaceous bush, due to the presence of quinazoline alkaloids, is widely used in indigenous medicine for cold, cough, bronchitis, and asthma.

Lupinine is a quinolizidine alkaloid present in several *Leguminosae* (*Lupinus luteus*, *L. hispanicus*) and *Chenopadiaceae* (*Anabasis aphylla*). Lupinine hydrochloride is a mildly toxic acetylcholinesterase inhibitor; it has an inhibitory effect on acetylcholine receptors. The resolution of racemic lupinine, by transesterification with vinyl acetate in the presence of AK lipase, provides (+)-lupinine with 89% ee (Figure 14.31).[110]

In order to obtain both enantiomers of the tetrahydroisoquinoline alkaloid 8,9-bis(methyloxy)-1,2,3,5,6,10b-hexa-hydropyrrolo[2,1-*a*]isoquinoline (crispine A), discrimination of the stereogenic center in the δ position against the hydroxyl group was carried out using kinetic resolution catalyzed by lipase PS (*B. cepacia* lipase). The decanoylation of the racemic mixture of its analogue led to an (*S*)-ester up to 96% ee (Figure 14.31). Enzymatic hydrolysis of the corresponding ester made it possible to obtain (*S*)-alcohol with similar optical purity.[111] The obtained crispine A and another isoquinoline alkaloid, containing a guanidyl group (crispine E), display high biological activity against SKOV3, KB, and HeLa human cancer cell lines. They are isolated from the plant *Carduus crispus* has long been used in Chinese folk medicine for the treatment of colds, stomach problems, and rheumatism.[112]

Crispine A was also obtained by the deracemization reaction of catalyzed monoamine oxidase from *A. niger* (MAO-N). Monoamine oxidases (MAOs) are flavin-dependent enzymes that catalyze the oxygen oxidation of amines to the corresponding imines or imine ions. The combination of this reaction with chemical reduction of imine leads – within several cycles of enantioselective oxidation and non-stereoselective reduction – to the accumulation of the slower reacting amine enantiomer. In this case, the reaction was carried out in the presence of ammonia-borate ($NH_3 \cdot BH_3$) as a reducing agent. Under optimized chemo-enzymatic deracemization reaction conditions, the biologically active (*R*)-enantiomer of crispine A in 97% ee was obtained (Figure 14.31).[38]

Since the first reaction using monoamine oxidase in the chemo-enzymatic deracemization of the alkaloid – racemic nicotine,[113] significant progress has been observed in this field. The catalytic properties of MAO-N were applied in the asymmetric synthesis of the indole alkaloids (*R*)-eleagnine and (*R*)-harmicine while MAO-N-5 was used in the deracemization of the hemlock neurotoxin coniine. A variant with MAO-N-11, an oxidase with opposite enantioselectivity, was also developed and applied in the synthesis of (*S*)-1-phenyltetrahydroisoquinoline, a building block for the urinary antispasmodic drug solifenacin (Figure 14.32).[114]

An interesting method of enzymatic resolution is the use of the berberine bridge enzyme (BBE, EC 1.21.3.3), which is a bicovalent flavoenzyme found in various alkaloid-producing

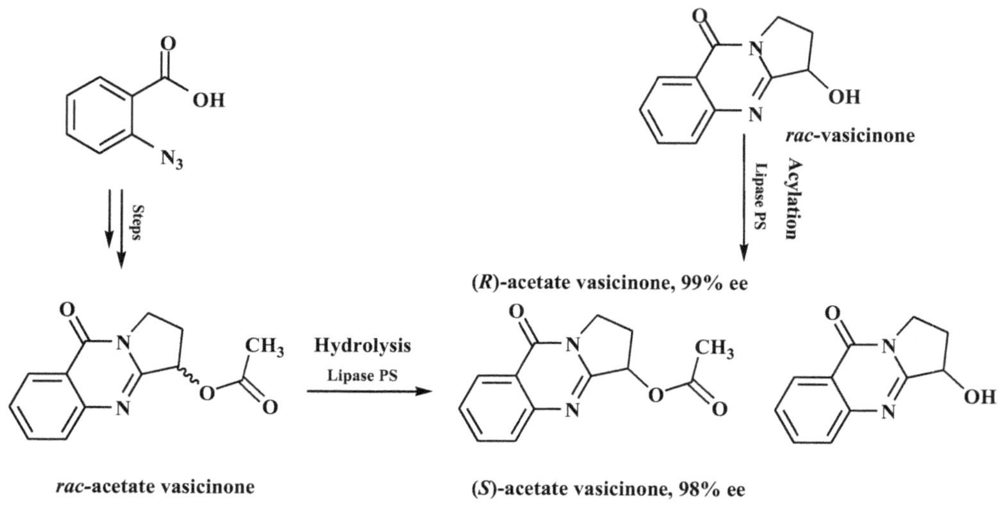

FIGURE 14.30 Kinetic resolution of acetyl vasicinone.

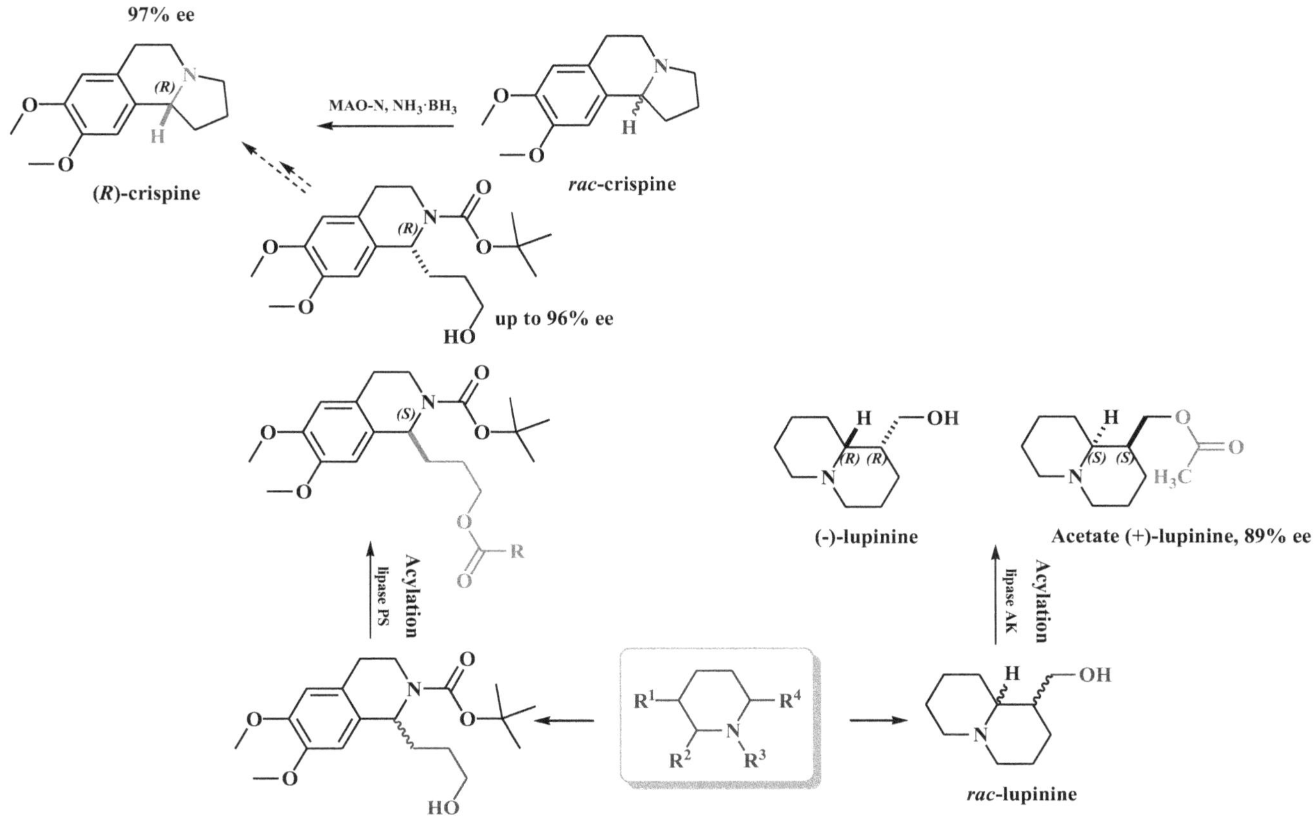

FIGURE 14.31 Kinetic resolution of lupinine and crispine A.

plants, in particular from the *Papaveraceae* (poppy) and *Fumariaceae* (fumitory) families. BBE catalyzes the stereospecific conversion of (*S*)-reticuline to (*S*)-scoulerine, thus forming the so-called berberine bridge (C8) by joining the isoquinoline ring with the benzyl ring system. This cyclization reaction has no known equivalent in organic chemistry; it consists of intramolecular oxidative carbon–carbon bond formation, at the expense of molecular oxygen as a terminal electron acceptor.[74,115,116] The BBE is a fairly stable enzyme capable of working in the presence of various organic solvents and at relatively high substrate concentrations. Transforms several unnatural substrates, racemic reticuline derivatives with high enantioselectivity. As a result of the reaction on an analytical scale, besides berbine-type cyclization (*S*)-products with over 97% ee and unreacted (*R*)-benzylisoquinolines in small quantities, regioisomeric 11-hydroxyberbine derivatives are produced (Figure 14.33).[117]

Another example of the biocatalytic kinetic resolution of alkaloids is the deracemization of tetrahydroberberrubines, which were separated by kinetic glycosylation and enantioselective sulfation. The reaction catalyzed by whole cells of the fungus *Gliocladium deliquescens* NRRL 1086 gave a product glycosylation with the *S* configuration, which was formed much faster than the (*R*)-isomer. Upon long-term incubation (120 hours), the minor isomer was converted to the sulfate ester. The sulfation reaction was found to be more selective than the glycosylation, which allows for complete resolution of both enantiomers to (*S*)- and (*R*)-isomers with a yield of 48% and 43%, respectively, and in greater than 99% ee for each enantiomer. Based on the research, it was found that effective glycosylation occurs only when the hydroxyl group of the substrate is in position 9 of the berbinescaffold.[118,119]

The alkaloids described above, berbines and tetrahydroberberrubines, containing an isoquinoline skeleton and a stereogenic center at position C-14, have been demonstrated to possess various biological activities such as anti-inflammatory, antitumor, antimicrobial, analgesic activity, anticonvulsant, antianxiety, antidepressant, and antipsychotic activities in the central nervous system (CNS), antiarrhythmic, antihemorrhagic, and hypotensive activities in the cardiovascular system.[120]

In asymmetric alkaloid synthesis, the use of enzymes in one of the synthesis steps can lead to the formation of new C-C or C-N bonds in a highly selective manner, in a shorter and more practical methodology. Aldolases are one of the enzymes that create C-C binding in chiral alkaloid synthesis, belonging to the class of lyase, catalyzing the reaction of aldol cleavage, in glycolysis the fructose-1,6-diphosphate (FDP) aldolase substrate is a six-carbon fructose-1,6-bisphosphate (F-1,6-BP) molecule.

D-fructose-1,6-diphosphate (FDPA) was used for highly stereoselective aldol reactions an route to (-)-D-1-deoxynojirimycin, (+)-D-1-deoxynojirimycin, and (+)-D-fagomine. They are useful for the treatment of carbohydrate-dependent metabolic disorders because of their selective inhibition of glycosidases, and were isolated from plants of genus Mows and strains of

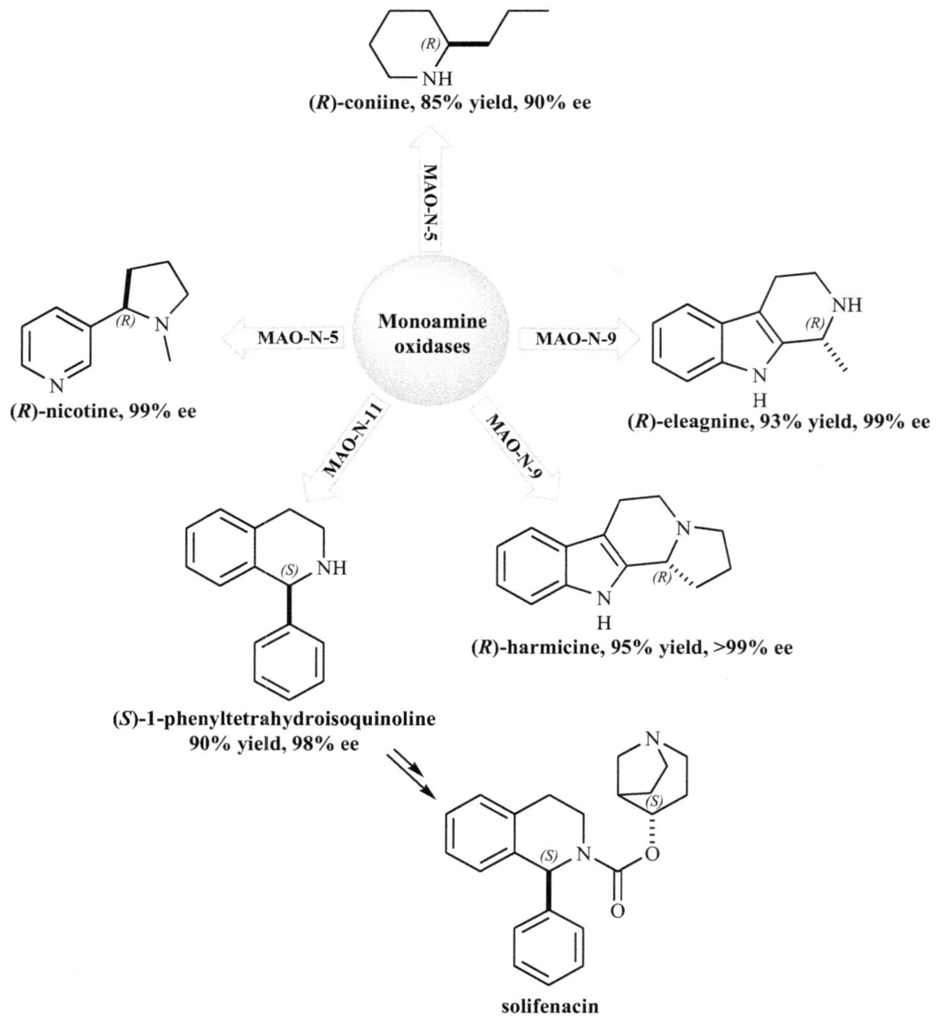

(*R*)-coniine, 85% yield, 90% ee
MAO-N-5
Monoamine oxidases
MAO-N-5
(*R*)-nicotine, 99% ee
MAO-N-9
(*R*)-eleagnine, 93% yield, 99% ee
MAO-N-11
MAO-N-9
(*R*)-harmicine, 95% yield, >99% ee
(*S*)-1-phenyltetrahydroisoquinoline 90% yield, 98% ee
solifenacin

FIGURE 14.32 Chemo-enzymatic deracemization of the alkaloid.

Bacillus (Figure 14.34).[121,122] In these chemo-enzymatic syntheses, aldolase catalyzes the coupling of DHAP with an aldehyde (sp^2 prochiral center) containing an azide substituent or with an amine-protected amino (Cbz-amine). After catalytic hydrogenation of the dephosphorylated linear aldol product, a free amine is formed, which cyclizes into the iminocylitol ring via intramolecular reductive amination.[121–123] The stereochemistry of the new C-C bond formed is completely controlled by the enzyme, the carbanion (or the enolate *si* face) generated via enzymatic abstraction of the H_s at C-3 of DHAP always attacks the *si* face of the aldehyde carbonyl group.[121]

Aldolases have also found application in the asymmetrical synthesis of polyhydroxylated pyrrolizidine alkaloids, i.e., 7-*epi*-alexine, (+)-3-*epi*-australine, and (+)-australine.[124] The synthesis of structurally related pyrrolizidines, which represent the stereoisomers of naturally occurring hyacinthacin A_1 and A_2 alkaloids, was performed using the step of coupling a proline derivative with DHAP catalyzed by 1-ramnulose-1-phosphate aldolase (RhuA) from *E. coli*. The reaction proceeded with excellent diastereoselectivity. Hyacinthacines A_1 and A_2 were isolated from the bulbs of *Muscari armeniacum* (Hyacinthaceae) and demonstrated to be good inhibitors against rat intestinal lactase, rat epididymis α-l-fucosidase, and amyloglucosidase from *Arpergillus niger* (Figure 14.34).[125] 1-fuculose-1-phosphate aldolase (FucA) as a biocatalyst is also successfully used in the production of some closely related pyrrolizidines (Figure 14.34).[126]

ω-Transaminases (ω-TA) which generate C-N bond formation, are also increasingly used in asymmetric synthesis. The natural function of these enzymes is to catalyze the transamination reaction between amino acids and α-keto acids. The use of ω-transaminases in the synthesis of piperidine alkaloids has been reported (Figure 14.35). Reductive amination of δ-diketones catalyzed by various ω-transaminases (*Chromobacterium violaceum*, *Bacillus megaterium*, (*R*)-*Arthrobacter*, *Aspergillus terreus*, and *Hyphomonas neptuniuoccurs*) only at the sterically less demanding ω-1 ketone moiety giving optically pure amino ketones. *Cis*-2,6-disubstituted piperidines as single stereoisomers are formed as a result of spontaneous cyclization of amino ketones to imines, which are catalytically hydrogenated in the next step.[127] Regioselective asymmetric mono-amination of 1,5-diketones catalyzed by ω-transaminase was used to the chemo-enzymatic synthesis of dihydropinidine and *epi*-dihydropinidine, by using (*R*)- and (*S*)-selective ω-transaminases.[128]

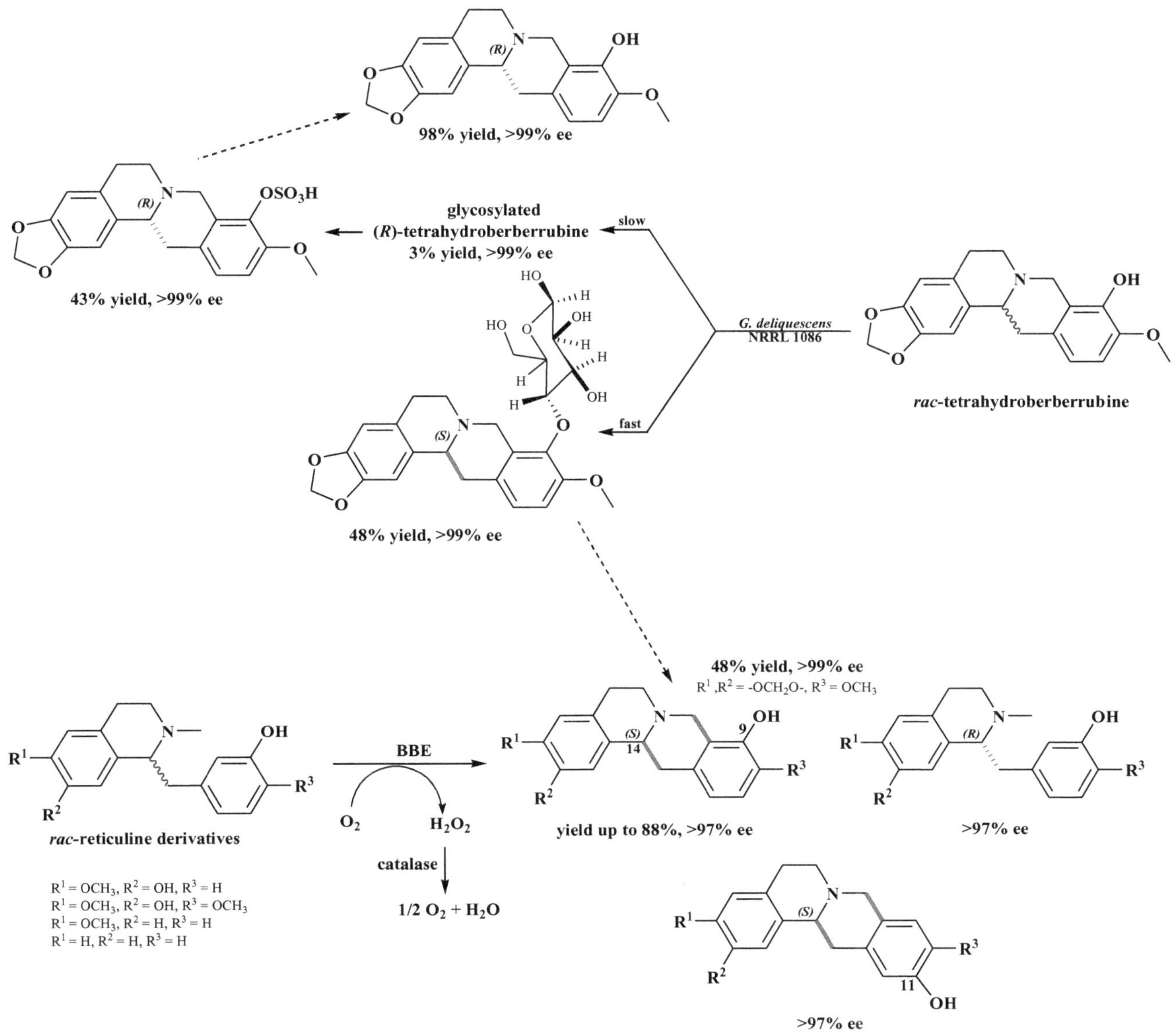

FIGURE 14.33 Kinetic resolution of alkaloids using berberine bridge enzyme.

A combined enzymatic transamination reaction and subsequent reduction of imine has also been used in the synthesis of 1,3-disubstituted tetrahydroisoquinoline.[129]

(*S*)-Norcoclaurinesynthase (4-hydroxyphenylacetaldehyde hydrolyase) belongs to the lyase class, it is an enzyme that catalyzes the reaction of 4-hydroxyphenylacetaldehyde and 4-(2-aminoethyl)benzene-1,2-diol to (*S*)-norcoclaurine. This enzyme participates in benzylisoquinoline alkaloid biosynthesis. The effective biosynthesis of (*S*)-norcoclaurine, the recombinant (*S*)-norcoclaurinesynthase (NCS) enzyme from *Thalictrum flavum*, was performed in a one-pot, two-step process from the cheap tyrosine and dopamine substrates. Key biotransformation steps included oxidative decarboxylation of tyrosine by stoichiometric amounts of sodium hypochlorite to produce 4-hydroxyphenylacetadehyde, followed by the addition of enzyme and dopamine in the presence of ascorbate. Under optimized conditions, enantiomerically pure (*S*)-norcoclaurine (93% ee) was obtained with a yield greater than 80% (Figure 14.36). The process thus developed was the first example of green Pictet–Spengler synthesis.[130]

The potential of norkoclaurine synthase (NCS) from *Coptis japonica* NCS2 was investigated using a new and efficient fluorescence-based assay using almost forty amines/aldehydes. Preparative biotransformations gave (*S*)-tetrahydroisoquinoline, including compounds having no natural counterparts, in good to excellent yield (56–77%) and with an excellent ee > 95% (Figure 14.36).[131]

Strictosidine synthase (STR) is another enzyme involved in asymmetric Pictet–Spengler reactions. The natural function of this enzyme is the condensation of tryptamine and the monoterpenoid aldehyde secologanin to form 3α(*S*)-strictosidine. The biocatalytic application of STR relates, among others, to the chemo-enzymatic synthesis of (+)-nacycline,

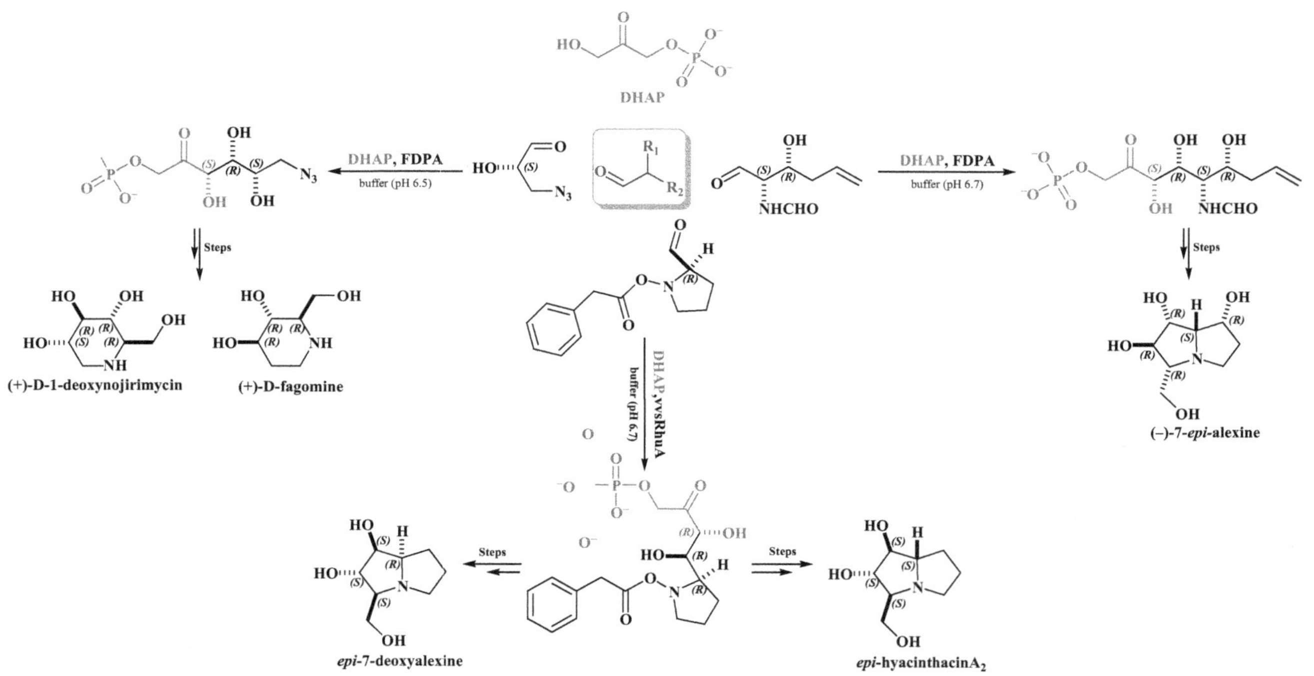

FIGURE 14.34 Synthesis of alkaloids by aldolases.

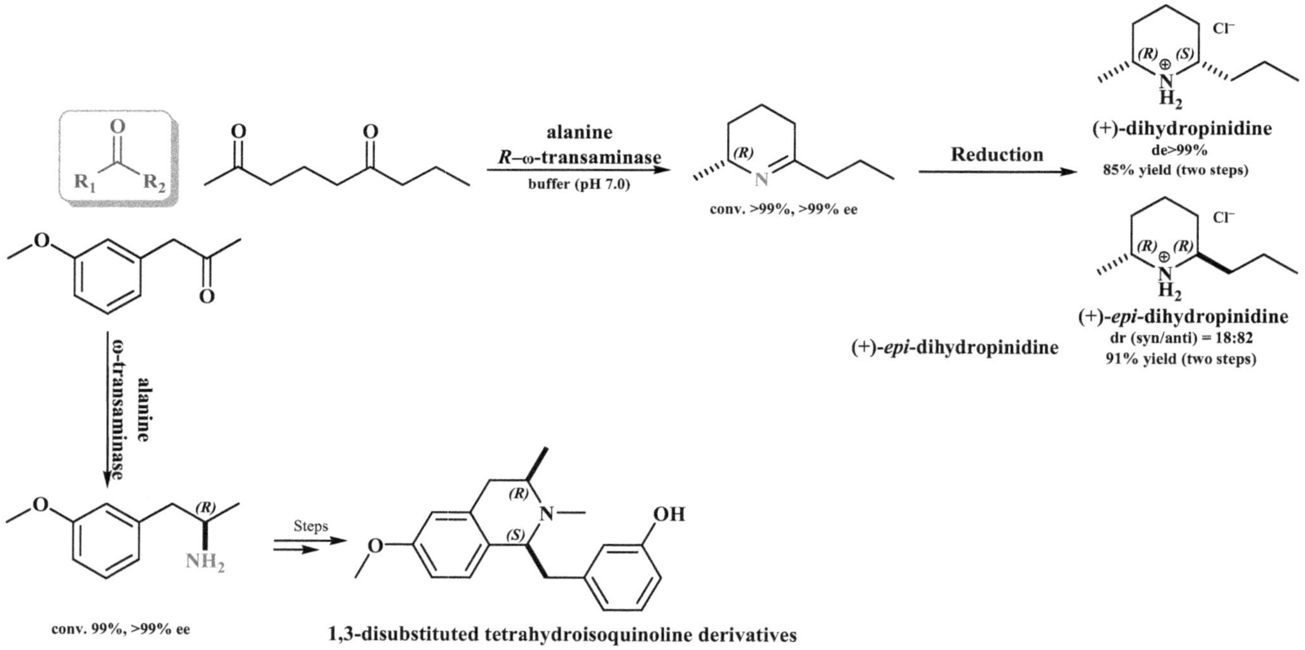

FIGURE 14.35 Synthesis of alkaloids by ω-transaminases.

(-)-tetrahydroalstonine and the production of piperazino[1,2-*a*] indole alkaloid (Figure 14.37).[74,132,133]

14.5 Enzymatic Synthesis and Biotransformation of Terpenoids

Terpenes are secondary metabolites of plant origin. They are also called terpenoids or isoprenoids and are one of the largest groups of natural compounds. Terpenoids are volatile substances that generally give plants and flowers their flavor and fragrance. They are widely found even in the leaves and fruits of higher plants, conifers, citrus, and eucalyptus, and are an important component of essential oils. The terpenes group consists of isoprene (C5) units, i.e., 2-methylbut-1,3-diene molecules.

The division of terpenes according to the structure is made on the basis of determining how many isoprene units the carbon chain of a given molecule consists of, which is why the terpenoids are classified as follows (Figure 14.38):

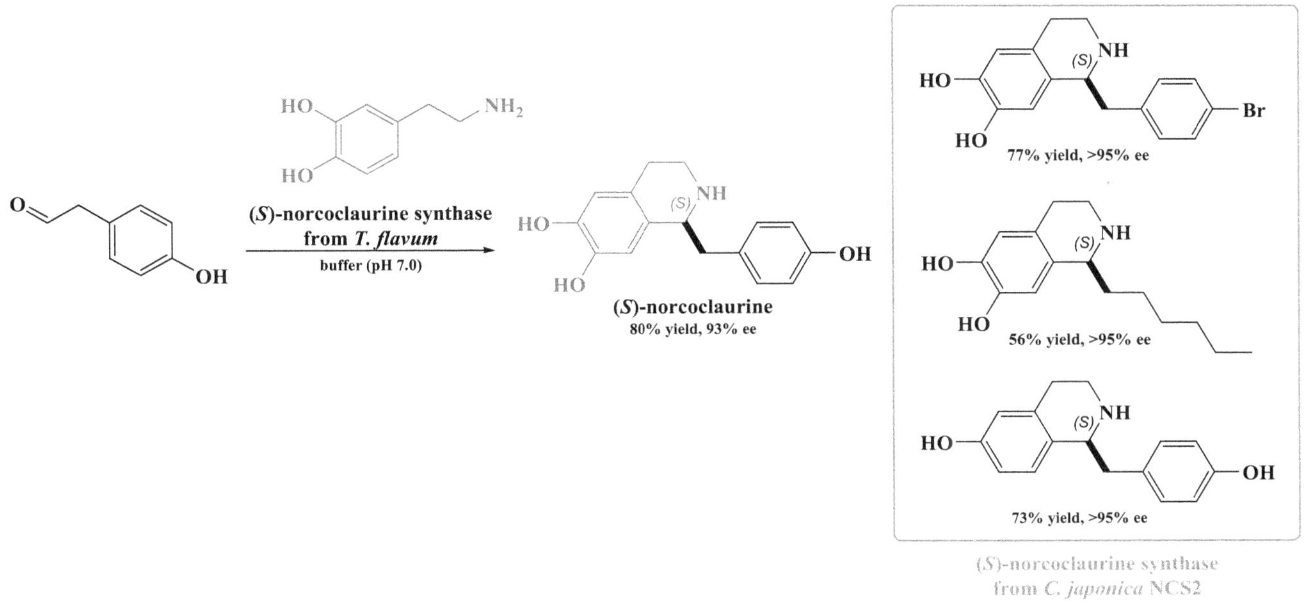

(S)-norcoclaurine synthase from T. flavum
buffer (pH 7.0)
(S)-norcoclaurine
80% yield, 93% ee
77% yield, >95% ee
56% yield, >95% ee
73% yield, >95% ee
(S)-norcoclaurine synthase from C. japonica NCS2

FIGURE 14.36 Synthesis of alkaloids by (*S*)-norcoclaurine synthase.

piperazino[1,2-a]indole alkaloid
Strictosidine synthase buffer (pH 7.0)
Strictosidine synthase
buffer (pH 7.0)
Steps
(+)-nacycline
(-)-tetrahydroalstonine

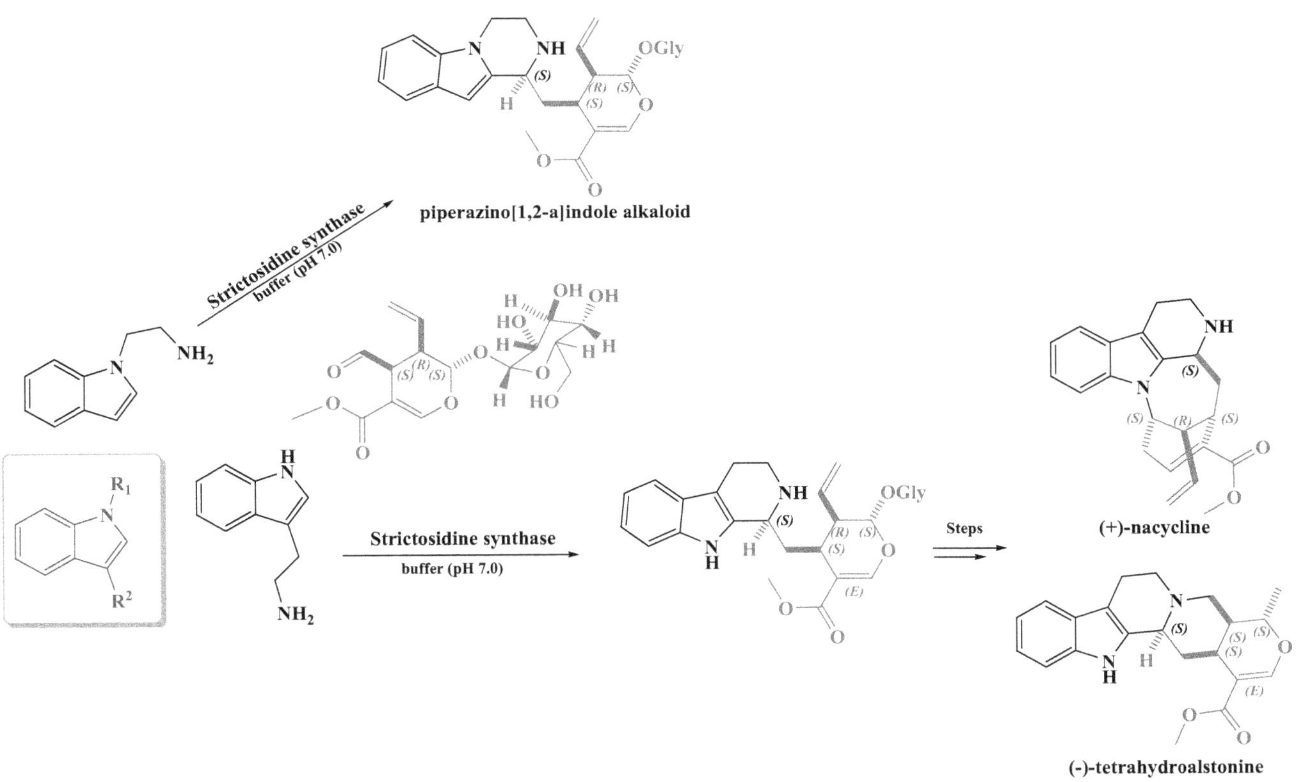

FIGURE 14.37 Synthesis of alkaloids by strictosidine synthase.

- Hemiterpenoids (C5) consist of a single isoprene unit. The only hemiterpene is the isoprene itself, but oxygen-containing derivatives of isoprene such as isovaleric acid and prenolare also included in this class.
- Monoterpenoids (C10) consist of two isoprene units. The monoterpenes can be further grouped into linear or acyclic and cyclic ring-containing, for example, geranyl pyrophosphate, eucalyptol, limonene, citral, camphor, and pinene.
- Sesquiterpenoids (C15) consist of three isoprene units. This group includes, for example, artemisinin, bisabolol and fernesol, flower oil, or cyclic compounds such as eudesmol (eucalyptus oil).

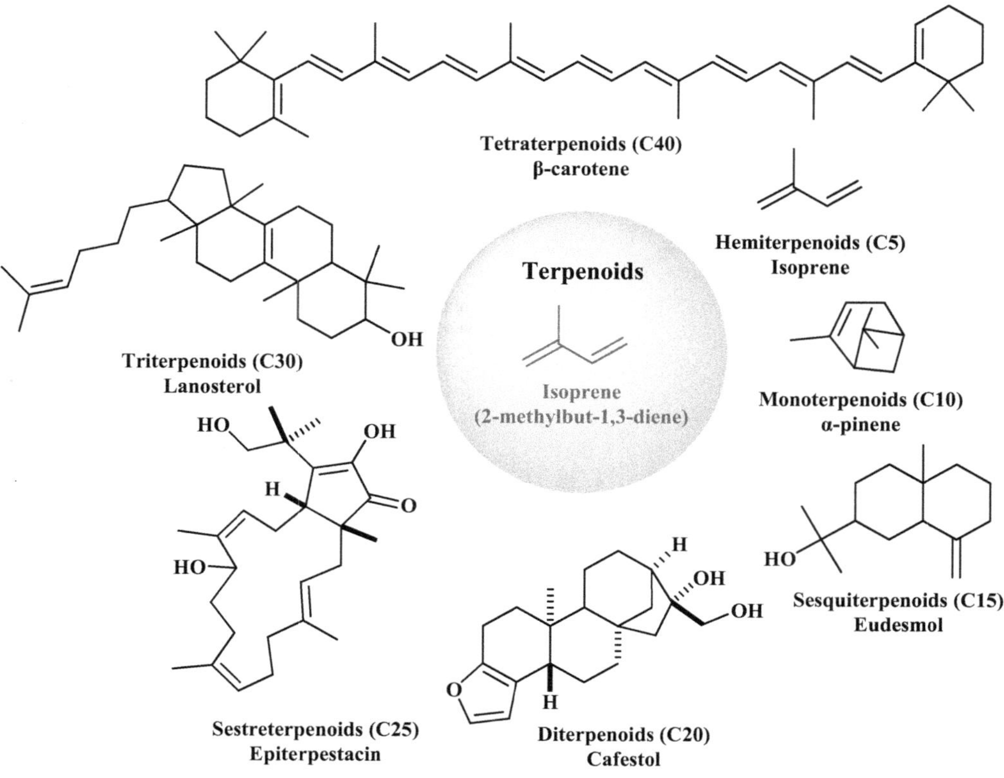

FIGURE 14.38 The molecular structure of terpenoids.

- Diterpenoids (C20) consist of four isoprene units and are derivatives of geranylgeranyl pyrophosphate. Some peculiar diterpenes are cembrene, taxadien, and cafestol. Biologically important natural compounds based on the diterpene unit include retinol, retinal, and phytol.
- Sestreterpenoids (C25) – composed of five isoprene units, which include, for example, epiterpestacin.
- Triterpenoids (C30) consist of six isoprene units, for example, lanosterol and squalene found in wheat germ and olives.
- Tetraterpenoids (C40) consist of eight isoprene units. They may be acyclic like lycopene, monocyclic like gammacarotene, and bicyclic like alpha- and beta-carotenes.[134]

Terpenes have drawn increasing commercial attention due to their wide use in the pharmaceutical, chemical, and agriculture industries. They exhibit several beneficial effects from a biological perspective, including cancer, and also have antimicrobial, antifungal, antiparasitic, antiviral, antiallergenic, antispasmodic, antihypercholesterolemic, antidiabetic, anti-inflammatory, and immunomodulatory properties. Terpenes can be used as natural insecticides and antimicrobial agents and as building blocks for the synthesis of many high-value compounds. The biotransformation of terpenes, as well as the kinetic resolution of racemic mixtures, allows the production of enantiomerically pure terpenes, flavors, and fragrances as well as therapeutic agents.[135]

EKR of terpenes most often relates to smaller natural monoterpenoid molecules. The chirality of monoterpenoids is important in fragrances and flavors because the perception of odor and flavor depends on the absolute conformation of the isomers. Carvone, for example, in the *R* configuration, has a spearmint smell and its mirror image of caraway, limonene has an orange odor if it is an (*R*)-isomer and turpentine odor if it is an isomer of the opposite configuration [135].

Rhodococcus erythropolis DCL14 cells allow the separation of the racemic mixture of limonene-1,2-epoxide, only *cis*-limonene-1,2-epoxide is selectively converted to diol by epoxide hydrolase. Reactions were carried out in two-phase systems, the conversion degree of limonene-1,2-epoxide reached 43%, the diastereomeric excess of *trans*-limonene-1,2-epoxide was greater than 99%. In manufacturing, limonene is used as a fragrance, cleaner (solvent), and as an ingredient in household cleaning products, cosmetics, and personal-hygiene products (Figure 14.39).[136]

The same cells, *R. erythropolis* DCL14, selectively in the biphasic system *n*-dodecane:water oxidize (-)-*trans*-carveol to (-)-carvone, giving a greater than 98% diastereomeric excess when the (-)-carveol conversion was 59% (Figure 14.39). *R*-(-)-Carvone is used for air-freshening products and, like many essential oils, oils containing carvones are used in aromatherapy and alternative medicine. *S*-(+)-Carvone has shown a suppressant effect against high-fat diet-induced weight gain in mice.[137,138]

Lavandulol, a constituent of essential oil extracted from plants of the genus *Lavandula*, is prized as an additive in the perfumery industry and also acts as a defensive pheromone in the red-lined carrion beetle. The asymmetric esterification of the racemic primary alcohol lavandulol was achieved using

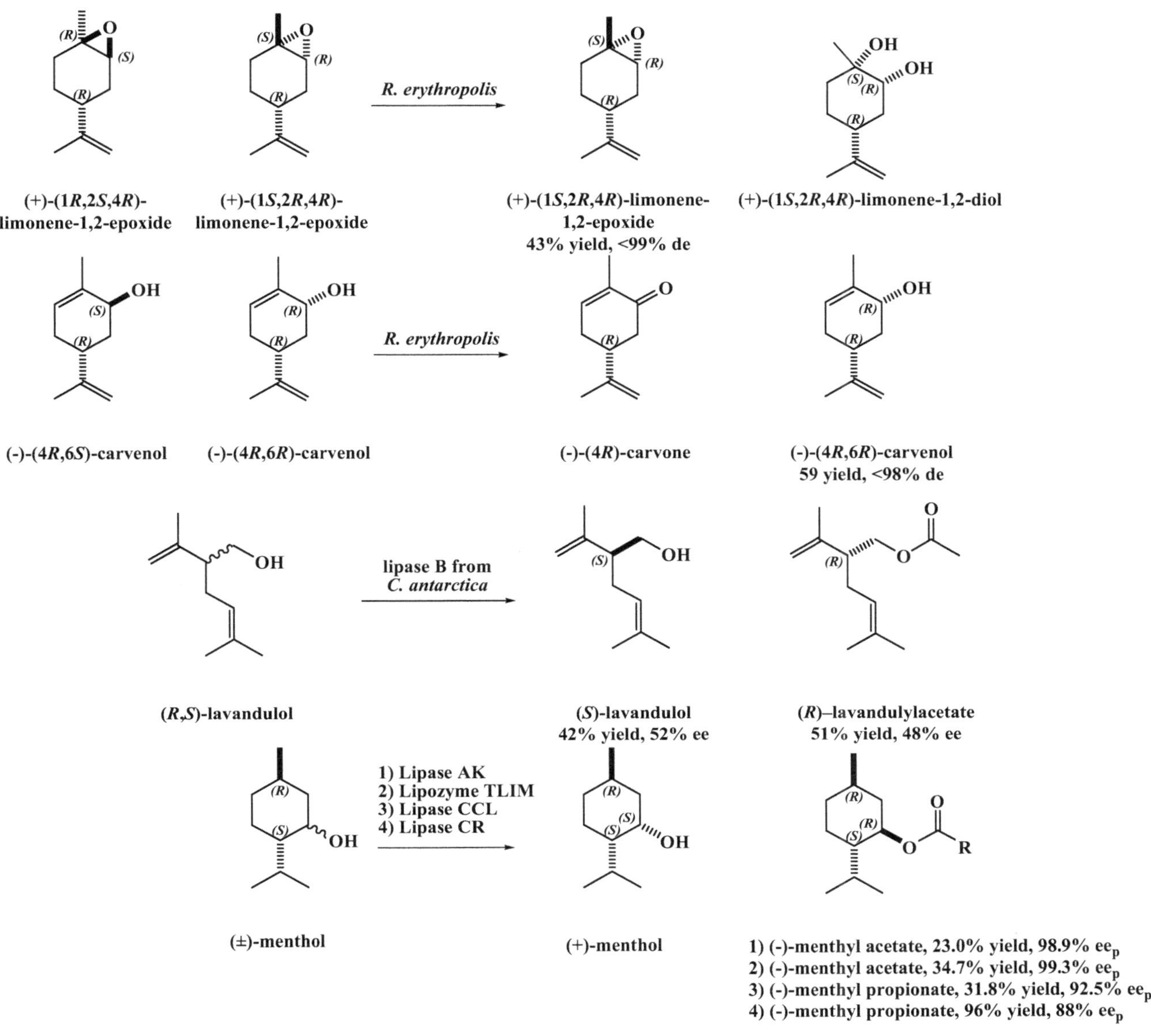

FIGURE 14.39 Kinetic resolution of monoterpenoids.

lipase B from *C. antarctica* and acetic acid as acyl donor. The bioconversion of *rac*-lavanduiol by *C. antarctica* yielded (*S*)-lavandulol with 42% yield and 52% ee and (*R*)-lavandulyl acetate with 51% yield and 48% ee (Figure 14.39).[139]

(-)-Menthol is one of the most important flavor compounds and it is used extensively as a food additive. Racemic menthol is a cheap commodity produced by Haarman and Raimer process but the desired organoleptic properties are related only to (1*R*,3*R*,4*S*) isomer. EKR has already proved to be a successful method for the resolution of *rac*-menthol, and the use of enzymes such as lipases has been extensively reported. Lipase AK exhibited high enantioselectivity (ee$_p$ 98.9%) for the resolution of racemic menthol. The best results were obtained in green solvent cyclopentyl methyl ether with vinyl acetate as the acyl donor.[140] A lipase from *Thermomyces lanuginosus* (Lipozyme TL IM) exhibited high enantioselectivity for kinetic resolution of (±)-menthol in methyl tert-butyl ether. In these conditions, (−)-menthyl acetate with 99.3% enantiomeric excess was obtained, whereas the conversion was 34.7% with the reaction time of 12 h.[141] The resolution of DL-menthol with propionic acid by *Candida cylindracea* lipase (CCL) in organic solvent reaction systems and a reverse micelles system of sodium 1,4-bis(2ethylhexyl)sulfosuccinate (AOT) were studied. The enantioselectivity (ee$_p$ = 92.5%) in the two systems was relatively high, although the conversion was moderate.[142] The resolution of (±)-menthol by the immobilized *Candida rugosa* (CR) lipase-catalyzed leads to (-)-menthyl propionate with a yield higher than 96% and over 88% enantiomeric excess (Figure 14.39).[143]

Enzymes contained in microorganisms transform natural terpenoids into new active biomolecules through the insertion of oxygen into C-H and C-C bonds, the introduction of hydroxyl groups into remote positions of terpenes, selective cleavage of the side chains of tetracyclic terpenoids, regioselective glycoside transfer, selective ring cleavage through a Baeyer–Villiger-type oxidation, carbon-skeleton rearrangements involving a methyl group migration, addition of oxygen to alkenes, hydrolysis or amide bond formation, transfer of acyl or sugar units from one substrate to another, hydrogenation, hydration, and elimination of small units, racemization reactions, epimerization, isomerization, and formation of C-C, C-O bonds, C-N, and C-S. Most often, however, during biotransformation, the process of reducing carbonyl groups and hydroxylation occurs.[144] Some selected examples of biotransformation sesquiterpenoids, diterpenoids, and tetraterpenoids are presented below.

One of the terpenoids subjected to microbial biotransformation was ambrox, a product of oxidation decomposition of ambrein sesquiterpene by the action of seawater, air, and sunlight. Ambrox fermentation from *Fusarium lini* leads to mono-, di-, and trihydroxylated metabolites (1α-hydroxyambrox, 1α,11α-dihydroxyambrox, 1α,6α-dihydroxyambrox, 1α,6α,11α-trihydroxyambrox). Enantioselective α-hydroxylation catalyzed by monooxygenases, through the transfer of one oxygen atom to the organic substrate, occurs at the C-1, C-6, and C-11 giving optical isomers with (*S*)-configuration on newly formed stereogenic centers. The incubation of ambrox with *R. stolonifer* allows obtaining hydroxyl derivatives at the C-3 or C-3 and C-6 position with *S* or *R* configuration (3β-hydroxyambrox, 3β,6β-dihydroxyambrox), the oxidation product of 3-oxoambrox as well as and its cleaved product (tetranor-12,8-diol-labdane). 3-oxoambrox, 3β-hydroxyambrox, and tetranor-12,8-diol-labdane were also obtained in microbiological transformation using the following catalysts: *Cephalosporium aphidicola* (wild-type), *A. niger* (IFO 4049), and *Aspergillus cellulose* (IFO4040). The incubation of ambrox with *C. elegans* and *Curvularialunata* resulted in the formation of two metabolites 3-oxoambrox and 3β-hydroxyambrox.[145] The spectrum of ambrox metabolites was obtained in a biotransformation reaction with cell suspension cultures of *Actinidia deliciosa* (Kiwifruit): 3-oxoambrox, 3β-hydroxyambrox, 1α-hydroxyambrox, 3β,6β-dihydroxyambrox, 1α,6β-dihydroxyambrox, and 1α,3β-dihydroxyambrox. Microorganisms enantioselectively hydroxylate the C-1 and C-3 to the (*S*)-isomer, and the C-6 positions to the isomers with opposite configuration (Figure 14.40).[146]

Sclareolide, an ambrox derivative that differs by the presence of a ketone group at the C-12 position, as a result of incubation with *C. elegans*, is transformed into new metabolites: 3-oxosclareolide, 3β-hydroxysclareolide, 2α-hydroxysclareolide, 2α,3β-dihydroxyepisclareolide, 1α,3β-dihydroxyepisclareolide, and 3β-hydroxyepisclareolide (Figure 14.41). Sclareolide exhibits good phytotoxic and cytotoxic activities against human cancer cells.[145]

Microbiological oxidation of (+)-isolongifolen-4-one, another sesquiterpenoid, using a standard two-stage fermentation technique in the presence of *A. niger* gave three metabolites: (7*R*)-12-hydroxyisolongifolen-4-one, (7*S*)-13-hydroxyisolongifolen-4-on, (11*R*)-11-hydroxyisolongifolen-4-one, while metabolites (10*R*)-10-hydroxyisolongifolen-4-one, and (9*R*)-9-hydroxyisolongifolen-4-one were formed by *F. lini* (NRRL 68751). Similarly, *C. aphidicola* (IMI 68689) provided (9*R*)-9-hydroxyisolongifolen-4-one, while *Rhizopus stolonifer* (ATCC 10404) afforded (7*R*)-12-hydroxyisolongifolen-4-one and (7*S*)-13-hydroxyisolongifolen-4-on. It was

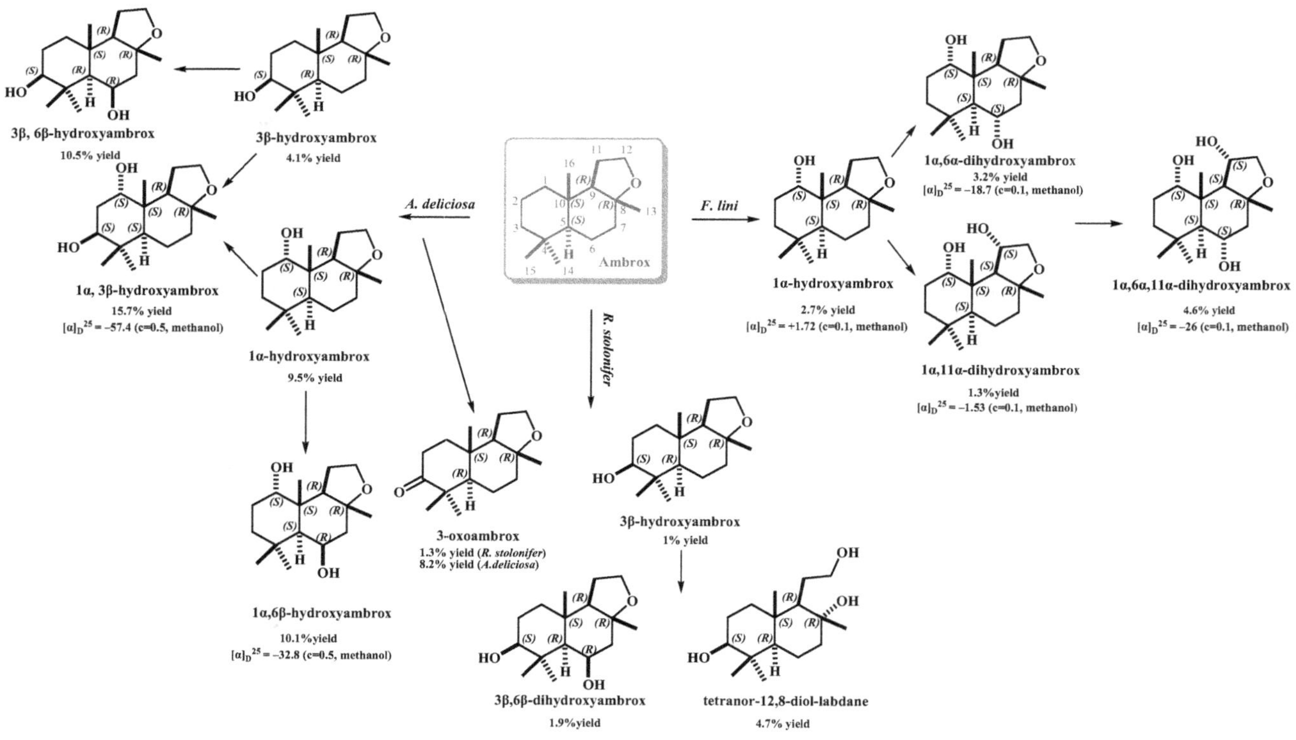

FIGURE 14.40 Biotransformation of ambrox.

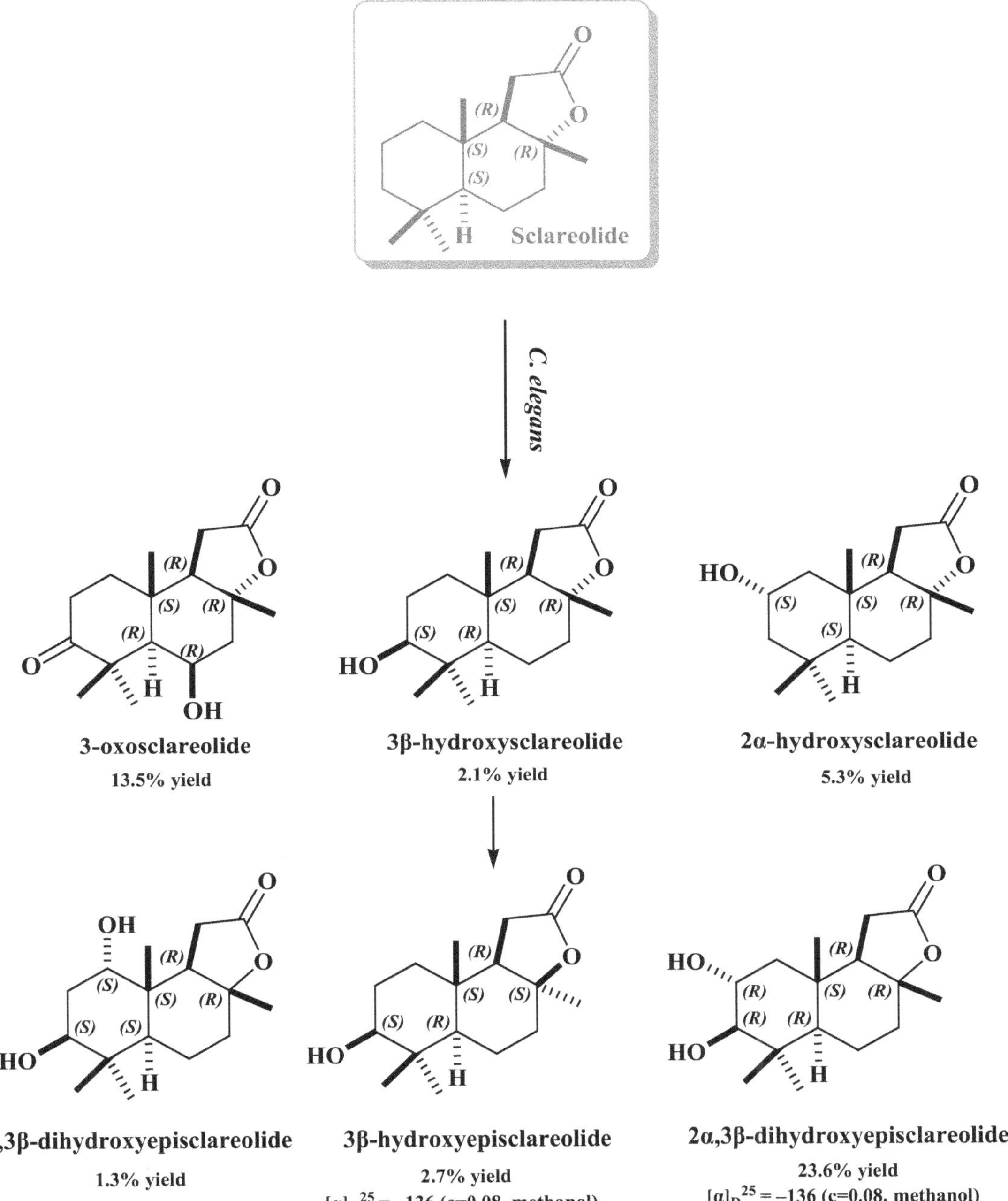

FIGURE 14.41 Biotransformation of sclareolide.

found that (7*S*)-13-hydroxyisolongifolen-4-on and (9*R*)-9-hydroxyisolongifolen-4-one showed potent tyrosinase inhibitory activity (Figure 14.42).[147]

Biotransformation of caryophyllene oxide using *Botrytis cinerea* afforded 15 different metabolites, among which we can distinguish products obtained as a result of stereoselective epoxidation at the C-8/C-13 and hydroxylation at the C-7 (Figure 14.43). The epoxidation at the double bond gives (4*R*,5*R*,8*R*)-4,5:8,13-diepoxycaryophyllane and (4*R*,5*R*,8*S*)-4,5:8,13-diepoxycaryophyllane as the main biotransformation products. The obtained epoxides could be the precursors of empirical alcohols (8α-hydroxymethyl caryophyllane oxide and 8β-hydroxymethyl caryophyllane oxide), which are formed by the reductive opening of the 8,13-epoxide. The hydroxylation process concerns the methyl group and the carbon atom at the C-7 position, enabling the following metabolites to be obtained: 15-hydroxymethylcaryophyllene oxide, 14-hydroxymethylcaryophyllene oxide, and 7ß-hydroxycaryophyllene oxide. 7ß-hydroxycaryophyllene oxide after oxidation gives (4*R*,5*R*)-4,5-epoxycaryophyll-8(13)-en-7-one.[148] The C-2 hydroxylation product can be got if the biotransformation is carried out in the presence of *Catharanthus roseus* (2β-hydroxycaryophyllene oxide and 2β-hydroxy-4,5-epoxycaryophyllan-13-ol), while *F. lini* allows obtaining the compound by

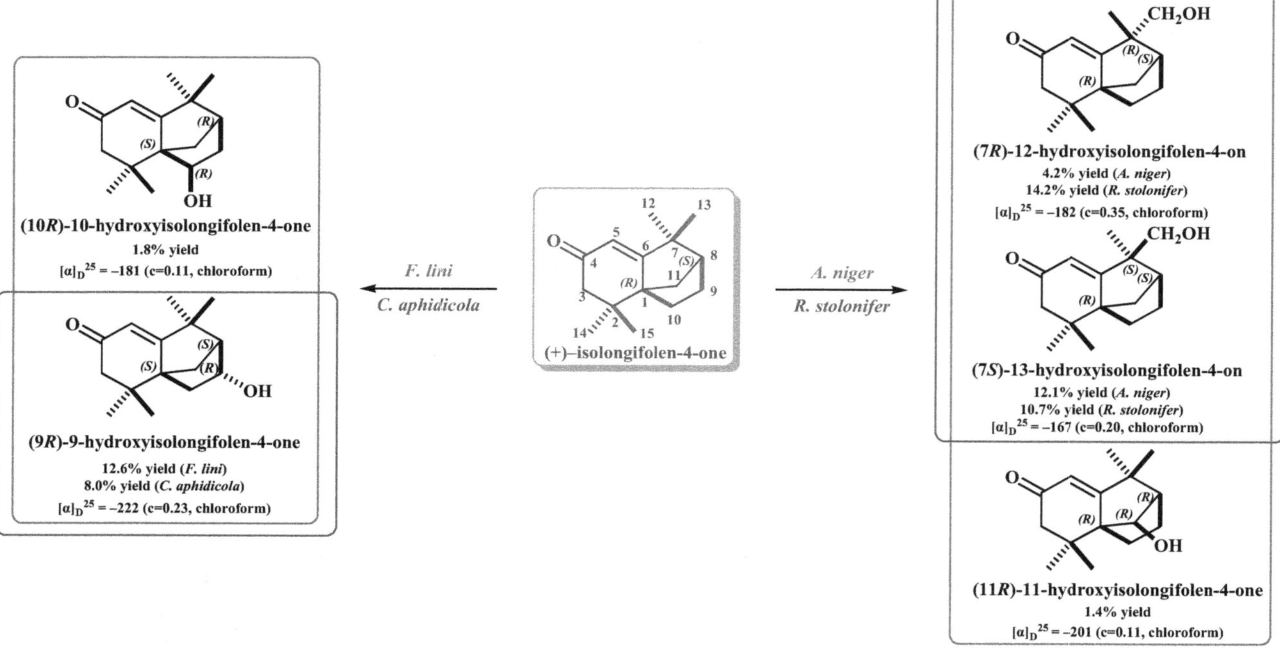

FIGURE 14.42 Biotransformation of (+)-isolongifolen-4-one.

C-3 hydroxylation and anti-Markonikov hydration of the C-8/C-13 bond (3β,8α-dihydroxymethylcaryophyllan oxide).[149,150] *C. roseus* and *C. aphidicola* leads to the epoxide hydrolysis product by the enzyme epoxide hydrolase – (1*R*,4*R*,5*R*,9*S*)-4,5-dihydroxycaryophyllan-8(13)-ene. In addition to this compound, (1*S*,4*R*,5*R*,8*S*,9*S*)-clovane-5,9-diol is also formed in biotransformation by *C. aphidicola*. The incubation with *Gibberella fujikuroi* and *A. niger* yielded metabolites clovane-5α,9β-12-triol and caryolane-5α,6β,13β-triol.[149,150]

Caryophyllene oxide was reported to have strong anti-inflammatory activity. It also exhibits potent antimutagenic properties.

Ent-pimara-7,15-dienes belong to diterpenes subjected to microbiological biotransformation, they undergo the epoxidation of the 7,8-double bond, followed by rearrangement to afford 7-oxo derivatives and allylic hydroxylation at the C-6, C-9, or C-14. Incubation of 18-hydroxy-9-epi-*ent*-pimara-7,15-diene with the fungus *G. fujikuroi* gave the compounds 18-hydroxy-7α,8α-epoxy-9-epi-*ent*-pimara-15-ene, 18-hydroxy-7-oxo-*ent*-pimara-15-ene,6β,18-dihydroxy-7α,8α-epoxy-9-epi-*ent*-pimara-15-ene, 9β,18-dihydroxy-7α,8α-epoxy-*ent*-pimara-15-ene, and 6β,14α,18-trihydroxy-9-epi-*ent*-pimara-7,15-diene (Figure 14.44).[151,152]

The incubation of 2β-hydroxy-*ent*-13-epi-manoyl oxide with *G. fujikuroi* gives in good yield 2β,6β-dihydroxy-*ent*-13-epi-manoyl oxide, 2β,12β-dihydroxy-*ent*-13-epi-manoyl oxide, and 2β,20-dihydroxy-*ent*-13-epi-manoyl oxide, confirming that although *ent*-13-epi-manoyl oxide is a final metabolite of a biosynthetic branch in this fungus (Figure 14.45). The products of this biotransformation are mainly the result of hydroxylation at the C-6, C-12, or C-20.[153]

The biotransformation of 13-epi-*ent*-pimara-9(11),15-diene-19-oic acid by *G. fujikuroi* leads to the oxidation of carbon at the C-1 to the carbonyl bond, followed by hydroxylation at the C-2 or C-12 (1-oxo-2α,9α-dihydroxy-13-epi-*ent*-pimara-11,15-dien-19-oic acid, 1-oxo-2β,9α-dihydroxy-13-epi-*ent*-pimara-11,15-dien-19-oic acid, and 1-oxo-12β-hydroxy-13-epi-*ent*-pimara-9(11),15-dien-19-oic acid), and the product formed by Baeyer–Villiger oxidation to afford the lactone (13-epi-*ent*-pimara-9(11),15-dien-1,19-dioic acid 1,2-lactone) (Figure 14.46). Baeyer–Villiger monooxygenases (BVMOs) catalyzing the lactone formation reaction is a flavoprotein which employs atmospheric oxygen as the natural oxidant.[154]

Glomerellacingulata reduces the carboxyl group of *ent*-pimara-8(14),15-dien-19-oic acid to the corresponding primary alcohol, while *Mucor rouxii* causes isomerization of the endocyclic double bond and oxidation at the C-7 (Figure 14.47).[155]

Dehydroabietic acid (DHA) is one of the major tricyclic diterpenoid constituents of pine resin, exhibits a broad spectrum of biological action. Several activities like antiulcer, antimicrobial, anxiolytic, antiviral, antitumor, anti-inflammatory, and cytotoxic have been reported.[152] This compound, subjected to microbial biototransformation, is mainly oxidized to the carbonyl group and alcohol by hydroxylation at the C-1, C-2, C-7, C-16 or C-17 positions (Figure 14.48). *C. roseus* transforms the substrate into 16-hydroxydehydroabietic acid, follows the glycosylation by putative glycosyltransferase into dehydroabietic 17-*O*-glucoside.[156] Stereoselective hydroxylation at the C-1 position is possible if two fungi are applied: *Trametes versicolor* and *Phlebiopsisgigantea*.[157] This hydroxylation at the C-1 position, has only been previously reported in cultures of two different *Fusarium* species and *A. niger*.[158,159] The C-7 and/or C-16 hydroxylations produced in the incubation of dehydroabietic acid with *T. versicolor* and *P. gigantea*. After 13 days, four compounds

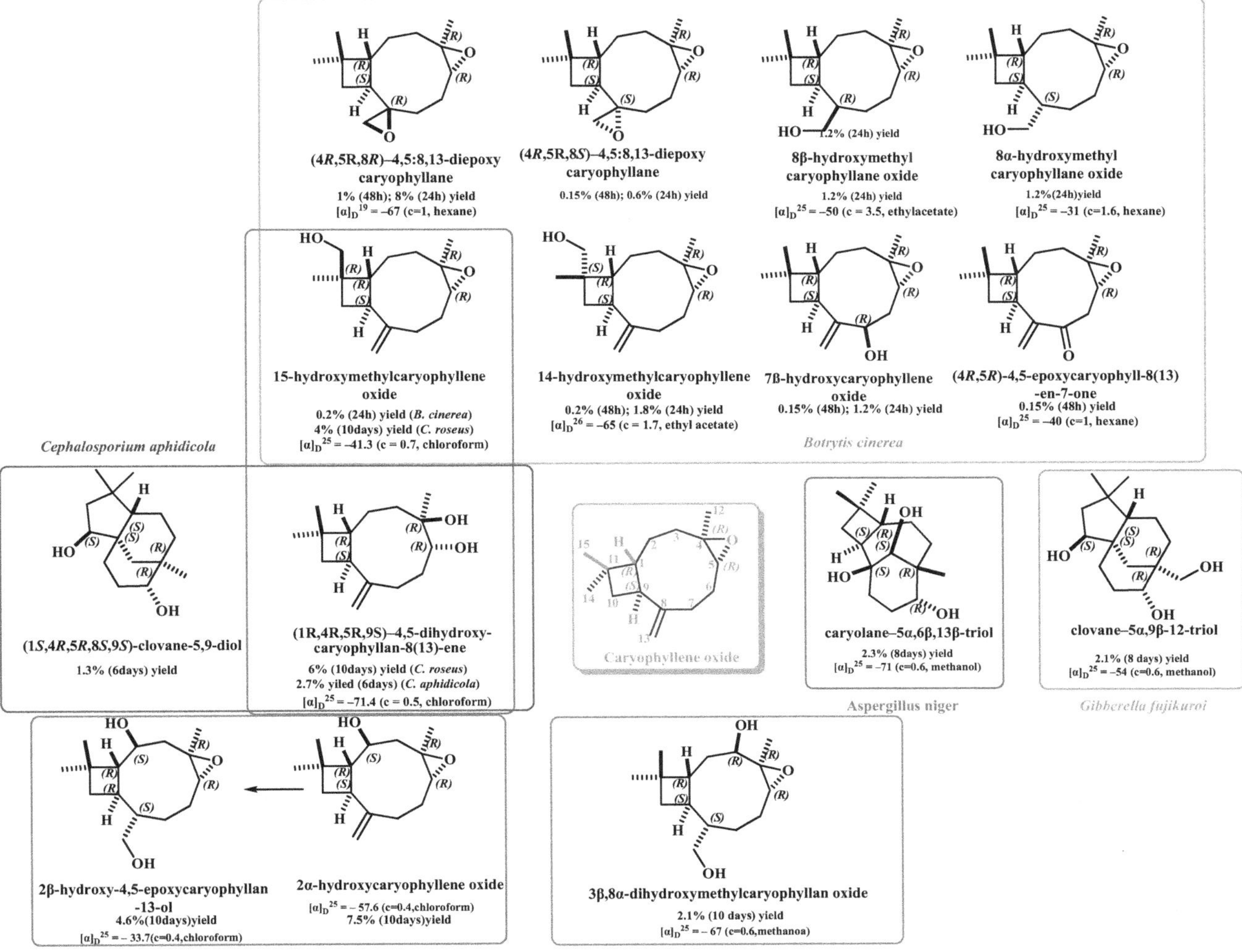

FIGURE 14.43 Biotransformation of caryophyllene oxide.

were identified in *P. gigantea* cultures: 1β-hydroxy-DHA, 1β,7α-dihydroxy-DHA, 1β,16-dihydroxy-DHA, and tentatively 1β-hydroxy-7-oxo-DHA. In *T. Versicolor* cultures, 1β,16-dihydroxy-DHA, 7β,16-dihydroxy-DHA, 1β,7β,16trihydroxy-DHA, 1β,16-dihydroxy-7-oxo-DHA, 1β,15-dihydroxy-DHA, and 1β,7α,16-trihydroxy-DHA were identified after 9 days of incubation.[157] Hydroxylation at C-7 has been reported in studies with the aerobic bacteria *Alcaligenes sp.*, *Pseudomonas sp.* and *Pseudomonas abietaniphila*.[160,161] Oxidation of the hydroxyl group at the C-7 and/or C-3 position to the carbonyl group is observed in the presence of aerobic bacteria *Flavobacterium resinovorum* and *Moraxella sp.* (HR6). Whereas *Mucor circinelloides* and *Mortierella isabellina* regio- and stereoselectively convert the substrate to 2α-hydroxydehydroabietic acid.[162]

Kaurene diterpenoids are a very important class of natural products that are widespread in the plant kingdom. Due to their construction, Kaurenes belong to tertracyclic diterpenes, they consist of a perhydrophenanthrene unit (A, B, and C rings) fused with a cyclopentane unit (D ring) formed by a bridge of two carbons between C-8 and C-13 (Figure 14.49). *ent*-Kaurene and many of their natural derivatives have important biological activities such as antimicrobial, antiparasitic, insect antifeedant, cytotoxic, antifertility, hypotensive, anti-HIV have been associated with them.[163]

Biotransformation of *ent*-kaur-16-ene derivatives (candidiol (15α,18-dihydroxy-*ent*-kaur-16-ene), 15α,19-dihydroxy-*ent*-kaur-16-ene, *ent*-7α,15β-dihydroxy-kaur-16-ene, candicandiol (7α,18-dihydroxy-*ent*-kaur-16-ene) and epicandicandiol (7β,18-dihydroxy-*ent*-kaur-16-ene) allows obtaining a number of hydroxylated derivatives. For example, incubation of *ent*-7α,15β-dihydroxy-kaur-16-ene with *G. fujikuroi*yielded *ent*-7α,15β,19-trihydroxy-kaur-16-ene, *ent*-7α,11α,15β-trihydroxy-kaur-16-ene,*ent*-7α,11α,15β,16β,17-pentahydroxy-kaurane,and *ent*-7α,11α,13,15β-tetrahydroxy-kaur-16-ene. While its derivative eubotriol (*ent*-7α,15β,18-trihydroxy-kaur-l6-ene) in the presence of the same microorganism gives *ent*-7α,11α,15β,18-tetrahydroxy-kaur-16-ene, *ent*-7α,15β,18,19-tetrahydroxy-kaur-16-ene, *ent*-7α,17,18-trihydroxy-15β,16β-epoxykaurane, and *ent*-7β,17α,18α,19β-tetrahydroxy-kaur-15-ene. *G. fujikuroi* selectively introduced hydroxyl groups at positions C-11, C-17, C-18 and C-19. *Ent*-7α,17,18-trihydroxy-15β,16β-epoxykurane

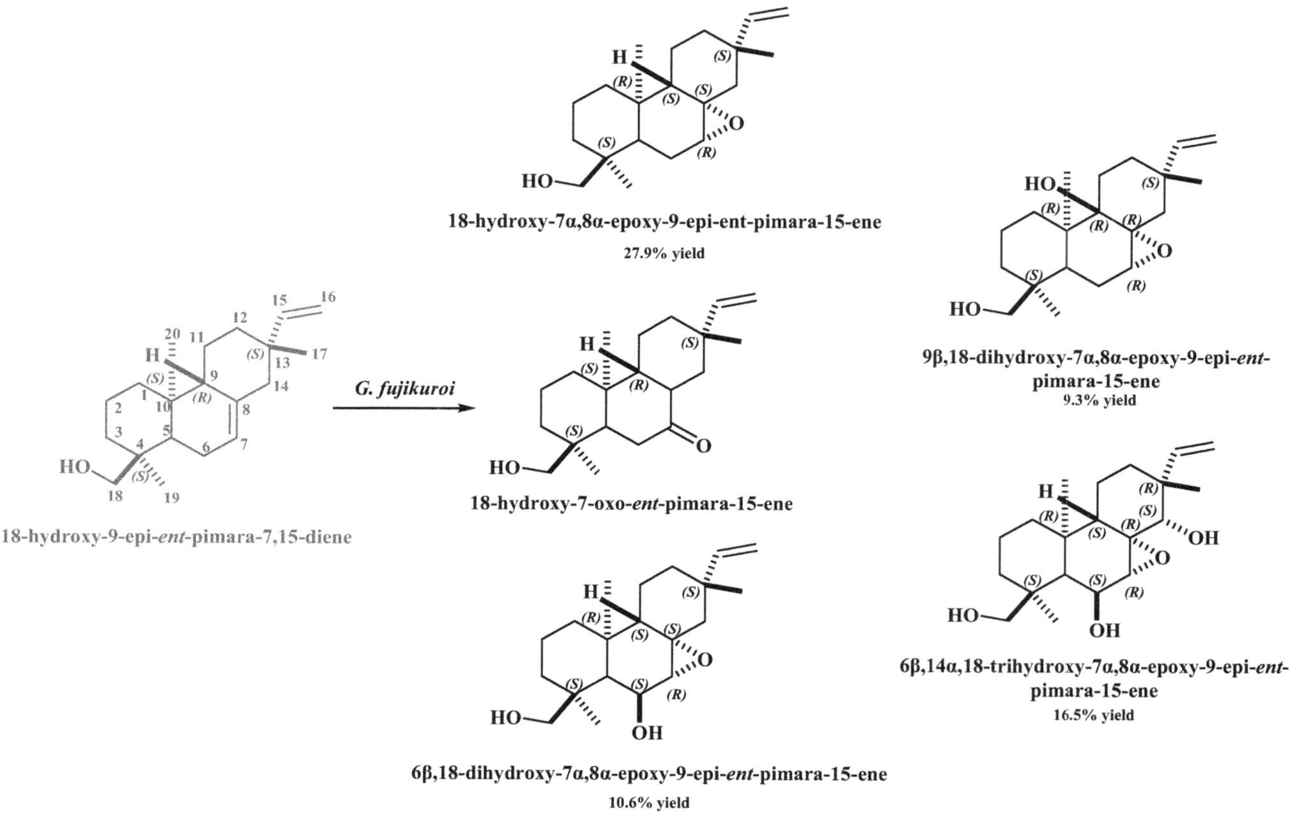

18-hydroxy-7α,8α-epoxy-9-epi-ent-pimara-15-ene
27.9% yield

9β,18-dihydroxy-7α,8α-epoxy-9-epi-*ent*-pimara-15-ene
9.3% yield

G. fujikuroi

18-hydroxy-9-epi-*ent*-pimara-7,15-diene

18-hydroxy-7-oxo-*ent*-pimara-15-ene

6β,14α,18-trihydroxy-7α,8α-epoxy-9-epi-*ent*-pimara-15-ene
16.5% yield

6β,18-dihydroxy-7α,8α-epoxy-9-epi-*ent*-pimara-15-ene
10.6% yield

FIGURE 14.44 Biotransformation of 18-hydroxy-9-epi-*ent*-pimara-7,15-diene.

was obtained by enzymatic epoxidation of the exocyclic double bond to give the corresponding epoxides, followed by opening of these in the medium (Figure 14.49).[164]

For the formation of ent-7α,17,18,19-tetrahydroxy-kaur-15-ene a mechanism was proposed in which took place enzymatic abstraction of hydrogen at C-15 in with the formation of a carbonium ion, followed by migration of the double bond to the 15,16-position and neutralization of the cation at C-17 by a hydroxyl group.[165]

The microbiological transformation of candidiol (15α,18-dihydroxy-*ent*-kaur-16-en) and 15α,19-dihydroxy-*ent*-kaur-16-en by *Mucor plumbeus* yielded the same hydroxylation products at C-3α, C-6α or C-11β and exocyclic double bond epoxidation product except for 9β,15α,19-trihydroxy-ent-kaur-16-ene, which is a hydroxylation product of only 15α,19-dihydroxy-*ent*-kaur-16-en. This means that a change in the spatial orientation of the hydroxymethylene group at C-4, from equatorial in 15α,18-dihydroxy-*ent*-kaur-16-ene to axial in 15α,19-dihydroxy-*ent*-kaur-16-ene, does not affect the way in which these *ent*-kaurenes bind to the oxidative enzymes.[166]

However, the difference in the spatial orientation of the hydroxyl group at the C-7 of two *ent*-kaurene derivatives: candicandiol (7α,18-dihydroxy-*ent*-kaur-16-ene) and epicandicandiol (7β,18-dihydroxy-*ent*-kaur-16-ene) affect the binding of these compounds at the active center of oxidative enzymes affording a different hydroxylation pattern in the A and B rings (Figure 14.50). The following derivatives are formed in *M. plumbeus*-catalyzed biotransformation: 9β and3α hydroxylated derivative for candicandiol and epicandicandiol, respectively. In addition to these products, 16,17-dihydroxylated compounds are also obtained, which are formed by epoxidation of the exocyclic double bond followed by epoxy ring-opening.[164]

Kauren-19-oic acid is one of the intermediate compounds involved in the biosynthesis of diverse kaurene diterpenes, including gibberellins, which represent an important group of growth phytohormones. Kauren-19-oic acid is a compound having diverse biological functions such as antihepatotoxic activity, antimicrobial activity, anti-inflammatory activities, larval growth inhibition.[152,163] (-)-Kaur-16-en-19-oic acid with *F. oxysporum* leads to two metabolites: 16α-hydroxy kauran-19-oic acid and 2β,16α-dihydroxy kauran-19-oic acid.[167] *Ent*-kaur-16-en-19-oic acid with *R. stolonifer* after 7 days of incubation was transformed into *ent*-7β-hydroxy-kaur-16-en-19-oic acid and *ent*-12α-hydroxy-kaur-9(11),16-dien-19-oic acid, after 15 days, *ent*-16α,17-dihydroxy-kauran-19-oic acid was also obtained (Figure 14.51).[168]

Triterpenoids are another large group of naturally occurring molecules that are readily subjected to enzymatic modification to obtain new biologically active compounds. Triterpenoids exhibit low toxicity and were demonstrated to have a variety of biological activities, such as anti-HIV, antimalarial, cytotoxicity to tumor cell lines, anti-inflammatory, and antiatherosgenic

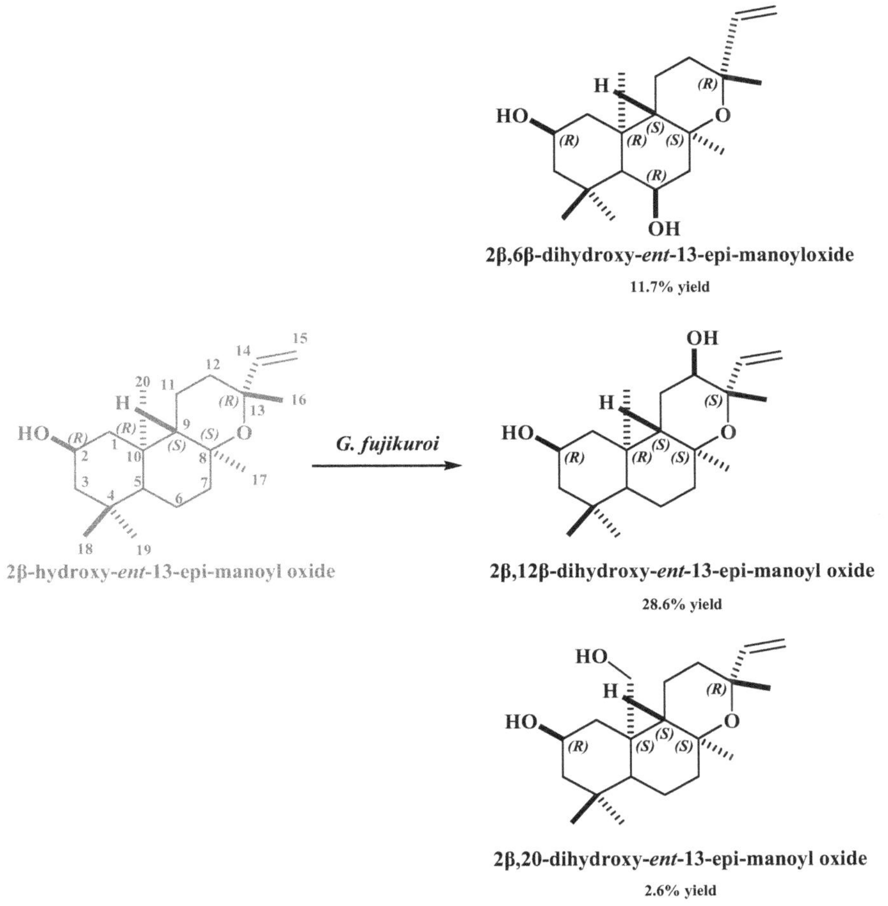

2β,6β-dihydroxy-*ent*-13-epi-manoyloxide
11.7% yield

2β-hydroxy-*ent*-13-epi-manoyl oxide

G. fujikuroi

2β,12β-dihydroxy-*ent*-13-epi-manoyl oxide
28.6% yield

2β,20-dihydroxy-*ent*-13-epi-manoyl oxide
2.6% yield

FIGURE 14.45 Biotransformation of 2β-hydroxy-*ent*-13-epi-manoyl oxide.

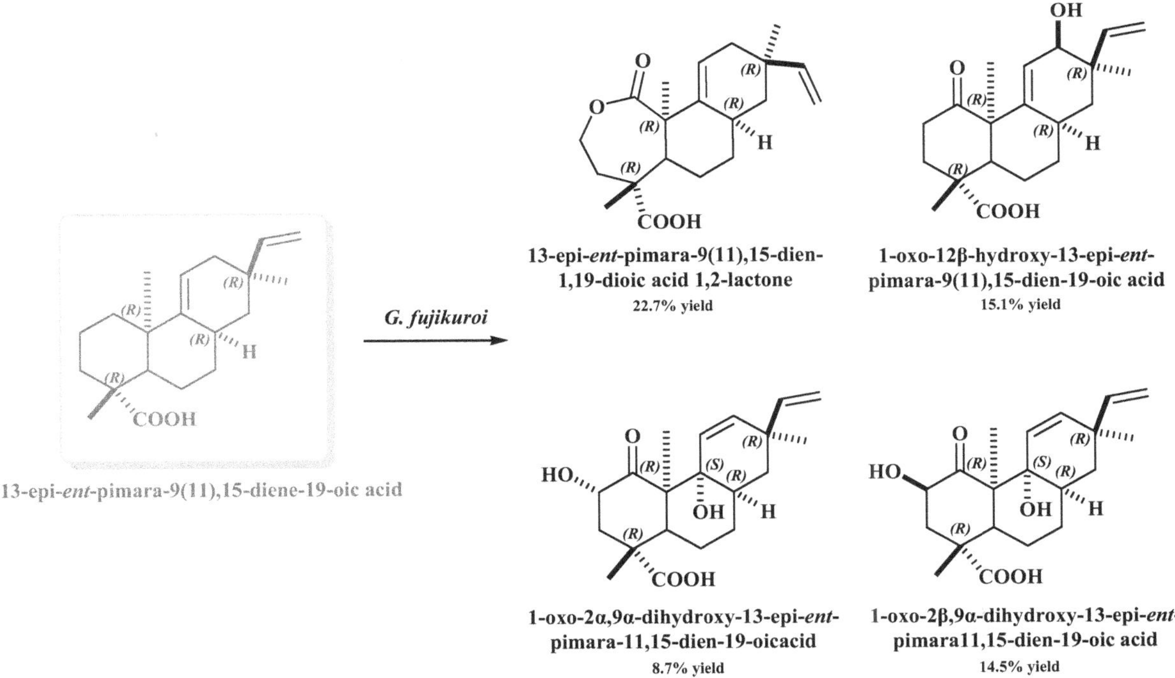

FIGURE 14.46 Biotransformation of 13-epi-*ent*-pimara-9(11),15-diene-19-oic acid.

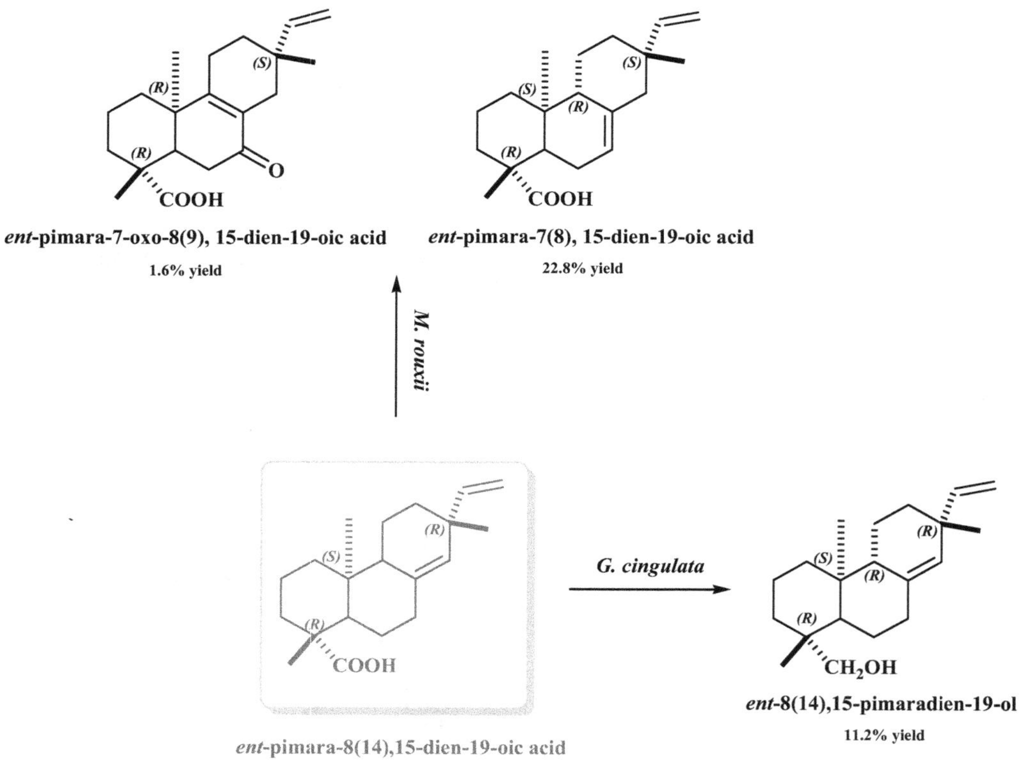

FIGURE 14.47 Biotransformation of *ent*-pimara-8(14),15-dien-19-oic.

properties, and antihyperlipidemic activities have a remarkable activity against LDL oxidation. Microbial cell cultures are capable of performing specific chemical transformations in triterpenoids, such as rearrangement, hydroxylation, oxidation, reduction, hydrolysis, epimerization, and isomerization, with high regio- and stereoselectivity.[163,169]

Cycloastragenol ((20*R*,24*S*)-3β,6α,16β,25-tetrahydroxy-20,24-epoxycycloartan), belonging to the tetracyclic triterpenoid, is a sapogenin of astragaloside IV, ta major bioactive constituent of *Astragalus* plants. Astragaloside IV exhibits various pharmacological properties such as anti-inflammatory, antiviral, antiaging, and antioxidant. Cycloastragenol has been considered as a new generation antiaging agent, it can retard the onset of cellular aging by progressing telomerase activity. In addition, it regulates the immune system by inducing the release of IL-2 and elevates the antiviral function of human CD8+ T lymphocytes.[169]

The microbial transformation of cycloastragenol by the fungi *Cunninghamella blakesleeana* NRRL 1369 and *Glomerellafusarioides* ATCC 9552 provided hydroxylated metabolites together with products formed by cyclization, dehydrogenation, and Baeyer–Villiger oxidation resulting in a ring cleavage. The bacteria *Mycobacterium* sp. NRRL 3805 yielded a single oxidation product, namely, 3-oxo-cycloastragenol. *C. blakesleeana* efficiently transformed cycloastragenol into a complicated rearrangement product, i.e., ring cleavage and methyl group migration, (20*R*,24*S*)-3β,6α,6β,19,25-pentahydroxy-ranunculan-9(10)-ene.[170] (20*R*,24*S*)-epoxy-1α,3β,6α,16β,25-pentahydroxycycloartane, (20*R*,24*S*)-epoxy-3β,6α,11β,16β,25-pentahydroxycycloartane, (20*R*,24*S*)-epoxy-3β,6α,12β,16β,25-pentahydroxycycloartane, and (20*R*,24*S*)-16β,24;20,24-diepoxy-3β,6α,12β,25-tetrahydroxycycloartane were monohydroxylated products of cycloastragenol having hydroxyl groups at C-1, C-11, and C-12 positions. (20*R*,24*S*)-epoxy-3β,6α,16β,25-tetrahydroxycycloarta-11(12)-ene was the microbial Δ^{11}-dehydrogenation product derived from *C. blakesleeana*. (20*R*,24*S*)-16β,24;20,24-diepoxy-3β,6α,12β,25-tetrahydroxycycloartane was a result of the cyclization reaction of cycloastragenol, a rare feature of cycloartanes, whereas 3,4-seco cycloastragenol was the molecule resulting from the cleavage of ring A. This compound is formed next to (20*R*,24*S*)-epoxy-3β,6α,11β,16β,25-pentahydroxycycloartane by the incubation with *G. fusarioides* ATCC 9552 (Figure 14.52).[171]

The biocatalysis of cycloastragenol by cultured whole cells of three strains of filamentous fungi, namely *C. elegans* AS 3.1207, *Syncephalastrum racemosum* AS 3.264 and *Doratomycesstemonitis* AS 3.1411 produced 15 metabolites.[172] *C. elegans* preferred to catalyze hydroxylation reactions, particularly on 28- and 29-CH_3 ((20*R*,24*S*)-3β,6α,16β,25,28-pentahydroxy-20,24-epoxy-cycloartane,(20*R*,24*S*)-3β,6α,16α,25,29-pentahydroxy-20,24-epoxy-cycloartane). Hydroxylation on methylene was also observed as minor reactions (C-2 for (20*R*,24*S*)-2α,3β,6α,16β,25-pentahydroxy-20,24-epoxy-cycloartane; C-12 for (20*R*,24*S*)-3β,6α,12α,16β,25-pentahydroxy-20,24-epoxy-cycloartane), which showed α-stereo selectivity. In addition, *C. elegans* could also catalyze carbonylation ((20*R*,24*S*)-3β,6α,16β,25-

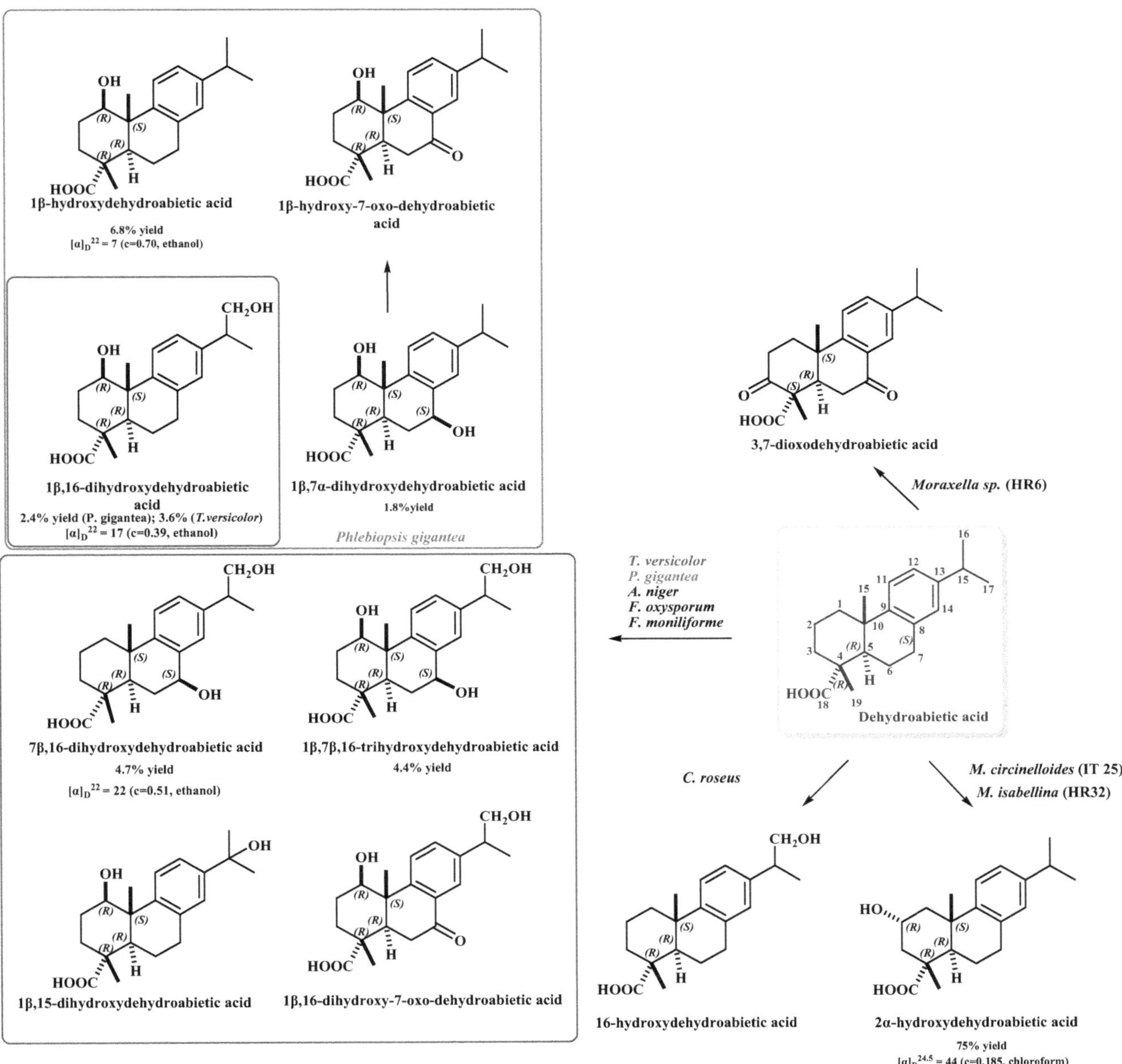

FIGURE 14.48 Biotransformation of dehydroabietic acid.

tetrahydroxy-20,24-epoxy-cycloartan-28-aldehyde), rearrangement ((20*R*,24*S*)-3β,6α,16β,19,25-pentahydroxy-ranunculan-9(10)-ene), and the rare 16,24-cyclization reaction ((20*R*,24*R*)-3β,6α,25,28-tetrahydroxy-16β,24:20,24-diepoxy-cycloartan). *S. racemosum* displayed a strong capacity to catalyze the complicated rearrangement reaction to produce the unusual ranunculane skeleton. Moreover, *n*-butylated, isopentenylated and acetylated derivatives of ((20*R*,24*S*)-3β,6α,16β,19,25-pentahydroxy-ranunculan-9(10)-ene were also produced. *S. racemosum* could also catalyze hydroxylation ((20*R*,24*S*)-3β,6α,12α,16β,25-pentahydroxy-20,24-epoxy-cycloartane, carbonylation ((20*R*,24*S*)-3β,6α,16β,25-tetrahydroxy-20,24-epoxy-cycloartan-11-one, dehydrogenation ((20*R*,24*S*)-3β,6α,16β,25-tetrahydroxy-20,24-epoxy-cycloartan-11(12)-ene, reactions and realized the ring expansion reaction to produce the unusual 9(10)α-homo-19-nor-cycloartane triterpene skeleton (neoastragenol (20*R*,24*S*)-3β,6α,16β,25-tetrahydroxy-20,24-epoxy-9(10)a-homo-19-nor-cycloartane. *D. stemonitis* AS 3.1411 transformed cycloastragenol to two carbonylated metabolites, (20*R*,24*S*)-6α,16β,25-trihydroxy-20,24-epoxy-cycloartan-3-one,(20*R*,24*S*)-6α,16β,25,30-tetrahydroxy-20,24-epoxy-cycloartan-3-one and (20*R*,24*S*)-3β,6α,6β,19,25-pentahydroxy-ranunculan-9(10)-ene (Figure 14.53).[172]

Pentacyclic triterpenoids are widely distributed in many medicinal plants, such as birch bark (betulin), plane bark (betulinic acid), olive leaves, olive pomace, mistletoe sprouts

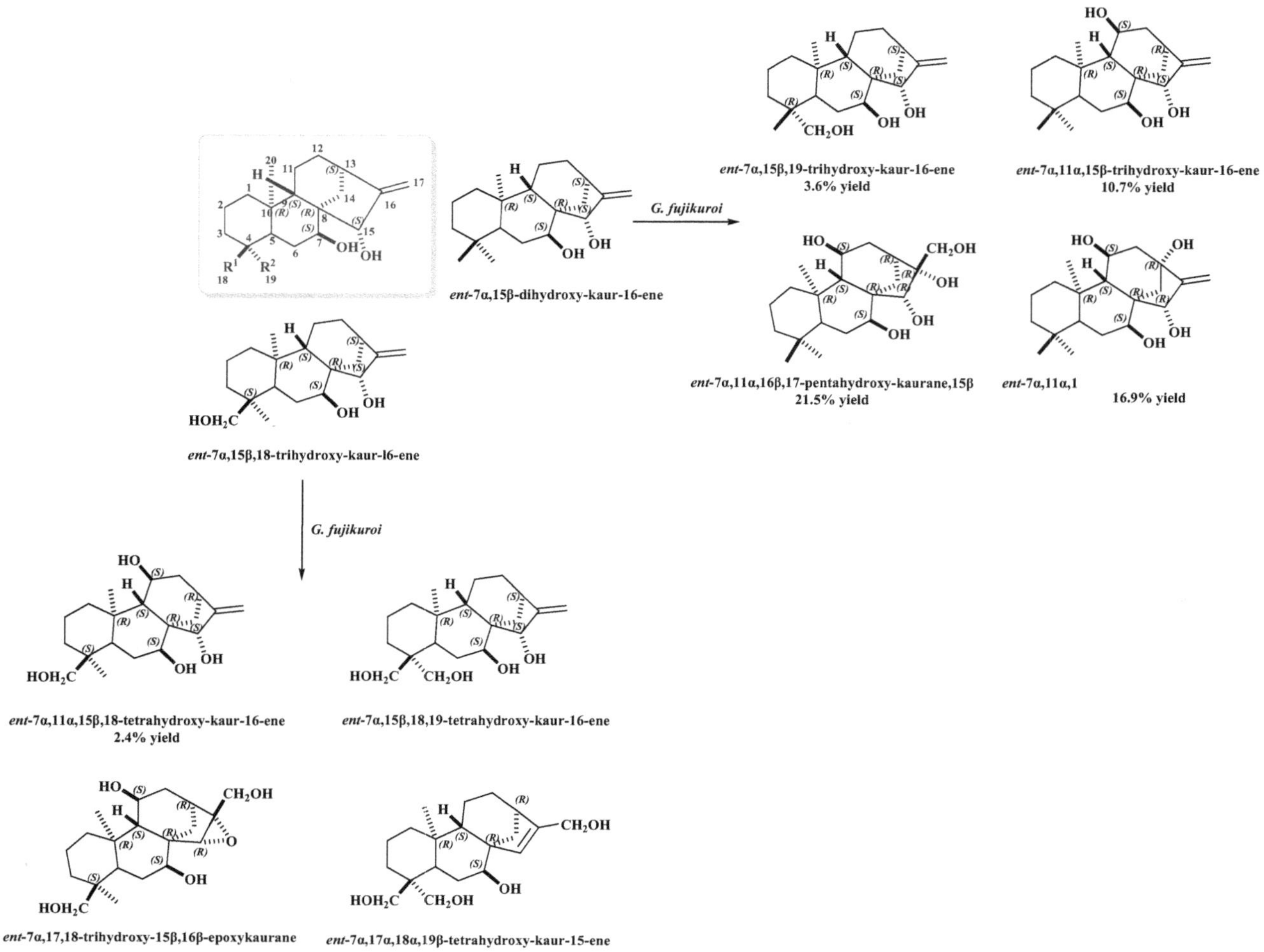

FIGURE 14.49 Biotransformation of *ent*-kaur-16-ene derivatives.

and clove flowers (oleanolic acid), and apple pomace (ursolic acids).[169]

One of the triterpenoids belonging to the lupane-type pentacyclic triterpene is betulinic acid (3β-hydroxy-lup-20(29)-en-28-oic acid). As a result of biotransformation, betulinic acid is bioconverted to the more polar metabolite 28-*O*-β-D-glucopyranosyl 3β-hydroxy-lup-20(29)-en-28-oate in the presence of *Cunninghamella* species NRRL 5695 and hydroxylated derivatives at the C-7 or/and C-1: 3β,7β-dihydroxylup-20(29)-en-28-oic acid and 1β,3β,7β-trihydroxy-lup-20(29)-en-28-oic acid, in the presence of *Mucor mucedo* UI- 4605 and *C. elegans* ATCC 9244.[173,174] In contrast, the fungus *Colletotrichum* sp. selectively oxidizes the hydroxyl group in the C-3 position and hydroxylation in the C-15 position leads to 3-oxo-15α-hydroxylup-20(29)-en-28-oic acid.[175] *B. megaterium* ATCC 13368 resulted in the production of four metabolites, which were identified as 3-oxo-lup-20(29)-en-28-oic acid, 3-oxo-11α-hydroxy-lup-20(29)-en-28-oic acid, 1β-hydroxy-3-oxo-lup-20(29)-en-28-oic acid, and 3β,7β,15α-trihydroxy-lup-20(29)-en-28-oic acid.[176]

Betulin, a derivative of betulinic acid, extracted from white birch (*Betula platyphylla* Sukastshev var. *Japonica*) is transformed in the fermentation process with *Chaetomium longirostre* into 4,28-dihydroxy-3,4-seco-lup-20(29)-en-3-oic acid (Figure 14.54).[177]

The triterpenoid ursolic acid, from the medicinal plant *Morinda lucida*, has been shown to be an antimalarial inhibitor of *Papilloma virus*. The incubation of ursolic acid from the soil fungus *Umbelopsis* is abelline gives three metabolites, which are different hydroxylated lactones (3β-hydroxy-urs-11-eno-28,13-lactone, 3β,7β-dihydroxy-urs-11-eno-28,13-lactone, and 1β,3β-dihydroxy-urs-11-eno-28,13-lactone. The endophytic fungus *U. isabellina* can hydroxyate the C12–C13 double bond at position 13 of ursolic acid and form a five-member lactone effectively. In the meantime, this fungus can also introduce the hydroxyl group at C-1 or C-7 of ursolicacid.[178] Microbial transformation of ursolic acid by filamentous fungus *S. racemosum* CGMCC 3.2500 leads to hydroxylated derivatives at the C-1, C-7, and C-21, lactone and oxidation product of the C-21 hydroxyl group (3β,7β,21β-trihydroxy-urs-12-en-28-oic acid, 3β,21β-dihydroxy-urs-11-en-28-oic acid-13-lactone, 1β,3β,21β-trihydroxy-urs-12-en-28-oic acid, 3β,7β,21β-trihydroxy-urs-1-en-28-oic acid-13-lactone, and 21-oxo-1β,3β-dihydroxy-urs-12-en-28-oic acid).[179]

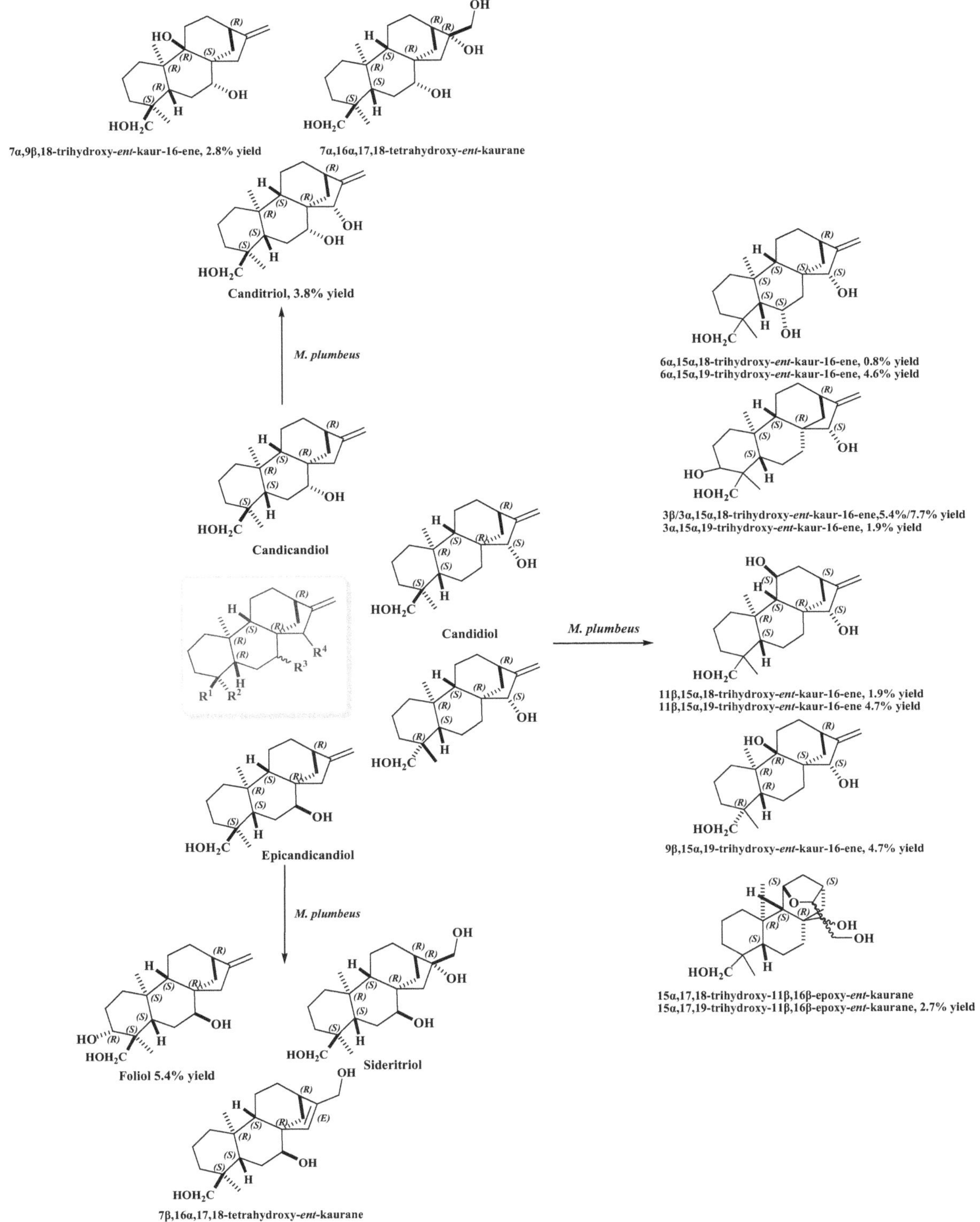

FIGURE 14.50 Biotransformation of candidiol, candicandiol and epicandicandiol.

The bioconversion of ursolic acid by *B. megaterium* CGMCC 1.1741 yielded five metabolites: 3-oxo-urs-12-en-28-oic acid, 1β,11α-dihydroxy-3-oxo-urs-12-en-28-*O*-β-D-glucopyranoside, 1β,11α-dihydroxy-3-oxo-urs-12-en-28-oic acid, 1β-hydroxy-3-oxo-urs-12-en-28,13-lactone, 1β,3β,11α-trihydroxyurs-12-en-28-oic acid. *B. megaterium CGMCC 1.1741* enables oxidation of the hydroxyl group in C-3, hydroxylation of carbon atoms at C-1, C-11, lactonization, and glycosylation (Figure 14.55).[180]

Incubation of 3-oxo oleanolic acid with *Streptomyces griseus* ATCC 13273produced two polar metabolites, 3-oxo-olean-12-en-28,29-dioic acid and 24-hydroxy-3-oxo-olean-12-en-28,29-dioic

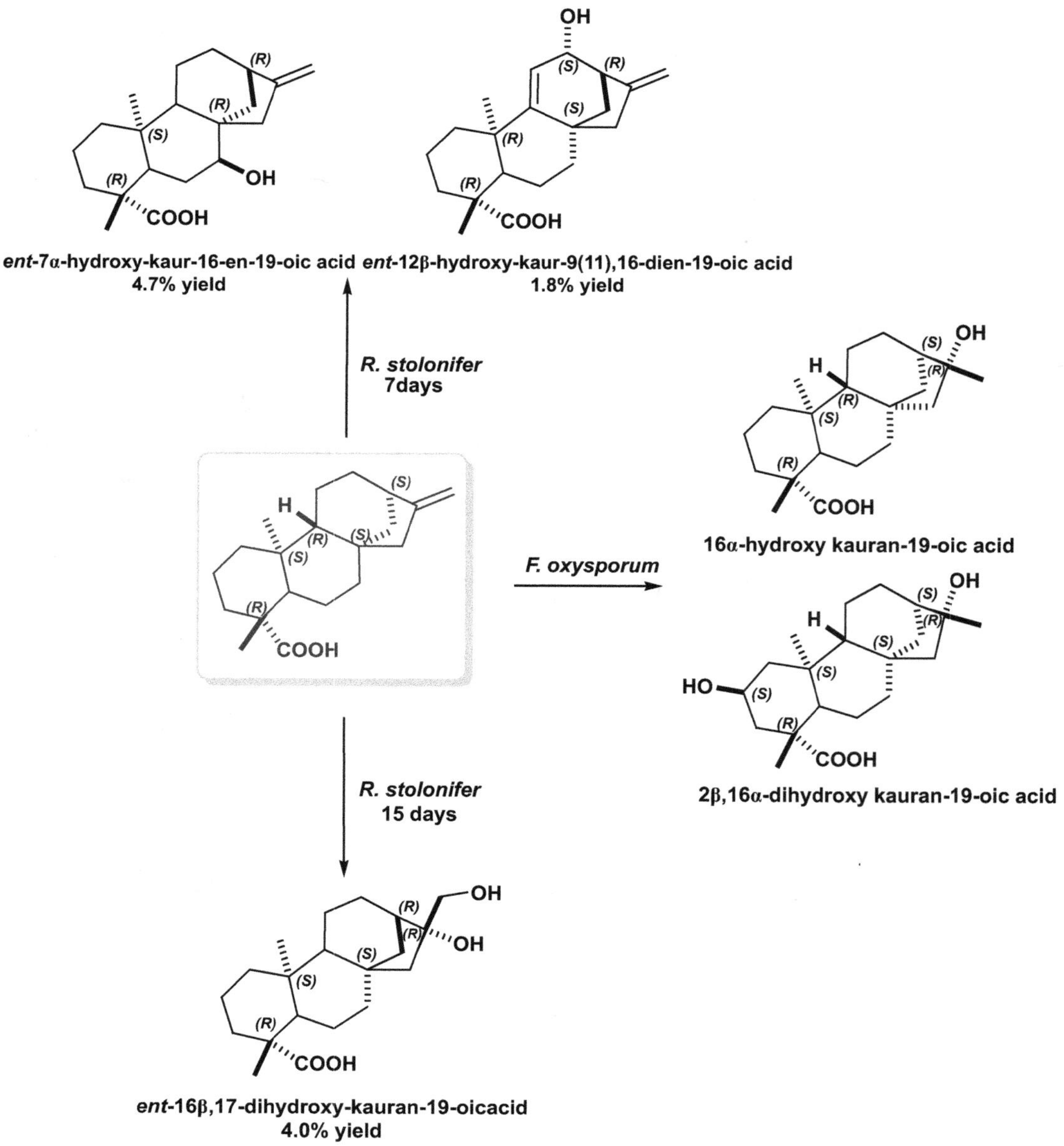

ent-7α-hydroxy-kaur-16-en-19-oic acid
4.7% yield

ent-12β-hydroxy-kaur-9(11),16-dien-19-oic acid
1.8% yield

R. stolonifer
7days

F. oxysporum

16α-hydroxy kauran-19-oic acid

2β,16α-dihydroxy kauran-19-oic acid

R. stolonifer
15 days

ent-16β,17-dihydroxy-kauran-19-oicacid
4.0% yield

FIGURE 14.51 Biotransformation of kauren-19-oic acid.

acid. On the other hand, the incubation of 3-oxo oleanolic acid with *A. ochraceus* CICC 40330 afforded a different polar metabolite 28-*O*-β-D-glucopyranosyl 3-oxo-olean-12-en-28-oate.[181] The biotransformation of 3-oxo-oleanolic acid by *Absidia glauca* CGMCC leads to three metabolites, one of which is similar to bioconversion ursolic acid is a lactone (1β-hydroxy-3-oxo-olean-11-eno-28,13-lactone), the other two are hydroxylated derivatives at positions C-1, C-11, and C-21 (1β,11α-dihydroxy-3-oxo-olean-12-en-28-oic acid, and 1β,11α,21β-trihydroxy-3-oxoolean-12-en-28-oic acid) – Figure 14.56.[182]

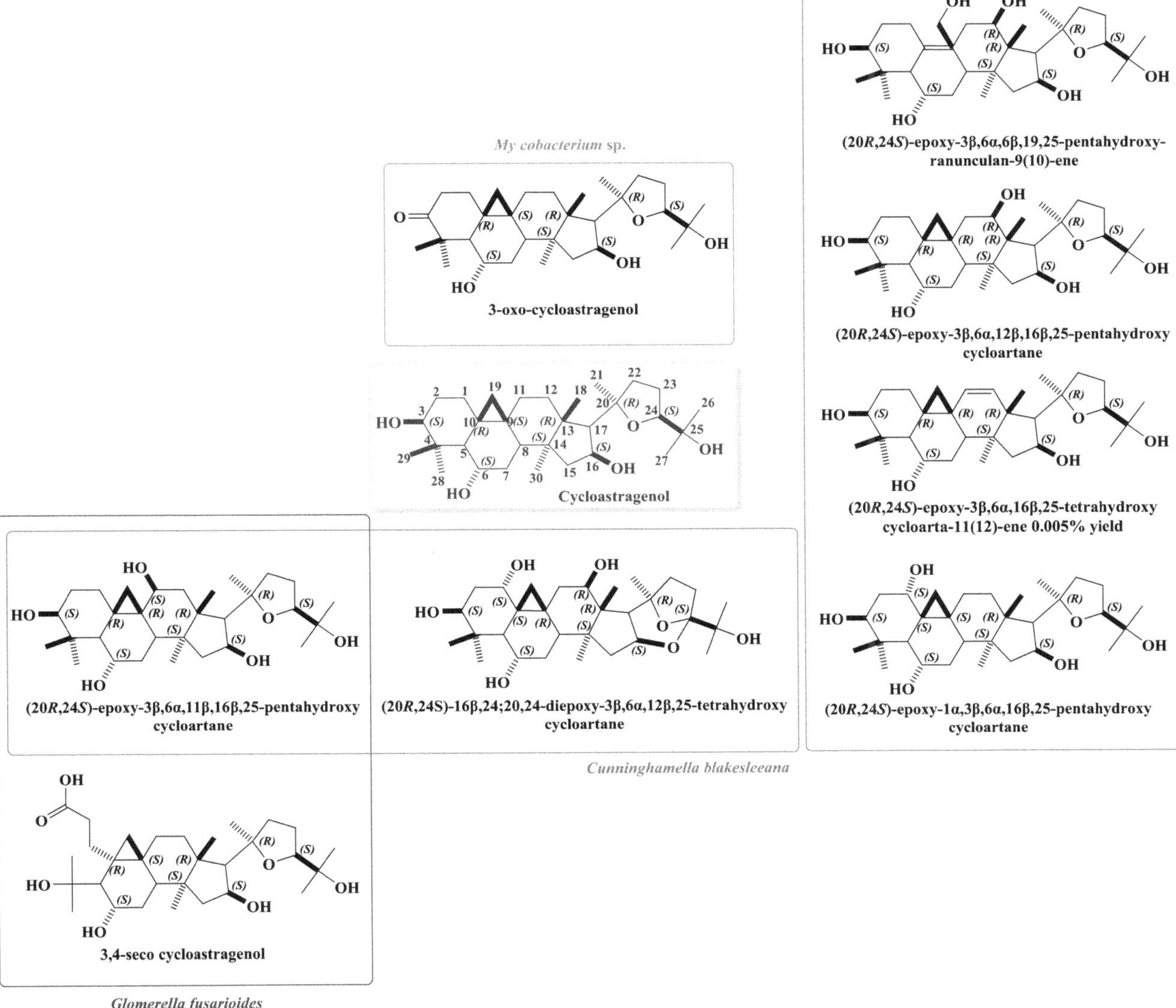

FIGURE 14.52 Biotransformation of cycloastragenol.

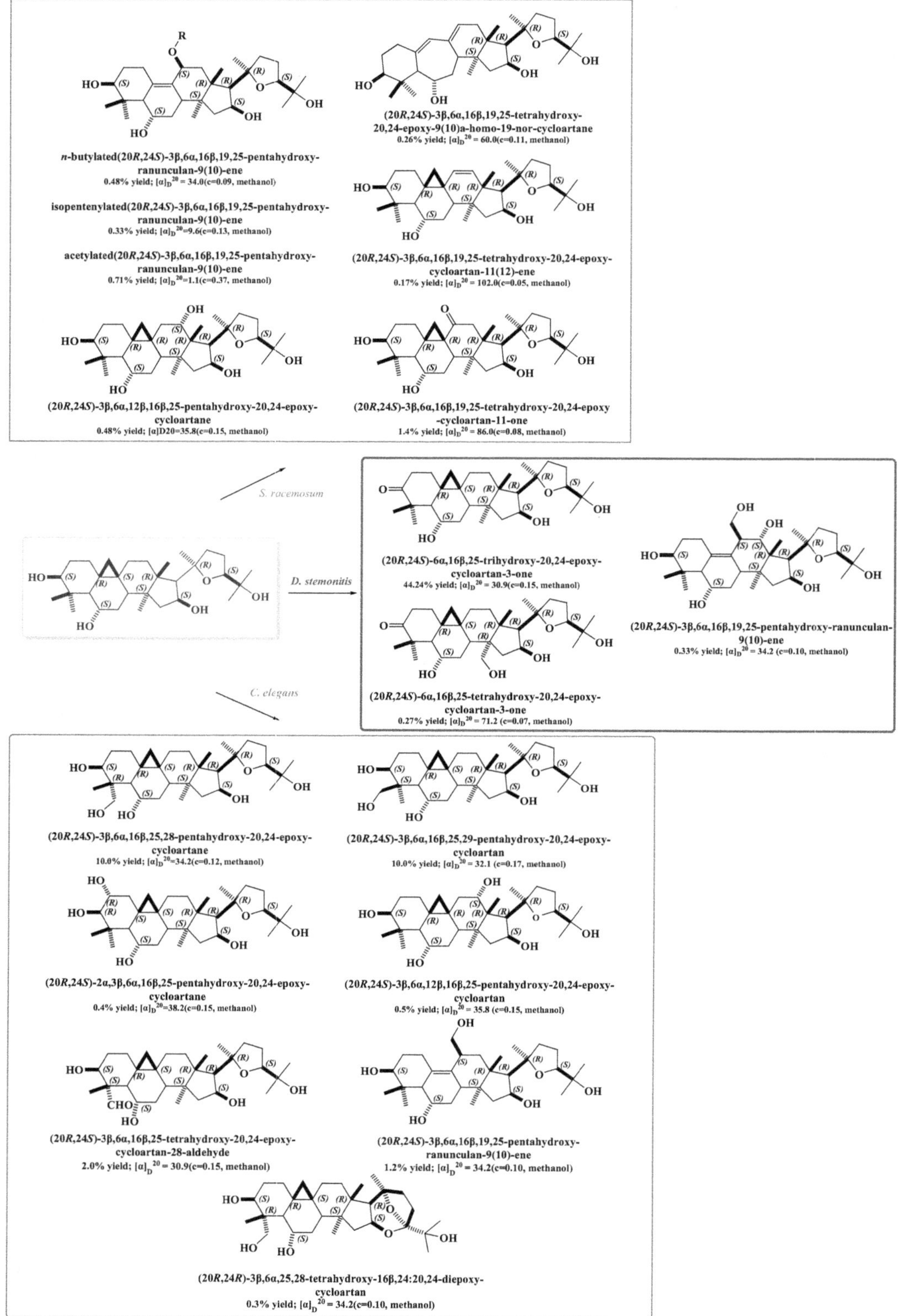

FIGURE 14.53 Biocatalysis of cycloastragenol.

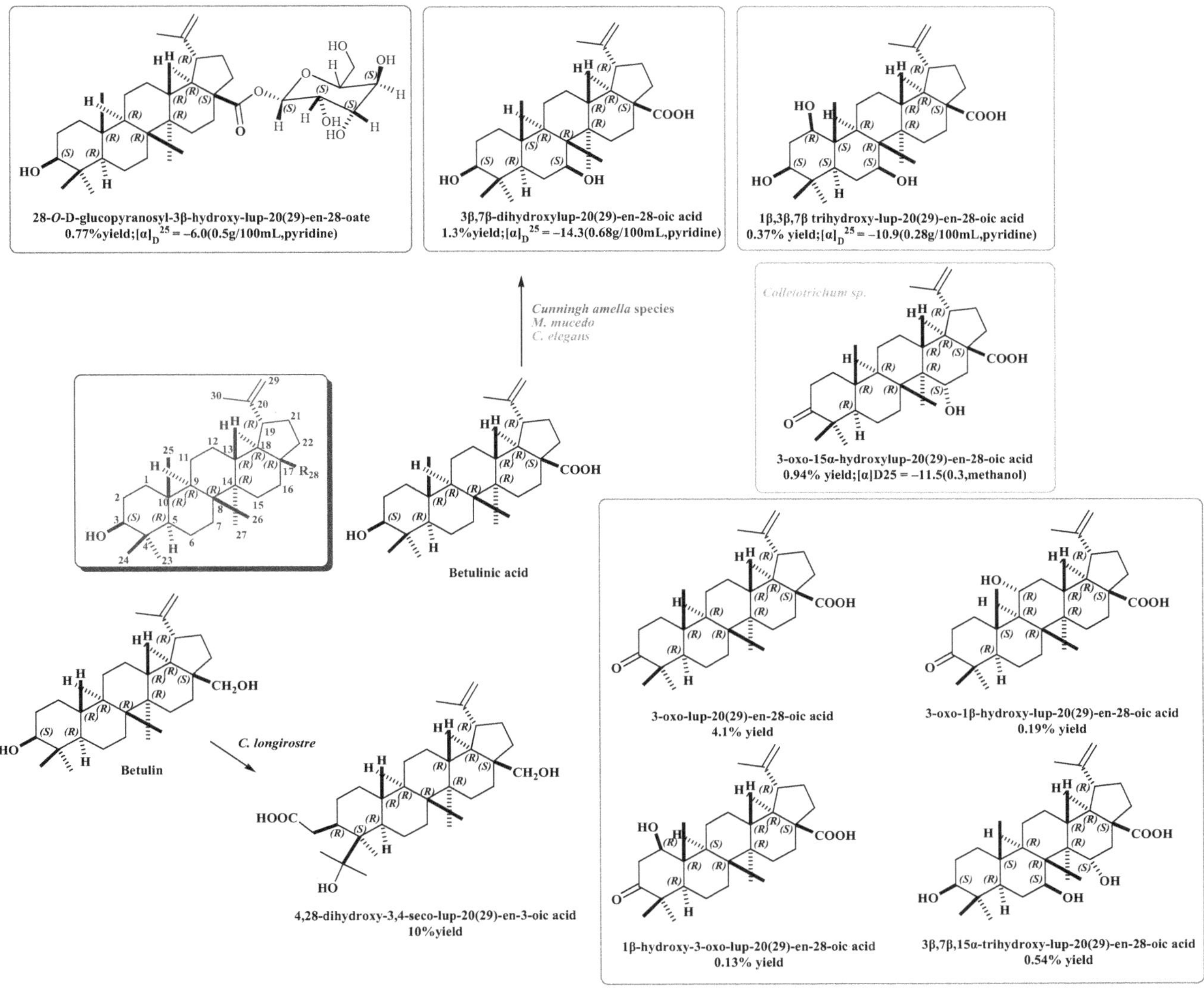

FIGURE 14.54 Biotransformation of betulinic acid and betulin.

S. racemosum

3β,21β-dihydroxy-urs-11-en-28-oic acid-13-lactone
[α]D20 = 55.3 (c=0.00023, ethanol)

3β,7β,21β-trihydroxy-urs-11-en-28-oic acid-13-lactone
$[\alpha]_D^{20}$ = 17.3 (c=0.00023, ethanol)

3β,7β,21β-trihydroxy-urs-12-en-28-oic acid
$[\alpha]_D^{20}$ = 32.0 (c=0.00018, ethanol)

1β,3β,21β-trihydroxy-urs-12-en-28-oic acid
$[\alpha]_D^{20}$ = 26.0 (c=0.00062, ethanol)

21-oxo-1β,3β-dihydroxy-urs-12-en-28-oic acid
$[\alpha]_D^{20}$ = 34.0 (c=0.00042, ethanol)

Ursolic acid

U. isabellina

3β-hydroxy-urs-11-eno-28,13-lactone

3β,7β-dihydroxy-urs-11-eno-28,13-lactone
$[\alpha]_D^{20}$ = 5.3 (c=0.75, methanol)

1β,3β-hydroxy-urs-11-eno-28,13-lactone
$[\alpha]_D 20$ = 70.6 (c=0.11, methanol)

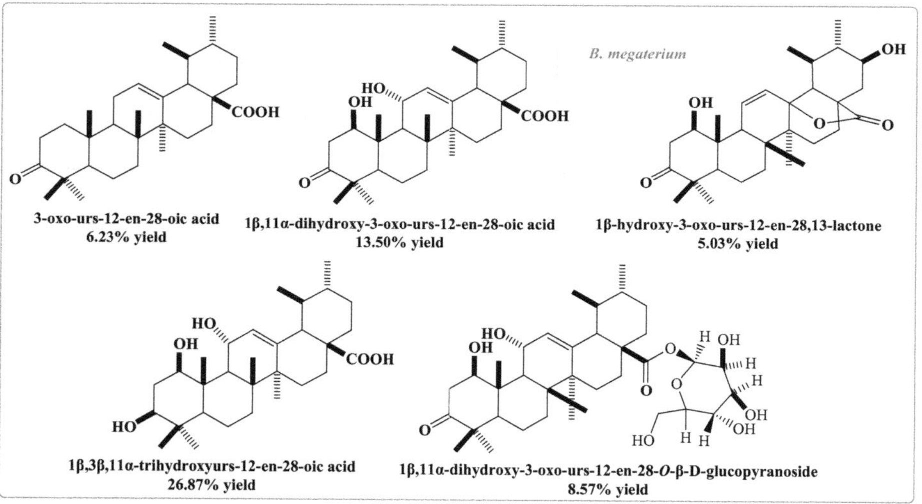

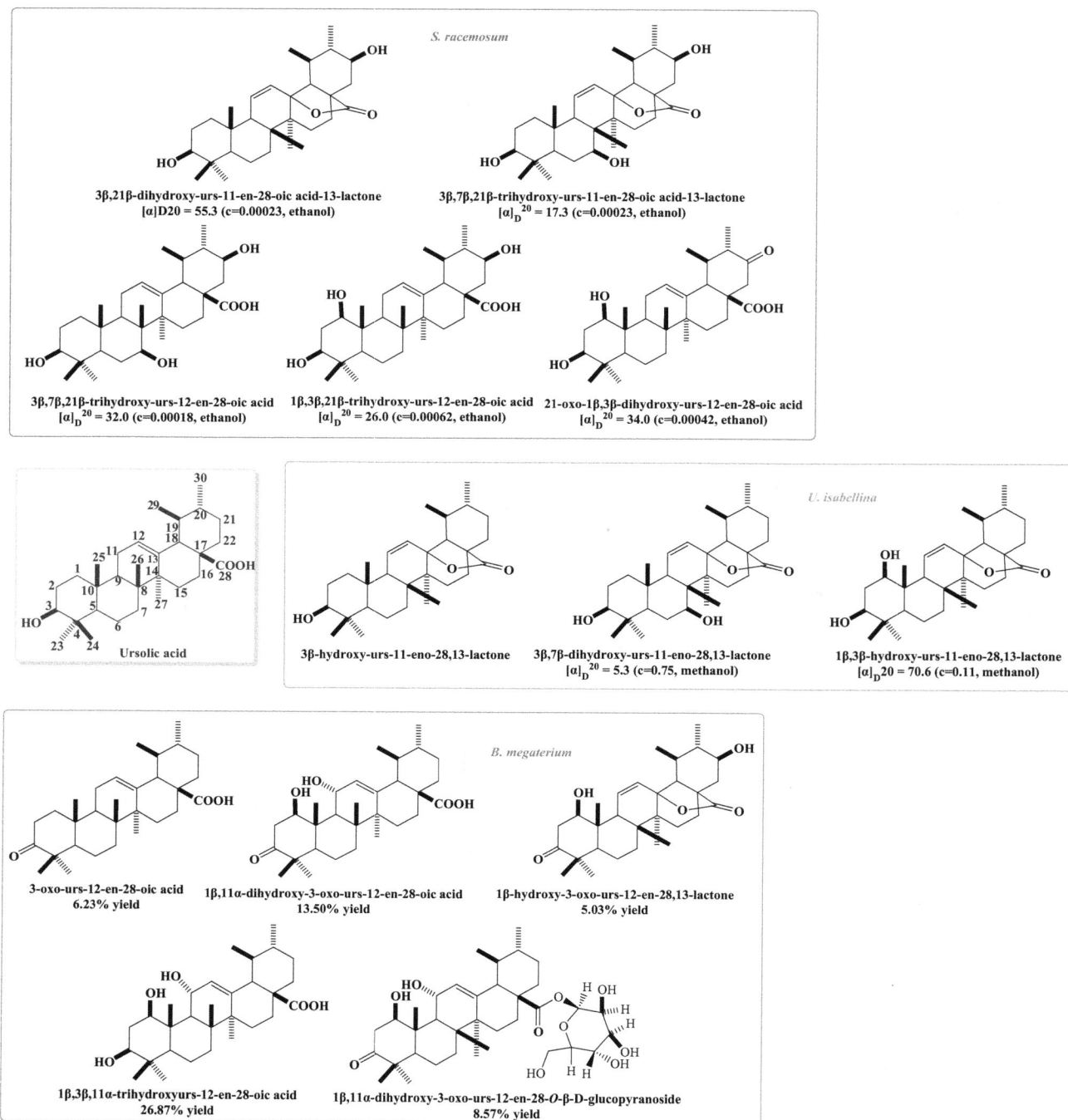

FIGURE 14.55 Biotransformation of ursolic acid.

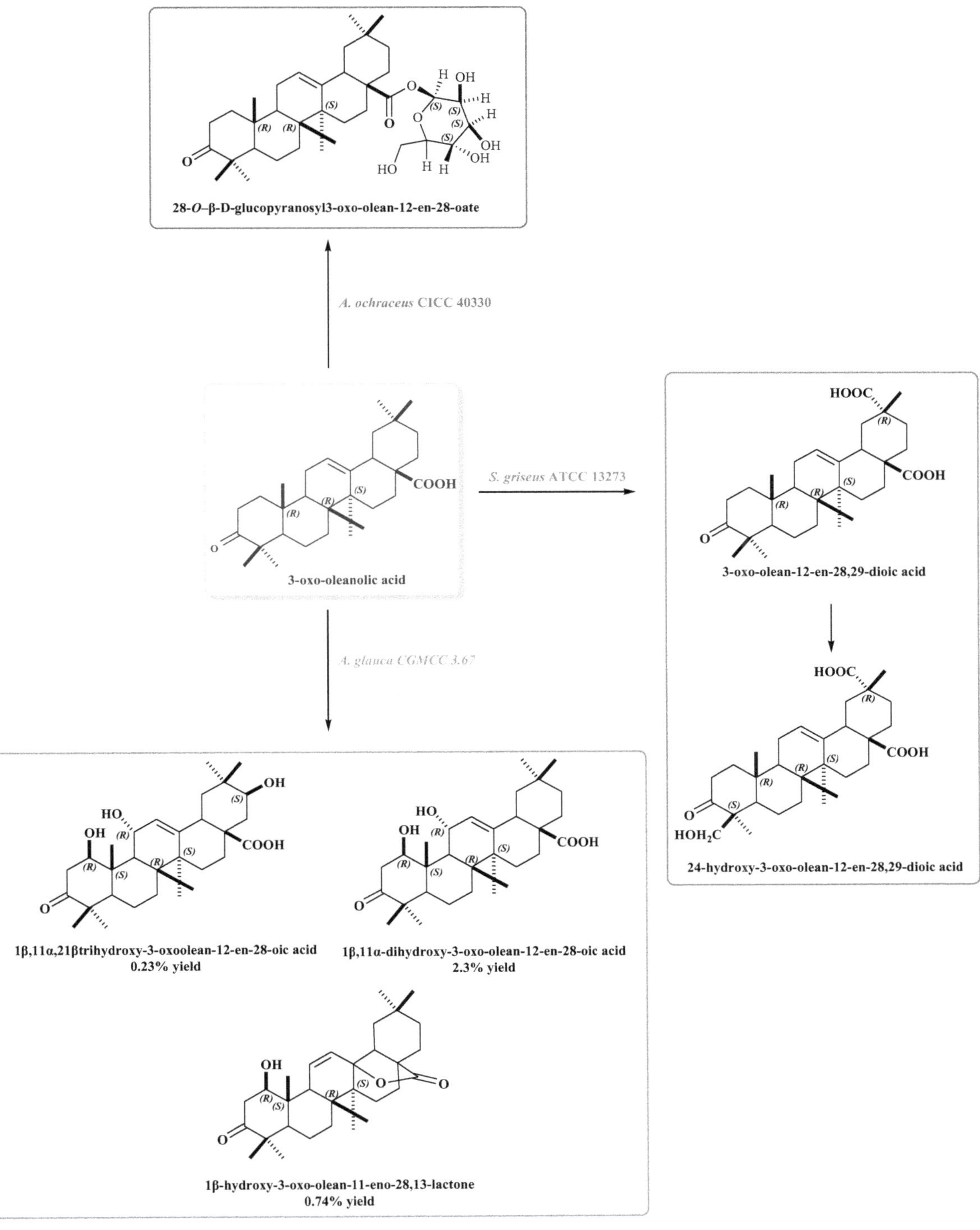

FIGURE 14.56 Biotransformation of 3-oxo oleanolic acid.

REFERENCES

1. R. A. Sheldon, The E factor 25 years on: the rise of green chemistry and sustainability, *Green Chem.*, 2017, 19, 18–43.
2. P. T. Anastas, J. C. Warner, *Green chemistry: theory and practice*, New York, Oxford University Press, 1998.
3. H. Sun, H. Zhang, E. L. Ang, H. Zhao, Biocatalysis for the synthesis of pharmaceuticals and pharmaceutical intermediates, *Bioorg. Med. Chem.*, 2018, 26, 1275–1284.
4. J.-H. Liu, B.-Y. Yu, Biotransformation of bioactive natural products for pharmaceutical lead compounds, *Curr. Org. Chem.*, 2010, 14, 1400–1406.
5. A. R. Alcántara, Biotransformations in drug synthesis: a green and powerful tool for medicinal chemistry, *J. Med. Chem. Drug. Des.*, 2017, 1, 1–7.
6. L. A. Nguyen, H. He, Ch. Pham-Huy, Chiral drugs: an overview, *Int. J. Biomed. Sci.*, 2006, 2, 85–100.
7. J.-M. Choi, S.-S. Han, H.-S. Kim, Industrial applications of enzyme biocatalysis: current status and future aspects, *Biotechnol. Adv.*, 2015, 33, 1443–1454.
8. E. Garcia-Urdiales, I. Alfonso, V. Gotor, Enantioselective enzymatic desymmetrizations in organic synthesis, *Chem. Rev.*, 2005, 105, 313–354.
9. R. N. Patel, Biocatalysis for synthesis of pharmaceuticals, *Bioorg. Med. Chem.*, 2018, 26, 1252–1274.
10. T. Classen, J. Pietruszka, Complex molecules, clever solutions – Enzymatic approaches towards natural product and active agent syntheses, *Bioorg. Med. Chem.*, 2018, 26, 1285–1303.
11. R. J. Ouellette, J. D. Rawn, *Organic chemistry study guide*, Elsevier, 2015.
12. J. Gawroński, K. Gawrońska, *Stereochemia w syntezie organicznej*, PWN, Warszawa, 1988.
13. I. Z. Siemion, *Biostereochemia*, PWN, Warszawa, 1985.
14. F. A. Carey, R. J. Sundberg, *Advanced organic chemistry*, Springer, New York, 2007.
15. V. Gotor, I. Alfonso, E. Garcia-Urdiales, *Asymmetric organic synthesis with enzymes*, Hardcover Wiley-VCH Verlag GmbH & Co. KGaA, 2008.
16. K. Faber, *Biotransformations in organic chemistry*, Springer, Berlin, 2018.
17. S. E. Milnera, A. R. Maguire, Recent trends in whole cell and isolated enzymes in enantioselective synthesis, *Arkivoc*, 2012, 321–382.
18. J.-F. Cai, Z. Guan, Y.-H. He, The lipase-catalyzed asymmetric C–C Michael addition, *J. Mol. Catal., B Enzym.*, 2011, 68, 240–244.
19. S. C. Lemoult, P. F. Richardson, S. M. Roberts, Lipase-catalysed Baeyer-Villiger reactions, *J. Chem. Soc. Perkin. Trans.*, 1, 1995.
20. Y. Ding, X. Xiang, M. Gu, H. Xu, H. Huang, Y. Hu, Efficient lipase-catalyzed Knoevenagel condensation: utilization of biocatalytic promiscuity for synthesis of benzylidene-indolin-2-ones, *Bioprocess. Biosyst. Eng.*, 2016, 39, 125–131.
21. R.-Ch. Tang, Z. Guan, Y.-H. He, W. Zhu, Enzyme-catalyzed Henry (nitroaldol) reaction, *J. Mol. Catal., B Enzym.*, 2010, 63, 62–67.
22. Z. Guan, L.-Y. Li, Y.-H. He, Hydrolase-catalyzed asymmetric carbon–carbon bond formation in organic synthesis, *RSC Adv.*, 2015, 5, 16801–16814.
23. I. Izquierdo, M. T. Plaza, M. Rodríguez, J. Tamayo, Chiral building-blocks by chemoenzymatic desymmetrization of 2-ethyl-1,3-propanediol for the preparation of biologically active natural products, *Tetrahedron: Asymm.*, 1999, 10, 449–455.
24. R. Chênevert, G. Courchesne, D. Caron, Chemoenzymatic enantioselective synthesis of the polypropionate acid moiety of dolabriferol, *Tetrahedron: Asymm.*, 2003, 14, 2567–2571.
25. M.-J. Kim, Y. Ahn, J. Park, Dynamic kinetic resolutions and asymmetric transformations by enzyme-metal Combo catalysis, *Bull. Korean Chem. Soc.*, 2005, 26, 515–523.
26. Ch. Li, X.-W. Feng, N. Wang, Y.-J. Zhou, X.-Q. Yu, Biocatalytic promiscuity: the first lipase-catalysed asymmetric aldol reaction, *Green Chem.*, 2008, 10, 616–618.
27. J. S. Yadav, S. Nanda, P. T. Reddy, A. B. Rao, Efficient enantioselective reduction of ketones with *Daucus carota* root, *J. Org. Chem.*, 2002, 67, 3900–3903.
28. E. Burda, W. Hummel, H. Gröger, Modular chemoenzymatic one-pot syntheses in aqueous media: combination of a palladium-catalyzed cross-coupling with an asymmetric biotransformation, *Angew. Chem. Int. Ed.*, 2008, 47, 9551–9554.
29. R. N. Patel, C. G. McNamee, A. Banerjee, J. M. Howell, R. S. Robison, L. J. Szarka, Stereoselective reduction of β-keto esters by *Geotrichumcandidum*, *Enzyme Microb. Technol.*, 1992, 14, 731–738.
30. R. N. Patel, L. Chu, R. Chidambaram, J. Zhu, J. Kant, Enantioselective microbial reduction of 2-oxo-2-(1′,2′,3′,4′-tetrahydro-1′,1′,4′,4′-tetramethyl-6′-naphthalenyl)acetic acid and its ethyl ester, *Tetrahedron: Asymm.*, 2002, 13, 349–355.
31. A. Romano, R. Gandolfi, P. Nitti, M. Rollini, F. Mollinari, Acetic acid bacteria as enantioselective biocatalysts, *J. Mol. Catal. B: Enzymol.*, 2002, 17, 235–240.
32. M. Landwehr, L. Hochrein, C. R. Otey, A. Kasrayan, J.-E. Bäckvall, F. H. Arnold, Enantioselective α-hydroxylation of 2-arylacetic acid derivatives and buspirone catalyzed by engineered cytochrome P450 BM-3, *J. Am. Chem. Soc.*, 2006, 128, 6058–6059.
33. K. Hofstetter, J. Lutz, I. Lang, B. Witholt, A. Schmid, Coupling of biocatalytic asymmetric epoxidationwith NADH regeneration in organic–aqueous emulsions, *Angew. Chem. Int. Ed.*, 2004, 43, 2163–2166.
34. D. V. Rial, D. A. Bianchi, P. Kapitanova, A. Lengar, J. B. Van Beilen, M. D. Mihovilovic, Stereoselective desymmetrizations by recombinant whole cells expressing the Baeyer–Villigermonooxygenase from *Xanthobacter* sp. ZL5: anew biocatalyst accepting structurally demanding substrates, *Eur. J. Org. Chem.*, 2008, 7, 1203–1213.
35. A. Kerridge, A. Willetts, H. J. Holland, Stereoselective oxidation of sulfides by cloned naphthalene dioxygenase, *Mol. Catal. B: Enzymol.*, 1999, 6, 59–65.
36. A. N. Phung, M. T. Zannetti, G. Whited, W.-D. Fessner, Stereospecific biocatalytic synthesis ofpancratistatinanalogues, *Angew. Chem., Int. Ed.*, 2003, 42, 4821–4824.
37. W. Adam, W. Boland, J. Hartmann-Schreier, H.-U. Humpf, M. Lazarus, A. Saffert, C. R. Saha-Möller, P. Schreier, α-Hydroxylation of carboxylic acids with molecular oxygen catalyzed by the *R* oxidase of peas (*Pisum sativum*): a novel biocatalytic synthesis of enantiomerically pure

(*R*)-2-hydroxy acids, *J. Am. Chem. Soc.*, 1998, 120, 11044–11048.
38. K. R. Bailey, A. J. Ellis, R. Reiss, T. J. Snape, N. J. Turner, A template-based mnemonic for monoamine oxidase (MAO-N) catalyzed reactions and its application to the chemo-enzymatic deracemisation of the alkaloid (±)-crispine A, *Chem. Commun.*, 2007, 3640–3642.
39. M. De Mancilha, R. De Conti, P. J. S. Moran, J. A. R. Rodrigues, Bioreduction of α-methyleneketones, *Arkivoc*, 2001, 85–93.
40. C. Stueckler, T. C. Reiter, N. Baudendistel, K. Faber, Nicotinamide-independent asymmetric bioreduction of C=C-bonds via disproportionation of enones catalyzed by enoate reductases, *Tetrahedron*, 2010, 66, 663–667.
41. M. Hall, C. Stueckler, B. Hauer, R. Stuermer, T. Friedrich, M. Breuer, W. Kroutil, K. Faber, Asymmetric bioreduction of activated C=C bonds using *Zymomonasmobilis* NCR enoate reductase and old yellow enzymes OYE 1–3 from yeasts, *Eur. J. Org. Chem.*, 2008, 9, 1511–1516.
42. P. D. de María, G. de Gonzalo, A. R. Alcántara, Biocatalysis as useful tool in asymmetric synthesis: an assessment of recently granted patents (2014–2019), *Catalysts*, 2019, 9, 802–844.
43. M. Xian, S. Alaux, E. Sagot, T. Gefflaut, Chemoenzymatic synthesis of glutamic acid analogues: substrate specificity and synthetic applications of branched chain aminotransferase from *Escherichia coli*, *J. Org. Chem.*, 2007, 72, 7560–7566.
44. A. A. Orden, J. H. Schrittwieser, V. Resch, F. G. Mutti, W. Kroutil, Controlling stereoselectivity by enzymatic and chemical means to access enantiomerically pure (1*S*,3*R*)-1-benzyl-2,3-dimethyl-1,2,3,4-tetrahydroisoquinoline derivatives, *Tetrahedron: Asymm.*, 2013, 24(12), 744–749.
45. W.-D. Fessner, Ch. Walter, *Bioorganic Chemistry, Enzymatic C-C bond formation in asymmetric synthesis*, Springer, Berlin, Heidelberg, 2005.
46. M. Brovetto, D. Gamenara, P. S. Méndez, G. A. Seoane, C-C bond-forming lyases in organic synthesis, *Chem. Rev.*, 2011, 111, 4346–4403.
47. W. D. Fessner, G. Sinerius, A. Schneider, M. Dreyer, G. E. Schulz, J. Badia, J. Aguilar, Diastereoselective enzymatic aldol additions: L-rhamnulose and L-fuculose 1-phosphate aldolases from *E. coli*, *Angew. Chem. In. Ed. Engl.*, 1991, 30, 555–558.
48. A. S. Demir, M. Pohl, E. Janzen, M. Müller, Enantioselective synthesis of hydroxy ketones through cleavage and formation of acyloin linkage. Enzymatic kinetic resolution via C–C bond cleavage, *J. Chem. Soc., Perkin Trans.*, 2001, 633–635.
49. J. Steinreiber, K. Fesko, Ch. Reisinger, M. Schürmann, F. van Assema, M. Wolberg, D. Mink, H. Griengl, Threonine aldolases—an emerging tool for organic synthesis, *Tetrahedron*, 2007, 63, 918–926.
50. H. Yu, X. Chen, Aldolase-catalyzed synthesis of β-D-Galp-(1→9)-D-KDN: a novel acceptor for sialyltransferases, *Org. Lett.*, 2006, 8, 2393–2396.
51. F. S. Sariaslani, J. P. N. Rosazza, Biocatalysis in natural products chemistry, *Enzyme Microb. Technol.*, 1984, 6, 242–253.
52. A. N. Panche, A. D. Diwan, S. R. Chandra, Flavonoids: an overview, *J. Nutr. Sci.*, 2016, 5, 1–15.
53. R. J. Nijveldt, E. van Nood, D. E. C. van Hoorn, P. G. Boelens, K. van Norren, P. A. M. van Leeuwen, Flavonoids: a review of probable mechanisms of action and potential applications, *Am. J. Clin. Nutr.*, 2001, 74, 418–425.
54. Ø. M. Andersen, K. R. Markham, *Flavonoids: chemistry, biochemistry and applications*, CRC, Taylor & Francis; Boca Raton, FL, 2006.
55. A. E. Nibbs, K. A. Scheidt, Asymmetric methods for the synthesis of flavanones, chromanones, and azaflavanones, *Eur. J. Org. Chem.*, 2012, 3, 449–462.
56. H. Cao, X. Chen, A. R. Jassbi, J. Xiao, Microbial biotransformation of bioactive flavonoids, biotechnology advances, *Biotechnol. Adv.*, 2015, 33, 214–223.
57. J. Luo, Q. Liang, Y. Shen, X. Chen, Z. Yin, M. Wang, Biotransformation of bavachinin by three fungal cell cultures, *J. Biosci. Bioeng.*, 2014, 117, 191–196.
58. H. Matsuda, S. Kiyohara, S. Sugimoto, S. Amdo, S. Nakamura, M. Yoshikawa, Bioactive constituents from Chinese natural medicines. XXXIII. Inhibitors from the seeds of *Psoralea corylifolia* on production of nitric oxide in lipopolysaccharide-activated macrophages, *Biol. Pharm. Bull.*, 2009, 32, 147–149.
59. A. K. Ibrahim, M. M. Radwan, S. A. Ahmed, D. Slade, S. A. Ross, M. A. ElSohly, I. A. Khan, Microbial metabolism of cannflavin A and B isolated from *Cannabis sativa*, *Phytochemistry*, 2010, 71, 1014–1019.
60. M.C. Henderson, C.L. Miranda, J.F. Stevens, M.L. Deinzer, D.R. Buhler, In *vitro* inhibition of human P450 enzymes by prenylated flavonoids from hops, *Humulus lupulus*, *Xenobiotica*, 2000, 30, 235–251.
61. C. L. Miranda, G. L. Aponso, J. F. Stevens, M. L. Deinzer, D. R. Buhler, Prenylated chalcones as inducers of quinone reductase in mouse Hepa 1c 1c 7 cells, *Cancer Lett.*, 2000, 149, 21–29.
62. US Patent and Trademark Office, United States Patent Number: 5, 679, 716, 1997.
63. W. H. M. W. Heratha, D. Ferreiraa, I. A. Khana, Microbial transformation of xanthohumol, *Phytochemistry*, 2003, 62, 673–677.
64. W. H.M.W. Heratha, D. Ferreiraa, S. I. Khan, I.A. Khan, Identification and biological activity of microbial metabolites of xanthohumol, *Chem. Pharm. Bull.*, 2003, 51(11), 1237–1240.
65. T. Tronina, A. Bartmańska, J. Popłoński, E. Huszcza, Transformation of xanthohumol by *Aspergillus ochraceus*, *J. Basic Microbiol.*, 2013, 53, 1–6.
66. T. Izumi, T. Hino, A. Kasahara, Enzymatic kinetic resolution of flavanone and *cis*-4-acetoxyflavan, *J. Chem. Soc. Perkin. Trans.*, 1992, 1265–1267.
67. T. Todoroki, A. Saito, A. Tanaka, Lipase-catalyzed kinetic resolution of (±)-*cis*-flavan-4-ol and its acetate: synthesis of chiral 3-hydroxyfavanones, *Biosci. Biotechnol. Biochem.*, 2002, 66(8), 1772–1774.
68. T. Izumi, K. Suenaga, Enzymatic resolution of flavanone oximes, *J. Heterocycl. Chem.*, 1997, 34, 1535–1538.
69. T. Izumi, S. Murakami, Enzymatic resolution of *trans*-dihydro-3-hydroxo-2-phenyl-4*H*-1-benzopyran-4-one (*trans*-flavanon-3-ol) by lipase, *J. Heterocycl. Chem.*, 1995, 32, 1125–1127.

70. M. Kawasaki, H. Kakuda, M. Goto, S. Kawabata, T. Kometani, Asymmetric synthesis of 2-substituted chroman-4-ones using lipase-catalyzed kinetic resolutions, *Tetrahedron: Asymm.*, 2003, 14, 1529–1534.
71. E. Kostrzewa-Susłow, M. Dymarska, A. Białońska, T. Janeczko, Enantioselective conversion of certain derivatives of 6-hydroxyflavanone, *J. Mol. Catal., B Enzym.*, 2014, 102, 59–65.
72. M. F. Roberts, M. Wink (ed.), *Alkaloids: biochemistry, ecology, and medicinal applications*, Plenum Press, New York, 1998.
73. S. W. Pelletier, *Alkaloids: chemical and biological perspectives*, ed. S. W. Pelletier, Wiley, New York, 1983.
74. J. H. Schrittwiesera, V. Resch, The role of biocatalysis in the asymmetric synthesis of alkaloids, *RSC Adv.*, 2013, 3(39), 17602–17632.
75. K. Laumen, M. Schneider, Enantioselective hydrolysis of *cis*-1,2-diacetoxycycloalkanedimethanols: enzymatic of chiral building blocks from prochiral *meso*-substrates, *Tetrahedron Lett.*, 1985, 26, 2073–2076.
76. G. Guanti, L. Banfi, E. Narisano, R. Riva, S. Thea, Enzymes in asymmetric synthesis: effect of reaction media on the PLE catalysed hydrolysis of diesters, *Tetrahedron Lett.*, 1986, 27, 4639–4642.
77. B. Danieli, G. Lesma, M. Mauro, G. Palmisano, D. Passarella, First enantioselective synthesis of (-)-akagerine by a chemoenzymatic approach, *J. Org. Chem.*, 1995, 60, 2506–2513.
78. B. Danieli, G. Lesma, M. Mauro, G. Palmisano, D. Passarella, An efficient chemo-enzymatic approach to (+)-meroquinene, *Tetrahedron: Asymm.*, 1990, 1, 793–800.
79. B. Danieli, G. Lesma, M. Mauro, G. Palmisano, D. Passarella, A highly enantioselective synthesis of (-)-antirhine by chemo-enzymatic approach, *Tetrahedron*, 1994, 50, 8837–8852.
80. P. Celestini, B. Danieli, G.O. Lesma, A. Sacchetti, A. Silvani, D. Passarella, A. Virdis, trans-6-Aminocyclohept-3-enols, a new designed polyfunctionalized chiral building block for the asymmetric synthesis of 2-substituted-4-hydroxypiperidines, *Org. Lett.*, 2002, 4, 1367–1370.
81. G. Lesma, S. Crippa, B. Danieli, D. Passarella, A. Sacchetti, A. Silvani, A. Virdis, Concise asymmetric synthesis of (-)-halosaline and (2*R*,9a*R*)-(1)-2-hydroxy-quinolizidine by ruthenium-catalyzed ring-rearrangement metathesis, *Tetrahedron*, 2004, 60, 6437–6442.
82. T. Momose, N. Toyooka, M. Jin, Asymmetric twin-ring differentiation by lipase-catalyzedenantiotoposelectivereaction of the ring-crossed *meso*glycol: asymmetric synthesis of a highly functionalized piperidine from the conjoined twin piperidine system, *Tetrahedron Lett.*, 1992, 33, 5389–5390.
83. T. Momose, N. Toyooka, M. Jin, Bicyclo[3.3.1]nonanes as synthetic intermediates. Part 21.1 Enantiodivergent synthesis of the *cis,cis* 2,6-disubstitutedpiperidin-3-ol chiral building block for alkaloid synthesis, *J. Chem. Soc., Perkin Trans.* 1997, 2005–2014.
84. T. Momose, N. Toyooka, Asymmetric synthesis of the alkaloid 2,6-disubstituted piperidin-3-ols, (-)-cassine and (+)-spectaline, *Tetrahedron Lett.*, 1993, 34, 5785–5786.
85. R. Chênevert, M. Dickman, Enzyme-catalysedhydrolysis of N-benzyloxycarbonyl-*cis*-2,6-(acetoxymethyl)piperidine. A facile route to optically active piperidines, *Tetrahedron: Asymm.*, 1992, 3, 1021–1024.
86. R. Chênevert, M. Dickman, Enzymatic route to chiral, nonracemic *cis*-2,6- and*cis,cis*-2,4,6-substituted piperidines. Synthesis of(+)-dihydropinidine and dendrobatealkaloid (+)-241D, *J. Org. Chem.*, 1996, 61, 3332–3341.
87. R. Chênevert, G. M. Ziarani, M. P. Morin, M. Dasser, Enzymatic desymmetrization of *mesocis*-2,6- and*cis,cis*-2,4,6-substituted piperidines. Chemoenzymatic synthesisof (5*S*,9*S*)-(+)-indolizidine 209D, *Tetrahedron: Asymm.*, 1999, 10, 3117–3122.
88. G. Lesma, A. Colombo, N. Landoni, A. Sacchetti, A. Silvani, A chemoenzymatic-RCM strategy for the enantioselective synthesis of new dihydroxylated 5-hydroxymethyl-indolizidines and 6-hydroxymethyl-quinolizidines, *Tetrahedron: Asymm..*, 2007, 18, 1948–1954.
89. B. Danieli, G. Lesma, D. Passarella, A. Silvani, Highly enantiopure C1-symmetric*cis*-piperidine-3,5-dimethanol monoacetatesby enzymatic asymmetrization, *J. Org. Chem.*, 1998, 63, 3492–3496.
90. B. Danieli, G. Lesma, D. Passarella, A. Silvani, N. Viviani, An efficient chemoenzymatic access to chiral3,7-diazabicyclo[3.3.1]nonane derivatives, *Tetrahedron*, 1999, 55, 11871–11878.
91. B. Danieli, G. Lesma, D. Passarella, A. Sacchetti, A. Silvani, A. Virdis, Total enantioselective synthesis of (-)-cytisine, *Org Lett.*, 2004, 6(4), 493–496.
92. B. Danieli, G. Lesma, D. Passarella, A. Sacchetti, A. Silvani, Chiral diamines for asymmetric synthesis: an efficient RCM construction of the ligand core of (-)- and (+)-sparteine, *Tetrahedron Lett.*, 2005, 46, 7121–7123.
93. T. Hudlicky, H. Luna, J. D. Price, F. Rulin, Microbial oxidation of chloroaromatics in the enantio divergent synthesis of pyrrolizidine alkaloids: trihydroxyheliotridanes, *J. Org. Chem.*, 1990, 55, 4683–4687.
94. T. Hudlicky, J. Rouden, H. Luna, S. Allen, Microbial oxidation of aromatics in enantiocontrolled synthesis. Rational design of aza sugars (endo-nitrogenous). Total synthesis of (+)-kifunensine, mannojirimycin, and other glycosidase inhibitors, *J. Am. Chem. Soc.*, 1994, 116, 5099–5107.
95. C. R. Johnson, A. Golebiowski, H. Sundram, M. W. Miller, R. L. Dwaihy, Synthesis of (+)-l-deoxygalactonojirimycin and a related indolizidine, *Tetrahedron Lett.*, 1995, 36, 653–654.
96. M. G. Banwell, X. Ma, N. Asano, K. Ikeda, J. N. Lambert, Chemoenzymatic syntheses of (-)-1-deoxymannojirimycin (DMJ) and its naturally occurring 6-*O*-α-L-rhamnopyranosyl glycoside, *Org. Biomol. Chem.*, 2003, 1, 2035–2037.
97. T. Hudlicky, Chemoenzymatic synthesis of complex natural and unnatural products: morphine, pancratistatin, and their analogs, *Arkivioc*, 2006, 7, 276–291.
98. U. Rinner, T. Hudlicky, Synthesis of Amaryllidaceae constituents-an update, *Synlett.*, 2005, 365–387.
99. L. Ingrassia, F. Lefranc, V. Mathieu, F. Darro, R. Kiss, Amaryllidaceae isocarbostyril alkaloids and their derivatives as promising antitumor agents, *Transl. Oncol.*, 2008, 1, 1–13.
100. G. Butora, T. Hudlicky, S. P. Fearnley, A. G. Gum, M. R. Stabile, K. Abboud, Chemoenzymatic synthesis of the

morphine skeleton via radical cyclization and a C_{10}-C_{11} closure, *Tetrahedron Lett.*, 1996, 37, 8155–8158.
101. H. Leisch, A. T. Omori, K. J. Finn, J. Gilmet, T. Bissett, D. Ilceski, T. Hudlicky, Chemoenzymatic enantiodivergent total syntheses of (+)- and (-)-codeine, *Tetrahedron*, 2009, 65, 9862–9875.
102. L. V. White, B. D. Schwartz, M. G. Banwell, A. C. Willis, A chemoenzymatic total synthesis of (+)-clividine, *J. Org. Chem.*, 2011, 76, 6250–6257.
103. B. D. Schwartz, M. G. Banwell, I. A. Cade, A chemoenzymatic total synthesis of the Amaryllidaceae alkaloid narseronine, *Tetrahedron Lett.*, 2011, 52, 4526–4528.
104. T. Hudlicky, U. Rinner, D. Gonzalez, H. Akgun, S. Schilling, P. Siengalewicz, T. A. Martinot, G. R. Pettit, Total synthesis and biological evaluation of *Amaryllidaceae* alkaloids: narciclasine, *ent*-7-deoxypancratistatin, regioisomer of 7-deoxypancratistatin, 10b-*epi*-deoxypancratistatin, and truncated derivatives, *J. Org. Chem.*, 2002, 67(25), 8726–8743.
105. R. Chênevert, P. Morin, Synthesis of (-)-lobeline via enzymatic desymmetrization of lobelanidine, *Bioorg. Med. Chem.*, 2009, 17, 1837–1839.
106. B. Das, P. Madhusudhan, Enantio seleetive synthesis of (*S*)- and (*R*)-mappicines and their analogues, *Tetrahedron*, 1999, 55, 7875–7880.
107. K. Shinya, M. Tanaka, K. Furihatat, Y. Hayakawa, H. Seto, Structure of carquinostatin A, anew neuronal cell protecting substance produced by *Streptomyces exfoliates*, *Tetrahedron Lett.*, 1993, 34, 4943–4944.
108. T. Choshi, Y. Uchida, Y. Kubota, J. Nobuhiro, M. Takeshita, T. Hatano, S. Hibino, Lipase-catalyzed asymmetric synthesis of desprenyl-carquinostatin A and descycloavandulyl-lavanduquinocin, *Chem. Pharm. Bull.*, 2007, 55(7), 1060–1064.
109. A. Kamal, K. V. Ramana, M. V. Rao, Chemoenzymatic synthesis of pyrrolo[2,1-b]quinazolinones: lipase-catalyzed resolution of vasicinone, *J. Org. Chem.*, 2001, 66, 997–1001.
110. Ch. Rusconi, N. Vaiana, M. Casagrande, N. Basilico, S. Parapini, D. Taramelli, S. Romeo, A. Sparatore, Synthesis and comparison of antiplasmodial activity of (+), (-) and racemic 7-chloro-4-(*N*-lupinyl)aminoquinoline, *Bioorg. Med. Chem.*, 2012, 20, 5980–5985.
111. E. Forró, L. Schönstein, F. Fülöp, Total synthesis of crispine A enantiomers through a *Burkholderiacepacia* lipase-catalysed kinetic resolution, *Tetrahedron: Asymm.*, 2011, 22, 1255–1260.
112. Q. Zhang, G. Tu, Y. Zhao, T. Cheng, Novel bioactive isoquinoline alkaloids from *Carduus crispus*, *Tetrahedron*, 2002, 58, 6795–6798.
113. C. J. Dunsmore, R. Carr, T. Fleming, N. J. Turner, A chemo-enzymatic route to enantiomerically pure cyclic tertiary amines, *J. Am. Chem. Soc.*, 2006, 128(7), 2224–2225.
114. D. Ghislieri, A. P. Green, M. Pontini, S. C. Willies, I. Rowles, A. Frank, G. Grogan, N. J. Turner, Engineering an enantioselective amine oxidase for the synthesis of pharmaceutical building blocks and alkaloid natural products, *J. Am. Chem. Soc.*, 2013, 135, 10863–10869.
115. A. Winkler, A. Łyskowski, S. Riedl, M. Puhl, T. M. Kutchan, P. Macheroux, K. Gruber, A concerted mechanism for berberine bridge enzyme, *Nat. Chem. Biol.*, 2008, 4, 739–741.
116. A. Winkler, K. Motz, S. Riedl, M. Puhl, P. Macheroux, K. Grube, Structural and mechanistic studies reveal the functional role of bicovalentflavinylationin Berberine bridge enzyme, *J. Biol. Chem.*, 2009, 284, 19993–20001.
117. J. H. Schrittwieser, B. Groenendaal, V. Resch, D. Ghislieri, S. Wallner, E. M. Fischereder, E. Fuchs, B. Grischek, J. H. Sattler, P. Macheroux, N. J. Turner, W. Kroutil, Deracemization by simultaneous bio-oxidative kinetic resolution and stereoinversion, *Angew. Chem. Int. Ed.*, 2014, 53, 3731–3734.
118. H.-X. Ge, J. Zhang, C. Kai, J.-H. Liu, B.-Y. Yu, Regio- and enantio-selective glycosylation of tetrahydroprotoberberines by *Gliocladiumdeliquescens* NRRL1086 resulting in unique alkaloidal glycosides, *Appl. Microbiol. Biotechnol.*, 2012, 93, 2357–2364.
119. H.-X. Ge, J. Zhang, Y. Dong, K. Cui, B.-Y. Yu, Unique biocatalytic resolution of racemic tetrahydroberberrubine via kinetic glycosylation and enantio-selective sulfation, *Chem. Commun.*, 2012, 48, 6127–6129.
120. V. L. D. Emidio, M. M. F. Ivana, N. G. Diego, *The alkaloids: chemistry and biology*. Academic: New York, 62, 2005.
121. R. L. Pederson, M.-J. Kim, C.-H. Wong, A combined chemical and enzymatic procedure for the synthesis of 1-deoxynojirimycin and 1-deoxymannojirimycin, *Tetrahedron Lett.*, 1988, 29, 4645–4648.
122. C. H. Von Der Osten, A. J. Sinskey, C. F. Barbas, R. L. Pederson, Y. F. Wang, C. H. Wong, Use of a recombinant bacterialf Fructose-1,6-diphosphate aldolase in aldol reactions: preparative syntheses of1-deoxynojirimycin, 1-deoxymannojirimycin,1,4-Dideoxy-1,4imino-arabinitol, and fagomine, *J. Am. Chem. Soc.*, 1989, 111, 3924–3927.
123. T. Ziegler, A. Straub, F. Effenberger, Enzyme-catalyzed synthesis of 1-deoxymannojirimycin, 1-deoxynojirimycin, and 1,4-dideoxy-1,4-imino-D-arabinitol, *Angew. Chem., Int. Ed. Engl.*, 1988, 27, 716–717.
124. A. Romero, Ch.-H. Wong, Chemo-enzymatic total synthesis of 3-epiaustraline, australine, and 7-epialexine, *J. Org. Chem.*, 2000, 65, 8264–8268.
125. J. Calveras, J. Casas, T. Parella, J. Joglar, P. Clapés, Chemoenzymatic synthesis and inhibitory activities of Hyacinthacines A_1 and A_2stereoisomers, *Adv. Synth. Catal.*, 2007, 349, 1661–1666.
126. X. Garrabou, L. Gómez, J. Joglar, S. Gil, T. Parella, J. Bujons, P. Clapés, Structure-guided minimalist redesign of the L-fuculose-1-phosphate aldolase active site: expedient synthesis of novel polyhydroxylated pyrrolizidines and their inhibitory properties against glycosidases and intestinal disaccharidases, *Chemistry*, 2010, 16(35), 10691–10706.
127. R. C. Simon, B. Grischek, F. Zepeck, A. Steinreiber, F. Belaj, W. Kroutil, Regio- and stereoselective monoamination of diketones without protecting groups, *Angew. Chem. Int. Ed. Engl.*, 2012, 51(27), 6713–6716.
128. R. C. Simon, F. Zepeck, W. Kroutil, Chemoenzymatic synthesis of all four diastereomers of 2,6-disubstituted piperidines through stereo selective monoamination of 1,5-diketones, *Chemistry*, 2013, 19(8), 2859–2865.
129. A. A. Orden, J. H. Schrittwieser, V. Resch, F. G. Mutti, W. Kroutil, Controlling stereoselectivity by enzymatic and chemical means to access enantiomerically pure (1*S*,3*R*)-1-benzyl-2,3-dimethyl-1,2,3,4-tetrahydroisoquinoline derivatives, *Tetrahedron: Asymm.*, 2013, 24(12), 744–749.

130. A. Bonamore, I. Rovardi, F. Gasparrini, P. Baiocco, M. Barba, C. Molinaro, B. Botta, A. Boffi, A. Macone, An enzymatic, stereoselective synthesis of (*S*)-norcoclaurine, *Green Chem.*, 2010, 12, 1623–1627.
131. T. Pesnot, M. C. Gershater, J. M. Ward, H. C. Hailes, The catalytic potential of *Coptis japonica* NCS2 revealed–development and utilisation of a fluorescamine-based assay, *Adv. Synth. Catal.*, 2012, 354, 2997–3008.
132. J. Stöckigt, B. Hammes, M. Ruppert, Construction and expression of a dual vector for chemo-enzymatic synthesis of plant indole alkaloids in *Escherichia coli*, *Nat. Prod. Res.*, 2010, 24, 759–766.
133. F. Wu, H. Zhu, L. Sun, C. Rajendran, M. Wang, X. Ren, S. Panjikar, A. Cherkasov, H. Zou, J. Stöckigt, Scaffold tailoring by a newly detected Pictet-Spenglerase activity of strictosidine synthase: from the common tryptoline skeleton to the rare piperazino-indole framework, *J. Am. Chem. Soc.*, 2012, 134(3), 1498–1500.
134. S. Jan, N. Abbas, *Himalayan phytochemicals: sustainable options for sourcing and developing bioactive compounds*, 1st Edition, Elsevier, 2018.
135. C. C.C.R. de Carvalho, M. M. R. da Fonseca, Biotransformation of terpenes, *Biotechnol. Adv.*, 2006, 24, 134–142.
136. C. C. C. R. de Carvalho. F. van Keulen, M. M. R. da Fonseca. Biotransformation of limonene-1,2-epoxide to limonene-1,2-diol by *Rhodococcuserythropolis* cells: an introductory approach to selective hydrolysis and product separation, *Food Technol. Biotechnol.*, 2000, 38, 181–185.
137. C. C. C. R. de Carvalho, M. M. R. da Fonseca, Influence of reactor configuration on the production of carvone from carveol by whole cells of *Rhodococcuserythropolis* DCL14, *J. Mol. Catal. B Enzyme.*, 2002, 19–20, 377–387.
138. C. C. C. R. de Carvalho, F. van Keulen, M. M. R. da Fonseca, Modelling the biokinetic resolution of diastereomers present in unequal initial amounts, *Tetrahedron: Asymm.*, 2002, 13, 1637– 1643.
139. H. Cross, R. Marriott, G. Grogan, Enzymatic esterification of lavandulol – a partial kinetic resolution of (*S*)-lavandulol and preparation of optically enriched (*R*)-lavandulyl acetate, *Biotechnol. Lett.*, 2004, 26, 457–460.
140. A. Belafriekh, F. Secundo, S. Serra, Z. Djeghaba, Enantioselective enzymatic resolution of racemic alcohols by lipases in green organic solvents, *Tetrahedron: Asymm.*, 2017, 28, 473–478.
141. H. D. Yan, Q. Li, Z. Wang, Efficient kinetic resolution of (±)-menthol by a lipase from *Thermomyceslanuginosus*, *Biotechnol. Appl. Biochem.*, 2017, 64(1), 87–91.
142. Z. Lü, Y. Chu, Y. Han, Y. Wang, J. Liu, Enzymatic esterification of DL-menthol with propionic acid by lipase from *Candida cylindracea*, *Biotechnol.*, 2005, 80(12), 1365–1370.
143. S. Bai, Z. Guo, W. Liu, Y. Sun, Resolution of (±)-menthol by immobilized *Candida rugosa* lipase on superparamagnetic nanoparticles, *Food Chem.*, 2006, 96, 1–7.
144. N. Sultana, Z. S. Saify, Enzymatic biotransformation of terpenes as bioactive agents, *J. Enzyme. Inhib. Med. Chem.*, 2013, 28(6), 1113–1128.
145. M. I. Choudhary, S. G. Musharraf, A. Sami, Atta-Ur-Rahman, Microbial transformation of sesquiterpenes, (-)-Ambrox and (+)-Sclareolide, *Helv. Chim. Acta*, 2004, 87, 2685–2694.
146. A. Nasib, S. G. Musharraf, S. Hussain, S. Khan, S. Anjum, S. Ali, Atta-Ur-Rahman, M. I. Choudhary, Biotransformation of (-)-Ambrox by cell suspension cultures of *Actinidia deliciosa*, *J. Nat. Prod.*, 2006, 69, 957–959.
147. M. I. Choudhary, I. C. Muhammed, G. M. Syed, M. T. H. Khan, D. Abdelrahman, M. Pervaz, F. Shaheen, Atta-Ur-Rahman, Microbial transformation of (+)-isolongifolen-4-one, *Helv. Chim. Acta*, 2003, 86, 3450–3460.
148. R. Duran, E. Corrales, R. Hernandez-Galan, I. G. Collado, Biotransformation of caryophyllene oxide by botrytis cinereal, *J. Nat. Prod.*, 1999, 62, 41–44.
149. M. I. Choudhary, Z. A. Siddiqui, S. A. Nawaz, Atta-Ur-Rahman, Microbial transformation and butyrylcholinesterase inhibitory activity of (-)-caryophyllene oxide and its derivatives, *J. Nat. Prod.*, 2006, 69, 1429–1434.
150. M. I. Choudhary, Z. A. Siddique, S. K. Saifullah, S. G. Musharraf, Atta-Ur-Rahman, Biotransformation of caryophyllene oxide by cell suspension culture of *Catharanthus roseus*, *Z. Naturforsch.*, 2006, 61, 197–200.
151. B. M. Fraga, M. G. Hernández, P. González, M. C. Chamy J. A. Garbarino, The biotransformation of 18-hydroxy-9-*epi*-*ent*-pimara-7,15-dieneby *Gibberella fujikuroi*, *Phytochemistry*, 2000, 53, 395–399.
152. M. Rico-Martínez, F. G. Medina, J. G. Marrero, S. Osegueda-Robles, Biotransformation of diterpenes, *RSC Adv.*, 2014, 4, 10627–10647.
153. B. M. Fraga, P. Gonzá, G. Melchor, N. Herná, S. Sergio, The biotransformation of the diterpene 2-hydroxy-*ent*-13-epimanoyl oxide by *Gibberellafujikuroi*, Ph 294, *Pytochemistry*, 2003, 62, 67–70.
154. B. M. Fraga, R. Guillermo, M. G. Hernández, M. C. Chamy, J. A. Garbarino, *J. Nat. Prod.*, 2009, 72, 87–91.
155. M. E. Severiano, M. R. Simao, T. S. Porto, C. H. G. Martins, R. C. S. Veneziani, N. A. J. C. Furtado, N. S. Arakawa, S. Said, D. C. R. de Oliveira, W. R. Cunha, L. E. Gregorio, S. R. Ambrosio, Anticariogenic properties of *ent*-pimaranediterpenes obtained by microbial transformation, *Molecules*, 2010, 15, 8553–8566.
156. S. T. Häkkinen, P. Lackman, H. Nygrén, K. M. Oksman-Caldentey, H. Maaheimo, H. Rischer, Differential patterns of dehydroabietic acid biotransformation by *Nicotiana tabacum* and *Catharanthus roseus* cells, *J. Biotechnol.*, 2012, 157, 287–294.
157. T. A. Van Beek, F. W. Claassen, J. Dorado, M. Godejohann, R. Sierra-Alvarez and J. B. P. A. Wijnberg, Fungal biotransformation Products of dehydroabietic acid, *J. Nat. Prod.*, 2007, 70, 154–159.
158. A. A. Tapia, M. D. Vallejo, S. C. Gouiric, G. E. Feresin, P. C. Rossomando and D. A. Bustos, Hydroxylation of dehydroabietic acid by *Fusarium* species, *Phytochemistry*, 1997, 46, 131–133.
159. S. C. Gouiric, G. E. Feresin, A. A. Tapia, P. C. Rossomando, G. Schmeda-Hirschmann, D. A. Bustos, 1β,7β-Dihydroxydehydroabietic acid, a new biotransformation product of dehydroabietic acid by *Aspergillus niger*, *World J. Microbiol. Biotechnol.*, 2004, 20, 281–284.
160. D. J. Smith, V. J. J. Martin, W. W. J. Mohn, A cytochrome P450 involved in the metabolism of abietane diterpenoids by *Pseudomonas abietaniphila* BKME-9, *J. Bacteriol.*, 2004, 186, 3631–3639.

161. J. F. Biellman, G. Branlant, M. Gero-Robert, M. Poiret, Degradation bacterienne de l'acidedehydroabietique par *Flavobacterium resinovorum*, *Tetrahedron*, 1973, 29, 1227–1236.
162. K. Mitsukura, T. Imoto, H. Nagaoka, T. Yoshida, T. Nagasawa, Regio- and stereo-selective hydroxylation of abietic acid derivativesby *Mucor circinelloides* and *Mortierella isabelline*, *Biotechnol. Lett.*, 2005, 27, 1305–1310.
163. H. N. Bhattia, R.A. Kheraa, Biotransformations of diterpenoids and triterpenoids: a review, *J. Asian Nat. Prod. Res.*, 2014, 16, 70–104.
164. B.M. Fraga, M.G. Hernandez, P. Gonzalez, The microbiological transformation of same *ent*-7α,15β-dihydroxykaurene derivatives by *Gibberellafujikuroi*, *Phytochemistry*, 1991, 30, 2567–2571.
165. B. M. Fraga, L. Alvarez, S. Suárez, Biotransformation of the diterpenes epicandicandiol and candicandiol by *Mucor plumbeus*, *J. Nat. Prod.*, 2003, 66, 327–331.
166. B. M. Fraga, I. de Alfonso, V. Gonzalez-Vallejo, R. Guillermo, Microbial transformation of two 15α-hydroxy-*ent*-kaur-16-ene diterpenes by *Mucor plumbeus*, *Tetrahedron*, 2010, 66, 227–234.
167. K.Y. Ho, S.H. Hyun, H.S. Kim, S.W. Lee, K. Dong-Hyun, J.J. Lee, Microbial transformation of bioactive diterpenoids from *Acanthopanax koreanum* by *Fusiariumoxysporum*, *J. Microbiol. Biotechnol.*, 1992, 2, 92–97.
168. E.A. Silva, J.A. Takahashi, M.A.D. Boaventura, A.B. Oliveira, The biotransformation of *ent*-kaur-16-en-19-oic acid by *Rhizopus stolonifera*, *Phytochemistry*, 1999, 52, 397–400.
169. S. A. A. Shah, H. L. Tan, S. Sultan, M. A. B. M. Faridz, M. A. B. M. Shah, S. Nurfazilah, M. Hussain, Microbial-catalyzed biotransformation of multifunctional triterpenoids derived from phytonutrients, *Int. J. Mol. Sci.*, 2014, 15, 12027–12060.
170. M. Kuban, G. Öngen, E. Bedir, Biotransformation of cycloastragenol by *Cunninghamella blakesleeana* NRRL 1369 resulting in a novel framework, *Org. Lett.*, 2010, 12, 4252–4255.
171. M. Kuban, G. Öngen, I. A. Khan, E. Bedir, Microbial transformation of cycloastragenol, *Phytochemistry*, 2013, 88, 99–104.
172. W.-Z. Yang, M. Ye, F.-X. Huang, W.-N. He, D.-A. Guo, Biocatalysis of cycloastragenol by filamentous fungi to produce unexpected triterpenes, *Adv. Synth. Catal.*, 2012, 354, 527–539.
173. P. Chatterjee, J.M. Pezzuto, S.A. Kouzi, Glucosidation of betulinic acid by *Cunninghamella* species, *J. Nat. Prod.*, 1999, 62, 761.
174. S. A. Kouzi, P. Chatterjee, J. M. Pezzuto, M. T. Hamann, Microbial transformations of the antimelanoma agent betulinicacid, *J. Nat. Prod.*, 2000, 63, 1653–1657.
175. D. Z.L. Bastos, I. C. Pimentel, D. A. de Jesus, B. H. de Oliveira, Biotransformation of betulinic and betulonic acids by fungi, *Phytochemistry*, 2007, 68, 834–839.
176. P. Chatterjee, S. A. Kouzi, J. M. Pezzuto, M. T. Hamann, Biotransformation of the antimelanoma agent betulinicacid by *Bacillus megaterium* ATCC 13368, *Appl. Environ. Microbiol.*, 2000, 66, 3850–3855.
177. T. Akihisa, Y. Takamine, K. Yoshizumi, H. Tokuda, Y. Kimura, M. Ukiya, T. Nakahara, T. Yokochi, E. Ichiishi, H. Nishino, Microbial transformations of two lupane-type triterpenes and anti-tumor-promoting effects of the transformation products, *J. Nat. Prod.*, 2002, 65, 278–282.
178. S.-B. Fu, J.-S. Ynag, J.-L. Cui, X. Feng, D.-A. Sun, Biotransformation of ursolicacid by an endophytic fungus from medicinal plant *Huperzia serrata*, *Chem. Pharm. Bull.*, 2011, 59(9), 1180–1182.
179. S.-B. Fu, J.-S. Yang, J.-L. Cui, D.-A. Sun, Biotransformation of ursolic acid by *Syncephalastrumracemosum* CGMCC 3.2500 and anti-HCV activity, *Fitoterapia*, 2013, 86, 123–128.
180. Ch. Zhang, S.-H. Xu, B.-L. Ma, W.-W. Wang, B.-Y. Yu, J. Zhang, New derivatives of ursolic acid through the biotransformation by *Bacillus megaterium* CGMCC 1.1741 as inhibitors on nitric oxide production, *Bioorg. Med. Chem. Lett.*, 2017, 27, 2575–2578.
181. Y.-Y. Zhu, L.-W. Qian, J. Zhang, J.-H. Liu, B.-Y. Yu, New approaches to the structural modification of olean-type pentacylic triterpenes via microbial oxidation and glycosylation, *Tetrahedron*, 2011, 67, 4206–4211.
182. N. Guo, Y. Zhao, W.-S. Fang, Biotransformation of 3-oxo-oleanolic acid by *Absidia glauca*, *Planta. Med.*, 2010, 76, 1904–1907.

15

The Prelude of Green Syntheses of Drugs and Natural Products

Leonardo Xochicale-Santana, C. C. Vidyasagar, Blanca M. Muñoz-Flores, and Víctor M. Jiménez Pérez

CONTENTS

15.1 Introduction ... 289
15.2 Mechanosynthesis of Drugs ... 289
15.3 Pharmaceutical Cocrystals by Mechanochemical ... 293
15.4 Ultrasound Synthesis of Drugs ... 295
15.5 Green Synthesis of Natural Products ... 297
15.5.1 History of Natural Products ... 297
15.5.2 Primary and Secondary Metabolites ... 298
15.5.3 Historically Important Natural Products ... 298
15.5.4 Development of New Medicines ... 298
15.5.5 Sustainable Synthesis of Natural Products ... 299
15.5.6 Grinding Method ... 299
15.5.7 Microwave-Assisted Method ... 300
15.5.8 Using Green Catalyst Method ... 300
15.5.9 Solvent-Free Method ... 301
15.5.10 Water as Greener Solvent ... 302
References ... 303

15.1 Introduction

Nowadays, there is tremendous concern about the negative environmental impact due to the synthesis of drugs and natural products from the pharmacy industry. It is well known that oil refining, bulk, and fine chemicals, plastics as well as the pharmacy industry generate large amounts of chemical wastes per year.[1] Thus, green chemistry has become a strong priority and philosophy all over the world where the chemists are reaching for new and renewable solvents and chemicals, and methodologies with energy efficiency.[2] Green synthetic methods include mechanosynthesis, sonosynthesis, electrosynthesis, and microwave (MW) assisted synthesis. However, the disadvantage of microwave-assisted synthesis of drugs or natural products is not yet scalable. Thus, mechanosynthesis and ultrasound have been the object of growing attention from both academia and the pharmacy industry.[3,4] For several decades, synthetic chemists have developed a wide variety of chemical reactions for the formation of chemical bonds such as the C–C or C-heteroatom bond, to develop synthetic methodologies and strategies for the synthesis of drugs of great importance. However, a drawback of these reactions is that many of them involve the use of toxic substances for the formation of bonds or strict reaction conditions, due to the use of transition metals as catalysts or the formation of more reactive 3intermediates from reagents toxic. Today, the term "green chemistry" has been used to give an environmentally friendly approach to chemical processes with the aim of reducing or eliminating the use of hazardous substances in these processes. Therefore, chemists have developed techniques and methods based on the 12 principles of green chemistry established by Paul Anastas and John Warner in the 1990s,[5] whose purpose is to establish the criteria to develop and design more sustainable chemical processes. In recent years, techniques have been developed that allow the formation of chemical bonds in an effective and efficient manner, maintaining the objective of a "green synthesis"; for this, the use of techniques such as mechanosynthesis, ultrasound, and microwaves, and even methodologies that involve the use of low-toxicity reagents have allowed access to more sustainable methods. Therefore, the purpose of this chapter is to describe these methodologies as well as their scope in drug synthesis.

15.2 Mechanosynthesis of Drugs

Mechanosynthesis is a technique based on solid-state reactions, using grinding as a process. An advantage of this technique is to avoid the excessive use of solvents to carry out a successful chemical reaction. The development of this technique as a synthetic method has been excellently supported by obtaining results over the last few years in a wide variety of applications,[6] as a method for obtaining leading drug structures. For example, Colacino and co-workers prepared disubstituted hydantoins using an eco-friendly methodology,[7] applying the mechanosynthesis technique as an easy-handling tool. Hydantoins are

DOI: 10.1201/9781003089162-15

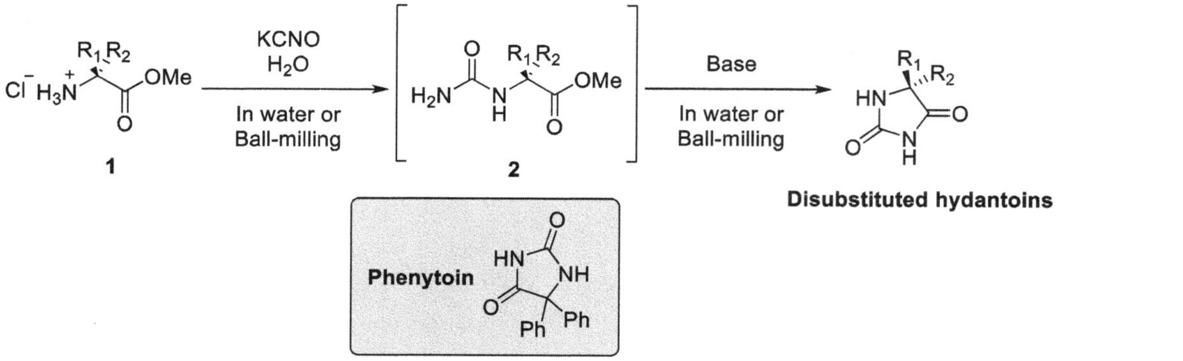

SCHEME 15.1 Hydantoin synthesis using an eco-friendly methodology by Colacino's group.

a group found in a wide variety of drugs, hydantoin derivatives are used as anticonvulsants,[8] muscle relaxants, antidiabetics,[9] antiandrogenic, or antifungal activities.[10] The application of this tool entails obtaining the disubstituted hydantoins in excellent yields from amino ester hydrochloride (**1**), in addition to avoiding the use of toxic substances to prepare said structures.[11] This methodology was applied to the "green synthesis" of the antiepileptic drug Phenytoin (Scheme 15.1).

On the other hand, a report by Fullenwarth and co-workers describes the methodology for the synthesis of N-substituted hydantoins from 1-Amino hydantoin hydrochloride (**3**) and aldehydes (**4**) to obtain active pharmaceutical ingredients by mechanosynthesis (Scheme 15.2).[12] An advantage of this method, samples that it is possible to obtain this type of structures avoiding the use of solvents and bases, accessing a process eco-friendly and low cost. This methodology allows to synthesize drugs such as Nitrofurantoin (Furantin©), Dantrolene (Dantrium©), and their structurally related derivatives.

In a separate report, Colacino and co-workers reported a simple and environmentally friendly methodology to prepare 3,5-disubstituted hydantoin-mediated mechanochemical Lossen rearrangement using benzhydroxamic acid (**5**) and amino esters hydrochloride (Scheme 15.3).[13] Highlighting a one-pot multistep fashion reaction and solvent-free.

Other research groups have reported the use of other techniques and methods for the preparation of these structures; a clear example is that reported by Parrot who makes use of microwave-assisted synthesis,[14] or Waser who uses a hypervalent iodine cyanation reagent (cyanobenziodoxolone, CBX) as a source of electrophilic carbon in dark and metal-free,[15] observing an efficient synthesis, and eco-friendly (Scheme 15.4).

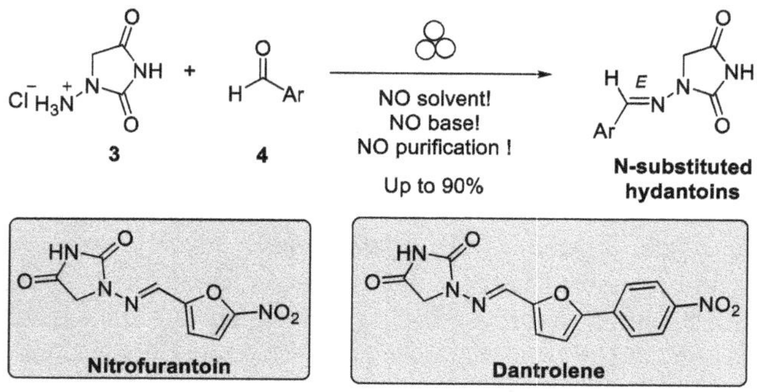

SCHEME 15.2 Synthesis of N-substituted hydantoins1-Amino hydantoin hydrochloride and aldehydes in a solvent-free protocol.

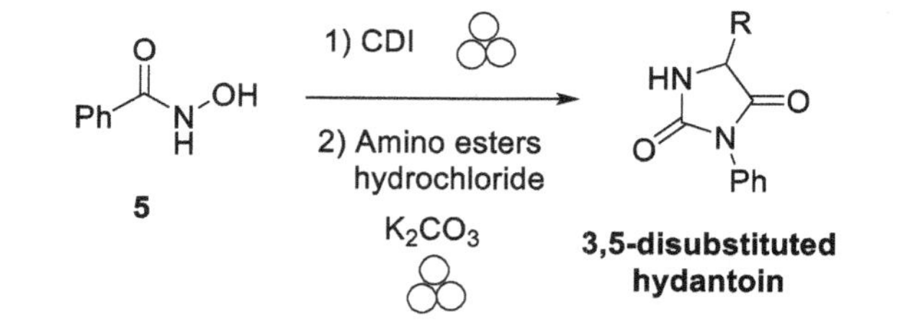

SCHEME 15.3 Lossen rearrangement reaction by mechanochemical.

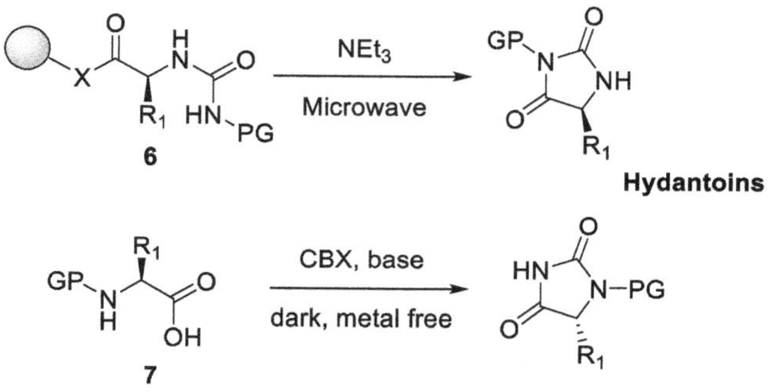

SCHEME 15.4 Other eco-friendly methods for obtaining hydantoins.

Previously, a variety of methodologies have been described for the synthesis of the heterocycle of hydantoins through mechanosynthesis; however, the mechanosynthesis is not limited to the construction of structures of this nature, and this tool can also be used to carry out specific reactions in important intermediaries of some drugs; an example is reported by Maciá *et al.*, describing a reduction of the carbonyl group with $NaBH_4$ focusing a simplified and fast synthetic pathway for the eco-friendly synthesis of Fluoxetine (Scheme 15.5),[16] proving that mechanosynthesis offers an opportunity for synthetic chemists in the process of conventional reactions at an industrial level to obtain active principles of great pharmaceutical interest.

Procainamide and Paracetamol can also be obtained using this tool; the first compound involves an amidation reaction and catalytic transfer hydrogenation to reduce the nitro group to amine (A), and the second one involves a reduction followed by an acylation reaction (B); in both cases, solvent-free (Scheme 15.6).[17] Mechanosynthetic pathways can be key

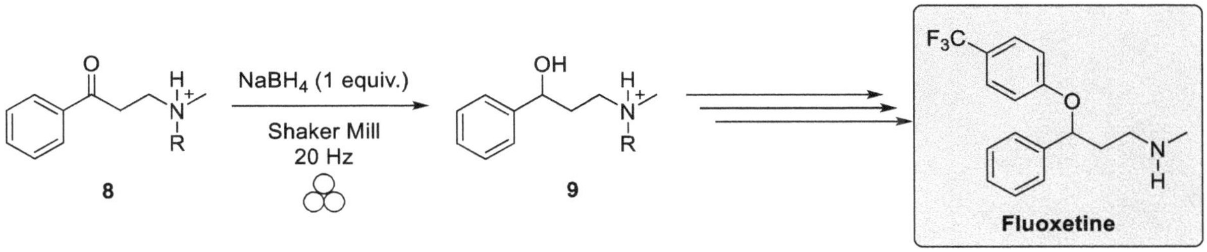

SCHEME 15.5 Eco-reduction of the carbonyl group by mechanochemical.

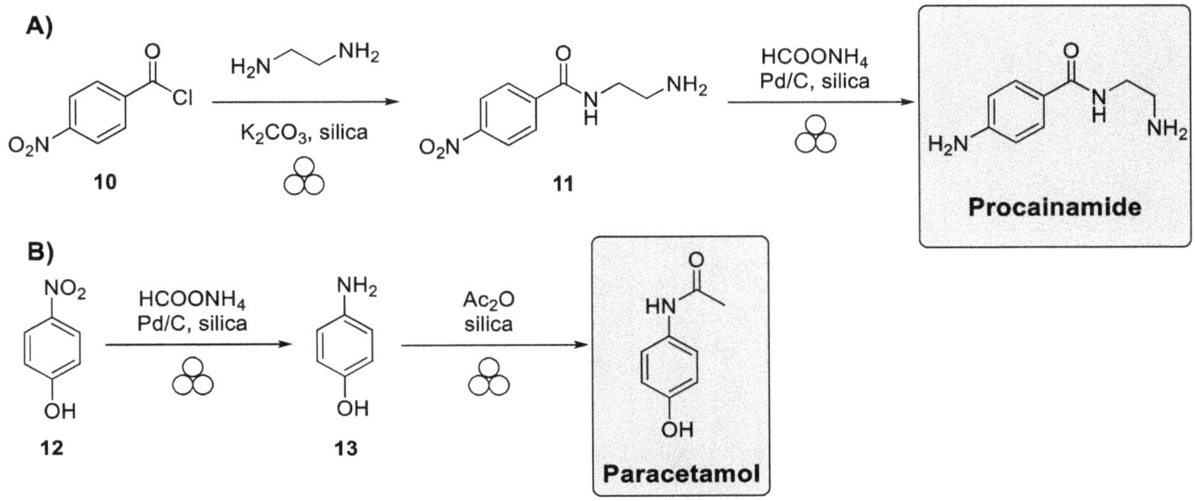

SCHEME 15.6 Reaction of amidation, acylation, and reduction of the nitro group by mechanochemistry.

methodologies for the preparation of pharmaceutically active compounds.

In some cases, not only the modification of conventional chemical reactions can be modified through the implementation of mechanochemistry, but even reactions that involve biological agents such as enzymes for a chemical reaction are susceptible to the application of this tool. For the synthesis of some peptides, it is necessary to use enzymes to carry out a specific type of reaction, for example, in the synthesis of the Leu-enkephalin, an endogenous opioid peptide neurotransmitter can be synthesized by use of enzymes[18] or through chemical routes.[19] In contrast, Lamaty and co-workers developed a mechanochemical procedure for the synthesis of this peptide in an efficient and eco-friendly manner as shown in Scheme 15.7.[20]

Recently, Lamaty and co-workers report a potent and selective 5-HT_7 receptor antagonist using a mechanosynthesis approach; this protocol offered huge advantages such as reduction of reaction time, limitation of the use of toxic solvents, no purification, and is eco-friendly compared to classical synthesis.[21] The synthetic route involved an opening of the epoxide by mechanochemistry followed by sulfonation of the primary amine via mechanochemical (Scheme 15.8).

The 2,3-Dihydroquinazolin-4(1H)-ones have traditionally been prepared by the action of metal catalysts such as AgOAc, Fe_3O_4, and CuBr.[22–24] However, a report recently published by Saha describes the synthesis of these molecular scaffolds through rapid, eco-friendly, and energy-efficient scalable methods by grinding in a mortar pestle as well as mechanochemically milling using only p-TSA as catalyst (Scheme 15.9).[25] These structures are highly demanded by the pharmaceutical industry because many drugs for the treatment of cancer, anti-inflammatories, diuretics, anticonvulsant, and antihypertensive activities contain this scaffold in their structure.

Benzoxaboroles are another heterocycle of great importance in medicine; these compounds have antifungal activity. Adamczyk-Woźniak *et al.* were the first to demonstrate that 3-aminobenzoxaboroles were active in various strains of fungi, in comparison with their phenylboronic acid analogues.[26,27] Tavaborole is the first representative of a novel class of antifungal drugs, sold under the brand name Kerydin®. In the literature, we can find some reports for the synthesis of this type of structure, for example, Adamczyk-Woźniak and co-workers reported the synthesis of piperazine bis(benzoxaboroles) by mechanosynthesis between a substituted formylphenylboronic acid and piperazine with moderate-to-good yields (Scheme 15.10). In addition to conducting microbiological studies on different strains of fungi.[28]

C–C couplings have also benefited from this tool called mechanochemistry; this has made it possible to synthesize intermediates and/or structures of numerous pharmaceuticals and drug candidates in a more ecological way compared to their traditional synthesis. In this context, indazoles are part of several drugs of medicinal interest that have shown therapeutic applications such as anticancer drugs, contraceptives,

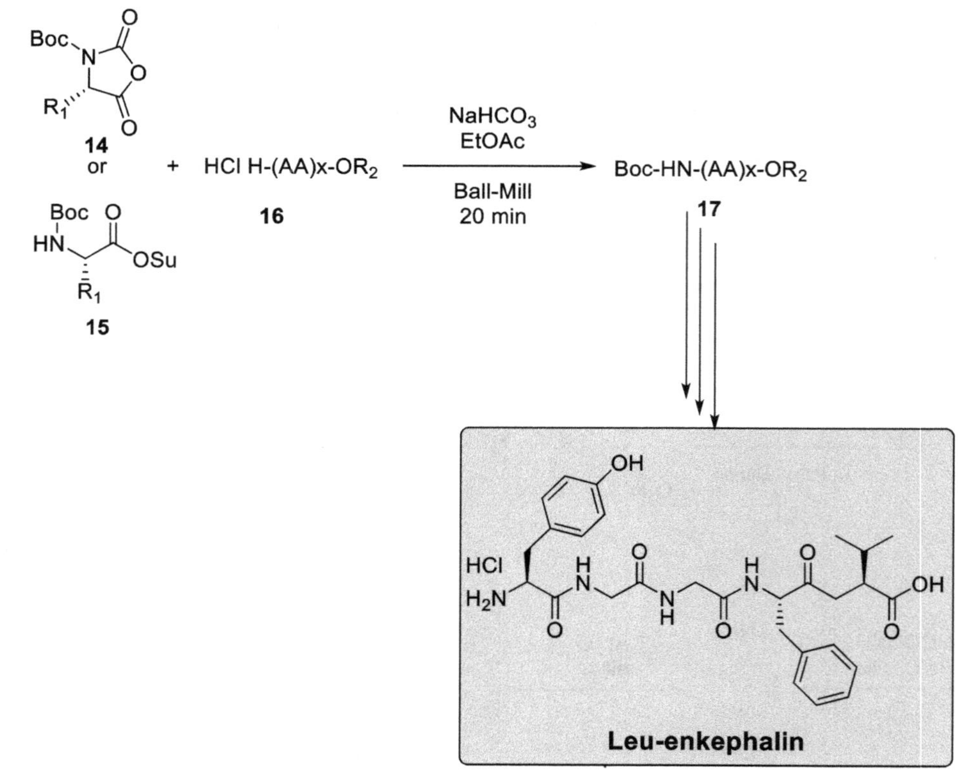

SCHEME 15.7 Synthesis of Leu-enkephalin without the intervention of enzymes.

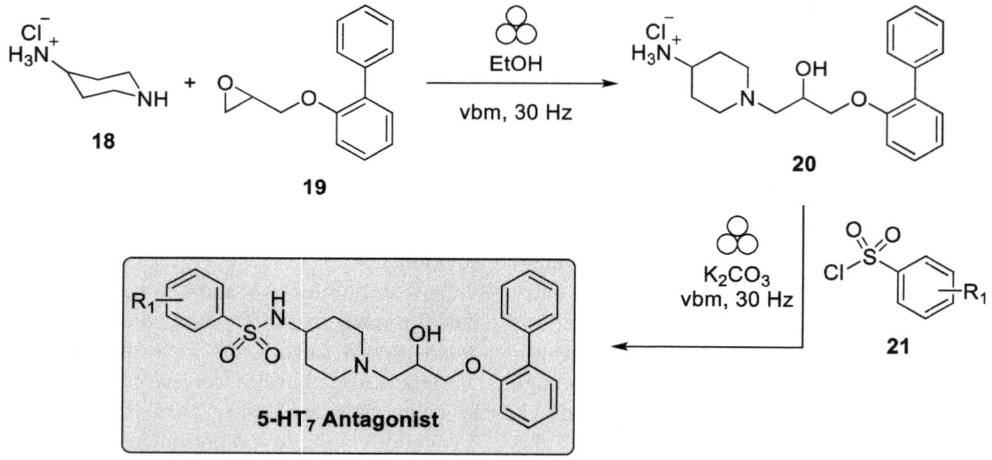

SCHEME 15.8 Eco-friendly synthesis of the 5-HT$_7$ antagonist using mechanosynthesis.

and protein inhibitors such as HIF-1a (hypoxia-inducible factor 1 alpha subunit). In various articles, it has been described that the functionalization of the pyrazole ring is carried out by using metal catalysts such as Rh (III),[29] Cu (I),[30] or palladium catalysts[31]; however, these reagents tend to be very expensive, in addition to using reaction conditions that make them more tedious. In contrast, a recent report published in 2019 by Su describes a methodology for the functionalization of the pyrrazole ring by a C–H/C–H coupling by mechanical chemistry using a palladium catalyst and a low-cost oxidizing agent with good product yields and wide tolerance of functional groups (Scheme 15.11).[32]

On the other hand, in 2020, Chettri *et al.* reported a Sonogashira cross-coupling reaction in a ball mill under the mechanochemical procedure and solvent-free conditions,[33] this reaction is a coupling reaction between terminal acetylenes with aryl or vinyl halides, to access various derivatives of isoxazolidine and isoxazoline (Scheme 15.12). It is important to mention that most of these synthesized compounds showed high biological activity in various cancer cell lines. The reaction conditions of this protocol allow having a simple, efficient, and eco-friendly methodology, whose attention may predominate among the community of organic chemists who are dedicated to developing "green chemistry."

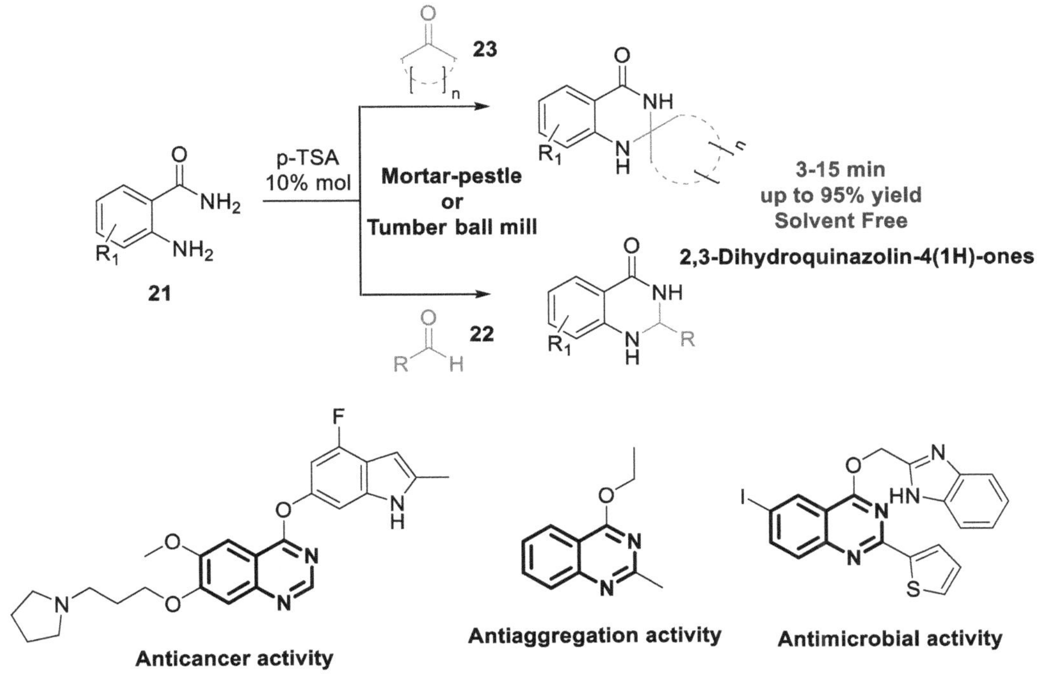

SCHEME 15.9 Synthesis of 2,3-Dihydroquinazolin-4(1H)-ones using mechanochemical approach.

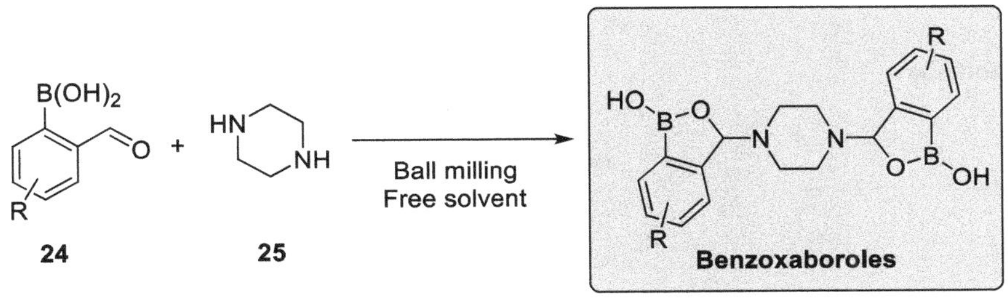

SCHEME 15.10 Benzoxaboroles synthesis using an eco-friendly methodology by Adamczyk-Woźniak's group.

15.3 Pharmaceutical Cocrystals by Mechanochemical

Cocrystals are defined by the Food and Drug Administration (FDA) as "crystalline materials composed of two or more different molecules, typically active pharmaceutical ingredient (API) and co-crystal formers ("coformers"), in the same crystal lattice."[34] There are two main methods for obtaining cocrystals: by solution and methods in solid states. The solution method consists of the combination of equimolar active pharmaceutical ingredient (API) and co-former in a solvent, and finally, when the solvent evaporates, the cocrystals are obtained. On the other hand, the solid methods consist of using mechanosynthesis by grinding in a mill or mortar.[35] Cocrystals allow modifying the biopharmaceutical properties of an API such as solubility, dissociation speed, physicochemical stability, and hygroscopicity.[36] Polymorphing is the ability of a compound to have more than one crystalline form. Some techniques for manipulating the crystalline form of APIs were previously mentioned; however, the newest technique is the formation of cocrystals, consisting of an API and one or more cocrystal forming agents. For the formation of cocrystals, it is important to consider some factors; among the most important ones are the functional groups of the API that allow molecular recognition by the cocrystallizing agent (co-former) through non-covalent interactions (heterosynthons and hydrogen bonding).[37] Among the most common co-forming agents, we can find carbohydrates, amino acids, alcohols, amines, and carboxylic acids.[21]

Most of the active principles of drugs are solid; this means that they are in a crystalline state. One of the problems of these drugs in the solid state is their low bioavailability due to various factors, including low solubility. There are techniques that modify the properties of compounds to increase solubility, among which are the formation of salts, solvates, and polymorphs, maintaining the solid state. In the last decade, the development of cocrystals through crystal-engineering principles using a mechanochemical approach has made it possible to improve the biopharmaceutical characteristics of drugs. An example of this is reported by Chadha *et al.* describing the development of the pharmaceutical cocrystal of Ambrisentan with syringic acid used to treat pulmonary arterial hypertension (Scheme 15.13). The developed cocrystal showed remarkable

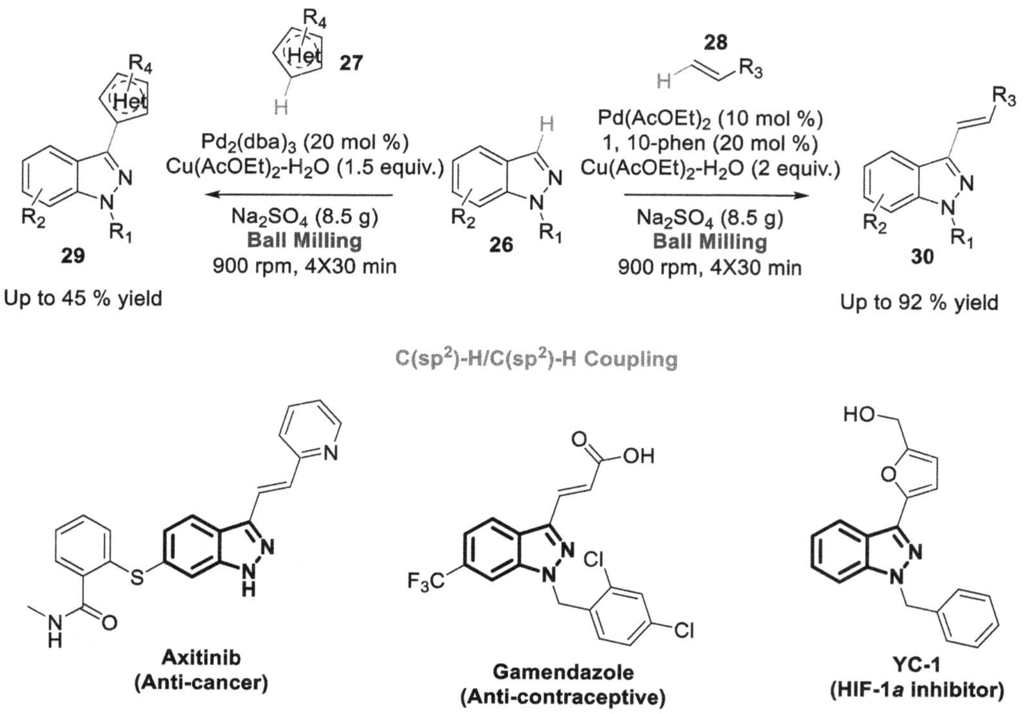

SCHEME 15.11 Coupling reaction using a mechanochemical approach for indazoles.

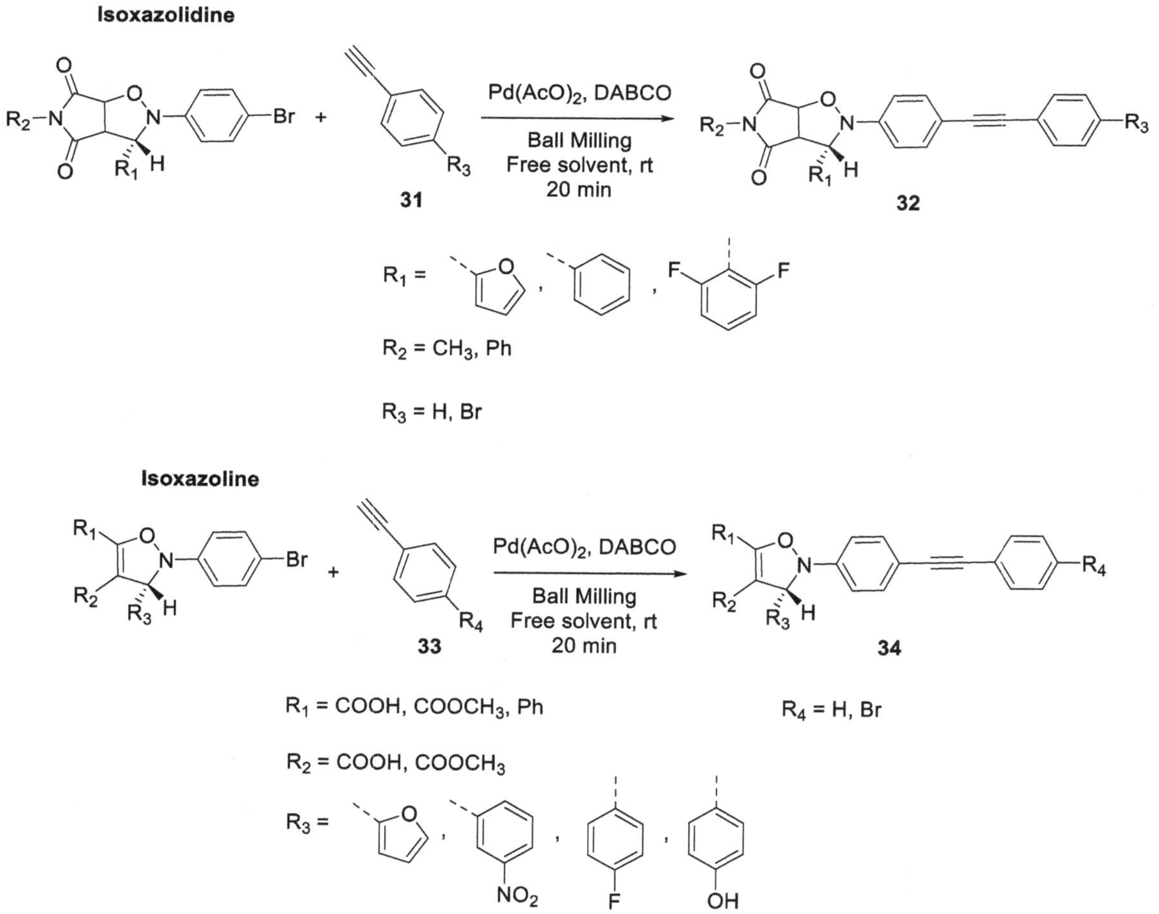

SCHEME 15.12 Sonogashira cross-coupling reaction in a ball mill under mechanochemical.

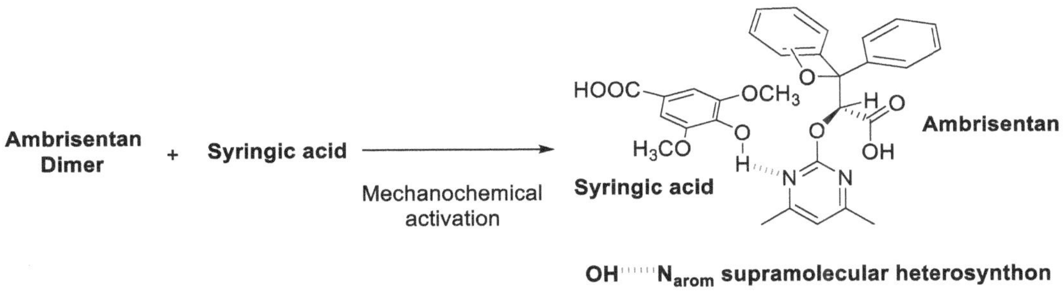

Ambrisentan Dimer + Syringic acid → Mechanochemical activation

Syringic acid · Ambrisentan

OH·····N_{arom} supramolecular heterosynthon

SCHEME 15.13 Cocrystal of Ambrisentan–syringic acid.

solubility and dissolution rate, compared to the drug in the pure form,[38] demonstrating that mechanosynthesis is a green alternative for the formation of cocrystals of APIs in solid state in the pharmaceutical industry.

We can cite other examples that allow us to visualize the scope of this tool for modifying the properties of APIs in the solid state. In this sense, a report by Ellena *et al.* described the synthesis of the co-drug involving the antimetabolite prodrug 5-fluorocytosine and the tuberculostatic drug isoniazid.[39] This cocrystal is a promising candidate for the treatment of fungal and bacteriological infections including for the treatment of cancer in patients undergoing therapy and, finally, to increase the physical stability of raw APIs. On the other hand, Marini *et al.* reported the mechanochemical synthesis of Bumetanide-4-aminobenzoic acid molecular cocrystals (Scheme 15.14). Bumetanide is a potent diuretic and natriuretic drug. They describe an easy, cheap, clean, and of course, green methodology.[40]

In addition to improving the solubility of an API by the formation of a cocrystal by means of mechanochemistry, other research groups have studied various properties of pharmaceutical cocrystals such as relative humidity,[41,42] competition between co-formers,[43] or addition of additives.[44] As described with the previous examples, the use of mechanosynthesis in the formation of cocrystals has allowed access to these crystalline materials more efficiently and with an environmentally friendly approach, thus making it a sustainable alternative to conventional solution synthesis. Furthermore, the formation of cocrystals has made it possible to improve the properties of the APIs, mainly the physicochemical properties. However, the stability of cocrystals toward formulation with excipients is little investigated as mentioned by Erxleben in an article published in 2020.[45] This report shows that the excipients used to incorporate the active principle can interact in different magnitudes with the APIs, observing a competition of non-covalent

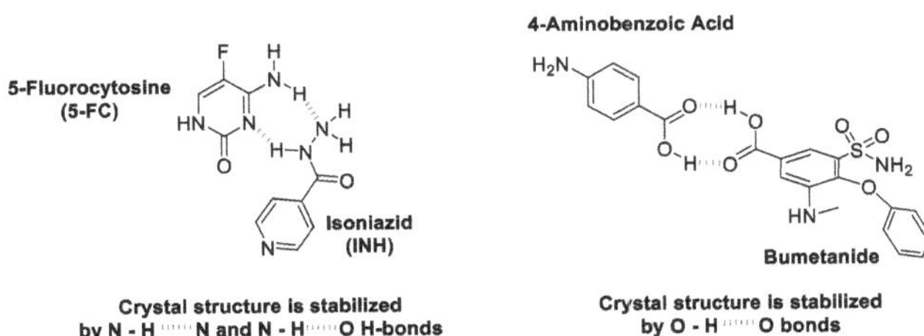

SCHEME 15.14 Cocrystals 5-FC-INH and Bumetanide-4-Aminibenzoic acid.

interactions (donor and acceptor groups of H bonds that can compete with the co-former for API H-binding sites) between the API, co-former, and excipient, whose result can interfere with the biopharmaceutical stability of the API.

In summary, we can say that mechanosynthesis has impacted in various ways in the pharmaceutical area; one of them is the formation of cocrystals for the modification of the biopharmacological properties of APIs. On the other hand, the development of new synthetic strategies for the maintenance of APIs is under development because there is a great variety of reactions that even with the implementation of mechanical chemistry, it is not possible to achieve the desired results, and this is due to the great diversity of functional groups in chemistry and their respective reactivity; hence, for there to be a chemical reaction, it is necessary to break that energy gap and manifest the reactivity of the functional groups, which is one of the limitations of mechanosynthesis, for now.

15.4 Ultrasound Synthesis of Drugs

Ultrasound is another method for the activation of chemical reactions; this technique can be used as an alternative to supplying energy to the system (heat, radiation) improving the reaction rate and chemical yields. The supply of energy occurs through an indirect and complex process called cavitation. This alternative is considered a green method, due to the low energy required to reach drastic reaction conditions in the medium that could not be achieved in a conventional way (high temperatures and pressures). In this sense, the application of this method for the synthesis of drugs with pharmacological activity is scarce in the literature, but that does not mean that this technique is out of reach for chemical processes; that is why the development of strategies using this tool would allow access to cheaper and more sustainable processes than conventional methodologies for the synthesis of drugs of pharmaceutical interest. For example, the application of ultrasound in the synthesis of 1,3,4-oxadiazole drugs for anti-cancer activity is reported by Jha and co-workers.[46] They reported a multi-compound reaction (three-component) of a series of 2-(N-heterocycle) substituted 1,3,4-oxadiazoles (Scheme 15.15). Moreover, cytotoxicity studies evidenced that product **O-1** which contained *p*-OH phenyl group of oxadiazole ring exhibited the best activity against two human cancer cell lines HT29 and HepG2 among all the different drugs synthesized. Ultrasound conditions not only improved the chemical yields but also accelerated the reaction times.

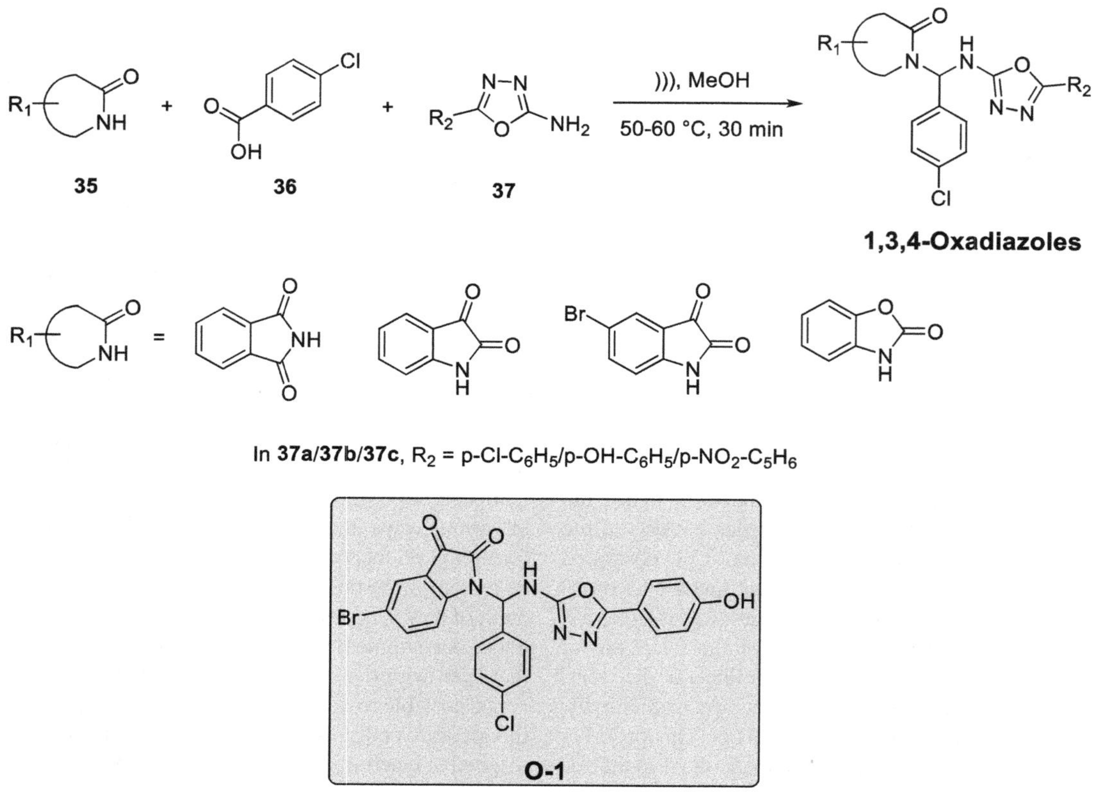

In **37a/37b/37c**, R_2 = p-Cl-C_6H_5/p-OH-C_6H_5/p-NO_2-C_5H_6

SCHEME 15.15 Synthesis of 1,3,4-oxadiazoles by ultrasound.

Thioesters are a functional group that is found in a great variety of organic compounds, this since the structure is used for the construction of heterocyclic compounds for the development of thio drugs with pharmaceutical importance.[47,48] However, there are very few reports for the synthesis of these compounds by ultrasound; one among one of them was reported by Siqueira and co-workers[49]; they describe a rapid method for the synthesis of thiols from mercaptobenzooxazole and benzoyl chloride, in excellent yields and short reaction times in comparison with conventional thermal methods (Scheme 15.16).

Peptides are molecules formed by the union of several amino acids through peptide bonds. Peptides differ from proteins due to their size; their mass is between 10,000 and 12,000 Daltons. These molecules have a broad spectrum of bioactivity such as antibacterial, antitumor, and antidiabetic activity. Furthermore, peptides can be linked to different heterocycles that allow them to improve bioactivity. That is why the pharmaceutical industry has an interest in obtaining these compounds; however, the most common methods for obtaining these species are usually expensive, due to the use of toxic reagents, use of catalysts, and handling of strict conditions. However, in 2018, Hooshmand *et al.* reported a methodology for obtaining pseudopeptides bound to Rhodamine-combined aliphatic amines, aromatic amines, carbon disulfide, aldehydes, anhydrides, and isocyanides through of one-pot reaction of six components in water using ultrasound irradiation (Scheme 15.17).[50] This methodology allows obtaining these compounds in an easy way, avoiding the excessive use of toxic substances and excluding purification methods, in addition to being an eco-friendly method.

Similarly, pyrazole and pyrimidine derivatives are structures that present important pharmacological properties such as antibacterial,[51] anti-inflammatory,[52] analgesic, antitumoral, and antifungal activities. Some of the methods used for the preparation of pyrazoles and pyrimidines are based on cycloadditions,[53,54] condensations of aldehydes or ketones,[55,56] and annulations,[57,58]; however, these methods use metal catalysts that are usually expensive, in addition to using toxic solvents, and have strict reaction conditions and long reaction times.

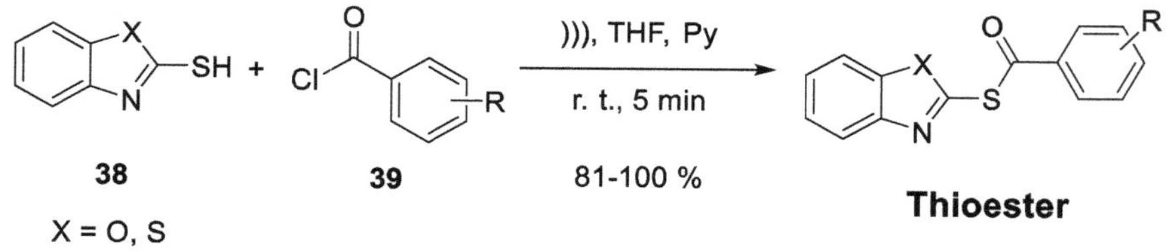

SCHEME 15.16 Synthesis of thioester.

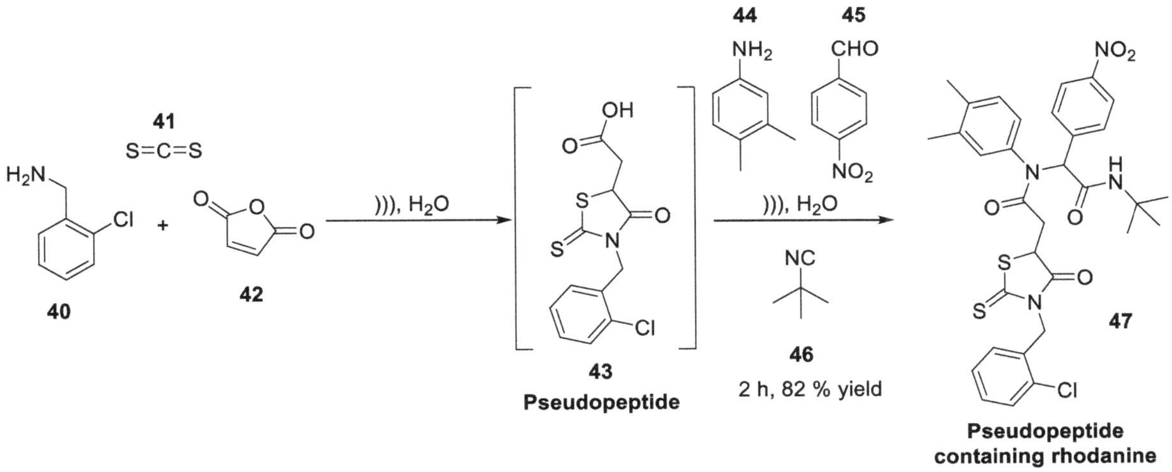

SCHEME 15.17 One-pot synthesis of pseudopeptide and derivates.

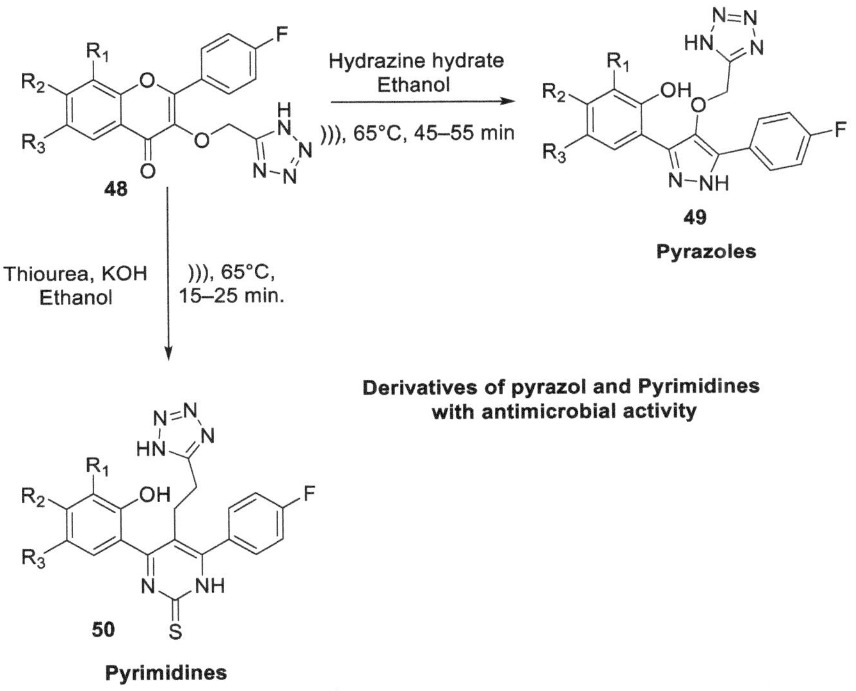

SCHEME 15.18 Synthesis of pyrimidines and pyrazole assisted by ultrasound.

In this sense, a report by Gill *et al.* describes a green method using ultrasound radiation to obtain these structures, avoiding the use of metallic catalysts and toxic solvents, in addition to reducing the reaction time and good yields (Scheme 15.18).[59]

15.5 Green Synthesis of Natural Products

15.5.1 History of Natural Products

Natural products, such as plants, minerals, and animals, have been commonly used to treat several diseases since ancient times. There are records dating back thousands of years that indicate people having used certain drugs for medical purposes. Medicinal plants and microorganisms have long been the main source of medicines. Plants contain a vast array of natural products with a wide range of structures. In contrast to the "main metabolites," which are required for plant growth and development, these products are referred to as "secondary metabolites." Secondary metabolites were once thought of as "waste products" that had no physiological purpose for the plant. However, it became clear about 30 years ago, with the advent of the field of chemical ecology, that these natural products play an important role in the relationship between plants and their biotic and abiotic climate. Natural products are compounds that occur naturally and are end products of primary metabolites; they are often special compounds for specific species.[60] They are plant chemicals that have protective or disease-preventive properties but are non-nutritive. They are non-essential nutrients, which means that the human body does not need them for survival. Alkaloids, tannins, saponins, steroids, flavonoids, terpenoid, and cardiac glycosides are the most common phytochemicals (secondary metabolites).[61]

15.5.2 Primary and Secondary Metabolites

Primary metabolism refers to the biosynthesis and breakdown of proteins, fats, nucleic acids, and carbohydrates, which are important to all living organisms. The substances involved in these pathways are referred to as "primary metabolites." The process by which an organism biosynthesizes compounds known as secondary metabolites (natural products) is known as "secondary metabolism" and it is often found to be peculiar to an organism or to be an example of a species' uniqueness. Secondary metabolites are developed because of the organism adapting to its surroundings or as a potential defensive mechanism against predators to help in the organism's survival.[62] The biosynthesis of secondary metabolites is derived from the fundamental processes of photosynthesis, glycolysis, and the Krebs cycle, which result in biosynthetic intermediates and, eventually, secondary metabolites, often known as natural products.[63] Although the number of building blocks is small, it can be seen that the possibilities are endless.[64]

15.5.3 Historically Important Natural Products

Heterocyclic compounds are abundant in nature and are essential for life in a variety of ways. Some heterocyclic compounds are essential to our survival because they are the building blocks of life. This category also contains essential life-saving medicines, as well as components of coloring matter and perfumery.[65] Many early medicines were based on conventional medical methods, which were then accompanied by clinical, pharmacological, and chemical research. The synthesis of the anti-inflammatory agent acetylsalicylic acid **(51)** (aspirin) from the natural product salicin **(52)** isolated from the bark of the willow tree *Salix alba* L. is perhaps the most popular and well-known example to date.[66] The isolation of many alkaloids from *Papaver somniferum* L. (opium poppy) led to the discovery of morphine **(53)**, a commercially valuable medication first published in 1803 (Figure 15.1). In the 1870s, crude morphine from the plant *P. somniferum* was boiled in acetic anhydride to produce diacetylmorphine (heroin), which was quickly converted to codeine (painkiller). Poppy extracts were used medicinally by the Sumerians and Ancient Greeks in the past, while the Arabs described opium as addictive. The (foxglove) *Digitalis purpurea* L. was first discovered in Europe in the 10th century, but it was not until the 1700s that the active constituent digitoxin **(54)**, a cardiotonic glycoside, was discovered to boost cardiac conduction and therefore cardiac contractibility. Digitoxin **(54)** and its analogues have long been used to treat congestive heart failure, but they can have long-term side effects and are being phased out in favor of other treatments for"heart failure." The anti-malarial drug quinine **(55)** isolated from the bark of *Cinchona succirubra* Pav. ex Klotsch and approved by the US FDA in 2004, had been used for centuries to treat malaria, flu, indigestion, mouth and throat infections, and cancer.[67]

The bark was first used to cure malaria in the mid-19th century when the British started cultivating the plant all over the world. Pilocarpine **(56)**, an L-histidine-derived alkaloid present in *Pilocarpus jaborandi* (Rutaceae), has been used as a prescription medication in the treatment of chronic open-angle glaucoma and acute angle-closure glaucoma for over a century.[68] The FDA licensed an oral formulation of pilocarpine **(56)** in 1994 to relieve dry mouth (xerostomia), a side effect of radiation therapy for head and neck cancers, and to stimulate sweat glands to test sodium and chloride concentrations (Figure 15.1). Sjogren's syndrome is an infectious condition that affects the salivary and lacrimal glands. In 1998, the oral preparation was approved for the treatment of Sjogren's syndrome.[69]

15.5.4 Development of New Medicines

The global pharmaceutical industry is worth approximately 1.1 trillion dollars. Natural products account for about 35% of all medicines. Plants (25%), microorganisms (13%), and animals (about 3%) all contributed directly or indirectly to the development of these medicines. Natural-derived products are a valuable resource for pharmaceutical firms seeking to create new drugs around the world. They are used as (i) a direct source of therapeutic agents (both pure drugs and medicinal products), (ii) a source of raw material for the manufacture of complex, semi-synthetic drugs, (iii) prototypes for lead molecule design, and (iv) taxonomic markers for drug discovery. Natural products or their derivatives account for about a third of the world's

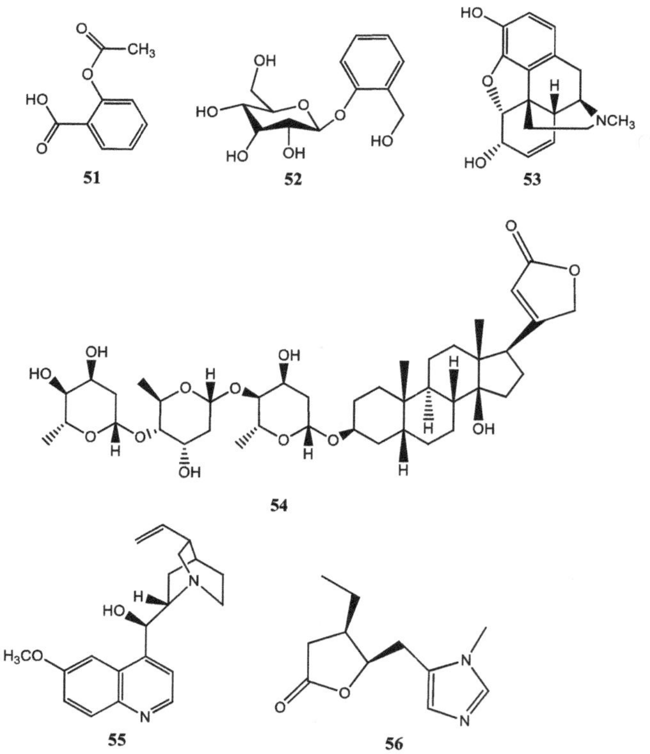

FIGURE 15.1 Acetylsalicylic acid (51), salicin (52), morphine (53), digitoxin (54), quinine (55), and pilocarpine (56).

best-selling medicines. Between 1983 and 1994, 39% of the 520 new drugs approved by the FDA were natural products or derived from natural products, and about 60–80% of antibiotics and anti-cancer drugs are derived from natural products. In addition, 13 new medicines based on natural ingredients were introduced to the market. Ixabepilone, retapamulin, trabectedin, and the peptides exenatide and ziconotide are some of these. In the biopharmaceutical industry, natural product drug discovery continues to play an important role in the clinical production of new therapies.

15.5.5 Sustainable Synthesis of Natural Products

Chemistry now plays an important part in improving and sustaining our quality of life. The chemical industry's race has a negative impact on the natural world and public well-being. The word "Green Chemistry" was coined by the US EPA (Environmental Protection Agency), in 1990, to describe chemical processes that minimize the use of hazardous or unsafe compounds in synthesis and manufacturing. The pharmaceutical industry and research laboratories that develop and synthesize organic molecules have made a lot of money. Green chemistry seeks to reduce the release of toxic by-products while increasing the yield of the desired compound without causing environmental damage.[70] Green chemistry, according to Paul Anastas and John C. Warner (1998), is the application of a series of concepts that restricts or avoids the use of dangerous compounds in reaction system design, waste avoidance, atom-economy maximization, use of cleaner solvents, reduced derivatization, and energy-efficient design. The development of sustainable synthetic methods for the construction of complicated cyclic target molecules with minimum environmental impacts is a challenging aim for academia and industry. New approaches to heterocycle synthesis have had immense repercussions both on organic and inorganic chemistry in the past two decades.[71] Heterocyclic sub-structuring also comes from natural materials, recycled fuels, agrochemicals, pharmaceuticals, and macromolecules. Synthesis approaches have increasingly evolved from classic condensation procedures to click reactions to modern domino multicomponent procedures. In addition, the development of modern heterocycle synthesis methods was a significant research issue for green and sustainable chemists. Further, more green aspects that would be aimed at synthesis will have:

- Minimum number of steps
- Mild operating conditions
- One-pot synthesis
- Increased yield

The largest class of bioactive organic compounds is heterocyclic compounds. The aim of green chemistry is to overcome the adverse environmental consequences caused by different dangerous chemicals used in synthetic pathways.[72] The examples below illustrate an extensive use of green chemistry in heterocyclic synthesis. Compounds that contain at least atoms other than carbon are heterocyclic. The largest class of bioactive organic compounds are heterocyclic compounds. Green chemistry seeks to overcome the harmful impact of multiple toxic substances on the atmosphere of synthetic paths. Examples in green chemistry demonstrate that heterocyclic compounds are extensively used in the synthesis.

15.5.6 Grinding Method

Advancing green chemistry made for those who applied chemistry in the medical, educational, and scientific sectors a host of challenges. The beginning of green chemistry is seen as a breakthrough in the fact that synthetic materials and pathways are used to minimize environmental degradation and human health harm. Green chemistry is better used in science to synthesize chemicals with benign, moderate, non-toxic, reproducible, and efficient catalysts. A. K. Bose and team reported that kilo-scale preparation of the target compound is an essential step in the progress of the procedure. A protocol for performing a few exothermic reactions on a wide scale was created involving water-based biphasic reaction media. A solvent-free, green-chemistry technique for Biginelli reaction using p-toluene sulfonic acid as a catalytic agent is used to demonstrate the energy-efficient and speedy preparation of dihydropyrimidinones. Water-based biphasic reactions were carried on a wide scale under vigorous agitation with immiscible organic reagents (for a few exothermic reactions); a quantity change of the water (or inclusion of crushed ice) will regulate the final temperature of the reaction mixture. Water-insoluble solid products that crystallized out in a short time were thus obtained in an essentially pure condition and with a very high yield. This protocol was energy-efficient and promised high nuclear efficiency (Scheme 15.19).[73]

It is well known that for many heterocyclic compounds, nitriles are commonly used as intermediates. Following our

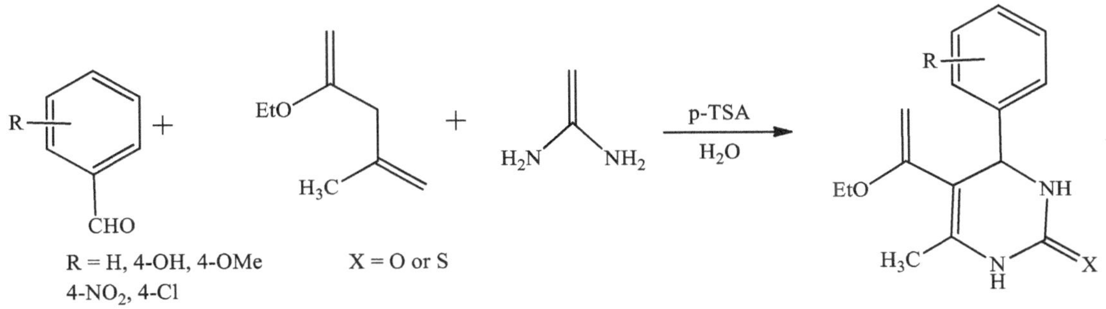

R = H, 4-OH, 4-OMe 4-NO_2, 4-Cl

X = O or S

SCHEME 15.19 The century-old Biginelli reaction.

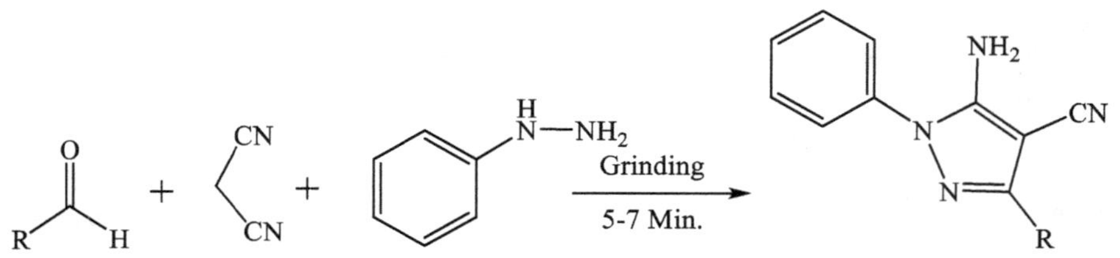

SCHEME 15.20 Synthesis of poly-substituted amino pyrazoles by simple grinding method.

interest in the synthesis of heterocyclic compounds of biological significance, they have synthesized a variety of new pyrazole derivatives by simply grinding the aromatic aldehydes, malononitrile and phenyl hydrazine. These catalyst-free reactions were efficient and successful and gave various other advantages including rapid reaction, fast experimental work-up, and no toxic by-products. The solution presented herein to pyrazole systems prohibits the use of catalyst, organic solvent toxicity. This protocol is a promising green path for this form of compound to synthesize (Scheme 15.20).[74]

Harjyoti Thakuria group has produced a simple, solvent-free, and highly efficient protocol for the synthesis of quinoxaline derivatives without the use of a catalyst. The thermal analysis of the hydrated crystals of the two derivatives has been completed. Powder X-ray diffraction and FT-IR study of hydrated and dehydrated crystals were also performed to demonstrate the function of water in stabilizing the overall solid-state structural network of these molecules (Scheme 15.21).[75]

15.5.7 Microwave-Assisted Method

A microwave technique aided the synthesis of organic compounds. In relation to grinding and traditional methods such as reaction time reduction, high returns, more atomic economy, this approach offers significant advantages. This technology is more environmentally conscious and presents a big solution to green chemistry. In recent years, microwave irradiation is an additional source of thermal energy to traditional heat. Many active reactions were revealed with great efficiency and significantly increased reaction rates. Auto-concentrating, temperature-controlled, microwave flash heating in containers has recently proved to be very effective in the acceleration of organic reactions and was commonly used in a parallel synthesis. However, the preparation of 6-aryl-2,4-diamino-1,3,5-triazines by the microwave technique has not been identified, despite the effective application of microwave irradiation in organic synthesis (Scheme 15.22).[76]

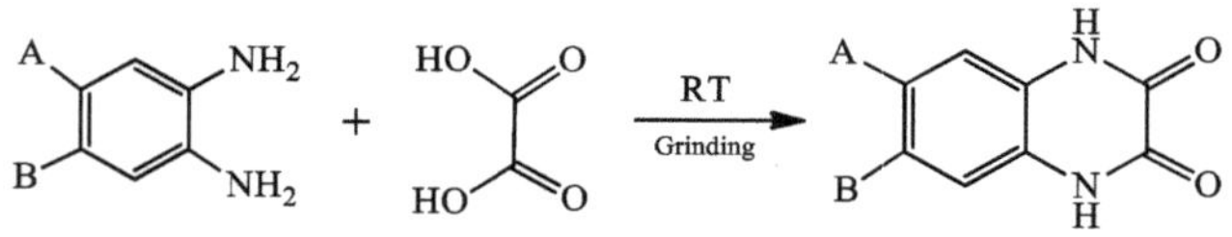

SCHEME 15.21 1,4-dihydro-quinoxaline-2,3-dione by grinding method.

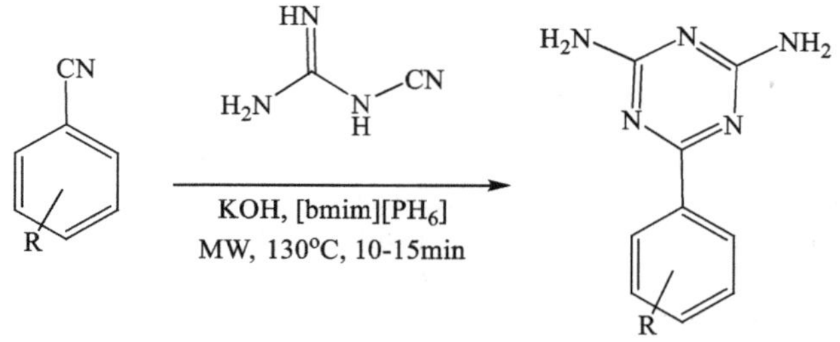

SCHEME 15.22 Cyclization of dicyandiamide with various aryl nitriles using ionic liquid [bmim][PF_6] as an eco-friendly solvent.

Chalcones and coumarins are natural plant components of importance and exhibit a broad variety of biological and medicinal activities. Synthetization of biphenyl chalcone and coumarin derivatives by Suzuki combination by using PEG400 as a solvent with microwave irradiation was achieved in an environmentally benign way. This methodology's salient attribute includes fast response time, decent-to-excellent results, and influential tolerance for various functional categories. The Suzuki coupling allowed the synthesis of a series of new chalcone and coumarin derivatives. In natural products and analogues, using palladium acetate as a precursor, PEG as solvent with microwave irradiation, it can be introduced with this method. Moreover, several function groups with our typical reaction conditions that guarantee a decent-to-excellent production of the target product are one of the most important features of the present methodology (Scheme 15.23).[77]

15.5.8 Using Green Catalyst Method

Among the difficulties faced by chemists are the discovery and development in selecting alternate reaction conditions and solvents for significantly improving selectivity, conservation of energy, and even less dangerous waste generation, of non-hazardous, environmentally friendly chemical processes.[78,79] This

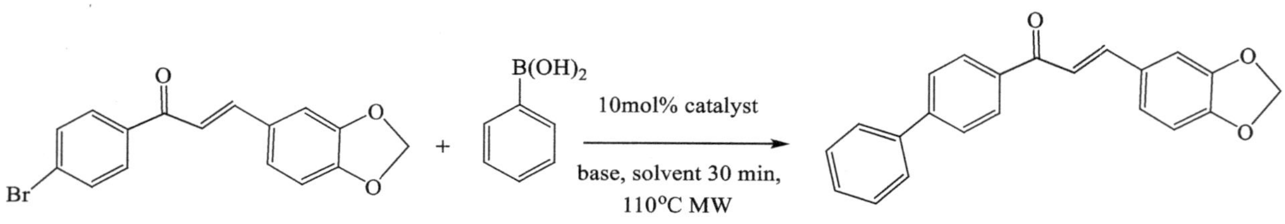

SCHEME 15.23 Suzuki reaction of chalcone with phenylboronic acid in different conditions.

includes chemical goods that are not desirable or inherently safe. Organic reactions, such as clay, phosphates, gold, animal bone, and other natural catalysts, have been published in the literature. As a catalyst, the authors record a solvent-free one-pot cyclocondensation reaction of substituted arilaldehydes, diketone/ketoester, and urea with good results (Scheme 15.24).[80]

In this connection, under solvent-free conditions, the Suresh Patil group report the synthesis of DHPMs using pineapple juice as the natural catalyst. According to a literature review, pineapple juice as a catalyst for Biginelli reactions has not been previously recorded. This catalyst produced higher yields for various aromatic aldehydes in addition to its clean and simple nature. Recently, fluorescent sensors with triazole were developed to detect platinum ions selectively in aqueous solutions and a fluorescent sensor for Cd^{2+} and Zn^{2+} was synthesized with the use of pyrenyl adjacent triazole-based Calyx arena. The discovery of new fluorescent organic compounds with high versatility, which has been subject to an intensive analysis for more than a decade, with the fast expansion of the use of organic fluorescent materials for electroluminescence (EL), dye lasers, filters, probes, and phototherapeutic agents and the fluorescent properties of triazole derivatives (Scheme 15.25).[80]

A comfortable and speedy method of synthetization of the nitrogen heterocycle derivatives is defined with the use, on a commercially available basis, of chitosan in 2% acetic acid at 60–65°C, of a single pot, three-component reaction of the replaced aromatic aldehydes, dicarbonyl and 2 aminobenzothiazole/3 amino-1,2,4-triazole/urea/thiourea. Chitosan was used for this multicomponent reaction as an important biodegradable and reusable green catalyst. In a fresh reaction, the chitosan catalyst is reusable ten times. In the first few runs, catalytic activity was maintained but in the following cycles, it marginally decreased (Scheme 15.26).[81]

15.5.9 Solvent-Free Method

When it comes to organic reactions, it is normal to believe that they take place in a solvent medium. This principle has a straightforward rationale. That is, if the reactants are in a homogeneous solution, they can interact effectively, which makes stirring, shaking, or other forms of agitation easier, allowing the reactant molecules to come together quickly and continuously. Furthermore, if uniform heating or cooling of the mixture is needed, it is relatively simple to do so in a solution. However, the position of a solvent in an organic reaction is much more complicated than simply providing a homogeneous environment for many reactant collisions to occur. A solvent can dramatically increase or decrease the speed of a reaction.[82] The rate of a reaction can be influenced by changing

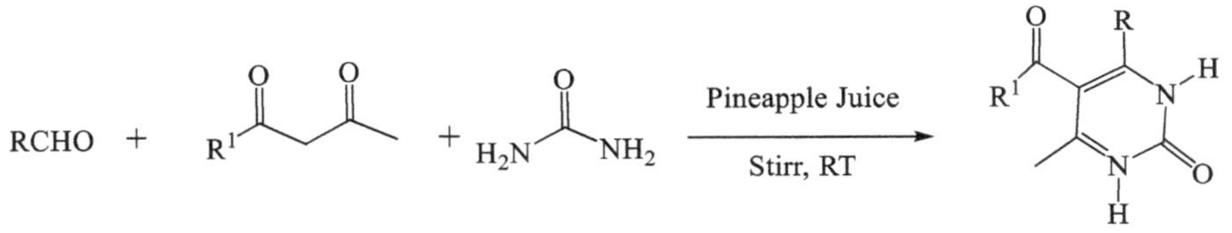

R = -H, -Ph, o-ClC_6H_4, p-OHC_6H_4, p-$OCH_3C_6H_4$, p-OH m-$OCH_3C_6H_3$
-CH = CHC_6H_4, o-$NO_2C_6H_4$, -C_4H_3O(furfural)
R^1 = OEt, Me

SCHEME 15.24 Synthesis of dihydropyrimidinones.

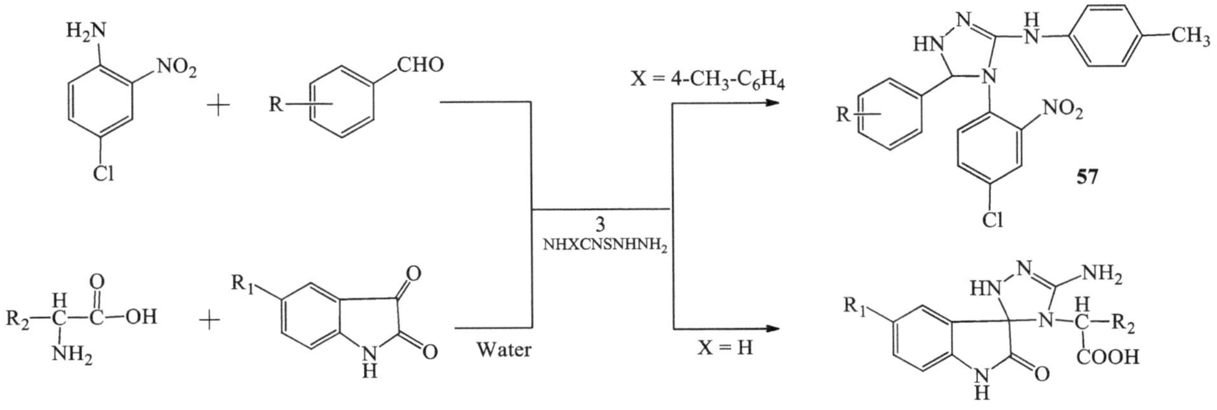

R = 4-OH, 3-OH-4-OCH_3, 3,4-dimethyl 4-Cl, 3-OH, 2,4-dimethyl, 2-OH, 3-OCH_3, 4-OCH3, 3-4-dimethoxy.
R_1 = -Cl, -NO_2, H
R_2 = -CH_3, -C_3H_7S, -$C_4H_5N_2$, -C_9H_8N, -C_4H_9, -C_3H_7, -CH_2OH, -C_7H_7

SCHEME 15.25 Synthesis of triazole derivatives 57 and 58.

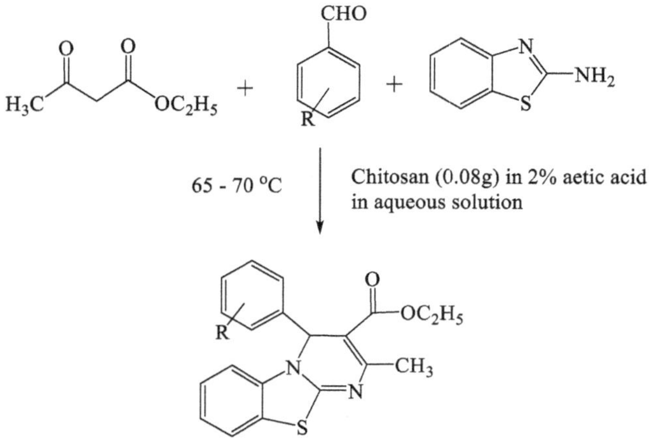

SCHEME 15.26 Preparation of 4H-Pyrimido [2,1-b]benzothiazole derivatives.

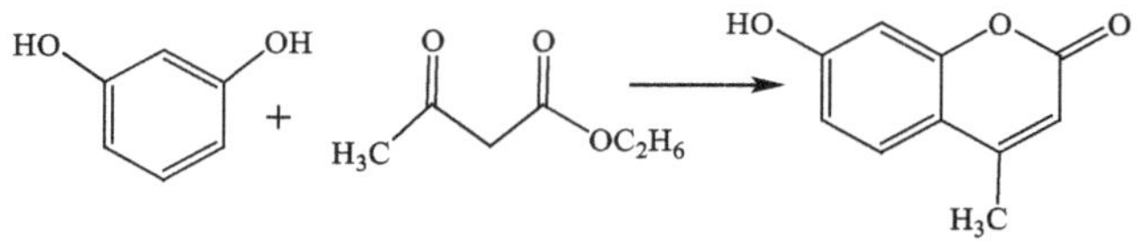

SCHEME 15.27 Pechmann reaction of resorcinol with ethyl acetoacetate to produce 7-hydroxy-4-methylcoumarin.

the solvent, and it can be strong enough to alter the reaction's trajectory. This could result in different product yields and ratios. As a result of the solvation of the reactants, products, transition-state, or other intervening species, a solvent could be strongly and inextricably linked to the mechanism of an organic reaction. Electrostatic, steric, and conformational effects, among other things, are responsible for certain close interactions between the solvent and the reaction partners. Despite its heavy presence, the solvent does not usually become part of the substance (except in solvolysis reactions) and is recovered intact after the reaction is completed. Even then, conducting a reaction in the absence of a solvent should not be considered or expected.[83]

As the Souad Bouasla group reported, the Pechmann reaction over heterogeneous solid acid catalysts in a free solvent media under microwave irradiation is described as a suitable methodology for the synthesis of coumarin derivatives. In the Pechmann condensation, resorcinol, phenol, and ethyl acetoacetate were selected as model reactants. Several materials – Amberlyst-15, zeolite, and a sulfonic-acid-functionalized hybrid – were tested for catalytic activity. Firstly, resorcinol and ethyl acetoacetate were employed as the model reactants in the Pechmann condensation under microwave reaction. Resorcinol (1,3-dihydroxybenzene) was chosen due to its high reactivity (Scheme 15.27).[84]

Benjaram M. Reddy reported that the Pechmann reaction catalyzed by silica-gel-assisted sulfuric acid (H_2SO_4/silica gel) at 120°C in high yields under solvent-free reaction conditions with short reaction times is defined as a remarkable acceleration in the synthesis of substituted coumarins. This approach outperforms other methods for coumarin synthesis in terms of product yield, ease of use, and environmental benefits due to the absence of poisonous catalysts and solvents (Scheme 15.28).[85]

Under solvent-free microwave irradiation conditions, the group demonstrated a novel technique based on the Pechmann condensation, catalyzed by FeF_3 as an efficient eco-friendly catalyst, for the synthesis of substituted coumarins. The related coumarins were obtained in moderate to high yields. One-pot synthesis, experimental simplicity under solvent-free microwave irradiation, high yields obtained in short reaction times, and simple and fast product isolation are all advantages of this process. The bulk of the compounds had strong efficacy against a variety of bacteria and fungi, with inhibition areas that were almost identical to those of normal medicines. As a result, a new class of compounds has been discovered that has antimicrobial potency equal to certain already used commercial bactericides/fungicides (Scheme 15.29).[86]

15.5.10 Water as Greener Solvent

Diels and Alder in 1931 announced the first described example of an organic reaction, the cycloaddition of furan and maleic anhydride in water, which was confined only to the Diels–Alder reaction. Previously, the use of aqueous media in organic reactions was reduced due to the limited solubility of organic substrate in the water solvent. Such researches led to the creation of high-temperature water (HTW) as a more environmentally friendly technique for organic reactions. Because of its lower dielectric constant, water acts as a "pseudo-organic solvent" under near-critical and supercritical conditions. As a result, the solvating ability of water for organic compounds at room temperature is equal to that of ethanol or acetone.[87]

Chemical reactions mediated by water have sparked a lot of interest in synthetic chemistry.

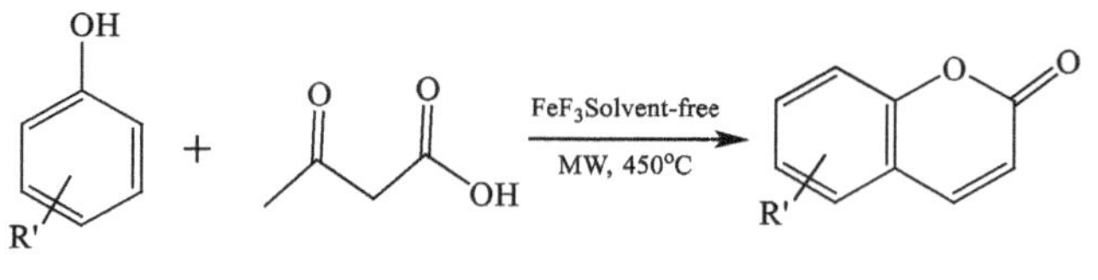

SCHEME 15.29 FeF_3-catalyzed Pechmann reaction.

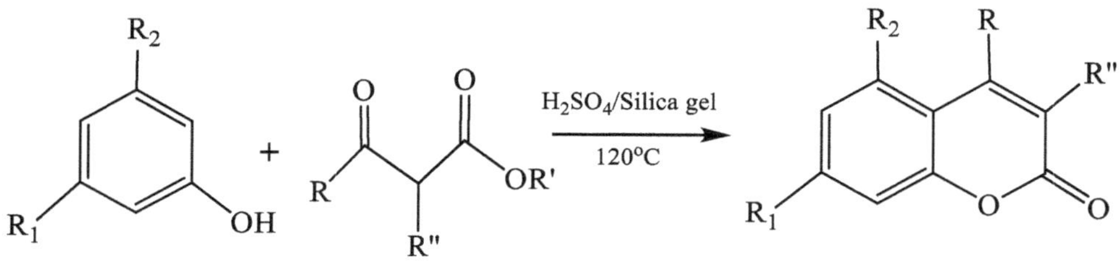

SCHEME 15.28 Pechmann condensation reaction of phenols with -keto esters to substituted coumarins.

- It is cleaner, less expensive, non-flammable, and widely available.
- Water's peculiar and unusual physical properties, such as its high specific heat, high surface tension, high dielectric constant, large cohesive capacity, amphoteric nature, and ability to form H-bonds, can help chemical reactions speed up and be more selective.
- Because of its ability to form tight hydrogen bonds, it has a high surface tension (three times that of liquid ammonia), which can aid in the aggregation of reactants.
- Its ability to participate in electron-transport reactions, as both biological and synthetic reactions demonstrate. In a water solvent, heterogeneous and homogeneous catalysts may be recycled.

REFERENCES

1. Sheldon, R. A. *Green Chem.*, 2007, 9, 1273–1283.
2. Bryan, M. C. Dunn, P. J. Entwistle, F. Gallou, S. G. Koenig, J. D. Hayler, M. R. Hickey, S. Huges, M. E. Kopach, G. Moine, P. Richardson, F. Roschangar, A. Steven, F. Weiberth, *J. Green Chem.*, 2018, 20, 5082–5103.
3. de los Santos Castillo-Peinado, L., M. D. Luque de Castro, *J. Pharm. Pharm.*, 2016, 68, 1249–1267.
4. Ying, P., J. Yu, W. Su. *Adv. Synth. Cat.*, 2021, 363, 1246–1271.
5. Anastas, P. T.; Warner, J. C. *Green chemistry: Theory and practice*, Oxford University Press: New York, 1998, p. 30.
6. Muñoz-Batista, M. J., Rodriguez-Padron, D., Puente-Santiago, A. R., and Luque, R., *ACS Sustain. Chem. Eng.*, 2018, 6, 8, 9530–9544.
7. Konnert, L., Reneaud, B., Marcia de Figueiredo, R., Campagne, J.-M., Lamaty, F., Martinez, F., and Colacino, E., *J. Org. Chem.* 2014, 79, 10132–10142.
8. Kutt, H., Harden, C. L., *Handb. Exp. Pharmacol.* 1999, 138, 229–265.
9. Iqbal, Z., Ali, S., Iqbal, J., Abbas, Q., Qureshi, I. Z., Hameed, S., *Bioorg. Med. Chem. Lett.*, 2013, 23, 488–491.
10. Marton, J., Enisz, J., Hosztafi, S., Timar, T., *J. Agric. Food Chem.*, 1993, 41, 148–152.
11. Bucherer, H. T., and Steiner, W., *J. Prakt. Chem.*, 1934, 140, 291.
12. Colacino, E., Porcheddu, A., Halasz, I., Charnay, C., Delogu, F., Guerra, R., and Fullenwarth, J., *Green Chem.*, 2018, 20, 2973–2977.
13. Porcheddu, A., Delogu, F., De Luca, L., and Colacino, E., *ACS Sustain. Chem. Eng.*, 2019, 7, 12044–12051.
14. Coursindel, T., Martinez, J., and Parrot I., *J. Chem. Educ.*, 2010, 87, 6, 640–642.
15. Declas, N., Le Vaillant, F., and Waser, J., *Org. Lett.*, 2019, 21, 524–528.
16. Solà, R., Sutcliffe, O. B., Banks, C. E., and Maciá, B., *Sustain. Chem. Pharm.*, 2017, 5, 14–21.
17. Portada, T., Margetić, D., and Štrukil, V., *Molecules*, 2018, 23, 3163.
18. Kullmann, W., *Biochem. Biophys. Res. Commun.*, 1979, 91, 693–698.
19. Rasmussen, G. J.; Bundgaard, H., *Int. J. Pharm.*, 1991, 76, 113–122.
20. Bonnamour, J., Métro, T.-X., Martinez, J., and Lamaty, F., *Green Chem.*, 2013, 15(5), 1116.
21. Canale, V., Frisi, V., Bantreil, X., Lamaty, F., and Zajdel, P., *J. Org. Chem.*, 2020, 85, 16, 10958–10965.
22. Chen, D.-S., Dou, G.-L., Li, Y.-L., Liu, Y., and Wang, X.-S., *J. Org. Chem.*, 2013, 78, 5700–5704.
23. Zhang, Z.-H., Lü, H.-Y., Yang, S.-H., and Gao, J.-W., *J. Comb. Chem.*, 2010, 12, 5, 643–646.
24. Lin, Z., Qian, J., Lu, P., and Wang, Y., *J. Org. Chem.*, 2020, 85, 11766–11777.
25. Yashwantrao, G., Jejurkar, V. P., Kshatriya, R., and Saha, S., *ACS Sustain. Chem. Eng.*, 2019, 7, 13551–13558.
26. Borys, K. M., A. Matuszewska, D. Wieczorek, K. Kopczyńska, J. Lipok, I. D. Madura and A. Adamczyk-Woźniak, *J. Mol. Struct.*, 2019, 1181, 587–598.
27. Wieczorek, D., J. Lipok, K. M. Borys, A. Adamczyk-Woźniak and A. Sporzyński, *Appl. Organomet. Chem.*, 2014, 28, 347–350.
28. Borys, K. M., Wieczorek, D., Tarkowska, M., Jankowska, A., Lipokb, J., and Adamczyk-Woźniak, A., *RSC Adv.*, 2020, 10, 37187.
29. Cai, S., Lin, S., Yi, X., and Xi, C., *J. Org. Chem.*, 2017, 82, 512–520.
30. Ye, Y., Kevlishvili, I., Feng, S., Liu, P., and Buchwald, S.L., *J. Am. Chem. Soc.*, 2020, 142, 10550–10556.
31. Naas, M., El Kazzouli, S., Essassi, E. M., Bousmina, M., Guillaumet, G., *Org. Lett.*, 2015, 17, 4320–4323.
32. Yu, J., Yang, X., Wu, C., and Su, W.-K., *J. Org. Chem.*, 2019, 85, 1009–1021.
33. Chakraborty, B., and Chettri, E., *Results Chem.*, 2020, 2, 100037.
34. Guidance for Industry. Regulatory Classification of Pharmaceutical Co-crystals; U.S. Food and Drug Administration, 2018. https://www.fda.gov/downloads/Drugs/Guidances/UCM281764.pdf.
35. Delori, A., Friščićab, T., and Jones, W., *Cryst. Eng. Comm.*, 2012, 14, 2350–2362.
36. Herrera Ruiz, D., *Rev. Mex. Cienc. Farm.*, 2010, 41, 55–56.
37. Almarsson, Ö., and Zaworotko, M., *J. Chem. Comm.*, 2004, 7, 1889–1896.
38. Haneef, J., and Chadha, R., *CrystEngComm*, 2020, 22, 2507–2516.
39. Souza, M. S., Diniz, L. F., Vogt, L., Carvalho Jr., P. S., D'vries, R. F., and Ellena, J., *Cryst. Growth Des.*, 2018, 18, 5202–5209.
40. Bruni, G., Maietta, M., Berbenni, V., Mustarelli, P., Ferrara, C., Freccero, M., Grande, V., Maggi, L., Milanese, C., Girella, A., and Marini, A., *J. Phys. Chem. B*, 2014, 118, 9180–9190.
41. Pekar, K. B., Lefton, J. B., McConville, C. A., Burleson, J., Sethio, D., Kraka, E., and Runčevskj, T., *Cryst. Growth Des.*, 2021, 21, 1297–1306.
42. Eddleston, M. D., Thakuria, R., Aldous, B. J., and Jones, W., *J. Pharm. Sci.*, 2014, 103, 2859–2864.
43. Fischer, F., Joester, M., Rademann, K., and Emmerling, F., *Chem. Eur. J.*, 2015, 21, 14969–14979.
44. Kulla, H., Becker, C., Michalchuk, A. A. L., Linberg, K., Paulus, B., and Emmerling, F., *Cryst. Growth Des.*, 2019, 19, 7271–7279.
45. Aljohani, M., McArdle, P., and Erxleben, A., *Cryst. Growth Des.*, 2020, 20, 4523–4532.

46. Bhatt, P., Sen, A., and Jha, A., *Chemistry Select*, 2020, 5, 3347–3354.
47. Kharkar, P. M., Rehmann, M. S., Skeens, K. M., Maverakis, E., and Kloxin, A. M., *ACS Biomater. Sci. Eng.*, 2016, 2, 165–179.
48. Petit, E., Bosch, L., Costa, A. M., and Vilarrasa, J., *J. Org. Chem.*, 2019, 84, 11170–11176.
49. Duarte, A., Cunico, W., Pereira, C. M. P., Flores, A. F. C., Freitag, R. A., and Siquiera, G. M., *Ultrasonics Sonochem.*, 2010, 17, 281–283.
50. Shaabani, A., and Hooshmand, S. E., *Ultrasonics Sonochem.*, 2018, 40, 84–90.
51. Mugnaini, C., Sannio, F., Brizzi, A., Del Prete, R., Simone, T., Ferraro, T., De Luca, F., Corelli, F., and Docquier J.-D., *ACS Med. Chem. Lett.*, 2020, 11, 5, 899–905.
52. Szabó, G., Fischer, J., Kis-Varga, A., and Gyires, K., *J. Med. Chem.*, 2008, 51, 1, 142–147.
53. Onodera, S., Kochi, T., and Kakiuchi, F., *J. Org. Chem.*, 2019, 84, 6508–6515.
54. Yi, F., Zhao, W., Wang, Z., and Bi, X., *Org. Lett.*, 2019, 21, 3158–3161.
55. Lellek, V., Chen, C.-Y., Yang, W., Liu, J., Ji, X., and Faessler, R., *Synlett*, 2018, 29, 1071–1075.
56. Chu, X.-Q., Cao, W.-B., Xu, X.-P., and Ji, S.-J., *J. Org. Chem.*, 2017, 82, 1145–1154.
57. Jadhav, S. D., and Singh, A., *Org. Lett.*, 2017, 19, 5673–5676.
58. Zhang, J.-L., Wu, M.-W., Chen, F., and Han, B., *J. Org. Chem.*, 2016, 81, 11994–12000.
59. Dofe, V. S., Sarkate, A. P., Shaikh, Z. M., and Gill, C. H., *Heterocycl. Comm.*, 2018, 24, 59–65.
60. Dias, D. A., S. Urban and U. Roessner, *Metabolites*, 2012, 2, 303–336.
61. Aricò, F., *Front. Chem.*, 2020, 8, 1–2.
62. Barnes, J., *Focus Altern. Complement. Ther.*, 1998, 3, 78–78.
63. Dewick, P. M., *Medicinal natural products: A biosynthetic approach*, 3rd Edition, 2009.
64. Dewick, R., P. M. Perroy and S. Careas, *Medicinal Natural Products A biosynthetic approach*, 2009, 3rd Edition, Wiley.
65. Butler, M. S., *J. Nat. Prod.*, 2004, 67, 2141–2153.
66. Eagle, S. R. and S. C. Gad, in *Encyclopedia of Toxicology*, 3rd Edition, 2014.
67. Elderfield, R. C., *Science*, 1966, 151, 317.
68. Newman, D. J. and G. M. Cragg, *J. Nat. Prod.*, 2016, 79, 629–661.
69. Aniszewski, T., *Alkaloids-secrets of life: Alkaloid chemistry, biological significance, applications and ecological role*, 1st ed. Elsevier, 2007.
70. Ameta, K. L. and A. Dandia, *Green chemistry: Synthesis of bioactive heterocycles*, Springer, 2014.
71. Ilango, K., B. Baskar and S. Murugesan, *Green chemistry assisted synthesis of natural and synthetic compounds as anticancer agents*, Elsevier Inc., 2020.
72. Young, D. M., J. J. C. Welker and K. M. Doxsee, *J. Chem. Educ.*, 2011, 88, 319–321.
73. Bose, A. K., M. S. Manhas, S. Pednekar, S. N. Ganguly, H. Dang, W. He and A. Mandadi, *Tetrahedron Lett.*, 2005, 46, 1901–1903.
74. Bhale, P. S., S. B. Dongare and U. B. Chanshetti, *Res. J. Chem. Sci.*, 2014, 4, 16–21.
75. Thakuria, H. and G. Das, *J. Chem. Sci.*, 2006, 118, 425–428.
76. Peng, Y. and G. Song, *Tetrahedron Lett.*, 2004, 45, 5313–5316.
77. Vieira, L. C. C., M. W. Paixao, A. Correa, *Tetrahedron Lett.*, 2012, 53, 2715–2718.
78. Singh, R., S. Gulati, R. Suprita and S. Singh, *Int. J. Chem. Stud.*, 2017, 5, 479–485.
79. Patil, S., S. D. Jadhav and S. Y. Mane, *Int. J. Org. Chem.*, 2011, 1, 125–131.
80. Sachdeva, H., R. Saroj, S. Khaturia and D. Dwivedi, *Org. Chem. Int.*, 2013, *2013*, 1–19.
81. Sahu, P. K., P. K. Sahu, S. K. Gupta and D. D. Agarwal, *Ind. Eng. Chem. Res.*, 2014, 53, *2085–2091*.
82. Bougrin, K., A. Loupy and M. Soufiaoui, *J. Photochem. Photobiol. C Photochem. Rev.*, 2005, 6, 139–167.
83. Martins, M. A. P., C. P. Frizzo, D. N. Moreira, L. Buriol and P. Machado, *Chem. Rev.*, 2009, 109, 4140–4182.
84. Bouasla, S., J. Amaro-Gahete, D. Esquivel, M. I. López, C. Jiménez-Sanchidrián, M. Teguiche and F. J. Romero-Salguero, *Molecules*, 2017, 22, 2072.
85. Reddy, B. M., B. Thirupathi and M. K. Patil, *Open Catal. J.*, 2009, 2, 33–39.
86. Vahabi, V. and F. Hatamjafari, *Molecules*, 2014, 19, 13093–13103.
87. Dunn, P. *J. Chem. Soc. Rev.*, 2012, 41, 1452–1461.

16

Biosynthesis of Natural Products

Athar Ata, Samina Naz, and Kenneth Friesen

CONTENTS

16.1 Introduction 305
16.2 Biosynthesis of Homoisoflavonoids 305
16.3 Biosynthesis of Alkaloids 305
16.4 Biosynthesis of Terpenoids 313
Acknowledgment 316
References 316

16.1 Introduction

Natural product chemistry has provided a significant number of lead compounds to the drug-discovery program. Approximately 50% of the prescribed pharmaceuticals are of natural product origin [1, 2] and half of these come from plant origin. This huge contribution to the drug-discovery program is due to the enormous structural diversity present in natural product chemistry compared to combinatorial chemistry and genomic approaches [3]. An extensive research in combinatorial chemistry has yielded only one anticancer drug, sorafenib, compared to natural products which have provided 49% of 877 small molecules introduced as pharmaceuticals during 1981–2002 [4, 5]. One of the major problems in the drug-discovery process is the supply of potent bioactive compounds for detailed in-vivo evaluation of these compounds for clinical trials due to their presence in minor quantities in nature. Their synthesis on large scales seems to be difficult due to their complex structures and the presence of asymmetric centers in them. Presently, lead bioactive natural products are mainly provided for the aforementioned purposes by using traditional methods of extractions from natural sources. Plants are considered as a major contributor of lead bioactive natural products to the drug-discovery program and extraction of lead bioactive compounds from plants for a detailed in-vivo evaluation or clinical trial will damage rainforests and cause environmental problems. To avoid these problems and keep continuous the supply of lead bioactive natural products for clinical testing, it is absolutely necessary to develop biotechnological methods for the production of bioactive natural products. For developing these biotechnological methods, we need to determine the biosynthetic origin of potent bioactive natural products. These studies provide information on the key enzymes involved in the synthesis of pharmaceutical candidate molecules in nature [6]. These enzymes can be purified and their viability can be determined using biosynthesis experiments as assays. Cloning of these enzymes in a suitable vector, such as *Escherichia coli*, helps to overexpress them for developing biotechnological methods to produce bioactive compounds on a large scale [7]. In this chapter, we have described the biosynthesis of a few bioactive flavonoids, alkaloids and terpenoids summarized as follows.

16.2 Biosynthesis of Homoisoflavonoids

Homoisoflavonoids, an important class of bioactive natural products, are mostly found in nature [8]. These compounds are reported to be present in a few plant families including Hyacinthaceae, Liliaceae, Asparagaceae, Fabaceae, Agavaceae and Polygonaceae [9]. These compounds are reported to exhibit several bioactivities such as anti-oxidant, anti-diabetic, anti-inflammatory and anti-microbial activities, etc. [10–12].

We are involved in the identification of Glutathione *S*-Transferase inhibitors from plants. These compounds have applications in use as an adjuvant during cancer chemotherapy to overcome the drug-resistance problem of cancer chemotherapeutic agents during cancer chemotherapy [13–15]. Our anti-GST inhibition directed studies on *Caesalpinia bonduc* resulted in the isolation of two homoisoflavonoids, caesalpinianone (**1**) and 6-*O*-methylcaesalpinianone (**2**). Both of these compounds were significantly active in this bioassay with IC_{50} values of 16.5 and 17.1 μM, respectively [15]. These homoisoflavonoids are biosynthesized using phenylalanine, acetate and methionine to give 2′methoxychalcone (**3**) [16–18]. The latter compound undergoes cyclization to afford 3-benzylchroman-4-one (**4**). Our biosynthetic studies reveal that compound **2** is produced in nature by the hydroxylation of **4**. Methionine has methylated the C-6 hydroxyl group to afford **2**. The biosynthesis of **1** and **2** is outlined in Figure 16.1.

16.3 Biosynthesis of Alkaloids

Alkaloids are naturally occurring nitrogen-containing organic compounds. This class of natural products exhibits various biological activities. For instance, morphine (**5**) is used as a pain

DOI: 10.1201/9781003089162-16

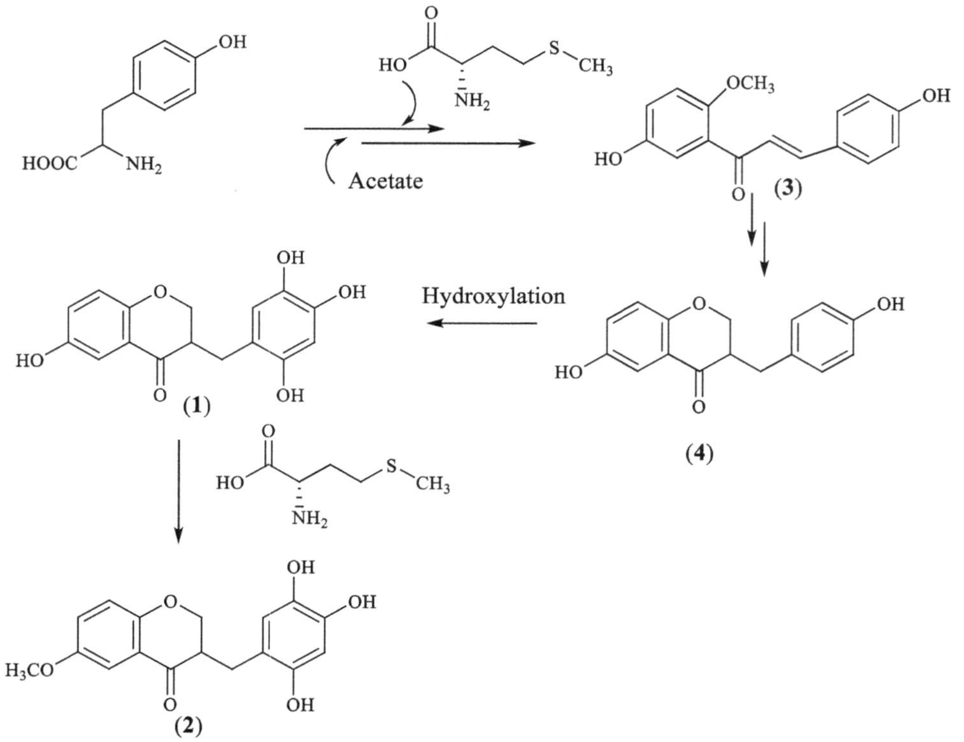

FIGURE 16.1 Biosynthesis of compounds **1** and **2**.

killer in clinics [19]. Alkaloids are derived from the primary metabolism of amino acids including Phe, Tyr, Trp, Orn and Lys [20]. It has been reported that nearly 20% of plant species have alkaloids and over 12,000 different alkaloids are reported in the literature. These compounds have huge structural and biosynthetic diversity compared to other natural products [21, 22]. Examples of Lys-derived alkaloids include piperidine (**6**), quinolizidine (**7**), indolizidine (**8**) and lycopodium (**9**) alkaloids (Figure 16.2). The synthesis of these alkaloids in nature involves decarboxylation of L-Lys into cadaverine by lysine decarboxylase (LDC) to afford 5-amino-pentanal, which is cyclized to Δ^1-piperideine (**10**) (Figure 16.3). Compound **10** is an important precursor of piperidine (**6**) which is also a precursor for the synthesis of quinolizidine (**7**), lycopodium (**9**). Piperline (**11**), a member of piperidine class of alkaloids, is biosynthesized on reaction with piperoyl CoA to afford compound **11** (Figure 16.3).

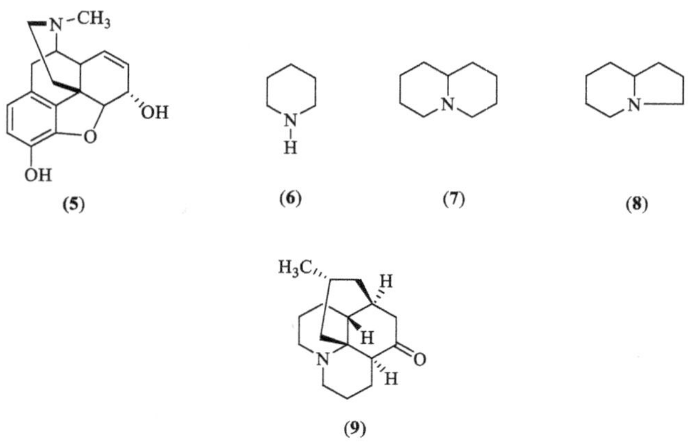

FIGURE 16.2 Structures of compounds **5–9**.

Lupinine (**12**), isolated from *Lupinus luteus* L. and other *Lupinus* species, is a member of quinolizidine-type alkaloid [23]. Compound **12** is biosynthesized from compound **10**. The latter compound in the presence of an acid produces imine and enamine which undergo aldol-type condensation to produce intermediate **12**. The hydrolysis of imine **12** affords compound **13** which on oxidative deamination of primary amine gives compound **14**. The intramolecular reaction of the amine with the aldehyde on **14** yields a Schiff base (**15**). This compound on reduction of double bond and aldehyde gives rise to lupinine (**17**) (Figure 16.4). Similarly, other quinolizidine alkaloids including sparteine and lupanine, etc., are synthesized in nature by using compounds **6** and **16** as intermediates.

Indolizidine alkaloids such as castanospermine (**23**) and swainsonine (**27**) are found in plants [24]. These compounds are also biosynthesized from L-Lys which upon oxidation of primary amine produces α-aminoadipic acid (**18**). This compound yields L-pipecolic acid (**19**). Compound **19** is a biosynthetic precursor to indolizidine alkaloids and yields compounds **23** and **27** after a series of biochemical reactions as outlined in Figure 16.5.

Lycopodium alkaloids are found in various plants and these compounds exhibit various biological activities including acetylcholinesterase inhibition activity [25]. Compounds with this bioactivity are used to treat early symptoms of Alzheimer's disease [25–27]. Lycopodine (**9**) is a member of this class of alkaloids [28]. These compounds are biosynthesized in nature from compound **10** which on reaction with acetoacetyl CoA gives compound **29**. The latter upon hydrolysis and

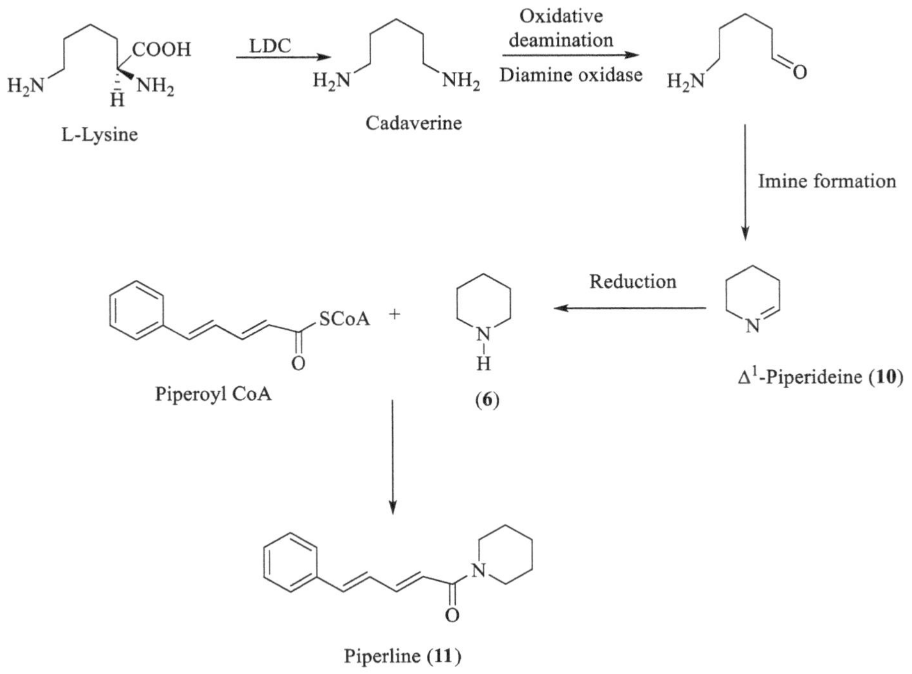

FIGURE 16.3 Biosynthesis of piperline (**11**).

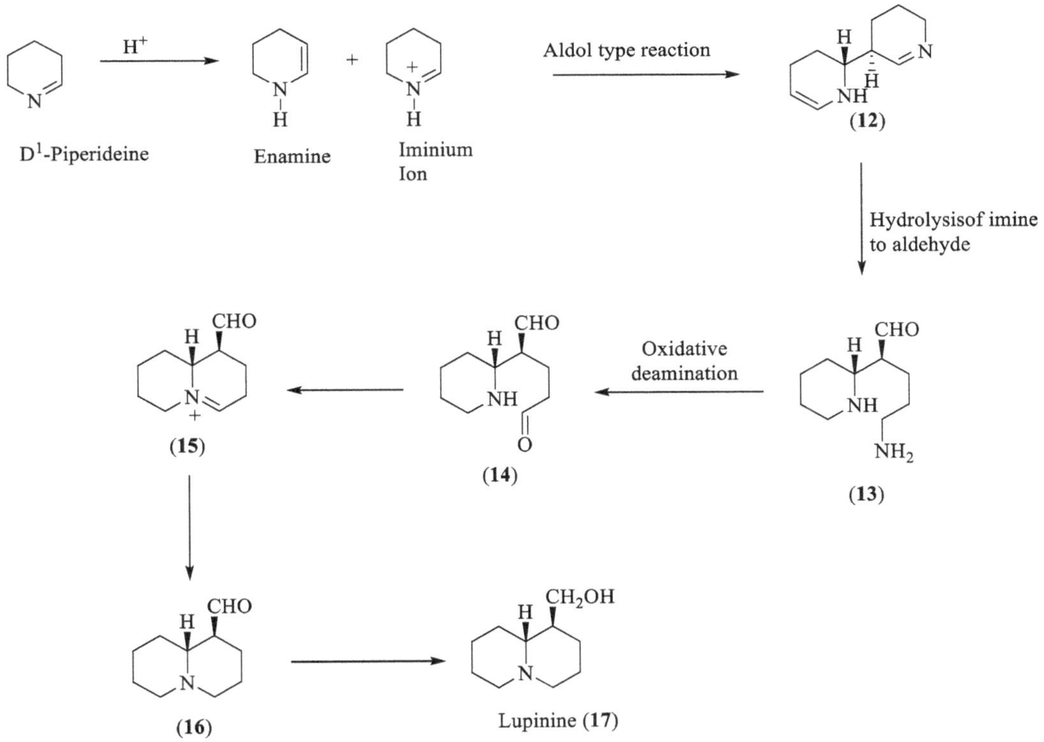

FIGURE 16.4 Biosynthesis of Lupinine (**17**).

decarboxylation yields pelletierine (**30**). Dehydrogenation of **30** yields compound **31** which undergoes a series of biochemical reactions including aldol reaction with **33** followed by hydrolysis, decarboxylation, dehydration and reduction affording compound **35**. All of these reactions are shown in Figure 16.6.

Compound **35** undergoes a series of biochemical reactions including enamine-type cyclization reactions and Schiff base formation affording compound **9**. All of these reactions are summarized in Figure 16.7. Further biochemical reactions incorporate other functional groups in other reported naturally occurring lycopodium alkaloids.

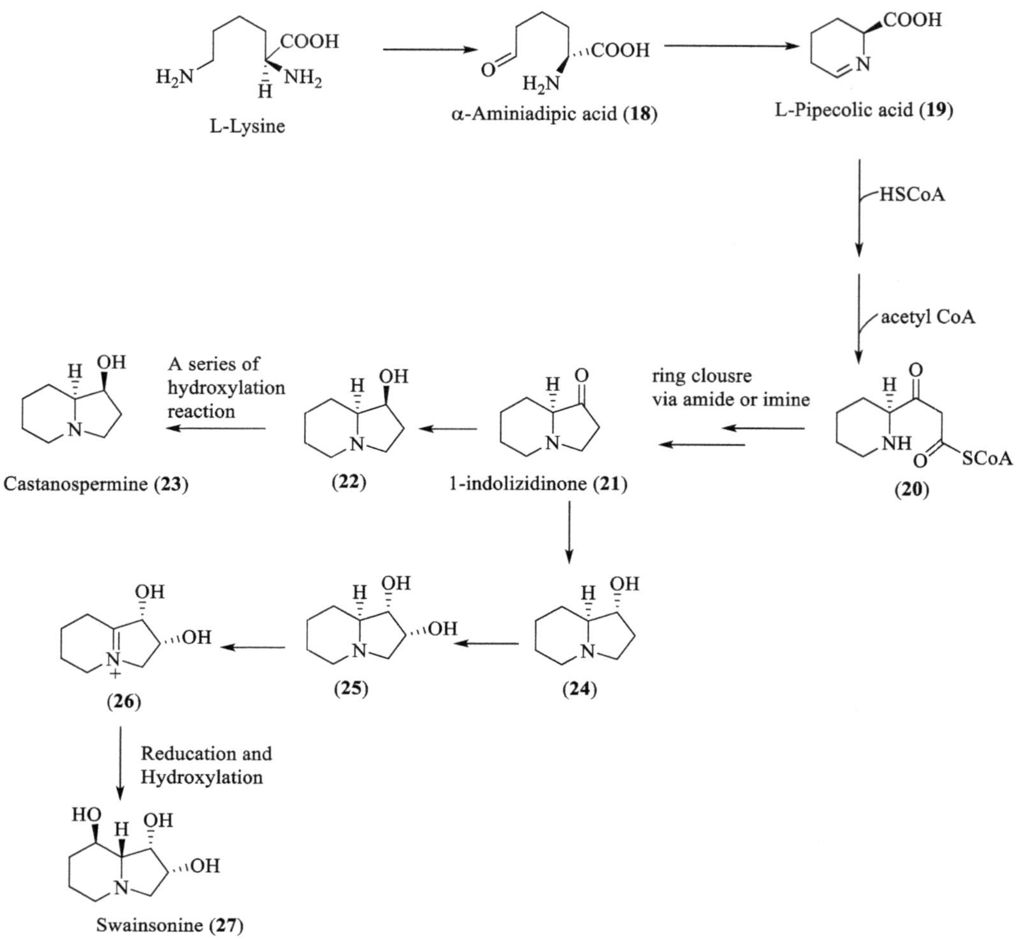

L-Lysine → α-Aminiadipic acid (18) → L-Pipecolic acid (19) —HSCoA→ —acetyl CoA→ (20) —ring clousre via amide or imine→ 1-indolizidinone (21) → (22) —A series of hydroxylation reaction→ Castanospermine (23)

1-indolizidinone (21) → (24) → (25) → (26) —Reducation and Hydroxylation→ Swainsonine (27)

FIGURE 16.5 Biosynthesis of compounds **23** and **27**.

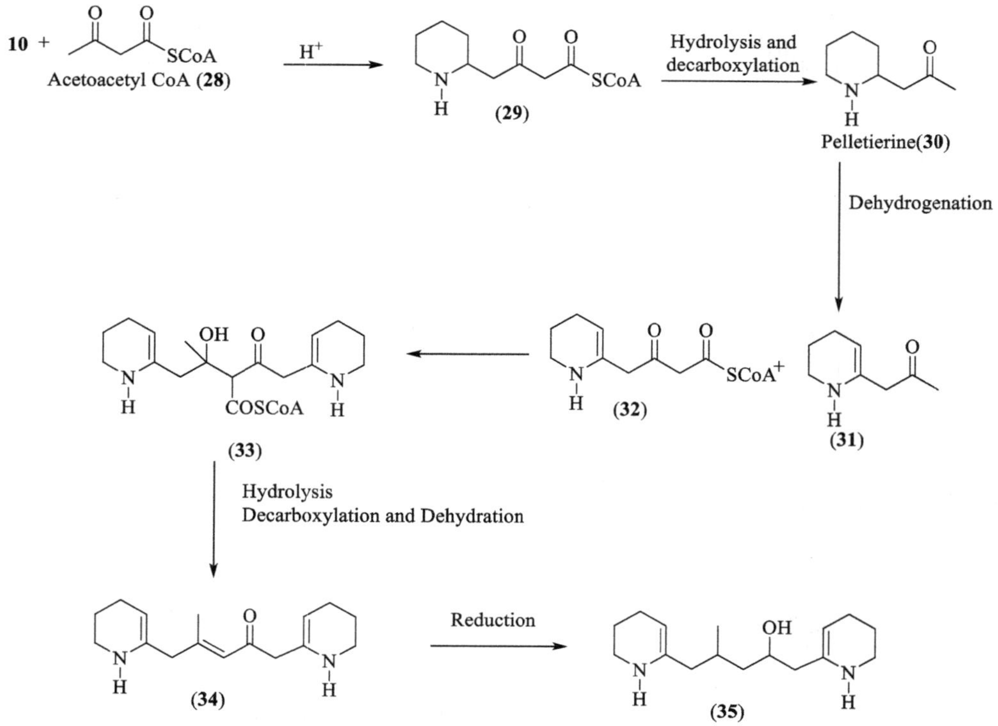

10 + Acetoacetyl CoA (28) —H^+→ (29) —Hydrolysis and decarboxylation→ Pelletierine(30) —Dehydrogenation→ (31) → (32) → (33) —Hydrolysis Decarboxylation and Dehydration→ (34) —Reduction→ (35)

FIGURE 16.6 Biosynthesis of compound **35**.

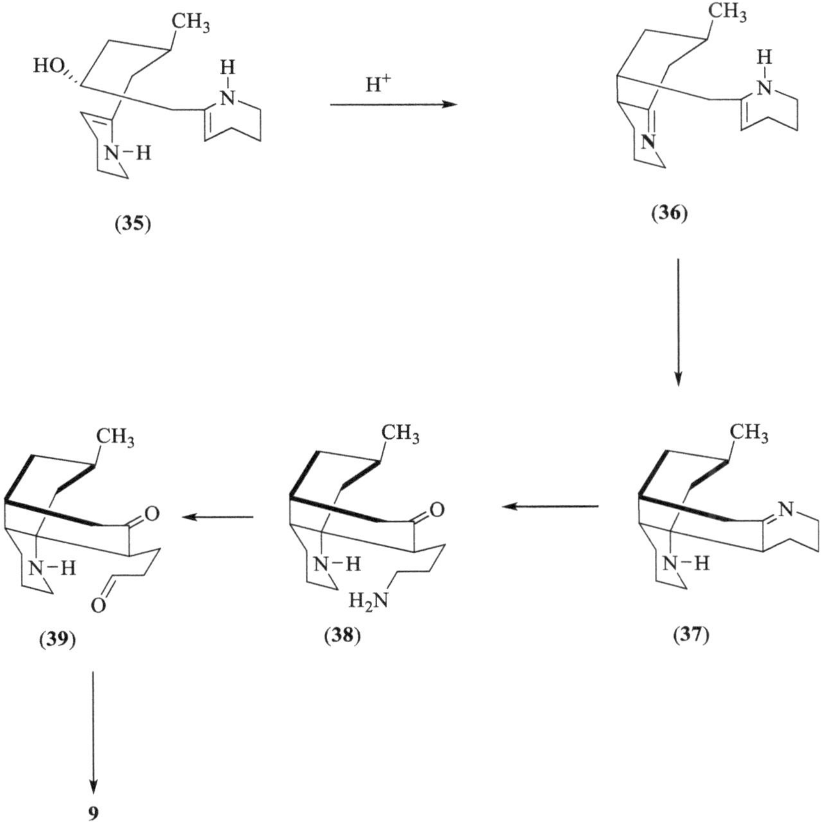

FIGURE 16.7 Biosynthesis of compound **9** from **35**.

Isoquinoline alkaloids are also distributed in various plants and they exhibit various biological activities [29–31]. For example, morphine (**5**) is used as an analgesic and papaverine (**40**) has muscle-relaxant activity [32–34]. This class of alkaloids is produced in nature from tyrosine (**40**) which upon decarboxylation and followed by hydroxylation reactions affords dopamine (**42**). Compound **40** also affords *p*-hydroxyphenyl acetaldehyde (**43**) using biocatalysts transaminase and decarboxylase. Reactions between compounds **42** and **43** yields isoquinoline alkaloid, norcocaurine (**44**). Further hydroxylation of compound **44** biosynthesizes norlaudanosoline (**45**) which undergoes *O*-methylation reaction to give papaverine (**46**). The enzymatic *O*-methylation and ether forming reactions on **45** yields thebaine (**47**), a precursor to morphine (**5**) (Figure 16.8). Hydrogenation and demethylation of **47** produce compound **5** [35, 36].

Tropane alkaloids are also an important class of alkaloids exhibiting various biological activities [37]. These compounds are mainly found in plants and are mainly derived from L-glutamic acid (**48**). Its decarboxylation leads to putrescine (**51**), a substituent of a natural product such as *p*-coumaroylputrescine (**52**). Putrescine (**51**) can also be further modified to *N*-methyl-Δ^1-pyrrolinium cation (**53**), which can be used in the synthesis of tropine (**54**), a precursor to atropine (**55**) as shown in Figure 16.9 [37].

Buxus alkaloids have a unique triterpenoid-steroidal pregnane-type structure with C-4 methyl groups, a 9β,10β-cycloartenol system and a degraded C-20 side chain [38–40]. These compounds have been reported to exhibit various biological activities including HIV, antimalarial, anti-acetylcholinesterase and anti-microbial activities, etc. [38–44]. Structurally, *Buxus* alkaloids are of the following two types (Figure 16.10):

1. Derivatives of 9β,10β-cyclo-4,4,14α-trimethyl-5-α-pregnane systems (**56**)
2. Derivatives of 9(10→19) *abeo* 4,4, 14α-trimethyl-5α-pegnane system (**57**)

Both of these types of *Buxus* alkaloids are reported to be differentiated with the aid of ^{1}H-NMR spectral studies. Alkaloids belonging to group **1** show a pair of AB doublets at δ 0.1–0.5 (J = 4.0 Hz) in their ^{1}H-NMR spectra. Compounds having basic structure **2** do not show these signals in their ^{1}H-NMR spectra but do exhibit signals for the C-11 and C-19 vinylic protons in the olefinic range of the spectrum. The presence of a 9(10→19) *abeo* diene system can also be inferred from absorption maxima at 238 and 245 nm with shoulders at 228 and 252 in their UV spectrum. These spectral features help in identifying two different classes of *Buxus* alkaloids [39].

Buxus alkaloids are assumed to be derived from cycloartenol-type terpenoids (**72**). These compounds are produced by the cyclization of squalene (**67**). The biosynthesis of precursor **67** starts with acetyl CoA (**58**) which on reaction with another molecule of acetyl CoA produces 3-hydroxy-3-methyl glutaryl SCoA (**60**). Compound **60** undergoes a series of reactions as outlined in Figure 16.11 to give isopentenyl pyrophosphate (**62**)

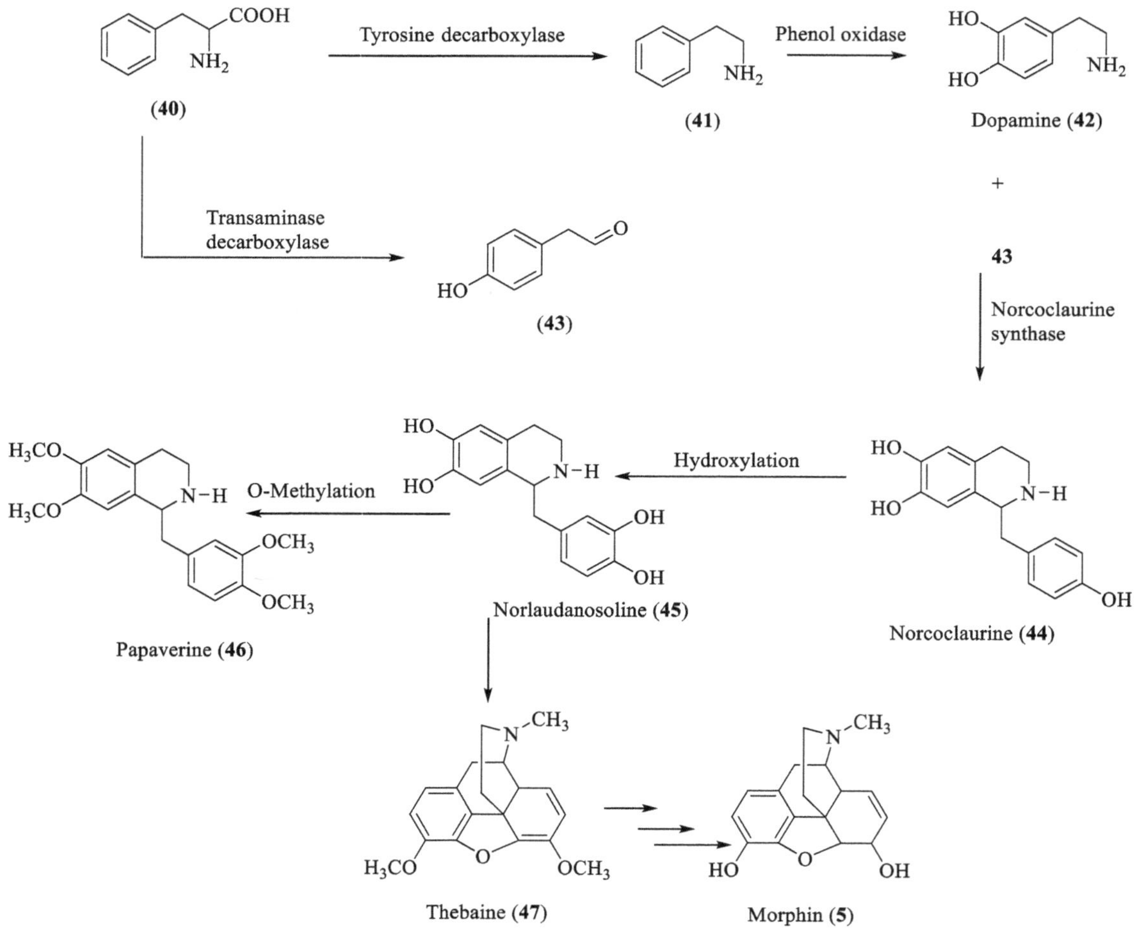

FIGURE 16.8 Biosynthesis of isoquinoline alkaloids (5, 44–47).

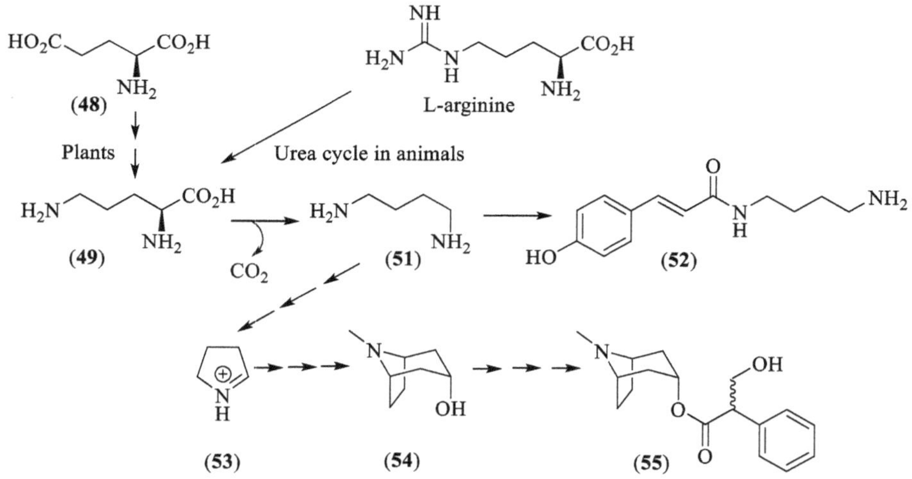

FIGURE 16.9 Biosynthesis of tropane alkaloid (55).

which is isomerized to produce dimethylallyl pyrophosphate (**63**). A nucleophilic attack of **62** on **63** yields geranyl pyrophosphate (**64**). The addition of another unit of **62** to **64** produces farnesyl pyrophosphate (FPP) (**65**). The coupling of two FPP units in a head-to-head manner synthesizes intermediate (**66**) which on rearrangement affords squalene (**67**) (Figure 16.11).

Squalene can be cyclized under oxidative, non-oxidative and basic oxidative conditions to afford intermediate (**70**). The cyclization of squalene under the aforementioned conditions is outlined in Figure 16.12.

Intermediate **70** produces lanosterol (**71**) which undergoes the formation of C-9 and C10 cyclopropane ring to afford

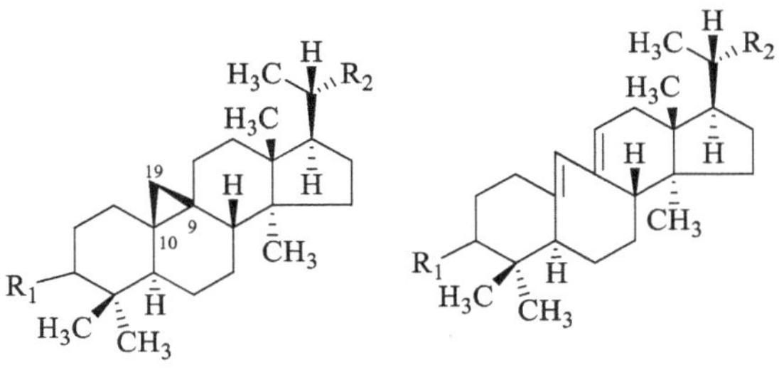
(56)
(57)
R1 and R2 are different amino or keto groups

FIGURE 16.10 Structures of *Buxus* alkaloids.

cycloartenol (**72**). The latter compound on oxidation yields compound **73** as shown in Figure 16.13.

Amination reaction at C-3 and C-20 or selective amination reaction at C-3 or C-20 of compound **73** affords *Buxus* alkaloids. All of these reactions are summarized in Figure 16.14. 9(10→19) *Abeo* diene system containing alkaloids may arise in nature from compound **74** by the decyclization of 9β,10β cyclopropane ring as shown in Figure 16.14. Similarly, 9(10→19) *abeo* $\Delta^{1\text{-}10}$ system containing alkaloids are produced by removing a proton from C-1 to initiate the opening of 9β,10β cyclopropane ring followed by the removal of C-11/hydroxyl group.

To confirm this hypothesis regarding the biosynthesis of *Buxus* alkaloids from cycloartenol, we have performed some biosynthetic experiments in our lab. For the incorporation of tritium at C-21 methyl group of compound **78**, we used chemo-enzymatic approach. In order to achieve this goal, compound **78** was fermented using the liquid culture of *Cunninghamella echinulata* to afford C-21 alcohol (**79**). Intermediate **79** was reacted with TMSI using a mixture of methanol and ^3H_2O as the solvent to afford compound **80** as shown in Figure 16.15 [45].

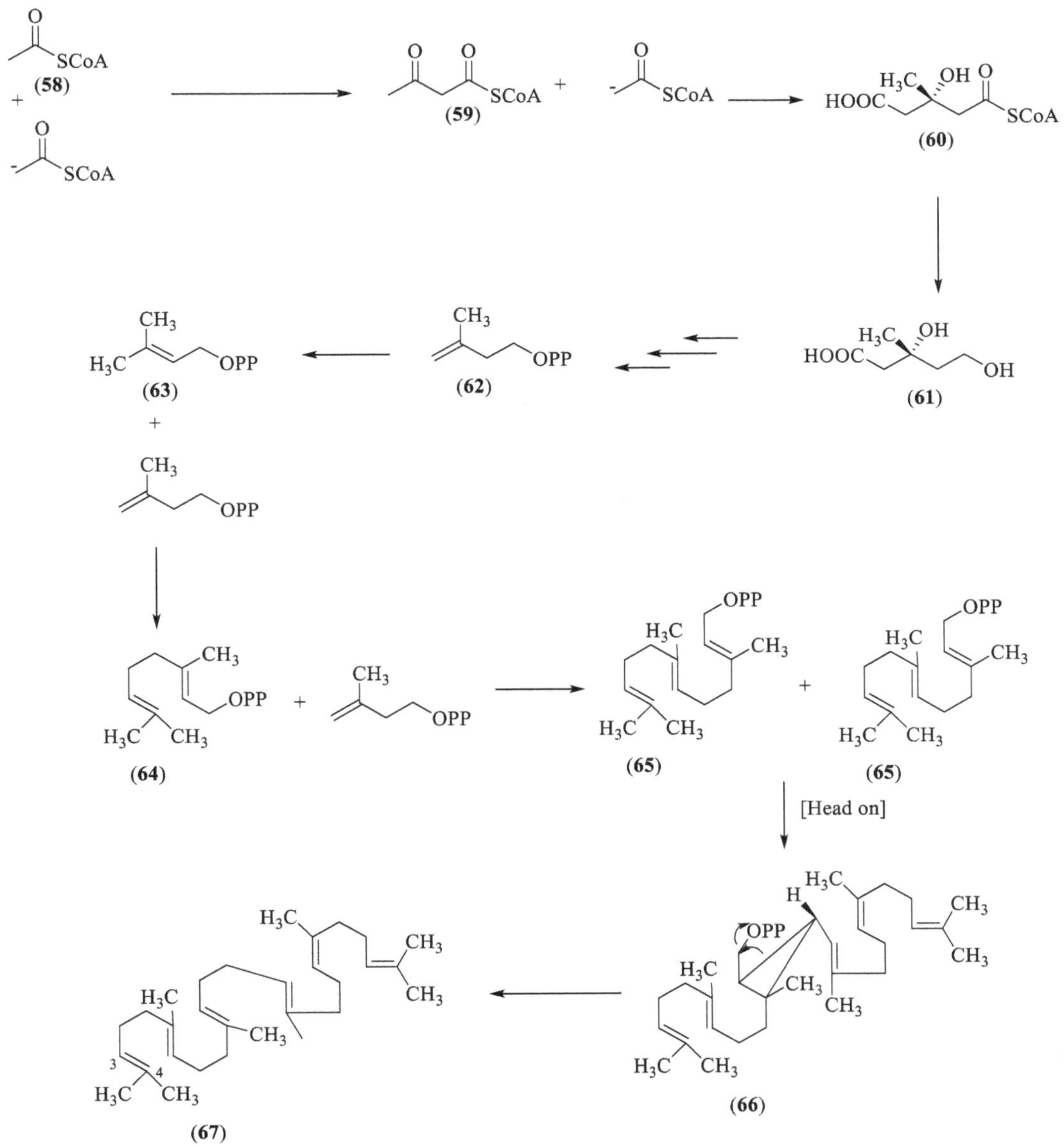
(58)
(59)
(60)
(61)
(62)
(63)
(64)
(65)
(65)
[Head on]
(66)
(67)

FIGURE 16.11 Biosynthesis of squalene (**67**).

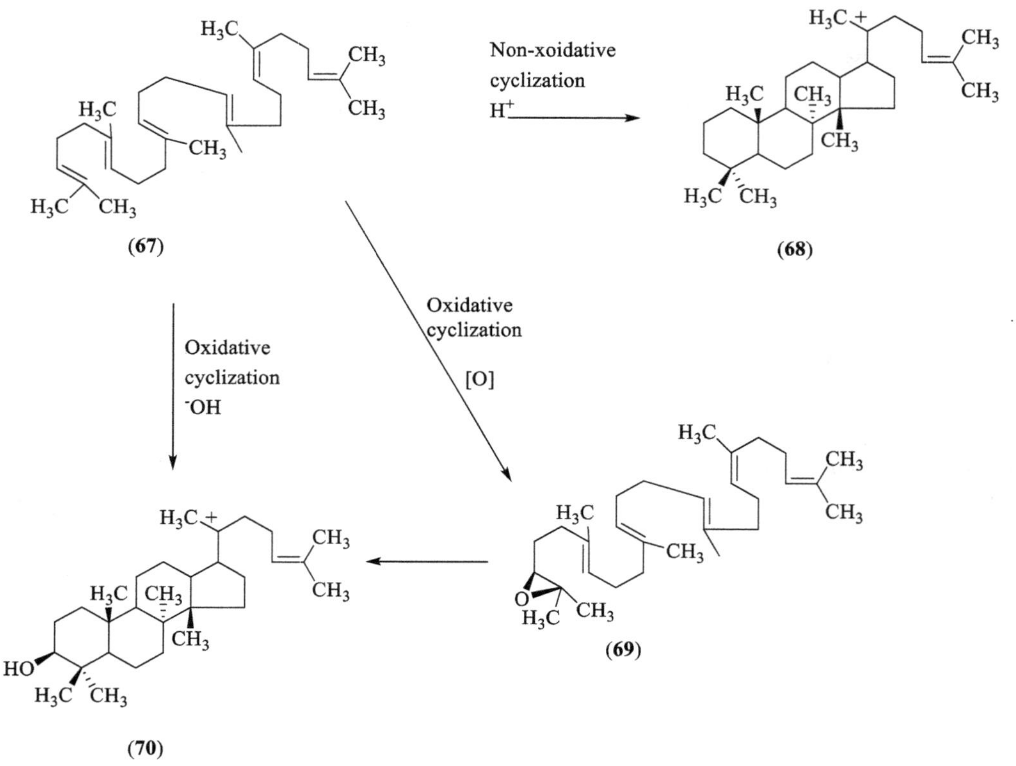

FIGURE 16.12 Cyclization of squalene (**67**).

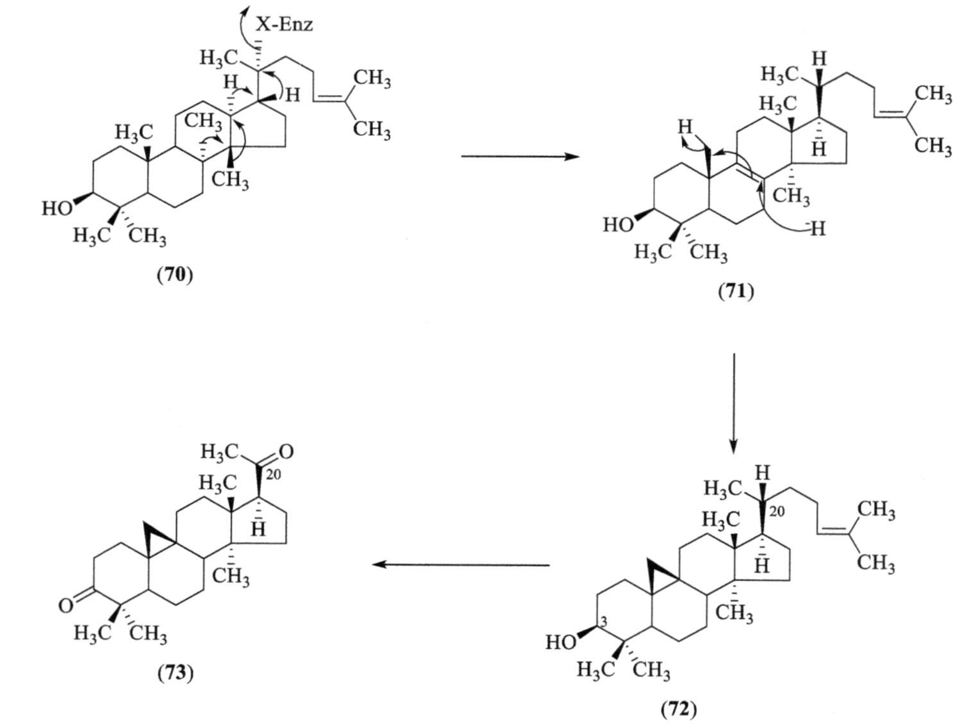

FIGURE 16.13 Biosynthesis of cycloartenol (**73**).

Compound **80** was fed to the fresh leaves of *Bulinus natalaensis* and an investigation of these leaves, after two weeks, afforded C-21 tritium labeled compounds **81**. During these studies, we have also identified compound (**81**) as a putative intermediate involved in the biosynthesis of *Buxus* alkaloid **82** as shown in Figure 16.16. Compound **82** has shown a potent anti-acetylcholinesterase activity [45]. Further studies on the biosynthesis of these compounds in order to elucidate the

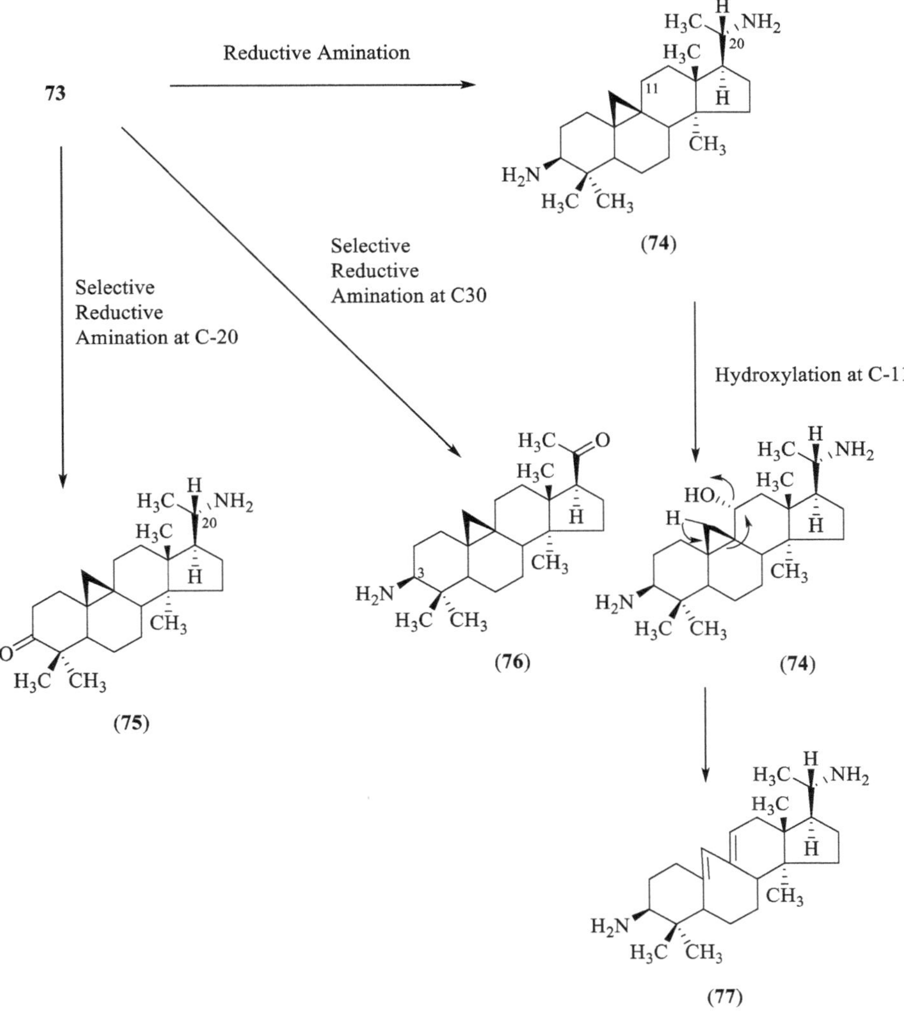

FIGURE 16.14 Biosynthesis of *Buxus* alkaloids from cycloartenol (**73**).

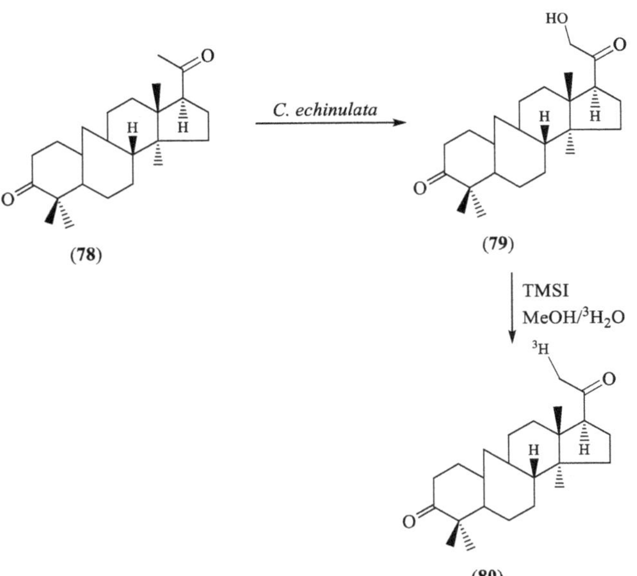

FIGURE 16.15 Chemo-enzymatic synthesis of compound **80**.

mechanism for the incorporation of an amino group at C-3 and C-20, the opening of 9β,10β cyclopropane ring to afford 9(10→19) *abeo* diene system and the substitution of other functional groups are currently underway in our lab.

16.4 Biosynthesis of Terpenoids

Terpenoids, an important class of natural products, have diverse structures. These compounds are derived from a 5-carbon containing compound, known as the isoprene unit (compounds **62** and **63**). These compounds are mainly classified as monoterpenes (containing two isoprene units), sesquiterpenes (having three isoprene units), diterpenes (having four isoprene units) and triterpenoids based on the number of carbon atoms derived from isoprene units. We have already shown the biosynthesis of cycloartenol, an example of triperpene. In this section, the biosynthesis of monoterpenes, sesquiterpenes and diterpenes has been discussed.

Monoterpenes are volatile compounds and play an important role in the perfume industry [46]. Geranyl pyrophosphate

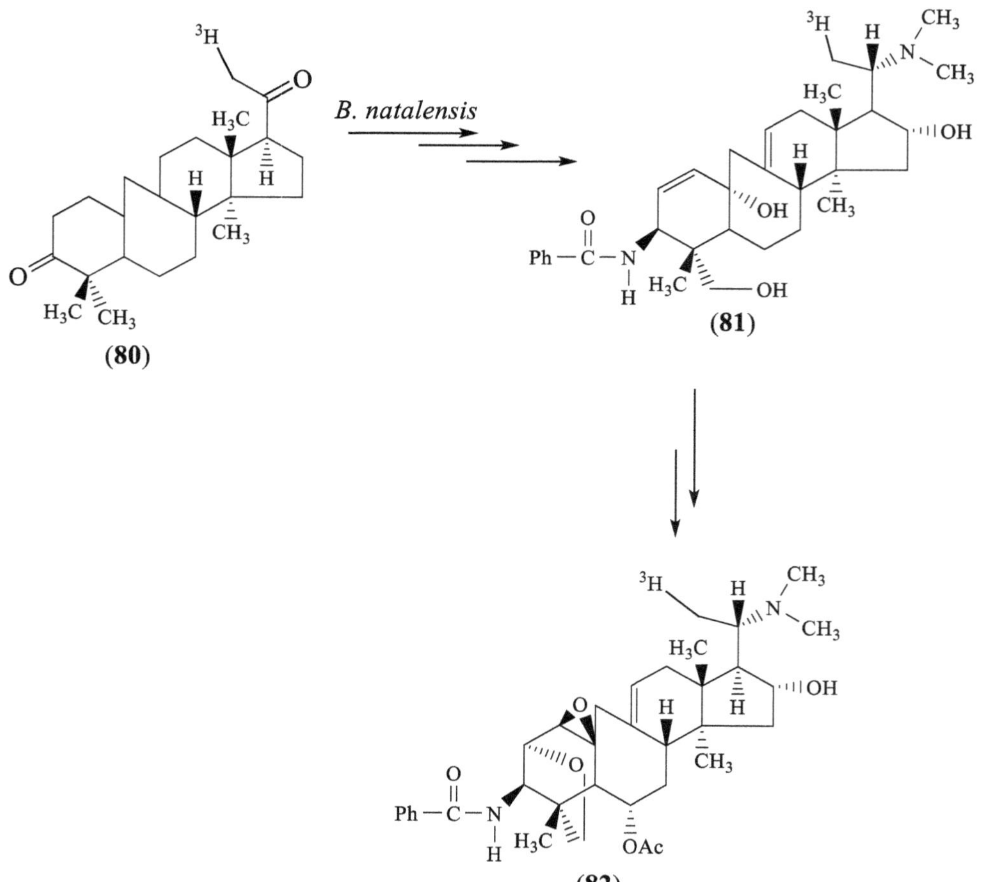

FIGURE 16.16 Biosynthesis of *Buxus* alkaloid **82**.

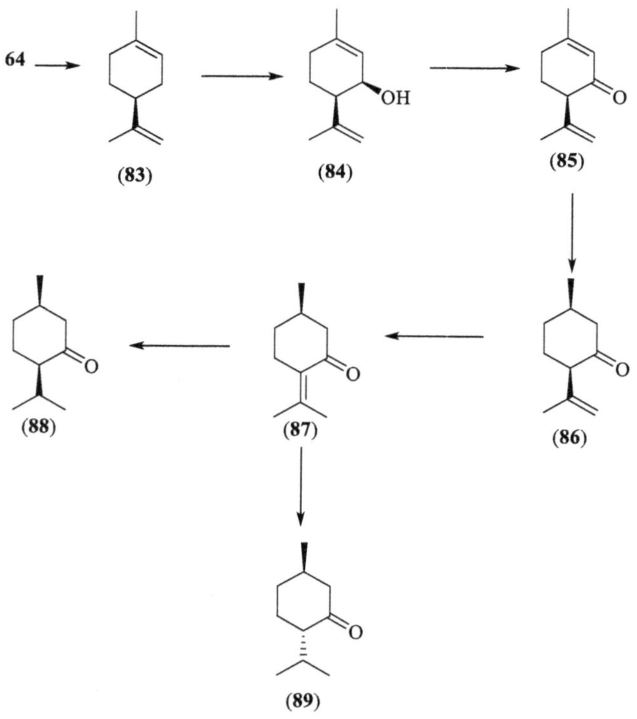

FIGURE 16.17 Biosynthesis of monoterpenes.

(**65**) is a biosynthetic precursor of monoterpenes. This undergoes cyclization to afford limonene (**83**) which on hydroxylation affords *trans*-isopiperitenol (**84**). The latter, on oxidation of hydroxyl group, gives isopiperitenone (**85**) which after the reduction of endocyclic double bond yields *cis* isopulegone (**86**). Isomerization of a double bond in **86** affords pulegone (**87**). The reduction of double a double in **87** produces both enantiomers of menthone (**88** and **89**) [47]. The biosynthesis scheme of monoterpenes is shown in Figure 16.17.

Sesquiterpenes are known to have various biological activities. For instance, artemisinin (**94**) exhibits antimalarial activity and its ether analogue is used in clinics [48]. These compounds are produced in nature from FPP (**65**) which is cyclized by an enzyme, cyclase to afford amorphan-4, 11-diene (**90**) which on hydroxylation reaction yields its hydroxy analogues (**91**). Further oxidation of primary alcohol gives aldehyde analogue (**92**). The oxidation of aldehyde affords carboxylic acid analogue (**93**) which on photooxidation produces a of artemisinin (**94**) [49]. This whole sequence of biosynthesis is shown in Figure 16.18.

Diterpenes are members of the terpenoidal class of natural products. These compounds are made up of four isoprene units. They are biosynthesized in nature using geranylgeranyl pyrophosphate (GGP) (**95**) as an intermediate [50]. Compound (**95**) is produced in nature by coupling FPP (**64**) with another unit of isoprene (**63**). There are a number of diterpene cyclase

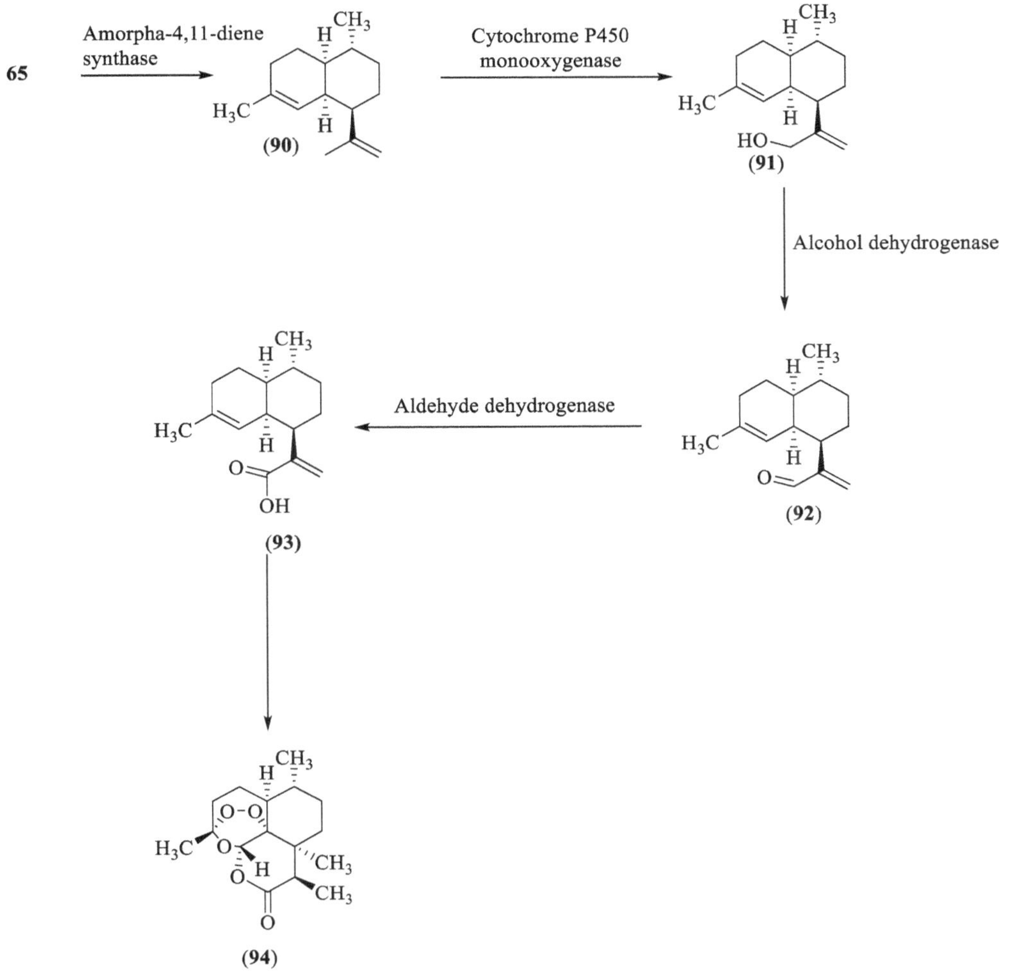

65 → Amorpha-4,11-diene synthase → (90) → Cytochrome P450 monooxygenase → (91) → Alcohol dehydrogenase → (92) → Aldehyde dehydrogenase → (93) → (94)

FIGURE 16.18 Biosynthesis of artemisinin.

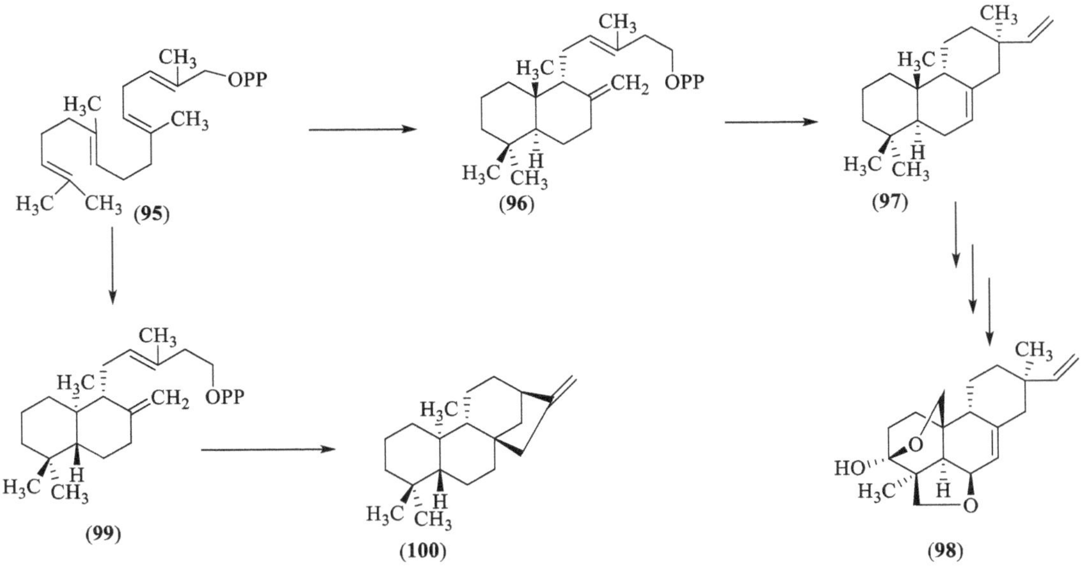

(95) → (96) → (97) → (98)

(95) → (99) → (100)

FIGURE 16.19 Biosynthesis of diterpenes **98** and **100**.

enzymes present in nature to fold/cyclize GGP (**95**) to give various classes of diterpenes [50]. For instance, momilactone (**98**) is produced from *syn*-primara-7,15-diene (**97**). GGP is cyclized to afford intermediate **96** which on further cyclization give compound **97**. Similarly, the cyclization of GGP (**95**) by ent-kaurene diterpene cyclase affords intermediate **99**, which on further cyclization yields ent-kaurene-type diterpenoid structure (**100**). All of these biosynthetic steps involved in the biosynthesis of diterpenes are shown in Figure 16.19 [51].

In summary, we have discussed the biosynthesis of various classes of natural products that might help to understand the key enzymology involved in the synthesis of these natural products in nature. This information can help biochemists to isolate key enzymes involved in the synthesis of natural products and use them for their in-vitro synthesis. Understanding enzymology and genes will help generate natural and unnatural bioactive compounds for the pharmaceutical industry.

ACKNOWLEDGMENT

Athar Ata would like to thank the Natural Sciences and Engineering Research Council of Canada for providing funding for his research program.

REFERENCES

1. Phillipson, D.J. 2007. *Phytochemistry* 68, 2960–2972.
2. Rates, S.M.K. 2001. *Toxicon* 39, 603–613.
3. Hasskarl, J. 2010. *Recent Results Cancer Res.* 184, 61–70.
4. Newman, D.J., Cragg, G.M. 2007. *J. Nat. Prod.* 70, 461–477.
5. Butler, M.S. 2008. *Nat. Prod. Rep.* 25, 475–516.
6. Watanabe, K. 2008. *Biosci. Biotechnol. Biochem.* 72, 2491–2506.
7. Cravens, A., Payne, J., Smolke, C.D. 2019. *Nat. Commun.* 10, 2142.
8. Lin, K.-G., Liu, Q.-Y., Ye, Y. 2014. *Planta Med.* 80, 1053–1066.
9. Abegaz, B.M., Kinfe, H.K. 2007. *Nat. Prod. Commun.* 2, 475–498.
10. Nirmal, N.P., Rajput, M.S., Prasad, R., Ahmad, M. 2015. *Asian Pac. J. Trop. Med.* 8, 421–430.
11. Chen, M.-H., Chen, X.-J., Wang, M., Lin, L.-G., Wang, Y.-T. 2016. *J Ethnopharmacol.* 181, 193–213.
12. Gupta, D., Bleakley, B., Gupta, R.K. 2008. *J. Ethnopharmacol.* 115, 361–380.
13. Ata, A., Udenigwe, C.C., Matochko, W., Holloway, P., Eze, M.O., Uzoegwu, P.N. 2009. *Nat. Prod. Commun.* 4, 1185–1188.
14. Udenigwe, C.C., Ata, A., Samarasekera, R. 2007. *Chem. Pharm. Bull.* 55, 442–445.
15. Iverson, C.D., Zahid, S., Li, Y., Shoqafi, A.H., Ata, A., Samarasekera, R. 2010. *Phytochem. Lett.* 3, 207–211.
16. Dewick, P.M. 1975. *Phytochemistry* 14, 983–988.
17. Dewick, P.M., Baz, W., Grisebach, H. 1970. *Phytochemistry* 9, 775–783.
18. Nguyen, A.-T., Fontaine, J., Malonne, H., Duez, P. 2006. *Phytochemistry* 67, 2159–2163.
19. Maier, C., Hildebrand, J., Klinger, R., Henrich-Eberl, C. Lindena, G. 2002. *Pain* 97, 223–233.
20. Bunsupa, S., Yamazaki, M., Saito, K. 2013. *Front. Plant Sci.* 3, 239.
21. Peng, J., Zheng, T.-T., Li, X., Liang, Y. Wang, L.-J., Huang, Y.-C., Xiao, H.-T. 2019. *Front. Pharmacol.* 10, 351.
22. Morales-Garcia, J.A., Revenga, M., Alonso-Gil, S., Rodriguez-Franco, M.I., Feildine A., Perez-Castillo, A., Riba, J. 2017. *Sci. Rep.* 7, 5309.
23. Saito, K., Murakoshi, I. 1995. Chemistry, biochemistry and chemotaxonomy of lupine alkaloids in the Leguminosae, In *Studies in Natural Products Chemistry*, Vol. 15, *Structure and Chemistry (Part C)*, Ed. Rahman, A.U. Elsevier Science Publishers, Amsterdam, pp. 519–550.
24. Harris, C.M., Schneider, M.J., Ungemach, F.S., Hill, J.E., Harris, T.M. 1988. *J. Am. Chem. Soc.* 110, 940–94.
25. Kobayashi, J., Morita, H. 2005. The lycopodium alkaloids, In *The Alkaloids: Chemistry and Biology*, Ed. Cordell, G.A. Academic Press, New York, pp. 1–57.
26. Berkov, S., Codina, C., Viladomat, F., Bastida, J. 2008. *Bioorg. Med. Chem. Lett.* 18, 2263–2266.
27. Wen, H., Lin, C., Que, L., Ge, H., Ma, L., Cao, R., Wan, Y., Peng, W., Wang, Z., Song, H. 2008. *Eur. J. Med. Chem.* 43, 166–173.
28. He, J., Wu, X.D., Liu, F., Liu, Y.-C., Peng, L.-Y., Zhao, Y., Cheng, X., Luo, H.-R., Zhao, Q.-S. 2014. *Nat. Prod. Bioprospect.* 4, 213–219.
29. Zhang, Y., Li, M., Li, X., Zhang, T., Qin, M., Ren, L. 2018. *Front. Pharmacol.* 9, 602.
30. Marasco, D., Vicidomini, C., Krupa, P., Cioffi, F., Huy, P.D.Q., Mai Suan Li, M.S., Daniele Florio, D., Broersen, K., DePandis, M.F., Roviello, G.N. 2021. *Chem. Biol. Interact.* 334, 109300.
31. Eliwa, D., Albadry, M.A., Ibrahim, A.-R.S., Kabbash, A., Meepagala, K., Khan, I.A., El-Aasr, M., Ross, S.A. 2021. *Phytochemistry* 183, 112598.
32. El-Sayed, K. 2000. *Phytochemistry* 53, 675–678.
33. Hagel, J., Facchini, P. 2013. *Plant Cell Physiol.* 54, 647–672.
34. Hegazy, M., Mohamed, T. ElShamy, Mahalel, U., Reda, E., Shaheen, A., Tawfik, W., Shahat, A., Shams, K., Abdel-Azim, N., Hammoud, F. 2015. *J. Adv. Res.* 6, 17–33.
35. Runguphan, W., Glenn, W., Conor, S.E. 2012. *Chem. Biol.* 19, 674–678.
36. Hagel, J.M., Facchini, P. 2010. *Nat. Chem. Biol.* 6, 273–275.
37. Dewick, P.M. 2009. *Medicinal Natural Products: A Biosynthetic Approach*, 3rd ed. John Wiley & Sons: New York, p 8.
38. Choudhary, M.I., Atta-Ur-Rahman, Shamma, M., Freyer, A.J. 1986. *Tetrahedron* 42, 5747–5752.
39. Ata, A., Andresh, B.J. 2008. *Buxus* steroidal alkaloids: chemistry and biology, In *Alkaloids*, Vol. 66, Ed. Cordell, G.A. Elsevier Science Publishers, San Diego, CA, pp. 191–213.
40. Babar, Z.U., Ata, A., Meshkatalsadat, M.H. 2006. *Steroids* 71, 1045–1051.
41. Ata, A., Iverson, C., Kalhari, K.S., Akhter, S., Betteridge, J., Meshkatalsadat, M.H., Orhan, I. Sener, B. 2010. *Phytochemistry* 71, 1780–1786.

42. Ata, A., Naz, S., Choudhary, M.I., Atta-Ur-Rahman, Sener, B., Turkoz, S.Z. 2002. *Z Naturforsch C* 57, 21–28.
43. Lam, C.W., Wakeman, A., James, A., Ata, A., Gengan, R.M.., Ross, S.A. 2015. *Steroids* 95, 73–79.
44. Yu-Ling, W., Wei, W., Yong-Nan, S., Zhi-Peng, A., Han-Chuan, M., Luo-Sheng, W., Ying, L., Ming-Hua, Q.J., Hong, Z. 2020. *Phytomedicine* 68, 153187.
45. Matochko, W.L., James, A., Lam, C.W., Kozera, D.J., Ata, A., Gengan, R.M. 2010. *J. Nat. Prod.* 73, 1858–1862.
46. Banthorpe, D.V., Charlwood B.V., Francis, M.J.O. 1972. *Chem. Rev.* 72, 115–155.
47. Turner, G., Gershenzon, J., Croteau, R.B. 2000. *Plant Physiol.* 124, 655–664.
48. Krishna, S., Bustamante, L., Haynes, R.K., Staines, H.M. 2008. *Trends Pharmacol. Sci.* 29, 502–507.
49. Wen, W., Yu, R. 2011. *Pharmacogn Rev.* 5, 189–194.
50. Heskes, A.M., Sundram, T.C.M., Boughton, B.A., Jensen, N.B., Hansen, N.L., Crocoll, C., Cozzi, F., Rasmussen, S., Hamberger, B., Hamberger, B., Staerk, D., Møller, B.L., Pateraki, I. 2018. *Plant J.* 93, 943–958.
51. Okada, K., Kawaide, H., Miyazaki, S. 2016. *Sci. Rep.* 6, 25316.

Problems and Answers

1. Give the IUPAC name of the below compounds.

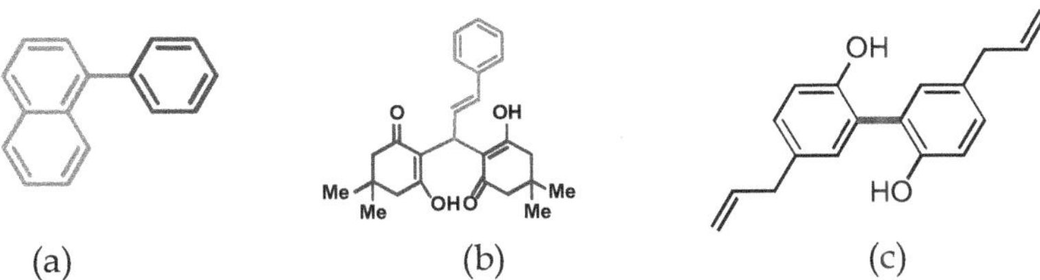

2. a. Describe the 12 principles of green chemistry.
   b. List the main physicochemical characteristics that a solvent must fulfill to be considered as a "green solvent".
3. Glycerol is considered a green biobased solvent and it is becoming more available, as it is a by-product in the production of biodiesel starting from raw materials. Given that lipases are one of the most used biocatalysts both in academia and industry, could you envision why there are not many cases reported using lipases in biotransformations using glycerol as a biosolvent?
4. A patent from Galactic S.A, Belgium (Moreau, B.; Richard, G.; Wathelet, J.-P.; Paquot, M. BE1019914A3, 2013) reports the use of immobilized lipase of *Candida antarctica* in the separation of the two isomers of racemic ethyl or butyl lactates *via* acylation of its hydroxyl moiety with acetic anhydride. For this process, could you clearly assign which is the green biosolvent (**clue: see Scheme 1**).
5. Supercritical CO_2 (sCO_2) is a fluid state of carbon dioxide where it is held at or above its critical temperature and critical pressure. It behaves as a supercritical fluid above its critical temperature (31.0 °C) and critical pressure (72.8 atm), expanding to fill its container like a gas but with a density like that of a liquid. It is generally considered a green solvent because of its great stability, low toxicity and environmental impact (can be easily removed by simply opening the reactor). Additionally, the solubility of many extracted compounds in sCO_2varies with pressure, allowing selective extractions of many bioactive compounds. sCO_2 has been also used in biotransformations, but there is a decisive parameter which has hampered its broad application. Could you indicate which would it be?
6. Water is generally considered the greenest possible solvent, as well as the prototypical solvent for biocatalysis. Anyhow, could you mention some issues associated with its use in biotransformations and postulate some solutions? (clue: Frontiers in Microbiology 2015, 6, 1257. DOI; 10.3389/fmicb.2015.01257)
7. What are the limitations in the use of glycerol as a solvent?
8. What is the Paal–Knorr reaction? What compounds can be obtained from this reaction? What solvents can be used in the green Paal–Knorr reaction?
9. Give three examples of syntheses of biologically active compounds carried out in DES.
10. List three methods of obtaining the thiazole derivatives that can be carried out in ionic liquids.
11. Give one example of the green synthesis of nitriles, alcohols and amines.
12. Give three examples of green syntheses of pyrazole derivatives carried out in glycerol.
13. What are the most common enzymes utilized in the asymmetric synthesis of drug precursors?
14. The following reactions are most often used in the synthesis of chiral building blocks of drugs:
    a. reduction of prochiral carbonyl substrates,
    b. hydrolysis/acylation of prochiral compounds,
    c. hydrolysis/acylation of *meso* compounds,
    d. kinetic resolution of racemic compounds.
       A. Only a
       B. Only b and c
       C. Only d and a
       D. All answers (a–d)
15. Give two examples of the use of lipase in the synthesis of chiral drug precursors in green solvent?
16. Give two examples of the use of microorganisms in the synthesis of chiral drug precursors in green solvent?
17. The ^1H NMR for *N*-Benzol[1,3]dioxol-5-yl-formamide is shown below. The compound is pure and clean, but all signals appeared to be "duplicated". For example, the H of the formamide appears as a doublet at 8.50 and as a singlet at

8.31 (both integrating for 1 H). A similar effect can be observed for CH_2. Additionally, the aromatic region integrates for 6 H instead of 3. Propose a plausible explanation.

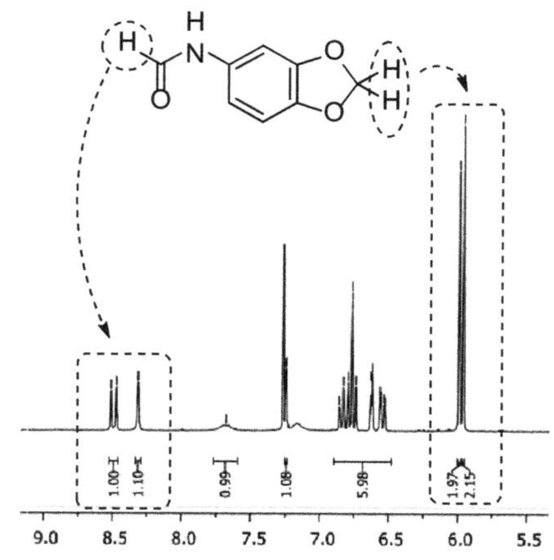

18. Even though when mixing 1 mol of acetic acid, 1 mol of butylamine and 1 mol of aniline, two possible products can be obtained, only a single product is observed. Draw the mechanism for the catalyst-free reaction and explain which compound will be formed and why.

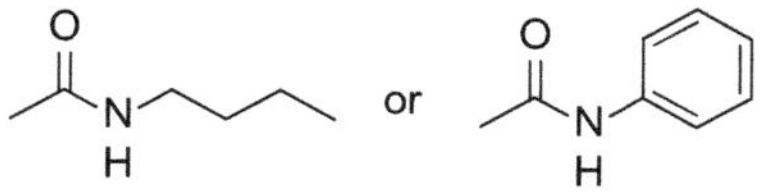

or

19. It should be clear by now that carboxylic acids have a dual reactivity, they are usually seen as electrophiles (Bronsted acids) since they can donate a proton. However, when reacting with more electrophilic species, they act as nucleophiles. So, imagine the activation of the acid in the next figure and propose a structure for the reaction product.

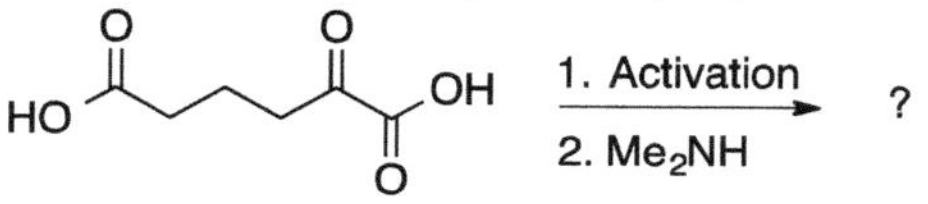

1. Activation; 2. Me_2NH → ?

20. Make a list of the more important characteristics for a good catalyst or additive for direct amidation reactions, having in mind the green chemistry principles.
21. Compound **74** was obtained in low yield. Propose an explanation and what do you expect for the same reaction performed with 2-amino phenol?

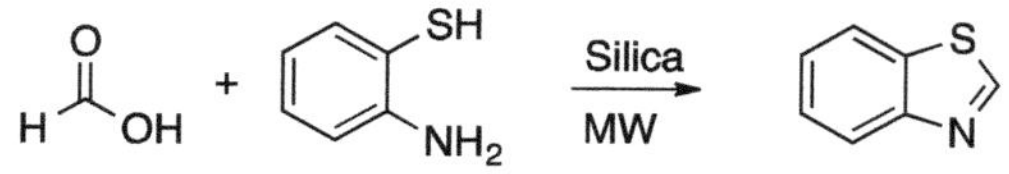

Silica, MW → **74** (34%)

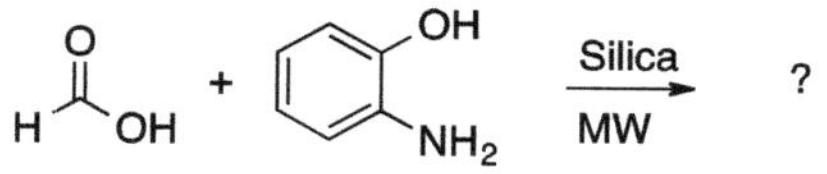

Silica, MW → ?

22. The chlorination of anthracene with NCS as the electrophilic chlorine source and phosphine sulfide as the catalyst, allows obtaining a new chlorinated compound. The profile of the molecular ion peak of the obtained compound is shown in the figure below. With this information, could you predict if the product obtained corresponds to the monochlorinated or the dechlorinated anthracene? Explain your answer.

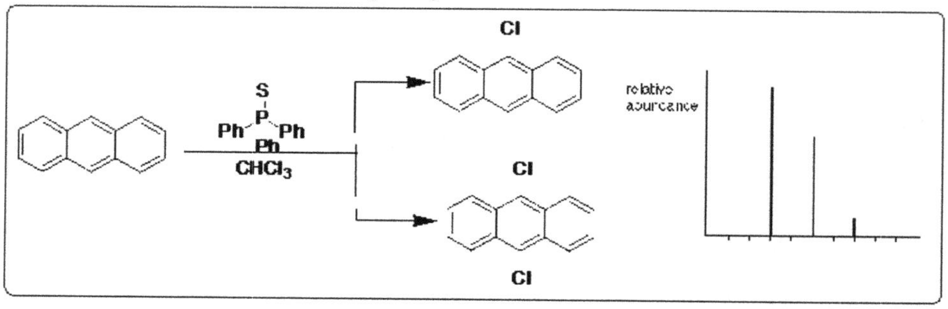

$Ph_3P{=}S$, $CHCl_3$

relative abundance

23. The fluorinated compound (*E*)-1-(2-fluorophenyl)ethenone *O*-methyl oxime was obtained from the reaction oxime ethers with a palladium catalyst and NFSI as the fluorinating agent. The product was obtained with excellent yield and purity. The characterization process was performed using ^{1}H, ^{13}C and ^{19}F NMR as well as high-resolution mass spectrometry.

    The carbon spectrum obtained is shown in the figure below, and it shows splitting in seven of the nine expected signals. Provide an explanation for this behavior.

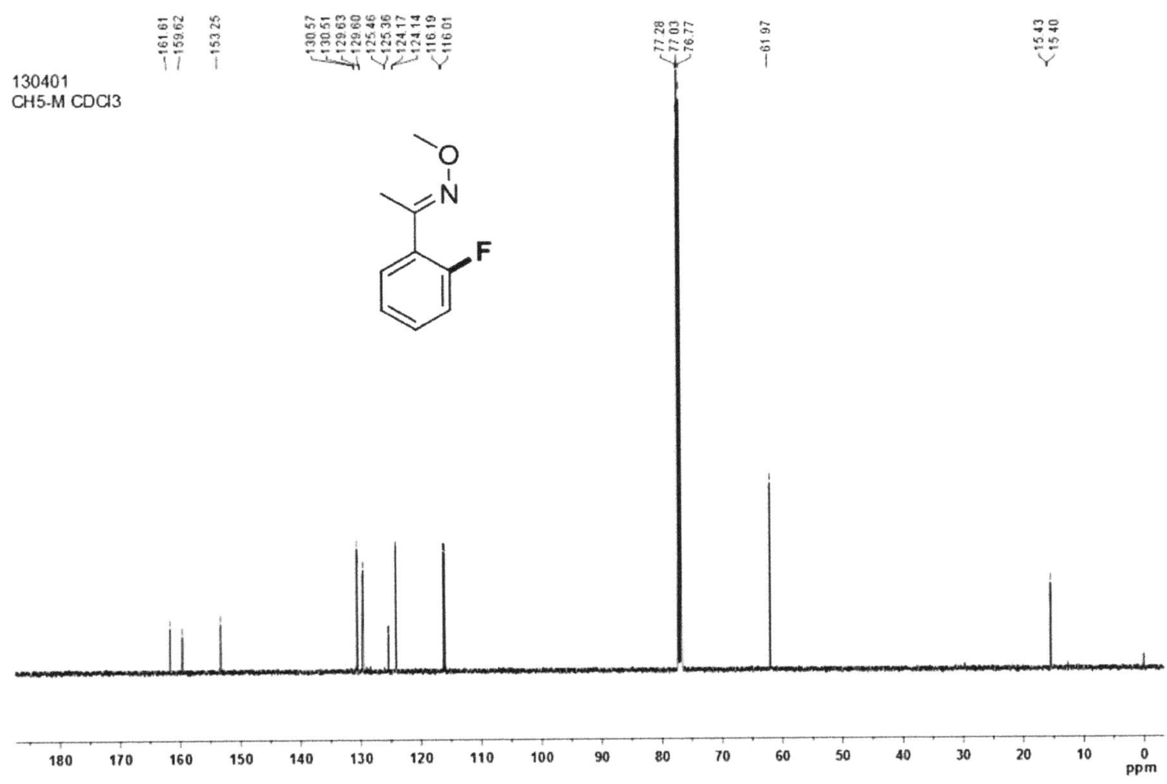

24. Which is the main product for the next reaction:

    H — Br-Cl / $FeBr_3$ → [ ]

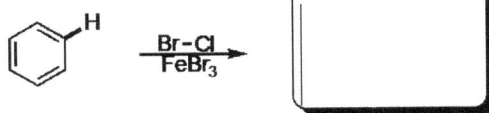

25. Explain the role of the Fe(III) halide in a halogenation reaction of an aromatic with Cl_2 or Br_2?
26. Classic aromatic halogenations using Cl_2 or Br_2 and a Lewis acid work very well. However, the same reaction does not work with iodine. Provide an explanation for this behavior and a comment about the optimal way to carry out iodinations.
27. Carrying out ortho-brominations on aromatic rings substituted with deactivating groups by a classic EAS is impossible, mention some synthetic procedures that allow carrying out this transformation.
28. Consider the high toxicity and low selectivity of the classic reagents $SOCl_2$, Cl_2 and *t*BuOCl that are used for chlorination reactions. Mention some examples of some reagents that allow obtaining the same reactivity but have better selectivity and that can be used under milder reaction conditions.
29. A compound with the molecular formula $C_{14}H_{14}O_2$ displays the following spectral data, ^{1}H NMR: **δ** 3.85 (s, 6H), 6.97 (d, 4H, $J = 8.8$ Hz), 7.49 (d, 4H, $J = 8.8$ Hz) ppm. ^{13}C NMR: **δ** 55.27, 114.18, 127.72, 133.56, 158.76 ppm. ^{13}C NMR DEPT-135 positive peaks at 55.27, 114.18, 127.72 ppm.

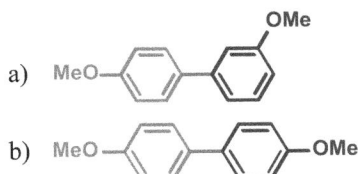

    a) MeO– (biphenyl) –OMe (meta)

    b) MeO– (biphenyl) –OMe (para)

c) 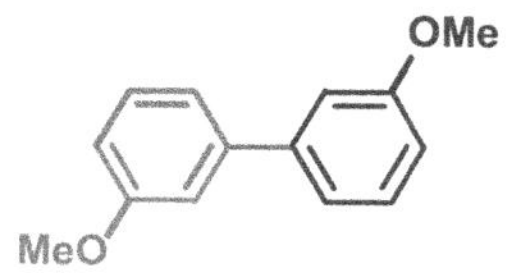 MeO OMe

30. A compound with the molecular formula $C_{17}H_{16}N_2O_3$ displays the following spectral data, ^{1}H NMR: **δ** 1.84–1.92 (m, 2H), 2.24–2.28 (m, 2H), 2.58 (s, 2H), 3.70 (s, 3H), 4.12 (s, 1H), 6.82 (d, 2H, J = 7.2 Hz), 6.93 (s, 2H), 7.05 (d, 2H, J = 7.9 Hz) ppm. ^{13}C NMR: **δ** 20.28, 26.92, 35.07, 36.83, 55.52, 58.97, 114.17, 114.58, 120.30, 128.67, 137.39, 158.41, 158.89, 164.57, 196.31 ppm.

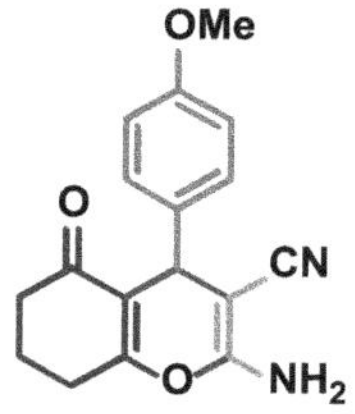

OMe O CN O NH2

(a)

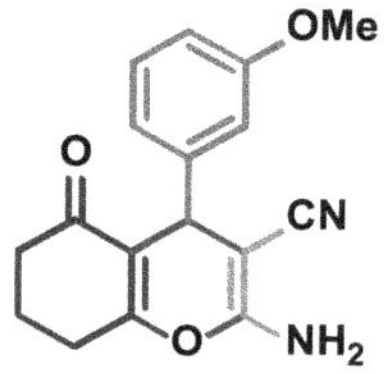

OMe O CN O NH2

(b)

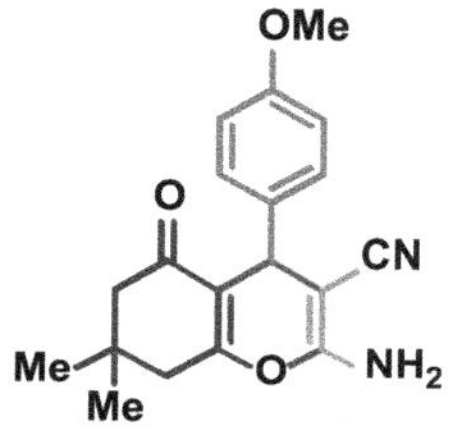

OMe O CN Me Me O NH2

(c)

31. What is the main difference between conventional and microwave heating?
32. Mention what type of solvents are suitable to carry out reactions under microwave irradiation.
33. What are the classifications of the materials based on their response to microwave?
34. Mention and explain the mechanisms that are involved in the heat by microwave.
35. Explain the effect of irradiation microwave in the selectivity of the reactions.
36. What is a photochemical reaction example? What is photochemical dissociation?
37. Ketones often decompose photochemically with the cleavage of the C-C bond alpha to the carbonyl. Can you suggest a detailed mechanism?
38. GN Lewis observed that in highly symmetric molecules, the fluorescence had a typical toothed structure, as in a saw, both in the absorption and in the emission spectrum, but with a different peak-to-peak distance? What was his rationalizing?
39. Benzyltrimethylsilane is photodecomposed by irradiation in the presence of naphthonitrile. Which products? Which mechanism?
40. Trans-stilbene is strongly florescent, while the cis isomer is a weak one. Why is that?
41. What is starch? Suggest a reaction mechanism for the synthesis of N-substituted pyrroles using starch.
42. Which is the best solvent in the synthesis of N-substituted pyrroles using starch? Explain the rule of substituted effects on the rate reaction in the synthesis of N-substituted pyrroles using starch? What are the advantages of this method?

43. Propose a consistent mechanism for obtaining pseudopeptide from reagents A, B and C.

S=C=S + H_2N ... Cl + ...))), H_2O →

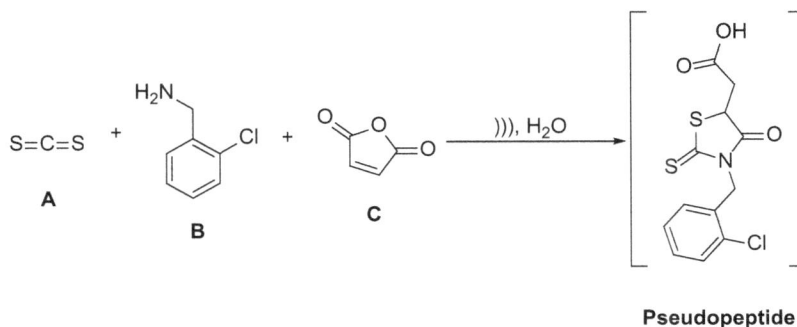

A B C

Pseudopeptide

44. Mention the main characteristics that a coformer must have for the synthesis of a pharmaceutical co-crystal.
45. In ultrasound-assisted synthesis, what parameters can be controlled? What influence does the solvent have when applying this type of radiation in a chemical reaction?
46. In the next nitrone cycloaddition, what reagent is needed?

MW 6 min

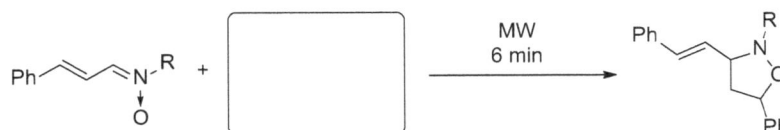

47. a. What are enantiotopic protons? How will you differentiate between them using ^1H NMR spectroscopy?

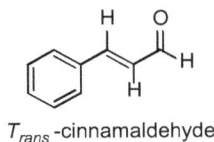

Trans-cinnamaldehyde

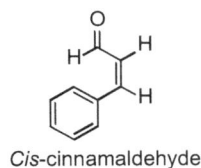

Cis-cinnamaldehyde

b. How will you differentiate between two isomers $HOCH_2C_6H_5CH_3$ and $C_6H_5CH_2CH_2OH$ as the following compounds are diastereoisomers of cinnamaldehyde?

48. a. What are pericyclic reactions? Discuss different types of pericyclic reactions. Discuss molecular rearrangement of the following compounds

(i) 80-110 °C

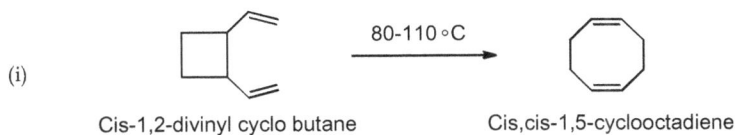

Cis-1,2-divinyl cyclo butane → Cis,cis-1,5-cyclooctadiene

(ii) 150-200 °C

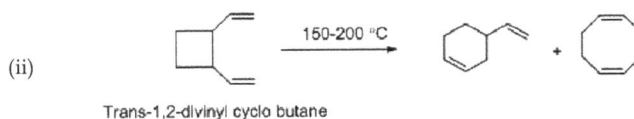

Trans-1,2-divinyl cyclo butane

(iii) Δ

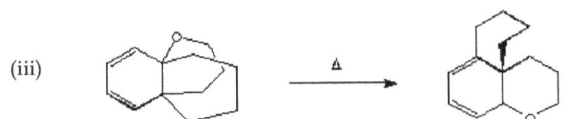

(iv)

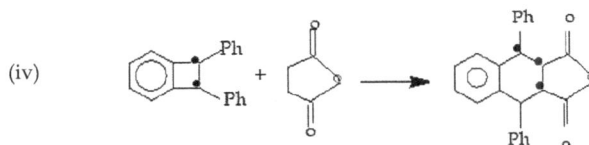

49. a. Reaction between aldehyde and ketone on base gives a compound with ^{1}H NMR spectrum δ_H (ppm):- 1.10 (9H.$_s$), 1.17 (9H,$_s$), 6.4 (1H, d.J = 15) and 7.0 (1H, d, J = 15).What is the structure with stereochemistry?
    b. When the compound reacts with HBr, it gives a compound B with the NMR spectrum δ_H (ppm):-1.08 (9H.$_s$), 1.13 (9H.$_s$), 2.71 (1H, dd.J, 1.9, 1.77), 3.25 (2H, dd.J, 10.0, 17.7) and 4.38 (1H, dd, J1.19, 10.0).
    c. Suggest a structure, assign the spectrum and give a mechanism of the formation of B.
50. Discuss the different stages of production of antibiotics. How can **β**-lactam be synthesized?
51. Discuss the biosynthesis of triglycerides (fats).
52. Discuss the biosynthesis of cholesterol.
53. Discuss the biosynthesis of bile acids.
54. a. Discuss the biosynthesis of amino acids.
    b. Discuss sources of carbon and nitrogen molecules of purines and pyrimidine ribionulcotides.
55. Give an example of the use of hydrolase in the biosynthesis of an antibiotic.

Answers to Problems

1. (a) (*E*)-2,2′-(3-phenylprop-2-ene-1,1-diyl)bis(3-hydroxy-5,5-dimethylcyclohex-2-en-1-one). (b) 1-phenylnaphthalene (c) 5,5′-diallyl-[1,1′-biphenyl]-2,2′-diol
2. a. 12 principles of green chemistry are as follows:
       1. Prevention, 2. economy of atoms, 3. less toxic intermediate chemicals, 4. safer end products, 5. reduction of the use of auxiliary substances, 6. reduction of energy consumption, 7. use of renewable raw materials, 8. reduction of unnecessary derivatization, 9. use of catalysts, 10. design for degradation, 11. development of analytical technologies for real-time monitoring, 12. minimization of the risk of chemical accidents.
   b. Main physicochemical characteristics that a solvent must fulfill to be considered as a "green solvent" are
      - Low vapor pressure
      - High boiling point
      - High biodegradability under environmental conditions
      - Lack of odor
      - Broad capability for dissolving as many chemicals as possible
      - Reasonably price
      - Derivation from renewable resources.
3. This is a "tricky" question. Lipases can perfectly work in glycerol; in fact, triglycerides (1,2,3-triacylglycerol) are the natural substrate for lipases, which, in aqueous media, are capable to catalyze its regio- and steroselective hydrolysis. Furthermore, in nearly anhydrous environments, they can catalyze the reverse reaction, that is, the regio- and steroselective acylation of glycerol, as schematically shown in the scheme below.

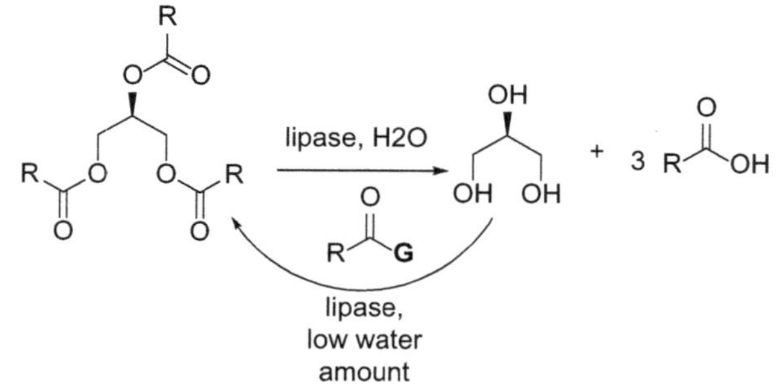

   lipase, H2O
   \+ 3 R–COOH
   lipase, low water amount

   Nevertheless, if we intend to use glycerol as a solvent for an acyl donor process, for example, the esterification of an alcohol with an acid, the number of molecules of glycerol will be much higher than that from the alcohol substrate, the main process catalyzed by the lipase will be the acylation of glycerol, leaving the target substrate intact. For this type of reaction, the use of GBSs (e.g., ethereal type), in which OH groups are blocked, would be highly beneficial.
4. In this process, ethyl (or butyl) lactate does play a double role: on the one hand, it is the substrate of the reaction, while, on the other hand, it is also the green biosolvent, as it is liquid at room temperature, allowing solubilization of the other reagent (acetic anhydride). So, this reaction could be considered a solventless biotransformation, therefore satisfying one of the norms of Green Chemistry: the best solvent is none.

5. Obviously, the high economic cost of a bioreactor (especially at the industrial scale) should be maintained at a high pressure to ensure CO_2 behaves as a supercritical fluid. A classical bioreactor using a green solvent would operate at an ambient pressure so that the equipment required is much cheaper.
6. The major disadvantage of water as a solvent lies within its high polarity. Therefore, hydrophobic reagents (synthetic ones) are poorly soluble. This drawback can be solved using water-soluble cosolvents, as illustrated all along this chapter. In fact, the use of these cost-effective and biodegradable water-miscible cosolvents represents an excellent alternative to enhance substrate loadings while leading to outstanding enzymatic enantio selectivities and productivities. Additionally, water-insoluble green cosolvents (2-MeTHF or ethyl acetate) can be used to generate the so-called two-liquid phase systems (2LPS), in which the water-immiscible solvent serves as substrate reservoir and product sink, so that amplified work-up is reached at the same time, in which the product is directly obtained from the organic phase upon a simple distillation.
7. a. High viscosity of glycerol – it should be fluidified with a cosolvent or the reactions can be proceeded at temperatures higher than 60 °C because then, its viscosity is much lower.
   b. Possibility of reacting with substrates due to the presence of three hydroxyl groups which can be mentioned as acidic sites in the molecule.
   c. Glycerol can obtain complexes with metal catalysts resulting in unwanted side products and/or unreactivity of catalysis, because of enough length of molecule the presence of a donor atom.
8. Paal–Knorr reaction is a reaction between a diketone and an amine that produces substituted pyrrole derivatives.

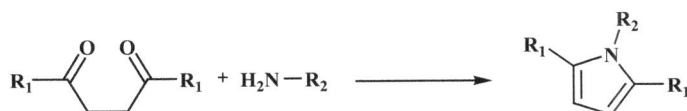

   The reaction can be carried out in green solvents, for example, in glycerol or DES (ChCl: urea).
9. a. Synthesis of pyrrole derivatives in Paal–Knorr reaction (ChCl: urea)

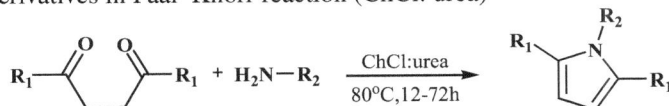

   ChCl:urea, 80°C,12-72h

   b. Synthesis of oxazole derivatives from phenacyl bromide derivatives and amide (ChCl: urea with ultrasounds)

   ChCl:urea, RT,12-17min, ultrasounds

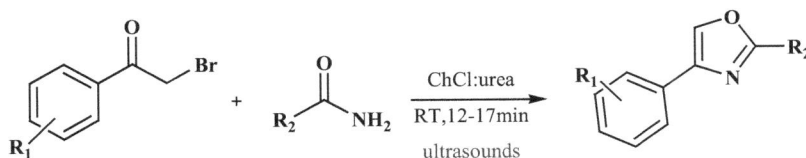

   b. Synthesis of 2-aminoimidazoles from α-chloroketones and guanidine derivatives (glycerol: choline chloride (ChCl) or urea: ChCl)

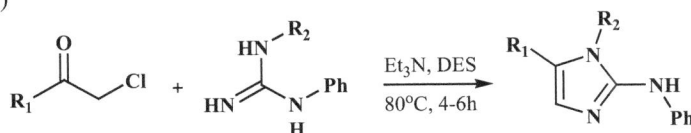

   Et_3N, DES, 80°C, 4-6h
10. a. Synthesis of 2,4-diarylthiazoles from arylthioamides and α-bromoacetophenones in $[BMIM]BF_4$, under ultrasound irradiation at room temperature.

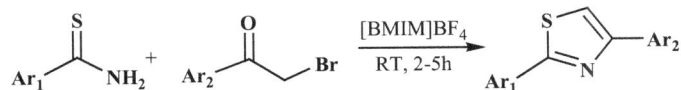

   $[BMIM]BF_4$, RT, 2-5h

   b. The synthesis of thiazolidinones from hetero/aromatic amines, 2-mercaptoacetic acid and carbonyl compounds in $[BMIM]BF_4$ or [MOEMIM]TFA.

   IL, 80-180min

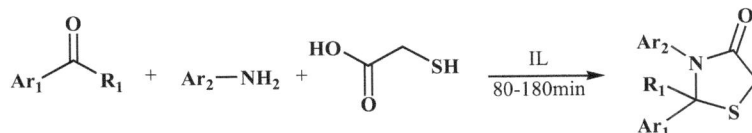

c. Synthesis of 1,3-dithiazolylbenzene derivatives from 1,3-benzenedithioamide and α-bromoacetophenones in $[BMIM]BF_4$.

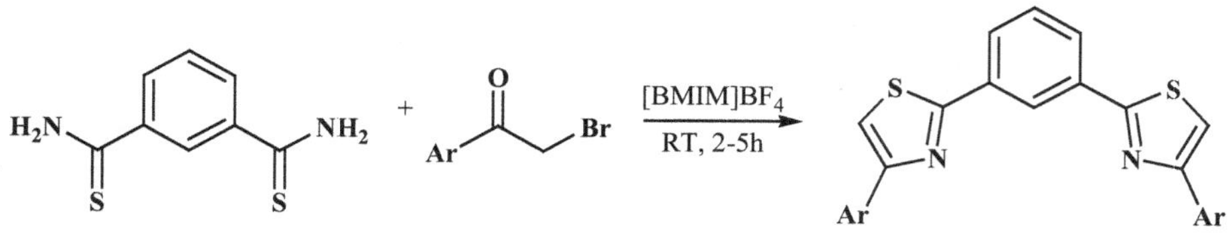

$[BMIM]BF_4$, RT, 2-5h

11. a. Synthesis of nitriles – Reaction of aromatic aldehydes with hydroxylamine hydrochloride using glycerol as the solvent.

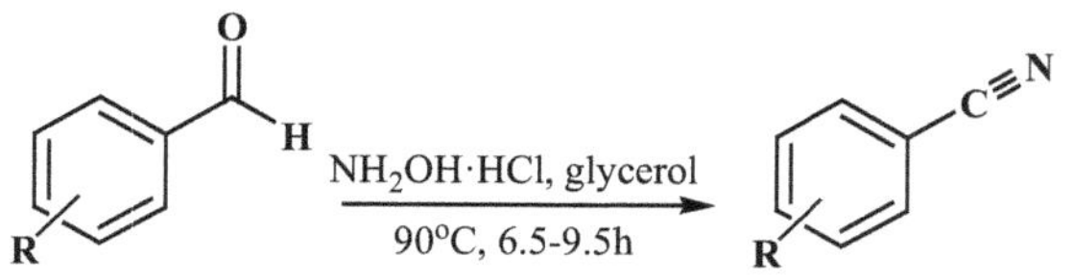

$NH_2OH \cdot HCl$, glycerol, 90°C, 6.5-9.5h

b. Synthesis of alcohols – reduction reactions of carbonyl compounds with glycerol as a solvent and a hydrogen source.

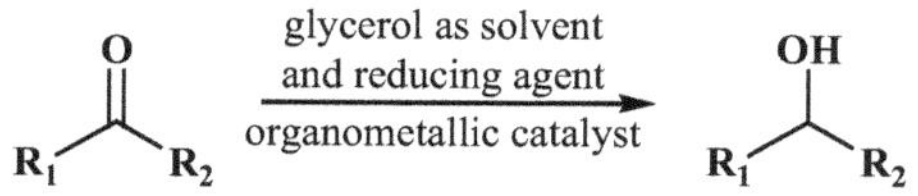

glycerol as solvent and reducing agent, organometallic catalyst

c. Synthesis of amines – the reduction of nitro compounds with glycerol as solvent and hydrogen source in the presence of magnetic ferrite nickel nanoparticles (Fe_3O_4-Ni) as the catalyst.

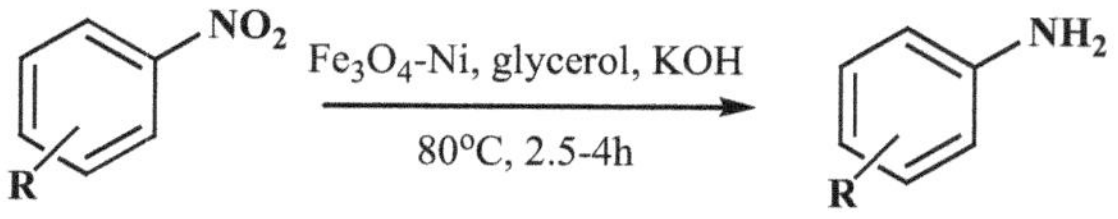

Fe_3O_4-Ni, glycerol, KOH, 80°C, 2.5-4h

12. a. Synthesis of pyrano[2,3-*c*]pyrazole derivatives in a multi-component reaction between aromatic aldehydes, malononitrile, hydrazine hydrate or phenyl hydrazine and 1,3-dicarbonyl compounds.

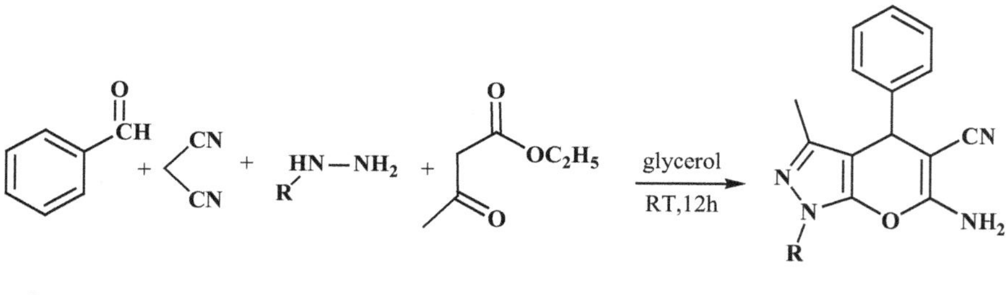

glycerol, RT,12h

b. Synthesis of pyrano[2,3-*c*]pyrazoles in two-step multicomponent reaction of arylhydrazine, β-oxo ester, alkene and formaldehyde.

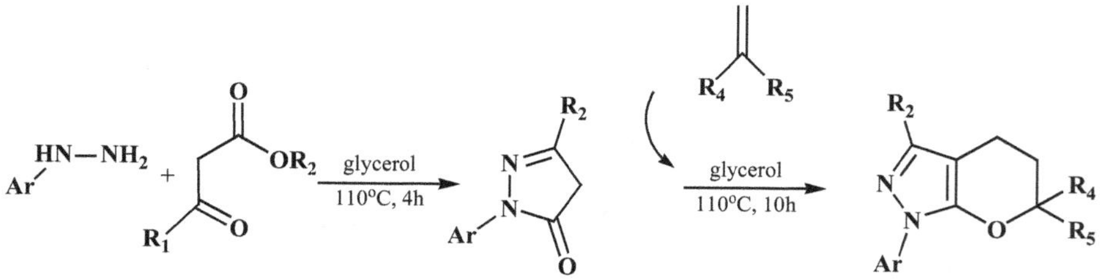

glycerol, 110°C, 4h; glycerol, 110°C, 10h

c. Synthesis of 4-arylselanylpyrazolesfrom α-arylselanyl-1,3-diketones and arylhydrazines under N_2 atmosphere.

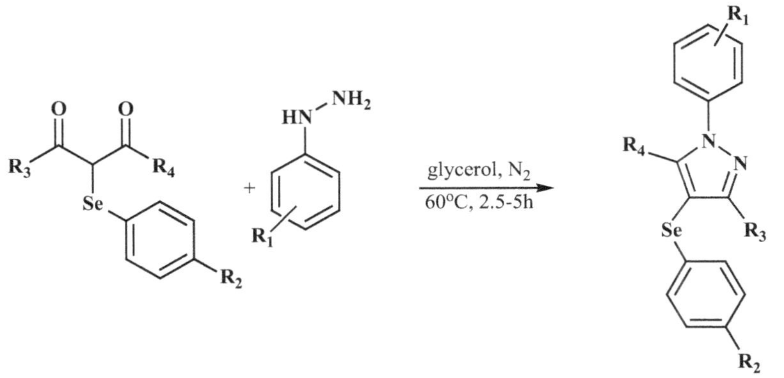

glycerol, N_2
60°C, 2.5-5h

13. Hydrolases and oxidoreductases.
14. D
15. a. Enantioselective desymmetrization ofdiacetate of *meso*-2-(2-propynyl)cyclohexane-1,2,3-triol by*Candida antarctica* lipase in buffer solution:

CAL
buffer pH 7.0, 30°C

80% yield, >99% ee

Aquayamicin

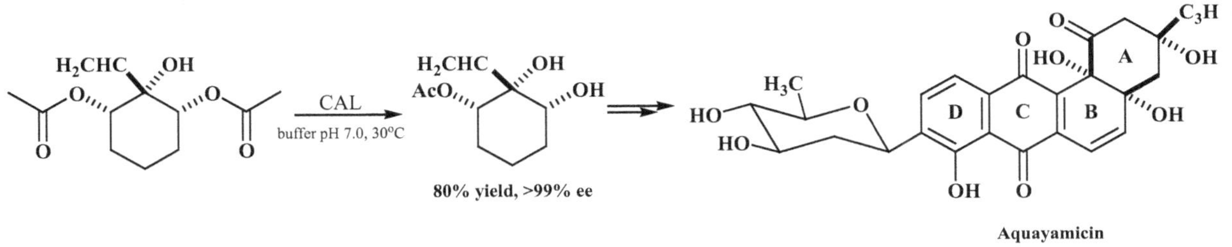

b. Enantioselectivedesymmetrization of the prochiral diethyl 3-[3',4'-dichlorophenyl] glutarate by *Candida antarctica* lipase in buffer solution:

CALB
buffer

80% yield, >99% ee

NK1/NK2 receptor

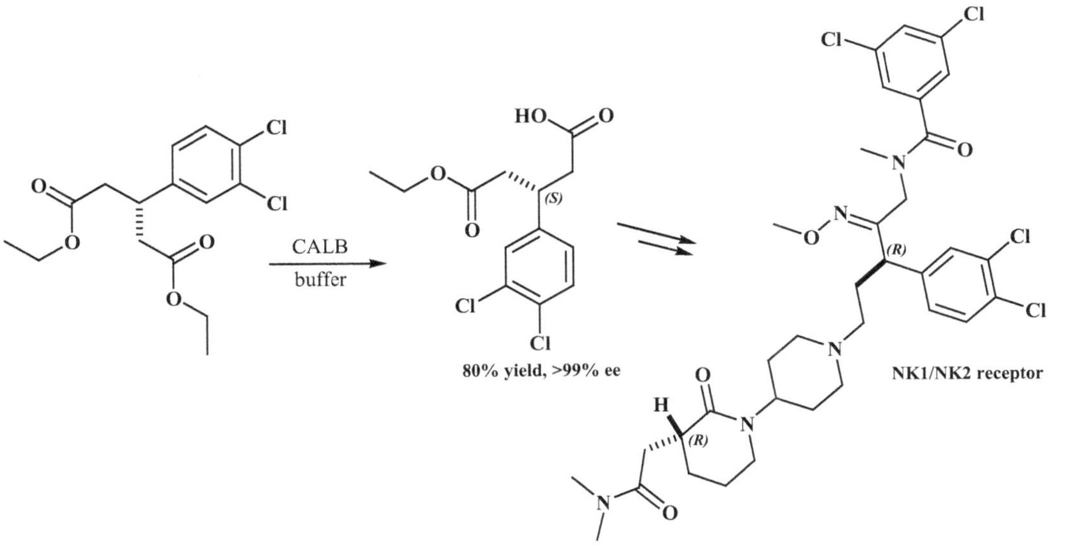

16. a. Diastereoselective microbial reduction of α-chloroketone by three strains of *Rhodococcus sp.* in buffer solution:

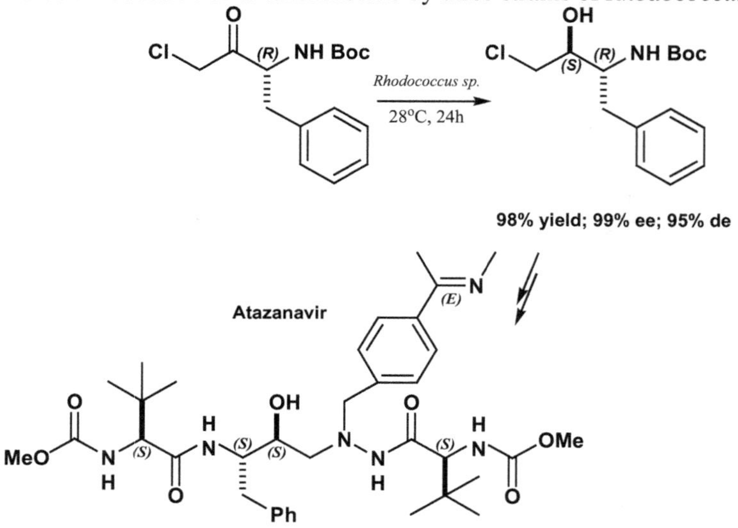

Rhodococcus sp.
28°C, 24h

98% yield; 99% ee; 95% de

Atazanavir

b. Enantioselective microbial reduction of 6-oxobuspirone by *Pseudomonas putida* and *Hansenula polymorpha* in buffer solution:

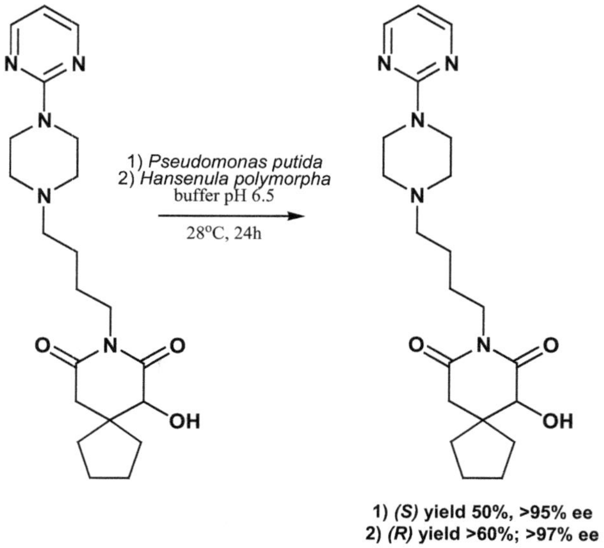

1) *Pseudomonas putida*
2) *Hansenula polymorpha*
buffer pH 6.5
28°C, 24h

1) *(S)* yield 50%, >95% ee
2) *(R)* yield >60%; >97% ee

18. Even though the compound is pure, it exists as a mixture of two different rotamers due to the delocalization of the lone pair of the nitrogen into the carbonyl group, forming a partial C-N double bond. This delocalization is responsible for the limited rotation around the C-N bond and its structural rigidity, resulting in two different rotamers that can be differentiated by the ^{1}H NMR

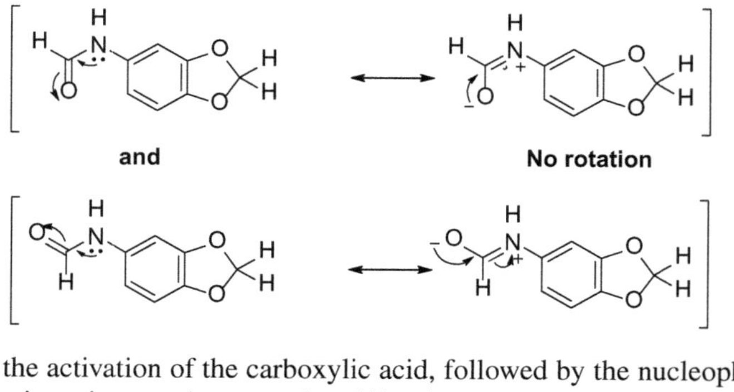

and

No rotation

19. The mechanism starts with the activation of the carboxylic acid, followed by the nucleophilic attack of the N to the carbonyl group. Since aliphatic amines are better nucleophiles than aromatic amines, in this case, the product of the reaction will be the amidation of acetic acid with butylamine.

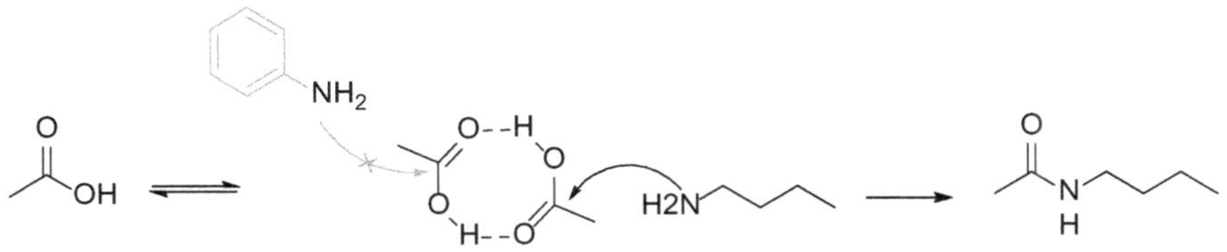

20. A plausible product for this reaction is shown in the next figure. The traditional activation of carboxylic acids is accomplished using electrophiles, so the less electrophilic acid should react with the more electrophilic activator; in other words, the more nucleophilic acid should be easily activated. That means the effect of the electron-withdrawing group will cause a more electrophilic character for the oxo-acid; consequently, this function is already activated. If the activation step is performed with a strong electrophile, the reaction should occur in the less acidic group.

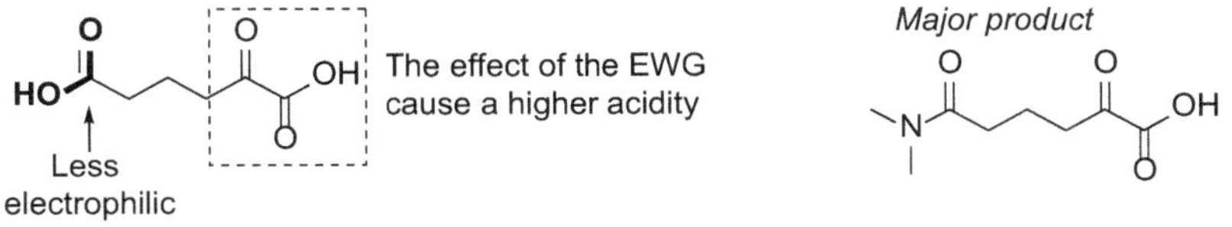

Less electrophilic

The effect of the EWG cause a higher acidity

Major product

21. A good activator or catalysts for direct activation should have the following characteristics: a. non-toxic; b. low molecular weight; c. highly electrophilic; d. metal-free or using eco-compatible metals.
22. Probably the reaction proceeds by an amidation followed by dehydration promoted by the nucleophilic attack of sulfur on the electrophilic carbonyl. The last step is promoted by heating, but oxidation of thiol is in competition; so that may explain the low yield.

    The same reaction with 2-aminophenol should produce the amide and no cyclic product since oxygen is less nucleophilic than nitrogen and is less nucleophilic than sulfur. Consequently, the dehydration promoted by the nucleophilic attack should be less favored.

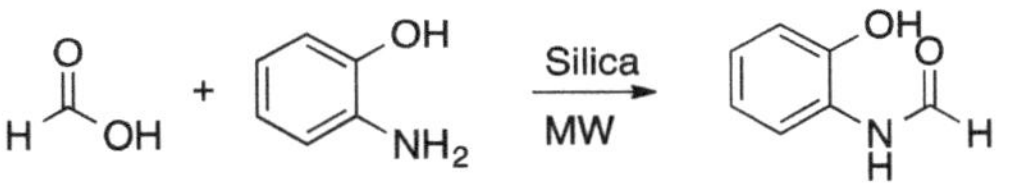

Silica, MW

23. It is possible to identify the reaction product. If you have three lines in the molecular ion region (M^+, M+2 and M+4) with gaps of 2 m/z units between them and with peak heights in the ratio of 9:6:1, the compound contains 2 chlorine atoms.

    The lines in the molecular ion region (at m/z values of M^+, M+2 and M+4) result from the various combinations of chlorine isotopes that are possible. The relative abundance must be calculated using the relative abundance of each isotope.

$$R+35+35=M^+$$

$$R+35+37=M+2$$

$$R+37+37=M+4$$

24. The signals of ^{13}C NMR are split because the product has one fluorine in its structure and the ^{19}F and ^{13}C are coupling.

    Fluor has only one isotope ^{19}F and its nuclear spin is −1/2. Any ^{13}C near to the fluorine (within a few bonds) will be split into a doublet in the ^{13}C NMR spectra because the ^{19}F nucleus has two spin states: +1/2 and −1/2.

    Standard proton-decoupled ^{13}C spectra for compounds containing both fluorine and protons present split, since the coupling C-H is avoided, but the coupling C-F is still present. The signal/noise ratio may affect the signal appearance; however, in modern spectrometers, it is possible to perform F-C decoupling experiments, thus increasing the intensity for carbons close or linked to fluorine atoms.
25. El main product is bromobenzene

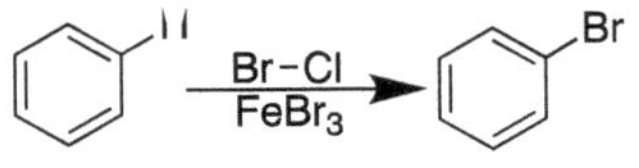

Br-Cl, $FeBr_3$

It is necessary to have in mind the electronegativity values for bromine (2.96) and chlorine (3.16); with this information, it is possible to conclude that bromine has a positive partial charge and chlorine a negative partial charge; that is the reason why Br^+ will be the electrophile that reacts with the aromatic ring.

26. The main role of iron halide in an electrophilic substitution reaction is the polarization of the X-X bond allowing to obtain a more electrophilic species (X^+), which is a much better electrophile than X_2, allowing the reaction to proceed.
27. Iodine is the less electrophilic of the halogens and, as a consequence, normal iodination is harder to accomplish than bromination or chlorination; besides, iodine adds to the aromatic reversibly generating HI, which is a strong reducing agent and reacts with the iodinated compound producing I_2 and the aromatic reagent. The addition of oxidizing agents (stronger than the aromatic) will produce the oxidation of HI to I_2 without affecting the iodinated organic compound, thus shifting the equilibrium to the desired product.

    The right way to do the iodination is to use catalytic amounts of an oxidizing agent; common examples are HNO_3, HIO_3 and HgO, among others.
28. Useful alternatives to access the ortho-substituted aromatics using transition metal catalysis; examples of this are the use of Rh or Pd catalysts published by Glorius and Fabis, respectively. In these reactions, the amide or the oxime plays as a directing group, making the reaction feasible with very good selectivities.

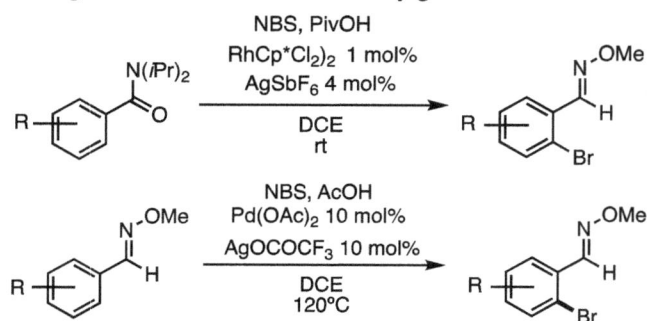

NBS, PivOH
$RhCp^*Cl_2)_2$ 1 mol%
$AgSbF_6$ 4 mol%
DCE
rt

NBS, AcOH
$Pd(OAc)_2$ 10 mol%
$AgOCOCF_3$ 10 mol%
DCE
120°C

29. 1-chloro-1,2-benziodoxol-3-one (CBDO), a known reagent that showed excellent reactivity with heteroaromatics and activated carbocycles. The reaction is performed at room temperature and the reagent can be easily regenerated.

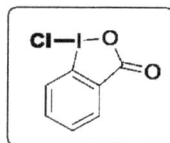

    The Palau'chlor or CBMG is a guanidine-based chlorinating reagent which exhibits very good activity for the chlorination of heteroaromatics; it is inexpensive, can be prepared on the decagram scale and is useful under soft reaction conditions.

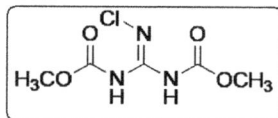

30. DBE (Double bond equivalent) = C + 1 − (H/2) − (X/2) + (N/2)

$$= 14 + 1 - (14/2)$$
$$= 15 - 7$$
$$= 8$$

As we know, a benzene ring requires four DBEs.

So, from the given options, we concluded that there are two benzene rings.

From the ^{1}H NMR data, the presence of a peak at 3.65 ppm confirms the –OMe group. So, there are two –OMe groups.

Also confirmed is the ^{13}C NMR peak at 55.27 ppm, which confirms the presence of the –OMe group.

According to another data from the ^{1}H NMR spectra, there are two peaks at 6.97 (d, 4H, J = 8.8 Hz) and 7.49 (d, 4H, J = 8.8 Hz). So, we can conclude that the molecular structure should be symmetrical as below.

Along with this, it confirms the ^{13}C and DEPT-135 peaks. Else, all the carbons show positive signals corresponding to either –CH or –CH_3 group.

So, the molecular formula is option (b).

MeO–C6H4–C6H4–OMe

DBE (Double bond equivalent) = C + 1 – (H/2) - (X/2) + (N/2)
= 17 + 1 – (16/2) + (2/2)
= 18 – 8 + 1
= 11

From the given molecular options, a benzene ring requires four DBEs, –CN group requires two DBEs in a ring, two double bonds require two DBEs, the carbonyl group requires one DBE and two adjacent rings require two DBEs.

So, the molecular unit can be

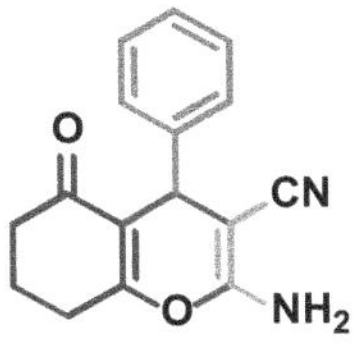

O CN NH2 O

From the ^1H NMR spectra, there are two peaks at 6.82 (d, 2H, J = 7.2 Hz) and 7.05 (d, 2H, J = 7.9 Hz). So, we can conclude that the benzene ring should be symmetrical as below:

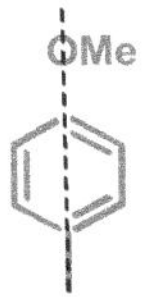

OMe

From the ^1H NMR data, the presence of a peak at 3.70 ppm confirms the –OMe group. So, there are two –OMe groups.

Also confirmed is the ^{13}C NMR peak at 55.52 ppm, which confirms the presence of the –OMe group.

The –NH_2 group comes at 6.93 ppm.

The three –CH_2 groups come from 1.84 to 2.58 ppm which confirms the 13C NMR also. The ^{13}C NMR peak at 196.31 ppm confirms the presence of –C=O group.

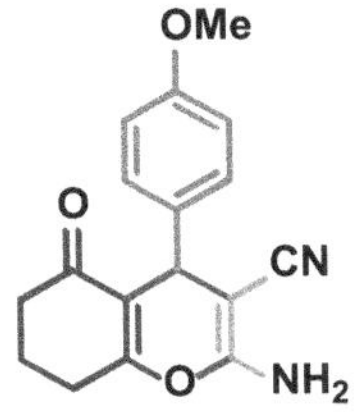

OMe O CN NH2 O

So, the molecular formula is option (a)

31. Under conventional conditions, organic reactions are heated using traditional heat transfer equipment such as sand baths, oil bath and heating jackets. However, rather slow and temperature gradient can develop within the sample. Also, local overheating can lead to product, substrate and reagent decomposition. In contrast, in microwave heating, the microwave energy is introduced into the chemical reactor remotely and direct access by the energy source to the reaction vessel is obtained. The microwave radiation passes through the walls of the vessel and heats only the reactants and solvent, not the reaction vessel itself.
32. The larger the dielectric constant, the greater the interaction with the solvent. So, solvents such as water, DMF, methanol, acetone, ethyl acetate, acetic acid, chloroform and dichloromethane are heated when they are irradiated with microwaves, while solvents such as toluene, hexane, CCl_4 and diethyl ether do not interact; in other words, they are more transparent to microwaves.
33. The materials can be classified into three categories based on their interactions with microwaves. The first one corresponds to high dielectric materials which lead to strong absorption of microwaves and consequently to rapid heating of the medium. They are the most important class of materials for microwave-induced synthesis, for example, polar solvents or aqueous solutions. The second class is composed of materials that are transparent to microwaves,

exhibiting only small interactions with the penetrating microwaves; examples are borosilicate glass, ceramic, Teflon and fused quartz among others. The last category is composed of materials that reflect microwaves; there is no, or only small, coupling of energy into the system. In this case, the temperature increases in the material only marginally, for example, metals.

34. The main different mechanisms that are involved in the heating of materials are:
    a. Dipolar polarization mechanism: It corresponds to one of the interactions of the electric field component with the sample. It is important that the substrate possesses a dipole moment to generate heat when it is irradiated by microwaves, since the dipole is sensitive to external electric fields and will attempt to align itself with the field through rotations.
    b. Conductions mechanism: This phenomenon is due to the second major interaction of the electric field component with the sample. It takes place when there are ions or ionic species free in the dissolution, which in the presence of an electric field try to orient themselves. The result is an instantaneous localized superheat.
35. The literature shows many different examples where stereoselectivity of the reaction is affected by microwave actions. This phenomenon is observed mainly when a competitive reaction is involved because the ground states in both are the same. However, the mechanism that occurs under more polar transitions state, could be favored by the actions of microwave irradiation.
36. There is an example of photosynthesis is where plants convert carbon dioxide and water into glucose and oxygen using solar energy. In humans, vitamin D production is completed by sunlight exposure. Bioluminescence is an example where an enzyme in the abdomen catalyzes a light-producing reaction in fireflies.

    Photodissociation is a chemical reaction where photons may break down a chemical compound. It is defined as one or more photons interacting with a single target molecule. There is no limited photodissociation to visible light.
37. Ketones have two types of transitions, an np* and a pp*. The latter is higher in energy and easy to see in the UV spectrum, as is the lower singlet. In the spectrum, one cannot see the triplets (a forbidden transition) but one may play with energy transfer sensitizers and quenchers until to bracket their energy.
38. They were the S0-S1 and S1-S0 transitions in each case with vibrational structures S_{00}-S_{1n} and $S_{10}S_{0n}$, respectively; the first figure indicates the electronic level and the second one, the vibrational level.
39. A $PhCH_2SiMe_3$ +NpCN – $PhCH_2SiMe_3^{+}$. + $NpCN^{-}$. $PhCH_2$.
40. Cis stilbene suffers from some sterical hindering to perfect coplanarity.
41. Glucose is one of the most numerous homopolysaccharides that is manufactured by plants. A reaction mechanism for the synthesis of *N*-substituted pyrroles using starch is as follows:

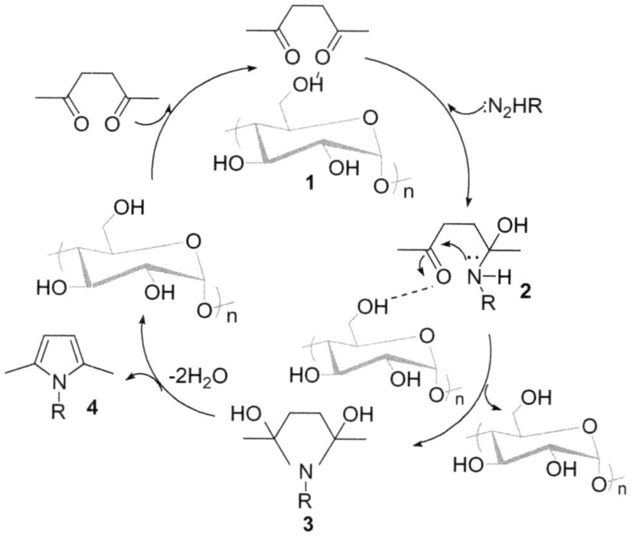

42. The best solvent is without solvent. The electron-donating groups on the phenyl ring increase the yield and reduce the reaction time relative to electron-withdrawing groups. The major advantages of this procedure are high atom economy, solvent-free reaction, reusability of the catalyst, short reaction time, no heating required and its waste-free nature as H_2O is the only by-product.

43.

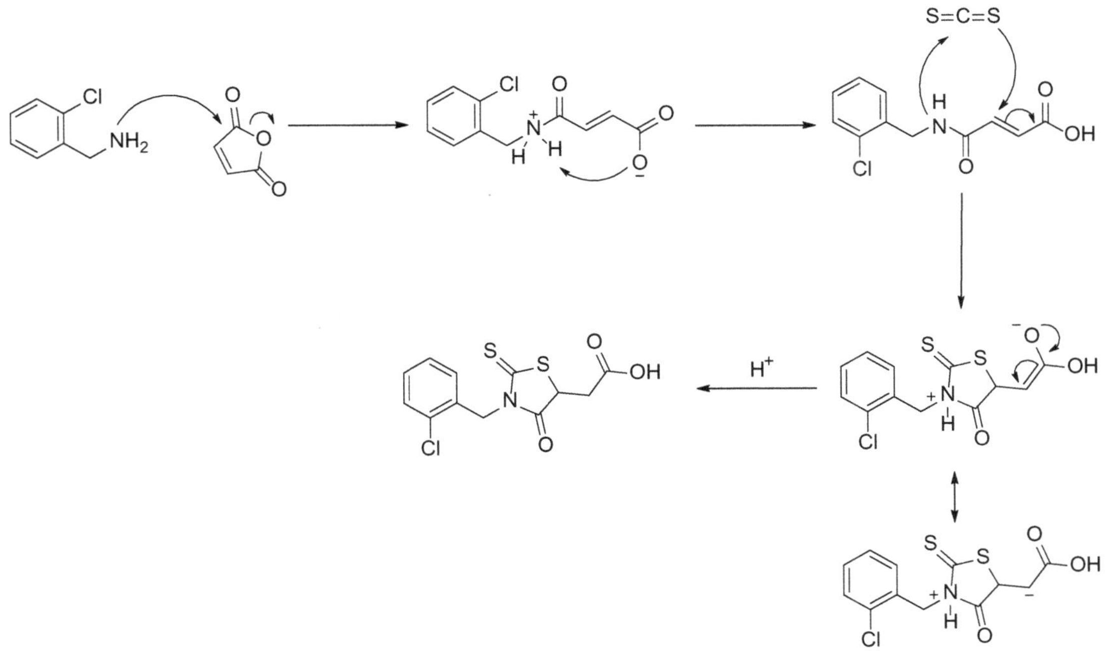

44. Coformers are selected based on hydrogen bonding. In general, coformers are organic acids, although some have other ionization properties. Co-crystals depend on interactions such as hydrogen bonds, **π-π** interactions and Van der Waals interactions; however, hydrogen bonding is the most frequent.

45. Frequency: the increase in frequency leads to a decrease in the production and intensity of cavitation in liquids.
 Solvent: Cavitation produces minor effects in viscous or high surface-tension liquids.
 Temperature: The increase in temperature allows cavitation to be carried out at lower acoustic intensities.
 External pressure: The increase in external pressure leads to an increase in the intensity of the destruction of cavitation bubbles.
 Intensity: The increase in ultrasound intensity strengthens the effects produced.

46.

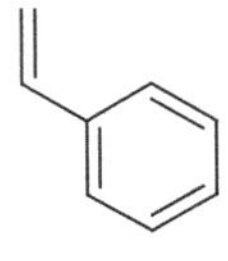

47. Enantiotopic protons have the same chemical shift in the vast majority of situations but if they are placed in a chiral environment (e.g., chiral solvent), they will have different chemical shifts.
 a. These two sterioismers can be differentiated with the help of coupling constants. The process of cis double bond will exhibit coupling constants in the range of 9–12 Hz while protons of trans double bond show coupling constants in the range of 15–18 Hz.

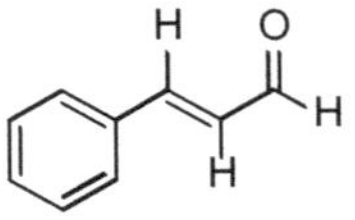

Trans-cinnamaldehyde

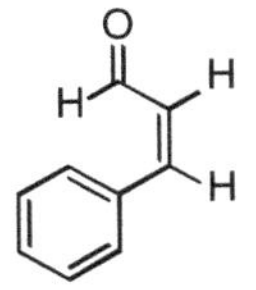

Cis-cinnamaldehyde

b. $HOCH_2C_6H_5CH_3$ compound has two singlets: one for CH_2 and the other for CH_3. The four phenyl H_s will be multiplet, whereas C_6 H_5 CH_2 CH_2OH five phenyl groups H_s multiplet and two triplet for the two methylene groups. In both cases, a broad singlet for both H and OH.

48. a A pericyclic reaction is a concerted reaction that proceeds through a cyclic transition state which is insensitive to polar solvents and catalysts but requires light or heat. It has the following properties: a) concentrated, b) reversible, c) single transition structure (TS) with no intermediates, d) mechanisms explained through the analysis of frontier molecular orbitals (FMOs), e) forward or reverse reactions provide the same analysis. The reaction is completely sterospecific; that is, a single stereoisomer of the reactant forms a single stereoisomer of the product. As the reactions proceed through cyclic transition states, the reactions are called pericyclic reactions. In this process, half of the molecular orbitals having lower energies than the isolated p orbitals are called **bonding molecular orbitals** and the other half of the orbitals having higher energies than the isolated p orbitals are called **antibonding molecular orbitals**. The two molecular orbitals which are of occupied molecular orbitals of highest energy is termed as **highest occupied molecular orbital (HOMO)** and the unoccupied molecular orbital of lowest energy is termed as **lowest unoccupied molecular orbital (LUMO)**. **Frontier molecular orbital** (FMO) allows the interpretation of a molecular interaction to be restricted to an analysis of the interactions between the highest occupied and lowest unoccupied molecular orbitals (HOMOs and LUMOs) of the reacting partner.

There are three types of pericyclic reactions: (i) electrolcyclic reactions, (ii) cycloaddition reactions and (iii) sigmatropic reactions.

i. Electrolcyclic reactions: In that reactions ð and π bonds are interconnected such examples of 1,3 butadiene becomes a ð in cyclobutane and the next ð bond of the cyclobutene becomes a π bond in 1,3 butadiene. Two features determine the course of the reactions: the number of π bonds involved and whether the reaction occurs in the presence of heat (thermal conditions) or light.

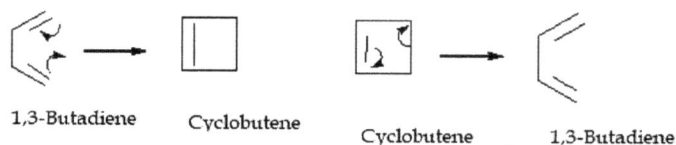

1,3-Butadiene → Cyclobutene   Cyclobutene → 1,3-Butadiene

ii. Cycloaddition reactions: The cycloaddition is a reaction between two compounds with π bonds to form a cyclic product with two new σ bonds. The reactions are concentrated and sterospecific. The reactions can be initiated by heat or light which are also identified by the number of π electrons in the two reactants.

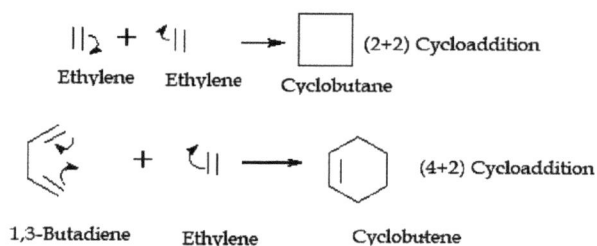

Ethylene + Ethylene → Cyclobutane (2+2) Cycloaddition

1,3-Butadiene + Ethylene → Cyclobutene (4+2) Cycloaddition

Cycloaddition reactions may take place either across the same face (called superfacial) or across the opposite face (called antarafacial) of planes in each reacting system.

iii. Sigmatropic reactions: These are uncatalyzed reactions where a ð bond migrates along with the atom or group attached to a new position within a π framework.

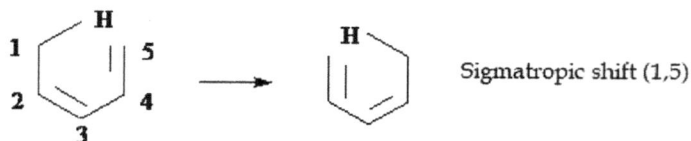

H 1 5 2 4 3 → H Sigmatropic shift (1,5)

Here, one end of ð moves over five atoms.

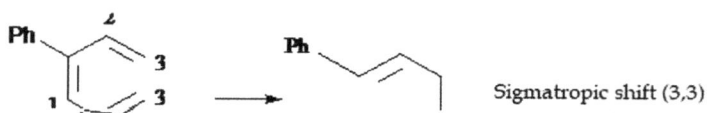

Ph 2 3 1 3 → Ph Sigmatropic shift (3,3)

Here both ends of a ð bond migrate over the three atoms.

b. i. Here, cis 1,2-divinyl cyclobutane rearranges not via the normal pathway (i.e., chair-like transition state) as a steric reason but rearranges via a boat-like six-membered transition state to give *cis,cis* 1,5-cyclooctadiene.

80-110°C

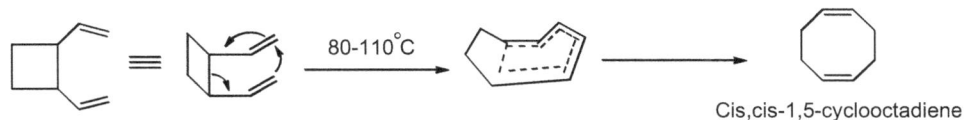

Cis,cis-1,5-cyclooctadiene

ii. This is a molecular arrangement reaction of *trans* isomer of 1,2–divinyl cyclobutane where steric reason the compound does not proceed through a cyclic transition state but the reaction actually proceeds by a non-concentrated pathway involving dissociates into allylic radicals to give products 4-vinylcyclohexene and *cis, cis* -1,5-cyclooctadiene.

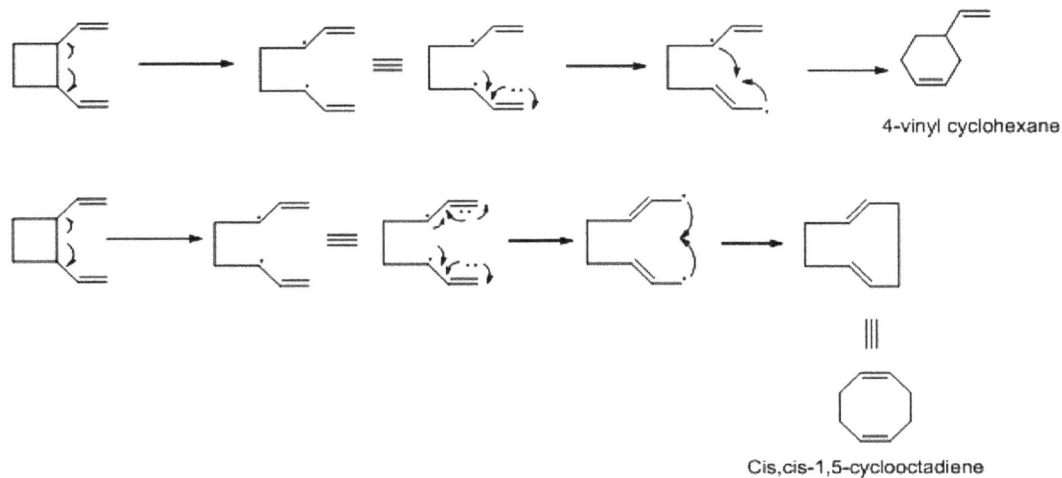

4-vinyl cyclohexane

Cis,cis-1,5-cyclooctadiene

iii. The reaction is 1,5 sigmatropic carbon-shift process.

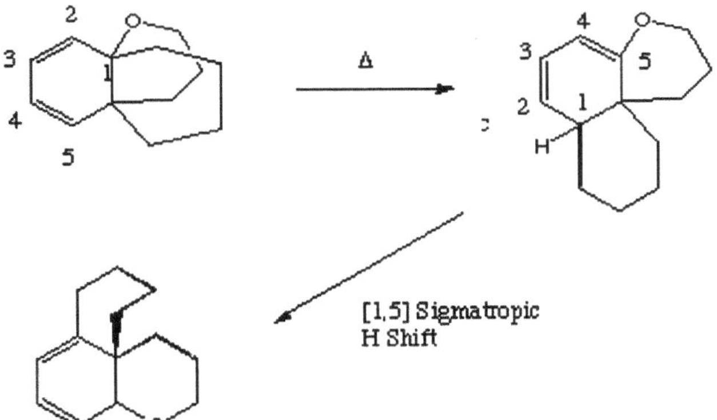

iv. In this case, conrotory ring-opening of the cyclobutane ring (to a 4π electron system) occurs to give a non-aromatic intermediate which undergoes a Diels–Alder reaction with maleic anhydride in the endo sense to give the aromatic product.

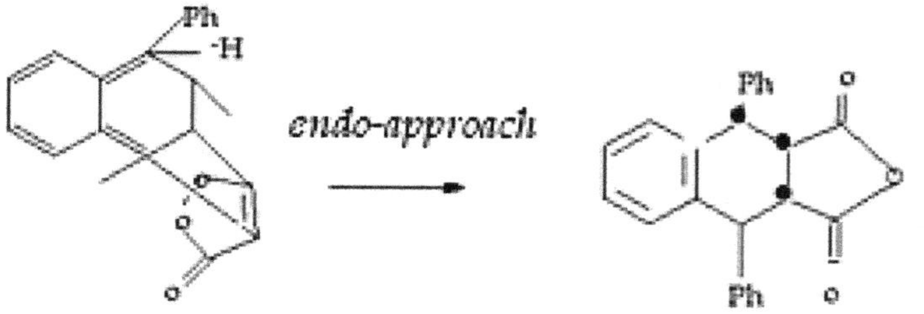

Ph H Ph Ph Δ Conrotatory H Ph H

Ph H endo-approach Ph O O O Ph O

49.

Base (A) $C_{11}H_{20}O$ HBr (B) $C_{11}H_{21}BrO$

-H -H2O J 15 Hz 7.0 6.4 (A)

HBr Br (B)

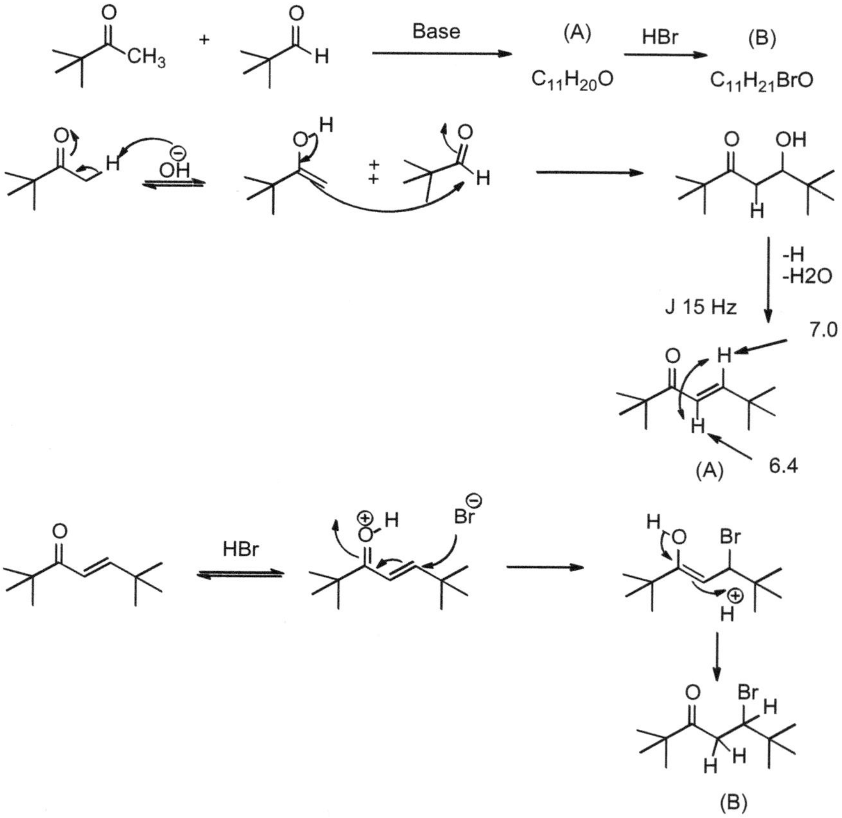

Compound A is *trans* isomer and compound B is one type of BX system: AB are diastereotopic CH_2 group (J_{AB} =17.7) and H is CHBr proton (J_{ax} = 10; Jbx = 1.9).

50. Production of antibiotics can be done using three methods:

a. Natural microbial production using fermentation technology, for example, penicillin
b. Semi-synthetic production (post-production modification of natural antibiotics), for example, ampicillin
c. Synthetic production of antibiotics in the laboratory, for example, quinoline

Synthesis of 4-substituted **β**-lactams through the Kinugasa reaction utilizing calcium carbide by Abolfazl Hosseini, Peter R. Schreiner synthesis of **β**-lactams. The reaction is a one-pot synthesis of lactams from inexpensive calcium carbide and nitrone derivatives. Calcium carbide is activated by TBAF·3H $_2$O in the presence of CuCl/NMI.

(A. Hosseini, P. R. Schreiner, *Org. Lett.*, 2019, *21*, 3746-3749.)

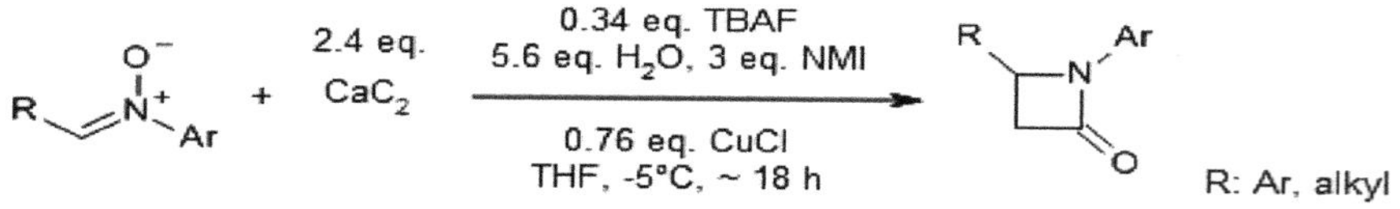

2.4 eq. CaC_2 — 0.34 eq. TBAF, 5.6 eq. H_2O, 3 eq. NMI / 0.76 eq. CuCl, THF, -5°C, ~ 18 h

R: Ar, alkyl

51. Animals can synthesize and store large quantities of triacylglycerols which are synthesized within the body in various organs, that is, in the liver, kidney, lactating mammary glands, aorta and intestinal mucosa to be used later as fuel enough to supply the body's energy needs for 12 hours. These triacylglycerols are stored in adipose tissue, but plants also manufacture triacylglycerols as an energy-rich fuel, stored especially in fruits, nuts and seeds. It consists of the following stages: (i) Activation of glycerol to glycerol-3-phosphate. It can be obtained from two important sources. (ii) Activation of fatty acids to fatty acyl-CoA. Fatty acids are activated by the respective thiokinase in the presence of ATP and coenzyme A.

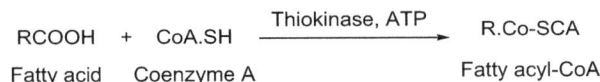

i. The formation of glycerol-3-phosphate consists of two stages (a) and (b):

a. In the first stage, the free glycerol which is formed by the hydrolysis of glycerol phosphate or glycerol-containing lipids may be phosphorylated by means of ATP in the presence of a glycerokinase.

CH_2OH — HOH — CH_2OH (Glycerol) $\xrightarrow[\text{ATP, } Mg^{2+}]{\text{Glycerol kinase}}$ CH_2OH — CHOH — CH_2OP (L-Glycerol-3-Phosphate)

b. In the second case, glycerol-3-phosphate is formed in the tissues by the reduction of glyceraldehyde-3-phosphate or dihydroxyacetone phosphate produced in the pentose phosphate pathway or glycolysis.

CH_2OH — CO — CH_2OP (Dihydroxyacetoxyphosphate) $\xrightarrow{\text{Reductase}}$ CH_2OH — CHOH — CH_2OP (Glycerol-3-Phosphate) $\xleftarrow{\text{Reductase}}$ CHO — CHOH — CH_2OP (Glyceraldehyde-3-Phosphate)

ii. In this stage activation of fatty acids to fatty acyl-CoA. Fatty acids are activated by the respective thiokinase in the presence of ATP and coenzyme A.

$$\underset{\text{Fatty acid}}{\text{RCOOH}} + \underset{\text{Coenzyme A}}{\text{CoA.SH}} \xrightarrow{\text{Thiokinase, ATP}} \underset{\text{Fatty acyl-CoA}}{\text{R.Co-SCA}}$$

Glycerol-3-phosphate is then esterified with 2 moles of fatty acyl-CoA to form 1, 2-diacyl glycerol phosphate (phosphatidate) via 1-acyl glycerol-3-phosphate (lysophosphatidate). The enzyme (glycerophosphate acyl transferase) for this reaction is not very specific for the definite chain length of the activated acids which reacts more swiftly, however, with C_{16}–,$_{17}$– and C_{18}– fatty acids. This fact provides one possible explanation for the predominance of C_{16}– and C_{18}– fatty acids in neutral fats. In that reaction phosphatidate, a common intermediate is formed in the synthesis of triacylglycerols (fats) and phosphoglycerides.

CH_2OH — CHOH — CH_2O–P(=O)(–O⁻)–O⁻ (Glycerol-3-Phosphate) $\xrightarrow{R_1\text{–C(=O)–S.CoA} \;\to\; \text{CoA}}$ CH_2O–C(=O)–R_1 — CHOH — CH_2O–P(=O)(–O⁻)–O⁻ (Lysophosphatidate)

$\xrightarrow{R_2\text{–C(=O)–S.CoA} \;\to\; \text{CoA}}$ CH_2O–C(=O)–R_1 — CHO–C(=O)–R_2 — CH_2O–P(=O)(–O⁻)–O⁻ (Phosphatidate)

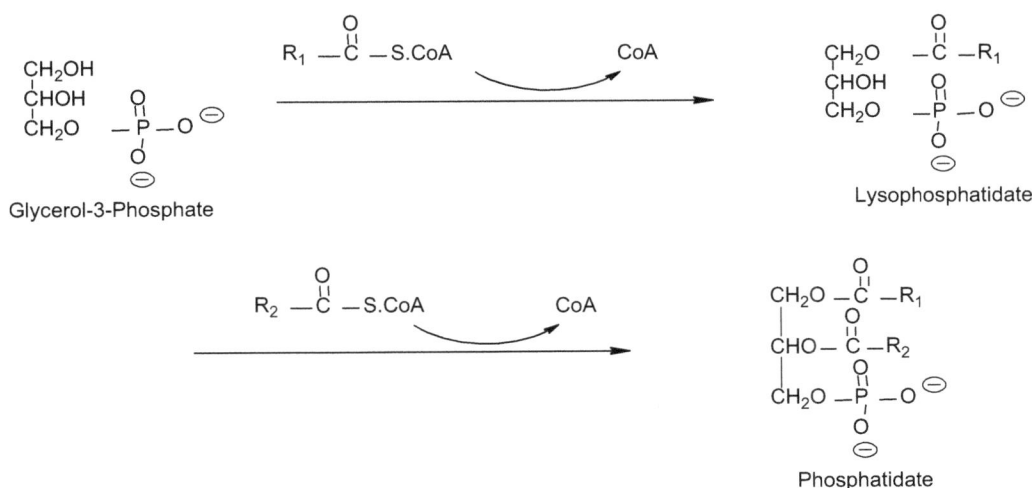

In the next step, the phosphatidate is dephosphorylated by means of a phosphatase and is then esterified with another mole of acyl-CoA to form the so-called neutral fat, the triglyceride (triacylglycerol).

Phosphatidate $\xrightarrow[\quad]{H_2O \;\rightarrow\; P}$ Diacylglycerol $\xrightarrow[\quad]{R_3\text{–}CO\text{–}S.CoA \;\rightarrow\; CoA}$ Triacylglycerol

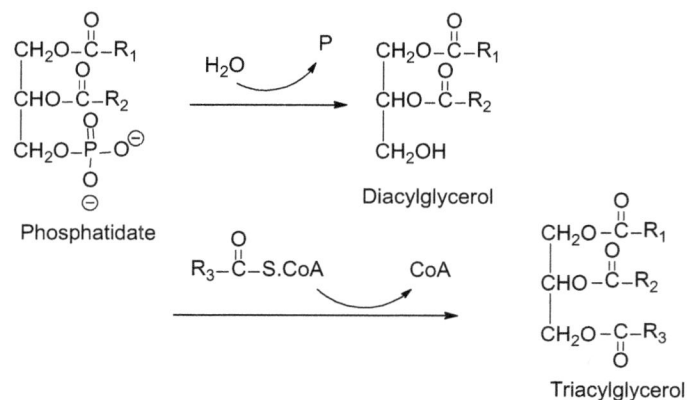

52. Cholesterol is a cyclopentaperhydrophenanthrene ring system that possesses both hydrophilic and hydrophobic regions in the structure. which is found in animals so-called animal sterol. It is generally derived from acetyl CoA (18 moles), ATP (36moles) and NADPH (16 moles). Cholesterol takes place in the endoplasmic reticulum of hepatic cells from acetyl-CoA, which is mainly from pyruvate derived from an oxidation reaction formed from pyruvate in mitochondria whose membrane is impermeable to acetyl-CoA. Hence, acetyl-CoA condenses with oxaloacetate to form citrate which is permeable to the mitochondrial membrane and hence enters into the cytoplasm where it decomposes back to acetyl-CoA and oxaloacetate in the presence of and CoAby the action of an enzyme called *ATP-citrate lyase* and then cholesterol has been biosynthesized in five stages (Figure 1)

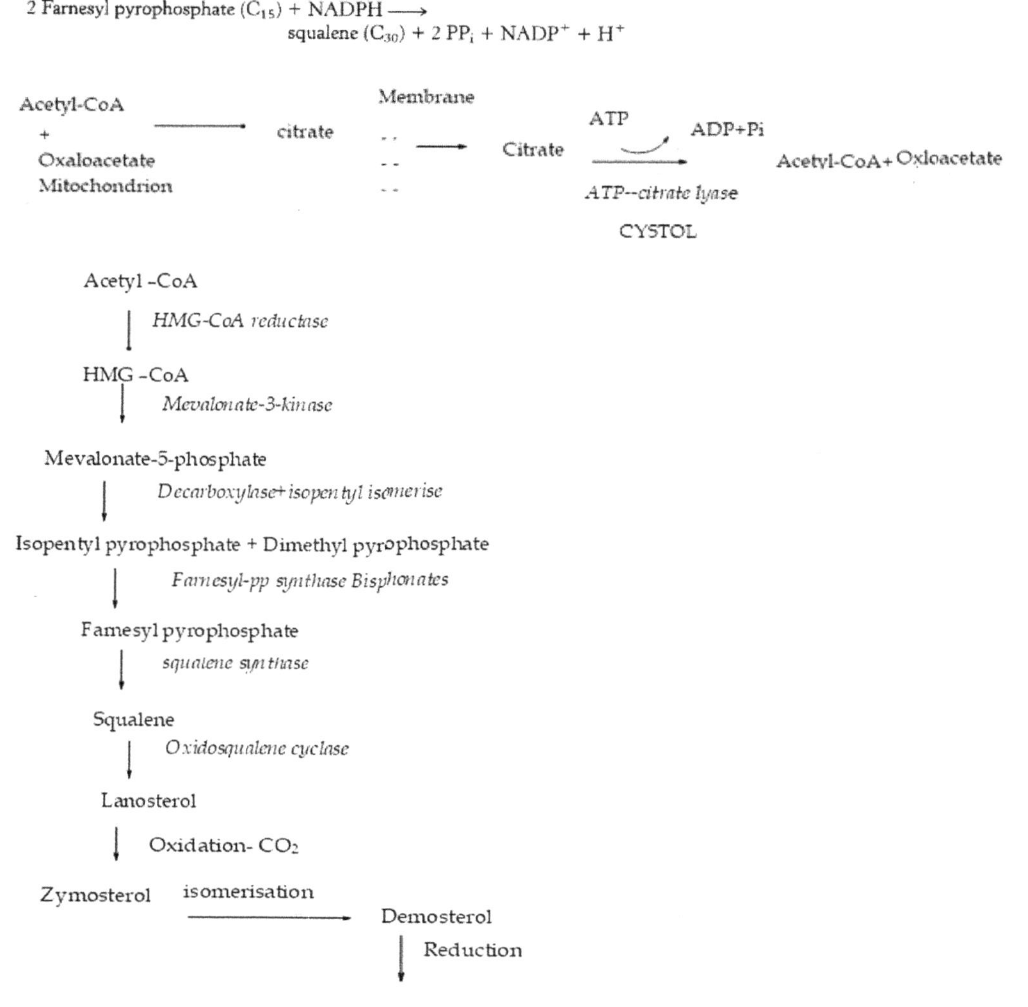

FIGURE 1 Total Biosynthesis of Cholesterol

1. Synthesis of HMG (**β**-hydroxy-**β**-methyl glutaryl CoA), 2) formation of mevalonate, 3) Production of isopentylpyrophosphate (IPP) and dimethyl pyrophosphate (DPP) units, 4) synthesis of squalene and 5) conversion of squalene into cholesterol.
2. Two moles of acetyl CoA condenses to form acetoacetyl CoA which is added to another molecule of acetyl CoA to form 3-hydroxy-3-methylglutaryl CoA (HMG CoA). The whole process occurs in cystol.
3. HMGCoA present in endoplasmic reticulum catalyzes the reduction of *HMG-CoA reductase* to mevaonate acid in the presence of NADPH which is then in presence of *mevalonate 3-kniase* and ATP to mevalonate 5-phosphate. The production of mevalonate is the rate-limiting and irreversible step in cholesterol synthesis.
4. Mevalonate 5-phosphate catalyzed by *phosphomevalonate kinases* to mevalonate-5-pyrophosphate which is in presence of mevalonate-5-pyrophosphate decarboxylase into 3-isopentyl-pyrophosphate (IPP) in presence of ATP and also the liberation of carbon dioxide. IPP in presence *of isopentyl pp-isomerse* is also converted into dimethylpyrophosphate (DPP).
5. IPP and DPP condense in the presence of *fanesyl-ppsynthase* Biosphosphonates to form 10-carbon geranylpyrophosphate (GPP) where another molecule of IPP in the presence of *fanesyl-ppsynthase Biosphosphonates* to form 15 carbon atom farnesyl pyrophosphate (FPP). Now, two molecules of FPP in the presence of squalene synthase from squalene. The process is a reductive tail-to-tail condensation of two molecules of farnesyl pyrophosphate catalyzed by the endoplasmic reticulum enzyme *squalene synthase*.

The process for conversion from squalene to cholesterol is shown in Figure 1. The next step for the conversion from squalene to lanosterol is the oxygenase–catalyzed cyclization where the oxygen required is derived from gaseous oxygen (1) (2a). Generally, compound II is formed by the rearrangement of hydrogen proton and the compound is converted into III by the migration of the methyl group from C_{14} to C_{13}. Compound III is converted into IV with the migration of the methyl group from C_8 to C_{14}. These are examples of 1,2 shift. Next, the conversion of lanosterol into cholesterol involves the removal of three methyl groups (at the 4, 4 and 14 positions) probably by the oxidation of –COOH followed by

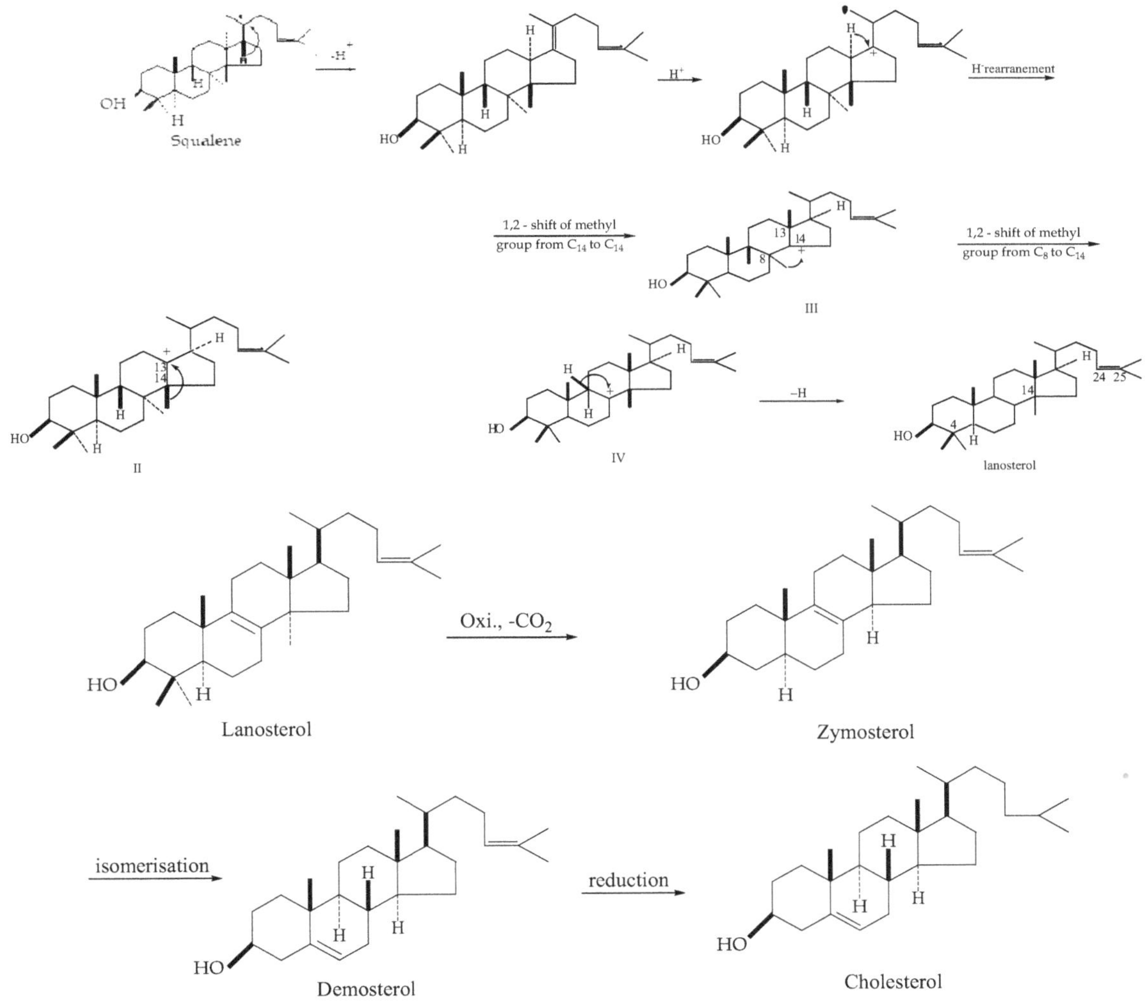

Squalene -H⁺ H⁺ H rearranement

1,2 - shift of methyl group from C_{14} to C_{13} III 1,2 - shift of methyl group from C_8 to C_{14}

II IV -H lanosterol

Lanosterol Oxi., -CO_2 Zymosterol

isomerisation Demosterol reduction Cholesterol

FIGURE 2 (a) Conversion of lanosterol from Squalene, (b) Conversion of Lanosterol into Cholesterol decarboxylation, shifting of the double bond at the position (C_8 and C_9) to C_5 and C_6 and the reduction of C_{24} and C_{25} double bond to get cholesterol (Figure 2b)

53. Bile acids are steroid acids found predominantly in the bile of mammals and other vertebrates. The biosynthesis of bile acids such as monohydroxy cholanoic acid or lithocholic acid ($C_{23}H_{38}(OH)COOH$) and dihydroxycholanic acid or deoxycholic acid ($C_{23}H_{37}(OH)_2COOH$ is formed from cholesterol in the liver. Cholesterol 7 alpha-hydroxylase is the rate-limiting enzyme in the synthesis of bile acid from cholesterol catalyzing the formation of 7 α-hydroxy cholesterol by 7α-*hydrolase* (CYP7A1) and the following process is as follows:

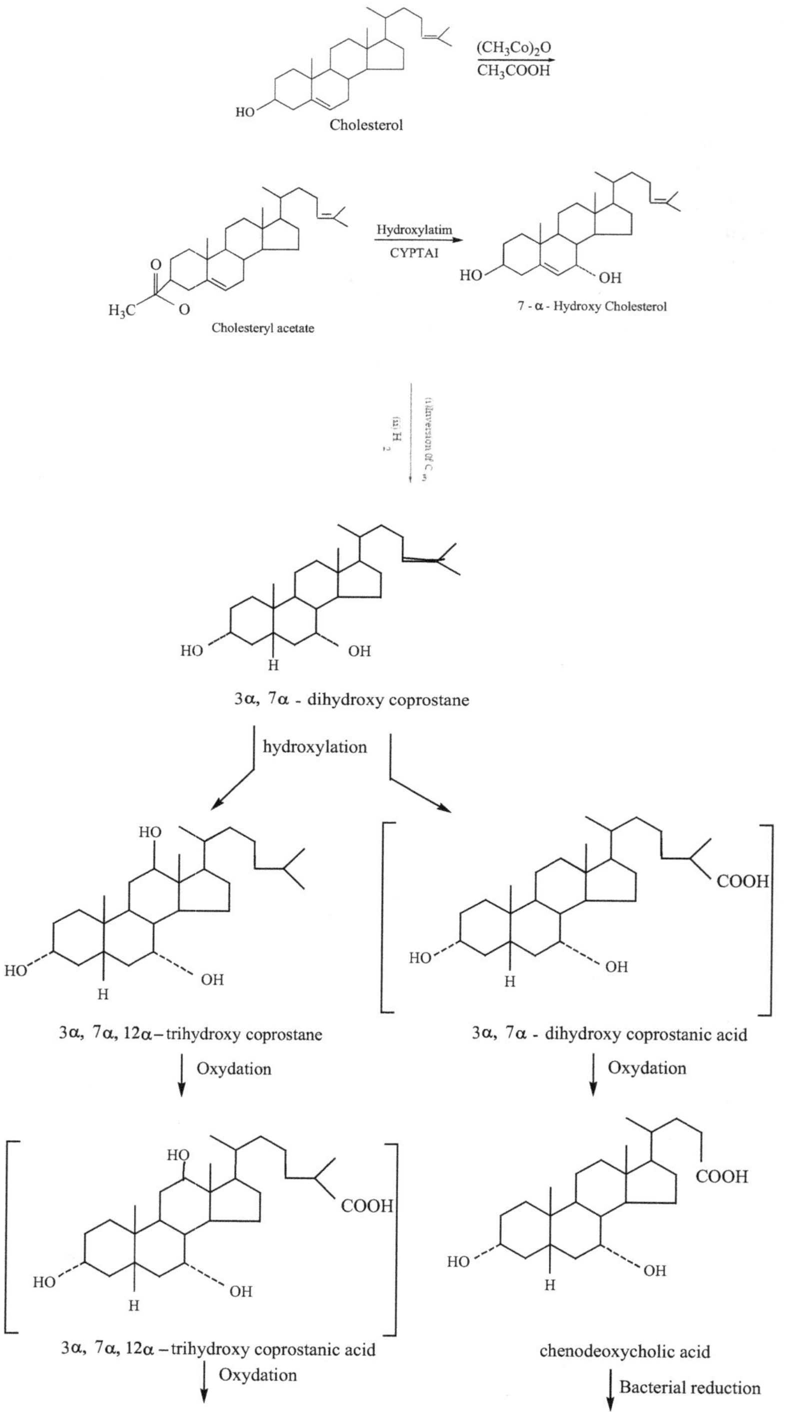

HO Cholesterol

$(CH_3Co)_2O$ / CH_3COOH

O H$_3$C O Cholesteryl acetate

Hydroxylatim / CYPTAI

HO OH 7 - α - Hydroxy Cholesterol

HO H OH

3α, 7α - dihydroxy coprostane

hydroxylation

HO HO H OH

3α, 7α, 12α–trihydroxy coprostane

Oxydation

COOH HO H OH

3α, 7α - dihydroxy coprostanic acid

Oxydation

HO COOH HO H OH

3α, 7α, 12α –trihydroxy coprostanic acid

Oxydation

COOH HO H OH

chenodeoxycholic acid

Bacterial reduction

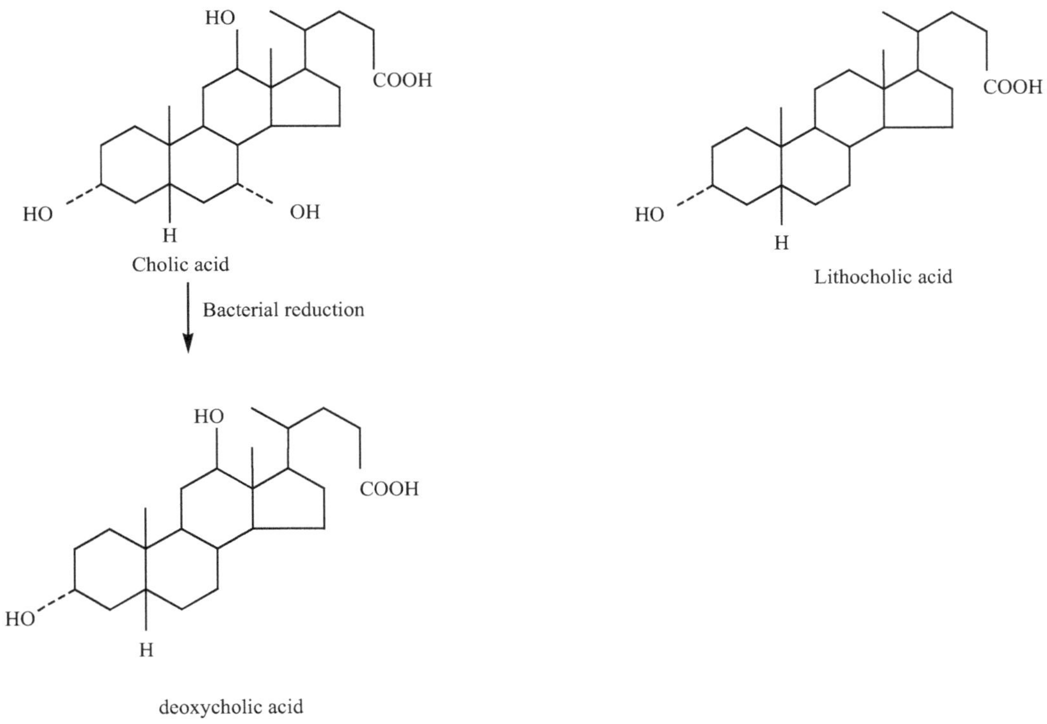

HO COOH HO OH H Cholic acid

Bacterial reduction

HO COOH HO H deoxycholic acid

COOH HO H Lithocholic acid

Synthesis and interconversions of bile acid

54. a. Amino acids are the building blocks for proteins and are also key metabolic intermediates in living cells. It is of general understanding that plants, as well as fungi, synthesize all amino acids required for the protein synthesis. There are two types of amino acids: essential and nonessential amino acids. Essential amino acids are not synthesized by man which includes histidine, isoleucine, leucine, lysine, methionine, phenylalanine, threonine, tryptophan and valine. They must come from food. Nonessential amino acids can be made by humans, plants and many microorganisms, which include alanine, arginine, asparagine, aspartic acid, cysteine, glutamic acid, glutamine, glycine, proline, serine and tyrosine. Nonessential amino acids support tissue growth and repair, immune function, red blood cell formation and hormone synthesis.

1. **Biosynthesis of glutamate and aspartate family amino acids**

   The precursor of these amino acids family is 2-oxoglutaric acid which is synthesized from acetic acid by way of the Krebs cycle which is as shown in Figure 3. But in the aspartate family, the precursor is oxaloacetic acid which is formed from acetic acid as shown in Figure 4.

2. **Biosynthesis of phenyl alanine, tyrosine and tryptophan**

   These are aromatic amino acids where benzene rings can be derived in living organisms by two methods: (i) acetate method (ii) carbohydrate method which is known as the **Shikimic acid route**. It is an important biochemical metabolite in plants and microorganisms. It is a 3,4,5 trihydroxy-1-cyclohexen-carboxylic acid. The Shikimic acid pathway is essentially used to synthesize basic amino acids, alkaloids and other aromatic metabolites by plants and lower organisms. In this process, D-glucose is the starting material which is converted into phosphoenolpyruvic acid anderytrose-4-phosphate. The latter two compounds react to form 3-deoxy-D-arbn oheptulosonicacid-7-phosphate which is converted to Shikimic acid which ultimately changes into chroismic acid to prephenic acid (also known anionic form prephanate) is an intermediate biosynthesis of amino acids. which comes by [3,3]-sigmatropic Claisen rearrangement of chroismic acid shown in Figure 5. This prephenic acid is the route of biosynthesis of phenyl alanie and tyrosine (Figure 6). But chorismic changes into in presence of glutamine and aromatic amino acid (anthranilic acid) which in a different biosynthetic pathway is converted into tryptophan (Figure 7).

b. Purine nucleotides are essential cellular constituents that are involved in energy transfer, metabolic regulation and the syntheses of DNA and RNA. Both RNA and DNA contain two major purine bases, adenine and guanine, and two major pyrimidines. When a purine or a *pyrimidine* is attached to the C–1 position of a sugar, the corresponding structure is called a *nucleoside*. From isotopes experiments, it has defined that origin of various atoms of carbon and nitrogen are as follows:

HOOC–C(=O)–CH2·CH2COOH
2-oxoglutaric acid
NH2-NADPH
HOOC–CH2·CH2·CH(NH2)COOH
L-glutamic acid
NH2ATP
NADPH, ATP
H2NOC–CH2·CH(NH2)COOH
L-glutamine
OHC–CH2·CH(NH2)COOH
L-glutamic-4-semialdehyde
-H2O
Spontaneous cyclization
COOH
N
3,4-dihydro-2H-pyrrole-2-carboxylic acid
NADPH
N
H
COOH
L-proline

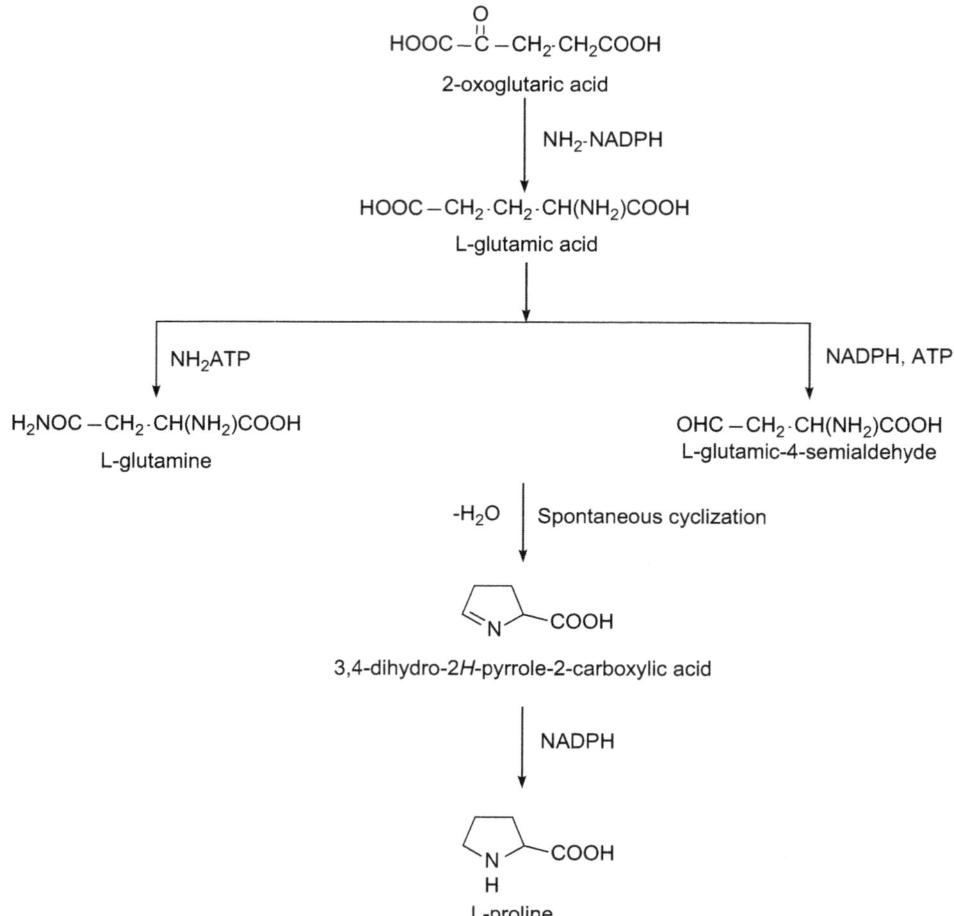

FIGURE 3 Biosynthesis of Glutamate Family.

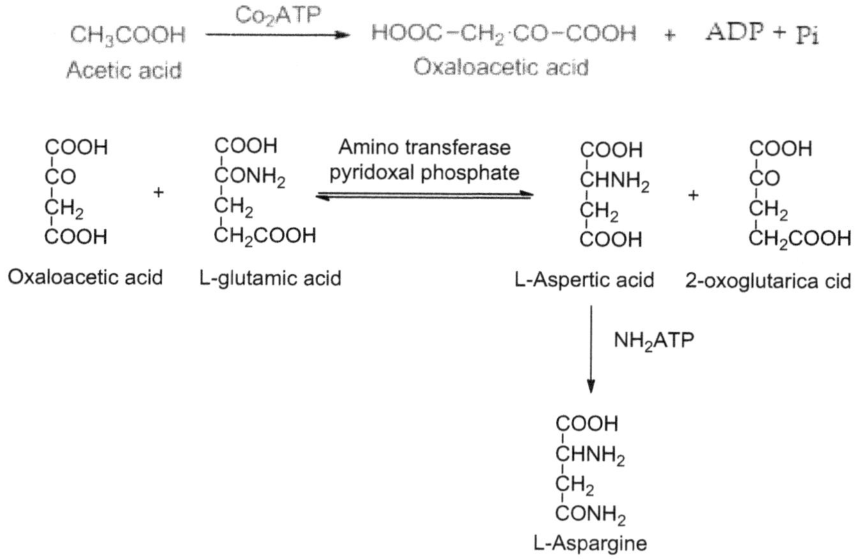

CH3COOH
Acetic acid
Co2ATP
HOOC–CH2·CO–COOH
Oxaloacetic acid
+ ADP + Pi
COOH CO CH2 COOH
Oxaloacetic acid
COOH CONH2 CH2 CH2COOH
L-glutamic acid
Amino transferase
pyridoxal phosphate
COOH CHNH2 CH2 COOH
L-Aspertic acid
COOH CO CH2 CH2COOH
2-oxoglutarica cid
NH2ATP
COOH CHNH2 CH2 CONH2
L-Aspargine

FIGURE 4 Biosynthesis of Aspartate Family.

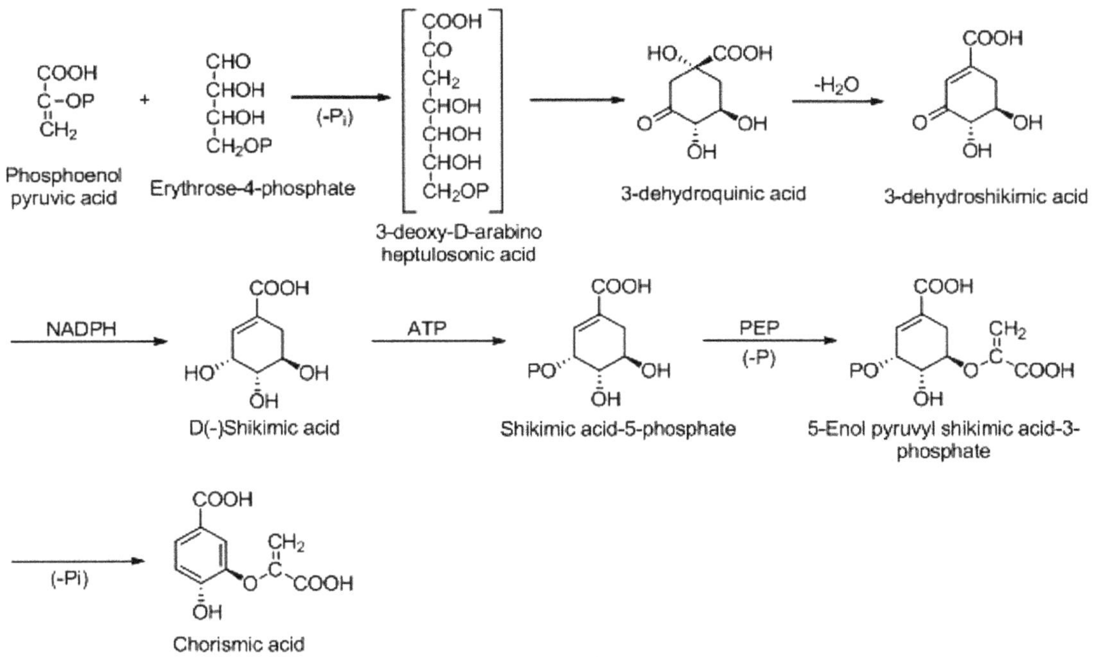

Phosphoenol pyruvic acid
Erythrose-4-phosphate
(-Pi)
3-deoxy-D-arabino heptulosonic acid
3-dehydroquinic acid
-H2O
3-dehydroshikimic acid
NADPH
D(-)Shikimic acid
ATP
Shikimic acid-5-phosphate
PEP
(-P)
5-Enol pyruvyl shikimic acid-3-phosphate
(-Pi)
Chorismic acid

FIGURE 5 Synthesis of Chorismic Acid.

Chorismic acid
Prephenic acid
-H2O, -CO2
-H2O
Phenylpyruvic acid
p-Hydroxyphenylpyruvic acid
L-glutmic acid
NADPH/ 02
L-Phenyl alanine
L-Tyrosine

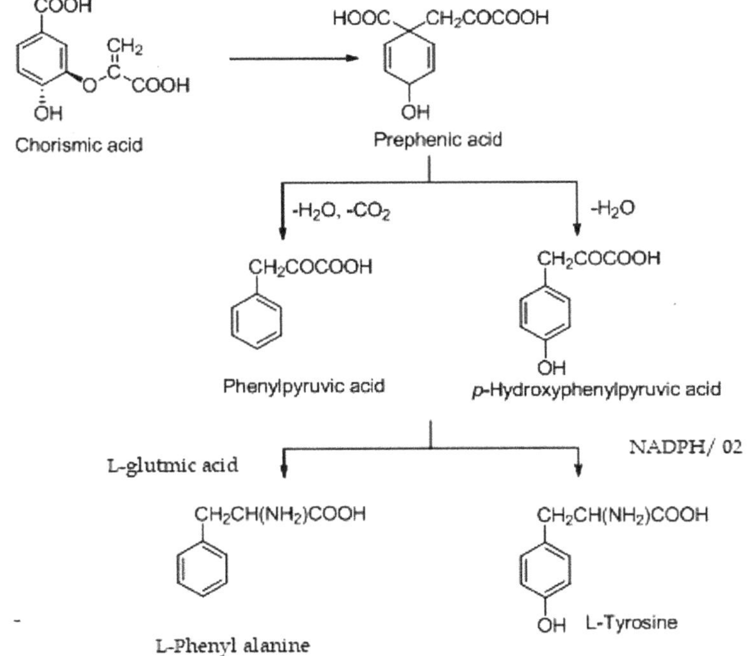

FIGURE 6 Synthesis of Phenyl Alanine and Tyrosine.

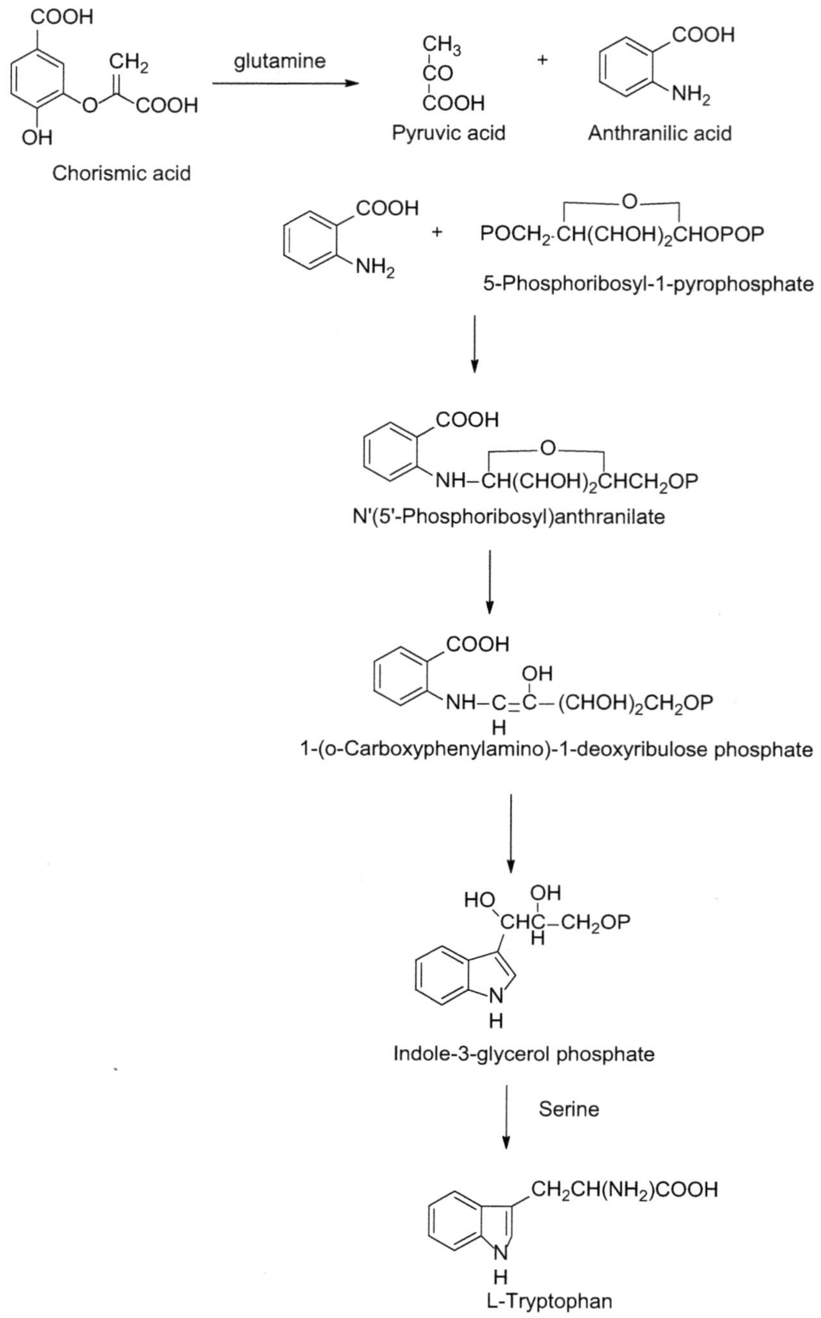

COOH
CH_2
O COOH
OH
Chorismic acid

glutamine →

CH_3
CO
COOH
Pyruvic acid

\+

COOH
NH_2
Anthranilic acid

COOH
NH_2
\+
O
$POCH_2 \cdot CH(CHOH)_2CHOPOP$
5-Phosphoribosyl-1-pyrophosphate

↓

COOH
O
$NH-CH(CHOH)_2CHCH_2OP$
N'(5'-Phosphoribosyl)anthranilate

↓

COOH
OH
$NH-C=C-(CHOH)_2CH_2OP$
H
1-(o-Carboxyphenylamino)-1-deoxyribulose phosphate

↓

HO OH
$CHC-CH_2OP$
H
N
H
Indole-3-glycerol phosphate

↓ Serine

$CH_2CH(NH_2)COOH$
N
H
L-Tryptophan

FIGURE 7 Synthesis of Tryptophan.

SOURCE OF CARBON AND NITROGEN IN PURINE NUCLEUS

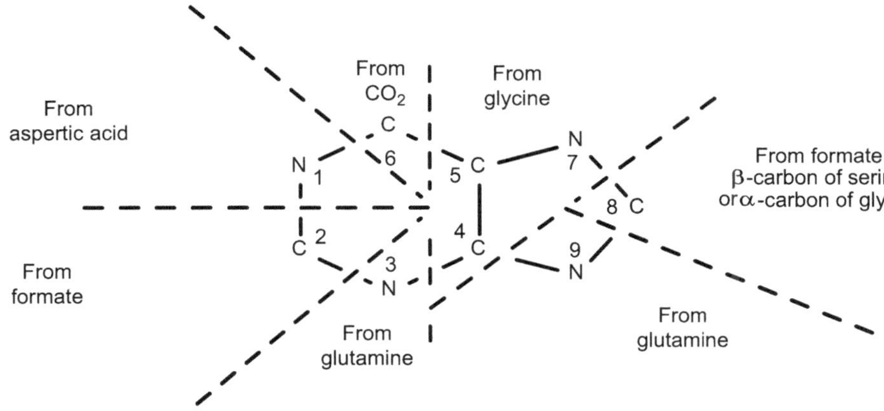

From aspertic acid
From CO_2
From glycine
From formate, β-carbon of serine, or α-carbon of glycine
From formate
From glutamine
From glutamine

In a nucleoside, the anomeric carbon is linked through a glycosidic bond to the N9 of a purine or the N1 of a pyrimidine; examples of nucleosides include cytidine, uridine, adenosine, guanosine, thymidine and inosine. Pyrimidines are synthesized in the body from simpler compounds such as carbon dioxide (C_2), ammonia (N_2) and the rest of the atoms are from aspartic acid (N_3, C_4 to C_6)

ORIGIN OF DIFFERENT ATOMS IN THE PYRIMIDINE NUCLEUS

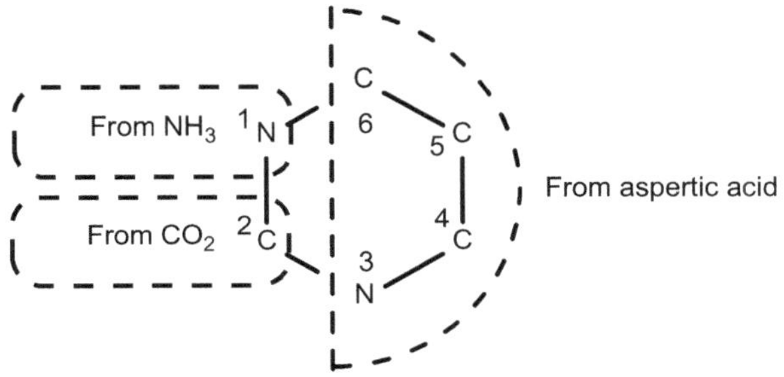

From NH_3
From CO_2
From aspertic acid

55. Biosynthesis of amoxicillin by penicillin G acylase:

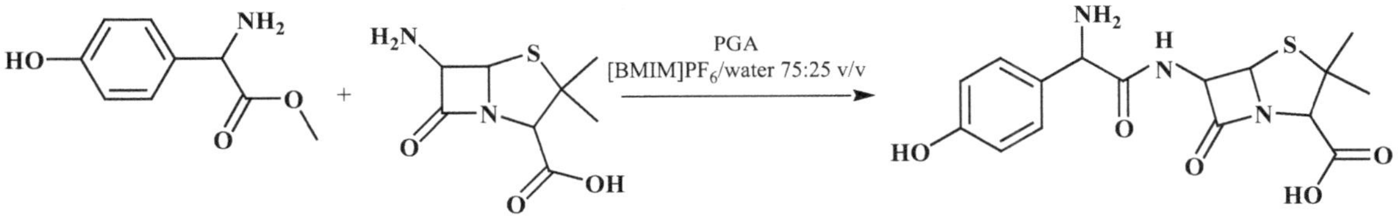

PGA
$[BMIM]PF_6$/water 75:25 v/v

Index

Note: Page numbers followed by n indicate notes, page numbers in *italics* indicate figures and page numbers in **bold** indicate tables.

(+)-isolongifolen-4-one, 266, *268*

1-chloro-1,2-benziodoxol-3-one (CBDO), 45, 330
1-fuculose-1-phosphate aldolase (FucA), 260
^{1}H NMR, 25, 319–320, 321–322, 323, 324, 328, 330–331
1-methoxyethyl-3-methylimidazolium trifluoroacetate ([MOEMIM]TFA), 202, 206
1-phenyl-2,5-dimethyl-1H-pyrrole, **193**
1β,3β-dihydroxysclareolide, 138, *139*
[1H-Indeno[1,2-b]quinoxalin-11-one], 141, *141–142*
1,1′-[bis(di-tert-butylphosphino)ferrocene] dichloropalladium(II) [PdCl2(dtbpf)], 164
1,2-dihydro-α-santonin, 138, *139*
1,2,3 triazoles, 158, *158*
1,2,3,4-tetrahydropyrimidene-5-carbonitrile derivatives, 69, *69*
1,3 butadiene, 334
1,3-benzenedithioamide, 205, 326
1,3-diiodo-5,5-dimethylhydantoin (DIH), 50
1,3-dithiazolylbenzenes, 205, *206*, 326
1,3,4-oxadiazole drugs, 295, *296*
1,4-diketones, 201–202
1,5-sigmatropic carbon-shift process, 335
2-amino phenol, 320, 329
2-aminobenzothiazoles, 205, *208*
2-aminoimidazoles, 203, *204*, 325
2-arylbenzothiazoles, 203, *205*
2-heteroaryl-5-methoxyindole, 66, *66*
2-mercaptoacetic acid, 206, 325
2-oxo-4-substituted aryl-azetidine derivatives of benzotriazole, 70, *70*
2-oxoglutaric acid, *248*, *342*
2-phenylchromane, 247, *249*
2-propylquinoline-4-carbohydrazide hydrazone derivatives, 67, *68*
2α,7α-dihydroxysphaerantholide, *140*, 140
2*H*-chromene derivatives, *8*, 202
2*H*-chromene, 7, *7–8*
2*H*-pyrans, 7, *7*, 91
2,3-butanediol, 157, *157*
2,3-dihydro-4-quinolone, 141–142
2,3-dihydro-8-nitro-4-quinolone, 142, *142*
2,3-dihydroquinazolin-4(1H)-ones, 291, *293*
2,4-diarylthiazoles, 205, *206*, 325
2,5-dimethyl-1-phenyl-1H-pyrrole, **192**
3-arylidene-2,3-dihydro-8-nitro-4-quinolones, 142, 143, *145–146*
3-hydroxy-3-methylglutaryl coenzyme A (HMG CoA), 180
3-oxo oleanolic acid, 275–276, *281*
3-oxosclareolide, 138, *139*
3-phenylchromane, *249*, 249
3,4-epoxy-α-santonin, 138, *139*
3β,6α-dihydroxysclareolide, 138, *139*
3β-hydroxysclareolide, 138, *139*
4-arylselanylpyrazoles, *203*, 327
4-vinylcyclohexene, 335
4α,5α-epoxy-7α-hydroxyfrullanolide, 140, *140*
4*H*-chromenes, 7, *7–8*, 64–65, *65*, *178*
4*H*-pyrans, 7, *7*, 15, *16*, **17**, 201
5-fluorocytosine, 295, *295*
5-HT_7 receptor antagonist, 291, *292*
5-hydroxysclareolide, 138, *139*
7α-hydroxyfrullanolide, 139, *140*
7β-hydroxysclareolide, 138, *139*
7-deoxypancratistatin, 255, *256*
9-hydroxysclareolide, 138, *139*
9-methoxy mappicine
  kinetic resolution, 256, *257*
10-β-glucosyloxyvincoside lactam, 140, *141*
10-hydroxystrictosamide, 140, *141*
13-acetyl-7α-hydroxyfrullanolide, *140*, 140
13-epi-*ent*-pimara-9(11),15-diene-19-oic acid, 268, *271*
^{13}C NMR, 321–322, 329, 330–331
16,17-dihydro-10-β-glucosyloxyvincoside lactam, 140, *141*
17α-hydroxy-11,13-dihydrofrullanolide, *140*, 140
18-hydroxy-9-epi-*ent*-pimara-7,15-diene, 268, *270*
21-hydroxybetulin, 148, *148*

α-aminoadipic acid, 306, *308*
α-aminophosphonates, 210, *211*
α-bromoacetophenones, 205, 207–208, 325, 326
α-chloroketones, 203, 325, 328
α-glucosidase, 141, 143, 145
α-glycosylation, *221*, 222
α-pinene, 91, *264*
α-santonin, 138, *139*
β-carotene, *264*, 264
β-keto esters, *14*, 14, 107, *245*
β-lactams, 99, 137, 213, 231, 336
γ-lindane
structure, 84, *85*
γ-valerolactone (GVL), 90, 91, 107, *157*, 157–158
Δ^1-piperideine, 306, *307*
p-cymene, *90*, 91, 107
ω-transaminases (ω-TAs), 247, 260, *261*

A

Absidia glauca, 276, *281*
Acacia concinna, 15
acetaldehyde, *90*, 247
acetic acid, 106, 320, 328, *342*
acetone, 98, 165, 166
acetyl CoA, *308*, 309, *338*, 339
acetylcholinesterase, 138, 258, 306, 309, 312
acetylsalicylic acid (aspirin), 298, *298*
Ačkar, Đ., 191
Acosta-Guzmán, Paola, xi, 41–73
acridines, 158, *158*
ACS, 166
  Green Chemistry Institute (GCI), 24, 195
Actinidia deliciosa (kiwifruit), 266, *266*
active compounds, 59–71
active methylene nitriles, 15, **17**
active pharmaceutical ingredients (APIs), vii, 89, 98, 164–166, 195, 241, 293, 295
acylation, **94**, 243, *244*, *251*, *257–259*, *291*
acyloin, 247, *248*
Adamczyk-Woźniak, A., 291, *293*
ADBA (2-amino-2,3-dimethylbutyramide), 107
adenosine triphosphate, 337, 339
  ATP-citrate lyase, 338, *338*
ADBN (2-amino-2,3-dimethylbutyronitrile), 107
Adhatoda vasica, 258
Aeropyrumpernix, 101
agricultural products, 165–166, *167*
Albini, Angelo, xi, 75–88
Alcaligenes sp., 269
Alcaligenes xylosoxidans, *248*
Alcántara, Andrés R., xi, 89–117
Alcazar, J., 128
alcohol dehydrogenases (ADHs), 101, 245, *245*
alcohols
  esterification, 324
  green synthesis (one example), 326
  lipase-catalyzed acylation of ~ in 2-MeTHF, 98–100
aldehydes, vii, *126*, 142, *142*, 306, *307*, 314, *315*, 324
aldol reactions, *159*, *210*, 243, *244*, *248*, *307*
aldolases, 247, 259–260
aliphatic
  ~ acids, 34
  ~ amines, 328
  ~ carboxylic acids, 33
alkaloids
  biosynthesis, 305–313, *314*
  chemo-enzymatic deracemization, *260*
  definitions, 252–253, 305
  enzymatic synthesis (strategies), 253
  kinetic resolution, *261*
  molecular structure, *253*
  synthesis from halogenodiols, *256*
  synthesis by ω-transaminases, 260, *261*
alkenes, 245, 326
alkylamines, 37–38
Almendros, M., 101
alternative synthesis routines for pollution prevention (EPA), 195
Alzheimer's disease, 66, 139, 177, 306
Amano PS lipase, **95**, 100, 214, *252*, 256, *257*
Amaryllidaceae pancratistatin, 255
Ambrisentan, 293, *295*

ambrox, 266
  biotransformation, *266*
amidation, *291*, 328, 329
  2-halo-arylacetc acids, 37
  amino and azido acids, *33*
  carboxylic acids, *34*
  microwave-assisted ~, 37
amide synthesis, *24*
  formation of ammonium carboxylate, *24*
  mechanistic considerations, 24–25
  using stoichiometric amounts of boron reagents, *32*
amides, 23–24, *36*, 213, *308*, 330
  also denoted as "carboxamides", 23
  applications, 23
  chemical properties, 23
  electrochemical synthesis, *30–31*
  enzymatic reactions in ionic liquids, *214*
  microwave-assisted synthesis, *27*
  non-conventional heating methods, 26–27
  ortho-iodination, *51*
  primary, secondary, tertiary, 23, *23*
  solvothermal synthesis, *26*
  tauromerization, *27*
amination, *219*, *248*
  ~ in water, 44, *45*
amines, 100, 156, **193**, 209, 306, *307*
  greener synthesis, *211*, 326
  hetero/aromatic ~, 206, 325
  meta-chlorination of ~ amines, *48*
amino acids, 32, 218, 260, 306
  biosynthesis, 341, *342–344*
  enantio-enriched α ~, *27*
  essential versus nonessential ~, 341
amiodarone, 41, *42*
ammonia, 25
  potential hydrogen carrier, 154
ammonia (NH_3) synthesis, 152–154, 169
  ~ over supported nanocatalyst, *153*
ammonia-borate ($NH_3 \cdot BH_3$), 258
ammonium carboxylate
  formation in amide synthesis, *24*
amorphan-4, 11-diene, 314, *315*
amoxicillin, 213, *214*, 345
amphiphilic copolymers, 103, *104*
Anabasis aphylla, 258
Anastas, P. T., vii, 1, 241
Andrea Ojeda-Porras, xii, 23–40
Andrographis paniculata, 101
anilines, *123–124*, 156, 191, *199*, 320
animals, 8, 241, 245, 298, 337
  urea cycle, *310*
anions, 177, 196
anisole, 50, 98
Anisopappuschinensis, 18
Anthemis xylopoda, 37
anthraquinone, 154, *154*
antiaggregation activity, 155, *293*
antibacterial agents, 7, 64–70 *passim*, 138–143 *passim*, 177, *178*, 180, *184*, 197–207 *passim*, *230*, 231, 249, 296; *see also* antimicrobial potency
antibiotics, *184*, 213, 299
  production methods, 336
antibonding molecular orbitals, 334
antidiabetic drugs (ADDs), 63
antimicrobial potency, 6, 219, 259, 264, 268–270, *293*, *297*, 302, 305, 309
antioxidants, 6, 7, 37, 64, 66, 67, 99, 177, *178*, 207, 222, 249, 272
antiparasitic effects, *8*, 140, 219, 264, 269
aquayamicin, 215, *217*, *327*
aqueous micellar solutions, 163–165
Arabpourian, K., xi, 191–194
Arias, K. S., 104, 106
aripiprazole, 41, *42*
Armstrong, H. E., 41
Arnica angustifolia, 138
aromatic aldehydes, **17**, 179, 326
aromatic amines, 27, 328
  formylation, 25, *26*
  meta-chlorination, *48*
aromatic compounds
  halogenation (greener methods), 41–56
aromatic hydrazine, 144, *146–147*
aromatics, 321
  activated and deactivated, 53
artemisinin, *87*, 314, *315*
arthritis, 214
  osteoarthritis, 207
  rheumatism, 258
aryl aldehydes, 177, 179
aryl boronic acids, 124, 126, 180–182, *187*, 209, *210*
aryl bromide, 42, 121, 125, *129*, 181
aryl halides, 120, 121, *122*, *123*, 126, 128, 129, 180–182, *187*, 209, 292
aryl hydrazines, 326, 327
aryl iodide, 5, 181
aryl ketones, 215, *218*
aryl lithium, 121, 125
aryl thioamides, 205, 325
aryl umpolung, 53, 54
Arylex™, 166, *167*
Asakawa, K., 12
aspartate family, 341, *342*
Aspergillus cellulose, 266
Aspergillus flavus, 249, *250*
Aspergillus niger, 138, *140*, *246*, 251, 255, 258, 260, 266, *268*, *269*, *273*
Aspergillus ochraceus, 250, 276, *281*
Aspergillus terreus, 217, *218*, *248*, 260
aspertic acid, 243, *342*, *345*
aspirin, 59
  synthesis promoted by microwaves, *60*
asthma, 214, 219, 249, 258
Astragalus plants, 272
AstraZeneca, 100, 161
Ata, Athar, xi, 137–150, 305–317
atazanavir, 216, *218*, *328*
atorvastatin, 23, *24*
Aureobasidium pullulans, 217, 219, *219–220*, *245*
azoles
  definition, 70
azomethine ylide, 142, *143*, *144*

B

B. circullans, 107
Bacillus lichen, 67
Bacillus megaterium, 260, 274, 275, *279–280*
Bacillus subtilis, 65, 101, *247*
Baeyer, A., 2
Baeyer-Villiger addition, 243
Baeyer-Villiger mono-oxygenases (BVMOs), 101, 268
Baeyer-Villiger oxidation, 245, *246*, 266, 268, 272
baker's yeast, 12–14, 102, 107, 177, 215–216, 217, *218*, 220, *221*, 256, *257*, 258
ball milling, 178, 292, *294*
Baran, P. S., 45
Barbieri, V., 66
barbituric acid derivatives, 38, *38*
Basak, A., 10
batch reactors, 4, 106, 119
bavachinin, 249, *250*
Behbahani, Farahnaz K., xi, 191–194
benzaldehyde, 221, *248*
benzaldehyde lyase (BAL), 101, *248*
benzamide, *38*, *197*
benzene, 41, 84
benzene ring, 330–331, *331*
benzimidazoles, 202, *204*, *209*
benzodiazepines, *204*, 207, *209*
benzofurans, 158, *158*, 219
benzohydrazide analogues, 141, *142*
benzohydroazides, 141, *142*
benzoic acid, 108, *197*
benzothiazoles, 205, *205*
benzoxaboroles, 291, *293*
benzoxazinones, *208*
berberine bridge enzyme (BBE), 258–259
Betula platyphylla, 274
betulin, 148, *148*, 274, *279*
betulinic acid, 274, *279*
Bhanage, B. M., 25
BHMF [2,5-bis-(hydroxymethyl)furan], 103–104, *104*, *106*, 106–107
biaryl derivatives, *126*, 129, 181, 182, *184*, *187*
  substrate scope, **186–187**
biaryls, 163, 177, 179, 180
Biginelli reaction, 4, 102, 299, *299*, 301
bile acids, 340–341
bioactive compounds, *37*, 39, 50
bioactive heterocyclic compounds, 137–150
bioactive heterocyclic natural products, 145, 148, *148*
biocatalysis, 89–117, 196–197, 210–222, 241, *241*, 247
  definition, 196
biocatalysts, 8
  advantages, 210
biocatalytic cascades, **97**
  using 2-MeTHF as (co)solvent, 101–102
bioflavonoids, 247–252
  molecular structure, *249*
biomass, 15–16, 18, 91, 157, 232–233
bioreduction, 18, **97**
  unsaturated ester (by ene-reductase YqiM from *Bacillus subtilis*), *247*
  unsaturated nitro-compound (by *Zymomonas mobilis* NCR-reductase), *247*
  using 2-MeTHF as (co)solvent, 101
biosolvents, 89–117
  most common, *90*
biosurfactants, 9, *9*
biphenyl dioxygenase (BPDO), 245

biphenyls
  single-atom catalyst synthesis, *168*
bis(indolyl)methanes, 212–213
bleach (NaOCl), 44
blood-brain barrier (BBB), 61
[BMIM]BF_4, 198–221 *passim*, 325–326
Bodenstein, M., **85**
bonding molecular orbitals, 334
Boni Protect, 217, 219
boron, 31–32, *33*, 52
  amide synthesis, *32*
boronic acids, 32, 120–123, 125, 128
  amidation of amino and azido acids catalyzed by ~, *33*
  amide coupling mechanism, *34*
  brominated ~, 44
  bromination catalyzed by Ag, *46*
  (hetero)aryl ~, 126
borylation, 121, *121–123*, *126–127*, 130
boscalid, 165, *166*
Bose, A. K., 299
Botrytis cinerea, 267
Bouasla, S., 302
branched chain aminotransferase (BCAT), *248*
breast cancer, 66–67, 143, 164, 213, 218
Breslow, R., 2
brimonidine, 41, *42*
bromides, 122, 126
bromination, 41, 42–44, *44–46*, *206*, 330
  with N-Br hindered reagents, *44*
  ortho-selective ~, *43*
  ~ in water, 44, *45*
bromine, 121, 145, 330
  electrophilic ~, 44
bromo-dragonfly, 41, *42*
bromobenzene, 128, 329, 329
Brønsted acids, 42, 59, 107, 196, 320
Bruton's tyrosine kinase (BTK), 180
Buchwald, S. L., 121, 122
Buchwald-Hartwig reaction, *5*, 119, 179, *183*
buffer solution, 327, 328
Bulinus natalaensis, 312, *314*
bulk chemical synthesis, 152–155
bumetanide-4-aminobenzoic acid, 295, *295*
Burkholderia cepacia (BCL; lipase PS), 99, 214, 250, *251*, 253, *254*, *258*
butanediol, 102, 105
butylamine, 320, 328
Buxus alkaloids, 309, *311*, *313–314*

C

C5-C6 carbohydrate feedstocks, 103, 106
C5-C6 sugars, *90*, 91
cafestol, *264*, 264
caffeic acid (CA), *6*, *212*, 212
Cahn-Ingold-Prelog (CIP) system, 242
calcium carbonate ($CaCO_3$), 59
Camptotheca acuminate, 143
camptothecin, *87*, *138*, 143, *253*, 256
cancer, 64–70 *passim*, 137, 140, 141, 177, *178*, 191, 197–207 *passim*, 220, 264, 291–299 *passim*, 305
  antitumor effects, *8*, 101, 143, 180, *184*, 211, 212, 219, 249, 259, 268
cancer cell lines, 158, 258, 292, 295
cancer cells, 214, 231, 266
candicandiol, 270, *275*
Candida antarctica, 99, 155, 256, *265*, 265
Candida antarctica lipase B (CAL-B), *9*, 10, 91, **92–96**, 108, *108*, *212*, 212, 213, *214*, *216*, *244*, 256, *257*, 319, 327
Candida cylindracea lipase (CCL), 265
Candida maltose, 219
Candida rugosa (CR), 265
Candida rugosa lipase (CRL), 9, **93**, 99, 100, 108, 211, *244*
candidiol (15α,18-dihydroxy-*ent*-kaur-16-en), 270
  biotransformation, *275*
cannflavin A, 249, *250*
Canty, A. J., 125
carbohydrates
  acylations, 99
  biomass for renewable energy production via HTC, 232–233
  chemical and enzymatic syntheses, 229–232
  medicinally valuable intermediates and drugs, 229–232
  utilization, 229–239
carbon atoms, 341, *345*
carbon bonds
  C-C, 3, 119, 120, 126, 157, 159, 160, 179, 209, 212, *213*, 221, 243, 247, 253, 259, 260, 265, 266, 289
  C-F, 329
  C-H, **17**, 43, 156, 159, 245, 265, 292, 329
  C-N, 3, 52, 179, 253, 259, 260, 266
  C-O, 3, 266
  C-P, 3
  C-S, 3, 212, *213*, 266
carbon dioxide, 168, *168*, *345*
  CO_2-expanded liquids (CXL), 99
carbon double bonds
  C=C, 245
  C=O, 247
  C=N, 328
carbon nanotubes (CNTs), 18, 156, *156*
carbon quantum dots (CQDs), 18
carbonyl compounds, **17**, 206, 209, 217, 245, 325, 326
carbonyl group, 208, 215, 243, 245, 328, 331
  eco-reduction, *291*
carboxylic acids, 106, 251, 314, *315*, 328
  amidation using silica as a heterogeneous catalyst, *34*
  direct amidation (greener developments), 23–40
  direct amidation (proposed mechanism), *25*
  dual reactivity, 320
Carduus crispus, 258
carvone, 264, *265*
caryophyllene oxide, 267, 268, *269*
castanospermine, 306, *308*
catalysis
  pillar of green chemistry, 151; *see also* biocatalysis
catalyst-free reactions, 25–28, 320
catalysts
  amount, **185**
  desirable characteristics, 329
  recycling, *165–166*, 166, 168–169, 332
  small particle ~, 151–176
catalytic cycle, *161*
catalytic systems, 36–39
Catharanthus roseus, 267, 268, *269*, *273*
cations, 196, 210, 270
  metallic ~, 18, 169
cavitation, 295, 333
CBMG, 45, 330
cellulose, *90*, 91, 108, 232
CEM, 60, 204
central nervous system (CNS), 60, 61, 63, 259
Cephalosporium aphidicola (wild-type), 266, *268*, 268
Chadha, R., 293, 295
Chaetomium longirostre, 274, *279*
chalcones, *249*, 300, *300*
chelating agents, 168, *169*
CHEM21 solvent selection guide, 91, 102, 105
chemical synthesis, 81
  amphiphilic copolymers, *104*, *106*
  drugs from carbohydrates, 229–232
chemoenzymatic approach, *102*, 148, 311
chemoenzymatic synthesis, *12*, *103*, *105–106*, 253
chemoselective reduction
  levulinic acid, *157*, 157
chemoselectivity, 27–28, *28*, 33–34, 36, 61, 156, *156*, 168
Chen, X-Y., 53
Chen, Z. G., 99
Chettri, E., 292
chiral alkaloids
  enzymatic synthesis, 252–262
chiral bioflavonoids
  enzymatic synthesis, 247–252
chiral compounds, 4, 213
chiral drugs, 10
  use of microorganisms (examples), 319, 328
chitin deacetylation, 207, *209*
chitosan, 207, 301, *302*
chlorides, 23
  aryl ~, 126
  (hetero)aryl ~, 126, *127*
chlorination, 41, 42, 44–50, 320, 321, 330
  classic reagents, *48*
  with $Fe(NTf_2)_3$ as catalyst, *47*
  with iodine reagent, *47*
  photocatalytic ~, *49*
  with Palau'chlor, *47*
  regioselective ~, 49, *50*
chlorine, 329, 330
chloroquine, 41, *42*
cholesterol
  biosynthesis, *338*, 339–340
  irradiation of 7-dehydrocholesterol, 86, *86*
cholic acid, *341*
choline chloride (ChCl), 107, *198*, 202, 203, 209, 214, *216*
chorismic acid
  biosynthesis, 341, *343–344*
"chromanes", 63–65
chromene system, 63
  synthetic analogs, 64
Chromobacterium violaceum, 260
Cinchona succirubra, 298
Cinnamom zeylanicum, 18
Cinnamomum camphora, 18
cis-1,2-divinyl cyclobutane, *323*, 335
cis isopulegone, *314*, 314

cis-cinnamaldehyde, *323*, *333*
cis,cis 1,5-cyclooctadiene, *323*, 335
citalopram, 10, *12*
citrate, 338, *338*
citronellol, 85, *86*
Claisen condensation, 3, 247
Clark, J. H., 33, *34*
click chemistry, 3–4, 231
climate change (global warming), 152, 157
Clostridium ani, 65
cobalt (Co), 50, *51*, 106
coconut juice (*Cocos nucifera*), *16*
  "abbreviated as ACC", 15
cocrystals
  FDA definition, 293
  methods for obtaining ~, 293
coenzyme A (CoA), 180, 306, *307–308*, 309, *311*, 337, *338*, 339
coformers, 293, 295
  main characteristics for synthesis of co-crystals, 333
Colacino, E., 289–290, *290*
Colletotrichum sp., 274
conductions mechanism, 332
conrotatory ring-opening, 335, *336*
continuous flow chemistry
  advantages, 119
continuous flow device, 86
continuous flows
  cross-coupling, 119–135
contraceptives, 291, *294*
copolymers, **92**, 103, *104*, 105, *106*
copper, 4, 157, *157*, 168, 207
  Cu(I)-catalyzed azidealkyne 1,3-dipolar cycloaddition (CuAAC), 3
copper acetylacetonate ($Cu(acac)_2$), 209, *211*
copper nanoparticles
  green synthesis, *37*, 37
Coptis japonica, 261, *263*
Corteva Agriscience, 165–166
coumarin-maltol hybrids, 67, *68*
coumarins, 67–69, 158, *158*, 300, *302*
covalent organic frameworks (COFs), 129
CPME, 98, 100, 101
crispine A, 258, *259*
  crispine E, 258
cross-coupling, 44, 159–163, *183*
  mechanistic components (catalytic cycle), *161*
  multiple-step ~ with recycling of catalyst, *165*
  palladium-mediated, 179–181
  selected series, *160*; *see also* Suzuki-Miyaura cross-coupling reaction
cucumber juice (*Cucumis sativus*), 15, *15*
Cunninghamella bainieri, 138, *139*
Cunninghamella blakesleeana, 138, *139*, 140, *141*, *148*, 272, *277*
Cunninghamella echinulata, 138, *139–140*, 250, *251*, 311, *313*
Cunninghamella elegans, 249, *250*, 266, *267*, 272, 274, *278–279*
Cunninghamella lunata, 138, *140*, 140
Cunninghamella species, 274, *279*
Curvularia lunata, 266
cyanobenziodoxolone (CBX), 290, *290*
cyanohydrin (mandelonitrile), *102*, *103*, *105*, 221
cyclization, 316, *342*
cycloaddition, 2, 3, 334
cycloartenol, 309, 311, *312–313*
cycloastragenol, 272
  biocatalysis, *278*
  biotransformation, *277*
cyclobutane, 334, 335
cyclodextrin, *221*, 222
cyclodextringlucanotransferase (CGTase), *221*, 222
cyclooxygenase (COX), 10, 59
cyclopropane rings, 310–311, 313
Cyrene™, 108, *108*
cytotoxic concentration (CC), 60

D

D-fructose-1,6-diphosphate (FDPA), 259
D-glucose, **97**, *197*, *210*
D-threonine aldlolase (DTA), *248*
dantrolene (dantrium©), 290, *290*
Das, P., 126
Daucus carota, *245*
DBM (solvent), 103, 105
DBU (1,8-diazabicyclo[5.4.0]undec-7-ene), 67, 103, *104*, 105, 207
Dean-Stark trap, 25, 33
decarboxylation, 307, *308*, 309
deep eutectic solvents (DESs), *34*, 89, 195–214 *passim*
  general formula, 196
  syntheses of biologically active compounds (examples), 325
degree of substitution (DS), 99
dehydration, *308*, 329
dehydroabietic acid (DHA), 268
  biotransformation, *273*
dehydrogenases, 13, **97**, 245
  application in asymmetric synthesis, *245*
  classified as oxidoreductases, 12
dehydrogenation, 307, *308*
demosterol, *338–339*
desymmetrization
  definition, 242
  ~ of ester, *216*
  ~ of pyrimidine acyclonucleoside, *216*
diabetes mellitus, 63, 98, 207, 264, 290, 296, 305
  type two ~, 141, 145, 218
diacetate of *meso*-2–(2-propynyl)cyclohexane-1,2,3-triol, 214–215, *217*
diamines, 210, *307*
diastereoselectivity, 58, *243*, 328
Díaz-Álvarez, A. E., 209
diazotization, *123–124*, 130
dibromo compound, 206–207
Diego Gamba-Sánchez, xi, 23–40
Diels-Alder reaction, 2, 3, *158*, 159, 232, 302, 335
Digitalis purpurea (foxglove), 298
digitoxin, 298, *298*
dihydrolevoglucosenone (cyrene), *90*, 90, 108, *108*
dihydropyrimidinones (DHPMs), *69*, 301, *301*
dihydroxyacetone phosphate (DHAP), 247, *248*, 260, 337
diltiazem, 23, *24*
diketones, vii, 177, 179, 201–202, *199*, 260, 327
dimethoxymethane (DME), 202
dimethyl pyrophosphate (DPP), 339
dimethylacetamide (DMAc), 108
dimethylformamide (DMF), 5, 10, 39, 46, 60, 61, 67, 101, 108
dimethylsulfoxide (DMSO), 101, 108
diols, *16*, 105, 214
  kinetic resolution, *215*, *254*
dioxygen molecule (O_2), 80, *80*
diphenylphosphine palladium (II), 128–129
dipolar polarization, 58, 332
dipolarophiles, 142, *142–144*
direct amidation
  carboxylic acids, 23–40
  catalyzed by zeolite, activated alumina balls, starbon® acid, *36*
  under MW irradiation using silica gel as solid support, *35*
  using mesoporous silica SB-15 as catalyst, *35*
dispiroheterocyclic compounds, 141–143, *144*
dispiropyrrolidine oxindole derivatives, 142, *143*
dispirothiapyrrolizidines, 142, *144*
diterpenes, 314–316, 315
diterpenoids, *264*, 264, 266
diversity-oriented synthesis (DOS), vii, 6
Dominguez de Maria, P., 101, 108
Doratomyces stemonitis, 272, *273*, *278*
double bond equivalent (DBE), 330–331
Drewsen, V., 2
drug
  substrates in synthesis of potential ~, 207–210, *211*
drug discovery, 3, 305
drug molecules, *183*
drugs synthesis, 59–63
  from carbohydrates (chemical and enzymatic syntheses), 229–232
  chiral building blocks (reactions most often used), 319, 327
  green syntheses (prelude), 289–297
  greener synthesis, 195–227
  mechanosynthesis, 289–293, *294*
  new medicines, 298–299
  ultrasound synthesis, 295–297
Dubey, A., 70
Dunn, Peter, 2
dynamic KR (DKR)
  effectiveness requirements, 242, *243*

E

ee values, 102, 107, 214, 221
Einstein, A., 88n1
Einstein (Avogadro number of photons), **85**
electroluminescence (EL), 301
electromagnetism, 58, 59
electron transfer, 80, *81*
electron-donating groups, 332
electronics waste, 168, *169*
electron-withdrawing group (EWG), 42, 329, *329*, 332
electrophilic aromatic substitution (EAS), 41, 42–43, 51, 45, 50, 52, 53, 321
  halogenations (general mechanism), *41*
electrophilic fluorine, 53, *54*
electrophilic substitution, vii, 47, 330
Ellena, J., 295

enamine, 306, *307*
enantiomers, 8, 242, 243
enantiopure cascade, *103*, *105*
enantiopure precursor of zeposia®, 102, *103*
enantioselective desymmetrization, *217*, 327
enantioselective enzymatic desymmetrization (EED), 213, 215, *216*
enantioselective microbial reduction, 219, 328
enantioselectivity, 2, 10, *13*, 209, 214, 241, 243, 258, 265
ene-reductase YqiM, **97**, *247*
energy transfersensitization, 80–81
Englezou, G., 103, 105
enolizable 3-acyl quinolin-2-one, 144, *146–147*
ent-7α,17,18,19-tetrahydroxy-kaur-15-ene, 270
ent-kaur-16-ene derivatives, 269–270, *274*
ent-kaurene, 269–270, 316
ent-pimara-7,15-dienes, 268
ent-pimara-8(14),15-dien-19-oic, 268, *272*
environmental factor (E factor), vii, 1–2, 34, 161–163, 162, 241
  amide synthesis from *N*-protected amino acids, 30, *31*
enzymatic desymmetrization (EED), 8
enzymatic hydrolysis, 214, 243
enzymatic kinetic resolution (EKR), 102, 242, *243*, 250, 251, 253, 255, 256, 265
  terpenes, 264; *see also* KR
enzymatic synthesis
  amphiphilic copolymers, *104*, *106*
  chiral alkaloids, 252–262
  chiral bioflavonoids, 247–252
  drugs from carbohydrates, 229–232
  multi-step processes, 101
  polyesters (in 2-MeTHF), *105*
enzymatic-catalyzed enantioselective hydrolysis of ketoprofen monochloroethyl ester and trifluroethyl ester, *11*
enzyme-catalyzed hydrolysis
  using 2-MeTHF as (co)solvent, 100–101
enzymes, 291
  asymmetric synthesis, 243–247, *248*
  classification, 243, *244*
  stereoselective property, 243
eosin Y, 28, *29*
epicandicandiol, 270, *275*
epidermal growth factor receptor (EGFR), 180
epilepsy, 100, 290
Epilobium angustifolium, 145, 148
epimerization, 32, *243*
epiterpestacin, *264*, 264
epoxidation, 3, 267, 270
Erxleben, A., 295
Escherichia coli, 65, 67, **97**, 101, 107, 140, 143, 214, 218, 230, 260, 305
eslicarbazepine, **96**, 100
essential oils, 262, 264
esterase synthesis
  ~ of paracetamol, *13*
  ~ of thromboxane synthetase inhibitor, *13*
esterases, 12, *13*, 243
  catalytic properties, 215
esterification, 9, 106
  reactions catalyzed by lipases, *212*
ethanol, 2, 10, 25, 143, 182
  EtOH, 107, 108, *142*, *145–147*, *179*, *184*, **185**
ethyl acetate (EtOAc), 5, 91, 325
eucalyptol, *90*, 91, 263
exciplexes, *81*, 83
excited states, *81*, 88n1
exothermic catalytic reaction energy pathway, *152*

F

F. moniliforme, *273*
Fabis, F., 43
farnesyl pyrophosphate (FPP), 310, 314, 339
fatty acid mixture (FAM), 104, 106–107
fatty acids, *90*, 106, 337
fatty acyl-CoA, 337
FDA, 230, 231, 293, 298, 299
ferric oxide (Fe_2O_3), 156
Ficher syntheses, 65–66
Filipan, M., 99–100
fine chemistry, 159, 165, 169
flavanones, *249*, 249
  biotransformation, *252*
Flavobacterium resinovorum, 269
flavonoids, *6*
  molecular structure, *249*
flow chemistry equipment, 4, *5*
flow chemistry reactions, 4–6
  advantages, 4
  definition, 4
  types, 4–5
flow chemistry reactors, *4–5*
fluconazole, 41, *42*
fluorination, 41, 42, 51–53, *54*
  ~ using catalytic directing auxiliary, *55*
fluoro analogues of 3-arrylidene-2,3-dihydro-8-nitro-4-quinolones, 143, *145*
fluoxetine, 290, *291*
fondaparinux, 229, *230*
food, 152–153, 165–166, 191
formaldehyde, 202, 326
formic acid (HCOOH), 25, 168, *168*
free baker's yeast (FBY), 217–218, *219*
Friedel-Crafts reaction, 5, *6*, 14–15, 18, 251
Friesen, Kenneth, xi, 305–317
frontier molecular orbitals (FMOs), 334
fructose-1,6-diphosphate (FDP), 259
fruits, 7, 107, 191, 247, 249, 262, 337
FTIR [Fourier transform infrared spectroscopy], 4, 300
Fullenwarth, J., 290
fungi, 8, 9, 12, 25, 67, 137, 138, 140, 249, 268, 272, 341
  antifungal activity, 7, 64, 66, 70, 180, *184*, 191, 201–203, 214, 264, 290, 291, 296
fungicides, 151, 165, *166*, 211, 217, 302
furan, 103, 104, *178*, 207, 232, 250, 302
furan derivatives, 105, 121, *123*, 129, 201–202
Fusarium lini, 266, *266*, 267, *268–269*
Fusarium oxysporum, 270, *273*, *276*

G

GABA (γ-aminobutyric acid), 213
Galactic SA, 319
Gamba-Sánchez, Diego, xi, 41–56
gamma-valerolactone (GVL), 90, 91, 107, *157*, 157–158
gangliosides, 231, *231*
Gastrodiaelata, 99
Gaussian curve, 75, *76*
Geotrichum candidum, 217, *218*, *245*
geranyl pyrophosphate, 310, 313–314
geranylgeranyl pyrophosphate (GGP), 314, 316
geranylpyrophosphate (GPP), 339
Gerardy, R., 107
Gibberella fujikuroi, 268, *269–271*, *274*
Gibbs energy change, 151, *152*
Gill, C. H., 297
Glaser, John A., xi, 151–176
Gliocladium deliquescens, 259
Globo H, 231, *231*
Glomerella cingulata, 268, *272*
Glomerella fusarioides, 272, *277*
Glorius, F., 43
Gluconobacter asai, *245*
glucose, *90*, 332
glucose dehydrogenase (GDH), **97**, 101, *108*
glutamate, 341, *342*
glutamic acid, *90*, 178
glutamine, *342*, *344–345*
glutathione *S*-transferase (GST), 140, 141, 305
glycans, 229, *230*
  mucin-attached ~, *231*
glyceraldehyde-3-phosphate, 337
glycerol, 2, 90, 107, 108, 196, *197*, *199*, *203–205*, 207, *209–211*, 319, 324, 326
  limitations as solvent, 325
glycerol-3-phosphate, 337
glycerol-based solvents (GBSs), 107, 324
glycine, 247, *345*
glycolysis, 259, 337
glycosidases, 100–101
glycosylation, **96**, 140, 259
gold nanoparticles, 155, 156, 158
gram-negative bacteria, 143, 231
gram-positive bacteria, 138, 139, 143, 231
graphene oxide (GO), 18, 44
green biosolvents
  biocatalysis, 89–117
green catalysts, 7–18
  method, 300–301, *302*
green chemistry (1990–), 1–7, 210, 299
  problems and answers, 319–324, 324–345
  twelve principles, vii, 1, 7, 57, *58*, 151, 229, 241, *241*, 289, 324
green nanoparticles, 17–18
green organic synthesis
  potential drugs, 195–227
green syntheses
  prelude (drugs and natural products), 289–304
greener alternative
  microwave in synthesis of organic compounds, 57–73
greener developments
  direct amidation of carboxylic acids, 23–40
greener methods for halogenation of aromatic compounds, 41–56
greener organic transformations
  literature survey, 177–179
  by plant-derived water extract ashes, 177–190

grinding method, 299–300
Grotthuss-Draper law, 3, 78
Guajardo, N., 108
guanidine, 45, 203, 325, 330
Gustafson, J. L., 49

H

H. lanuginosa lipase, *255*
H-Cube®, *125–126*
Halder, Bipasa, vii, xi, 177–190
Hall, C. D., 59
Hall, D. G., 32
halogenation, 124
  aromatic compounds, 41–56, 321
halogens, *42*
Hansenula polymorpha, 219, 220, *328*
Hantzsch reaction, 4, 101
Hartwig, J. F., 52
Heck reaction, 3, 119, 159, *160*, 179, *183*
HeLa cancer cells, 69, 249, 258
  cervical cancer, 67
hemiterpenoids, 263, *264*
heteroannulated 8-nitroquinolines, 143, *144–146*
heteroarenes, 121, *122*
heteroatoms, 158, 289
heterocycles, *208*
  microbiological reduction, *220*
heterocyclic compounds, 7, 141–148, 299
  bioactive heterocyclic natural products, 145–148
  dispiroheterocyclic compounds, 141–143, *144*
  heteroannulated 8-nitroquinolines, 143, *144–146*
  pyrazoloquinolines, 143–145, *146–147*
  quinoxaline analogues, 141, *142*
heterocyclic natural products (bioactive), 145, 148
  microbial reactions, 137–141
heterogeneous catalysis, 151
  advantages, 155
heterogeneous catalyst, 17–18
Hevea brasiliensis hydroxynitrile lyase (*Hb*NHL), 221
highest occupied molecular orbital (HOMO), 80, 334
high-temperature water (HTW), 302
HIV-AIDS, *8*, 66, 67, 141, 177, 191, 197, 216, 269, 270, 309
HIV-associated neurocognitive disorder (HAND), 60
Hiyama reaction, 119, 159, *160*, 179, *183*
HMF [5-(hydroxymethyl)furfural], 103–104, *104*, 105, 106
HMG (β-hydroxy-β-methyl glutaryl), 339
homoisoflavonoids, 305, *306*
Hooshmand, S. E., 296
Hosamani, K. M., 67
Hosseini, A., 336
hot compressed water (HCW), 232
Hu, Y. D., 99
huajiasimuline, 137, *138*, 143
Huisgen's 1,3-dipolar azide-alkyne cycloaddition (AAC), 3
human cytomegalovirus (HCMV), 256
human epidermal growth factor receptor 2 (HER2), 67
Huttunen, K. M., 63
hydantoins, 289–290, *290*
hydrazine, 143, *145*, 326
hydrogen
  energy-carrying potential, 154
  valuable commodity in high demand, 156
hydrogen bonds, 303, 333
hydrogen bromide (HBr), 324, *336*
hydrogen peroxide (H_2O_2), 2
hydrogen peroxide synthesis, 154–155, 169
  ~ over Pd-based nanocatalyst, *155*
hydrogenation, *156*, *157*, 166, 168, 309
hydrogen-bond acceptors (HBAs), 61, 196, *198*
hydrogen-bond donors (HBDs), 61, 196, *197*
hydrolases, 100–101, 210, 243, *244*, 255, *345*
  advantages, 211
  enzymes utilized in synthesis of drug precursors, 319, 327
hydrolysis, 9, 12, *15*, *54*, **96**, 106, *244*, *251*, *257*, *308*
  reactions catalyzed by lipases in green medium, *212*
  regio- and stereoselective ~, 324
hydrometallurgy, 168–169
hydrothermal carbonization (HTC), 229, 232–233
hydrothermal carbons (HCs), 232
hydroxy analogues, 314, *315*
hydroxyl group, 208, 210, 214, 243, 265
hydroxylation, 137, *210*, *250*, *306*, *308*, 309, *310*, *313*
  ~ of vitamin D3, *221*
Hyphomonas neptuniuoccurs, 260
hypoxia-inducible factor 1 alpha subunit (HIF-1a), 292, *294*

I

ibuprofen, 5, *6*, 10, 32, 60, *62*
Ielo, Laura, xi, 89–117
imidazole derivatives, 202–203, *204*
imines, 258, 261, 306, *307–308*
immersion well apparatus, *82*
immobilized baker's yeast (IBY), 218, *219*
indazoles, 291
  coupling reaction, *294*
indigo, *2*
indole alkaloid, 140, 255
indoles, 65–66, 158, *158*
indolizidine, 253, 306, *306*
indomethacin, 32, 60, *62*
inductive heating (IH), 4
inflammation, 2, 6, 7, 10, *11*, 38, 59–60, 64, 66, 67, 99, 101, 138, 141, 143, 177, 180, *184*, 197–212 *passim*, 259, 264, 268, 270, 272, 291, 296, 298, 305
influenza, 215, 229–230, 298
iodination, 41, 42, 50–51, *52*, 321, 330
  ~ of phenylacetic acids, *51*
iodine, *169*, 169, 321, 330
iododediazotization, *123–124*, 130
ionic liquids (ILs), 2, 89, 195, 196, 205, 209–210, 221
  common organic cations, *196*
  greener synthesis of thiazole derivatives, *206*, 207
  synthesis of amides, 37, *38*
Irannejad, H., 70
irinotecan, 137, *138*
iron, 46, 50, 153–154
iron halide, 321, 330
iron (III) phosphate ($FePO_4$), 191–192, **193**
iron trifluoride (FeF_3), 302, *302*
irradiation time, **85**
isatin, 142, *143–144*, 199
isatoribine, **96**, 100
isomers, 3, 264, 323
isoniazid (INH), 69, *69*
isoniazid, 295, *295*
isopentylpyrophosphate (IPP), 339
isopiperitenone, *314*, 314
isoprene, 262–263, *264*
isoprene units, 313, 314
isoprenoids
  same as terpenoids (*qv*), 262
isoquinoline, 121, 258, 259, 309, *310*
isoxazolidine, 292, *294*
isoxazolines, *208*, 292, *294*
Izumoring process, 232

J

Jain, S., 69
Jamison, T. F., 5
Janus kinase (JAK), 180, *183*
Jassem, A. M., 66
Jensen, K. F., 122
jet fuels, 157, *157*
Jiménez Pérez, Víctor M., xi, 289–304

K

Kanerva, L. T., 99
Kappe, C. O., 128–129, 130
Kato, K., 10
kauren-19-oic acid, 270, *276*
kaurene diterpenoids, 269
Kekule, A., 41
keto acids, 218, *219*, 247
  amidation, 27, *28*
keto esters, *14*, 14, 218, *219*, *245*
ketones, 101, 107, 108, 215, 216, *245*, 247, 324, 332
  amination by *Aspergillus terreus* ω-transaminase, *248*
  reduction by *Thermoanaerobium brockii*, *14*
kinetic resolution (KR), 8, 100, 102, 104, 213, 242
  acetyl vasicinone, *258*
  alkaloids, *261*
  lipase catalyzed ~, *215*
  lupinine and crispine A, *259*
  methods, *243*
  monoterpenoids, *265*; *see also* enzymatic KR
Kluyveromyces marxianus, 215–216
Knoevenagel condensation, 3, 14, 38, 69, 243
Knoevenagel-Michael-type reaction, vii, 182
Kołodziejska, Renata, xi, 195–227, 241–287
Krebs cycle, 298, 341
Kumada reaction, 3, 119, 159, *160*, 179, *183*
Kupczyk, Daria, xi, 195–227

L

L-glutamic acid, 309, *342–343*
L-lysine, 177, 306, *307*, *308*
L-pipecolic acid, 306, *308*
L-proline, *210*, *342*
L-serine, 100, 107, *108*
Lăcătuş, M. A., 104, 106
lactic acid, *90*, 90
lactide, 103, 105
Lactobacillus α-amylases, 191
Lactobacillus reuteri, 216
Lamaty, F., 291
lamps (used for photochemical synthesis), *82*
lanosterol, *264*, 310, *338–339*
Lavandula genus, 264
lavandulol (essential oil), 264–265, *265*
LEDs, 83, 84, **85**, 87
Leguminosae, 258
Leishmania donovani, 213
Len, Christophe, xi, 119–135
leu-enkephalin, 291, *292*
levoglucosenone (LGO), *90*, 108
levulinic acid, *90*, *157*, 157
Lewis, G. N., 322
Lewis acids, 5, 32, 42, 49, 107, 144, 196, 321
life cycle assessment (LCA), 91
light, *77*, 78, 80, **85**, 88n2, 334
 UV ~, 3, 332
lignin, *6*, *90*, 232
lignocellulosic biomass, 89, *90*, 157–158
Lilium lancifolium, 99
lily polysaccharide (LP), 99
limonene, *90*, 91, 264, *265*, *314*, 314
lineal alcohols
 lipase-catalyzed acylation, **91–98**
linear isomer, 144, 145, *147*
lipase-catalyzed acylation, 9, *9*, 99–100
 ~ of alcohols in 2-MeTHF, 98–100
 ~ with carbonates, 100
 ~ of lineal alcohols, **91–98**
lipases, 8–12, **93**, 100, *212*, 213, 243, 265, 319, *324*
 advantages, 211
 application in asymmetric synthesis, *244*
 catalytic action, *9*
 synthesis of chiral drug precursors (examples), 319, 327
 uses, 8
Lipińska, T., *66*
lithiation, *121–123*, *127*, 130
Liu, C., 207
Liu, Y., 98
lixiviation, 168–169
López-Belmonte, M. T., 10
Loupy, A., 34
lowest unoccupied molecular orbital (LUMO), 80, 334
LSZ102 (drug candidate), *164*, **164**
Luong, H., 44
lupinine, *253*, 258, *259*, 306, *307*
Lupinus hispanicus, 258
Lupinus luteus, 258, 306
luteinizing hormone-releasing hormone (LHRH), 98
lyases, 247
 application in asymmetric synthesis, *248*
 use as solvents, 101
lycopodium, *306*, 306–307
lysine decarboxylase (LDC), 306

M

m-chloroperenzoic acid, *140–141*
Maciá, B., 290
magnetic ferrite nickel nanoparticles (Fe_3O_4-ni), 209
Maillard reaction, 232, 233
Maity, Himadri Sekhar, vii, 1–21
malaria, 41, 66, 87, 141, 143, 177, 197, 270, 274, 298, 309, 314
maleic anhydride, 302, 335
malononitrile, 143, *146*, 177, 179, 326
mammals, 137, 138, 340
Mandal, B., 30
Mannich reaction, 3, 4
mappicine
 kinetic resolution, 256, *257*
Marder, K., 61
Marini, A., 295
McKenna, C. E., 63
mechanosynthesis, 289–293, *294*
medicinal chemistry, 159
 carbohydrates, 229
melatonin, 29, *30*
menthol, **93**, 98, *265*, 265
menthone
 enantiomers, *314*, 314
meso 1,3-cyclohexanedicarboxylic acid diester, 215, *217*
mesolobelanidine
 desymmetrization, 256, *257*
mesoporous solids as heterogeneous catalysts, 32–36
mesoreactor, 119, 128
metabolites, 6, *90*, 91, 138, 140, 196, 247, 250, 266, 267, 274–276
 primary and secondary ~, 297, 298
metal hydride salts, 103, 106
metal-organic frameworks (MOFs), 129
methanol, 2, 10, 25, 59, 99, 125, 168, 196, *250–251*, *266–267*, *278–280*, 331
methyl-polyethylene glycol (mPEG), 103, 105
methyltetrahydrofuran (2-MeTHF), 90, 91–108, 325
 biocatalytic cascades, 101–102
 bioreductions, 101
 chemical-physical properties, 91
 enzymatic synthesis of polyesters, *103*, *105*
 enzyme-catalyzed hydrolysis using ~ as (co)solvent, 100–101
 examples of biotransformation, **91–98**
 lipase-catalyzed acylation of alcohols, 98–100
 recent examples of biocatalysis, 102–107
 use as (co)solvent, 101
Michael addition, 3, 29, 34, 69, 243
Microbacterium campoquemadoensis, 220, *220*
microbial lipases, 8–9
microbial reactions
 heterocyclic natural products, 137–141
microflow system, *126*
microorganisms, 8, 9, 67, 70, 137, 196, 241, *245*, 249, 251, 265, 266, 297, 298
 use in synthesis of chiral drug precursors (two examples), 319, 328
microreactors, 87, 125, 126
 advantages and disadvantages, 119
 same as "millireactors" and "minireactors", 119
microwave (MW), vii, 143, *320*, *329*
microwave effects, 58–59
microwave heating, 178
 versus conventional heating, 331
 mechanisms, 332
 response of materials (classification), 331–332
 solvents, 331
microwave irradiation, 4, 34, 36
 advantages, 57
 direct amidation, *35*
 greener alternative (synthesis of organic compounds), 57–73
microwave-assisted method, 144, 300
microwave-assisted reactions, 119, 204, *205*, *208*
microwave-assisted synthesis, 3, 290
 amides, *27*
 disadvantage, 289
Miele, Margherita, xi, 89–117
migraine, 65, 199, 214
Mohan, S. B., 69
molecules
 versatile, small, biologically active, 6
momilactone, *315*, 316
Monguchi, Y., 124
monoamine oxidases (MAOs), *246*, 258, *259*, *260*
monoglycosides, *221*
 diglycosides, *221*
monooxygenases (MO), 101, 220, 245, 266, 268
 application in asymmetric synthesis, *246*
monosaccharides, 229
 biosynthesis, *313*, 313–314
monoterpenoids, 263–264, *264*
 kinetic resolution, *265*
montelukast, 220, *220*
Montilla Arevalo, R., 101
Moraxella sp., 269, *273*
Morinda lucida, 274
morphine, 298, *298*, 305–306, *306*, 309, *310*
Mortierella isabellina, 269, *273*
MTBE, 99, 100, 101
Mucor circinelloides, 269, *273*
Mucor hiemalis, 220
Mucor javanicus, 10
Mucor mucedo, 274, *279*
Mucor plumbeus, 138, *139*, 270, *275*
Mucor racemosus, 219
Mucor ramannianus, 249
Mucor rouxii, 268, *272*
multi-component reactions (MCRs), vii, 3, 4, 15, 18, **98**
Muñoz-Flores, Blanca M., xi, 289–304
Muscari armeniacum (Hyacinthaceae), 260
Mycobacterium smegmatis (MsAcT), *221*, 221–222
Mycobacterium sp., 272, *277*
Mycobacterium tuberculosis, 70

N

N-acetylation, 209, *211*
N-acylation, 23, 28
N-arylation, 209, *211*

N-benzol[1,3]dioxol-5-yl-formamide, 319–320
N-bromosuccinimide (NBS), 42, 44, *46*, *206*, 206
N-butylpyrrolidinone (NBP), *90*, 91
N-fluorobenzenesulfonimide (NFSI), 52, 321
N-formyl α-amino acid derivatives, 37, *38*
N-formylation of amines, 25, 37
N-methyl-2-pyrrolidone (NMP), 98, 108
N-methyl-Δ^1-pyrrolinium cation, 309, *310*
N-substituted pyrroles, 191–194, 332
  synthesis mechanism, *193*
N,N′-dimethylperhydrodiazepine-2,3-dithione (Me2dazdt), 169
Nag, Ahindra, ix, xii, 1–21, 177–190
Nagaki, A., 125
nanocatalysts
  applications, *153*
  comparison with single-atom catalysts, *167*
  heterocyclic systems, *158*
  Pd-based ~, *155*
  syntheses, *167*
nanocatalyzed organic transformations, 155–169
  applications to agricultural products, 165–166, *167*
  catalyst recovery, 168–169
  cross-coupling reaction, 159–163
  future research, 168, 169
  recycling of aqueous micellar solutions, 163–165
  single-atom catalysts, 166–168, 169
nanomicelles, 165, 166
nanoparticles (NPs), 18, **91**, **96**, *152*, 156, *159*
nanoscience, 36, 151–152, *153*
nanotechnology, vii, 36
naphthalene dioxygenase (NDO), 245
naproxen, 10, 60
naringenin, *249*
Nasrollahzadeh, M., 37
natural deep eutectic solvents (NDES), 107, 195, 196, *197–198*, 208, 210, 212
natural products, *183*
  biosynthesis, 305–317
  definition, 247
  green catalyst method, 300–301, *302*
  green syntheses (prelude), 297–303
  greener synthesis, 241–287
  grinding method, 299–300
  history, 297–298
  lipase-catalyzed *O*-acylation, **94**
  microwave-assisted method, 300
  modification, 99
  new medicines, 298–299
  solvent-free method, 301–302
  sustainable synthesis, 299
  water as greener solvent, 302–303
Nauclea latifolia, 140
Naz, Samina, xi, 137–150, 305–317
NCR-reductase, *247*
Negishi reaction, 3, 119, 179, *183*
Neurospora crassa, 219
Nguyen, Remi, xii, 119–135
nicotinamide adenine dinucleotide
  NAD, 218
  NAD+, 12
  NADH, 12, *14*, *108*, *245*, 255
nicotinamide adenine dinucleotide phosphate (NADPH), 12, *14*, 101, *108*, 245, *338*, 339, *342–343*
  NADP+, 12, **97**
nicotine, 253, 258, *260*
nicotinic acid, *198*, 253
nitriles, 3, 15, **17**, 208, *210*, *221*, 299, *300*
  green synthesis (example), 326, *326*
nitro group, 214, *291*
nitrocompounds, 5, *211*
nitrofurantoin (furantin©), 290, *290*
nitrogen, *23*, 143, 218, 301, 328, 329
nitrogen atoms, *208*, 341, *345*
nitrones, 323, 336
nitrostyrene, 156, *156*
NMR, 12, 25
no observed adverse effect level (NOAEL), 91
noble metals, 155, 169
non-noble metal nanoparticles, *104*, 104, 106
non-steroidal, anti-inflammatory drug (NSAID), 59, 60, 63, 65
  synthesis of novel ~ bis-conjugates with acetaminophen, *61*
norcoclaurine synthase (NCS), 44, 46–47, *47*, 49, *50*, 261, *263*, 309, *310*, 320
norlaudanosoline, 309, *310*
Novartis, 100, 164
Novozym 435 (catalyst), **94**, **96**, *102–106*, 107, 108, 213–215, *216*
nucleophilic attack, 28, 32, 38, 144, 192, 250, 310, 328, 329
nucleophilic fluoride, *53*, 53
nucleophilic substitution, *5*, 144, 207–208, *210*, *211*
nucleoside phosphorylases (NPs), 101
nucleosides, 341, 345

O

o-phenylene diamine, 141, 192
octanediol, 102, 105
olanzapine, 4
  continuous flow synthesis, *5*
olaparib, 41, *42*
OMe, **93**, **98**, *183–184*, 321–322, 330–331
one-pot reaction, 3, 4, 69, 177
one-pot synthesis, *66*, 144, 179, 299, 336
one-pot three-component reactions (3-CRs), 177
onion-like carbon (OLC), 18
Organ, M. G., 122
organic additives, 28–31
organic chemistry
  main goal, 57
organic compounds, 232
  synthesis (microwave as greener alternative), 57–73
organic reactions, 331
organic solvents, 162–163
organic transformations, 179, **180**; *see also* nanocatalyzed organic transformations
organoboron, 120, 160
organocatalysis, 28, 48–49, 50, *52*, *54*
ortho-brominations, 321, 330
oseltamivir (tamiflu), 215, *217*, 229, *230*, 231
osteoporosis, 63, 249, 250
ovarian cancer, 137, 218
oxaloacetate, 338, *338*
oxazole derivatives, 207, 325
oxidases, 245, *246*
oxidoreductases, 210, 215, 220, *244*, 255, 327
  types, 245
oxindoles, 37, 199–200
  microwave-assisted synthesis, *38*
oxygen, *87*, 218, 329
  photochemical reactions, 80
oxygen atoms, *208*
oxygenases, *244*, 245, 339
ozanimod (zeposia®), 102, *103*, *105*

P

p-toluene sulfonic acid (p-TSA), 291, *293*, 299, *299*
Paal-Knorr reaction, 191, 192, 198
  definition, 325
Pace, Vittorio, xii, 89–117
Paggiola, G., 107
palladium (Pd), 5, 37, *51*, 53, 155, 160, *161*, 167, 168, *169*, **185**, 292, 321, 330
  immobilized ~ on polymer monolith, *127*
  nanocatalysis in water, 165
  single-atom catalyst, 166
  Suzuki-Miyaura cross-coupling, 119–135
palladium on alumina (Pd/Al_2O_3) catalyst, 154
palladium complex (ArPdXLn), 120
palladium nanoparticles, 126, 128
  dendrimer-encapsulated ~, *128*
palladium-mediated cross-coupling reaction, 179–181
panthenyl monoacyl esters (PMEs), 212
Papaver somniferum (opium poppy), 298, 298
Papaveraceae (poppy) family, 259
papaverine, 309, *310*
paracetamol, 12, 290, *291*
Park, H. J., 10
Parkinson's disease, 66, 177
Parmar, Virinder S., xii, 119–135
Parrot, I., 290
Patil, S., 301
Patil, S. A., 64
Pawluk, Hanna, xii, 241–287
Pechmann reaction, 302, *302*
pelletierine, 307, *308*
penicillin, 23, *24*, 137, *138*
  production method, 336
penicillin G acylase (PGA), 213, *345*
Penicillium expansum, **93**, 99
Penicillium raistrickii, 249, *250*
pentacyclic triterpenoids, 273–274
peptides, 23, 31, 32, 39, 296
per-*O*-acetylated thymidine, 211–212, *212*
Perez, M., 107
Perez-Sanchez, M., 107
perfluoroalkoxyalkane (PFA)
  ~ capillary tube, 121, 122
  ~ reactor coil, 6
perfume, 23, 85, 264, 298, 313
pericyclic reactions, 323, 334
Peris, E., 102
permitted daily exposure (PDE), 91
pH, **96**, 179, 181, 214, *219–220*, 250, *254*, *256*, *263*, 328

pharmaceutical cocrystals, 293–295
pharmaceutical industry, vii, 23, 197, 229, 241, 289, 291, 296, 298, 299, 316
Phaseolus aureus, 15
phenacyl bromides, *205*, 207–208, *208*, *210*, 325
phenolic compounds, 6, *6*
phenyl alanine, 341, *343*
phenyl hydrazine, 144, *146–147*, 326
phenylacetic acids, 50
  iodination with Pd(II) catalysts, *51*
phenylalanine dehydrogenase, 218, *219*
phenylboronic acid, 128, 129, 181, *300*
Phenytoin, *290*
Phlebiopsis gigantea, 268–269, *273*
PhMe (solvent), 102, 103, 105
phosphatidate, 337–338
phosphatidylcholine (PC), 100, 107, *108*
phosphatidylserine (PS), 100, 107
phosphine, 166, 167
phospholipase, **95**, 107, *108*
phosphors, 82, **85**
photocatalysis, 49, *49*
photochemical reactions, 75–88
  choice of apparatus, 81–83
  concentrating mirror, *82*
  conditions, *84*
  cost issue, 83–87
  satisfactory ~, 78–80
  via singlet and triplet states, *76*
photochemistry, vii, 3, 49, 87
  liquid phase, 81
  vocabulary, *76*
photoinitiated processes, 80, *80*
  generation of radicals, 81
photoreactions
  bimolecular ~, *79*
  unimolecular ~, *78*
photoredox catalysis, 28, *29*, 87
Pichia membranifaciens, 250, *251*
Pichia pastoris, 101, 218
pig liver esterase (PLE), 12, *13*, 215, *217*, 243, 253, *254*
pilocarpine, 298, *298*
Pilocarpus jaborandi, 298
pineapple juice, 301, *301*
piperidine, 177, *253*, 255, 260, 306, *306*
piperidine building blocks, 254, *255*
piperline, 306, *307*
plant-derived water extract ashes, 177–190
  transformations, 179, **180–181**, *182*
plants, 8, 241, 245, 297, 298, *310*, 337, 341
  as biocatalysts, 14–15
platinum, 168, 301
  nanoparticles, 156
  Pt-$SrTiO_3$ perovskite nanocuboids, 156–157, *157*
pollution, vii, 1, 43, 49, 54
poly-L-glutamine (PGN), 229
poly(ADP-ribosyl) transferase (PARP), 142
polyesters
  enzymatic synthesis, *103*, *105*
polyethylene (PE), 156–157
  nanocatalytic degradation, *157*
polyethylene glycol (PEG), 2, 50, *52*, 195, 300
Polygonum cuspidatum, 99
polylactides, 103, 105
polymers, 3, 128
  furan-derived ~, 105
polypropylene, **93**, 98, 156
ponatinib, 23, *24*
porcine pancreas lipase (PPL), 10, 212, *213*, *244*, 250, *251*, 253, *254*, 255, *256*
pregablin, 2, 213
primary amides, 27
  mechanochemical synthesis, *31*
primary amines, 27, 34
procainamide, 290, *291*
process mass intensity (PMI), 100
  LSZ102 drug candidate, **164**
processes on excited state surface (PES), *77*, 81
prochiral compounds, 215, *216*, 242, *244*
  microbial reduction, *218*
prochiral keto esters, 107, *219*
prochiral olefin
  epoxidation by isolated MO, *246*
prochiral carbon atoms, 242, *242*
proteins, 23, 253, 296, 341
Proteus vulgaris, 67
provenge®, 231
Prunus amygdalus hydroxynitrile lyase (*Pa*HNL), 221
Pseudomonas abietaniphila, 269
Pseudomonas aeruginosa, 67, 140, 143, 231
Pseudomonas cepacia, **95**, 98, 99, 100, 214, 250
Pseudomonas fluorescens lipase (PFL), 99, 101, *244*, *248*, 250, *255*, 255
Pseudomonas putida, 219, *220*, *255*, 255, *256*, 328, *328*
Pseudomonas stutzeri lipase, **92**, 99, 107
pseudopeptides, 323
  one-pot synthesis, 296, *297*
Psoralea corylifolia, 249
pulegone, *314*, 314
Pulicaria glutinosa, 18
purine, *253*, 341, 345
purine nucleoside phosphorylase (PNPase), 101
pyran derivatives, 201–202
pyran ring system, 201
pyrano[2,3-*c*]pyrazoles, 202, *203*, 326
pyrazole derivatives, 202, *203*
  green synthesis in glycerol (examples), 326–327
pyrazole ring, 292
pyrazoles, 158, *158*, 296, *297*, *300*, 300
pyrazolo[3,4-b]quinoline, 143–144
pyrazolo[4,3-c]quinoline, 143–145
pyrazolo[4,3-c]quinoline-3-one, 137, *138*
pyrazoloquinolines, 143–145, *146–147*
Pyrex glassware, 82, 83
pyridine, 143, *146*
pyridoxal phosphate (PLP), 247, *248*
pyrimidines, 69, 121, 296, *297*, 341, 345
  heterocyclic compounds, 38
  origin of atoms, *345*
pyrrole derivatives, 198–200, *200*, 325
pyrroles
  n-substituted ~, 191–194
pyrrolidine, *142*, 143
pyruvate, 247, *248*, 338
pyruvate decarboxylase (PDC), **97**, 101
Pythium oligandrum, 219, *220*

Q

QLM lipase, 256, *257*
quantum yield (Φ)
  definition, 3
quinazoline, *253*, 258
quinine, 298, *298*
quinoline, 53, *54*, 66–67, 121, 197–198, *199*, *253*, 336
quinolizidine, 253, 258, 306, *306*
quinolones, *67*, 141–142, 143–146, 148
quinoxalines, 141, 158, *158*, 197–198, *199*, 300, *300*

R

racemic α-cyclogeraniol, **93**, 98
racemic flavonol derivatives
  kinetic resolution, *251*
racemization, 27, 32, 37, 100, 242
re face, *242*, *245–246*
reacting oxygen species (ROS), 80
reagents, *32*, 41, 44–46, *47–48*, 52, 53, 321, 323, 325, 330
Rebolledo, F., 100
Reddy, B. M., 302
reductases, 245
  asymmetric synthesis, *247*
reduction reactions, *210*
reductive amination, *313*
reductive elimination, 180
regioselective acylation, 91
  ~ of nucleosides and analogs, 99
regioselective enzymatic hydrolysis, 100
regioselective reactions, 144
regioselectivity, 47, 52, 53, 211–212, *212*, 214
renewable energy, 229, 232–233
resorcinol, *302*
respiratory syncytial virus (RSV), 199
resveratrol, *6*, *221*, 222
Rhizomucor miehei, 99
Rhizomucor miehei lipase, *10*
Rhizomucor species, 9
Rhizopus circinans, 140, *140–141*
Rhizopus delemar, 213
Rhizopus miehei, 10
Rhizopus stolonifer, 138, *139*, 219, 266, *268*, 270, *276*
Rhizopus species, 9
rhodamine, 296, *297*
rhodium (Rh), 292, 330
Rhodococcus erythropolis, 264, *265*
Rhodococcus sp., 216, *218*, 245, 328
Rhyzopus oryzae, 217, *218*
ring-opening polymerizations (ROPs), 103, 105
rinskor™, 166, *167*
Ritter, T., 52
room-temperature ionic liquids (RTILs), 99
rose oxide, 85, *86*, *90*, 91
ruthenium (Ru), 154
ruthenium-catalyzed azide-alkyne cycloaddition (RuAAC), 3–4

S

Saha, S., 291
Saiganesh, R., 42

Sajiki, H., 124
salicin, 298, *298*
Salix alba (willow), 298
Salmonella typhi, 65
Salvadorapersica, 18
Sanford, M. S., 52
Sangani, B. C., 65
Sapindus trifolistus, 15
sarcosine, 142, *143*
savolitinib, 161, *161*
saxagliptin, 218, *219*
Schiff base, 60, *62*, 178, 306
Schmidt, S., 101
Schreiner, P. R., 336
sclareolide, 138, *139*, 266, *267*
SCoA, 309, *311*
scopolamine, 253
secondary amines
  formylation, 25, *26*
Secundo, F., 98
SELECT (safety, environmental, legal, economy, control, throughput) criteria, 161
selective estrogen receptor-degrader (SERD), 164
sensitized electron transfer activation, *81*
serine hydrolases (class), 9
serotonin, 10, 32
sesquiterpenes, 314
sesquiterpenoids, 263, *264*, 266
sestreterpenoids, *264*, 264
Shah, N. K., 64–65
Sharpless, K. B., 2, 3–4
Shi, X., 230
Shikimic acid, 341, *343*
si face, *242*, *245*, 260
sialyl-Tn (STn), 231, *231*
sigmatropic reactions, 86, 334
SiliaCat DPP-Pd, 128–129
SiliaCat DPPP-Pd catalyst, *129*
silica, 32–33, 82, 320, 329
  amidation of carboxylic acids, *34*
silica gel, 34, *35*
silica SBA-15, 34
silica sulfuric acid, 192, **193**, 302, *302*
silicon dioxide (SiO_2), 155, 156, *156*
silver nanoparticles (AgNPs), 156, *158–159*
Simeó, Y., 91, *93*, 99
Simone, Michela I., xii, 229–239
Singh, P. P., 28
single-atom catalysts (SACs), 166–168, 169
  comparison with nanocatalysts, *167*
  synthesis of biphenyls, *168*
Siquiera, G. M., 296
small particle catalysts, 151–176
  ammonia synthesis, 152–154, 169
  bulk chemical applications, 152–155
  hydrogen peroxide synthesis, 154–155, 169
  nanocatalyzed organic transformations, 155–169
  nanoscale applications, 151–152, *153*
sodium bromide (NaBr), *44*
sodium hydroxide, 2, 6, 63
sodium hypochlorite (NaClO), *44*, 261
solar energy, *82*, 84, **85**, 332
solid state, 81, 161, 166, 289, 293, 295, 300
solvate ionic liquids (SILs), 210
solvent selection, 102, 105, 160, 161
solvent-free method, 301–302
solvents, 89
  best ~, 332
Sonogashira reaction, 3, 15, 119, 159, *160*, 179, *183*, 292, *294*
sorbiterrin A molecule, 159, *159*
Sorensen, E. J., 53
Sorocenol B, *158*, 158–159
soybean (*Glycine max*), 18
space-time yield (STY), 102
Spencer, Percy, 57
Sphaeranthus indicus, 139
sphingosine-1-phosphate (S1P), 102
squalene, 309, 310, *311*, *338–339*
strontium titanate ($SrTiO_3$), 156
Staphylococcus aureus, 67, 138, 140, 143, 231
Saccharomyces cerevisiae, 101, 216, 217, 220
starbon® acid, 35, *36*
starch, 332
  catalyst reusability, **192**
  synthesis of n-substituted pyrroles, 191–194
Stark-Einstein law, 3
stereoisomers, 85, 333, 334
stereoselective
  ~ biotransformation, 242
  ~ epoxidation, 267
stereoselectivity, 211, 332
Stille reaction, 3, 119, 159, *160*, 179, *183*
stimuli multistep reaction sequences, 4
stoichiometry, *32*, 103, 106, 151, 154, 164, 261
Streptococcus agalactiae, 138, 140, 143
Streptococcus pneumoniae, 65
Streptomyces chromofuscus, 100, 107
Streptomyces exfoliates, 258
Streptomyces griseus, 275, *281*
strictosamide, 140, *141*
strictosidine synthase (STR), 261, *263*
structure-activity relationships (SARs), 7, 60, 65, 137, 140, 148
Studzińska, Renata, xii, 195–227, 241–287
Su, W-K., 292
Sulfolobus solfataricus, 101
supercritical carbon dioxide ($scCO_2$), 2, 319, 325
supercritical fluids, 89, 232
superheating, 4, 58, 59
supported ionic-liquid-like phases (SILLPs), 102
surfactants, 162–166 *passim*, 168
sustainability, 156, *167*, 169
  synthesis of natural products, 299
Sutherland, A., 50
Suzuki reaction, 31, 119, 159, *160*, 300, *300*
Suzuki-Miyaura cross-coupling reaction (SMCR), 3, 15, 119–135, *161–162*, 163–168 *passim*, 177, 179, 180, 182, *183*
  future perspectives, 130–131
  heterogeneous ~, 120, *120*, 124–130
  homogeneous ~, 120, *120*, 121–124, 130
  mechanism, 120, *120*
  ortho-chemoselective ~, *124*
  substrate scope, *122–125*, *127–130*
swainsonine, *253*, 306, *308*
syn-primara-7,15-diene, *315*, 316
Syncephalastrum racemosum, 272, *273*, 274, *278*, *280*
synthetic chemistry, 151, 160
syringic acid, 293, *295*

T

tamiflu (oseltamivir), 215, *217*, 229, *230*, 231
tandem reaction, 3
Taxol, 137, *138*, 218, *219*
TBAF, 126, *336*
temperature, 232, 333
terpenes, *9*, 91
  uses, 264
  same as "terpenoids", 262
terpenoids
  biosynthesis, 313–316
  classification, 313
  enzymatic synthesis and biotransformation, 262–281
  molecular structure, *264*
tert-butyl hydroperoxide (TBHP), 202
tetrabutylammonium bromide (TBAB), 28–29, *30*
tetrahydro-7*H*-pyrido[*a*]carbazoles, 66, *66*
tetrahydrobenzo[*b*]pyran derivatives, 177–179, *178*, 179, *182*, 182
  substrate scope using WETSA, **181**
tetraketones, 177–179, *178*, 179, *182*, 182
  optimization of reaction condition, **180**
  substrate scope, **180**
tetraterpenoids, *264*, 264, 266
Thakuria, H., 300
Thalictrum flavum, 261, *263*
thalidomide, 241
thebaine, 309, *310*
Thermoactinomyces intermedius, 218
Thermoanaerobium brockii, 13
thermocatalytic Haber-Bosch (HB) shortcomings, 153
thermocatalytic process, 154
Thermomyces lanuginosus, 99, 212, 265
Thermus thermophilus, 101, 107
THF, 60, 91, 98, 102, 103, 105, 126
thiamin diphosphate (ThDP), **97**, *248*
thiaproline, 142, *144*
thiazole derivatives, 203–207, 325–326
thiazolidinones, 206, 325
thiazolobenzimidazoles, 202, *204*
thiobarbituric acid derivatives, 38, *38*
thioester, 296, *296*
thiokinase, 337, *337*
Thomsen-Friendreich (TF), 231
Thomsennouveau (Tn), 231
thymine dimer, 3, *3*
tin, 43, *44*, 52
titanium dioxide (TiO_2), 156
toluene, *29*, *38*, 59, 100, 163
toluene dioxygenase (TDO), 245, 255, *256*
topological polar surface areas (TPSA), 61
topotecan, 137, *138*
TPGS surfactant, 164–166
Trametes versicolor, 268, *273*
trans-1,2-divinyl cyclo butane, 323, 335
trans-cinnamaldehyde, 323, 333
trans isomer, 335, *336*
trans-isopiperitenol, *314*, 314
transaminases (TAs), 247, *248*
transesterification, 98, 99, 106, 214

transferases, *244*, 247
definition, 221
transglycosylation, 107, 231
transition metals, 43, 50, 53, 154, 155, 160, 179, *183*, 330
nanoparticles, 36
transition structure (TS), 334
transmetalation, *161*, 180, 181
transphosphatidylation, **95**, 100, 107, *108*
triacylglycerol, 337–338
triazinas, 70–71
triazoles, 3, 4, 301, *301*
trichloro triazine (TCT), 29–30
amide bond formation using DES, *31*
mechanochemical synthesis of primary amides, *31*
triethylamine, 165, 166
trifluoroacetic acid, 50, *252*
trifluoromethylsulfonyloxy (Tf), *257*
triglycerides (fats), 324
biosynthesis, 337–338
triiodothyronine, 41, *42*
trileptal® (Novartis), 100
trimethyl orthoformate (TMOF), 5
trisubstituted methane derivatives (TRSMs), vii
triterpenoids, *264*, 264, 270, 272, 274
tritium, 311, 312
tropane, *253*, 309, *310*
tryptophan, 341, *344*
tuberculosis, 7, *8*, 66, 69, 70, 141, 180, 202, 207, 295
tumor-associated carbohydrate antigens (TACAs), *231*, 231–232
two-liquid phase systems (2LPS), 325
tyrosinase, 177, *178*, 267
tyrosine, 261, 309
biosynthesis, 341, *343*

U

ultrasound, 1, 119, 126, 178, 198, *199*, 207, 295–297, 323, 325, 333
Umbelopsis, 274
U. isabellina, 274, *280*
UN FAO, 165
urea, *197*, 209, 214, *216*, 325
uridine phosphorylase (UPase), 101
ursolic acid, 274–275, *280*
US Environmental Protection Agency (EPA), 195, 299

V

Vaccaro, L., 107
Van der Waals interactions, 67, 333
van Pelt, S., 99
vancomycin, *230*, 231
vancomycin aglycon, 231
vasicinone, 258, *258*
vegetables, 7, 195, 247
Verboom, W., 126
Vibrio cholerae, 65
Vidyasagar, C. C., xii, 289–304
vinyl acetate, **92**, **94**, 98, *244*, 250, *257*
vitamins
A, 86, 87
C (ascorbic acid), 99, *197*
D, 332
D3 (cholecalciferol), 86, *86*, 220, *221*
volatile organic compounds (VOCs), 195

W

Warner, J. C., vii, 1, 241
Waser, J., 290
waste avoidance, 164, 197, 241
waste feedstock (green catalyst), 15–16, **17**
waste materials, 16–17, 289
water, 89, 162, 163, 164, 319
advantages as solvent, 195
disadvantage, 325
as greener solvent, 2–3, 302–303
in/on ~ reactions (interface technique), 2
transesterification reactions, *221*
water extract ashes, 177–190
water extract of banana peel ash (WEB), 15
water extract of banana stem ash (WEBSA), 180–181, 182, **185**, *187*
SMCR, *184*
water extract of rice straw ash (WERSA), 15
water extract of tamarind seed ash (WETSA), vii, 179, *179*, **180–181**, 182
plausible mechanism, *182*
water-glycerol combination, 216–217
Whiting, A., 24
Wittig olefination, 86
Wolfson, A., 107
Woźniak, Alina, xii, 241–287

X

Xanthobacter sp., *246*
xanthohumol, 250, *251*
Xochicale-Santana, Leonardo, xii, 289–304
Xue, D., 45

Y

Yamamoto, H., 32
Yarrowialipolytica lipase (YLL), 98
You, Q. D., 26–27
Young, S. C., 61
Yu, J-Q., 47, 50
Yuan, R., 37

Z

zanamivir (relenza), 229, *230*, 230
zanthosimuline, 137, *138*, 143
zeolite, *36*, 157, *157*, 177
zeposia®, 102, 102
Zhu, S. J., 107
Zymomonas mobilis, *247*
Zymosterol, *338–339*

For Product Safety Concerns and Information please contact our EU representative GPSR@taylorandfrancis.com
Taylor & Francis Verlag GmbH, Kaufingerstraße 24, 80331 München, Germany

www.ingramcontent.com/pod-product-compliance
Lightning Source LLC
Chambersburg PA
CBHW080714220326
41598CB00033B/5422

* 9 7 8 0 3 6 7 5 4 4 0 8 9 *